Thermal, Mechanical, and Hybrid Chemical Energy Storage Systems

Thermal, Mechanical, and Hybrid Chemical Energy Storage Systems

Edited by

Klaus Brun
Director Research & Development, Elliott Group, Jeannette, Pennsylvania, United States

Timothy Allison
Director Research & Development, Machinery Department, Southwest Research Institute, San Antonio, Texas, United States

Richard Dennis
Technology Manager for Advanced Turbines and Supercritical Carbon Dioxide Power Cycle Programs, U.S. Department of Energy's National Energy Technology Laboratory (NETL), Morgantown, West Virginia, United States

Academic Press is an imprint of Elsevier
125 London Wall, London EC2Y 5AS, United Kingdom
525 B Street, Suite 1650, San Diego, CA 92101, United States
50 Hampshire Street, 5th Floor, Cambridge, MA 02139, United States
The Boulevard, Langford Lane, Kidlington, Oxford OX5 1GB, United Kingdom

Notices

Knowledge and best practice in this field are constantly changing. As new research and experience broaden our understanding, changes in research methods, professional practices, or medical treatment may become necessary.

Practitioners and researchers must always rely on their own experience and knowledge in evaluating and using any information, methods, compounds, or experiments described herein. In using such information or methods they should be mindful of their own safety and the safety of others, including parties for whom they have a professional responsibility.

To the fullest extent of the law, neither the Publisher nor the authors, contributors, or editors, assume any liability for any injury and/or damage to persons or property as a matter of products liability, negligence or otherwise, or from any use or operation of any methods, products, instructions, or ideas contained in the material herein.

Library of Congress Cataloging-in-Publication Data
A catalog record for this book is available from the Library of Congress

British Library Cataloguing-in-Publication Data
A catalogue record for this book is available from the British Library

ISBN 978-0-12-819892-6

Publisher: Brian Romer
Acquisitions Editor: Lisa Reading
Editorial Project Manager: Joanna Collett
Production Project Manager: Prasanna Kalyanaraman
Cover Designer: Christian J. Bilbow

Typeset by SPi Global, India

Contents

8. Applications of energy storage

Pablo Bueno, Craig Turchi, Joseph Stekli, Hitesh Bindra, Brendan Ward, James Underwood, and David Voss

9. Path to commercialization

David Voss, James Underwood, Jason Kerth, David K. Bellman, Kevin Pykkonen, and Kenneth M. Bryden

Contributors

Numbers in parenthesis indicate the pages on which the authors' contributions begin.

Miles Abarr (65, 293), Carbon America, Arvada, CO, United States

Timothy Allison (1), Southwest Research Institute, San Antonio, TX, United States

Bahar Anvari (139), ABB, Greenville, SC, United States

Huashan Bao (569), Department of Engineering, Durham University, Durham, United Kingdom

David K. Bellman (513), All Energy Consulting, Houston, TX, United States

Hitesh Bindra (463), Kansas State University, Manhattan, KS, United States

Gareth Brett (293), Highview Power, London, United Kingdom

Eugene Broerman (139), Southwest Research Institute, San Antonio, TX, United States

Kenneth M. Bryden (513), Iowa State University, Ames, IA, United States

Pablo Bueno (463), Southwest Research Institute, San Antonio, TX, United States

Giuseppe Casubolo (65), SQM, Antwerp, Belgium

Terry Creasy (139), Texas A&M University, College Station, TX, United States

Sebastian Freund (65, 293, 451), Energiefreund Consulting, Munich, Germany

Konor L. Frick (65), Idaho National Laboratory, Idaho Falls, ID, United States

Jeffrey Goldmeer (293), General Electric, Schenectady, NY, United States

Scott Hume (293, 451), Electric Power Research Institute, Charlotte, NC, United States

Emmanuel Jacquemoud (293), MAN Energy, Zurich, Switzerland

Jamshid Kavosi (139), Texas A&M University, College Station, TX, United States

Jason Kerth (293, 513), Siemens Energy, Houston, TX, United States

James Klausner (249), Department of Mechanical Engineering, Michigan State University, East Lansing, MI, United States

Klaus Krueger (139), Voith, Heidenheim, Germany

Rainer Kurz (293), Solar Turbines, San Diego, CA, United States

R.B. Laughlin (27), Department of Physics, Stanford University, Stanford, CA, United States

Xiaojun Li (139), Gotion Inc., Fremont, CA, United States

Zhiwei Ma (1), Department of Engineering, Durham University, Durham, United Kingdom

Christos N. Markides (293), Imperial College London, London, United Kingdom

Anoop Mathur (65), Terrafore Technologies, Minneapolis, MN, United States

Josh D. McTigue (65, 293), National Renewable Energy Laboratory, Golden, CO, United States

Michael A. Miller (249), Department of Materials Engineering, Southwest Research Institute, San Antonio, TX, United States

Jeff Moore (293), Southwest Research Institute, San Antonio, TX, United States

Mohamad Naraghi (139), Texas A&M University, College Station, TX, United States

Alan Palazzolo (139), Texas A&M University, College Station, TX, United States

Mark Pechulis (569), Elliott Group, Jeannette, PA, United States

Joerg Petrasch (249), Department of Mechanical Engineering, Michigan State University, East Lansing, MI, United States

Brian Pettinato (569), Elliott Group, Jeannette, PA, United States

Kevin Pykkonen (513), Carbon America, Arvada, CO, United States

Nima Rahmatian (249), Department of Mechanical Engineering, Michigan State University, East Lansing, MI, United States

Kelvin Randhir (249), Department of Mechanical Engineering, Michigan State University, East Lansing, MI, United States

Douglas Reindl (65), University of Wisconsin-Madison, Madison, WI, United States

Richard Riley (293), Highview Power, London, United Kingdom

Aaron Rimpel (139), Southwest Research Institute, San Antonio, TX, United States

Joshua Schmitt (569), Southwest Research Institute, San Antonio, TX, United States

Eric Severson (139), University of Wisconsin, Madison, WI, United States

Sarah Simons (569), Southwest Research Institute, San Antonio, TX, United States

Michael Simpson (293), Imperial College London, London, United Kingdom

Natalie R. Smith (1, 293), Southwest Research Institute, San Antonio, TX, United States

Joseph Stekli (463), EPRI, Palo Alto, CA, United States

Brittany Tom (569), Southwest Research Institute, San Antonio, TX, United States

Craig Turchi (463), NREL, Golden, CO, United States

James Underwood (463, 513), Solar Turbines Incorporated, San Diego, CA, United States

Amy Van Asselt (65), Lafayette College, Easton, PA, United States

David Voss (463, 513), Solar Turbines Incorporated, San Diego, CA, United States

Zhiyang Wang (139), Vycon Energy, Cerritos, CA, United States

Brendan Ward (463), Kansas State University, Manhattan, KS, United States

Karl Wygant (293), Hanwha Power Systems America, Houston, TX, United States

Editors biography

Dr. Klaus Brun is the Director of Research & Development at Elliott Group where he leads a group of over 60 professionals in the development of turbomachinery and related systems for the energy industry. His past experience includes positions in product development, applications engineering, project management, and executive management at Southwest Research Institute, Solar Turbines, General Electric, and Alstom. He holds ten patents, has authored over 350 papers, and published four textbooks on energy systems and turbomachinery. Dr. Brun is a Fellow of the American Society of Mechanical Engineers (ASME) and won an R&D 100 award in 2007 for his Semi-Active Valve invention. He also won the ASME Industrial Gas Turbine Award in 2016 and 11 individuals ASME Turbo Expo Best Paper awards. Dr. Brun has chaired several large conferences including the ASME Turbo Expo and the Supercritical CO2 Power Cycles Symposium. Dr. Brun is a member of the Global Power Propulsion Society Board of Directors and the past chair of the ASME International Gas Turbine Institute Board of Directors, the ASME Oil & Gas Applications Committee, and ASME sCO2 Power Cycle Committee. He is also a member of the API 616 Task Force, the ASME PTC-10 task force, the Asia Turbomachinery Symposiums Committee, and the Supercritical CO2 Symposium Advisory Committee. Dr. Brun is currently the Executive Correspondent of Turbomachinery International Magazine and an Associate Editor of several journal transactions.

Dr. Timothy Allison is the Machinery Department Director at Southwest Research Institute where he leads an organization that focuses on R&D for the energy industry. His research experience includes analysis, fabrication, and testing of turbomachinery and systems for advanced power and oil & gas applications including high-pressure turbomachinery, centrifugal compressors, expanders, gas turbines, reciprocating compressors, and test rigs for bearings, seals, blade dynamics, and aerodynamic performance. Dr. Allison holds two patents, has authored two book chapters, and has published over 70 articles on various turbomachinery topics. He received the best tutorial/paper awards from the ASME Oil & Gas and Supercritical CO2 Power Cycle Committees in 2010, 2014, 2015, and 2018. He is a past chairman of the ASME Oil & Gas Applications Committee, a member of the Supercritical CO_2 Symposium and Thermal-Chemical-Mechanical Energy Storage Workshop Advisory Committees, and an Associate Editor for the ASME Journal of Engineering for Gas Turbines & Power.

Mr. Richard Dennis is currently the Technology Manager for Advanced Turbines and Supercritical Carbon Dioxide Power Cycle Programs at the U.S. Department of Energy's National Energy Technology Laboratory (NETL). These programs support US University, industry and U.S. national laboratory research, development and demonstration projects. Mr. Dennis has worked at NETL since 1983. Rich has a BS and MS in Mechanical Engineering from West Virginia University.

Foreword

The total worldwide energy storage capacity has been doubling every six months for the last three years. This is a trend that is primarily driven by the need to provide electrical backup capacity for renewable energy sources with high variability, primarily wind and solar energy. For a range of environmental, political, and economic reasons, this trend will continue for the foreseeable future. Historically, large-scale energy storage for electricity generation was accomplished using pumped hydro plants which require large elevated water reservoirs. However, over the last ten years most newly installed energy storage projects have been Lithium-Ion battery based. Although batteries provide an economical means to store chemical energy, they have several practical shortcomings making them less than ideal for large utility-scale applications.

Thus the relevance of nonbattery energy storage has dramatically increased over the recent past due to the rapid introduction of utility-sized power supply sources that provide drastically fluctuating, irregular, and difficult to forecast power to the electric grid. These large MW- or even GW-scale electricity providers, primarily alternative energy wind turbine farms and photovoltaic solar power plants, supply inherently unreliable and unpredictable power. This electricity cannot be utilized in conventional baseload or dispatch modes. Grid-scale energy storage for electricity generation that can be quickly dispatched can address the irregularity and unpredictability of these sources allowing for reliable and steady power to end-users. Fundamentally, electric energy storage provides a means of short-term and long-term capacitance in the power grid to smooth irregular supply source to match demand cycles without wasteful plant operation curtailment or the requirement for artificial and wasteful power sinks such as load banks.

The storage of energy in its most elemental forms is a natural process that has occurred over billions of years since the beginning of time. There are many energy conversion to storage processes that occur but one of the most relevant for all forms of life has been sunlight conversion by photosynthesis. Plants convert sunlight into long cellulose chains, which primarily contain carbon and some hydrogen. This stored energy is utilized directly as feed by animals or further transformed over millions of years into other forms of stored energy such as oil and natural gas. Historically the most utilized stored energy is common tree wood which has been used for heating, cooking, and building materials. Over the last several thousand years mankind has taken advantage of this stored energy since when stone-age humans lit the first fire. Ancient societies became more organized in their consumption of energy and consequently increased their conversion rate of naturally stored energy. This process then accelerated drastically with the beginning of the industrial age due to the growing need for consumption of energy to meet modern requirements for food, transportation, clothing, shelter, heating, etc. When the first heat engines were invented to produce mechanical energy the rate of use of stored energy increased drastically. Due to a low specific energy density, wood was found to be inadequate for steam engines and humans began to utilize other sources of stored energy in the form of coal and eventually oil and gas. Today's industrial society relies on these hydrocarbon-based energy carriers for most of its energy and transportation needs. But these long-term carriers of stored energy are not inexhaustible, and they also contribute to various forms of air pollution and greenhouse gas-based climate change.

Other energy sources are needed in the future to assure a reliable supply of affordable electricity, heat, and mechanical power to the consumers. A wide range of renewable generation technologies exist, including hydropower, geothermal, biomass, wind, solar photovoltaic, and solar thermal sources. The fastest-growing renewable energy sources utilized for electricity production are solar and wind. Solar generation without energy storage produces power in direct response to immediate solar radiation and therefore varies significantly by short-time events (cloud pass, weather) and daily/annual cycles. Wind is generated by solar heating of the earth and therefore utilizes stored thermal energy. However, the resulting winds are highly variable due to local surface geometries and atmospheric instabilities.

An effective energy storage system that enables high penetration of variable renewables has many challenges. The foremost challenge is the staggering scale of energy required to fuel the global energy demand. Multiple studies of dispatch scenarios spanning various countries have shown that the amount of storage required to achieve high penetration of renewables >85% ranges widely for each application; the storage requirement is potentially on the order of the total daily energy demand. While there is uncertainty and variation in this number, the correct answer will be much higher than the current global energy storage capacity of 0.4% of daily demand. Energy demand is also rapidly increasing; worldwide energy consumption for electricity has grown by 239% in the past several decades to reach the current value of 26.6 terawatt-hours. This demand is projected to increase exponentially at 2.1% per year, thus approximately doubling again in the next three decades. Even at relatively low renewable penetration, energy storage can be combined with baseload fossil/nuclear resources to maximize efficiency, load following, and reliability of these assets. In these cases, there is a challenge and opportunity to synergistically minimize storage cost by integrating various energy storage technologies into a wide variety of power generation processes. In addition to scale and integration, energy storage systems must be cost effective, efficient, safe, secure, reliable, and sustainable for various applications combining a wide range of generation technologies and thermal and electrical demands.

Energy can be stored in many forms from kinetic to potential to chemical. Currently the most common form of energy storage is by batteries. However, batteries are electrochemical devices that have several inherent shortcomings that make them less suitable for large-MW and bulk long-term energy storage. These limitations are driven mostly by cost and performance and are further exacerbated by short life expectancy, limited supply of materials for manufacturing (specifically expensive rare-earth metals needed for electrodes), and environmental impact during disposal. Consequently, electrochemical batteries may not provide a long-term solution for utility-scale energy storage. Other forms of energy storage include a near infinite array of technology options including mechanical potential, mechanical kinetic, cold thermal, hot thermal, various chemical options, and compressed gases. It is beyond any book to cover all these options so this book, by no means an exhaustive text, will focus on technology options that exist commercially or have demonstrated the potential to be technically feasible and commercially viable in the near- to mid-term as alternatives to conventional batteries.

This book provides a comprehensive overview of thermal, mechanical, and hybrid chemical energy systems that are utilized or are currently being developed for the electric generation and utility power industries. In the first section of this book (Chapters 1 and 2), an introduction of general energy storage concepts and a high-level technology discussion is provided. Specifically, Chapter 1 is an overview of energy storage requirements and technologies while Chapter 2 is a special chapter on thermal-mechanical energy storage authored by Nobel Prize Laureate Professor Robert Laughlin of Stanford University. Subsequent Chapters 3–6 comprise the second section of the book and focus on specific energy storage technologies such as thermal, mechanical, chemical, and heat engine-based storage. These chapters provide a detailed discussion of each of these technology areas with significant engineering insight related to their application in the electric power industry. The third section of the book (Chapters 7, 8, and 9) discusses the applications, commercial considerations, and the different types of energy storage services. This includes a detailed discussion of all major energy storage applications and the type of utility services that they are optimized for. Chapter 9 also looks at commercial considerations for the installation and operation of different energy storage technologies, including plant total costs, round-trip efficiencies, and operation and maintenance costs. Finally, Chapter 10 provides an overview of novel and advanced concepts for energy storage that are currently being considered for future research and development. Research requirements and recommendations to advance some of these technologies are also summarized.

The topics covered in this book were selected to provide engineers, scientists, and other practitioners interested in the energy storage industry with a comprehensive technology and application overview. Each chapter provides sufficient background material to stand alone and can be used on its own, although an attempt was made to avoid duplication throughout this book.

We, the editors, are indebted to the chapter authors. They are all subject matter experts in their field who were selected from the scientific and engineering community based on their relevant contributions to the field. They represent a broad and diverse range of expertise and have volunteered numerous hours to provide their contributions; we thank them wholeheartedly!

Klaus Brun
Timothy Allison
Richard Dennis

Acknowledgments

We would like to thank Andrea Barnett for her tireless efforts and assistance while putting this book together.

Nomenclature

a	speed of sound [m/s]
A	amplitude
AF	amplification factor
b	impeller exit width
C	flow heat capacity ($\dot{m}c_P$)
c	flow velocity in absolute reference frame
c	specific heat capacity for a solid or incompressible fluid
$C_{effective}$	effective damping $= C_{xx} - K_{xy}/\omega$
c_v, c_P	specific heat capacity at constant volume and specific heat at constant pressure, respectively
C_{xx}	direct damping coefficient
C_{xy}	cross-coupled damping coefficient
D	diameter
D_2	impeller tip diameter
e	energy per unit mass
E	elastic modulus (Young's modulus)
f	frequency
G	shear modulus (modulus of rigidity)
h	enthalpy
$\hat{h}$	convective heat transfer film coefficient
h	height
H	head
i, I	irreversibility per unit mass and total irreversibility (entropy), respectively
I	electrical current
J	polar moment of inertia
K	stiffness
k	isentropic exponent
k_{xx}	direct stiffness coefficient
k_{xy}	cross-coupled stiffness coefficient
L	length
m	mass
$\dot{m}$	mass flow rate
Ma	Mach number
MW	molecular weight

n	polytrophic exponent
N	rotational speed
N_c	critical speed
Nu	Nusselt number
P	power
p	pressure
pe	potential energy per unit mass (gz, where z represents elevation)
ρ	density
Pr	Prandtl number of the fluid
q, Q	heat transfer per unit mass, total heat transfer
Q	volume flow rate
r_v	specific volume ratio
R	gas constant for a specific gas
$\overline{R}$	universal gas constant
Re	Reynolds number
SQ	std. flow
s	entropy
T	temperature
u	internal energy per unit mass
$\hat{U}$	overall heat transfer coefficient
U_2	impeller tip speed
V	voltage
v	flow velocity in stationary reference frame
v_d	specific volume at discharge
v_i	specific volume at inlet
w	flow velocity in rotating reference frame
W	weight
W	work
X	reactive impedance
Z	total impedance
ΔP	pressure drop
ε	heat exchanger effectiveness
Φ	exergy
η	efficiency
ρ	density
ψ	head coefficient
μ	absolute (dynamic) viscosity
φ	phase angle
ϕ	flow coefficient
υ	kinematic viscosity
Γ	torque

θ	angular displacement
δ	ratio of specific heats (c_P/c_v), Fluid thermal conductivity
α	absolute flow angle
β	relative flow angle

Abbreviations

CSR	critical speed ratio
FFT	fast Fourier transfer
HP	horsepower
ke	kinetic energy per unit mass ($v^2/2$, where v represents velocity)
MCOS	maximum continuous operating speed
PF	power factor

Subscripts

1, 2, 3	property at defined point
I, II	first law (or energy) and second law (or exergy) basis, respectively
C, H	heat exchanger cold and hot fluids, respectively
C, T	compressor, turbine, respectively
f	saturated liquid
fg	difference in property for vaporization from liquid to vapor
g	saturated vapor
H	heat source
o	dead state
p	polytropic
r	rejected heat
R, S	heat rejected and supplied, respectively
S	state point that would be reached in an isentropic process
s, d	suction, discharge
S	isentropic
th	thermal efficiency (refers to energy transformations within the working fluid)

Over dot

$\dot{\square}$	rate or time derivative
$\longrightarrow$	vector
$\sim$	matrix

Chapter 1

Introduction to energy storage

Timothy Allison[a], Natalie R. Smith[a], and Zhiwei Ma[b]
[a]Southwest Research Institute, San Antonio, TX, United States, [b]Department of Engineering, Durham University, Durham, United Kingdom

Chapter outline

Significant global integration of renewable energy sources with high variability into the power generation mix requires the development of cost-effective, efficient, and reliable grid-scale energy storage technologies. Many energy storage technologies are being developed that can store energy when excess renewable power is available and discharge the stored energy to meet power demand when renewable generation drops off, assisting or even displacing conventional fossil- or nuclear-fueled power plants. The development and commercialization of these technologies is a critical step for enabling a high penetration of renewable energy sources.

Many mature and emerging energy storage technologies utilize combinations of thermal, mechanical, and chemical energy to meet storage demands over a variety of conditions. These systems offer the potential for better scalability than electrochemical batteries. Energy storage demands are complex and the resulting solutions may vary significantly with required storage duration, charge/discharge duty cycle, geography, daily/annual ambient conditions, and integration with other power or heat producers and consumers. This introductory chapter provides details regarding the needs that motivate development

Thermal, Mechanical, and Hybrid Chemical Energy Storage Systems
https://doi.org/10.1016/B978-0-12-819892-6.00001-0

efforts for new thermal, mechanical, and chemical energy storage technologies; discusses fundamental thermodynamic principles that govern energy storage; and describes the opportunities and challenges for successful development and commercialization of these technologies.

1.1 Motivation for energy storage

Energy storage systems help to bridge the gap between power generation and demand and are useful for systems with high variability or generation-demand mismatch. The increasing introduction of renewable power sources into the generation mix results in power availability that is highly variable and poorly matched with demand profiles, thus increasing the high turndown and ramping requirements for baseload power plants that are poorly equipped for this service.

1.1.1 Worldwide power generation mix and trends

In 2018 the world consumed approximately 26,641 TWh of electric power [1], produced by a combination of sources illustrated in Fig. 1. Based on these data, fossil-based sources accounted for 64.2% of generation, supplemented by 10.2% nuclear power. The remaining ~25% was produced by renewable sources including hydroelectric (15.8%), wind (4.8%), solar (2.2%), and geothermal/biomass (2.4% combined). Notably, although wind and solar sources are still a relatively low percentage of the overall energy mix, they are the fastest-growing categories globally and particularly for OECD (Organization for Economic Cooperation and Development) member countries. From 2017 to 2018, the IEA [2] reports overall declines in electricity production in OECD

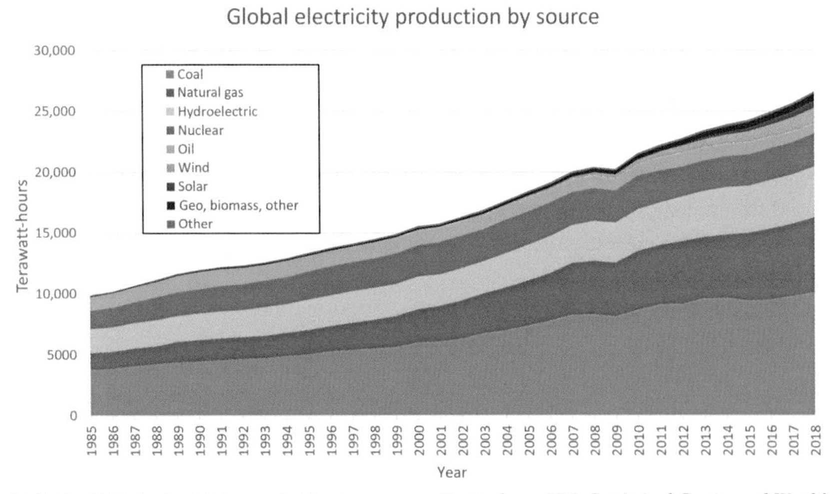

FIG. 1 Global electricity production by source. *(Data from "BP Statistical Review of World Energy," 68th ed., 2019.)*

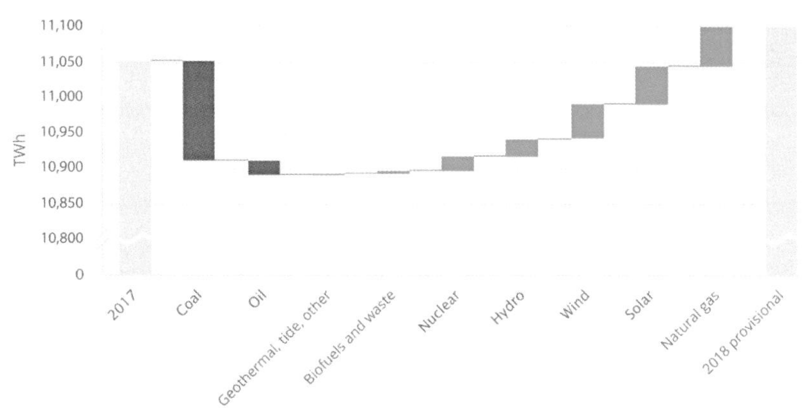

FIG. 2 OECD electricity production variation, 2017–18 provisional [2].

countries from combustible fuels (particularly coal and oil) that are substantially offset by 19.8% and 7.0% growth in solar and wind production, respectively, as shown in Fig. 2 (red bars represent a decrease in production, green bars an increase).

In local regions, more dramatic changes can be seen. California's electricity production profile (Fig. 3) shows that coal-based electricity in that location has declined to negligible amounts. Natural gas power plants constitute the largest source of electrical power at about 46%, but renewables have grown rapidly in the past decade, combining for 21% growth from 2017 to 2018. In 2018

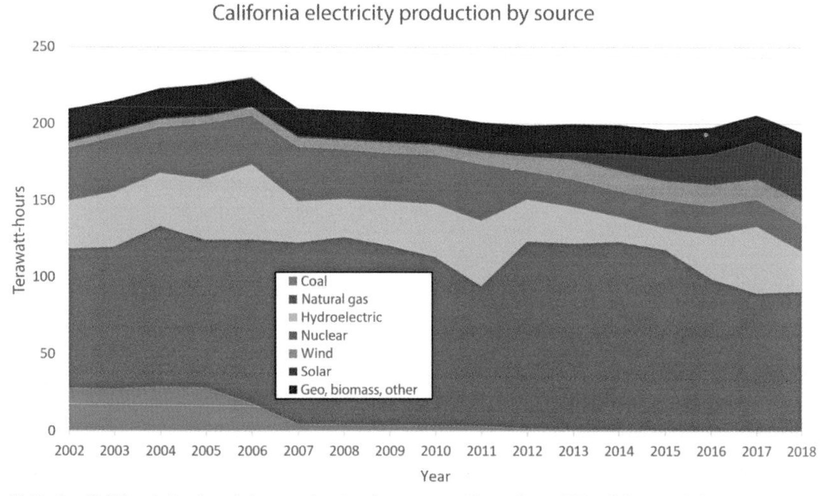

FIG. 3 California's electricity production by source. *(Data from "Total System Electric Generation," California Energy Commission, 2019. https://ww2.energy.ca.gov/almanac/electricity_data/total_system_power.html [Accessed 18 October 2019].)*

renewable sources including solar, wind, hydro, geothermal, and biofuels were a close second to natural gas, providing 44% of California's electricity production [3].

Renewable energy sources are expected to continue their rapid growth in the future. The Paris Agreement established in 2016 set a goal of limiting global temperature rise to less than 2 degrees Celsius, to be achieved in part by lowering greenhouse gas emissions and increasing energy production from renewable sources. This agreement has been signed by 186 states and the European Union, representing nearly 97% of global greenhouse gas emissions. In alignment with this agreement, REN21 [4] reports that 162 countries have established national targets for substantially increasing the share of renewable power generation. The range of targets is illustrated in Fig. 4, with most countries resolving to achieve 50% renewable generation by 2035. In the United States, 6 states (Hawaii, California, New York, New Mexico, Maine, and Nevada) and Washington, D.C. have all passed laws to reach 100% clean or renewable energy production by target dates ranging from 2032 to 2050 [5].

1.1.2 Renewable variability and demand mismatch

Two significant challenges result from the rapid introduction of renewable resources into the energy mix. First, much of the capacity growth will be provided from solar and wind generators that have high variability. Second, the availability of renewable resources is also poorly matched with the power demand profile in a daily cycle.

Wind and solar power output can vary significantly by the minute, hour, and season. Wind speed varies due to weather patterns or diurnal effects. Likewise, solar power output will vary with storms, cloud passes, and ambient temperature/wind. An example of photovoltaic plant variability is shown in Fig. 5, where plant output variation is on the same order of magnitude as the average output. Fig. 5 also illustrates a typical daily cycle for solar plants, where solar

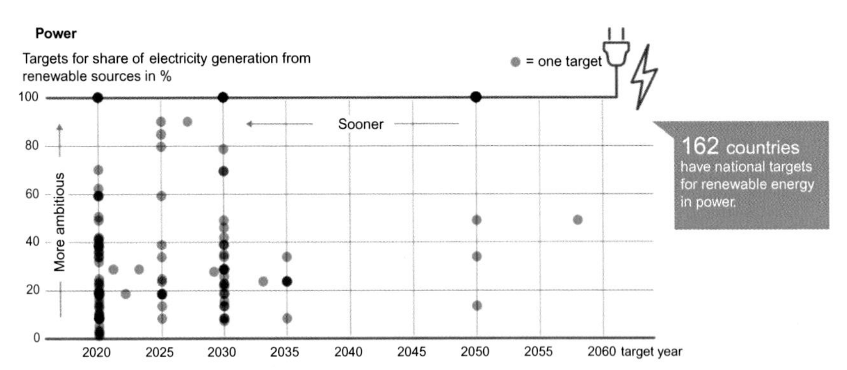

FIG. 4 Global targets for renewable electricity production [4]. Darker circles represent multiple policies.

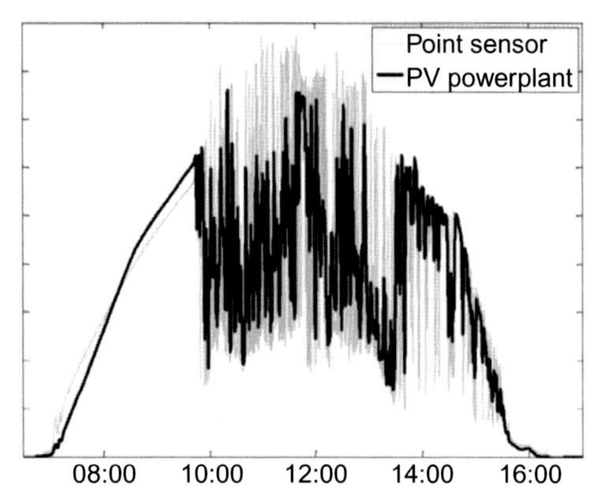

FIG. 5 Daily photovoltaic power plant output variability [6].

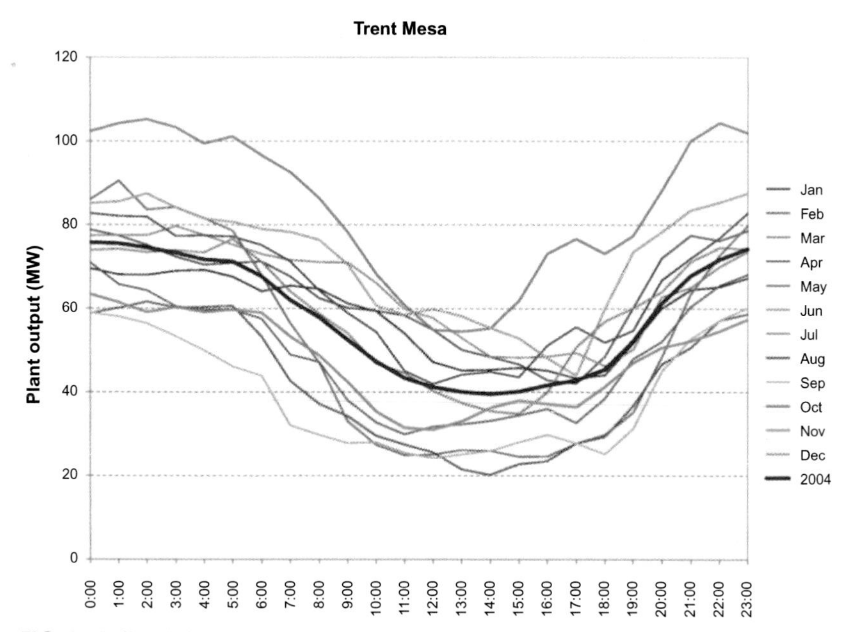

FIG. 6 Daily wind power plant output variability for trent Mesa WPP, 2004 [7].

input is only available when the sun is shining and peaks near midday. Data from wind power shows a similar trend, with hourly standard deviation even exceeding the average output during summer and winter months [7]. Average hourly wind power also varies significantly throughout the day, as illustrated in Fig. 6 by a typical daily cycle between peak and $\sim 50\%$ capacity for the Trent Mesa WPP [7]. Finally, similar amplitudes of seasonal variation also occur in wind and solar power plants. Summer capacity at Lake Benton WPP is just over

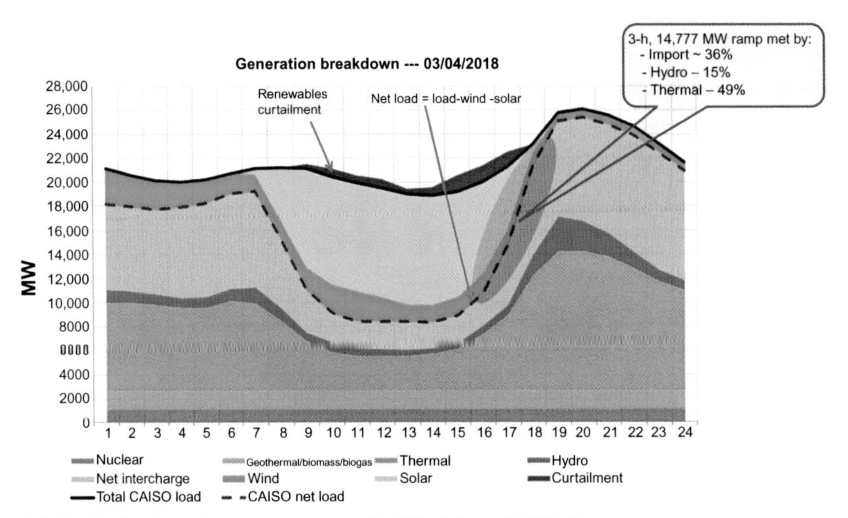

FIG. 7 Daily demand vs. power source; CAISO, March 42,019 [9].

half of winter capacity [7], and winter production from combined PV and CSP is approximately 1/3 or summer peak values [8].

The availability of wind and solar resources is also poorly matched with the typical electrical demand profile. A typical electrical demand curve is shown in Fig. 7 and illustrates small demand peaks in the morning and evening. The evening peak, in particular, occurs after solar production has dropped off, resulting in a fast ramp of baseload/import resources that must be brought online very quickly. Another challenge is highlighted during the middle of the day, when the renewable resource is very high and renewable resources (plus baseload generators operating at minimum output) exceed the power demand, requiring curtailment of renewables to achieve power balance. Historical monthly curtailment of wind and solar resources in California is shown in Fig. 8. Notably, every month since December 2016 has required curtailment of renewable resources. There is also a strong trend of growing curtailment from year to year, with peak curtailment in the spring and fall.

The high variability and resource-demand mismatch associated with renewable power sources impose significant ramp rate and turndown requirements on baseload power generators that were not necessarily designed for this service. Simple-cycle gas turbines are characterized by fast startup times and high turndown capabilities, but with poor efficiency and emissions performance at low load. Combined-cycle gas turbines and steam turbines incorporate large heat exchangers and have reduced ramp rate capabilities to minimize thermal stress, but have higher part-load efficiencies. A comparison of the range of turndown (expressed as minimum complaint load or MCL), ramp rate capabilities, and hot/cold startup times for various technologies is presented in Fig. 9. The figure describes large (>400 MW), midsize (<400 MW), and small (<100 MW)

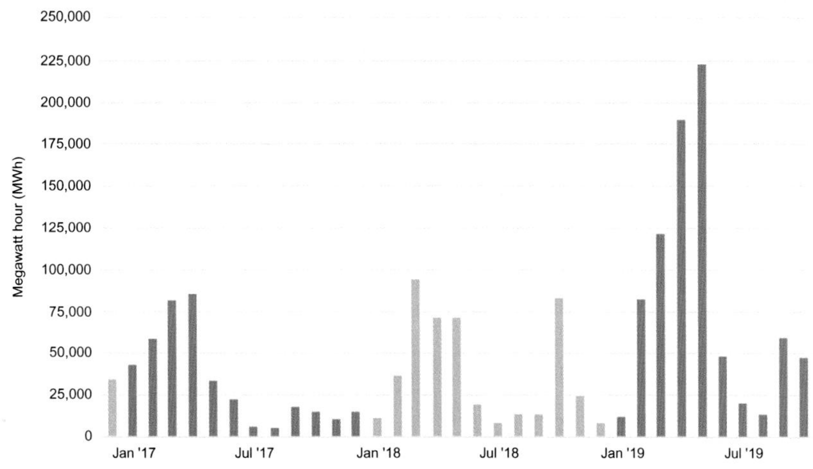

FIG. 8 Monthly wind and solar curtailment in California, Dec 2016–Nov 2019 [10].

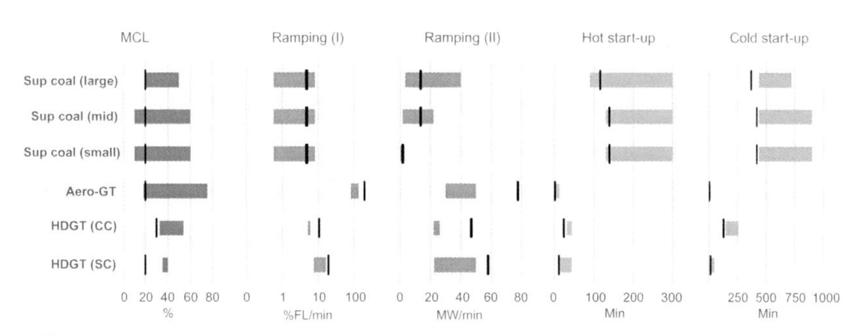

FIG. 9 Comparison of turndown, ramp rate, and startup times for various thermal power plants [11].

supercritical (sup) and subcritical (sub) steam-based coal plants, aeroderivative gas turbines, and heavy duty gas turbine (HDGT) combined-cycle (CC) and simple-cycle (SC) power plants. Ramp rates are expressed both as a percentage of full load (FL) per minute and MW per minute. In general, gas turbines are found to have a better ramping and startup performance than coal-based power plants, but poorer turndown (except for aeroderivative units). Simple-cycle gas turbines are the fastest starting and ramping systems, but have poorer efficiency and emissions performance. This study notes that significant performance improvements are necessary for all technologies to provide backup for a high penetration of renewables. Future operation of these conventional generators may include twice as many starts, 70%–100% faster ramp rates, 35%–70% faster starts, 35%–60% increased turndown, and lower emissions than achievable by the current state of the art [11].

1.1.3 Opportunities and challenges for energy storage

There is a rapidly growing need for energy storage technologies that can buffer or time-shift renewable resources and mitigate ramping/off-design requirements for conventional generators. These systems must function reliably, efficiently, and cost effectively at the grid scale. Energy storage systems that can operate over minute by minute, hourly, weekly, and even seasonal timescales have the capability to fully combat renewable resource variability and are a key enabling technology for deep penetration of renewable power generation. Energy storage technology can also improve grid resilience to overcome variability from nonrenewable power generation upsets.

Multiple commercial opportunities already exist for cost-effective energy storage systems. These include applications in front of or behind the electric meter for commercial and residential applications. Front-the-meter applications are more varied and include power quality (frequency regulation or load following), energy arbitrage (buy low, sell high), or deferral of upgrades to generation or transmission and distribution systems. Behind-the-meter storage is typically applied to improved local energy resiliency or to reduce power demand costs for high peak power consumers. These opportunities are driving forecasts for strong growth in energy storage systems, with market analysts predicting strong exponential growth worldwide (Fig. 10) based primarily on existing commercial electrochemical battery systems. Thermal, mechanical, or (nonbattery) chemical energy storage technologies compete with battery technologies for all of the previously listed commercial applications, but also enable additional applications for longer durations, higher power density, or involving hybridization with existing utility-scale heat and power resources.

How much storage is needed in the long term? The answer to this complex question depends on many factors including the depth of renewable penetration into the energy mix, the relative mix of wind/solar generators, grid size and diversity, geography and climate trends, degree of allowable energy

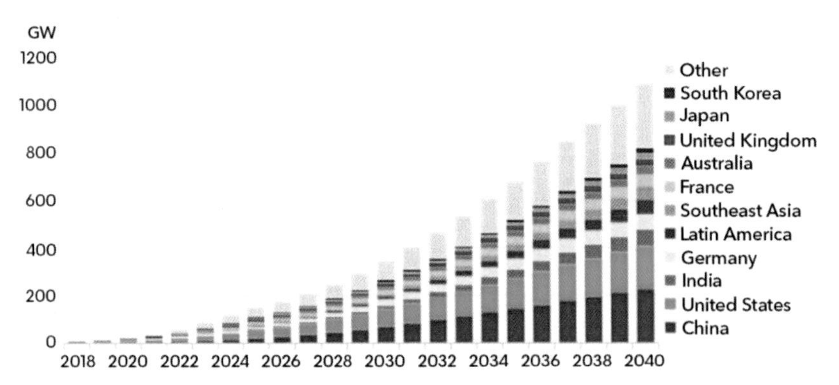

FIG. 10 Prediction of global cumulative energy storage installations [12].

curtailment, storage system performance capabilities, approach to utility load management, economic policy, etc. A recent review paper [13] highlighted a range of studies showing that the required storage capacity for 70%–100% renewable penetration of the required storage capacity was equal to 22%–2160% of the average daily demand—a very wide range. The authors conclude that an optimal solution balancing curtailment and storage may result in storage requirements approximately equal to the daily average demand.

The storage of energy in very large quantities introduces issues of proper location and safety. As an example of the required scale, a large city, such as Tokyo, has an average power demand of approximately 30–40 GW. Thus the daily energy demand is approximately 840 GWh. This amount of energy is equivalent to approximately 6500 battery banks like those manufactured by Tesla, Inc. for the Hornsdale Power Reserve in Australia; 35 of the world's largest pumped hydro facility in Bath County, Virginia; or 760 tanks of molten salt similar to those used in the Crescent Dunes concentrating solar power plant. More dramatically, the energy contained is equivalent to approximately 35 of the 20-kt nuclear weapons used during World War II! Although the energy amounts are not greater than what is already produced and consumed in a day, the collection and storage of this energy must be done in a safe and secure manner, ideally located for transmission from generators and to loads.

Multiple metrics are important for quantifying the cost and performance of energy storage systems for various applications. A summary of common metrics and their definitions is provided in Table 1. These metrics emphasize that significant details are required to fully characterize an energy storage system that may need to operate flexibly in response to grid demands, i.e., at different charge/storage/discharge profiles and different power rates. One key observation is that both power capital costs and energy capital costs are important and will scale differently for different systems. For example, a 2-h 100 MW Lithium-Ion battery storage system may have a significantly lower cost per kW than a 2-h pumped hydro system, but as energy increases to longer durations the pumped hydro system costs will increase much more slowly than the battery system. Thus meaningful cost evaluations must include both effects. Another important point is that the commercial viability of an energy storage system is typically a function of both performance and cost, i.e., a lower-cost system may be viable even with reduced performance or vice versa.

1.2 Basic thermodynamics of energy storage

Energy storage employs and exploits the true fundamentals of Thermodynamics. As such, it is appropriate to begin the discussion with first principles. This section will provide an overview of the first and second laws of thermodynamics to assist in the discussion of the thermodynamics and performance of various energy storage technologies presented throughout this book.

TABLE 1 Performance and cost metrics for energy storage systems.

Performance/cost metric	Typical units	Definition/explanation
Power rating	MW	Maximum output/discharge power allowed from system at nominal conditions. May be different than input/charge power rating
Power density	W/kg	Power rating divided by system weight. Emphasizes short-duration systems
Specific power	W/m^3	Power rating divided by system volume requirement. Emphasizes short-duration systems
Energy capacity or storage capacity	Wh	Maximum amount of stored energy that system can deliver, i.e., power rating multiplied by discharge time at rated power. Will be less than charging energy and stored energy due to system inefficiencies
Energy density	Wh/kg	Energy capacity divided by system weight. Emphasizes long-duration systems
Specific energy	Wh/m^3	Energy capacity divided by system volume requirement. Emphasizes long-duration systems
Charge efficiency	%	Total stored energy divided by total input energy for nominal charge profile
Discharge efficiency	%	Output energy divided by stored energy for nominal discharge profile
Round-trip efficiency or cycle efficiency	%	Output energy divided by input energy for nominal charge, storage, and discharge profile
Response time	Seconds—minutes	Various specific definitions, but generally time required to ramp discharge power up to rated power
Daily self-discharge	%/day	Percentage of energy capacity lost per day due to heat leaks, friction, chemical breakdown, system parasitics, or other energy losses
Lifetime	Years	Useful system life, may include major maintenance/overhauls
Performance degradation	%/year	Loss of system rated power or energy capacity due to degradation, fouling, etc.
Storage duration	Seconds—months	Time between charge and discharge events
Turndown	%	Lowest percentage of rated power that the system can be operated at

TABLE 1 Performance and cost metrics for energy storage systems—cont'd

Performance/cost metric	Typical units	Definition/explanation
Power capital cost	\$/W	System cost divided by power rating. Emphasizes short-duration systems
Energy capital cost	\$/Wh	System cost divided by energy capacity. Emphasizes long-duration systems
Operating and maintenance cost	\$	Operating and maintenance costs may be functions of time (\$/year), operating time (\$/Wh), or cycles (\$/cycle)
Siting requirements	–	Siting requirements other than power/energy density or specific power/energy may include safety, permitting, geographic, noise, environmental, and other constraints

1.2.1 First law of thermodynamics

The First Law of Thermodynamics, also referred to as the conservation of energy, governs the balance of energy for a defined system. This is defined in Eq. (1), where the total energy transferred into (E_{in}) or out of (E_{out}) the system must equal to the change in total energy of the system (ΔE_{system}) during a process. This indicates that energy cannot be created nor destroyed, it can only change forms.

$$E_{in} - E_{out} = \Delta E_{system} \tag{1}$$

The energy equation consists of forms of energy transfer on the left and changes in system energy on the right, as shown in Fig. 11.

The energy of a system is made up of macroscopic and microscopic forms of energy. Macroscopic energy is energy of a system relative to some reference frame; this includes kinetic energy and potential energy. Kinetic energy (*KE*) is the energy associated with a system's motion relative to a given reference frame. Potential energy (*PE*) is the energy associated with a system's elevation in a gravitational field relative to a given reference frame. Microscopic energy is independent of external reference frames and depends on the molecular structure and molecular activity of a system. The sum of all microscopic energies is referred to as internal energy (*U*) which includes sensible, latent, chemical, nuclear, electric, and magnetic energies. Sensible energy is associated with the kinetic energy (translation, rotation, spinning, and vibration) of the molecules in a system. Latent energy is associated with the phase of the system.

(Energy transfer into the system) − (Energy transfer out of the system) = (Change in energy of the system)

FIG. 11 Energy balance equates the energy transfer to and from the system to the change in energy of the system during a process.

Chemical energy is the internal energy associated with the atomic bonds of the molecules. Nuclear energy is the internal energy associated with the bonds in the nuclei of the atoms. System energies are often alternatively categorized based on how that energy can be transferred. Mechanical energy is energy that can be converted completely and directly to mechanical work, such as kinetic energy, potential energy, and the pressure of a flowing fluid. However, sensible and latent energy are thermal energy which cannot be converted directly to work. Finally, forms of system energy are point functions; they depend only on the state of the system. They are evaluated at the initial and final state of a process and do not depend on the process itself.

Energy can be transferred in three forms: work, heat, and mass. It is through these three energy transfers that energy crosses a system boundary during a process and the system energies change. Heat transfer (Q) is the transfer of energy between two systems due to a temperature difference. Work (W) is energy transfer associated with a force acting across some distance. These forms of energy transfer, heat and work, are path functions, that is, they depend on how the process is conducted. Polytropic, isentropic, isothermal, isobaric, etc. processes will all result in different amounts of energy transfer, even if they begin and end at the same state points. Finally, a flowing fluid contains energy just as the system contains energy, and thus, as a fluid enters or leaves a system, energy is transferred (E_{mass}). A flowing fluid contains the same energies as a stationary fluid, internal, kinetic, and potential, but it also consists of flow energy, which is the energy required to move a fluid mass into or out of a control volume. The flow energy term is pressure times volume (PV); however, typically for flowing systems, this term is grouped with internal energy and evaluated as enthalpy (H).

Substituting the forms of energy transfer and types of system energy into Eq. (1), the first law of thermodynamics can be written as Eq. (2) or Eq. (3), where N_{in} and N_{out} of the number of inlets and outlets to the system, respectively.

$$W_{in} + Q_{in} + E_{mass,in} - W_{out} - Q_{out} - E_{mass,out} = \Delta U + \Delta KE + \Delta PE \quad (2)$$

$$W_{in} + Q_{in} + \sum_{j=1}^{N_{in}} (H + KE + PE)_j - W_{out} - Q_{out} - \sum_{j=1}^{N_{out}} (H + KE + PE)_j = \Delta U + \Delta KE + \Delta PE \quad (3)$$

1.2.2 Second law of thermodynamics

The first law of thermodynamics provides a means to quantify energy and its changes. However, even based on empirical everyday observations, we know that not all energy is the same. For instance: a cup of warm coffee or tea will eventually become room temperature, your pen falls to the floor when bumped off the desk, gas leaks from a pressurized container. However, none of these processes will ever naturally occur in reverse. The second law of thermodynamics provides a means to describe the order or hierarchy of energy, and thus the natural direction of processes and their performance. For example, thermal energy is more chaotic, while mechanical energy is more ordered. These qualities effect how the energy can be used in energy transfer. Two additional thermodynamic quantities are defined for two law analysis: entropy and exergy.

Entropy is a thermodynamic property used to describe the amount of molecular chaos, randomness, or disorder a system contains. In a similar manner to energy, entropy can be transferred into (S_{in}) and out of (S_{out}) a system, and it can be evaluated as a system property for a given state point or change between state points of a process (ΔS_{system}). However, unlike energy, entropy is a nonconserved property, and thus, entropy can be generated during a process (S_{gen}). Eq. (4) describes the balance of entropy during a process.

$$S_{in} - S_{out} + S_{gen} = \Delta S_{system}. \tag{4}$$

Entropy can only be transferred by two mechanisms: heat and mass. Heat is a disorganized form of energy. As heat is transferred to a system, the disorder, and thus, entropy of the system increase. The opposite is true for heat rejection from a system, the disorder and entropy decrease. Entropy cannot be transferred by work. The entropy of a system is evaluated as a single property and it does not consist of different types like energy. The entropy generation term is zero for a nonphysical, idealized process, for example an isentropic process, which is adiabatic and internally reversible, but for all actual processes, the entropy generation term is always positive. Entropy generation enables the measurement of irreversibilities in a process. At the system level, increases in entropy due to entropy generation are realized as reduced efficiency. Considering Eq. (4) again, as the entropy generation term increases, the entropy transfer and/or the change in entropy of the system are directly affected. For the same amount of entropy transferred into or out of the system, with increased entropy generation, the system will not result in the same state points. Applied to energy storage, the implications of entropy generation are apparent in the fact that not all the energy stored during charge will be converted back to useful energy in discharge mode due to irreversibilities in the processes.

Exergy addresses the second law from the opposite perspective of entropy. Exergy is a measure of the useful work potential a system can deliver from a given state point. This is evaluated as the amount of useful work a system could produce through a reversible process from some given state point to the state of the surrounding environment. If a system ends at the conditions of the surrounding environment, then there is no further process that could generate energy from the system. Exergy can be transferred into (X_{in}) and out of (X_{out}) a system and it can be evaluated for a system at given state points at the start and end of a process (ΔX_{system}). Additionally, exergy can be destroyed during a process ($X_{destroyed}$) due to irreversibilities, such as friction, mixing, etc.

$$X_{in} - X_{out} - X_{destroyed} = \Delta X_{system}. \tag{5}$$

Exergy can be transferred by work, heat, and mass. The exergy of a system is evaluated as a difference between state points and include exergy from internal energy, flow energy, kinetic energy, and potential energy. The exergy destroyed term is zero for an internally reversible process, which is a nonphysical ideal case. For all actual processes, the exergy destroyed term is always positive. Exergy destroyed is proportional to entropy, and thus, exergy is always destroyed in real processes. This implies that a real process will never achieve the idealized work potential from a given energy source. In the exergy balance shown in Eq. (5), when exergy is destroyed not all of the exergy transferred into or out of the system contribute to changes in exergy of the system.

1.2.2.1 Materials for energy storage

Materials play a significant role in energy storage systems, especially for thermal energy storage (TES) and chemical energy storage.

1.2.3 Thermal energy storage materials

There are three general types of TES mechanism, sensible heat storage, latent heat storage, and sorption heat storage. Different materials are used by different mechanisms. The candidates of thermal energy storage materials should satisfy thermal, physical, chemical, economic, and environmental requirements, described as follows:

- Thermal requirements: high latent heat, high specific heat, high thermal conductivity, suitable phase change temperature;
- Physical requirements: high density, low density change after phase change, low supercooling degree, no phase separation;
- Chemical requirements: high chemical stability, no degradation, noncorrosive to the construction material, nontoxic, nonflammable, and nonexplosive;
- Economic requirements: cheap and abundant;
- Environmental requirements: nonpolluting, environment friendly.

1.2.3.1 Sensible heat storage materials

Thermal energy can be stored by simply changing the temperature of a material to higher level for heat storage or to lower level for cold storage. The amount of the stored energy can be calculated as the product of the specific heat capacity, the mass of the used material and the temperature difference. In the energy charging process of heating or cooling, phase change is not expected.

Typical sensible heat storage materials include water, thermal oil, molten salt, clay, brick, sandstone, steel, magnetite, etc. Different materials have different application temperature ranges, such as the application temperature of water is normally not expected to be higher than 95°C for heat storage and not lower than 0°C for cold storage. Other solid materials can be used for higher temperature utilization (>100°C).

Two of the key parameters of a sensible heat storage material that dominate its storage capability are the density and specific heat capacity; the higher value of the product of these two parameters leads to larger volumetric energy storage density with the unit of $J\ m^{-3}\ K^{-1}$.

1.2.3.2 Phase change materials

Different from sensible heat storage, latent heat storage involves a phase change (or phase transition) process of the used storage material, and there should be no change in chemical composition. This type of material is called phase change material (PCM). There is a jump of the enthalpy of PCM when the phase change occurs; the latent heat involved in the phase change process is much larger than the sensible heat. Theoretically, this phase change process happens at a constant temperature rather than in a temperature range like sensible heat storage process. Then the applied temperature range of the latent heat storage system is around the phase change temperature of the used PCM.

Majority of PCMs used for TES is solid-liquid phase change material due to its ubiquitous and negligible volume change comparing to that of liquid-vapor phase change. There are also some solid-solid phase change processes that can be used for TES, such as the phase transition process between crystalline phase and amorphous phase of a solid. One of few examples of solid-vapor phase change that is being used for TES is dry ice sublimation process, which is normally a one-off usage. Typical PCMs are listed as follows as the sequences of their temperature application ranges.

- Salty aqueous solutions, e.g., $CaCl_2$ aqueous solution, and organic aqueous solutions, e.g., ethylene glycol aqueous solution, are commonly used as PCMs for subzero degree TES.
- Ice-water phase change is widely used for cold energy storage at near 0°C.
- Paraffins, fatty acids, and salt hydrates are widely used for TES between 0 and 120°C.
- Sugar alcohols can be used for TES between 80 and 200°C.
- Solid-solid PCMs, such as FeS, Ag_2S, $LiSO_4$, can be used for TES from 100°C to 600°C.
- Molten salts, such as $NaNO_3$, KNO_3, KCl, can be used for TES from 200°C to 1000°C.

- Moreover, eutectic PCMs can be developed to deliberately achieve certain phase change temperature. A eutectic PCM is a composition of two or more pure materials, which melts and freezes congruently at a certain temperature like a pure PCM.

Various techniques have been developed to improve the performance of PCMs for TES [14]:

- Porous structure materials with high thermal conductivity, such as expanded graphite foam and metal foam, are used as promoter to enhance the thermal conductivity of PCMs.
- Nucleation agent, e.g., borax, is used to trigger heterogeneous nucleation to eliminate or to reduce supercooling degree of some PCMs, typically like salt hydrate. Moreover, mechanical impulsion, cold finger technology, and supersonic treatment have also been investigated to reduce the supercooling degree.
- Gelling or thickening agents can be used to prevent the phase separation of nonpure PCMs.

1.2.3.3 Sorption heat storage materials

Reactions between solid and gas (adsorption) and liquid and gas (absorption) can be used for TES, typically for temperatures lower than 200°C. Sorption heat storage has attracted intensive research interests in recent years. The adsorption/absorption process is an exothermic process while the desorption process is an endothermic process. The sorption heat involved in this reversible process is generally larger than sensible and latent heats. This method is also featured as minimum energy loss during the storage period since the thermal energy is stored not dependent on temperature but on the chemical adsorption/absorption potential. Therefore sorption heat storage has been recognized as the most promising technique for long-term TES [15]. There are different types of sorption reactions that have been investigated for TES and are listed as follows [16]:

- Physical adsorption. Porous structure materials such as zeolite, silica gel, activated carbon, and activated alumina have the ability to adsorb vapors like water. The adsorbate is bonded to the solid surface by van der Waals force. An open system using water moisture in the air as the adsorptive gas is one of the popular researches for seasonal solar TES.
- Chemical adsorption. This type of adsorption reaction involves the formation of new chemical substance. Typical chemical adsorption working pairs include $CaCl_2$, $MgSO_4$, $SrBr_2$, and so on as adsorbent and water as adsorptive gas, and $CaCl_2$, $SrCl_2$, $MnCl_2$, and so on as adsorbent and ammonia as adsorptive gas. For water-based chemical adsorption, composite materials of "salt in porous matrix" are investigated to solve the problems of salt deliquescence, swelling, and agglomeration; for ammonia-based adsorbent,

expanded graphite is widely used as supporting matrix to eliminate the swelling and agglomeration problems.
- Physical absorption. Some salty solutions are capable of absorbing/desorbing water, like aqueous solutions of LiCl, LiBr, NaOH, and so on. This reversible process can also be used for TES.

1.2.3.4 Chemical reaction materials (without sorption)

Different from chemical adsorption, the chemical reaction without sorption used for TES is a pure chemical reaction between solids/liquids and gases. The application temperature is normally ranged from 300°C to 1000°C. The materials have been classified according to their reaction families. Such as metallic hydrides (e.g., MgH_2), carbonates (e.g., $CaCO_3$), hydroxides (e.g., $Mg(OH)_2$), REDOX (e.g., BaO_2), and ammonium (e.g., NH_3), which can be decomposed by heating; the other type of material is organics, such as CH_4, which can have reaction with CO_2 or H_2O with heating.

1.2.4 Chemical energy storage materials

There is crossover between the TES and chemical energy storage. Abovementioned chemical adsorption/absorption materials and chemical reaction materials without sorption can also be regarded as chemical energy storage materials. Moreover, pure or mixed gas fuels are commonly used as energy storage materials, which are considered as chemical energy storage materials. The key factors for such kinds of chemical energy storage materials are as follows:

- High calorific value;
- Large density;
- Easy to store and transport;
- Compatible to the existing infrastructure;
- Easy to produce and high round-trip efficiency;
- Environment friendly.

Different chemical energy storage materials are listed as follows.

- Hydrogen. Hydrogen is the most important alternative fuel to fossil fuels because it is clean and affordable. Hydrogen can be produced by electrolyzing water. The production process is normally called power to hydrogen. The produced hydrogen can be stored in the forms of high pressure gas, adsorbed gas by solids of large surface area, metal hydrides, alanates, and other light hydrides [17].
- Methane. Power can be converted to methane through the reaction between hydrogen and CO_2. The storage of methane can use existing infrastructure; the volumetric energy storage density of methane is nearly four times as large as that of hydrogen [18]; the power to methane process is also accompanied with reduction of CO_2 emission.

- Ammonia. Ammonia has been recently evoked as an alternative fuel source as well as chemical energy storage material. Ammonia has been massively produced in agriculture sector; the conventional manufacturing process releases large quantities of CO_2. However, it can also be produced through renewable ways, like using hydrogen produced by water electrolysis and nitrogen from air. Ammonia can be converted back to power through fuel cell or combustion-based technology [19].

1.3 Introduction to energy storage technologies

The many forms of energy have resulted in a wide range of technologies that seek to store and convert energy, some of which are commercially mature and others that are currently under development. A graphical summary of mature and developing technologies is provided in Fig. 12, identifying nominal discharge times and operating scales for flywheels, various battery and supercapacitor technologies, and large-scale technologies including thermal, mechanical, and chemical storage concepts based on information presented in this book. This section provides an introductory summary of the various technologies; detailed descriptions are provided in the remaining chapters of the book.

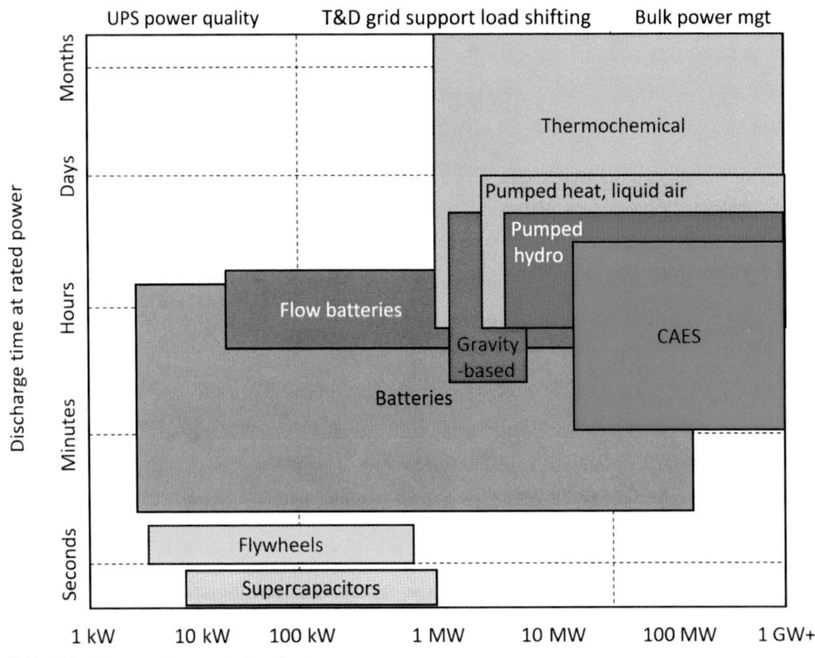

FIG. 12 Energy storage technologies.

The development and cumulative power generation capacity of various energy storage technologies across the world for the past several decades are illustrated in Fig. 13. This figure illustrates that pumped hydro comprises over 96% of global capacity, followed by thermal storage (primarily hot oil and molten salt) and electromechanical storage (primarily compressed air energy

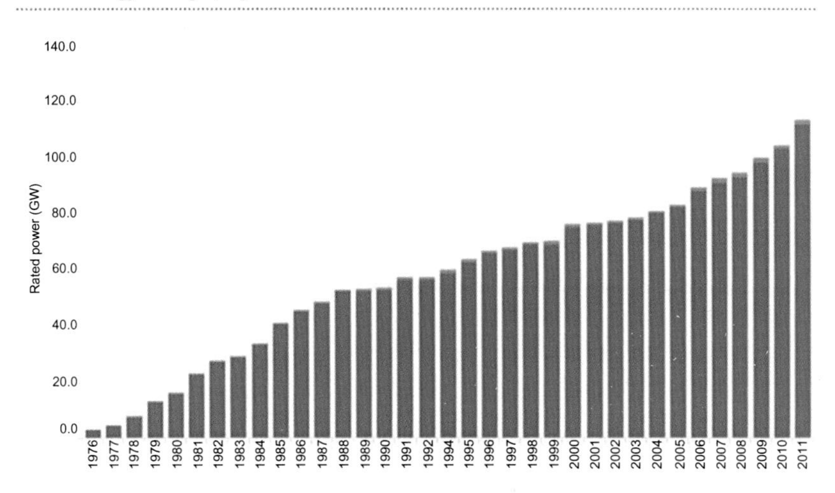

Global energy storage project installations—excluding PHS

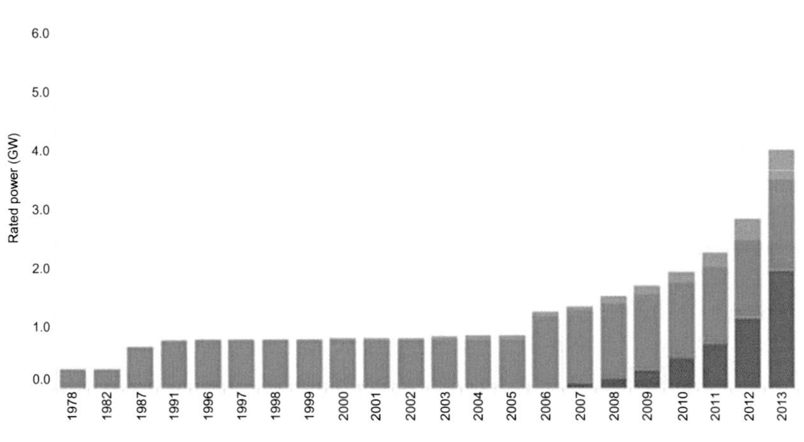

Electrochemical storage
Electromechemical storage
Hydrogen storage
Thermal storage
Pumped hydro storage

FIG. 13 Global growth of energy storage projects including (top) and excluding (bottom) pumped hydro [20].

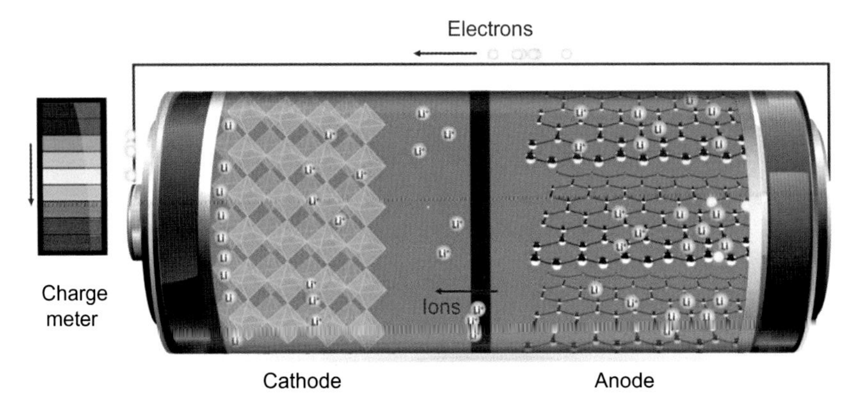

FIG. 14 Solid-state batteries generate charge by ion movement between electrodes [21].

storage and flywheels). Electrochemical battery storage systems have seen recent growth through 2013 and even more rapid growth in years since due to significant price declines.

Battery technologies store energy chemically and charge/discharge electricity via ion movement between electrodes as illustrated in Fig. 14. Although historically limited to small-scale applications, batteries have decreased dramatically in price in recent years and are considered for many large-scale applications, including 100+ MW applications for 1–4 h of storage. Lithium-Ion batteries are the current market leader with 80% market share [22], although many other technologies exist, including lead acid and sodium sulfur batteries that also target grid-scale applications. Batteries are advantageous in that they have high round-trip efficiencies of approximately 81%–87% and relatively low cost for high-power short-duration applications.

Despite these advantages, batteries suffer from a number of drawbacks that currently limit their widespread application to grid-scale energy storage. Most importantly, although battery costs have dropped significantly on a cost per kW basis, most applications have a short duration (the median is only 1.7 h). Battery technology costs will (approximately) scale proportionally with duration (duration is increased by adding parallel cells), so batteries are still prohibitively expensive for long-duration applications greater than about 2–6 h (depending on many factors). Lithium-Ion batteries require rare earth metals including lithium, cobalt, and others; there is significant disagreement in the literature about whether global reserves are adequate and/or can be sustainably accessed to scale battery production up to the necessary scales for supporting high renewable penetrations for decades and centuries to come [23]. Geopolitics are also a complicating factor for many countries, as these materials are sourced from relatively few locations worldwide. Even after raw materials have been sourced, processed, and manufactured into batteries, battery degradation reduces battery capacity and efficiency below initially specified values. The life of most utility-scale battery banks is limited to 10 years, with major maintenance required after 5–8 years [24]. There is also no methodology for recycling of lithium-ion battery materials with sufficient purity for reuse in batteries [25]. Finally, commercial battery systems are susceptible to thermal runaway and fire, requiring thermal management to avoid abuse/environmental hazards. Catastrophic fires have occurred at commercial utility-scale battery installations in the United States, Europe, Korea, Australia, and other countries [26], highlighting the need for improvements in battery system safety.

Flow battery technology is designed to scale better for long durations by storing chemical energy in electrolyte tanks rather than the electrodes. This architecture decouples power (controlled by electrode stack and electrolyte flow rate) from stored energy (electrolyte volume) and is a technology contender for grid-scale storage. Multiple chemistries exist, including vanadium redox, zinc bromine, and polysulfide bromine, and others. However, flow battery technology requires significant advances in order to meet the requirements of grid-scale storage. Existing flow batteries have low power density, requiring large volumes of costly electrolyte solutions. Additionally, the electrolyte solutions are generally corrosive and/or have poor chemical stability that results in precipitation and fouling [27, 28].

Large-scale thermal storage of energy for the grid has been pioneered in the 1980s by the concentrating solar power industry, initially using thermal oils and progressing to molten salts for systems with higher temperatures and efficiencies. Thermal storage is generally categorized into sensible and latent (phase change) heat storage, and is most commonly applied (for power generation) at high temperatures although low-temperature (ice or cold water) storage is also used for air conditioning or other cooling applications. Thermal storage typically relies on thermodynamic heat engine cycles for power generation, and heat addition may be obtained directly from existing heat sources such as solar or waste heat, or from electricity via resistive heating or other thermodynamic cycles (heat pumps or heat streams in other processes).

The earliest grid-scale energy storage technology is pumped hydroelectric storage, introduced to the grid in the 1930s. Significant capacity growth has continued since, and pumped hydro is still the dominant technology in energy storage on a capacity basis. For pumped hydro systems, electrical energy is converted to potential energy by pumping water from low to high elevation (Fig. 15), where it can be stored for long durations. The system is discharged by using the high-pressure water to drive a turbine and produce electrical power. Pumped hydro is cost advantageous over batteries for multihour storage durations, but has a high capital cost and can only be applied where suitable geography and permitting opportunities exist.

Flywheels also utilize potential energy and are described in Chapter 4, but store the energy in a high-speed rotating mass instead of changing the elevation of large volumes of water. The rotor typically operates in a vacuum environment to minimize parasitic drag losses. They offer high round-trip efficiencies

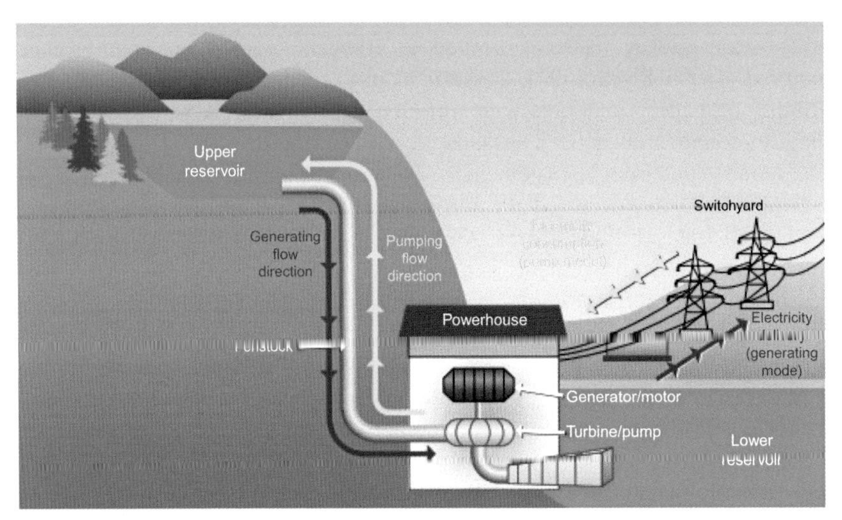

FIG. 15 Pumped hydro stores potential energy in water at different elevations.

of 90%–95% but have a high self-discharge rate and very short response time, typically limiting their application to power quality/frequency regulation use. Other potential energy storage systems under development include towers or elevated rail systems for large-scale energy storage using low-cost materials, e.g., masses of rock or concrete.

Hydrogen technologies are detailed in Chapter 5 and include a wide range of generation, storage, transmission, and electrical conversion systems. Hydrogen is an attractive storage medium due to its zero-carbon formulation and long-term stability enabling seasonal storage. Most existing hydrogen is formed by steam reforming using coal or natural gas, although electrolysis of water via renewable or nuclear power is being developed for a carbon-free solution. Hydrogen is already stored in large volumes in underground salt caverns, but poses challenges in compression and transportation due to its low mole weight (requiring significant compression power) and lower heating value than methane. Various hydrogen carriers are considered (ammonia, metal hydrides, sorbents, formic acid, methane, etc.). Power conversion with hydrogen and hydrogen products can be accomplished via combustion in a gas turbine or other process or electrochemically via fuel cells.

A variety of existing and developing heat engine-based storage systems exist that adapt existing industrial or power generation processes and machinery to energy storage. The oldest of these is a compressed air energy storage (CAES) system (Fig. 16, modified from [29]) that is charged by compressing air into underground solution-mined salt dome caverns. To discharge, the compressed air is released from the cavern through a turbo-generator. Existing systems increase power output by firing the air with natural gas (diabatic CAES). Newer precommercial concepts seek to improve round-trip efficiency and achieve zero-carbon operation by storing the heat of compression to preheat

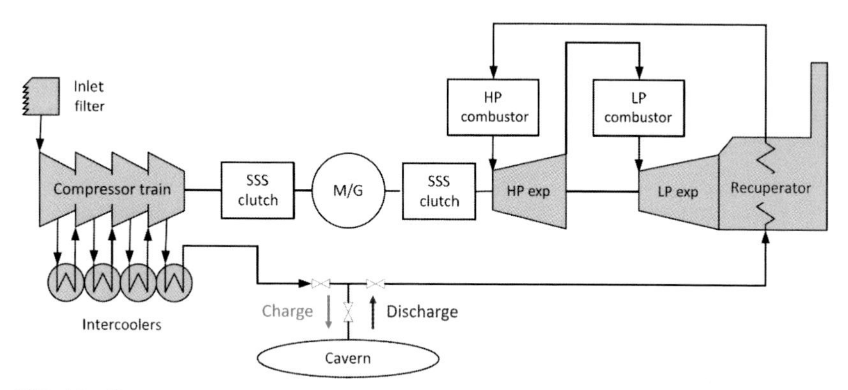

FIG. 16 Concept illustration of CAES system. *(Modified from Kerth, J. (2019). "Thermomechanical Energy Storage," Thermal-Mechanical-Chemical Electricity Storage Workshop and Roadmapping Session, San Antonio, TX.)*

expansion air during discharge mode and, for a zero-carbon solution, eliminating the combustors (adiabatic CAES). Due to cavern use, both diabatic and adiabatic CAES are inherently limited to areas with suitable geology. Liquid Air Energy Storage (LAES) is a noteworthy variation on CAES in that the air is liquefied for storage and heated (similar to CAES, diabatic and adiabatic variations exist) and expanded for discharge. Liquid air can be stored at relatively low pressure in commercial storage tanks, thus eliminating the geographic dependence of CAES. Pumped heat energy storage (PHES) systems store energy in hot (and possibly cold) thermal stores, which are charged by running machinery in a heat pump configuration and discharged by running a heat engine cycle [30]. Fig. 17 conceptually illustrates one implementation of this concept. PHES systems operating closed cycles decouple the machinery working fluid (typically pressurized air, argon, or carbon dioxide) from the thermal

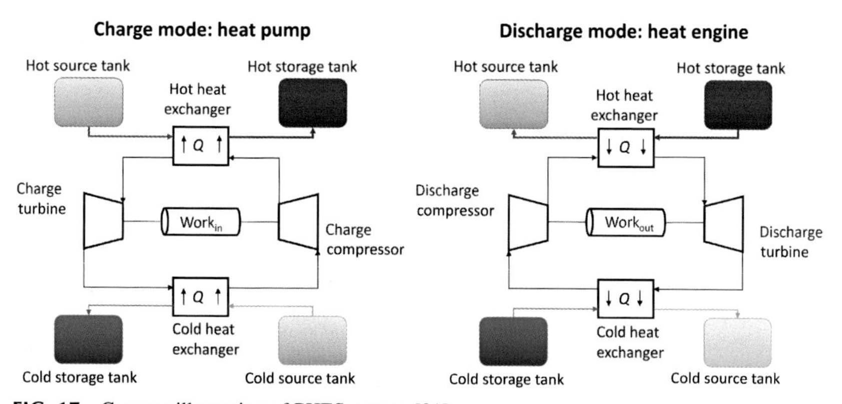

FIG. 17 Concept illustration of PHES system [31].

energy storage fluids (refrigerants, water, thermal oils, and molten salts) to minimize the cost of thermal fluid storage vessels.

The remainder of this book focuses on detailed descriptions of the large variety of thermal, mechanical, and chemical energy storage systems that also decouple generation capacity from storage capacity and have the potential for competitive economics and performance for grid-scale energy storage. These technologies are categorized broadly into thermal technologies (Chapter 2); mechanical technologies including pumped hydro, flywheels, and other gravitational storage concepts (Chapter 4); and hydrogen-derived thermochemical technologies (Chapter 5). The heat engine-based systems that incorporate thermal storage with thermodynamic cycles for power/heat generation are covered in Chapters 3 and 6, including compressed air energy storage, liquid air energy storage, and pumped heat energy storage. Chapters 7–9 focus on energy storage services, applications, and commercialization, and advanced storage concepts beyond the current state of the art are addressed in Chapter 10.

References

[1] "BP Statistical Review of World Energy," 68th ed., 2019.

[2] "Electricity Information: Overview," International Energy Agency, 2019.

[3] "Total System Electric Generation," California Energy Commission, 2019. https://ww2.energy.ca.gov/almanac/electricity_data/total_system_power.html [Accessed 18 October 2019].

[4] "Renewables 2019 Global Status Report," REN21 2019. https://www.ren21.net/wp-content/uploads/2019/05/gsr_2019_full_report_en.pdf [Accessed 4 November 2019].

[5] "100% Commitments in Cities Counties & States," Sierra Club, 2019. https://www.sierraclub.org/ready-for-100/commitments [Accessed 10 December 2019].

[6] R. Byrne, Energy Storage Overview, Sandia National Laboratories, 2016.

[7] Wan, Y.H., "Long-Term Wind Power Variability," NREL Technical Report TP-5500-53637, January 2012.

[8] Lew, D., "Impact of High Solar Penetration in the Wester Interconnection," Technical Report TP-5500-49667, NREL, 2010.

[9] C. Loutan, Briefing on Renewables and Recent Grid Operations, CAISO, 2018.

[10] "Managing oversupply", CAISO, 2019. http://www.caiso.com/informed/Pages/ManagingOversupply.aspx (Accessed 5 December 2019).

[11] M. Gonzalez-Salazar, T. Kirsten, Review of the operational flexibility and emissions of gas- and coal-fired power plants in a future with growing renewables, Renewable and Sustainable Energy Reviews 5 (278) (July 2017).

[12] P. Largue, $662 billion needed for energy storage market, Smart Energy International, August 2019. https://www.smart-energy.com/industry-sectors/storage/report-662-billion-needed-for-energy-storage-market/ (Accessed 29 October 2019).

[13] A.A. Solomon, M. Child, U. Caldera, C. Breyer, How much energy storage is needed to incorporate very large intermittent renewables? Energy Procedia 135 (2017) 283–293.

[14] P. Zhang, X. Xiao, Z.W. Ma, A review of the composite phase change materials: fabrication, characterization, mathematical modelling and application to performance enhancement, Appl. Energy 165 (2016) 472–510.

[15] N. Yu, R.Z. Wang, L.W. Wang, Sorption thermal energy storage for solar energy, Prog. Energy Combust. Sci. 39 (2013) 489–514.
[16] K. Edem N'Tsoukpoe, H. Liu, N. Le Pierres, L. Luo, A review on long-term sorption solar energy storage, Renewable and Sustainable Energy Review 13 (2009) 2385–2396.
[17] L. Schlapbach, A. Zuttel, Hydrogen-storage materials for mobile applications, Nature 414 (2002) 353–358.
[18] H. Blanco, W. Nijs, J. Ruf, A. Faaij, Potential of power-to-methane in the EU energy transition to a low carbon system using cost optimisation, Appl. Energy 232 (2018) 323–340.
[19] A. Valera-Medina, H. Xiao, M. Owen-Jones, W.I.F. David, P.J. Bowen, Ammonia for power, Prog. Energy Combust. Sci. 69 (2018) 63–102.
[20] "European Energy Storage Technology Development Roadmap, 2017 Update," EASE/EERA, 2017.
[21] Harman, S., Joyner, C., "How Lithium-ion batteries work," U.S. DOE, https://www.energy.gov/eere/articles/how-does-lithium-ion-battery-work (Accessed 4 January 2019).
[22] "U.S. Battery Storage Market Trends," U.S. Energy Information Administration, May 2019. https://www.eia.gov/analysis/studies/electricity/batterystorage/pdf/battery_storage.pdf [Accessed 15 December 2019].
[23] E.A. Olivetti, G. Ceder, G.C. Gaustad, X. Fu, Lithium-ion battery supply chain considerations: analysis of potential bottlenecks in critical metals, Joule 1 (2017) 229–243.
[24] Mongird, K., Viswanathan, V., Balducci, P., Alam, J., Fotedar, V., Koritarov, V., and Hadjerioua, B. "Energy Storage Technology and Cost Characterization Report," PNNL-28866, U.S. DOE, July 2019.
[25] "Is There Enough Lithium to Feed the Need for Batteries?" Green Journal, February 2018, https://www.greenjournal.co.uk/2018/02/is-there-enough-lithium-to-feed-the-need-for-batteries/ (Accessed 15 December 2019).
[26] Hering, G., "Burning Concern: Energy storage industry battles battery fires," S&P Global Market Intelligence, May 2019 https://www.spglobal.com/marketintelligence/en/news-insights/latest-news-headlines/51900636 (Accessed 15 December 2019).
[27] X. Yuan, C. Song, A. Platt, N. Zhao, H. Wang, H. Li, K. Fatih, D. Jang, A review of all-vanadium redox flow battery durability: degradation mechanisms and mitigation strategies, International Journal of Energy Research 43 (2019) 6599–6638. Wiley.
[28] X. Luo, J. Wang, M. Dooner, J. Clarke, Overview of current development in electrical energy storage technologies and the application potential in power system operation, J. Applied Energy 137 (2015) 511–536.
[29] J. Kerth, Thermomechanical Energy Storage, in: Thermal-Mechanical-Chemical Electricity Storage Workshop and Roadmapping Session, San Antonio, TX, 2019.
[30] R.B. Laughlin, Pumped thermal grid storage with heat exchange, J. Renew. Sust. Energy 9 (2017) 044103.
[31] B. Tom, Small-scale PHES demonstration, ARPA-E DAYS Program Kickoff, New Orleans, LA, 2019.

Chapter 2

Mass grid storage with reversible Brayton engines

R.B. Laughlin
Department of Physics, Stanford University, Stanford, CA, United States

Chapter outline

2.1 Introduction

The purpose of this chapter is to discuss the thermomechanical grid energy storage technology illustrated in Fig. 1. It is fundamentally a Carnot heat pump: a working fluid, such as air or argon, is alternately compressed and expanded by turbomachinery, just as would occur in a traditional gas turbine, except with heat transferred in or out through counterflow heat exchange rather than through combustion and exhaustion. In the limit of large heat exchanger size and perfect adiabatic compression/expansion, no entropy is created, so the machine is exactly reversible. In charge mode (shown in Fig. 1), energy taken from the grid is converted to heat, mixed with additional heat pumped uphill from a

Thermal, Mechanical, and Hybrid Chemical Energy Storage Systems
https://doi.org/10.1016/B978-0-12-819892-6.00002-2

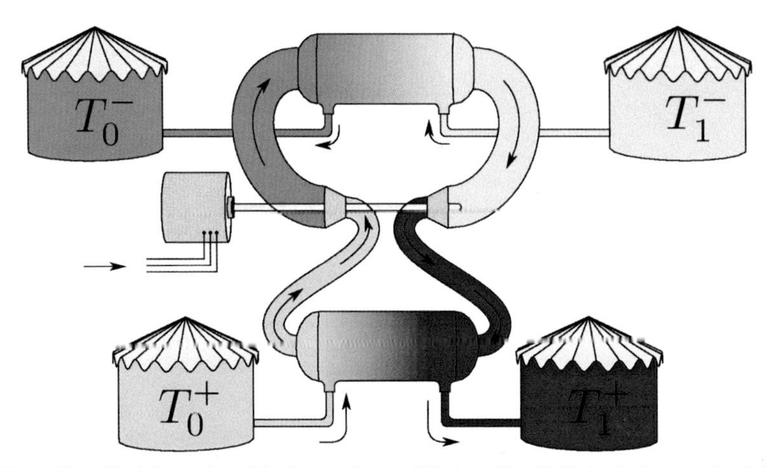

FIG. 1 Simplified illustration of the Brayton battery [1]. A working fluid circulating in a closed-loop Brayton engine exchanges heat by counterflow with two thermal storage fluids, one each for the high-pressure and low-pressure sides. The system is switched from charge mode (shown) to generation mode by reversing all mechanical motions, including the axle rotation sense and the fluid flow direction.

low-temperature reservoir, and deposited into a high-temperature reservoir. The amount of energy stored thus exceeds the amount taken from the grid. In generation mode, achieved by reversing the sense of all the mechanical motions, work is extracted from this stored heat, just as happens in a conventional Carnot cycle, and the remaining heat is allowed to fall back downhill into the low-temperature reservoir. The ideal round-trip storage efficiency is 1.

The importance of this technology lies not in its physical distinction from electrochemical batteries but in its potentially superior cost/efficiency calculus at storage times of 4 h or more. The round-trip efficiency of any real technology is of course not 1. Operation of the machinery creates entropy, and this entropy must be sloughed off into the environment as waste heat. The issue is rather the amount of investment required to reduce the entropy production to a certain level. The thermodynamics of the machine illustrated in Fig. 1 is not different from that of any battery. It is aptly called a "Brayton battery."

The technology nonetheless has some immediate advantages over electrochemistry beyond lowered cost. As shown in Fig. 1, it connects directly to the grid with no semiconductor interface. The grid is fundamentally mechanical in nature and thus works more simply and effectively with real mechanical generators and motors than it does with electronically simulated ones. The machine does not "lose" any energy but simply degrades some of it to waste heat that can be repurposed. The technology is also environmentally benign. There is no explosion danger, despite the enormous amounts of energy stored, and the storage media are not poisonous. A salt tank breach would result in energy dissipating harmlessly as heat and the nearby ground being covered with a layer of fertilizer, which could later be reclaimed. The cycle temperatures allow for cheap dry cooling.

2.2 The grid storage problem

The grid storage problem is succinctly explained with three graphs.

2.2.1 World energy budget

Fig. 2 shows the recent world energy budget history. It may be seen that (1) the demand for energy is insatiable and ever-increasing, (2) the heavy lifting is done by fossil fuels, and (3) the renewable energy remains so far largely irrelevant.

2.2.2 Renewable load leveling

Fig. 3 shows the electricity demand curve for the California Independent System Operator (ISO) for a typical day in spring. The large solar energy

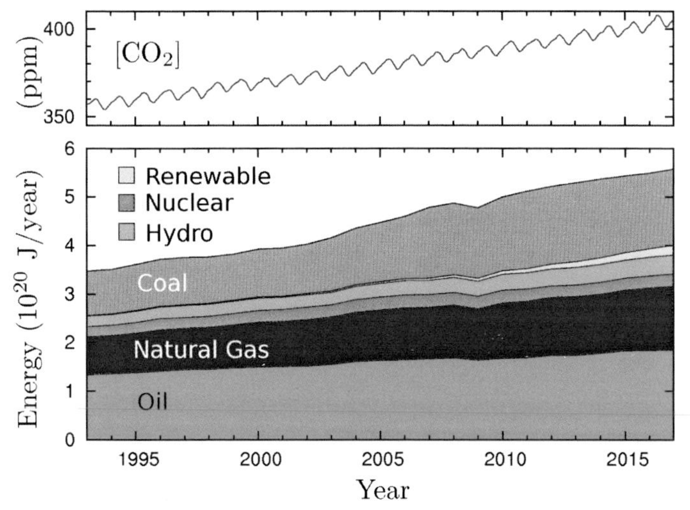

FIG. 2 World energy consumption from the BP Statistical Review of World Energy, converted with 1 TOE = 4.2×10^{10} J [2]. Total 2017 renewable contribution (*yellow*) shown in this graph is 2.05×10^{19} J year^{-1} out of 5.67×10^{20} J year^{-1}, or 3.6%. It breaks down (in multiples of 10^{19} J year^{-1}) to biofuel (mostly corn ethanol) 0.353, wind 1.064, solar 0.419, geothermal biomass, and other 0.211. The wind and solar contributions include a 38% thermal efficiency inflator, so the energy actually delivered to the grid from the sun and wind in 2017 was even smaller: $0.38 \times (1.064 + 0.419) \times 10^{19}$ J year^{-1} = 0.564×10^{19} J year^{-1} = 1565 TWH year^{-1} [2]. The *inset* at the top is the keeling curve, the carbon dioxide concentration measured at the top of Mauna Loa, Hawaii, in units of ppm (moles per million moles) of dry air [3–6]. *(Courtesy of NOAA.)* The 50 ppm growth over the time period shown corresponds to a CO_2 mass of $5.0 \times 10^{-5} \times (44/29) \times 4\pi r^2 p_0/g = 4.0 \times 10^{14}$ kg, where $r = 6.378 \times 10^6$ m is the radius of the earth, $p_0 = 1.01 \times 10^5$ Pa is the atmospheric pressure at sea level, and $g = 9.8$ m/s^2 is the acceleration due to gravity. This assumes the atmosphere to be thoroughly mixed. A crude estimate of the CO_2 mass generated by the coal, oil, and natural gas consumption shown above is $(44/14) \times 24$ years $\times 4.5 \times 10^{20}$ J year^{-1}/(4.2×10^7 J kg^{-1}) = 8.08×10^{14} kg. A more careful accounting for the three fuels gives 8.5×10^{14} kg. BP's own estimate is 6.8×10^{14} kg [2]. The emitted CO_2 not seen in the atmosphere is presumably absorbed into the oceans.

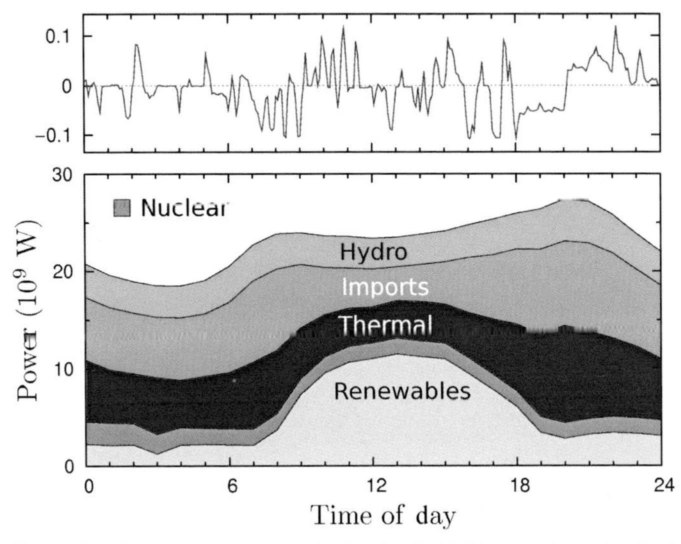

FIG. 3 Twenty-four-hour production source history for the California Independent System Operator for April 1, 2019, showing extreme variability of renewable sources, chiefly solar [7]. The load leveling from batteries (*inset top*) is too small to be visible on the lower graph. The magnitude is consistent with EIA estimates of CAISO battery capacity at 127 MW and 381 MWh in 2017 [8].

production bulge seen rising up at midday does not match the demand maximum, which occurs in early evening. It may be inferred from both this mismatch and the battery storage signal reported for the California ISO that day, also shown in Fig. 3, that (1) very long-distance transportation of large amounts of electric energy is cost impractical, (2) timing is central, and (3) mass energy storage, which could in principle shift the load, has a cost problem. These data say that there is a business model for batteries to regulate the grid phase but not for them to time shift the entire solar energy supply to satisfy peak demand.

2.2.3 Pricing

Fig. 4 shows a pricing graph for grid storage. This graph extremely crude (50% error bars), but it nonetheless reveals the core problem of conventional batteries: their "engines" (i.e., their electrodes, the places where electron motion converts to ion motion) are bundled together with their storage media, so that one cannot purchase more storage without also purchasing more engine. This causes the cost to rise out of control as the storage time increases. Pumped hydro, by contrast, separates the engine from the storage medium, with the latter having a very low cost. It, therefore, becomes superior to all conventional batteries as the storage time increases.

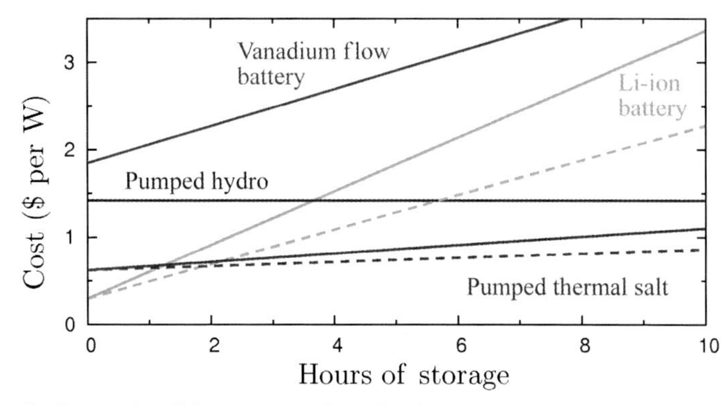

FIG. 4 Crude (error bar 50%) cost comparison of various energy storage technologies illustrating the scaling differences of engine cost per watt (intercept at 0 h of storage) and marginal cost per stored Joule (slope of line divided by 3600 s/h). The latter is irrelevant at short storage times but crucial for long ones. The present-day battery crossover time is about 4 h. The present-day Li-ion numbers come from Fu et al. [9] and are broken down to an intercept of \$0.294 W^{-1} and a slope of \$0.307 $(Wh)^{-1}$. The *dashed Li-ion line* assumes the same numbers but with the bare battery cost reduced from \$0.209 $(Wh)^{-1}$ to \$0.10 $(Wh)^{-1}$ as per optimistic predictions [10, 11]. The pumped hydro number of \$1.42 W^{-1} is a power-weighted average of existing facility construction costs taken from Key et al. [12], multiplied by 1.12 to adjust for inflation between 2010 and 2018. The present-day vanadium flow battery numbers of \$1.845 W^{-1} and \$0.212 $(Wh)^{-1}$ are from Lüth et al. [13], converted at €1 = \$1.1. The present-day pumped thermal salt line is broken down to an intercept of \$0.62 W^{-1} and a slope of \$0.048 $(Wh)^{-1}$. The intercept comes from Laughlin [1] and amounts to twice the estimated turbomachinery cost. The slope is the product of \$0.022 $(Wh\text{-}t)^{-1}$ assessed by Turchi et al. [14] for the storage cost salt towers times the thermal efficiency factor (1.6 + 0.6) implicit in Table 2 [1]. The dashed thermal salt line is an optimistic prediction obtained by lowering the \$0.22 $(Wh\text{-}t)^{-1}$ of Turchi et al. to \$12 $(Wh\text{-}t)^{-1}$, twice the cost of the nitrate salt priced at \$0.8 kg^{-1} [14]. The costs illustrated here are for off-the-shelf hardware only and are not adjusted for lifetime maintenance and operating costs (are not levelized). There is no Pb-acid battery line, even though this line properly lies slightly below the Li-ion line, because of Pb-acid lifetime issues [15]. The current Li-ion battery lifetime is not known but is estimated by Diorio et al. at 5 years beyond warranty, or 15 years [15].

2.2.4 Safety

There is also a safety issue. The solar bulge in Fig. 3 may be seen to be 7.8×10^{10} Wh × 3600 s/h × 8 h = 2.8×10^{14} J or 6.7×10^{4} tonnes of TNT. It is essential that sudden release of such amounts of energy, even by accident, be physically impossible.

2.3 Digression: Flow batteries

One electrochemical technology, the flow battery, separates the engine from the storage medium the way the Brayton battery does, and thus has the same scaling behavior. There are parallels between the engineering issues in two technologies, particularly the need to compromise among engine cost, electrical resistive loss, and viscous drag loss [19].

2.3.1 Thermodynamic reversibility

The thermodynamic nature of a battery is demonstrated by the variation of its terminal voltage with electrolyte charge state, shown for the specific case of a vanadium redox flow battery cell in Fig. 5. The slope of the curve at its midpoint is proportional to the Kelvin temperature T per the Nernst equation: $-\ \partial\phi/\partial\xi = 8\ k_B T/e = 0.206$ V. The underlying physical principle is that the vanadium ions form a dilute gas in the solvent that obeys mass action. The numbers in this case show that approximately 15% of the battery's energy is actually stored as heat sent into the electrolyte and retrieved on discharge.

Batteries have a capital cost trade-off that requires energy to be stored or retrieved at a finite rate. To linear order in this rate, entropy is created, and this entropy must then be sloughed off into the environment as waste heat. The electrical loss parameter of a battery is its internal resistance R_0, shown for the case of the vanadium flow battery in Fig. 5. The effect of this resistance on efficiency is shown in Fig. 6. The flow battery also has linear loss from viscous drag of the electrolyte passing through the electrode, which appears in Fig. 6 as a pumping loss. At the design point of 1.5×10^4 W, the pumping and electrical losses are roughly equal, indicating an engineering compromise between the two losses [19]. The total round-trip efficiency of this particular battery, including electrolyte pumping and AC/DC conversion losses, was just over 50% [16].

2.3.2 Membrane cost constraint

As shown in Fig. 6, the electrodes of the flow battery are made of graphite felt. This is necessitated by the very high cost of the ion-exchange membrane, a special material that allows H^+ ions to pass but no others. The bulk price of Nafion

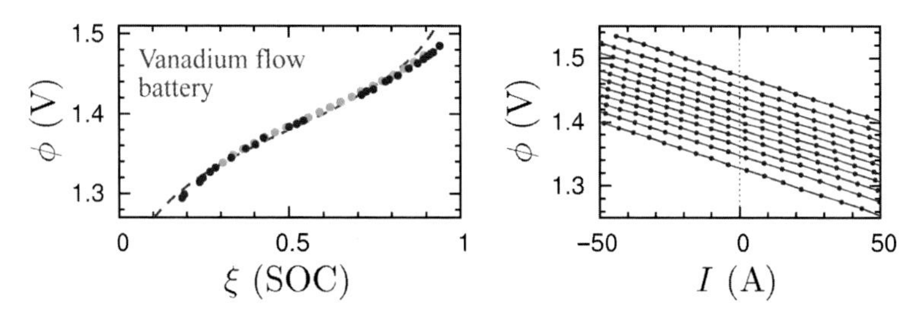

FIG. 5 *Left*: Cell voltage as a function of charge state ξ (SOC) for a vanadium flow battery reported by Bindner et al. [16]. The dashed line is a plot of the Nernst potential $\phi = \phi_0 + 2(k_B T/e)\ \ln[\xi/(1-\xi)]$, where $\phi_0 = 1.38$ V. Only the slope of this fit at the midpoint is meaningful. The offset ϕ_0 has a shift due to membrane effects and dense ion chemistry that is incompletely understood [17, 18]. *Right*: Cell current-voltage characteristics reported by Bindner et al. [16] for various values of ξ. The inferred cell internal resistance is $-\ \partial\phi/\partial I = 1.4 \times 10^{-3}\ \Omega$.

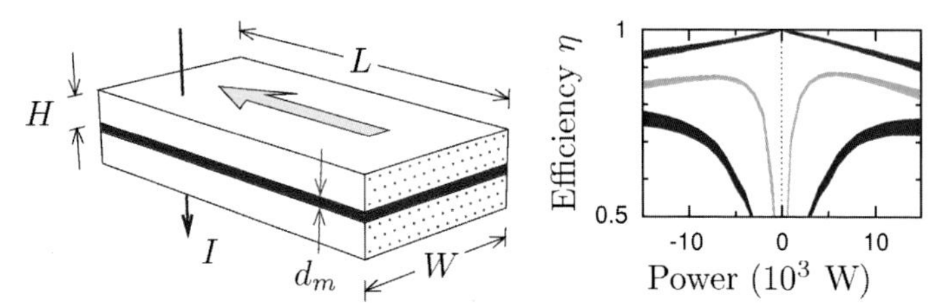

FIG. 6 *Right*: Breakdown of the efficiency losses for the vanadium flow battery reported by Bindner et al. [16] showing trade-off between electric and viscous resistance. Positive and negative power corresponds to discharge and charge, respectively. The *top (red) curve* includes only the resistive loss shown in Fig. 5. The *middle (green) curve* includes both this loss and that due to the power conversion electronics. The *bottom (blue) curve* includes all losses, including the power required for pumping the electrolyte through the graphite felt electrodes. Doubling these losses for a full storage cycle gives a total round-trip efficiency of just over 50% [16]. *Left*: Illustration of flow battery cell. An ion-exchange membrane sandwiched between carbon felt electrodes. The viscous drag loss is estimated by imagining felt fibers of radius r_0 to be oriented in the flow direction and surrounded by tubes of radius ℓ_0, as described in Table 1.

115, the typical membrane choice for this application, is cited by Bagotsky as \$700 m^{-2} [20], Viswanathan et al. as \$500 m^{-2} [21], Minke as \$500 m^{-2} [22], and Luth as \$230 m^{-2} [13]. Its current retail price is \$1300 m^{-2}. At \$500 m^{-2}, the battery described in Table 1 has a cost in membrane alone of \$0.6 W^{-1}. This unacceptably expensive, per Fig. 4 [13]. Thus, it is not economically feasible to increase the round-trip efficiency by making the electrode bigger. The only option is to make the electrode more porous, which is achieved by using felt.

2.3.3 Electrode entropy creation

When total current I travels through the cell, the rate of entropy creation (thermal loss power divided by the environmental dump temperature T_{dump}) is

$$\dot{S} = \dot{S}_{\text{visc}} + \dot{S}_{\text{ohmic}} = \frac{I^2}{T_{\text{dump}}}\left[\frac{2\mu}{\kappa_0}\left(\frac{40}{n_0 e}\right)\left(\frac{L}{WH}\right) + \frac{2}{3}\left(\sigma_H^{-1} + \sigma_f^{-1}\right)\left(\frac{H}{LW}\right)\right] \tag{1}$$

with parameters as in Table 1. Minimizing with respect to L, as shown in Fig. 7, gives the preferred aspect ratio

$$\frac{L}{H} = \left[\frac{2}{3}\left(\sigma_H^{-1} + \sigma_f^{-1}\right)\frac{k_0}{2\mu}\left(\frac{n_0 e}{40}\right)\right]^{1/2} \tag{2}$$

With this aspect ratio held at the optimal value, the total round-trip efficiency scales as $1/W$, as shown in Fig. 7.

2.3.3.1 Loss component: Viscous flow resistance

The viscous flow loss in Eq. (1) is the pdV work done in pushing the electrolyte through the graphite felt. The average flow velocity is related to the pressure

TABLE 1 Estimated vanadium flow battery parameters used to cross-check Figs. 5 and 6.

Quantity	Symbol	Value	
Graphite density	ρ_g	2200	kg m^{-1}
Felt density	ρ_f	100.0	kg m^{-3}
Felt conductivity	σ_f	285	Ω^{-1} m^{-1}
Fiber radius	r_0	5.0×10^{-6}	m
Electrolyte viscosity	μ	1.2×10^{-3}	Pa s
Electrolyte conductivity (H^+)	σ_{H^+}	28.0	Ω^{-1} m^{-1}
Vanadium ion density	n_0	6.02×10^{26}	m^{-3}
Vanadium ion mobility	μ_i	8.0×10^{-8}	m^2 V^{-1} s^{-1}
Membrane thickness	d_m	1.25×10^{-4}	m
Membrane conductivity	σ_m	10	Ω^{-1} m^{-1}
Felt electrode length	L	0.5	m
Felt electrode width	W	0.3	m
Felt electrode height	H	6.0×10^{-3}	m
Cylinder radius $\ell_0 = (\rho_g/\rho_f)^{1/2} r_0$		2.3×10^{-5}	m
Ideal permeability $k_0 = 0.12\ell_0^2$		6.35×10^{-11}	m^2
Cell electrical resistance	R_0	1.04×10^{-3}	Ω
Cell viscous resistance	R_v	1.8×10^{-3}	Ω

Notes: Bindner et al. [16] report three stacks of 40 cells apiece with a total membrane area of 18 m^2. The aspect ratio appears to be 3 × 5, but is not specified [16]. The membrane is assumed to be Nafion 115, which comes in standard widths that are multiples of 0.3 m [24]. The graphite felt is assumed to be Sigratherm GFA 5, which comes with standard 6 mm thickness [25]. A vanadium ion concentration of 1 mol L^{-1} is inferred from the stated tank volume of 6.5 m^3 and energy storage capacity of 120 kWh [16]. Industry standard electrolyte initialization is 0.9–1.6 mol L^{-1} $VOSO_4$ and 4 mol L^{-1} H_2SO_4 [26–28]. The H^+ conductivity measured under industry standard conditions [29] is suppressed because charging reduces the pH. The theoretical values of R_0 and R_v agree roughly with the experimental values seen in Fig. 6.

gradient by Darcy's law $\overline{v} = -(k/\mu)(\partial p/\partial x)$, where μ is the electrolyte viscosity and k is the hydraulic permeability. Values of the latter reported in the literature vary in the range 10^{-10} $m^2 < k < 10^{-9}$ m^2 [19, 23]. This is roughly consistent with the estimate of $k_0 = 6.3 \times 10^{-11}$ m^2 in Table 1, obtained by approximating the felt as parallel tubes as shown in Fig. 6 and solving the Navier-Stokes equation. The total current I supplied by a cell (or stack) in the steady state is $I = -n_0 e\overline{v}WH/40$, where n_0 is the vanadium ion density in the electrolyte, e

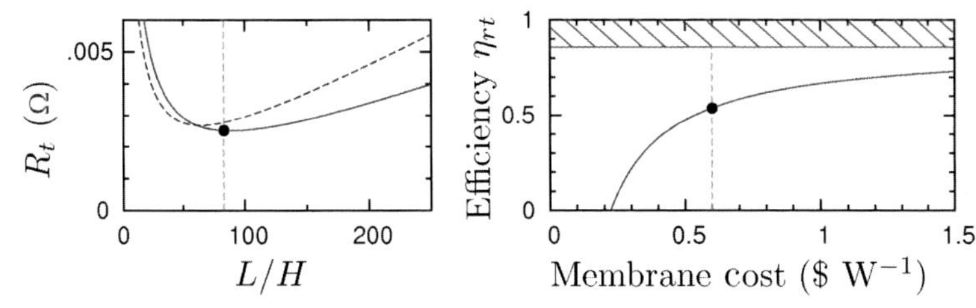

FIG. 7 Illustration of flow battery optimization [16]. *Left*: The *dashed line* is the sum of the theoretical losses $R_t = R_0 + R_v$ described in Eqs. (3), (4) as a function of the control parameter L/H, computed with the parameters of Table 1. The *solid curve* is the same quantity computed with the experimental values in Fig. 6. The *dot* shows the design point. *Right*: Round-trip efficiency η_{rt} (the square of the efficiency η in Fig. 6) computed with the optimal value of L/H versus membrane area LW, converted to a cost at \$500 m^{-2}, showing the trade-off between efficiency and cost per engine watt. The current is assumed to be held fixed. The hatched asymptote is the loss from the conversion electronics, taken from Fig. 6.

is the electron charge, and W, H, and L are the cell dimensions defined in Fig. 6. The factor of 40 accounts for passing the same electrolyte flow through all 40 cells in a stack in series, so as to minimize the loss due to the variation of ϕ with ξ seen in Fig. 5. The total power dissipated in each cell by viscous drag is thus $2\overline{v}(\partial p/\partial x)(LWH) = R_v I^2$, where

$$R_v = \frac{2\mu}{k_0}\left(\frac{40}{n_0 e}\right)^2 \left(\frac{L}{WH}\right) \tag{3}$$

The factor of 2 accounts for the two layers of felt. The parameters in Table 1 give $R_v = 1.8 \times 10^{-3}$ Ω. The very close agreement with the experimental value seen in Fig. 6 is fortuitous given the error bars in Table 1, but the rough agreement is meaningful. The corresponding total power loss from viscous drag is $120\, R_v I^2 = 1.73 \times 10^3$ W or 11.5% of the total power, in rough agreement with Fig. 6.

2.3.3.2 *Loss component: Electrical resistance*

The electrical resistive loss seen in Fig. 5 is almost entirely from hydrodynamic drag on the H^+ ions as they move through the electrolyte to pass from one side of the cell to the other. The theoretical value

$$R_0 = \frac{2}{3}\left(\sigma_{\text{H}^+}^{-1} + \sigma_f^{-1}\right)\left(\frac{H}{LW}\right) \tag{4}$$

obtained from the parameters in Table 1 is 1.04×10^{-3} Ω, which is 30% lower than the experimental value seen in Fig. 5. The H^+ conductivity σ_{H^+} is sensitive to the electrolyte pH, which is not specified and must be estimated from industry standards [16].

2.3.3.3 *Irrelevant losses*

The two other electrical resistance sources in the cell are negligible: the membrane resistance is $\delta R_m = d_m/(\sigma_m LW) = 5.8 \times 10^{-5}\ \Omega$, and the vanadium ion diffusive resistance is $\delta R_D = 5.12\ell_0^2/(\mu_i e n_0 HWL) = 3.9 \times 10^{-7}\ \Omega$. In the absence of the felt the latter would have been $(H/\ell_0)^2$ bigger, or $2.65 \times 10^{-2}\ \Omega$, so this loss is made negligible by the felt.

2.3.4 Battery as thermal engine

Fig. 8 shows how the flow battery may be reconfigured into a reversible thermal engine that stores energy as heat only. This arrangement is academic, in that the energy transferred is much larger than the energy stored, but the fact that the two technologies can be deformed into each other shows that they are physically equivalent. They are thus are properly compared on the basis of cost and safety only, not fundamentals.

The increase in the Nernst potential with charging seen in Fig. 5 is analogous to the increase of pressure that results from adiabatically adding particles to an ideal gas. The corresponding total internal energy per particle is $2/3(e\phi_0/k_B T) =$ 36 times larger in the electrochemical flow battery than in the Brayton battery, but this is misleading because energy in the Brayton battery is

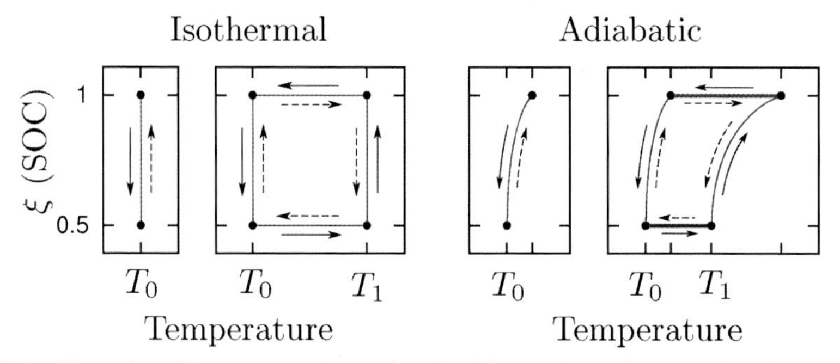

FIG. 8 Illustration of flow battery as heat engine. The *left panel* shows the conventional isothermal charge/discharge cycle. If, however, as shown in the *second panel*, one charges at a low temperature T_0 and discharges at a higher one T_1, then more energy will be extracted than is put in, due to the temperature dependence of the Nernst potential ϕ seen in Fig. 5. If one assumes (falsely) that the ionization state of the H_2SO_4 is independent of temperature and that $T_0 = 300°K$ and $T_1 = 600°K$ (under pressure), then this thermal energy "profit" is 0.07 of the total energy stored. The source of this extra energy is heat absorbed from the environment at T_1. If the operations are quasistatic so that total entropy is conserved, then the amount of heat dumped back into the environment at temperature T_0 satisfies the Carnot condition. Performing these operations in reverse order results in a loss of energy and a pumping of heat uphill from T_0 to T_1. On the right is shown the same set of operations with the apparatus disconnected thermally from the environment during charging and discharging. In this case, charging and discharging raises or lowers the system's temperature adiabatically. The heat transfers responsible for the useful energy gain or loss then occur during the heating and cooling steps. The heat capacity of the system has been made artificially small in this figure to exaggerate the temperature changes. In the case of the Fig. 5, an assumption of a $VOSO_4$ concentration of 1 mol L^{-1} gives a charging temperature rise of 2.3°K.

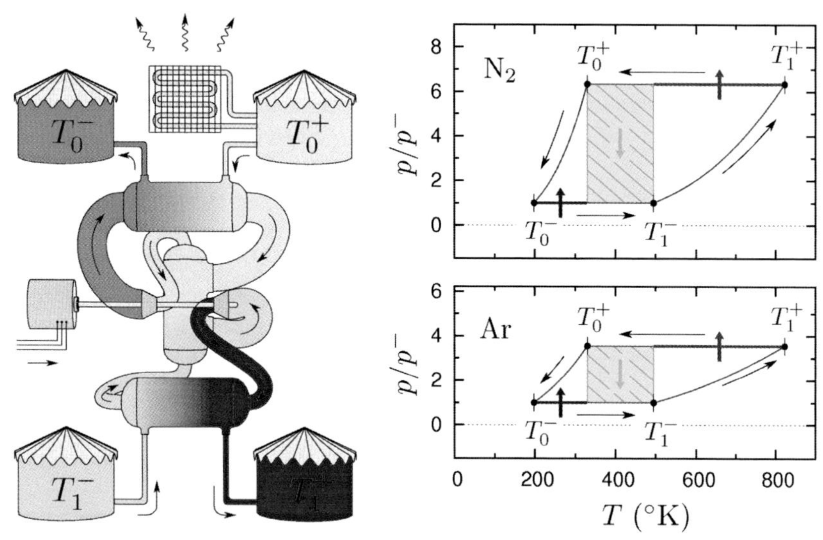

FIG. 9 *Left*: Illustration of a realistic Brayton battery. The four temperatures, listed in Table 2, are fixed by materials considerations and the cold-start requirement for initialization. This necessitates reversal of the roles of T_0^+ and T_1^- with respect to Fig. 1 and also the inclusion of a recuperator. The machine is switched from charge mode (shown) to generation mode by reversing the sense of all mechanical motions, including the working fluid flow and the axle rotation. *Right*: Pressure p versus temperature T for two potential choices of working fluid. The NIST equation of state tables is effectively polytropic over the entire range, with specific heat ratio γ of 1.38 and 1.67 for N_2 and Ar, respectively. The lower pressure p^- is arbitrary, but warrants being made as high as possible to minimize the overall cost per engine watt. The high-temperature storage fluid is molten $NaNO_3/KNO_3$ eutectic. The low-temperature storage fluid is anything that remains liquid over the relevant cryogenic temperature range.

transferred to a dense fluid for storage, whereas the energy of a flow battery remains diluted in the electrolyte. The energy storage density of the Brayton battery, reckoned with the sum of the hot and cold storage fluid volumes as per Fig. 9, is 2.04×10^8 J m^{-3}. The storage density the flow battery, reckoned with the sum of the two electrolyte volumes, is 0.67×10^8 J m^{-3}.

2.4 The Brayton battery

Fig. 9 shows a version of the Brayton battery that might actually be built [1]. The four temperatures, listed in Table 2, are constrained by (1) the liquid range

TABLE 2 Prototype Brayton battery parameters [1] used in generating Fig. 9.

T_0^-	T_0^+	T_1^-	T_1^+	T_{dump}	ξ	$\eta_c^{(p)}$	$\eta_t^{(p)}$
196°K	326°K	495°K	823°K	300°K	1.66	0.91	0.92

of the $KNO_3/NaNO_3$ eutectic storage medium, (2) the condition that T_0^+ be no lower than ambient, and (3) the working fluid equation of state. These choices require reversal of the roles of T_0^+ and T_1^- from those in Fig. 1 and the addition of a recuperator. The constraint on T_0^+ is necessary for the system to restart after an extended power failure. The salt can easily be remelted in this circumstance, but the cold side must be able to rerefrigerate itself.

The design based on these four temperatures turns out by coincidence to be nearly optimal from an engineering perspective, particularly as regards cost, materials properties, and turbomachinery performance. The one exception is the low-temperature storage fluid, which must be either a flammable hydrocarbon (gasoline) or a banned chlorofluorocarbon. This design also precludes the use of CO_2 as the working fluid because of the latter's low-temperature phase transition issues at elevated pressure, particularly snow-out at 216°K.

2.4.1 Molten nitrate salt technology

The use of molten $KNO_3/NaNO_3$ eutectic as the high-temperature storage medium is motivated by its success in concentrating solar power plants [30, 31]. The technology itself was first prototyped in the 1995 Solar-2 experiment in California, pictured in Fig. 10 [32, 33]. Andasol-1, the first commercial plant to use the technology, was commissioned in 2008 [31]. The ensuing 11 years saw the construction of 40 additional plants, listed in Table 3 [34]. Most of the plants in Table 3 were copies of the original Andasol design, but a few were of the more advanced salt tower receiver design. The SolarPaces database from which Table 3 was generated shows 13 additional molten salt plants under construction with total storage capacity of 1.93×10^4 MWh, and also 10 plants in advanced planning stages with 1.99×10^4 MW of storage, as of November 2019. The combined total of 5.86×10^4 MWh is 155 times the California 381 MWh battery capacity of Fig. 3 [8].

FIG. 10 *Left*: Andasol-I, the first commercial concentrating solar plant using molten nitrate salt storage [31]. The molten salt storage tank pair may be seen in the foreground with the parabolic trough solar field behind. *(Photo courtesy of Solar Millennium AG.)* Subsequent plants using this principle are listed in Table 3. *Right*: The original two-tank molten salt storage prototype built for the 1995 Solar Two retrofit of Solar One in Barstow, California [32, 33]. *(Photo courtesy of Sandia National Laboratories.)*

TABLE 3 List of operational concentrating solar power plants with molten nitrate salt storage.

Name	Country	Operational	Power (MW)	Storage (h)
Andasol-1	Spain	2008	50	7.5
Andasol-2	Spain	2009	50	7.5
Andasol-3	Spain	2011	50	7.5
Archimede	Italy	2010	5	8.0
Arcosol 50	Spain	2011	50	7.5
Arenales	Spain	2013	50	7.0
Ashalim	Israel	2019	121	4.5
Aste-1A	Spain	2012	50	8.0
Aste-1B	Spain	2012	50	8.0
Astexol-II	Spain	2012	50	8.0
Bokpoort	South Africa	2016	55	9.3
Casablanca	Spain	2013	50	7.5
Crecent Dunes	USA	2015	110	10.0
Delingha	China	2018	50	9.0
eLLO Solar	France	2019	9	9.0
Extresol-I	Spain	2010	50	7.5
Extresol-II	Spain	2010	50	7.5
Extresol-III	Spain	2012	50	7.5
Gemasolar	Spain	2011	20	15.0
Ilanga-I	South Africa	2018	100	5.0
Kathu	South Africa	2019	100	4.5
KaXu	South Africa	2015	100	2.5
La Africana	Spain	2012	50	7.5
La Dehesa	Spain	2011	50	7.5
La Florida	Spain	2010	50	7.5
Manchasol-I	Spain	2011	50	7.5
Manchasol-II	Spain	2011	50	7.5

Continued

TABLE 3 List of operational concentrating solar power plants with molten nitrate salt storage—cont'd

Name	Country	Operational	Power (MW)	Storage (h)
Noor-I	Morocco	2015	160	3.0
Noor-II	Morocco	2018	200	7.0
Noor-III	Morocco	2018	150	7.0
Qinghai Gonghe	China	2019	50	6
Shagaya	Kuwait	2019	50	9.0
Shouhang Dunhuang-I	China	2016	10	15.0
Shouhang Dunhuang-II	China	2018	100	11.0
Solana	USA	2013	280	6.0
Supcon Delingha-I	China	2018	10	2.0
Supcon Delingha-II	China	2018	50	7.0
Termesol 50	Spain	2011	50	7.5
Termosol-I	Spain	2013	50	9.0
Termosol-II	Spain	2013	50	9.0
Xina	South Africa	2018	100	5.0

Notes: These data agree with Pelay et al., where there is overlap [34].
Source: Reproduced from the SolarPaces database at the US National Renewable Energy Laboratory, as of November 2019.

Table 3 demonstrates that the finances of molten salt energy storage make sense, at least with an appropriate rate tariff. At an average cost of \$5 W^{-1}, the plants in Table 3 represent a total investment of \$14 billion [14]. The cost of the storage was roughly 15% of this total, or \$2.1 billion [14].

Nitrate salt is the medium of choice chiefly by virtue of being liquid over the range of interest ($230°C < T < 600°C$), having a low vapor pressure, and being cheap. It is not particularly distinguished either chemically or thermally. It is one of two classes of salt that has relatively low melting temperature, the other being hydroxides. Molten nitrates do corrode steels, but this corrosion is mild if electrochemical action is properly blocked and quite manageable over a plant lifetime [35, 36].

Molten nitrate salt has physical properties similar to those of liquid water [37–39]. It viscosity is between 1 and 5 times that of water, depending

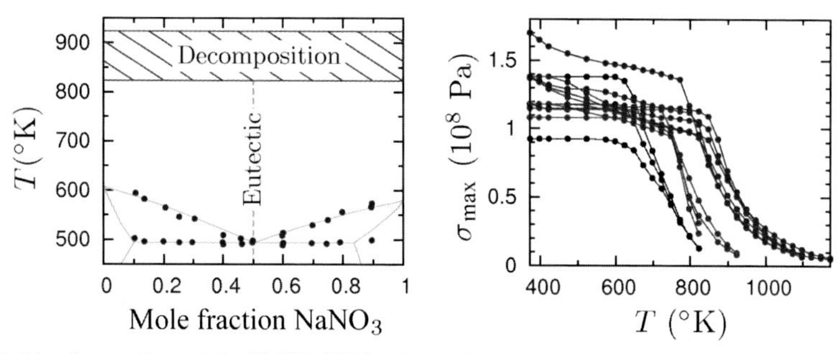

FIG. 11 Comparison of the $NaNO_3/KNO_3$ phase diagram [42–44] (*left*) with the ASME stress limit σ_{max} of steels [45] (*right*). The *black*, *red*, and *blue* data points correspond to carbon, low-alloy, and stainless steels (including inconels), respectively [1]. The decomposition boundary at 823°K (550°C) is the point, where the performance becomes difficult to predict, not where the salt begins to fail [46–48].

on the temperature. Its thermal conductivity is about the same as water's. Its heat capacity is close to the equipartition value of $3k_B$ per atom. Raising its temperature from T_1^- to T_1^+ stores energy at a density 9.15×10^8 J m^{-3}, whereas raising the temperature of water from 0°C to 100°C stores energy at 4.18×10^8 J m^{-3}.

It is also important that the liquid range of the salt correlates well with the performance window of steels. This is shown in Fig. 11. All steels develop creep strength issues at temperatures well below their melting points. For stainless steels, the "creep cliff" lies at about 873°K (600°C). This is the reason that modern supercritical steam plants work around this temperature [40]. The thermodynamic performance of the plant can be improved by raising this temperature, but the cost in extra steel required is problematic. It is possible to lower the eutectic melting point from 495°K (222°C) by about 100°K by adding ternary components, such as $CaNO_3$ or $LiNO_3$ [41], but the only advantage in doing so is to reduce the size of the recuperator. This benefit is outweighed by the high cost of the special salt. Use of these improved salts would negligibly affect the round-trip efficiency.

2.4.2 Entropy metric

The wide range of temperatures in the Brayton battery design mandates the use of the entropy creation as the efficiency metric, as opposed to energy loss. The two are equivalent in the end, for what matters is the amount of energy dumped into the environment as waste heat. However, direct accounting of Q_{dump} requires a complete set of experiments (or calculations) that close the cycle, which is awkward and time consuming. It is much simpler to add up the amounts of entropy created at each step in the cycle to get a total ΔS and then to assess the heat dump requirement by

$$Q_{\mathrm{dump}} = T_{\mathrm{dump}} \Delta S \tag{5}$$

The total entropy S of a system containing energy E at thermal equilibrium has the microscopic quantum definition

$$\begin{gathered} S = k_B \ln[\mathcal{D}(E)] \quad \mathcal{D}(E) = \sum_n \delta(E - E_n) \\ \mathcal{H}\Psi_n = E_n \Psi_n \quad \mathcal{H} = -\sum_j \frac{\hbar^2}{2m_j} \nabla_j^2 + \sum_{j<k} \frac{Z_j Z_k e^2}{|\mathbf{r}_j - \mathbf{r}_k|} \end{gathered} \tag{6}$$

in Gaussian units, where m_i and $Z_i e$ are the mass and charge of the ith particle either electron or ion and $\hbar$ is Planck's constant. The entropy is measured through its quasistatic energy derivative

$$\frac{\partial S}{\partial E} = \frac{1}{T} \tag{7}$$

The entropy constitutes the system's master equation of state.

The second law of thermodynamics dictates that $\partial S / \partial t > 0$. Since the entropy at the end of the storage cycle must equal the entropy at the beginning, all entropy created in the cycle must have been sloughed off into the environment per Eq. (5).

2.4.3 Turbomachinery entropy generation

The generation of entropy by the turbomachinery is illustrated in Fig. 12. When ν moles of fluid are compressed nonideally from pressure p^- to pressure p^+, they heat to a temperature $T_1^{++} > T_1^+$, where T_1^+ is the ideal adiabatic temperature. The temperature increase $T_1^+ \rightarrow T_1^{++}$ represents a friction loss that includes things such as turbulence, wetting drag, and blade leakage. Since the entropy creation along the adiabat is zero by definition, the entropy created by the nonideal compression is the constant-pressure integral

$$\frac{\Delta S}{\nu} = \int_{T_1^+}^{T_1^{++}} c_p \frac{dT}{T} = \left(\frac{\gamma}{\gamma - 1}\right) R \ln\left(\frac{T_1^{++}}{T_1^+}\right) = \left(\frac{\gamma}{\gamma - 1}\right) R \left[\frac{1}{\eta_c^{(p)}} - 1\right] \ln\left(\frac{T_1^+}{T_1^-}\right) \tag{8}$$

where R is the ideal gas constant. The increase $T_1^+ \rightarrow T_1^{++}$ also defines the adiabatic and polytropic efficiencies

$$\eta_c = \frac{\text{Ideal work done}}{\text{Actual work done}} = \frac{T_1^+ - T_1^-}{T_1^{++} - T_1^-} \quad \eta_c^{(p)} = \frac{\ln(T_1^+/T_1^-)}{\ln(T_1^{++}/T_1^-)} \tag{9}$$

The latter is a proxy for the former that describes theoretical infinitesimal stages that might be composed to achieve the compression. A similar analysis for the turbine yields for the total turbomachinery entropy creation rate

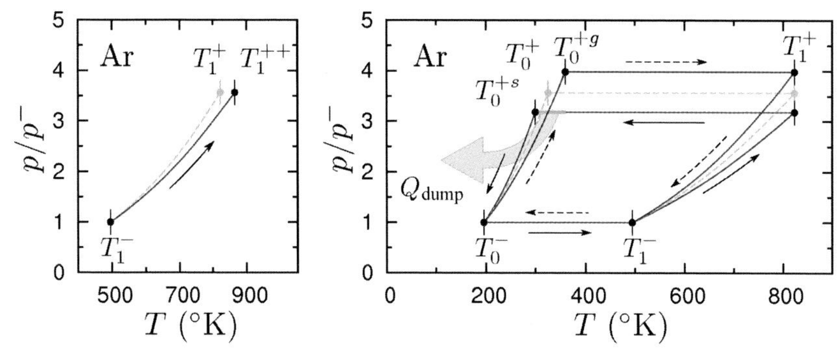

FIG. 12 *Left*: Close-up of the compression step in Fig. 9 showing the effects of nonideality. Compressing the working fluid (in this case Ar) heats it from T_1^- to a temperature T_1^{++} slightly higher than the adiabatic (entropy-conserving) value $T_1^+ = T_1^- \, (p^+/p^-)^{(\gamma-1)/\gamma}$. The shift from T_1^+ to T_1^{++} implies the entropy increase given by Eq. (8). It also defines the adiabatic and polytropic efficiencies, per Eq. (9). *Right*: Complete storage/generation cycle assuming no entropy creation in the heat exchangers. Closing the cycle requires (1) slightly modifying the compression ratios (3.18 and 3.98) for the storage and generation steps and (2) dumping heat Q_{dump} into the environment. The parameters assumed are those of Table 2. The shifted temperatures are $T_0^{+g} = 360^\circ$K and $T_0^{+s} = 300^\circ$K. The round-trip efficiency is $\eta_{\text{rt}} = [(T_1^+ - T_0^{+g}) - (T_1^- - T_0^-)]/[(T_1^+ - T_0^{+s}) - (T_1^- - T_0^-)] = 0.732$. This matches the value of 0.724 given by Eq. (12).

$$\dot{S} = \left(\frac{\gamma}{\gamma - 1}\right) R\dot{\nu} \left[\frac{1}{\eta_c^{(p)}} - \eta_t^{(p)}\right] \ln(\xi) \quad \left(\xi = \frac{T_1^+}{T_1^-} = \frac{T_0^+}{T_0^-}\right) \tag{10}$$

The corresponding rate of energy storage is

$$\dot{E} = \left(\frac{\gamma}{\gamma - 1}\right) R\dot{\nu} (T_0^- - T_1^-)[\xi - 1] \tag{11}$$

Since Eq. (5) gives $Q_{\text{dump}}/E_{\text{store}} = 2T_{\text{dump}}\dot{S}/\dot{E}$, the factor of 2 accounting loss in both charge and discharge, the fundamental limit of round-trip storage efficiency is

$$\eta_{\text{rt}} \le 1 - \frac{2T_{\text{dump}}}{T_1^- - T_0^-} \left[\frac{1}{\eta_c^{(p)}} - \eta_t^{(p)}\right] \frac{\ln(\xi)}{\xi - 1} \tag{12}$$

The full-cycle analysis of Fig. 12 is consistent with this expression.

Eqs. (10)–(12) facilitate quick assessments of potential design modifications. For example, Fig. 13 shows the effects of (1) employing cold-side intercooling, thus enabling the water-based antifreeze to be used as the cold storage fluid and (2) substituting a more sophisticated nitrate salt on the hot side. The result in both cases is a round-trip efficiency degradation 0.724 → 0.700. The first result shows that a water-based design with intercooling is a reasonable

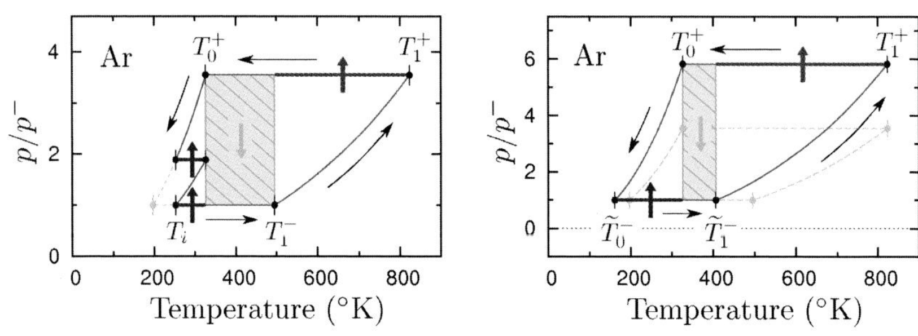

FIG. 13 Modifications of the design of Fig. 9. *Left*: The single expansion $T_0^+ \rightarrow T_0^-$ (in storage mode) is replaced with two successive expansions $T_0^+ \rightarrow T_i = (T_0^+ T_0^-)^{1/2} = 253°K$ (= −20°C), thus enabling conventional water antifreeze to be used as the cold storage fluid. Each mole of working fluid passing through the circuit generates exactly the same entropy as before, but the energy stored is smaller by the factor $[(T_1^+ - T_1^-) - 2(T_0^+ - T_i)]/[(T_1^+ - T_1^-) - (T_0^+ - T_0^-)] = 0.917$. The round-trip efficiency limit of Fig. 12 is thus degraded to $0.724 \rightarrow 0.700$. *Right*: Replacement of the simple $NaNO_2/KNO_2$ thermal storage fluid with a $NaNO_3/KNO_3/Ca(NO_3)_2$ ternary eutectic with a slightly lowered melting point [41]. Substituting the revised values of $\tilde{T}_1^- = 406°K$, $\tilde{\xi} = 2.027$, and $\tilde{T}_0^- = 161°K$ into Eq. (12) gives the round-trip efficiency degradation $0.724 \rightarrow 0.699$.

engineering alternative for suppressing fire danger. The second shows that substituting a special (more expensive) salt is counterproductive.

2.4.4 Heat exchanger entropy generation

The generation of entropy by the heat exchangers is design dependent, but the important scaling principles are not and may be prototyped with any design. For a tube-and-shell design with the parameters of Fig. 14, the entropy generation rate is

$$\begin{aligned}\dot{S} &= \dot{S}_{\text{visc}} + \dot{S}_{\Delta T} + \dot{S}_{\text{leak}} \\ &= \left(\frac{8\mu}{\pi a^4}\right)\frac{L}{N}\left(\frac{\dot{\nu}R}{p}\right)^2\left(\frac{T_B + T_A}{2}\right) + \frac{11}{48\pi\kappa}\left[\left(\frac{\gamma}{\gamma-1}\right)\dot{\nu}R\right]^2\frac{1}{NL}\frac{(T_B - T_A)^2}{T_B + T_A} \\ &\quad + \pi a^2\left(\frac{N}{L}\right)\left(\frac{2\Delta a}{a}\kappa_{\text{steel}}\right)\frac{(T_B - T_A)^2}{T_B T_A}\end{aligned} \tag{13}$$

where $N = (2\pi/\sqrt{3})(R/d)^2$ is the number of channels. As was the case with Eq. (1), this expression has optimal values of a, L, and N leading to

$$\frac{\dot{S}}{\dot{Q}_x} = 3\left[\frac{T_B + T_A}{2(T_B T_A)^2}\right]^{1/3}\left\{\frac{11\mu}{6\pi p^2}\frac{\kappa_{\text{steel}}}{\kappa}\left(\frac{2\Delta a}{a}\right)\left(\frac{\gamma-1}{\gamma}\right)^2\frac{\dot{Q}_x}{\Omega}\right\}^{1/3} \tag{14}$$

where

$$\dot{Q}_x = \left(\frac{\gamma}{\gamma-1}\right)R\dot{\nu}(T_B - T_A) \quad \Omega = a^2 NL \tag{15}$$

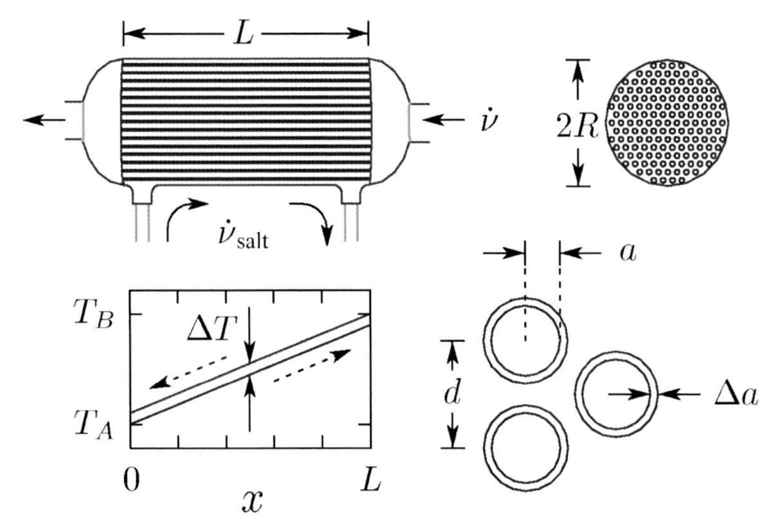

FIG. 14 Illustration of prototype shell-and-tube salt heat exchanger [49] showing the definitions of its various parameters. The number of channels is $N = (2\pi/\sqrt{3})(R/d)^2$.

This result, together related quantities generated by Eqs. (16)–(31), is shown in Fig. 15.

Fig. 15 includes a nominal design point reckoned chiefly by cost. Assuming a market price for stainless steel of \$5 kg^{-1} and an aggressive product delivery estimate of twice the cost of this steel, the cost per heat exchanger watt is 2 × \$0.05 W^{-1} × \$5 kg^{-1} = \$0.5 W^{-1}. The cost per watt delivered to the grid is 1.66 times this number, or \$0.83 W^{-1}, a value dangerously high, per Fig. 4. The entropy production rate at the nominal design point is extremely low. It would reduce the round-trip efficiency of Eq. (12) only by $1.66 \times 2T_{\text{dump}}\dot{S}/\dot{Q}_x = 9.7 \times 10^{-4}$.

The design point is not taken to even lower values of $M_{\text{steel}}/\dot{Q}_x$ in Fig. 15 because of manufacturing uncertainties associated with requisite submillimeter channels. These call for switching to a more advanced heat exchanger design such as that illustrated in Fig. 16. The parameters in Eq. (13) are easily modified to accommodate features of such designs, but accurate assessment of the entropy budget requires access to (proprietary) technical specifications.

Similar cost-efficiency trade-offs must be made for the low-temperature heat exchanger and recuperator. The costs are lower in these cases because (1) they handle less thermal power and (2) they can be made with less expensive carbon steel.

2.4.4.1 Loss component: Viscous flow resistance

The solution of the Navier-Stokes equation gives for laminar flow in a single channel of radius *a*

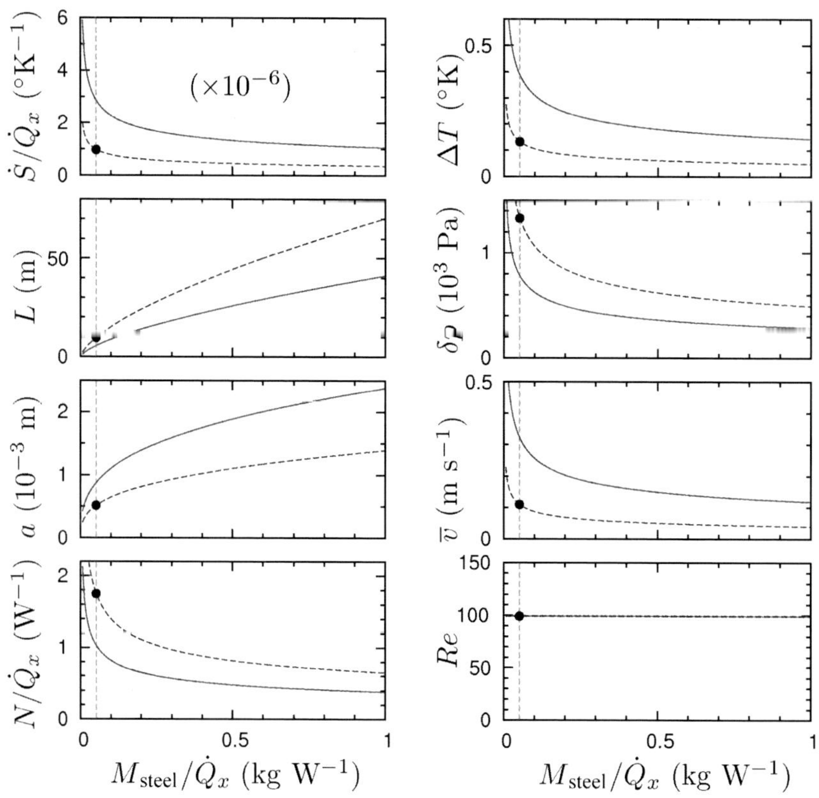

FIG. 15 Performance summary of the entropy-optimized heat exchanger of Fig. 14 as a function of total steel mass $M_{steel} = 2\pi(\Delta a/a)\Omega\rho_{steel}$. The entropy production rate, given by Eq. (14), assumes the values of L, a, and N shown, which are generated by Eqs. (15), (29), (29). The average flow velocity $\overline{v}$, approach temperature ΔT, and pressure drop $\delta p = L(\partial p/\partial x)$ are given by Eqs. (15), (17), (24). The optimization condition causes the Reynolds number $Re = 2m_\iota p\overline{v}a/(RT\mu)$ to be roughly 100, regardless of details. The working fluid is assumed to be Ar. The *solid* (*red*) and *dashed* (*blue*) *lines* correspond to pressures of $p = 1.0 \times 10^6$ and 5.0×10^6 Pa, respectively. The remaining parameters are listed in Table 4. The *dot* shows a nominal design point, as explained in the text.

$$\mu\left[\frac{\partial^2 v}{\partial r^2} + \frac{1}{r}\frac{\partial v}{\partial r}\right] = \frac{\partial p}{\partial x} \quad v = -\frac{a^2}{4\mu}\left(\frac{\partial p}{\partial x}\right)\left[1 - \left(\frac{r}{a}\right)^2\right] \tag{16}$$

The resulting average flow velocity

$$\overline{v} = \frac{2}{a^2}\int_0^a v\, r dr = -\frac{a^2}{8\mu}\left(\frac{\partial p}{\partial x}\right) = -\frac{1}{\pi a^2}\left(\frac{RT}{p}\right)\left(\frac{\dot{\iota}}{N}\right) \tag{17}$$

then gives a power dissipated per length of channel of

$$\frac{\partial \dot{W}}{dx} = -\pi a^2 \overline{v}\left(\frac{\partial p}{\partial x}\right) = \frac{\pi a^4}{8\mu}\left(\frac{\partial p}{\partial x}\right)^2 = \frac{8\mu}{\pi a^4}\left(\frac{RT\dot{\iota}}{pN}\right)^2 \tag{18}$$

TABLE 4 Parameters used to generate Fig. 15.

γ	1.667	(No units)	κ	3.27×10^{-2}	W m^{-1} °K^{-1}
m_ν	0.040	kg mol^{-1}	μ	4.17×10^{-5}	Pa s
ρ_{steel}	8.0×10^3	kg m^{-3}	κ_{steel}	18	W m^{-1} °K^{-1}
$\Delta a/a$	0.224	(No units)	R	8.314	J mol^{-1} °K^{-1}
T_A	495	°K	T_B	823	°K

Notes: The working fluid is Ar. T_A and T_B are the same as T_1 and T_1^+ **in Table 2.**

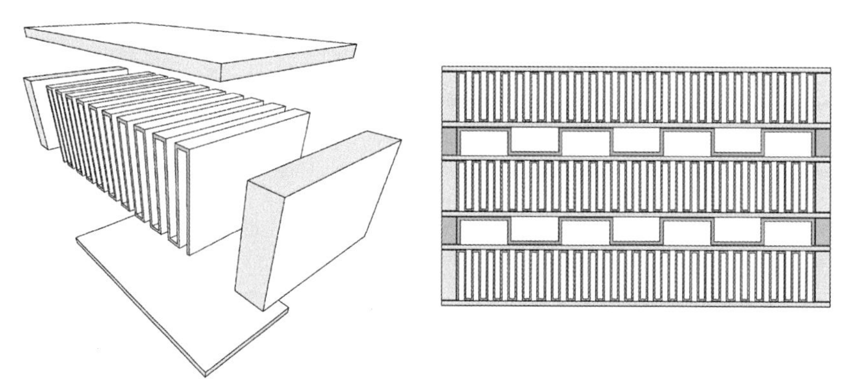

FIG. 16 Illustration of a plate-fin heat exchanger. A thin steel plate is plastically deformed to make a series of crenelations and then brazed onto steel end plates and parting sheets to form channels. The working fluid channels are drawn with a height of 3.0×10^{-3} m [50]. The salt channels are not optimized either for flow/maintenance issues or corrosion margins [36].

and thus a total entropy production due to viscous drag of

$$\dot{S}_{visc} = N \int_0^L \left(\frac{\partial \dot{W}}{\partial x}\right) \frac{dx}{T(x)} = \left(\frac{8\mu}{\pi a^4}\right) \frac{L}{N} \left(\frac{\dot{\nu} R}{p}\right)^2 \left(\frac{T_B + T_A}{2}\right) \tag{19}$$

2.4.4.2 Loss component: Approach temperature

The thermal impedance in Fig. 14 comes chiefly from the temperature difference δT between the working fluid interior and the channel wall. Solving the Graetz equation for a single channel of radius a

$$\kappa \left[\frac{\partial^2(\delta T)}{\partial r^2} + \frac{1}{r}\frac{\partial(\delta T)}{\partial r}\right] = \left(\frac{\gamma}{\gamma - 1}\right) \frac{\nu p}{T} \left(\frac{\partial T}{\partial x}\right) \tag{20}$$

with v defined by Eq. (16) gives

$$\delta T = \left(\frac{\gamma}{\gamma-1}\right)\left(\frac{\dot{\nu}R}{\pi\kappa N}\right)\left[\frac{3}{8}-\frac{1}{2}\left(\frac{r}{a}\right)^2+\frac{1}{8}\left(\frac{r}{a}\right)^4\right]\left(\frac{\partial T}{\partial x}\right) \tag{21}$$

The local channel entropy generation

$$\frac{\partial \dot{S}_{\Delta T}}{\partial x} = \frac{2\pi\kappa}{T^2}\int_0^a\left[\frac{\partial(\delta T)}{\partial r}\right]^2 rdr = \frac{11}{48\pi\kappa}\left[\left(\frac{\gamma}{\gamma-1}\right)\frac{\dot{\nu}R(T_B-T_A)}{NLT}\right]^2 \tag{22}$$

then gives a total entropy generation due to thermal impedance of

$$\dot{S}_{\Delta T} = N\int_0^L\left(\frac{\partial \dot{S}_{\Delta T}}{\partial x}\right)dx = \frac{11}{48\pi\kappa}\left[\left(\frac{\gamma}{\gamma-1}\right)\dot{\nu}R\right]^2\frac{1}{NL}\frac{(T_B-T_A)^2}{T_BT_A} \tag{23}$$

The formal approach temperature ΔT of Fig. 4 is related to the entropy creation rate by

$$\dot{S}_{\Delta T} = \frac{\dot{Q}_x\Delta T}{T_B-T_A}\int_{T_A}^{T_B}\frac{dT}{T^2} = \frac{\dot{Q}_x\Delta T}{T_BT_A} \quad \Delta T = \frac{11}{48\kappa\pi NL}\dot{Q}_x \tag{24}$$

with $\dot{Q}_x$ given by Eq. (15).

2.4.4.3 Loss component: Thermal leak

The entropy production due to thermal leak is $1/T_A - 1/T_B$ times the total heat flow conducted longitudinally through the steels and fluids. It equals

$$\dot{S}_{\text{leak}} = \pi a^2\left(\frac{N}{L}\right)\left[\kappa+\kappa_{\text{steel}}\left(\frac{2\Delta a}{a}\right)+\kappa_{\text{salt}}\right]\frac{(T_B-T_A)^2}{T_BT_A} \tag{25}$$

assuming that the hydraulic radii of the interior and exterior channels are the same. If they are not, then κ_{salt} must be multiplied by $(\sqrt{3}/2\pi)(d/a)^2-(1+\Delta a/a)^2$. With practical parameters, only the thermal leak through the steel matters.

2.4.4.4 Heat exchanger optimization

Using Ω, as defined by Eq. (15) to eliminate a from Eq. (13) gives

$$\begin{aligned}\dot{S} &= \dot{S}_{\text{visc}}+\dot{S}_{\Delta T}+\dot{S}_{\text{leak}}\\ &= \left(\frac{8\mu}{\pi\Omega^2}\right)L^3N\left(\frac{\dot{\nu}R}{p}\right)^2\left(\frac{T_B+T_A}{2}\right)+\frac{11}{48\pi\kappa}\left[\left(\frac{\gamma}{\gamma-1}\right)\dot{\nu}R\right]^2\frac{1}{NL}\frac{(T_B-T_A)^2}{T_BT_A}\\ &\quad+\frac{\pi\Omega\kappa_{\text{steel}}}{L^2}\left(\frac{2\Delta a}{a}\right)\frac{(T_B-T_A)^2}{T_BT_A}\end{aligned} \tag{26}$$

The constant-Ω optimization condition is then

$$N\left(\frac{\partial \dot{S}}{\partial N}\right)_{\Omega,L} = L\left(\frac{\partial \dot{S}}{\partial L}\right)_{\Omega,N} = 0 \quad \dot{S}_{\text{visc}} = \dot{S}_{\Delta T} = \dot{S}_{\text{leak}} = \dot{S}/3 \tag{27}$$

The relation $\dot{S}_{\text{visc}} = \dot{S}_{\Delta T}$ gives

$$NL^2 = \frac{p\Omega}{8}\left(\frac{\gamma}{\gamma-1}\right)\sqrt{\frac{11}{3\kappa\mu}\frac{(T_B - T_A)^2}{T_B T_A (T_B + T_A)}} \tag{28}$$

The relation $\dot{S}_{\Delta T} = \dot{S}_{\text{leak}}$ gives

$$\frac{N}{L} = \frac{11}{48\pi^2\Omega\kappa_{\text{steel}}}\frac{1}{\kappa}\left(\frac{a}{2\Delta a}\right)\left[\left(\frac{\gamma}{\gamma-1}\right)\dot{\nu}R\right]^2 \tag{29}$$

Combining Eqs. (28), (29), we then obtain

$$L = \left\{\pi^2\Omega^2\left(\frac{\gamma-1}{\gamma}\right)\frac{p\kappa_{\text{steel}}}{(\dot{\nu}R)^2}\left(\frac{2\Delta a}{a}\right)\sqrt{\frac{12\kappa}{11\mu}\frac{(T_B - T_A)^2}{T_B T_A (T_B + T_A)}}\right\}^{1/3} \tag{30}$$

Substituting back into Eq. (26) gives finally

$$\dot{S} = 3\left\{\frac{11\mu}{12\pi\Omega}\frac{\kappa_{\text{steel}}}{\kappa}\left(\frac{2\Delta a}{a}\right)\left(\frac{\gamma-1}{\gamma}\right)^2\frac{\dot{Q}_x^4}{p^2}\frac{(T_B + T_A)}{(T_B T_A)^2}\right\}^{1/3} \tag{31}$$

with $\dot{Q}_x$ defined per Eq. (15).

2.5 Steam technology precedents

Even though the Brayton battery uses a closed cycle and a compressor, it is more like a large steam power plant than an aircraft engine, due to the absence of internal combustion. The working fluid in both cases is a hot gas, the properties of which are universal except for molar density and internal heat capacity. There is a cost mandate in both cases to operate at the largest temperatures and pressures the materials can handle. In both cases, there are constraints from strength of steels, electrical conductivity, temperature sensitivity of magnetism, insulation, and lubricants that must be accommodated. And, most importantly, there is a similar list of subtle inefficiencies that need to be tracked down and minimized.

2.5.1 High pressure and power

There is an immense performance advantage to raising the background pressure p^- as high as possible. It reduces the heat exchanger entropy production, as indicated in Eq. (14), although this benefit diminishes when the channel size becomes small. The much more important effect is the increase in power produced by an engine of a given physical size. This lowers the cost per engine watt, potentially very steeply, as shown in Fig. 17. The scaling of power with pressure is also the key principle behind the supercritical CO_2 engine [52, 53], which operates between 7.4×10^6 and 2.2×10^7 Pa. The latter is also the operating pressure of modern supercritical steam plants [40].

Varying the total working fluid inventory of a closed-cycle Brayton engine up and down increases or decreases the power. The blade angles and rotation speed fix the working fluid velocity, so increasing the density while leaving the temperatures the same increases the number of moles per second passing a given point, thus increasing the power. The working fluid viscosity and thermal conductivity are also both independent of density, per Maxwell kinetics, so increasing the density also increases the Reynolds number. This results in more blade turbulence, which improves adiabatic efficiency, but otherwise has no effect.

The background pressure is, therefore, the appropriate control parameter for governing the power of a Brayton battery. As shown in Fig. 17, the total inventory can be easily varied up and down with minimal use of pumps by leveraging the pressure difference between the two sides of the circuit. Governing by means of a throttle in the flow circuit is not practical because it would generate

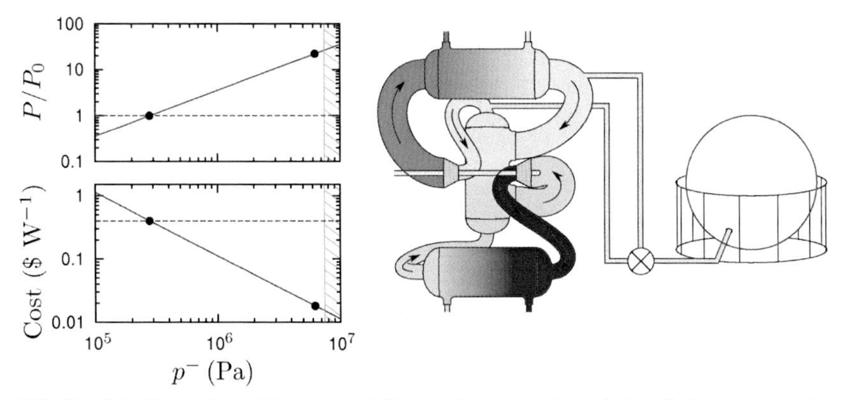

FIG. 17 *Left*: Comparison of the power of a Brayton battery engine with that of a large commercial gas turbine of the same physical size. Table 2 gives $P/\dot{\nu} = \gamma/(\gamma - 1)R(T_1^- - T_0^-)(\xi - 1) = 4.11 \times 10^3$ J mol^{-1}. Technical specifications for a PG9351(FA) gas turbine give $P_0/\dot{\nu} = 0.029$ kg mol^{-1} $\times$ 2.56×10^8 W/644 kg s^{-1} $= 1.153 \times 10^4$ J mol^{-1} [51]. Raising the background pressure of the former thus gives $P/P_0 = 0.357p^-/p_0$, where $p_0 = 1.0 \times 10^5$ Pa, or 1 bar. The *dots* correspond to (1) $P/P_0 = 1$ and (2) a value of p^- appropriate to a large commercial steam turbine [40]. The hatched region is the lower operating range of a supercritical CO_2 engine [52, 53]. An assumed bare cost per watt of the power gas turbine (without burners or blade cooling) of \$0.4 W^{-1} gives a cost for the Brayton battery engine operating at steam pressures of \$0.02 W^{-1}. *Right*: Illustration of power control by means of inventory. Working fluid is bled off of the high-pressure side of the loop into a storage tank, or fed into the low-pressure side of the loop from the tank, according to the need. The power delivered to the grid (or taken from it) is then proportional to the number of moles of gas circulating in the loop.

an immense amount of entropy. Governing by varying any of the four temperatures in Fig. 9 is also impractical because the heat capacities of the heat exchangers are too large. It is only practical to servo the flow of heat storage fluid to these temperatures and further stabilize them.

How pressure governing might work is illustrated by two concrete examples. One is a prototypical startup sequence: (1) initialize p^- to a very low value, (2) spin up the turbomachinery to grid speed by driving the generator/motor, (3) switch the machine onto the grid, and (4) slowly ramp up p^- to its production value. The other is the fail-safe response to a catastrophic loss-of-load accident in generation mode: blow out a relief valve and let the working fluid pressure drop quickly to a negligibly low value.

2.5.2 Motor/generator speed limitation

Elementary principles at work in the motor/generator constrain the machinery of the Brayton battery to rotate no faster than 3600 rpm (for 60 Hz, or 3000 rpm for 50 Hz). This is not a coincidence. The grid was effectively tailored to properties of naturally occurring magnetic materials and metallic conductors when it was originally created. Rotating significantly faster than the natural grid frequency results in intractable dissipative losses in the materials at the few percent level. These losses occur upstream of any semiconductor frequency conversion electronics (see Fig. 6).

This frequency constraint prevents the system from being small. For Ar working fluid at 823°K, an azimuthal blade tip speed of Mach 0.7 requires a blade radius of about 1 m.

As illustrated in Fig. 18, one cause for the speed limitation is the permeable magnetic shield that guides the magnetic flux lines around in a tight loop and prevents them from leaking out into the environment and causing external loss [58]. The permittivity of this shielding has an imaginary (lossy) part that grows

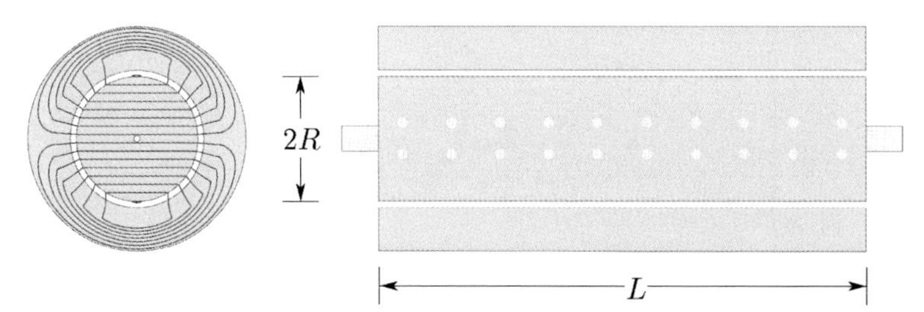

FIG. 18 Illustration of a large two-pole synchronous motor/generator [54]. The length scale is $R \simeq 0.5$ m [55, 56]. *Left*: Axial view showing magnetic field lines generated by the rotor. These cross the air gap, penetrate the stator coils, and return through a high-permeability shield. *Right*: Side view showing dimensions appropriate for very high powers (1.4 GW). Lengths longer than about 5 m are vibrationally problematic [57].

with frequency and needs to me aggressively minimized [59]. One cannot simply eliminate this lossy permeable material because doing so would prevent the magnetic field lines from concentrating properly through the stator coils. This is so even with permanent-magnet rotors. In theory, this magnetic "defocus" problem could be compensated by increasing the number of wire turns, but in practice the required decrease in wire diameter would unacceptably increase the coil resistive loss.

The other important factor is power per unit length of rotor, given by $P/L \leq (\pi B^2/\mu_0)\omega R^2$, where B is the operating magnetic field of the magnetic medium, μ_0 is the permittivity of free space, R is the rotor radius, and ω is the rotor angular frequency. This limit occurs because the torque is transmitted from the stator to the rotor through the Maxwell stress in the air gap. The rotor radius is further limited by the condition that the rim speed ωR not exceed about Mach 0.7 for the cooling gas used. For the case of air at 100°C and 60 Hz, this gives $R = 0.73$ m and $P/L < 2.82 \times 10^8$ W m^{-1}, assuming a value of $B = 0.75$ T. This estimate is crude and overlooks many important fine tunings (hydrogen cooling gas, B optimization, etc.), but it is roughly consistent with an overall active rotor length of about 5 m for a large (1.4 GW) commercial power generator.

The physical limit on P/L subsequently imposes an overall power limit on the motor/generator. The only way to increase the power is to increase L, and this lengthening eventually leads to a collision with the motor/generator's mechanical vibrations [57]. The maximum length L is design dependent, but it can be crudely estimated using the Euler-Bernoulli beam equation $\omega = \sqrt{E/\rho}(R/2)(\pi/L)^2$, where E and ρ are the rotor's Young's modulus and density, respectively. The values of E and ρ appropriate for silicon steel and the above values of R and ω give $L = 6.7$ m. As shown in Fig. 19, the total power limit is a significant problem in high-speed motor/generators [60, 61] but not for low-speed ones.

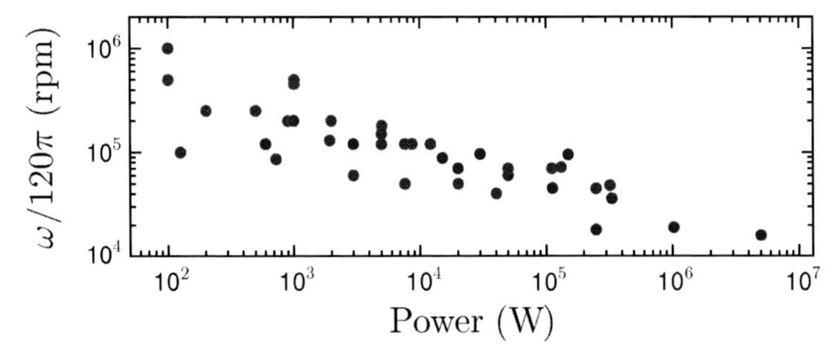

FIG. 19 Rated speed versus power of existing high-speed permanent-magnet motors taken from Borisevljevic (*red*) [60] and Shen et al. (*blue*) [61], showing the general tendency of maximum deliverable power to decline with speed. None of these are induction machines.

At the grid-matched frequencies, synchronous generators (and thus synchronous motors, for they are the same thing) are commodities. Industry standard efficiencies for such large synchronous machines are 99% at optimal loading [55, 56, 62]. The lost 1% is sloughed off as heat at about 130°C [62].

2.5.3 Bearings and seals

The large size of the turbomachinery and its relatively low rotation speed together mandate the use of oiled journal bearings with upstream seals [64–67]. An example of a bearing/seal arrangement from a large steam turbine is shown in Fig. 20 [40, 63].

The superiority of journal bearings for heavy applications is usually ascribed to their ability to take the loads [68, 69], but it is actually a complicated engineering decision involving physical size constraints, bearing lifetime, energy use, and cost. The decision is greatly simplified by the slow rotation speed of this application, which reduces the low friction loss advantages of other kinds of bearing. Losses from journal bearings in large steam plants are typically 0.5%, with the bulk of this being energy consumed in pumping the oil [40, 67]. Roller bearings are not used because they suffer from fatigue issues that cause them to fail unpredictably in high-load applications [69]. Foil bearings require high rotation speed to work and have a load limit of one-third that of journal bearings, making them bulky [68, 70]. Active magnetic bearings will work, but they are expensive and also have a load limit of one-third that of journal bearings [71].

The physics problem that oiled bearings address so brilliantly is that the resistance to compression and shear required for any substance to bear loads are properties of solids, not fluids. Fluids (including magnetic fields) can support large loads only if they are tricked into behaving like solids on short time scales. Lubricating oils facilitate this trick by combining (1) fluidity, (2) high stiffness (bulk modulus characteristic of a solid), and (3) high viscosity in their operating range. The latter comes from oil's special polymeric nature. When

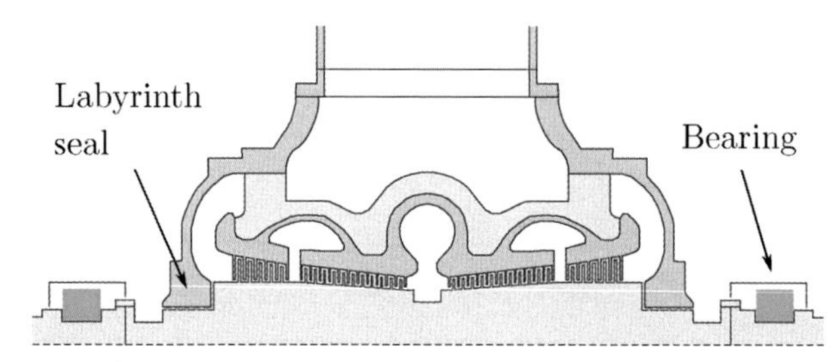

FIG. 20 Example of a large steam turbine cylinder showing relative locations of bearings and seals [40, 63]. This particular design flows working fluid in both direction through two sets matched turbines, thus neutralizing the thrust and minimizing losses in the thrust bearings. The bearing-to-bearing length scale is roughly 6.5 m [40].

axle rotation shortens the time scale, this viscosity becomes a proxy for the shear strength, and the load is supported. The details how this happens are implicit in solutions of the Reynolds bearing equation [69].

Sealing is a highly advanced science in steam technology, and involves a choices among a number of approaches, notably labyrinth seals and brush seals [40, 72–74]. The pressures across these seals and the temperatures they must withstand are exactly the same as those of the Brayton battery, so these sealing methods precise precedents. The one possible difference is that working fluid leaks might need to be reduced further in the Brayton battery for safety reasons. Any working fluid that escapes from a steam turbine condenses harmlessly into water, but Ar (and also CO_2) are asphyxiating gases and can be lethal if they leak and accumulate. This is one of the important arguments for using air as the working fluid rather than Ar.

2.5.4 Cooling

Both the motor/generator and the bearings of a steam power plant need to be kept cool. The heat rejection required to achieve this cooling is about 1.5% of the delivered power, so it need not be combined with the much larger dumping of entropy created by the fuel combustion, although it can be. These distinct cooling requirements also occur in the Brayton battery except that (1) the fuel entropy in that case is replaced by the entropy created by the turbomachinery and heat exchangers and (2) the motor/generator and bearing losses are doubled, due their occurring in both charge and discharge.

The heat issues of the motor/generator and bearings come from fundamentals of materials and are intractable. The magnetic permeability of motor/generator core material falls off slightly with temperature, although this effect is small [78]. Much more important is the wire insulation, which is a carbon-based polymer that begins to decompose at temperatures 423°K (150°C) [62, 79–81]. There is no noncarbon alternative for this insulation because only polymer materials are (1) flexible and (2) reliability noncrystalline. The latter is necessary for stopping shorting by avalanche. The maximum temperature the motor/generator can handle depends on the choice of insulation thickness, but is typically 403°K (130°C) [79–81]. Conventional oil begins to behave poorly at roughly 323°K (50°C) and becomes nonfunctional above 373°K (100°C) [67, 82–84]. Specialty synthetic lubricants resist decomposition to 423°K (150°C) and higher, but they are expensive and still suffer from rapid decline of viscosity with temperature.

A crucial difference between a steam power plant and the Brayton battery is the relatively high value of the latter's discharge temperature, denoted T_0^{+g} in Fig. 12. The immediate consequence is that the Brayton battery's waste heat is much easier to reject to ambient (should that be what is decided to do with it) than a steam plant, and dry cooling becomes economically feasible [75]. As illustrated in Fig. 21, the dry cooler is simply a cross-flow gas-gas heat

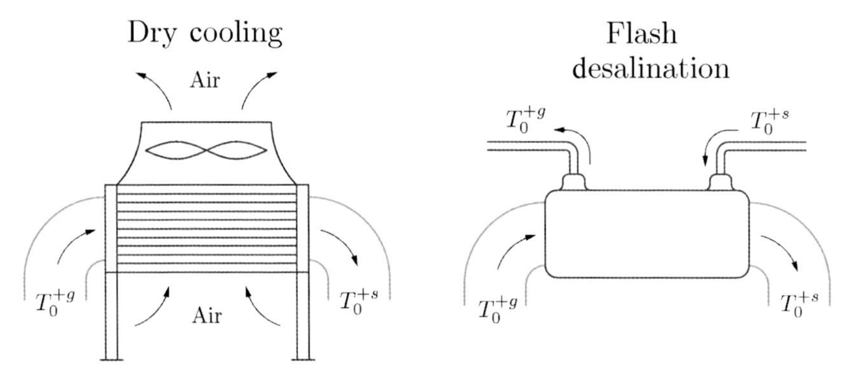

FIG. 21 Illustration of Brayton battery heat dump options. The relatively large value of the discharge temperature T_0^{+g} defined in Fig. 12 reduces the cost of dry cooling (*left*) and eliminates the need for wet cooling towers [75]. It also enables the waste heat to be transferred to a liquid (*right*) and used for other purposes, such as an organic Rankine bottoming cycle [76] or flash desalination [77].

exchanger with the working fluid on one side and ambient air on the other. With no need for the approach temperature between these two fluids to be small, this dry cooler can be physically much smaller than the recuperator illustrated in Fig. 9, and correspondingly cheaper.

The great size of the cooling apparatus of a steam plant of the same power comes about because, for efficiency reasons, the plant must expand its working fluid adiabatically to subatmospheric pressure, thus achieving a final temperature comparable to ambient [75, 85]. It then must condense the vapor to liquid and slough the heat of condensation as waste over a temperature difference that is zero formally, but is about 10°K in practice. This small temperature difference then calls for enormous heat exchanger structures, which for cost reasons are typically chosen to be wet cooling towers. The Brayton battery, by contrast, rejects its heat at about 373°K (100°C).

However, it is not necessary to just dump this heat. The high T_0^{+g} means that the heat has some entropic value remaining that can be extracted and repurposed. This is accomplished, as shown in Fig. 21, by replacing the dry cooler with a heat exchanger delivering a flow of liquid heated to a temperature T_0^{+g}. This fluid is then used for other things, for example, a flow of extremely hot water to a nearby industrial user, or input to an organic Rankine bottoming cycle [76]. The electric energy extraction by latter would amount to an overall round-trip storage efficiency increase of about 5%.

The most important potential use of this residual entropic value is desalination [86–88]. The value of T_0^{+g} turns out to coincide exactly with the optimal input temperature for multistage flash distillation of seawater [77, 89]. This technology, which is deployed in multiple locations in the Middle East [89], is simple and easy to build but notoriously energy inefficient. Reverse osmosis is, therefore, more widely deployed for cost reasons [90]. However, the value proposition reverses if the energy used is waste heat that is going to be discarded anyway. Assuming a flash distillation energy cost of 8.0×10^7 J m^{-3} [91, 92], a Brayton battery generating 10^9 W for 10 h/day would produce, as a side effect, 10^9 W $\times$ 0.3 $\times$ 3600 s h^{-1} $\times$ 10 h day^{-1}/8.0 $\times$ 10^7 J m^{-3} = 1.34 $\times$ 10^5 m^3 day^{-1} of freshwater. Jones et al. estimate the present world production of desalinated water to be 9.5 $\times$ 10^7 m^3 day^{-1} [93].

2.6 Reversible turbomachinery

The tight cost constraints of the Brayton battery implicit in Fig. 4 call for the use of a turbine that turns into a compressor, and vice versa, when the crankshaft rotation direction is reversed. This is illustrated in Fig. 22.

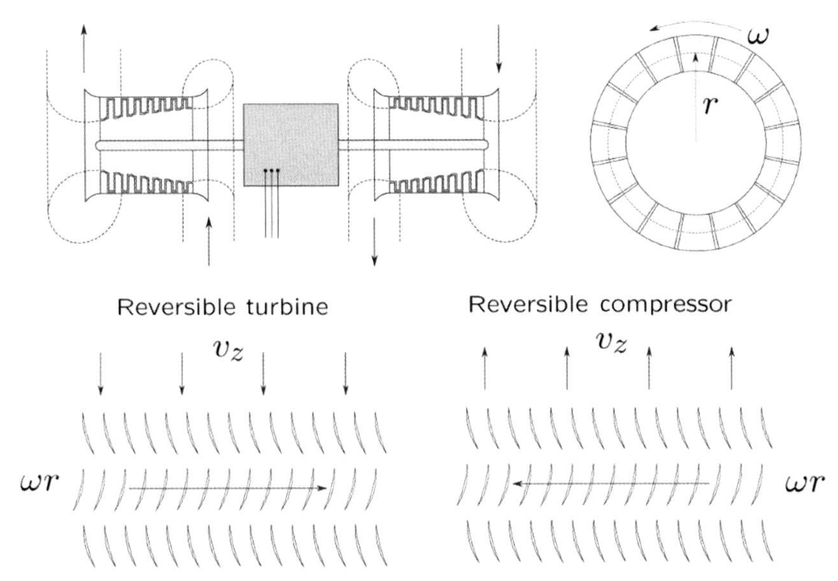

FIG. 22 Illustration of the reversible turbine/pair. The working fluid flow is the same as in Fig. 9. The *left and right sides* of the engine are mirror images of each other, up to modifications to the physical dimensions required to accommodate the different temperatures on the two sides. The *dashed lines* are conventional turbocompressor and turboexpander diffusers. The reversible motor/generator is placed in the middle for clarity. The arrangement differs from that in a conventional gas turbine in that (1) the number of turbine stages is large, so that the pressure loading on each is small and (2) the blades are compromise airfoils that are aerodynamically symmetrical. As discussed more fully in the text, the second aspect is made possible by the first: small loading on the turbine blades allows them to be thin and to have leading and trailing edges that are simultaneously tapered and blunt. The shapes drawn are double-circular-arc (DCA) with angles $\alpha = 0$ and $\beta = -0.165\pi$, as per Eq. (38) and the prototype parameters in Section 2.6.2.4. The maximum width is 0.05 times the chord. The solidity (chord divided by blade row pitch) is 2.

2.6.1 Stage loading

The blading compromises required to make such reversibility are only a small perturbation to conventional turbomachinery design philosophy, which presumes blades that are airfoils designed to work in one way only. As shown in Fig. 23, blades of modern gas turbine compressors are actually quite thin and so nearly symmetrical that it is difficult to distinguish by eye the leading edge from the trailing edge. This near-symmetry results from light loading of the stages, a feature required to suppress surge. A lightly loaded stage is optimized with the blade that is thin, and a thin blade must be relatively symmetric [94–96]. A relatively sharp leading edge is also desirable in compressors operating in the transonic range for the purpose of controlling bow shocks [97].

In contrast to blades of compressor stages, blades of turbine stages are typically heavily loaded, thick, and asymmetric [40, 98, 99]. This design feature is not actually necessary hydrodynamically, however. It is chiefly a means for reducing size and weight (critically important in an aircraft engine) and also for making enough room for internal blade cooling [100]. But the Brayton battery is stationary, so there is no need to minimize weight, and its operating temperatures are sufficiently low, per Figs. 9 and 11, that there is no need for blade cooling. The latter is actually undesirable because of the entropy it generates.

2.6.2 Velocity triangles

With thin blades, a rotor operating at fixed angular frequency ω, and no significant external load (i.e., the Brayton battery with large heat exchangers) both the axial velocity of fluid v_z and compression ratio $1 + \Delta p/p$ are fixed by the annular radius r and the edge blade angles α and β. A general design actually has four

FIG. 23 Blading of commercial gas turbine compressors is thin and nearly symmetrical. *Left*: Rotor of a Siemens SGT5 series power gas turbine. *(Courtesy of Siemens AG.) Right*: Stators of a General Electric J79 aircraft engine (progenitor of LM1500 stationary gas turbine) on display in the Deutsches Museum in Munich, Germany. *(Photo credit: Olivier Cleynen.)*

angles, two for the rotor and another two for the stator, but the principles are most easily illustrated with rotor and stator blades mirror images of each other, as shown in Fig. 22 [64, 101].

2.6.2.1 Minimal-loss condition

At the minimal-loss point, fluid exiting onc bladc row must match the entrance angle of the next blade row downstream in its rest frame. This is achieved when

$$\omega r = -v_z[\tan(\alpha) + \tan(\beta)] \tag{32}$$

2.6.2.2 Euler turbine equation

Conservation of energy causes the temperature change ΔT across a stage to obey

$$\frac{c_p}{m_\nu}\Delta T = \pm\omega^2 r^2 \tag{33}$$

where c_p and m_ν are the working fluid constant-pressure heat capacity per mole and mass per mole, respectively. The plus and minus signs refer to compressor and turbine mode. From the thermodynamic relations of the gas working fluid

$$\frac{\Delta p}{p} = \left(\frac{\gamma}{\gamma - 1}\right)\frac{\Delta T}{T} \quad c_p = \left(\frac{\gamma}{\gamma - 1}\right)R \quad v_s = \sqrt{\frac{\gamma RT}{m_\nu}} \tag{34}$$

where v_s denotes the speed of sound, one then obtains

$$\frac{\Delta p}{p} = \pm\gamma\left(\frac{\omega r}{v_s}\right)^2 \tag{35}$$

2.6.2.3 Half-reaction condition

The exact half-reaction condition, in which the fluid enthalpy loss/gain occurs half in the rotor and half in the stator, is

$$\omega^2 r^2 = \pm v_z^2[\tan^2(\beta) - \tan^2(\alpha)] \tag{36}$$

Dividing Eq. (36) by Eq. (32) gives

$$\omega r v = \pm v_z[\tan(\alpha) - \tan(\beta)] \tag{37}$$

Combining with Eq. (32) then gives

$$\alpha = 0.0 \quad \beta = -\frac{\omega r}{v_z} \tag{38}$$

2.6.2.4 Prototype values

A working rule of thumb in gas turbine design is that compressor loading should be no greater than $\Delta p/p = 0.2$ [102, 103]. Substituting this value into Eq. (35) gives $|\omega r|/v_s = 0.346$ for Ar ($\gamma = 5/3$). Setting the total Mach number to just slightly subsonic $(v_z^2 + \omega^2 r^2)^{1/2} = 0.7 v_s$ then gives $v_z/v_s = 0.609$, and thus $\alpha = 0$ and $\beta = -0.165\pi$. Fig. 22 is drawn with these values of α and β.

2.7 Summary

As Figs. 2 and 3 show, the grid storage problem remains unsolved today. The important barriers to achieving a solution are high cost and safety, not technical means. Energy is a primitive physical quantity that is universally convertible one form to another by means of machines one already knows how to build. These machines all produce entropy as they work, and they must slough this entropy off into the environment as waste heat, thus causing their round-trip storage efficiencies to be less than 1. Technical improvements that raise the efficiency are constantly being sought, but the core problem remains the machines' costs and their potential danger to health and the environment.

I have argued that the Brayton battery would be a workable solution to the problem in its most immediate form—renewable load leveling on the time scale of roughly one day. The technology would potentially fill the market niche presently occupied by pumped hydroelectric storage but at lower cost and without the latter's land use and safety issues. Like pumped hydroelectricity, it would separate the engine from the storage medium and thus control costs for long storage times. At grid-stabilization times of 8 h and longer, it would be economically superior to all electrochemistry, including flow batteries. Its energy storage density would be the same as that of a flow battery. It could be sited anywhere and, in particular, would not require use of underground caverns. Its mechanical interface with the grid would be superior to semiconductor inverter electronics for loss and stability reasons, but more fundamentally because the grid itself is mechanical. It would leverage engine expertise that reliably and cheaply powers the world's airliner fleet and one-fourth of the world's grid [2, 104].

The Brayton battery does not exist today as an off-the-shelf product one can purchase only because of the high cost of its engineering, not because of issues of principle or lack of scientific knowledge. Once designed, the technology should have the low manufacturing and maintenance costs shown in Fig. 4.

References

[1] R.B. Laughlin, Pumped thermal grid storage with heat exchange, J. Renew. Sustain. Energy 9 (2017) 044103.

[2] British Petroleum, BP Statistical Review of World Energy 2018, (June 2018).

[3] R. Showstack, Carbon dioxide tops 400 ppm at Mauna Loa, Hawaii, Eos 94 (2013) 192.

[4] R.F. Keeling, Recording the Earth's vital signs, Science 319 (2008) 1771.
[5] D.C. Harris, Charles David keeling and the story of atmospheric CO_2 measurements, Anal. Chem. 82 (2010) 7865.
[6] K.W. Throning, P.P. Tans, W.D. Kornhyr, Atmospheric carbon dioxide at Mauna Loa observatory: 2. Analysis of the NOAA GMCC Data, 1974–1985, J. Geophys. Res. 94 (1989) 8549.
[7] California Independent System Operator, Renewables Watch for Operating Day: Monday, April 1, 2019, (2019).
[8] U.S. Energy Information Administration, U.S. Battery Storage Market Trends, (May 2018).
[9] R. Fu, T. Remo, R. Margolis, U.S. Utility-Scale Photovoltaics-Plus-Energy Storage System Costs Benchmark, U.S. National Renewable Energy Laboratory, NREL/TP-6A20-71714, November 2018.
[10] S. Few, et al., Prospective improvements in cost and cycle life of off-grid lithium-ion battery packs: an analysis informed by expert elicitations, Energy Policy 114 (2018) 578.
[11] D. Feldman, R. Margolis, P. Denholm, Exploring the Potential Competitiveness of Utility-Scale Photovoltaics Plus Batteries With Concentrating Solar Power, 2015–2030, U.S. National Renewable Energy Laboratory, NREL/TP-6A20-66592, August 2016.
[12] T. Key, et al., Quantifying the Value of Hydropower in the Electric Grid: Final Report, Electric Power Research Institute, 1023144, February, 2013.
[13] T. Lüth, et al., Passive components limit the cost reduction of conventionally designed vanadium redox flow batteries, Energy Procedia 155 (2018) 379.
[14] C.S. Turchi, et al., CSP Systems Analysis—Final Project Report, U.S. National Renewable Energy Laboratory, NREL/TP-5500-72856, May 2019.
[15] N. DiOrio, A. Dobos, S. Janzou, Economic Analysis Case Studies of Battery Energy Storage With SAM, U.S. National Renewable Energy Laboratory, NREL/TP-6A20-64987, November 2015.
[16] H. Bindner, et al., Characterization of Vanadium Flow Battery, Risø DTU National Laboratory for Sustainable Energy, Risø-R-1753, October 2010.
[17] K.W. Knehr, E.C. Kumbur, Open circuit voltage of vanadium redox flow batteries: discrepancy between models and experiments, Electrochem. Commun. 13 (2011) 342.
[18] Z. Tang, et al., Monitoring the state of charge of operating vanadium redox flow batteries, ECS Trans. 41 (2012) 1.
[19] Q. Ye, et al., Design trade-offs among shunt current, pumping loss, and compactness in the piping system of a multi-stack vanadium flow battery, J. Power Sources 296 (2015) 352.
[20] V.S. Bagotsky, Fuel Cells: Problems and Solutions, Wiley, New York, NY, 2012 45 pp.
[21] V. Viswanathan, et al., Cost and performance model for redox flow batteries, J. Power Sources 247 (2014) 1040–1051.
[22] C. Minke, U. Kunz, T. Turek, Techno-economic assessment of novel vanadium redox flow batteries with large-area cells, J. Power Sources 361 (2017) 105.
[23] J. González-Garcia, et al., Characterization of a carbon felt electrode: structural and physical properties, J. Mater. Chem. 9 (1999) 419.
[24] Chemours Co, Nation N115, N117, N1110 Ion Exchange Materials, Product Bulletin P-12, February 2016.
[25] SGL Carbon GmbH, Flexible Carbon and Graphite Felts, May 2001.
[26] J.S. Lawton, et al., The effects of sulfuric acid concentration on the physical and electrochemical properties of vanadyl solutions, Batteries 4 (2018) 40.
[27] M.R. Mohamed, et al., Design and development of unit cell and system for vanadium redox flow batteries (V-RFB), Int. J. Phys. Sci. 7 (2012) 1010.

[28] F. Rahman, M. Skyllas-Kazacos, Solubility of vanadyl sulfate in concentrated sulfuric acid solutions, J. Power Sources 72 (1998) 105.
[29] Y.A. Gandomi, et al., Critical review—experimental diagnostics and material characterization techniques used on redox flow batteries, J. Electrochem. Soc. 165 (2018) A970.
[30] U. Herrmann, B. Kelly, H. Price, Two-tank molten salt storage for parabolic trough solar power plants, Energy 29 (2004) 883.
[31] Solar Millennium AG, The Parabolic Trough Power Plants Andasol 1 to 3, 2008.
[32] Sandia National Laboratories, Solar Two Demonstrates Clean Power for the Future, SAND-2000-0613, March 2008.
[33] H.E. Reilly, G.J. Kolb, An Evaluation of Molten-Salt Power Towers Including Results of the Solar Two Project, Sandia National Laboratories, SAND2001-3674, November 2001.
[34] U. Pelay, L. Luo, Y. Fan, D. Stitou, M. Rood, Technical data for concentrated solar power plants in operation, under construction and in project, Data Brief 13 (2017) 597.
[35] S.H. Goods, R.W. Bradshaw, Corrosion of stainless steels and carbon steel by molten mixtures of commercial nitrate salts, J. Mater. Eng. Perform. 13 (2004) 78.
[36] A. Kruizenga, D. Gill, Corrosion of iron stainless steels in molten nitrate salt, Energy Procedia 49 (2014) 878.
[37] G.L. Janz, et al., Molten salts: volume 3, nitrates, nitrites, and mixtures, J. Phys. Chem. Ref. Data 1 (1972) 581.
[38] M. Lasfargues, et al., Rheological analysis of binary eutectic mixture of sodium and potassium nitrate and the effect of low concentration of CuO nanoparticle addition to its viscosity, Materials 8 (2015) 5194.
[39] T. Bauer, et al., Material aspects of solar salt for sensible heat storage, Appl. Energy 111 (2013) 1114.
[40] A.S. Layzerovich, Steam Turbines for Modern Fossil Fuel Power Plants, CRC Press, Boca Raton, FL, 2007.
[41] R.W. Bradshaw, N.P. Siegel, Molten Nitrate Salt Development for Thermal Energy Storage in Parabolic Trough Solar Power Systems, Sandia National Laboratory, ES200854174, August 2008.
[42] D.J. Rogers, G.J. Janz, Melting-crystallization and premelting properties of $NaNO_3$-KNO_3 enthalpies and heat capacities, J. Chem. Eng. Data 42 (1982) 424.
[43] C.M. Kramer, C.J. Wilson, The phase diagram of $NaNO_3$-KNO_3, Thermochim. Acta 42 (1980) 253.
[44] X. Zhang, et al., Thermodynamic evaluation of phase equilibria in $NaNO_3$-KNO_3 system, J. Phase Equilib. 24 (2003) 441.
[45] ASME, 2007 Boiler and Pressure Vessel Code (With Addenda for 2008).
[46] E.S. Freeman, The kinetics of the thermal decomposition of potassium nitrate and of the reaction between potassium nitrite and oxygen, J. Phys. Chem. 79 (1957) 838.
[47] R.I. Olivares, The thermal stability of molten nitrite/nitrates salt for solar thermal energy storage in different atmospheres, Sol. Energy 86 (2012) 2576.
[48] P. Gimenez, S. Fereres, Effect of heating rates and composition on the thermal decomposition of nitrate based molten salts, Energy Procedia 69 (2015) 654.
[49] L. Chordia, High temperature heat exchanger design and fabrication for systems with large pressure differentials, Tech. Rep., Thar Energy LLC, DOE/NETL Final Report DE-FE0024012, March 2017.
[50] S.D. Sullivan, High-efficiency low-cost solar receiver for use in a supercritical CO_2 recompression cycle, Tech. Rep., Brayton Energy LLC, DOE EERE Final Report DE-EE0005799, April 2016.

[51] F.J. Brooks, GE gas turbine performance characteristics, GE Power Systems, GER-3567H, October 2000.
[52] Y. Ahn, et al., Review of supercritical CO_2 power cycle technology and current status of research and development, Nucl. Eng. Technol. 47 (2015) 647.
[53] K. Brun, P. Friedman, R. Dennis, Fundamentals and Applications of Supercritical Carbon Dioxide (sCO_2) Based Power Cycles, Woodhead Publishing, Duxford, UK, 2017.
[54] R.J. Zawoysky, K.C. Tornroos, GE generator rotor design, operational issues, and refurbishment options, GE Power Systems, GER-4212, August 2001.
[55] G.E. Power, Generators, GEA32229, November 2015.
[56] Siemens AG, SGen-3000W water-cooled generator series, PGGT-T10029-00-7600, n.d.
[57] E. Swanson, C. Powell, S. Weissman, A practical review of rotation machinery critical speeds and modes, Sound Vib. 39 (2005) 10.
[58] A. Hughes, Electric Motors and Drives, third ed., Newnes, Oxford, UK, 2006.
[59] W.T. McLyman, Transformer and Inductor Design Handbook, CRC Press, Boca Raton, FL, 2011.
[60] A. Borisevljevic, Limits, Modeling and Design of High-Speed Permanent Magnet Machines, Springer, New York, NY, 2013.
[61] J. Shen, X. Qin, Y. Wang, High-speed permanent magnet electrical machines—applications, key issues and challenges, CES Trans. Electric. Machines Syst. 2 (2018) 23.
[62] P. Breeze, Why generators are the unsung heroes of power plant efficiency, Power Eng. Int. (25 February) (2015).
[63] K. Nakamura, T. Tabei, T. Takano, Recent technologies for steam turbines, Fuji Electr. Rev. 56 (4) (2010) 123 (Fuji Electric Systems Co.).
[64] M.W. Boyce, Gas Turbine Engineering Handbook, Gulf Professional Publishing, Houston, 2002.
[65] T. Tanuma (Ed.), Advances in Steam Turbines for Modern Power Plants, Woodhead Publishing, Duxford, UK, 2016.
[66] D.M. Smith, Journal Bearings in Turbomachinery, Springer, New York, NY, 1969.
[67] M.J. Neale (Ed.), The Tribology Handbook, second ed., Butterworth Heinemann, Oxford, 1995.
[68] F. Stefani, et al., Comparative analysis of bearings for micro-GT: an innovative arrangement, in: P.H. Darji (Ed.), Bearing Technology, 2017.
[69] B.J. Hamrock, S.R. Schmid, B.O. Jacobson, Fundamentals of Thin Film Lubrication, second ed., CRC Press, Boca Raton, FL, 2004.
[70] C. DellaCorte, R.J. Bruckner, Remaining technical challenges and future plans for oil-free turbomachinery, J. Eng. Gas Turbines Power 133 (2010) 042502.
[71] G. Schweitzer, E.H. Maslen, Magnetic Bearings, Springer, New York, NY, 2009.
[72] R.E. Chupp, et al., Sealing in Turbomachinery, J. Propul. Power 22 (2006) 313.
[73] P.N. Jiang, et al., Influence of steam leakage through vane, gland, and shaft seals on rotordynamics of high-pressure rotor of a 1,000 MW ultra-supercritical steam turbine, Arch. Appl. Mech. 82 (2010) 177.
[74] T.S. Kim, K.S. Cha, Comparative analysis of the influence of labyrinth seal configuration on leakage behavior, J. Mech. Sci. Technol. 23 (2009) 2830.
[75] K. Birkinshaw, M. Masri, R.L. Therkelsen, Comparison of alternate cooling technologies for California power plants, 50002-079F, February 2002.
[76] S. Quoilin, et al., Techno-economic survey of organic Rankine cycle (ORC) systems, Renew. Sustain. Energy Rev. 22 (2013) 168.

[77] H.T. El-Dessouky, H.M. Ettouney, Y. Al-Roumi, Multi-stage flash desalination: present and future outlook, Chem. Eng. J. 73 (1999) 173.
[78] J. Chen, et al., Influence of temperature on magnetic properties of silicon steel lamination, AIP Adv. 7 (2017) 056113.
[79] A. Machimo, A. Nakayama, H. Hiwasa, Recent technologies for rotating machines, Fuji Electr. Rev. 56 (4) (2010) 129 (Fuji Electric Systems Co.).
[80] International Electrotechnical Commission, International Standard: Rotating Electrical Machines—Part 1: Rating and Performance., IEC 60034-1, April 2004.
[81] G.C. Stone, et al., Electrical Insulation for Rotating Machines, Wiley-IEEE, Hoboken, NJ, 2004.
[82] M.M. Khonsari, E.R. Booser, Applied Tribology: Bearing Design and Lubrication, Wiley-Interscience, Hoboken, NJ, 2001, 250 pp.
[83] S. Strzelecki, Operating characteristic of heavy loaded cylindrical journal bearing with variable axial profile, Mater. Res. 8 (2005) 481.
[84] R.X. Perez, D.W. Lawhon, Operator's Guide to General Purpose Steam Turbines, Wiley-Schrivener, Beverly, MA, 2016.
[85] J.L. Tsou, J. Maulbetsch, J. Shi, Power plant cooling system overview for researchers and technology developers, Electric Power Research Institute, Palo Alto, CA, 3002001915, May 2013.
[86] H. Rabiee, et al., Energy-water nexus: renewable-integrated hybridized desalination systems, in: K.R. Khalilpour (Ed.), Polygeneration With Polystorage for Chemical and Energy Hubs, Academic Press, London, UK, 2018.
[87] M. Shatat, S.B. Riffat, Water desalination technologies utilizing conventional and renewable energy sources, Int. J. Low-Carbon Technol. 9 (2014) 1.
[88] S.A. Kalogirou, Solar Energy Engineering, second ed., Academic Press, Oxford, UK, 2013, 431 pp.
[89] A.M. El-Ghonemy, Performance test of a sea water multi-stage flash distillation plant: case study, Alex. Eng. J. 57 (2018) 2401.
[90] M.W. Shahzad, et al., Desalination processes' efficiency and future roadmap, Entropy 21 (2019) 84.
[91] International Renewable Energy Agency, Water desalination using renewable energy., (March 2012).
[92] C. Gautier, Oil, Water, and Climate, Cambridge University Press, New York, 2008, 284 pp.
[93] E. Jones, et al., The state of desalination and brine production: a global outlook, Sci. Total Environ. 657 (2019) 1346.
[94] M. Schnoes, C. Voss, E. Nicke, Design optimization of a multi-stage axial compressor using throughflow and a database of optimal airfoils, J. Global Power Propul. Soc. 2 (2018) 516.
[95] E. Benini, Advances in aerodynamic design of gas turbine compressors, in: I. Gurrappa (Ed.), Gas Turbines, InTech, 2014.
[96] N. Gourdain, et al., High performance parallel computing flows in complex geometries: II. Applications, Comput. Sci. Discov. 2 (2009) 015004.
[97] L.C. Wright, Blade selection for a modern axial-flow compressor, in: B. Lakshminarayana, W.R. Britsch, W.S. Gearhart (Eds.), Fluid Mechanics, Acoustics, and Design of Turbomachinery, U.S. National Aeronautics and Space Administration, NASA-SP-304, 1974, 603 pp.
[98] R.A. Alexeev, et al., Turbine blade profile design method based on Bezier curves, J. Phys. C Conf. Series 891 (2017) 012254.
[99] D.E. Brandt, R.R. Wesorick, GE Gas Turbine Design Philosophy, GE Power Generation GER-3434D, September 1994.

[100] L.M. Wright, J.C. Han, Heat transfer enhancement for turbine blade internal cooling, J. Enhanc. Heat Transf. 21 (2014) 111.

[101] MIT Open Course Ware, Lecture 27—subjects: turbines; stage characteristics; degree of reaction, in: 16.50: Introduction to Propulsion Systems, Spring (2012).

[102] W.R. Britsch, W.M. Osborn, M.R. Laessig, Effects of diffusion factor, aspect ratio, and solidity on overall performance of 15 compressor middle stages, Tech. Rep., NASA Technical Paper 1523, September 1979.

[103] S. Lieblein, F.C. Schwenk, R.L. Broderick, Diffusion Factor for Estimating Losses and Limiting Blade Loadings in Axial-Flow Compressor Blade Elements, National Advisory Committee for Aeronautics, NACA RM E53D01, June 1953.

[104] U.S. Energy Information Administration, Electric Power Annual 2018, October, (2019).

Chapter 3

Thermal energy storage

Sebastian Freund[a], Miles Abarr[b], Josh D. McTigue[c], Konor L. Frick[d], Anoop Mathur[e], Douglas Reindl[f], Amy Van Asselt[g], and Giuseppe Casubolo[h]

[a]*Energiefreund Consulting, Munich, Germany,* [b]*Carbon America, Arvada, CO, United States,* [c]*National Renewable Energy Laboratory, Golden, CO, United States,* [d]*Idaho National Laboratory, Idaho Falls, ID, United States,* [e]*Terrafore Technologies, Minneapolis, MN, United States,* [f]*University of Wisconsin-Madison, Madison, WI, United States,* [g]*Lafayette College, Easton, PA, United States,* [h]*SQM, Antwerp, Belgium*

Chapter outline

Thermal, Mechanical, and Hybrid Chemical Energy Storage Systems
https://doi.org/10.1016/B978-0-12-819892-6.00003-4

3.1 Sensible heat liquid thermal energy storage

Sebastian Freund, Konor Frick

3.1.1 Liquid thermal energy storage

Sensible heat thermal energy storage is a technology using the change of internal energy of a liquid undergoing a temperature change without changing phase, and storing the heated or cooled liquid for a subsequent energy exchange in a tank. Of the numerous industrial applications of liquid thermal energy storage (TES), concentrated solar power plants (CSP) are an illustrative example. Solar heat collected during the day is stored in hot molten salt tanks. Later, the hot molten salt is used in a steam generator to drive turbines for making electricity on demand, and the cooled molten salt is awaiting to be heated again to continue the cycle. Current CSP plants have storage tanks that may be sized to allow for upwards of 15 h of generation [1]. The two dominant types of CSP plants with storage are either using parabolic trough mirrors heating a collector with thermal oil heat transfer fluid (HTF) to about 400°C and transferring this heat to either steam for generation or molten salt for storage, or use heliostat mirrors to heat a molten salt to 565°C in a central receiver, called power tower plants. Only the seamless integration of thermal storage made CSP plants competitive with and complementary to PV, wind, and even fossil power generation in the decade after 2010 (Fig. 1).

Two main types of liquid TES designs are in commercial use. Single tank thermocline and two-tank designs. Two-tank TES designs are the most commercially developed and readily understood systems for many applications. These designs have separate tanks that hold the hot and the cold fluid, respectively. An advantage of two-tank systems is the relative simplicity in their

FIG. 1 Molten salt TES tanks of Solana CSP plant in Arizona. The TES holds 135,000 t of salt in 12 tanks of 11,000 m^3 usable volume each and 40 m diameter. *(Courtesy of Solana Generating Station, Atlantica Sustainable Infrastructure.)*

FIG. 2 Two-tank molten salt TES of the Valle 1 CSP plant in San José del Valle, Cádiz, Spain, with tanks holding 28,500 t of salt and three salt/heat transfer fluid shell and tube heat exchangers in between. *(Image source: https://torresolenergy.com/wp-content/uploads/2017/12/torresol-energy-csp.jpg.)*

design and the reduced thermal degradation achieved. Single tank thermocline systems involve having the hot and cold fluid present within a single tank and relying on density differentials to thermally stratify the fluid, thus developing a thermocline between the hot and cold fluid. Alternatively, a floating separator baffle or a filler material in dual-media thermocline TES can be used to stabilize a thermocline; the latter technology is described in an extra section in this chapter (Fig. 2).

3.1.2 Two-tank TES in CSP

Two-tank thermal energy storage with molten salt has been widely used after the pioneering Solar Two project in the 1990s since the construction of a series of 50 MW parabolic trough CSP plants in Spain. The first one of what turned out to be a fleet of almost 40 similar plants was Andasol-1, in operation since 2008 and built by ACS Cobra and Sener for Endesa near Granada, Spain. Inside the solar heat collector elements of these plants, a synthetic heat transfer fluid, e.g., Dowtherm A, is heated by concentrated solar radiation to about 395°C and pumped through shell and tube heat exchangers to heat molten salt and store the heat in an "indirect" 2-tank TES. Andasol-1's tanks are 14 m tall with a diameter of 36 m to hold a total of 28,500 t of salt. With an average heat capacity of 1.56 kJ/kg-K and a temperature range of about 290°C in the cold to 385°C in the hot tank, the storage capacity is about 1000 MWh, allowing the plant to generate 50 MW for 7.5 h on storage only [2]. The heat is stored in salt rather than directly with the heat transfer fluid because of the high cost and the vapor pressure of synthetic oils, making large tank volumes too expensive compared to using salt.

This salt is a most economic and near-eutectic mixture of 60% sodium nitrate and 40% potassium nitrate, called Solar Salt. Both constituents can be mined from natural resources and purified or synthetically produced and are a major chemical commodity with more than a million tons annual world production. One manufacturer mines sodium nitrate from surface deposits in the Atacama Desert and produces potassium nitrate from mineral leaching processes including brines pumped from the ground in the same region in northern Chile [3]. Solar Salt costs range about $0.5 to 1.5 per kg, depending on market conditions, quantity and quality of the order, a typical indicative price (CIF main port) in 2019 was $800/t as communicated by Giuseppe Casubolo [3]. With a freezing point around 240°C, Solar Salt needs to be carefully protected from large heat losses through insulation and temperature-controlled electric heat tracing on all pipe and valve surfaces to avoid flow blockage. Ternary salt mixtures are available with freezing points as low as 120°C, such as "Hitec" or "Hitec XL," that additionally contain sodium nitride or calcium nitrate. These are used in industrial applications rather than for solar storage because of higher costs and lower thermal stability. Solar Salt has been successfully used up to 565°C for long duration in power tower CSP. Decomposition of the salt increases rapidly above this temperature range into products including alkali oxides, nitrides, oxygen, and nitrous oxides [4]. Above 565°C also the corrosivity of salt to steel increases, while the strength of most temperature-resistant steels degrades quickly. For these reasons, 565°C has been the preferred upper working temperature limit of solar power plants and Solar Salt storage.

Halogen salts with higher decomposition temperatures are available, such as chlorides and fluorides, but cost and material strength and corrosion issues have limited their commercial application to high-temperature niches including novel nuclear concepts. For even higher temperatures, certain types of glass have been proposed by researchers as heat transfer and storage fluids, besides molten metals and silicon.

For lower temperatures, mineral oil-based heat transfer fluids are suitable for storage temperatures up to about 300°C as they cost much less than synthetic fluids and have low vapor pressure, i.e., can be stored at atmospheric pressure. Some synthetic fluids are thermally stable to more than 400°C but need pressurization to prevent boiling. Common brand names of synthetic thermal fluids are Therminol and Dowtherm, but many other brands and unbranded same or similar mixtures of many different compositions tailored to operating range, thermal stability, and vapor pressure are on the market. Water is the HTF of choice if the temperature range permits, antifreeze additives and corrosion inhibitors are often added. Compared to molten salt TES, storage systems for water, thermal oil, or other HTFs are generally simpler in construction, design, and materials because of lower temperatures and no corrosiveness, but may require more attention to fire and spill prevention in case of oils.

For pumping hot heat transfer fluids, centrifugal pumps with adapted seals and bearings are used, positioned below the liquid level to provide net positive suction head. For salt, submersible pumps sitting just above the tank sump are used, the molten salt keeps the pump impeller, volute, and bearing hot, and a long shaft connects with the motor sitting on top of the tank; these specialty pumps are only made by a handful of manufacturers. To avoid temperature cycling of the pumps, separate hot and cold fluid pumps are installed in most TES systems. For large duties and redundancy, two (100%) or three (50%) pumps in parallel are used. VFDs are employed for speed control of the asynchronous pump motors to adapt the flow rate economically. Designing for gravity draining of salt piping avoids freezing salt plugs during standstill periods.

For temperatures below 400°C, storage tanks, even for molten salt, are made of pressure vessel grade fine grain carbon or chromoly steel plate, but for higher temperatures, as found in power tower CSP plants, stainless steel (e.g., Gr 347) is used for its high-temperature mechanical strength and corrosion resistance. Thick insulation of mineral or ceramic fiber is used to keep tank heat losses typically below 1°C/day. Molten salt storage tanks are equipped with tank heaters to prevent the salt from freezing should the temperature drop during prolonged standstill periods. Very long electric cartridge heaters are used that protrude radially from the perimeter above the bottom into the tank sump. Special attention is needed in proper design of the foundations to support the enormous weight of the tank yet allow for the tank bottom and the walls to glide and expand and shrink during temperature cycling without cracking. Several hot tank failures have been reported for the TES of power tower type CSP plants with 565°C hot storage temperature. These failures happened after only fractions of the design lifetime and led to expensive downtime for repairs. An advantage of two-tank TES design compared to single-tank storage is the possibility to pump the liquid from one tank into the other and keep one empty for repairs. For thermal oil, synthetic HTF and in many cases of lower temperature molten salt, nitrogen blanketing is used to fill the head space, ullage, of the tanks and heat exchangers to prevent air and moisture ingress, oxidation of the fluid, and potentially flammable mixture of oil vapors. Very large molten salt tanks and those for high temperature often have air in the ullage space for economical reasons and for stabilizing the nitrates by suppressing the decomposition to nitrides through higher oxygen partial pressure than under nitrogen. Storage tanks are no pressure vessels and can breathe through valves to keep the pressure inside almost equal to the atmosphere (within millibars) to protect the hull and roof from blowing up or collapsing when liquid volumes are exchanged or the gas volumes change its temperature. A gas balance line can connect the ullage space of both tanks to limit the loss of head gas during each cycle. However, changing temperatures between charged and discharged store and changes in ambient pressure still cause the required mass of gas to keep the pressure

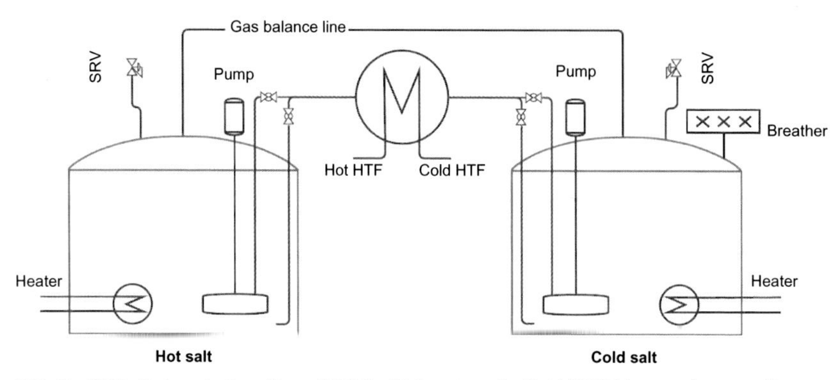

FIG. 3 PFD of a two-tank molten salt TES with heat transfer fluid (HTF) heat exchanger. *(Source: S. Freund.)*

within a narrow range to change periodically and a fraction of gas needs to be vented or actively removed with compressors during charge and replenished during discharge.

Shell and tube heat exchangers are predominantly used (with a few exceptions of welded plate and spiral-wound heat exchangers) for thermal oil-to-salt or salt-to-steam duty. Austenitic stainless steels are used for heat exchangers and thin-walled components especially for high salt temperatures; in order of increasing temperature and heat resistance, common grades include Gr. 316Ti, 321, and 347. Many alternative steel grades, even ferritics and lower alloys could be used with molten salts but there is little industry experience. Heat exchangers for lower temperatures, e.g., for oil and steam, can be made of chromoly or low-alloy carbon steels. In CSP applications, the heat exchangers between salt and oil, oil and steam, or salt and steam are designed for low approach temperatures, e.g., 10 K, and become often large enough to require several shells used in series to cover the temperature duty and two or more of these trains connected in parallel to handle the flow rates. Each heat exchanger shell can weight more than 100 t.

The project capital costs of two-tank TES systems including tanks, fluids, heat exchangers, and associated balance of plant equipment as well as structures can range from $60 down to 20/kWh, depending on the specifics, scale, markets, and temperature range. For smaller projects, less than about 200 MWh, costs may be higher than this range. Still, this storage technology is relatively inexpensive on a per unit of energy stored basis and scales easily in storage duration by increasing the tank size and inventory mass (Fig. 3).

3.1.3 Single-tank TES for district heating

In several European cities with vast district heating networks, single-tank TES in the form of large naturally stratified hot water tanks have been built to decouple hot water generation in central combined heat and power plants or electric boilers from the heating demand of the network. These tanks rely on gravity and the density difference between the hot and the cooler return water to establish a stable thermocline separating the warmer and cooler water to prevent self-discharge. To promote the thermocline, these tanks tend to be rather tall. Hot water inlets/exits are on top, while the return water enters from the bottom during discharge or leaves during charging. Unpressurized tanks with operating temperatures up to 98°C have been built as large as 50,000 m^3, 70 m tall or 40 m in diameter. Storage capacities can exceed 1500 MWh. Pressurized tanks for higher temperatures tend to be smaller and thinner and have been built for pressures up to 16 bar. The latest generation of single-tank TES for district heating water allows even water storage temperatures up to 120°C in a nominally unpressurized tank. This is achieved by dividing the tall tank through an internal roof into a large lower compartment housing the hot water and the thermocline with cold water at the bottom, and an upper chamber connected through risers to the cold water, which is filled with cold water and gas expansion volume. The water in the upper chamber exerts sufficient hydrostatic pressure to suppress boiling of the hot water below, a design known after its inventor as a "Hedbäck" storage and built by Bilfinger GmbH [5]. The following figure shows a photograph of such a tank just after construction in Kiel, Germany, in 2019. In all large tanks the walls are welded with grain-refined pressure vessel steel plates and insulated with mineral wool up to 0.5 m thick, enclosed by sheet metal weather protection. The project capital costs for such single-tank water TES systems are on the order of $10/kWh (Fig. 4).

3.1.4 TES with nuclear power

In an exemplary but rare application as compared to the established commercial use of two-tank TES for CSP and single-tank TES for water, thermal storage designs have also been put forth for use with nuclear reactors. A potential configuration of a two-tank TES with nuclear reactors is shown in Fig. 5 and the design basis and operation described; a single tank thermocline operation is not presented but would be effectively the same in this application. The purpose is the flexibilization of steam demand response of a nuclear reactor running at continuous power. Coming from the steam generator, steam is directed through a turbine bypass to an intermediate heat exchanger (IHX), heating the TES, and discharged to the main condenser. The TES system consists of two large storage tanks along with several pumps to transport the HTF between the tanks, the IHX, and a steam generator. Flow Bypass Valves are included in the discharge lines of both the Hot and Cold tanks to prevent deadheading the pumps when the Flow Control Valves are closing as the pumps may not be speed controlled. While the system could work with a large number of HTFs, the configuration proposed by Frick [6] uses Therminol-66 [6], which was chosen because it is readily available, can be pumped at low temperatures, and offers a wide thermal

FIG. 4 A 30,000 m^3 district heating water TES storing 1500 MWh at 115°C in a tank 70 m tall with 26 m diameter, note the hot and cold connection pipes at the center and bottom. *(Image source: Stadtwerke Kiel, https://www.energiefachmagazin.de/2019/Ausgabe-06/Jahresuebersichten-I/Energiespeicher?page=8.)*

stability over the range −3°C to 343°C. The anticipated normal operating range of the TES system would be 200–260°C, where Therminol-66 has a density of 840 kg/m^3 and a heat capacity of 2030 kJ/m^{3}K. Other benefits of using Therminol-66 include its safety data sheet classification as a nonhazardous material and that it would not be activated by absorbing tritium in the rare event of a contamination with radioactive primary water during simultaneous leaks in the steam generator and the IHX. The TES system is designed to allow the reactor to run continuously at 100% power over a wide range of operating conditions. During periods of excess capacity, bypass steam is directed to the TES unit, where it condenses on the shell side of the IHX. TES fluid is pumped from the Cold Tank to the Hot Tank through the tube side of the IHX at a rate sufficient to raise the temperature of the TES fluid to some set point. The TES fluid is then stored in the Hot Tank at constant temperature. Condensate is collected in a hot well below the IHX and drains back to the main condenser, or can be used for some other low pressure application such as chilled water production, desalination, or feed heating. Pressure relief lines connect the shell side of the

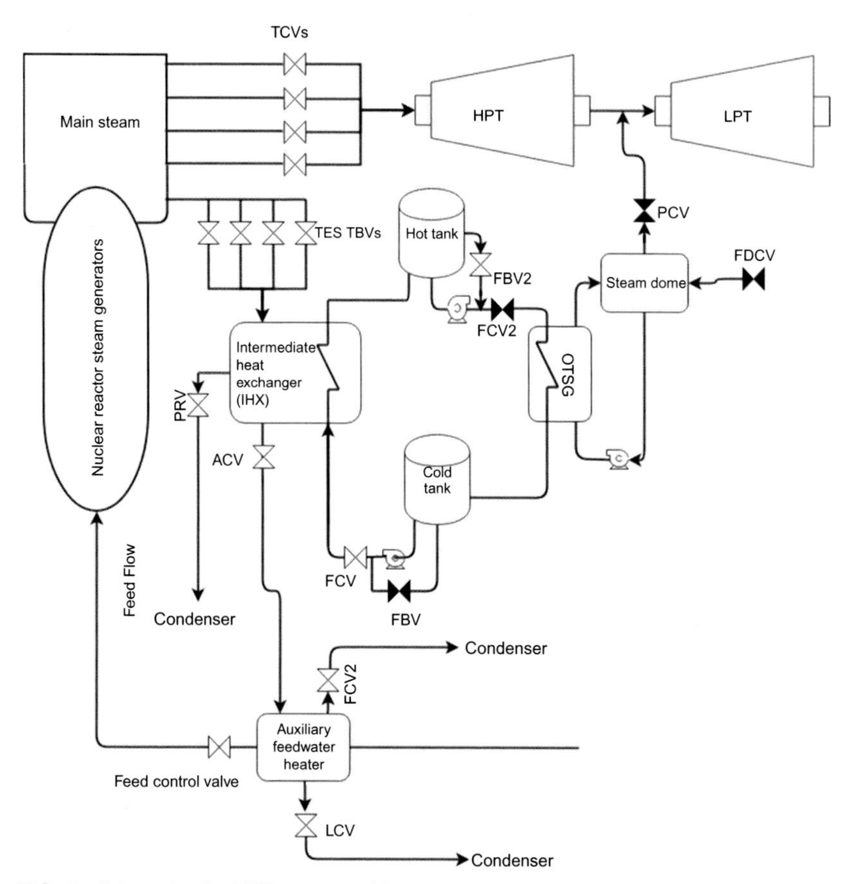

FIG. 5 Schematic of a PWR concept with a two-tank TES system, shown in discharge mode. *(Source: K. Frick, Modeling and Design of a Sensible Heat Thermal Energy Storage System for Small Modular Reactors, PhD Dissertation, North Carolina State University, Department of Nuclear Engineering, 2018. https://repository.lib.ncsu.edu/bitstream/handle/1840.20/34975/etd.pdf.)*

IHX with the condenser to prevent overpressurization of the heat exchanger during periods of low condensation rate. A nitrogen blanketing gas maintains a desired TES tank pressure.

Converse to the charging mode, during periods of peak demand, or when process steam is desired, the system is discharged by pumping TES fluid from the Hot Tank to the Cold Tank through the tube side of the Once Through Steam Generator (OTSG) producing a saturated liquid-vapor mixture. This two-phase mixture flows into a steam dome where it is separated into the gas and liquid phases. This saturated steam is then introduced into the turbine for electricity production or directed to some other process application through the Pressure Control Valve (PCV) at the exit of the steam dome. For operation as an electrical peaking unit integrated with a typical PWR, the steam is sent to the moisture separator/reheaters before entering the low pressure turbine. This conceptual TES application allows running a nuclear reactor more constantly and economical while allowing its electric output power and process steam generation to be more flexible, even replacing additional peaking power plants, thus potentially improving safety and decreasing costs through the addition of storage.

3.2 Solid thermal energy storage

3.2.1 Solid TES overview

Solid material thermal energy storage (TES) refers to sensible heat storage, where heat is brought to or removed from the material by a fluid through primarily convection and conduction. This intermittently heated and cooled solid TES structure is often referred to as a regenerator in the literature. There are two main categories of solid TES: fixed bed where the solid TES is stationary and a moving fluid exchanges heat with it, and a moving/fluidized bed where the solid TES is mobile along with the fluid that exchanges heat with it. Fixed bed solid TESs are a common arrangement and the focus of this chapter. TES systems employing solid storage media in a configuration such as fluidized particles may resemble liquid TES and are not considered much in this chapter. One such TES concept using fluidized sand with tubular heat exchangers and a hot and a cold silo is described by Hämmerle et al. [7]. With this in mind, the term "solid TES" will by default be referring to a fixed bed solid TES, and references to moving/fluidized solid TESs will be explicitly stated.

One fundamental difference between a solid TES and other thermal storage units such as two-tank liquid storage is the presence of a thermal gradient along the length of the storage material and heat transfer fluid. Details on solid TES thermal gradient behavior will be discussed in more detail in the Design Considerations section. Another key difference with solid TES compared to other TES types is that many solid materials can withstand higher temperatures than other storage materials. Therefore TES is particularly suited for high-temperature applications and has been used, e.g., for Cowper stoves in blast furnaces in steel making for more than a century with temperatures exceeding 1000°C.

While this book is focused on ways to store electricity in various forms, this solid TES section is specifically focused on how to store heat. Ultimately, these solid TES concepts can be incorporated into various power cycles to create an electricity energy storage system. Those system concepts will be discussed in other chapters.

3.2.1.1 Solid TES materials and structure

A solid TES is technically any TES where the storage material remains in the solid phase across the temperature range of its application. Therefore there are numerous materials that can be used in a solid TES, depending on the application, and also many ways to structure the TES design. This solid TES section will narrow the focus to three primary solid TES designs: tube-in-concrete TES, packed bed TES, and stacked brick TES. The purpose for narrowing the focus of potential solid TES designs is to give the reader more in-depth look into how solid TESs work, how they are designed, and the technical challenges involved with them. The three main TES technologies discussed in this section have enjoyed the most research and are considered by many to be the best options for high-temperature heat storage.

With this in mind, the reader can review the three different solid TES designs discussed later and recognize that different materials could be substituted into the different structures, and different structures could be utilized for any solid material. General requirements for the materials used in a solid TES are temperature resistance within the range of operating temperature, chemical and mechanical stability in conjunction with the working fluid, and chiefly low cost per unit of heat capacity. Many minerals, ceramics, and iron/steel fulfill these for different temperatures and working fluids. For further background on materials, the reader should refer to the ADELE project [8] where the German Aerospace Center (DLR) developed solid TESs and tested various kinds of inventory material including fireclay bricks, quartzite pebbles, and basalt gravel with pressure cycles to 65bar and temperature cycles beyond 400°C.

Some notable structures not discussed in this section include TES with flat plate channels between the storage material and solid TES rods where a fluid flows between them. The reader should refer to Xu et al. for more thorough description and explanation of these structures [9].

3.2.2 Tube-in-concrete

3.2.2.1 Description

Tube-in-concrete TES consists of tubes (or pipes) cast into a concrete. The tubes are usually made of some sort of steel, while the concrete recipe can vary widely. Fig. 6 shows an image of the most basic building block of a tube-in-concrete design where a single steel tube is cast into a concrete block. Fig. 7 gets more complicated by bending the tubes to create coils or other shapes.

The concrete recipe for tube-in-concrete can be any composite mixture of aggregate (e.g., gravel), binder (e.g., Portland cement), and additives (e.g., polypropylene fibers) which achieves the desired properties. Portland cement concrete is the most common form of concrete, but many other recipes have been proposed for specifically thermal storage applications [10–12]. For a TES, the concrete does not need to have the same strength properties that it does for building and pavement structures, which is what most concrete is used for. Instead, the goal is generally to get the most heat capacity and conductivity for as low as possible cost while meeting minimal structural requirements at the design temperature. It should also be noted that while many TES applications require

FIG. 6 Test article for a single tube-in-concrete TES (Bright Energy Storage Technologies).

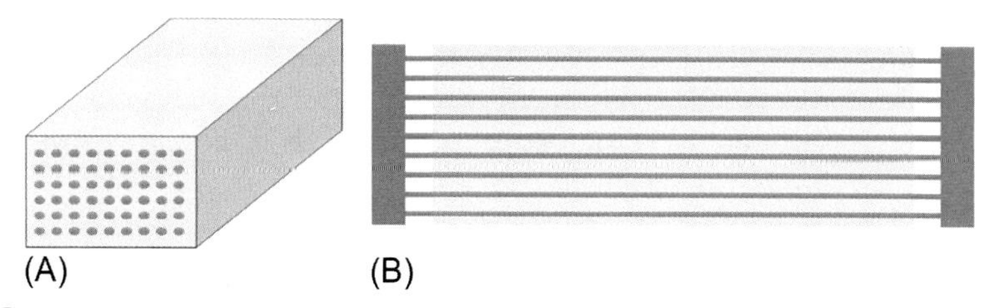

FIG. 7 Basic tube-in-concrete block design: (A) ISO view, (B) side view.

temperatures that well exceed 400°C, Portland cement concrete is unstable at temperatures over 400°C due to sequential dehydration and decomposition of its principal binding phases, gypsum, hydrated lime, and calcium silicate hydrate with increasing temperature. Concrete recipes designed specifically for high-temperature heat storage have been proposed that exhibit higher temperature resistance and improved thermal conductivity and specific heat.

While not technically concrete, another similar option is to pour fine particles such as sand or fly ash around tubes to create a TES. This has similar design features to tube-in-concrete TES, except there is no rigid contact between the thermal storage media and the tubes. This has the potential to lower cost, as no binder material is needed, but the main drawback is that air gaps between particles create low effective thermal conductivity for the TES system. Fin tubes can be used in this case (for an additional capital cost) to improve heat transfer and concepts have been developed for large, spiral-shaped heat stores with sand [13].

3.2.2.2 *Advantages and disadvantages*

Tube-in-concrete TES has several advantages over most other forms of TES technologies. The first is that it can be fabricated in modules in a factory

FIG. 8 Factory fabrication of tube-in-concrete TES modules (Bright Energy Storage Technologies).

(Fig. 8) before being shipped and stacked together onsite. This helps lower cost of fabrication and installation by leveraging economies of scale in a controlled manufacturing environment.

Another advantage of tube-in-concrete TES is that it can easily contain multiple working fluids within it, which may be beneficial for applications where heat is stored with one fluid, then extracted with another. The design only needs to add the other fluid passageways within the concrete, and the concrete acts as a heat transfer medium as well as storage medium between the fluids. Fig. 9 shows an image of a tube-in-concrete TES design that includes both natural gas exhaust and pressurized water/steam channels (Bright Energy Storage Technologies).

FIG. 9 Tube-in-concrete TES with channels for 2 fluid types (Bright Energy Storage Technologies).

Like some other solid TES concepts, tube-in-concrete TES utilizes a low-cost storage medium in the form of concrete, which is typically less than $100/ton. Tube-in-concrete TES has additionally been proven to be able to utilize fly ash, which is a by-product of coal burning. This not only provides a solution for cleaning up coal power plant waste, but it also helps lower the cost of tube-in-concrete TES since recycled fly ash is extremely cheap, and in some cases power plants may even pay someone to take their fly ash.

Although it is not an absolute truth, tube-in-concrete TES is typically better for higher pressure working fluids, as holding pressure within small tubes is generally cheaper than larger storage vessels that may be needed for other forms of solid TES. This is because as the mass of a cylindrical storage vessel increases proportionally to the square of the diameter of the pressure vessel. While this can be overcome in other TES designs by adding another heat exchanger and a low pressure heat transfer fluid, this hurts the thermodynamic performance by adding an additional heat transfer process, and likely with a fluid that does not match perfectly with the working fluid's temperature versus heat profile.

The main disadvantage of tube-in-concrete TES is that the steel tubes must be scaled up with longer duration storage. For tube-in-concrete TES, the tubes typically represent the largest cost piece of the system. As the storage duration gets extremely long, tube-in-concrete TES by itself may become uneconomical.

Another disadvantage of tube-in-concrete TES over other solid TES designs is that it generally has some of the poorest heat conduction when compared to other solid TES designs [14]. While this may seem quite bad, the optimal TES design must factor in its cost and performance together. For example, poor heat conduction can be overcome through more surface area, but at a cost. A detailed model that couples cost and performance together must be used to determine which TES design is best for a given application.

3.2.2.3 Technical challenges

The primary technical challenges for tube-in-concrete TES include decreasing thermal conductivity from increased temperature, thermal fatigue from coefficient of thermal expansion differences between the tube and concrete materials, and parallel flow instability potential for two-phase working fluid applications.

After fabrication of a tube-in-concrete TES and before using it at high temperature, the TES should be thoroughly dried, i.e., it should be heated up slowly for a prolonged period of time to remove much of its water content. The reason for this is that if the tube-in-concrete TES contains a lot of water and it is heated up quickly, it can cause the water within it to vaporize and build pressure in the pores, which can in turn lead to spalling (explosion) of the concrete.

Removing the water from the concrete has the downside of worsening the thermal properties of the concrete, particularly the thermal conductivity. Research by Bright Energy Storage Technologies has shown that even after a

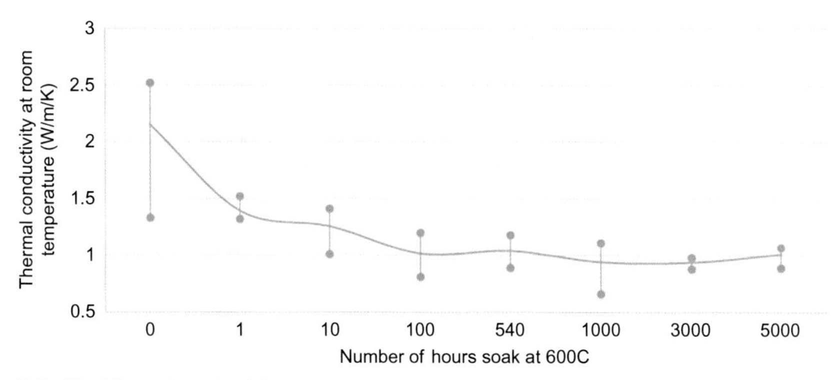

FIG. 10 Thermal conductivity (at room temperature) degradation from soaking concrete at high temperature (Bright Energy Storage Technologies).

thorough drying process, there can be further loss of conductivity of the concrete over the course of the first several hundred hours or more of use (Bright Energy Storage Technologies). Fig. 10 shows the degradation in thermal conductivity at room temperature for a concrete recipe held at 600°C for varying amounts of time. To conduct that test, concrete samples were brought to 600°C and held there for the given amounts of time before bringing the samples back to room temperature and measuring conductivity. The primary cause of this drop in conductivity is assumed to be due to further loss of water content in the concrete. As the concrete remains at high temperature, water within the concrete is forced into vapor and diffuses out of the concrete over time.

Other research has shown that the conductivity of concrete decreases when simply at a higher temperature, while the specific heat of concrete increases when at higher temperature [11, 15, 16].

Another technical challenge with tube-in-concrete TES is thermal fatigue. As can be seen, cracks may form near the tube/concrete interface, and propagate outward due to differences in thermal expansion between the steel tubing and the concrete as the tube-in-concrete article heats up and cools down. The only structural requirements of the tube-in-concrete TES are that it can hold its own weight, so cracking may be acceptable to a degree. Cracks may also hurt thermal conductivity by creating air gaps in the structure. Ultimately, the concrete recipe should be designed such that costs are minimized while load-carrying capacity and thermal conductivity requirements are met.

Parallel flow instability is a known challenge for two-phase flow applications where a fluid is either evaporating or condensing. As phase change occurs within a fluid, its density changes rapidly by an order of magnitude or more. As the density changes, the frictional resistance (pressure drop) also changes. For a heat exchange process like the one occurring in a tube-in-concrete TES, this may lead to instabilities in the flow distribution between tubes. If, for example, some tubes happen to have much higher heat transfer performance than others, during discharge of the TES those tubes will cool the local concrete next to themselves faster than other tubes in the TES. This in turn can lead to less vapor in those fluid channels, which leads to less flow resistance. The pressure balance would then lead to more mass flow in those tubes with higher heat transfer, which would in turn lead to even more relative cooling in those channels. This can lead to a runaway effect where some tubes end up cooling much faster than others, and hurting the overall performance of the tube-in-concrete TES.

Parallel flow instability in tube-in-concrete TES can be overcome by either designing the tubes to be highly symmetric within tight tolerances, and/or to put the main flow restriction in the manifolds where there is not a highly varying density as the block is heated or cooled.

3.2.2.4 Technology status

Tube-in-concrete TES is one of the more developed high-temperature TES technologies outside of molten salt TES. Some of the first work to develop tube-in-concrete TES was done by the German Aerospace Center. As of August 2012, they had designed and built a 20-m^3 tube-in-concrete test article and cycled it for approximately 13,000 h between 200°C and 400°C (roughly 600 cycles) in the water/steam test loop of the power plant of Endesa in Carboneras, Spain (Fig. 11) [17].

More recently, work has been done by EnergyNest out of Norway to develop their proprietary HEATCRETE concrete for tube-in-concrete TES (N. Hoivik, C. Greiner, J. Barragan, et al., Long-term performance results of concrete-based modular thermal energy storage system. J Energy Storage. 24 (2019) 100735.

FIG. 11 DLR: Cutaway image of the Züblin/DLR concrete embedded pipe solid TES prototype in the water/steam test loop of the Litoral power plant of Endesa in Carboneras, Spain. *(Courtesy of © Ed. Züblin AG.)*

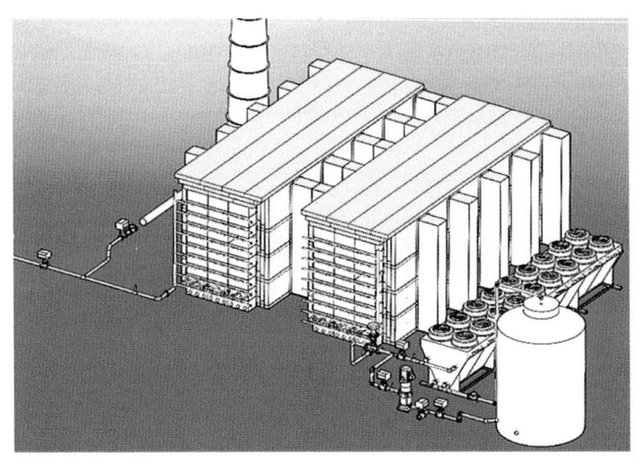

FIG. 12 Bright Energy Storage Technologies' 10 $MWh_{electric}$ tube-in-concrete TES pilot design (Bright Energy Storage Technologies).

doi:10.1016/j.est.2019.04.009) [11]. Over 20 months in 2017 and 2018, they demonstrated a 2×500 kWh_{th} tube-in-concrete TES at the Masdar Institute Solar Platform at temperatures up to 380°C.

Bright Energy Storage Technologies out of Colorado has been working on a higher temperature tube-in-concrete TES design for applications up to 600°C. They, along with the Electric Power Research Institute and AECOM, were awarded a DOE grant in late 2019 to demonstrate a 10-$MWh_{electric}$ (2.5 MW for 4 h duration) tube-in-concrete TES pilot plant [18]. Fig. 12 shows a schematic of their proposed pilot plant, which would be attached to a coal-fired steam plant that could be used to heat and cool the blocks with steam and water, respectively. Bright Energy Storage Technologies is primarily targeting steam power plant applications for the TES, such as natural gas, nuclear, coal, concentrated solar power, and geothermal power plants.

Similar to a Tube-in-Concrete solid TES, but in a very different arrangement, a sand-filled very long rectangular channel wound as a spiral has been published by Nielsen and Leithner [19]. Finned tubes run through this channel to transfer heat in and out of the sand. For improvement of conduction within the sand, liquid fillers such as oil or molten salt have been suggested. The advantage of the spiral arrangement is the compactness and the self-insulation by having the colder end on the outside and the hot end with pipe collector in the center.

3.2.3 Packed beds

3.2.3.1 Technology description

This solid TES type is in principle the simplest form, consisting of a vessel filled with a granular material with sufficient void fraction to allow the fluid to pass through. Packed beds are made of randomly oriented particles of storage material piled up in a vessel to contain and guide the flow of the working fluid. The working fluid passes through the porosity between the particles and transfer heat. Because of their large surface area when the particles are small, packed bed solid TESs have high heat transfer rates per unit volume and can have very small approach temperatures between the working fluid and the solid. In the process industry, packed beds for this reason are also used for catalytic chemical reactions and adsorption, e.g., dryers. While the wiggly and typically turbulent flow of the working fluid over the bed ensures good heat transfer and very thin boundary layers specifically for fine particles, packed beds at higher velocity create substantial pressure loss. This pressure loss limits the length of the packed bed for a required volume to store heat and, similar to other heat exchangers, necessitates the cross sectional area to grow. For a design with a steep thermal gradient and minimum inventory mass for a set approach temperature difference and maximum final temperature deviation, the ideal packed bed solid TES would be a long cylinder with very fine granulate. Pressure loss and practical considerations will lead to a compromise in design with larger diameter and coarser inventory material.

Axial conduction through a packed bed is typically very low because of the length between the warm and cold regions and the resistance between the particles' small contact area. Natural convection of the working fluid during standstill can be suppressed by orienting the packed bed such that the temperature gradient is vertically aligned with gravity with the higher temperature on top. Together with good insulation to the ambient, packed beds can store heat with minimal losses of heat and temperature (Figs. 13 and 14).

3.2.3.2 Applications

A large industrial application of packed bed solid TESs are recent Cowpers used for air preheating for blast furnaces in steel production in China using inch-size

FIG. 13 Packed bed made of alumina balls (http://refractoryonline.com/checker-brick/).

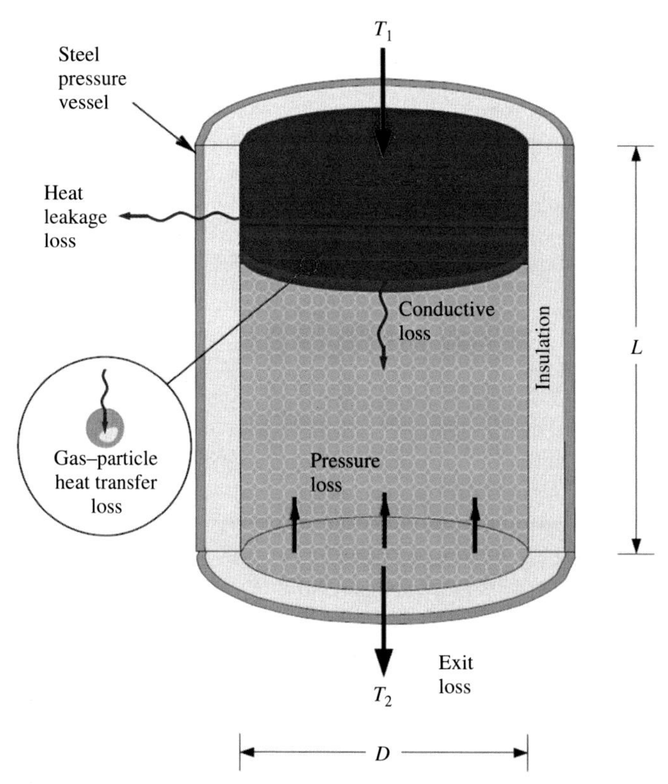

FIG. 14 Schematic representation of a packed bed thermal storage system with the main loss mechanisms. *(Source: J. McTigue. Analysis and Optimisation of Thermal Energy Storage, Department of Engineering, University of Cambridge, 2016.)*

ceramic balls instead of more typical checker bricks. Fired aluminum melting furnaces have used similar technology to recover heat from the flue gases at high temperature and preheat the combustion gas. Lower temperature applications include dryers for compressed air or other gases that have packed beds made of zeolite or silica gel have been used in drying or air conditioning processes that are operated intermittently with a moist cool gas and a warm dry gas for regeneration.

Several adiabatic compressed air energy storage (ACAES) concepts were published that included packed bed solid TESs with temperature-resistant mineral gravel or ceramic balls to store heat from compressed air. In this application, the pressure vessel becomes a major cost driver as pressures can be 65 bar or higher. Züblin and DLR in Germany successfully tested a section of a solid TES pressure vessel for 16 bar made of prestressed concrete with steel trusses in 2014 in Cologne [8]. Such solid TESs become more economical at lower pressures because of the cost of the pressure vessels at 65 bar may be uneconomical when made from concrete or steel even at small size. The use of solid TESs with packed beds of inexpensive heat-resistant minerals allows storing the heat of compression of ACAES processes at higher temperatures than, e.g., oil TES and, depending on pressure, less expensively than in molten salt TES.

A very different solid TES not requiring a pressure vessel has been built and tested by Siemens Gamesa with a 5 MWh prototype in 2016 and a 120-MWh heat storage pilot plant with a 1.4-MW steam turbine generator. This is a horizontally flowing packed bed solid TES filled with "volcanic rock" gravel that during discharge heats air used in an HRSG for steam production. For charging, the air is blown in a closed loop through an electric heating grid at a low, near atmospheric pressure, such that the containment can be made from internally insulated ordinary rectangular concrete walls (Fig. 15).

Isentropic Ltd in Fareham, UK has explored a solid TES design with a segmented packed bed of rather fine grit, in which the gas passes at any given time only through one or two segments instead of the entire bed length. The theory is that only a small fraction of the packed bed is actively exchanging heat, a thermal front of a steep gradient moving through the length of the solid TES, and opening segment by segment the gas bypasses the large volume of "inactive" bed material that largely adds pressure loss. By this, lower pressure loss is achieved and smaller granules with better heat transfer can be used and the inventory volume and cost reduced. The prototypes used internal rotating baffles for guiding air through the solid TES; the air was the working fluid in a pumped heat electricity system with storage temperatures between cryogenic and about 500°C (Fig. 16).

Liquids may also be used as the heat transfer fluid. For example, the DLR has built a packed bed test facility that uses molten salts as the working fluid. These packed beds are also unpressurized and potentially have applications in the concentrating solar power (CSP) industry. The combination of a packed bed with liquid can be considered a dual-media thermocline TES, a type which is

FIG. 15 Siemens Gamesa ETES prototype with 120 MWh_th packed bed TES in Hamburg, Germany. *(Source: Siemens Gamesa Renewable Energy.)*

FIG. 16 Isentropic Ltd segmented packed bed solid TESs for hot *(front)* and cold *(rear)* storage of a pumped heat electricity storage prototype in Fareham, UK (http://www.isentropic.co.uk).

described in detail in another section of this chapter. The molten salt TES vessel from DLR has a volume of 22 m^3 and is designed to enable different filler materials to be swapped in and out for testing. The nitrate molten salts operate up to a maximum temperature of 560°C and an electrical heater is used so that the salt can be circulated continuously (Fig. 17).

3.2.3.3 Technical challenges

Difficulties in packed beds may arise from temperature cycling as strain and resulting mechanical stress can damage the filling material and produce dust. Thermal expansion and settling of granular material in a silo lead to pressure distributions in the pressure vessel wall that are difficult to predict and deviate from a "hydrostatic" pressure, necessitating careful design of the vessel and any internal insulation. Depending on the size, size distribution, and settling, the porosity can change and the pressure loss of a working fluid for finer granulates in the vessel can become high, requiring larger cross sections and shorter bed lengths.

Pressure vessels are generally required for housing the packed bed solid TESs as the working fluid is often at an elevated pressure. Specifically for

FIG. 17 The DLR's molten-salt packed bed test facility. Molten salt tanks are in the bottom right of the picture, with the vertically orientated packed bed vessel on the left. *(Photo: by authors.)*

the very large solid TESs proposed in adiabatic compressed air energy storage concepts, the pressure vessels may become unprecedentedly large. Both steel and prestressed concrete vessels have been designed for this purpose. During the ADELE project, the German construction firm Ed. Züblin AG and the aerospace research center DLR developed solid TESs and tested a section of a concrete vessel wall [8]. In Cowper stoves with packed beds instead of more commonly used bricks, the pressures are lower (several atmospheres) and internally insulated concrete, masonry, or steel walls are used (Fig. 18).

Steel pressure vessels can be used instead of concrete for smaller size or when welding on site is possible. Inner insulation and cooling is needed for temperatures too high for the containment material. Suitable large steel pressure vessels are made by a few manufacturers as "heavy wall reactors" in dimensions of about 4.5 m in diameter and up to 300 mm wall thickness for refineries and other chemical processing plants. Typically creep resistant chromoly steel of higher yield strength (e.g., 16Mo3/A 387 Gr. P1, 13CrMo4-5/A 387 Gr. P12) is used, but for lower temperatures and pressures, inexpensive unalloyed pressure vessel steel (e.g., P355GH/A 516 Gr. 70) is suitable (Fig. 19).

3.2.3.4 Alternative packed bed designs

Packed beds are conventionally cylindrically shaped, with the flow axis aligned vertically in order to suppress buoyancy effects. Several alternative geometries have been proposed.

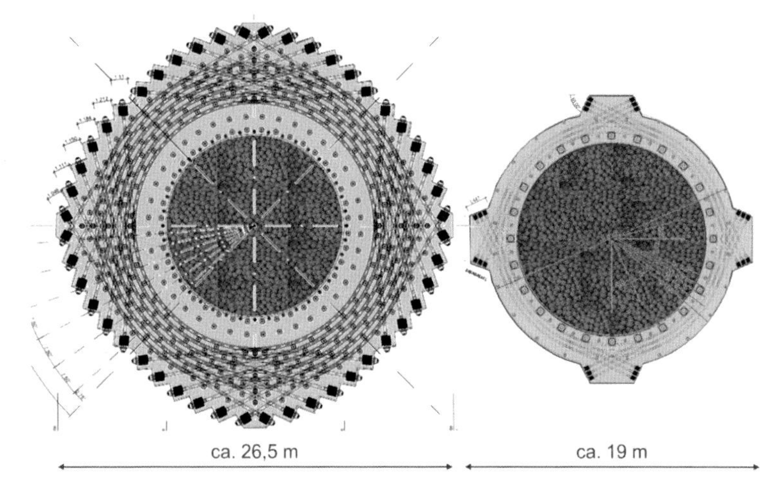

FIG. 18 Cross section of prestressed concrete pressure vessels for 65 bar *(left)* and 16 bar *(right)*. *(Source: ADELE-ING Engineering-Vorhaben für die Errichtung der ersten Demonstrationsanlage zur Adiabaten Druckluftspeichertechnik Öffentlicher Schlussbericht, DLR, 2018.)*

FIG. 19 A heavy wall reactor vessel for a refinery. *(Source: MAN Energy Solutions SE, DWE®.)*

Zangeneh et al. [20] developed a conical packed bed which has a larger diameter at the top than the bottom. This shape is intended to reduce the effect of thermal ratcheting, whereby particles are crushed due to their thermal expansion. The conical shape reduces the normal force on the particles and the sloped walls 'guide' them upwards as they expand. The study also pointed out that for hot storage systems the hottest region of the tank is at the top which has the lowest surface-area-to-volume ratio, thereby reducing heat leakage losses (Fig. 20).

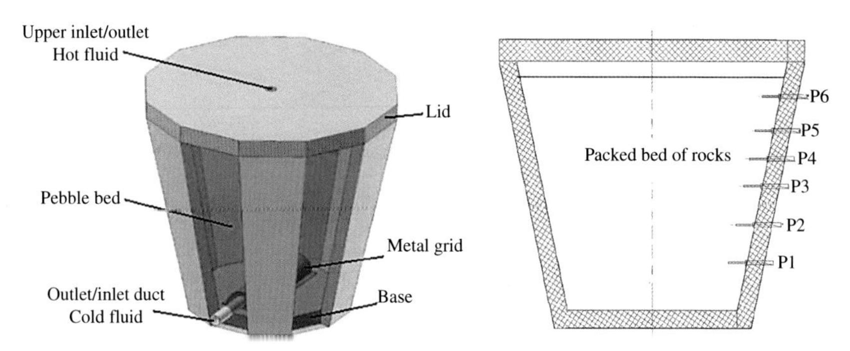

FIG. 20 Schematic of a conical packed bed from Zavattoni et al. [21].

Packed beds can be segmented into layers—effectively short packed beds—as first proposed by Crandall and Thatcher as a way to maintain a steeper thermal gradient and therefore higher energy densities. As described in a previous paragraph, the UK startup firm Isentropic Ltd. constructed segmented beds and noted that these systems could be controlled so that the heat transfer fluid only passes through the "active" segments, that is the segments where the thermal gradient exists and heat transfer thus occurs. The flow is diverted around the remaining segments, consequently reducing frictional resistance and pressure losses. Optimization studies found that segmentation affects the packed bed design. Using segmentation to reduce pressure losses allowed smaller particles to be used which improved heat transfer performance. Smaller particles also led to steeper thermal gradients and a higher energy density. The benefit of packed beds increases as the inlet pressure reduces, since pressure losses become more significant. Furthermore, segmentation preserves the shape of the thermal gradient during storage phases more effectively than an unsegmented bed, therefore reducing these conductive losses. However, due to additional complexity of segmentation and flow guidance with valves or baffles and the required extra volume, such packed beds have found no application outside of research and prototypes yet.

In a radial-flow packed bed, the heat transfer fluid enters an inner plenum before travelling along the radius rather than the vertical axis, and exiting via an outer plenum. These beds require internal structures to keep the particles in the correct location. Radial-flow packed beds have been used in the chemical industry, but only a few examples exist for energy storage. Since the area the fluid passes through increases along the radius, the fluid velocity decreases leading to lower pressure losses near the outer plenum. Bradley suggested that smaller particles could be placed here to improve heat transfer performance. One advantage of this geometry is that the hottest temperatures occur near the center of the bed rather than at the walls. Therefore the packing material itself is "self-insulating" and lower quantities of insulation are required. A techno-economic comparison of radial-flow packed beds with conventional packed beds suggested that both systems have similar efficiencies, but that the additional volume taken up by the plena lead to more costly containment and higher costs for radial-flow stores [22–24].

Combining sensible and latent storage media within a packed bed has been proposed. Phase change materials (PCMs) typically have higher energy densities than sensible materials and may reduce storage volumes. It has been suggested that PCMs can be used to reduce the fluid temperature variations that occur when the thermal gradient approaches the ends of the storage. The PCM is placed only at the ends of the packed bed and is used to stabilize the fluid exit temperature [20]. Multiple layers of phase change materials each with a different melting temperature may be used to approximate the shape of a thermocline. However, PCMs must be encapsulated to contain the material, and this is likely to lead to high costs. In addition, PCMs have relatively low thermal conductivities which negatively affects heat transfer rates (Fig. 21).

Packed beds may use fillers that have a variety of particle sizes. Pacheco et al. [26] investigated the compatibility of mixtures of crushed rock and sand with molten salts and thermal oils. Using a mixture of particles sizes reduces the void fraction and thereby leads to lower volumes of heat transfer fluid, potentially reducing the cost. In practice, having some distribution of particle sizes is unavoidable without increasing the cost of sourcing and processing the material.

3.2.4 Stacked bricks

3.2.4.1 Technology description

This type of solid TES is generally comparable to the packed bed solid TES, but instead of the loose fill of granular inventory, orderly stacked bricks with holes

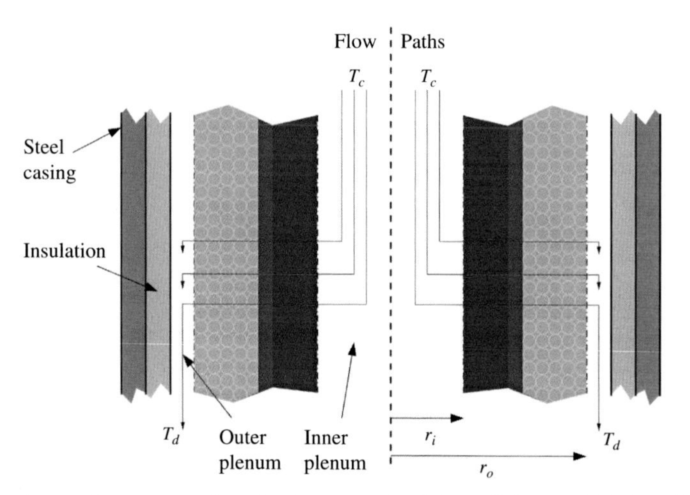

FIG. 21 Schematic cut-away of a radial-flow packed bed from McTigue and White [25].

FIG. 22 Checker bricks (http://refractoryonline.com/checker-brick/).

for channeling the working fluid are used to store the heat. The bricks are stacked such that their internal vertical channels line up to allow the gas to pass with minimal pressure loss and are often referred to as "Checker bricks."

The bricks feature frequent holes or honeycomb structures to increase the porosity and surface area for good heat transfer and have small wall thicknesses on the order of centimeters to reduce temperature losses from conduction path length. The bricks are made of refractory material for very high-temperature resistance beyond 1000°C without cracking or corrosion such as Cordierite, Corundum, Mullite, or Silicon Oxide (Fig. 22).

The bricks are typically housed inside long vertical cylindrical vessels, just as packed bed solid TESs. Their thermal performance is fundamentally similar and can be calculated the same way. With long continuous channels, the heat transfer coefficient is obtained from pipe flow correlations. Advantages over packed beds include that the material is meant for very high temperatures unlike most natural minerals, and the pressure loss can be low.

3.2.4.2 Applications

The main application for solid TESs with checker bricks is Cowper Stoves for blast furnaces in steel making, also known as Hot Stoves. These regenerators use the hot flue gases emanating from iron and coke conversion to heat the bricks with near atmospheric pressure. For cleaning the furnace exhaust gases and providing additional heat to the bricks, the waste gas from the blast furnace is mixed with air and combusted in the combustion chamber of the Cowper at temperatures up to 1500°C, before entering the solid TES section. During the discharge period, fresh air is driven by a fan at an elevated pressure of up to 10 bar across the bricks. After being heated to 1200°C, this air called hot blast is brought into the bottom of the blast furnace for iron ore smelting (Fig. 23).

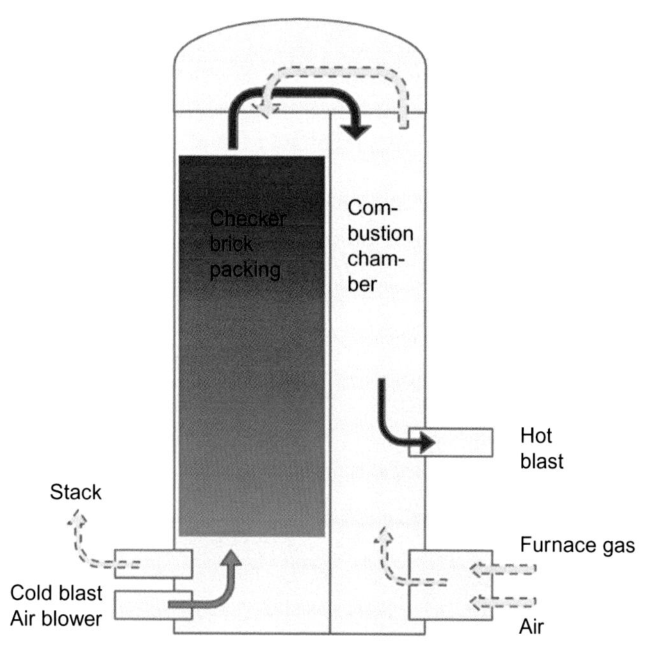

FIG. 23 Cowper stove solid TES schematic (discharging flow arrows in color, charging flow in gray dotted). *(Source: S. Freund.)*

Cowpers are typically constructed with a steel shell that is lined on the inside with firebrick for insulation. The dome on top that turns and distributes the hot gases across the storage material and the combustion chamber are also lined with layers of high-temperature resistant bricks to insulate against the intense heat. The height reaches up to 40 m with solid TES lengths of about 30 m and total vessel volumina up to 5000 m^3. Typically three units are installed and operated such that at any given time one or two are being charged with combusting gas while one is being discharged with fresh air to provide the hot blast to the bottom of the blast furnace. Cycle times are typically about 30 min. to 1 h and the energy stored may be up to 200 MWh with blast flow rates of 200 kg/s.

3.2.5 Analysis Methods

For all solid TESs except moving solids there is a moving temperature gradient throughout the TES, as described throughout this section. This means that any analysis method trying to approximate the performance of a solid TES with temperature gradients must be able to account for the dynamic (transient) behavior of that TES. This is different for moving solid TESs or two-tank fluid TESs where the dynamic behavior is negligible for high-level performance approximations, and the analysis becomes much more trivial. For this analysis methods discussion we will use the terms solid TES to mean a solid TES with a temperature gradient.

In general, analysis of a solid TES should include a simulation over time. While there may be ways to approximate the net behavior (e.g., the round-trip efficiency) of a solid TES without a time-domain simulation, any such methods would require a time-domain estimation of the TES behavior to be validated. For this section we will only be focusing on how to model solid TESs dynamically, and leave it to the reader to do further research and thinking on how one could approximate net solid TES behavior without a time-domain simulation.

3.2.5.1 Governing equations

Regardless of the modeling approach or type of solid TES, the governing equations needed to model a solid TES include:

- Energy conservation
- Mass conservation
- Momentum conservation
- Fluid state modeling
- Solid state modeling
- Heat transfer

The way this typically works is there needs to be a fluid channel where energy, mass, and momentum conservation are maintained throughout the fluid channel. The fluid state (temperature, pressure, enthalpy, etc.) is calculated via a fluid state model. A heat transfer model is then used to calculate the heat transfer between the fluid and the edges of the fluid channel where the solid TES will be located. The solid itself then also needs to include a way to approximate its state (e.g., specific heat versus temperature) during the simulation.

Many simplifications can be taken for these equations, such as:

- constant specific heat of the solid
- ideal gas law for some fluids
- neglecting changes in pressure or density to simplify the mass and momentum conservation equations

Nevertheless, all the aforementioned equations should be accounted for in any dynamic analysis of a solid TES.

3.2.5.2 3D/2D simulation

Computational fluid dynamics (CFD) code is a detailed way to model simple to complex fluid flows in typically 2D or 3D geometries. CFD can also be coupled to heat transfer finite element analysis (FEA) to provide all the necessary methodology to model a solid TES in 2D or 3D. While this method of solid TES analysis provides the highest fidelity analysis, it is very computationally expensive, and in most practical cases too slow to be very useful as of the writing of this book. It can, however, be quite useful as a tool to supplement a 1D model by spot checking and fitting a 1D model to a 2D or 3D model.

3.2.5.3 1D simulation

1D simulation is the most common method used to model solid TESs. This method provides the best trade-off between computational speed and accuracy, and enables the analyst to:

- Incorporate the solid TES into a larger 1D system model
- Calculate many design points to optimize solid TES designs and analyze design sensitivities
- Develop control systems
- Analyze the behavior of the solid TES under many types of operational conditions

The common way to create a 1D solid TES model is to discretize the solid TES along its length to account for changes in its temperature profile along the length. The fluid passages are then modeled with a control volume approach, where the control volumes fundamentally utilize the Reynold's Transport Theorem to be able to account for the three conservation laws (energy, mass, and momentum). The heat transfer into/out of the control volumes is then modeled in the radial direction toward the solid TES material(s). A useful derivation of lumped capacitance heat transfer models for common solid TES geometries was published by Xu et al. in 2012, which can be used for 1D TES modeling [9]. Heat transfer may also be modeled axially for the solid material along the TES length (between solid TES segments), but in most cases this is a small effect which is neglected in the model for simplicity. Such an approach is described further in the section Simplified Calculation Method.

Fig. 24 illustrates how a solid TES model could be split up into five segments. The red arrows represent heat flow from each fluid control volume to

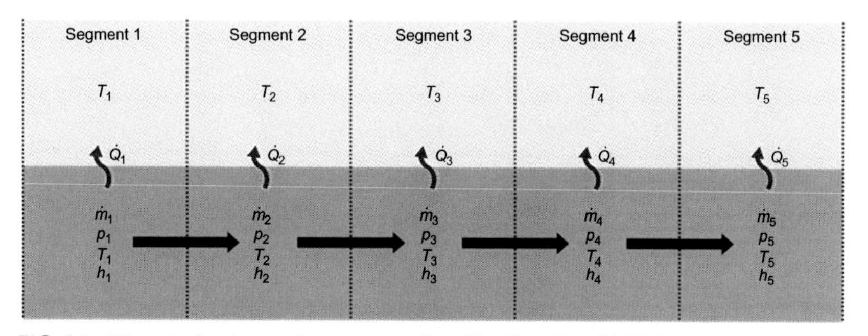

FIG. 24 Discretized volume scheme commonly utilized for 1D solid TES modeling.

the adjacent solid TES material of that control volume. The black arrows represent the fluid flow direction between control volume segments. Some but not all of the key fluid and solid state properties are also listed.

Many simplifications are possible in this 1D modeling scheme, but which of those simplifications are appropriate depends on the type of fluid and solid TES design being modeled. The reader should take caution when applying simplifications and make sure that they apply to the type of TES being modeled.

3.2.5.4 Packed bed 1D model example

A variety of heat transfer processes occur within packed beds, including convective heat transfer between the packing material and heat transfer fluid, conductive and radiative heat transfer between particles, conduction within particles, and heat leakage out of the storage. For cylindrical packed beds the dominant heat transfer mechanisms occur along the axis of fluid flow, but nonuniformities within the packed bed can also lead to radial effects and nonuniform fluid flow distributions. Other packed bed geometries may require three spatial dimensions to be modeled (Fig. 25).

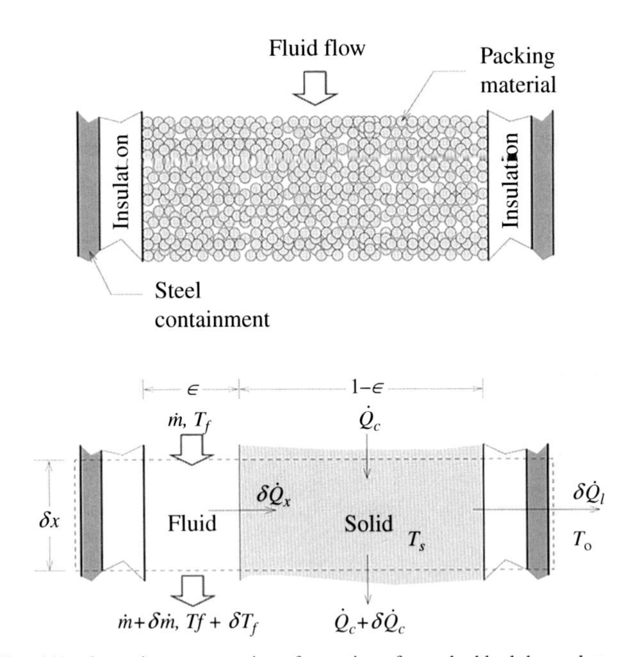

FIG. 25 *Top*, (A) schematic representation of a section of a packed bed thermal storage. *Bottom*, (B) simplified model of heat transfer modes. *(Figure adapted from White, Alexander J., Joshua D. McTigue, Christos N. Markides: Analysis and optimisation of packed-bed thermal reservoirs for electricity storage applications, Sage Eng. 2016, https://doi.org/10.1177/0957650916668447.)*

Despite the apparent complexity of packed bed thermal storage, a number of simplifying assumptions can be made that enable relatively straightforward solutions to the governing equations while capturing the dominant thermodynamic behavior. These simplifications include:

(1) Thermal gradients within particles may be ignored if the internal thermal resistance is small compared to the thermal resistance at the surface, i.e., the Biot number is small
(2) The governing equations are one-dimensional for a cylindrical packed bed if radial variations are insignificant. Two factors predominantly contribute to radial variations: heat leakage from the storage walls and variations in packing density. Heat leakage leads to radial temperature changes, but can be simply reduced by increasing the insulation thickness and decreasing the surface-area-to-volume ratio of the packed bed (larger stores). The void fraction (or packing density) is not constant along the radius, and the bulk packing density is disrupted by the presence of the wall. Void fractions of 1 occur at the wall, and then decrease to 0.3–0.4 as distance from the wall increases. The lower frictional resistance at the wall results in higher fluid velocities or 'bypass flow.' These variations typically occur within five particle diameters of the wall, and previous work has suggested these radial effects may be neglected if the ratio of the bed diameter to particle diameter is $D/d_p > 40$.
(3) In practice, particle sizes and shapes may vary. However, it is sufficient to assume that the particle heat transfer characteristics can be represented by an average equivalent diameter, which implies the particles are spherical and uniformly sized.
(4) The mass flow rate varies along the length of the bed due to density variations in the fluid. However, the thermal gradient moves relatively slowly compared to the fluid velocity meaning that the density at a given location does not change quickly. Therefore the momentum equation does not need to be solved or coupled to the energy equations.
(5) Solid and fluid properties (density, heat capacity) vary with temperature and can have a significant impact on the shape of the thermal front. For instance, solid heat capacities tend to increase with temperature, and this leads to hotter regions of the thermal front travelling more slowly. While it is not overly complicated to include temperature-dependent properties in packed bed models, assuming average values of the properties can usually provide reasonable results.
(6) Conduction occurs along the axis of fluid flow, and several mechanisms contribute to the *effective* thermal conductivity including conduction through the particle bulk, conduction between the surfaces of particles, radiation between particles, radiation between voids, conduction through the fluid film adjacent to the contact surface of particles, convection between the fluid and solid, and mixing of the fluid (turbulent transport). Various correlations have been developed for the effective

conductivity and estimates vary widely. However, models that include axial conductivity indicate that this term does not significantly affect the shape of the thermal gradient. On the other hand, axial conduction can contribute significantly to exergy losses (depending on particle size, temperatures, and operation). This is particularly relevant to the storage period when the packed bed is neither charging or discharging, and axial conduction acts to erode the steepness of the thermal gradient, thereby destroying exergy.

Assuming one-dimensional flow with no heat leakage or temperature variations within the particles leads to the following coupled energy equations for the fluid and the solid:

$$\varepsilon\frac{\partial}{\partial t}(\rho_f e_f) + \frac{\partial}{\partial x}(\rho_f u_o h_f) = (1-\varepsilon)S_v h(T_s - T_f) + k_{eff}\frac{\partial^2 T_f}{\partial x^2}$$

$$\rho_s c_s(1-\varepsilon)\frac{\partial T_s}{\partial t} = S_v h(1-\varepsilon)(T_f - T_s)$$

where ε is the void fraction, ρ is the density, e is the internal energy, $u_o = \dot{m}/A$ is the superficial fluid velocity (where $\dot{m}$ is the mass flow rate and A is the packed bed cross sectional area), h_f is the fluid enthalpy, $S_v = 6/d_p$ is the particle surface-area-to-volume ratio (where d_p is the particle diameter), T is the temperature, k_{eff} is the effective conductivity, c_s is the solid heat capacity, h is the heat transfer coefficient, and subscripts f and s correspond to the fluid and solid, respectively.

Noting that, for the fluid $h_f = e_f + p/\rho_f = c_p T_f$, the fluid equation may be rewritten as follows:

$$\rho_f c_p\frac{\partial T_f}{\partial t} + \rho_f u_o c_p\frac{\partial T_f}{\partial x} = (1-\varepsilon)S_v h(T_s - T_f) + k_{eff}\frac{\partial^2 T_f}{\partial x^2} + \varepsilon\frac{\partial}{\partial t}(p/\rho_f)$$

where p is the pressure and c_p is the fluid isobaric heat capacity. The unsteady pressure term is relatively small compared to the other terms and may be neglected.

Simple numerical schemes may be developed to solve equations x and y—numerical schemes are particularly straightforward if the conductive term is neglected. As an example of the effectiveness of this approach, Fig. 26 compares numerical results using these simplified equations (with no conduction) to experimental results for a packed bed operating between 251°C and 66°C with alumina packing and air as the heat transfer fluid. The figure shows the discharging cycle in terms of nondimensional parameters—the thermal gradients travel from right to left (Figs. 27 and 28).

The energy equations given previously are coupled and the simultaneous solution of these mathematically stiff equations can be computationally time consuming as small time steps and grid steps are required to achieve a stable, accurate result. Further simplifications may be made to the governing energy equations. The "one-equation" model may be derived by assuming that the fluid and solid temperatures are equal which leads to

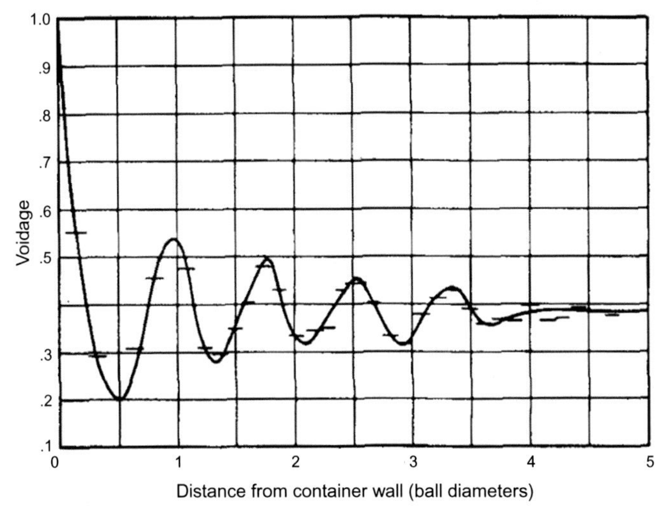

FIG. 26 Experimental results showing the variation in void fraction with distance from the wall for uniform spheres in a packed bed. *(Figure from Benenati and Brosilow.)*

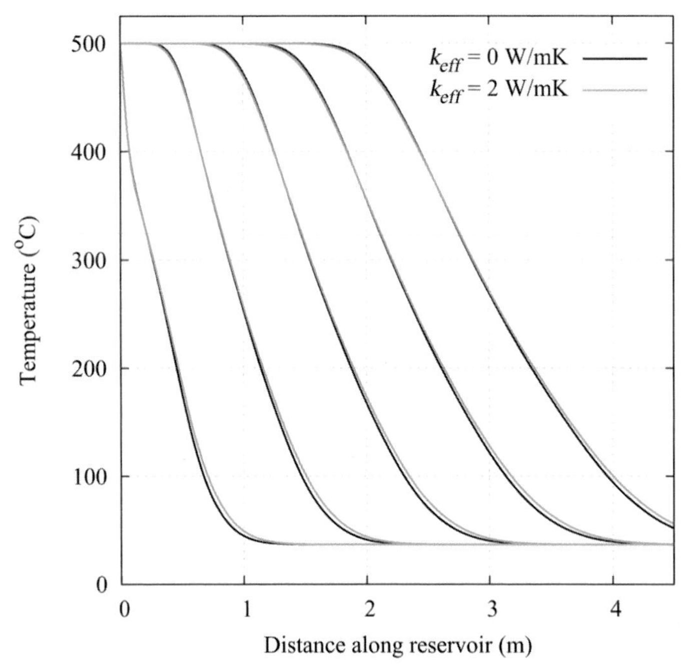

FIG. 27 Impact of effective conductivity on the thermal gradient shape within a packed bed. *(Figure from J. McTigue. Analysis and Optimisation of Thermal Energy Storage, Department of Engineering, University of Cambridge, 2016.)*

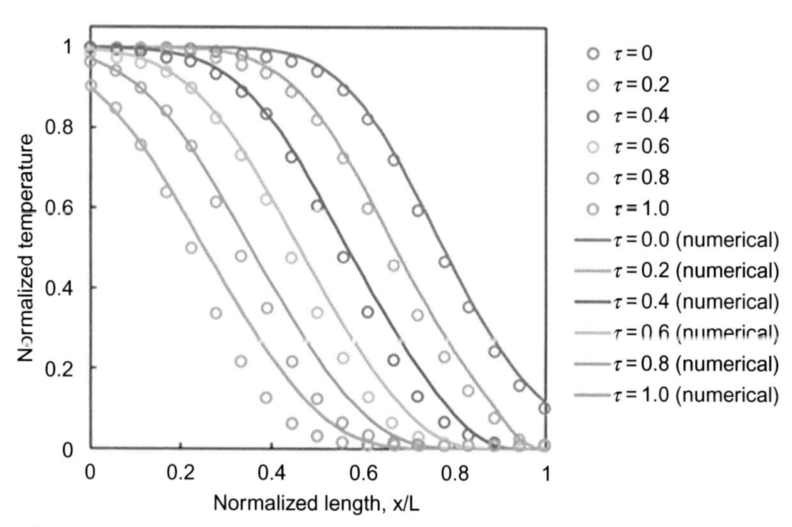

FIG. 28 Validation of numerical results against experimental results (Cascetta et al.) for a discharging sensible heat packed bed.

$$\left[\varepsilon\rho_f c_p + (1-\varepsilon)\rho_s c_s\right]\frac{\partial T}{\partial t} + \rho_f u_o c_p \frac{\partial T}{\partial x} - k_{eff}\frac{\partial^2 T}{\partial x^2} = 0$$

The $[\varepsilon\rho_f c_p + (1-\varepsilon)\rho_s c_s]$ term may be considered as an effective volumetric heat capacity $(\rho c)_{eff}$. Rewriting this equation in terms of the time derivative leads to

$$\frac{\partial T}{\partial t} + \frac{\rho_f u_o c_p}{(\rho c)_{eff}}\frac{\partial T}{\partial x} - \frac{k_{eff}}{(\rho c)_{eff}}\frac{\partial^2 T}{\partial x^2} = 0$$

From which numerical scheme can be developed.

An analytical solution to the governing equations was presented by Anzelius in 1926 and Schumann in 1929. The solid equation is written in terms of the time derivative while the fluid equation is written in terms of the spatial derivative. The unsteady fluid term $(\partial T_f/\partial t)$ can be effectively ignored if the unsteady accumulation of internal energy within the fluid is small compared to the other terms. This is a reasonable assumption for gaseous working fluids which have a low heat capacity per unit volume compared to the solid filler. In the absence of heat leakage, conduction, or pressure losses, the energy equations are then written as follows:

$$\frac{\partial T_f}{\partial x} = \frac{T_s - T_f}{\lambda}, \quad \frac{\partial T_s}{\partial t} = \frac{T_f - T_s}{\tau}$$

Where

$$\lambda = \frac{1}{(1-\varepsilon)S_v St}, \quad \tau = \frac{\rho_s c_s A}{c_p \dot{m} S_v St}$$

where St is the Stanton number. This form is commonly known as the Schumann equations. The equations are nondimensionalized and transformed leading to

$$\frac{\partial \theta_f}{\partial \xi} = \theta_s - \theta_f, \frac{\partial \theta_s}{\partial \eta} = \theta_f - \theta_s$$

where $\theta_{f,\, s} = (T_{f,\, s} - T_{out})/(T_{in} - T_{out})$ and $\xi$and η are dimensionless length and time variables given by

$$\xi = \frac{x}{\lambda}, \eta = \frac{t}{\tau}$$

Assuming solid and fluid properties do not vary with temperature allows for an analytical solution for the temperature difference between the fluid and the solid was first presented by Anzelius

$$\theta_f - \theta_s = \exp\left[-(\xi + \eta) I_0\left(2\sqrt{\xi\eta}\right)\right]$$

where I_0 is the zero-th order modified Bessel function of the first kind. Full analytical equations for $\theta_f$and $\theta_s$were developed by Schumann, but they require evaluation of an infinite sum of Bessel functions. Thus numerical methods may be solved more efficiently and with fewer assumptions than the analytical results.

3.2.5.5 Simplified calculation method

Following the considerations given previously and neglecting some secondary and 3D effects leads to a practical approach for solid TES calculations. A 1D transient finite volume method where the TES is discretized in the flow length and in the time domain has been used by many researchers and TES designers to calculate the temperature profiles. It was originally described by Willmott [27] to calculate the temperatures in a Cowper for a blast furnace that uses stacked checker bricks as inventory. A similar method is described by Stuckle et al. [28] for the application of an embedded tube concrete solid TES. The reduction of dimensions that greatly simplifies calculations can be achieved through two assumptions and by lumping the heat transfer in an individual particle or solid TES material wall segment. The first assumption is that of a radially uniform cross sectional solid and fluid temperature along the axial flow path of the working fluid. Second, axial conduction both within the fluid and the solid can be neglected. To use a mean solid temperature and neglect the thermal gradient within the particle or wall segment that would introduce another dimension, essentially a resistance is added to the inverse of the surface heat transfer coefficient. The heat transfer coefficient between the solid surface and the fluid comes from a Nusselt number correlation for the channel geometry and fluid properties. The added resistance includes the solid thermal conductivity, a mean conduction path length from the geometry of the material, and a factor accounting for the frequency of temperature profile reversals. For thin walls or small particles or for high solid conductivity or low heat transfer rates, this correction can become very small. With these simplifications, the governing equations for the fluid energy and the solid energy in time and axial space become the following:

$$\rho_f c_f \frac{\partial T_f}{\partial t} + \rho_f c_f \frac{\dot{V}_f}{A_{c,f}} \frac{\partial T_f}{\partial x} = 0 (\text{fluid}),$$

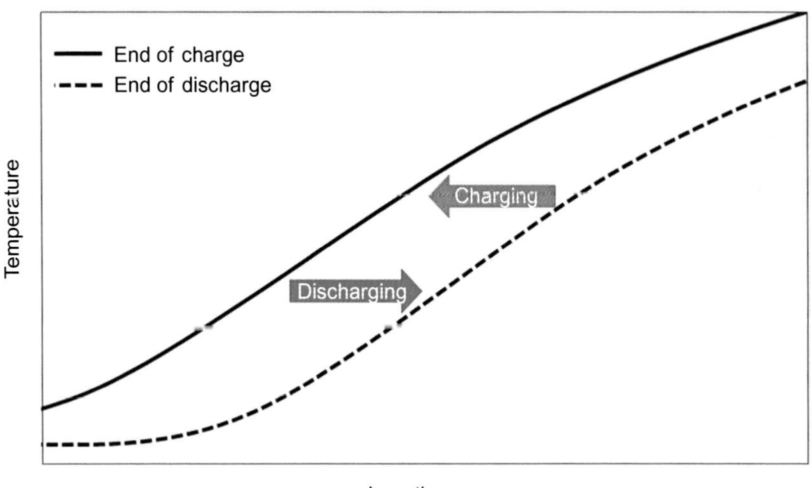

FIG. 29 Solid TES temperature gradients.

$$\rho_s c_s V_s \frac{\partial T_s}{\partial t} + \alpha_L A_L (T_s - T_{amp}) = \alpha_H A (T_f - T_s) \text{ (solid).}$$

Here subscript f denotes the fluid and s the solid, L refers to ambient losses, αH is the corrected heat transfer coefficient, with standard nomenclature for the rest. For the fluid, the change in temperature of a volume over time balances with the change in temperature along its path. For the solid, the change of a volume over time corresponds to the heat exchange with the fluid and the losses to the ambient. Calculations of these coupled 1D local differential equations in the time domain have been done with various discretization schemes and solvers (e.g., Finite Difference Methods) to find the temperature profiles. Stuckle et al. [28] used the Modelica process simulation software for their storage system and coded a discretized volume model of the simplified solid TES to run in the transient simulations with time steps on the order of seconds. With the measurement data from the prototype solid TES storing heat for steam at a power plant, they successfully validated their calculations. The boundary conditions for solving the coupled differential equations toward a solution at a certain time are first the initial temperature conditions in the inventory along the flow path and second the inlet temperature and flow rate of the working fluid, which can change over time. If initial conditions are not known, a guess can be made and calculations repeated for several cycles. The results after sufficient iterations within each time step to achieve a desired numerical accuracy and over many time steps to develop a realistic temperature profile that shows the characteristic curves of the solid and gas temperatures closely following each other, comparable to the curves in Fig. 29.

3.2.6 Design considerations

3.2.6.1 Temperature gradient effects

A unique phenomenon related to storing heat in a sensible, nonmoving solid TES is the dynamic temperature gradients in them. Fig. 29 shows a conceptual example of how the temperature gradients can move through a solid TES over the course of charging and discharging. This shifting of the temperature gradient poses multiple challenges for both prediction of system performance, as well as for operation and control of the system.

As the solid TES is charged, the temperature gradient shifts up and toward the cold end of the block. This leads to the cold end getting hotter over time, which leads to the fluid leaving the TES getting hotter over time. The same type of effect happens in the discharge process where the hot end of the block cools over time. With regards to control, this means that the rest of the system that is integrated with the TES needs to account for this varying outlet temperature over time. With regards to performance prediction, this means that the heat being absorbed by the block is varying over time, and any analysis of such a system needs to account for these dynamics.

One way to control the outlet temperature of a solid TES over time is to have a bypass stream that goes around the TES and mixes or exchanges heat with the outlet of the TES. For charge mode, this would involve bypassing hot fluid around the TES and mixing or exchanging heat with the fluid leaving the cold end in order to maintain a steady, albeit warmer exit temperature of the TES over time.

Variations in the working fluid outlet temperature over time may be undesirable and have an adverse effect on the system that the TES is integrated into. The magnitude of temperature variations can be controlled but impacts the performance of the storage, specifically the efficiency and energy density.

One way to begin to understand the influence of the outlet temperatures is to consider "cyclic" or "steady-state" operation of the storage. In this mode of operation, charge-discharge cycles are repeated under identical inlet conditions and operating durations until the behavior of the solid TES in the nth cycle is the same as the $(n-1)$th cycle, and the thermodynamic state is the same at the start of every charge cycle thereafter. This implies that the energy and entropy content of the TES bed is the same at the start of each charge cycle, that is, there is no net change in energy or entropy. We know, however, that charging and discharging the TES is not a reversible process due to frictional pressure losses, heat transfer between the fluid and TES inventory over a finite temperature difference, and other dissipative heat transfer processes such as conduction. Therefore there must be entropy rejection in each cycle to achieve a net zero change in entropy at the start of each new charge cycle. To do this, heat must be rejected by the TES for each cycle, which is the reason that some change in exit temperatures over the course of charging and discharging is fundamentally unavoidable, i.e., the charge outlet temperature will be higher than the cold inlet from discharging, and the discharge outlet lower than the hot inlet from charging.

The magnitude of the outlet temperature change is therefore linked to the irreversibilities within the storage and is intrinsically related to the steepness of the thermal gradient. (The thermal gradient thickness is related to the size of the temperature difference between the fluid and TES material, and thus to one source of entropy generation.) As a result, large outlet temperature variations typically correspond to larger losses and a lower round-trip efficiency. However, the solid TES designer has to balance this against the energy storage density. Higher energy storage density leads to lower round-trip efficiency, but also lower capital cost.

Fig. 30 shows the relation between outlet temperature change and energy content. Larger outlet temperatures correspond to the thermocline moving greater distances along the storage, and therefore a more cost-effective use of the TES material. This is partly facilitated by steeper thermal gradients, which then lead to larger losses. The trade-off between round-trip efficiency and energy capacity is shown in Fig. 31.

Consequently, when designing a TES several competing factors must be considered, including maximizing the utilization of the storage system (to minimize capital cost); maximizing the storage efficiency; and considering the constraints, requirements, and flexibility of the system that the storage is part of. Finally, it should also be evident that the thermal gradients act as a kind of 'memory' of the storage system. Therefore any variations in the charging or discharging cycle (such as changes to mass flow, inlet temperatures, or charging durations) will have not only a transient effect on the storage performance, but

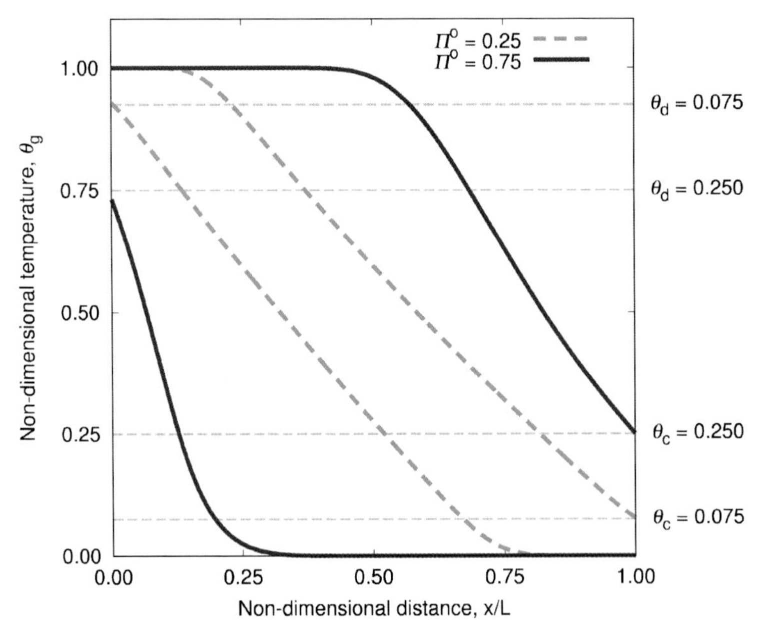

FIG. 30 Comparison of thermal gradient positions and shapes and the end of charge *(right)* and end of discharge *(left)*. Larger outlet temperatures lead to steeper fronts and larger energy capacities. *(From J.D. McTigue, A.J. White. A comparison of radial-flow and axial-flow packed beds for thermal energy storage. Appl. Energy 227 (2018) 533–541.)*

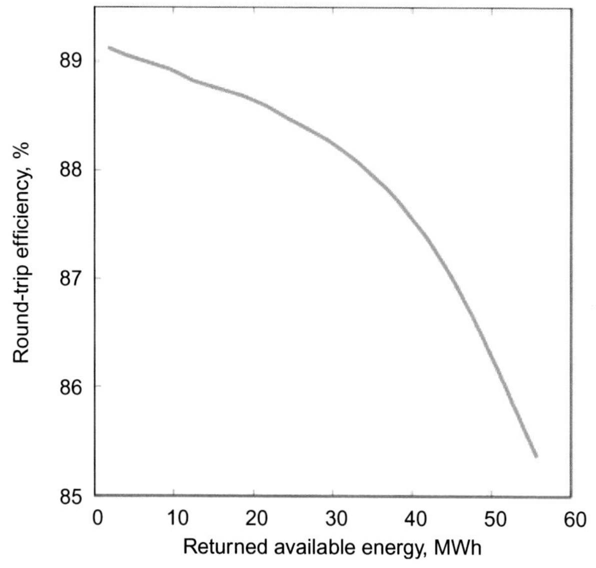

FIG. 31 Trade-off between efficiency and exergy density in cyclic operation for a packed bed. *(From J.D. McTigue, A.J. White. A comparison of radial-flow and axial-flow packed beds for thermal energy storage. Appl. Energy 227 (2018) 533–541.)*

also a longer term impact as the storage gradually returns to its steady-state behavior in subsequent charge-discharge cycles.

3.2.6.2 *Direct vs. indirect heat exchange*

Another consideration when designing a solid TES is whether to use a direct or indirect heat exchange process between the TES and the working fluid for the power cycle. Fig. 32 illustrates the difference between the two design approaches. For a direct heat exchange TES, the working fluid exchanges heat directly with the solid TES. For an indirect heat exchange TES the working fluid first exchanges heat with a heat transfer fluid within a separate heat exchanger, and the heat transfer fluid then exchanges heat with the solid TES.

The more typical design of the two is to do a direct heat exchange process. Direct heat exchange TES designs require only one heat exchanger to be purchased (the TES itself). Also, having less heat transfer processes can lead to less exergy loss due to temperature differences across the heat exchangers. Moreover, adding a heat transfer fluid to the design creates additional compatibility challenges including heat versus temperature matching at the desired pressures and material compatibility in particular.

That being said, sometimes there can be advantages to an indirect heat exchange design. Sometimes the working fluid and solid TES design are not compatible material wise, so a heat transfer fluid is needed to exchange heat

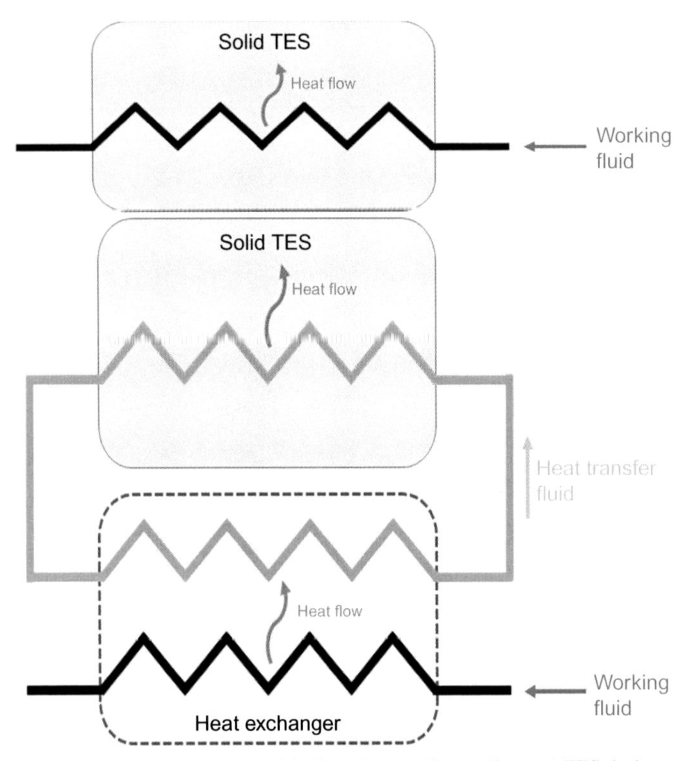

FIG. 32 Direct heat exchange *(top)* and indirect heat exchange *(bottom)* TES designs.

with the TES. An additional factor is that whatever heat exchanger is incorporated into the solid TES itself will need to scale with energy storage capacity. For long-duration energy storage (dozens of hours or more) this can become a capital cost challenge. If the working fluid operates under high pressure (as with most power cycles) then the working fluid heat exchanger will be costly. With the heat exchanger being a significant cost, there will be a point where it will not be cost effective to scale the working fluid heat exchanger with energy storage capacity. In other words, the longer the energy storage duration gets, the less likely a direct heat exchange TES design will be cost optimal.

3.2.6.3 2-Phase working fluids

When storing the latent energy of a working fluid, such as the condensation energy of steam, there are special considerations needed when using a solid TES. Since the solid TES can only store energy in sensible form, this means that a fairly large temperature difference must be induced in the working fluid saturated temperature point between charge and discharge. In other words, the charging pressure must be significantly higher than the discharging pressure of the 2-phase working fluid. Fig. 33 illustrates this point by showing the discharge process of a solid TES that has stored steam heat at 80 bar and is undergoing

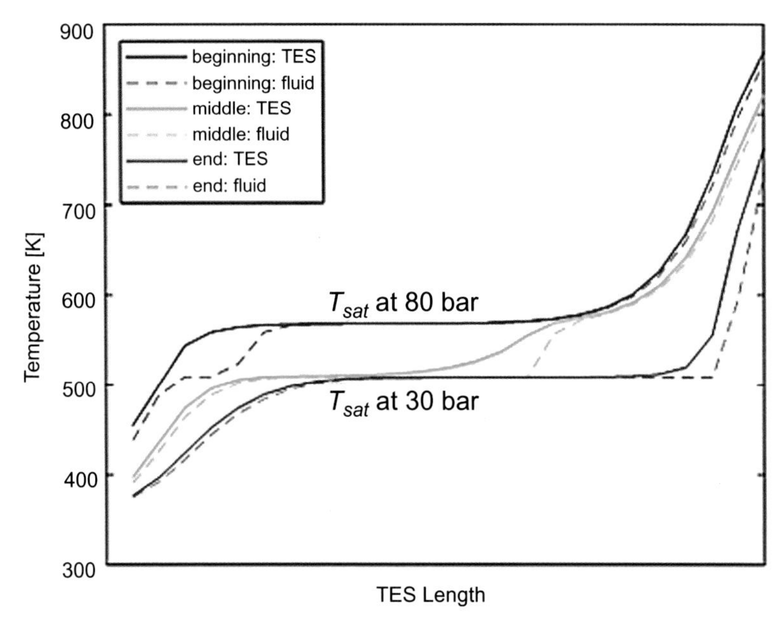

FIG. 33 Fluid and TES temperature gradients over the course of a discharge process for a solid TES used to store 2-phase steam energy.

discharge using 30 bar steam. The much higher latent plateau of steam at 80 bar compared to the latent plateau of steam at 30 bar allows for the solid TES mass to store the phase change energy in sensible form. The green line is midway through the discharge process and shows how the temperature gradients shift as the 30 bar steam uses the sensible heat to evaporate.

If the solid TES were instead a phase change TES that changed phase around the temperature of saturated steam at 80 bar, then both the charge and discharge pressures could be much closer together, because the latent heat of steam could go into changing the phase of the TES, and the temperature of the TES would stay constant there. TESs that use a phase change material are discussed in another section of this book.

3.3 Thermocline dual-media thermal energy storage

Anoop Mathur

Terrafore Technologies, Minneapolis, MN, United States

3.3.1 Motivation for using dual-media thermocline thermal energy storage

Storing sensible heat in thermally stratified layers in a mixture consisting of a solid filler such as granite and a liquid such as molten salt is a lower cost alternative to the conventional two-tank all liquid sensible heat thermal storage system. This storage system is referred to as the dual-media thermocline thermal energy storage (TES).

The conventional method is to store sensible heat in a liquid such as molten salt or hydrocarbon oil, referred to as heat transfer fluid (HTF), using two tanks. One tank called the hot tank stores the hot fluid, and a second cold tank stores the cold fluid. In this system HTF is drawn from the cold tank, is heated in a heat source such as concentrated solar receiver, and the resulting hot HTF is directed to the hot tank.

An alternative is to use a single hot tank with cold HTF as the lower layer and the hot HTF as the top layer. The hot liquid has lower density and 'floats' on top of the more dense cold liquid. The fluids are known to be thermally stratified. The separation of the fluids is not perfect and there is a transition zone called the thermocline zone, where a thermal gradient from hot operating temperature to cold operating temperature separates the hot and cold fluids. This thermocline zone moves down when storage is charged and moves up when it is discharged.

A tank is fully charged when the thermocline is at the bottom and the cold temperature exiting the tank is a few degrees higher than the cold operating temperature called the cold cut-off point. A tank is fully discharged when the thermocline is at the top and the hot temperature of HTF delivered is a few degrees lower than the hot operating temperature and is called the hot cut-off temperature. The usable energy is defined as the energy between charged state (when thermocline zone is at the bottom) and the discharged state (when the thermocline zone is at the top). Therefore the width or thickness of this thermocline zone is an important design parameter and depends on a number of design and operating factors and is discussed later (Figs. 34 and 35).

A key advantage of using a single thermocline tank is that the tank can be filled with solid filler material which can occupy over 60% of the volume depending on the particle shape and void fraction of the bed. The solid media

FIG. 34 Representation of thermocline temperatures from left, fully charged (hot, *red*) to right, fully discharged (cold, *blue*), at ideal conditions with thin thermocline layer (*gray*).

FIG. 35 Representation of thermocline temperatures from left, fully charged (hot, *red*) to right, fully discharged (cold, *blue*), after many cycles and thermocline degradation. Notice the growth of the thermocline layer (*gray*).

or filler material, typically particles of minerals such as, e.g., granite, dolomite, quartzite, or basalt, is significantly less costly than the HTF and has specific heat comparable to the HTF making the dual-media thermocline thermal energy storage a lower cost alternative to the conventional two-tank storage system. For example, for a typical concentrated solar power plant, a dual-media thermocline thermal storage using a combination of granite rock and a nitrate mixture of molten salt as HTF can be 30% less costly than the two-tank all-molten salt system. However, this system has two major concerns:

(1) Thermocline degradation. The thermocline zone, which is initially about 15% of the height of the tank, expands or thickens with cyclic charge and discharge of heat. Studies have shown that this zone can expand to as much as 60% of the height of the tank after ten charge and discharge cycles. The effect of this is that the usable thermal storage capacity decreases from 85% to 40%. The 40% usable capacity limit reached is a function of the configuration of the tank (height-to-diameter ratio), the cut-off points for temperature during discharge, properties of materials, size of solid particles, and the operation of storage system. The degradation of the thermocline is a physical phenomenon due to forced convection of HTF, fluid mixing during charge and discharge process, and some heat conduction between layers. Active intervention or control is required to restore the thermocline. A method to continuously maintain this is described later.
(2) Thermal ratcheting. This phenomenon occurs due to the difference in expansion coefficients of the filler and tank material. The tank material typically has higher expansion coefficient than the filler. When heat is added to the tank, the tank wall expands more than the filler material resulting in a gap which causes the filler material to settle lower to fill the gap. Next, when heat is discharged, the system is cooled and the tank is unable to contract completely, resulting in thermal stresses in the tank material, which may cause plastic deformation. If the final settling of the filler cannot prevent the same process in the subsequent heating and cooling cycles, the tank wall will slowly ratchet outwards and grow until it fails. The type and thickness of material used for lining inside the tank have been suggested to address this engineering design issue.

3.3.2 Dual-media thermocline thermal storage design considerations

A thermocline thermal storage system is one that uses a single tank to store energy. A thermal gradient separates the hot fluid which is the top layer from the cold fluid which is the bottom layer. A low-cost and inert filler material such as, e.g., granite gravel is used to displace the higher cost liquid making the overall cost of the storage system lower cost than an all liquid system.

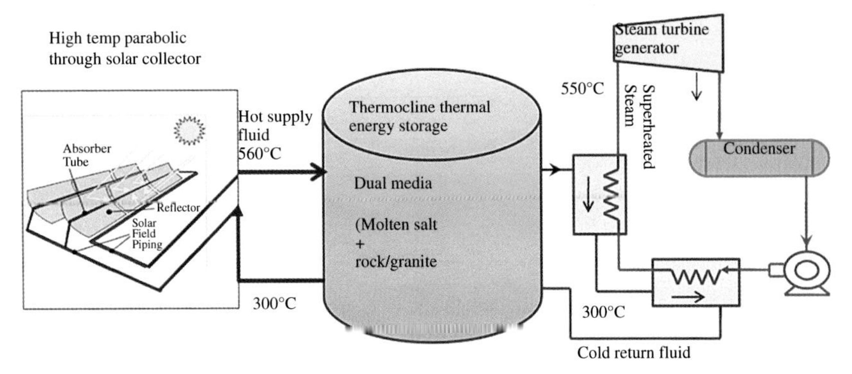

FIG. 36 Dual-Media Thermocline Tank with trough collector and steam Rankine cycle.

The following is a typical charge and discharge cycle. To charge the system, cold fluid is drawn from the bottom, heated in a heat source such as solar receiver, and is returned to the top of the tank. To discharge the thermal storage, hot fluid is drawn from the top, cooled as it passes through a heat exchanger to transfer heat to generate steam, and is returned to the bottom of the tank. This steam is used in turbine to make electricity (Fig. 36).

Solar One Thermocline Thermal Storage System. As an example, Solar One, a 10-MW Solar Central Receiver Pilot Plant constructed in 1981, stored thermal energy as sensible heat in a thermocline using dual media of rock and sand as filler media and a hydrocarbon oil Caloria HT-43 as the liquid. The operating temperature was between 218°C and 302°C. The thermocline storage concept worked well. It was able to establish a thermal gradient and maintain the gradient until it was used. Heat loss through the tank walls was acceptable. Solar One thermal storage was not a full-scale thermal storage but provided valuable information for designing a commercial scale system. At the upper limit temperature of Caloria Oil, the Rankine cycle conversion efficiency was only about 21%. At higher temperatures, hydrocarbon oils decompose and generate flammable vapors. More about concentrated solar thermocline thermal storage research can be found in literature, e.g., from Faas, 1986[a] and Pacheco, 2002[b]. The author, while at Honeywell, conducted a detailed charge/discharge analysis of the Solar One thermocline system and Babcock & Wilcox designed a tank with increasing thickness of tank from top to bottom. This is described in a 1977 report submitted to the DOE[c].

a. S. Faas, L. Thorne, E. Fuchs, N. Gilbertsen, 10 MWe Solar Thermal Central Receiver Pilot Plant: Thermal Storage Subsystem Evaluation—Final Report, SAND86-8212, Sandia National Laboratories, Livermore, CA, 1986.

b. J. Pacheco, S. Showalter, W. Kolb, Development of a molten-salt thermocline thermal storage system for parabolic trough plants, J. Sol. Energy Eng. 124 (2002) 153.

c. SAN 1109-9/7 Solar Pilot Plant Phase 1 Preliminary Design Report, Volume 5, Contract FY-76-C-03-1109.

Operating Temperature: Current thermal storage systems use molten salts (Solar Salt—a near eutectic mixture of sodium and potassium nitrates) which can be used up to 565ºC with Rankine conversion efficiencies of greater than 44%. The operating temperatures of the nitrate molten salt systems are between 285°C and 565°C. Thermocline systems with these operating temperatures have not yet been built or tested. To further increase the conversion efficiency, the next generation Concentrated Solar Power systems will be operating at temperatures >650°C. Molten eutectic mixtures of carbonates and chlorides are being considered for heat transfer and storage medium. At these temperatures, filler material compatibility can also become an issue. These issues are discussed later.

Tank Height to Diameter. The Solar One tank was designed for a thermal capacity of 182 MWth. It used 6170 metric tons of rock and sand and 906 cubic meters of Caloria. The diameter of the storage tank was 18.2 m and the height 13.3 m. The overall design and operability favor taller, smaller diameter tanks (high H/D aspect ratio). The tallest tanks that can practically be fabricated have a bed height of 16 m. Due to the ground load and attendant high cost of civil engineering of tall tanks, the height of the tank was limited to 13.3 m. Another way to increase the aspect ratio is to use multiple tanks in series. There are practical limits on the design, installation, operation, and maintenance of very large tanks, particularly thermocline tanks. A single large tank may have the lowest overall surface-to-volume ratio, minimizing the required insulation, heat trace, controls. On the other hand, a single tank may be so large that construction, startup, operation, maintenance may be very difficult. Multiple smaller tanks in series are also favored for the easier restoration of the thermocline when the fluid flow is actively controlled and routed based on the temperature.

Cut-off temperature. Note that during the charging cycle, the temperature of the cold salt entering the heat exchanger is constant until the gradient reaches the bottom of the tank. At that point the temperature coming out of the bottom of the tank starts to rise. Similarly during a discharge cycle, the hot salt coming off the top of the tank is nearly constant until the gradient enters the top. At that point the hot salt temperature starts to decay. Cut-off points for temperature are chosen by the designer for the charge and discharge. Typically, cut-off points are 15°C higher than the low operating temperature for the end of charge cycle and 15°C lower than the high operating temperature for the end of the discharge cycle depending on the operating temperatures. Cut-off temperatures affect the size of the storage tank or storage capacity, and turbine minimum temperature and are determined by design optimization.

Usable Storage Capacity. One way to quantify the practical storage capacity of a thermal storage system is to compare it to the amount of energy, the storage media could hold if the entire inventory were at its upper temperature and discharged completely to its lower temperature. This has been named the percent theoretical capacity. In a two-tank molten salt system, the percent theoretical capacity is typically about 85% because each tank has a heel of salt (about

1 m deep) which cannot be used for its storage capacity. In a thermocline storage system, the percent theoretical capacity is a function of the tank height and is typically about 69% because of the space required by the thermal gradient.

Thermal Ratcheting. The tank wall stress increases with subsequent cycles due to thermal growth and settling of the wall and filler material can be addressed by using sloped walls, screens to create layers within the tank, suitable size and geometry of the filler, bed-stabilizing structures, and a rugged tank design.

3.3.3 A solution to thermocline degradation: Terrafore's TerraKline™ technology

Storing thermal energy by stratifying the hot fluid layer on top of a cold fluid layer in a single tank packed with granite particles and then filling the volume between the particles with a fluid such as molten salt or oil can reduce the cost of thermal storage because large portion of the expensive molten salt or oil is replaced with lower cost filler mineral and a single tank is used. This storage system is commonly referred to as dual-media thermocline thermal energy storage and has been used experimentally in concentrated solar power plants. A technology barrier that so far has prevented widespread use of this thermal storage system is the thermocline degradation (widening of the transition zone) resulting from periodic charge and discharge. This reduces the availability of the initial thermal storage capacity (and/or requires periodic maintenance to restore the thermocline). Simulations have shown that the storage capacity can degrade by 40% of the initial capacity after just ten charge/discharge cycles, as shown in Fig. 38 on the left.

Terrafore has developed a method to eliminate the degradation of thermocline by using a proprietary technology. In this approach, the tank is designed with fluid collectors located at several levels inside the tank. During charging, the fluid enters from the top and is drawn out from the one of the collectors just below the thermocline, to leave the remaining fluid at the bottom undisturbed. The active collector and the flow rate drawn from the collector are controlled subject to process constraints such as maximum flow rate through the heat source and the current temperatures by a model predictive controller. During discharge, the process is reversed with the cold fluid entering from the bottom of the tank and the hot fluid drawn from a collector just above the thermocline. Using simulation models, Terrafore showed that this technology restores the thermocline diurnally, automatically during each cycle. Fig. 37 shows the distributors, sensors, and control required in a single tank. An alternative implementation, particularly for larger volume, is using multiple tanks in series with fluid distributors also between the tanks.

Fig. 38 shows the thermal profiles of the fluid inside a tank of an ordinary thermocline, with increasing degradation, where after 10 charge and discharge cycles only about 30% of the initial storage capacity is available, and one with the TerraKline technology, where the storage capacity does not degrade much after cycling.

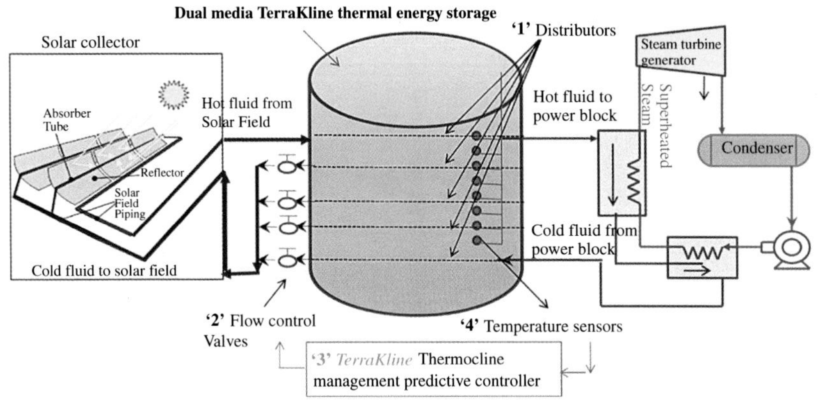

FIG. 37 Terrafore TerraKline™ thermocline storage system.

The thinner thermocline in this system increases the available storage capacity and thus reduces the cost of dual-media single tank even further when compared to a conventional two-tank TES system. Current two-tank TES systems with molten salt may cost about $30 per kWh depending on many specifics. An optimized single-tank thermocline TES is estimated by Terrafore to cost about 33% less or $20 per kWh and the company is open to working with CSP developers to demonstrate this technology on a pilot scale.

3.3.4 Packed-bed solid and fluid thermocline calculations

For design and simulation purposes, the temperatures of the solid and fluid inside a dual-media thermocline store need to be calculated. From a calculation perspective, this dual media thermocline is the same as a packed bed solid thermal store with a heat transfer liquid or gas as described in the analyses methods in a previous section and the calculation method described here is very similar with much the same outcome. The thermal performance of the store depends on the change of the temperatures over time, specifically the response of the outlet temperature to the inlet temperatures after multiple cycles. These can be calculated with transient mathematical models based upon reasonable simplifications and assumptions of the complex nature and detail of the real system. Assuming, for instance, radially homogenous temperatures in the tank, no losses, constant properties, and making the problem nondimensional by use of the thermal properties, length, and time scales, the following equation scheme of a simple, one-dimensional transient model was derived:

$$\frac{\partial \varphi}{\partial \tau} = -\left(\frac{1}{\lambda}\right)\frac{\partial \varphi}{\partial x} + \left(\frac{1}{Pe_{sx}}\right)\frac{\partial^2 \varphi}{\partial x^2} - \left(\frac{NTU}{\lambda}\right)(\theta - \phi)$$

$$\frac{\partial \theta}{\partial \tau} = \left(\frac{NTU}{\lambda}\right)(\theta - \phi) + \left(\frac{k_{sx}/k_{fx}}{Pe_{sx}}\right)\frac{\partial^2 \theta}{\partial x^2}$$

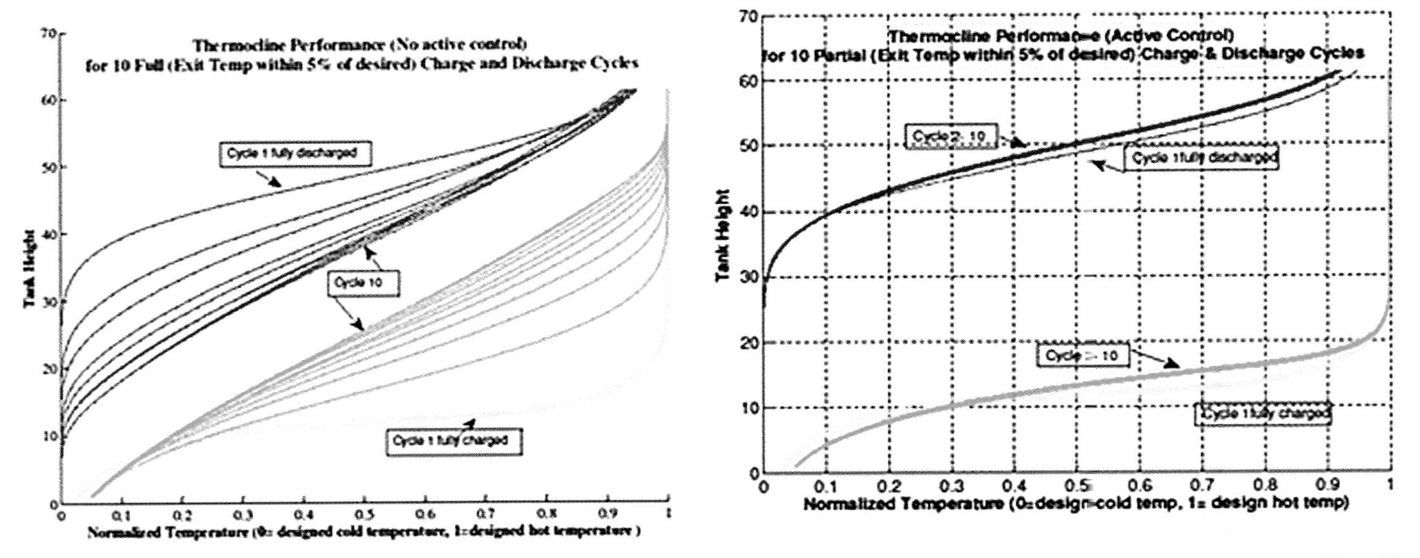

FIG. 38 *Left*, thermocline profiles at end of charge *(green)* and end of discharge *(red)* for 10 cycles without TerraKline technology. *Right*, same profiles with TerraKline.

Here the nondimensional variables are as follows

$$\theta = \frac{(T_f - T_C)}{(T_H - T_C)}, \varphi = \frac{(T_s - T_C)}{(T_H - T_C)}$$

$$x = \frac{z}{L}, Pe_x = \frac{GC_f L}{k_{fx}}, NTU = \frac{h_v L}{GC_f},$$

$$\lambda = \frac{\varepsilon \rho_f C_f}{(1 - \varepsilon)\rho_s C_s + \varepsilon \rho_f C_f}$$

$$\tau = \frac{GC_f t}{[1 - \varepsilon)\rho_s C_s + \varepsilon \rho_f C_f]} \left[\frac{1}{L}\right]$$

The boundary conditions of this model are the following for charging

$$x = 0, \theta = 0$$

$$x = 1, \frac{\partial \theta}{\partial x} = 0$$

$$x = 1, \frac{\partial \varphi}{\partial x} = 0$$

$$\tau \leq 0, 0 \leq x \leq 1, \varphi \,\&\, \theta \text{ specified profile}$$

and for discharging the storage

$$x = 1, \theta = 1$$

$$x = 0, \frac{\partial \theta}{\partial x} = 0$$

$$x = 0, \frac{\partial \varphi}{\partial x} = 0$$

$$\tau \leq 0, 0 \leq x \leq 1, \varphi \,\&\, \theta \text{ specified profile}$$

This set of equations to model the thermal storage cannot easily be solved analytically but, for example, with an explicit finite difference method. This numerical scheme converts the partial differential equations for the temperature into algebraic equations that can be solved sequentially to obtain the time history of the temperatures as a function of the spatial variable. The accuracy and efficiency of the two considered finite difference schemes are as follows: Center-spaced differencing is more accurate and less efficient while upwind differencing is less accurate and more efficient for easier convergence. The example is based on the dual-media model, with one spatial dimension, used by Ismail and Stuginsky[d] (Eqs. 3, 4, p. 762). The partial differential equations

---

d. K. A. R. Ismail, R. Stuginsky Jr. A parametric study of possible fixed bed models for pcm and sensible storage. Appl. Thermal Eng. 19(7) (1999) 757–788.

for the temperatures $\overline{T}(t,x)$ (of the fluid) and $\overline{\theta}(t,x)$ (average in the solid particles) are as follows

$$\varepsilon\rho_f c_f \frac{\partial \overline{T}}{\partial t} + Gc_f \frac{\partial \overline{T}}{\partial \overline{x}} = k_{fx} \frac{\partial^2 \overline{T}}{\partial \overline{x}^2} + h_v\left(\overline{\theta} - \overline{T}\right) \tag{1}$$

$$(1-\varepsilon)\rho_s c_s \frac{\partial \overline{\theta}}{\partial t} = k_{sx} \frac{\partial^2 \overline{\theta}}{\partial \overline{x}^2} + h_v\left(\overline{T} - \overline{\theta}\right) \tag{2}$$

where $0 \leq t$ and $0 \leq \overline{x} \leq L$. The wall heat transfer coefficient has been set to zero and $h_v = h_p a_p$.

The boundary conditions for Eq. (1) are $\overline{T}(t,0) = T_H$ and $\partial\overline{T}(t,L)/\partial\overline{x} = 0$ for $t \geq 0$. The boundary conditions represent a specified hot inlet temperature T_H, at $\overline{x} = 0$ and the lack of any temperature gradient at outflow, at $\overline{x} = L$. The initial temperature $\overline{T}(0,\overline{x})$ is assumed to be some colder temperature T_C, except at $\overline{x} = 0$ where $\overline{T}(0,0) = T_H$. The boundary conditions for Eq. (2) are $\partial\overline{\theta}(t,0)/\partial\overline{x} = 0$ and $\partial\overline{\theta}(t,L)/\partial\overline{x} = 0$ for $t \geq 0$.

The nondimensional variables to be used here are as follows

$x = \overline{x}/L, \tau = \frac{Gc_f t}{\left[(1-\varepsilon)\rho_s c_s + \varepsilon\rho_f c_f\right]L}, T = \frac{\left(\overline{T} - T_C\right)}{\left(T_H - T_C\right)}$, and $\theta = \frac{\left(\overline{\theta} - T_C\right)}{\left(T_H - T_C\right)}$.

This choice yields the following set of nondimensional equations

$$\lambda \frac{\partial T}{\partial \tau} = \frac{k_{fx}}{Gc_f L} \frac{\partial^2 T}{\partial x^2} - \frac{\partial T}{\partial x} + \frac{h_v L}{Gc_f}(\theta - T) \tag{3}$$

$$\gamma \frac{\partial \theta}{\partial \tau} = \frac{k_{sx}}{Gc_f L} \frac{\partial^2 \theta}{\partial x^2} + \frac{h_v L}{Gc_f}(T - \theta) \tag{4}$$

where $\lambda = \frac{\varepsilon\rho_f c_f}{(1-\varepsilon)\rho_s c_s + \varepsilon\rho_f c_f}$ and $\gamma = \frac{(1-\varepsilon)\rho_s c_s}{(1-\varepsilon)\rho_s c_s + \varepsilon\rho_f c_f}$. These nondimensional equations are a variation of the set of equations found in Ismail and Stuginsky[d]. They are based on a different choice of nondimensional time τ.

The nondimensional boundary conditions for Eq. (3) are $T(\tau,0) = 1$ and $\partial T(\tau,1)/\partial x = 0$ for $\tau \geq 0$. The initial fluid temperature $T(0,x)$ is assumed to be zero, except at $x = 0$ where $T(0,0) = 1$. The boundary conditions for the solid temperature in Eq. (4) are $\partial\theta(\tau,0)/\partial x = 0$ and $\partial\theta(\tau,1)/\partial x = 0$ for $\tau \geq 0$. The initial solid temperature $\theta(0,x)$ is assumed to be zero. Next, for the implementing the finite difference scheme, equations (3) and (4) are rewritten as

$$\lambda \frac{\partial T}{\partial \tau} = a \frac{\partial^2 T}{\partial x^2} - \frac{\partial T}{\partial x} + c(\theta - T) \tag{5}$$

and

$$\gamma \frac{\partial \theta}{\partial \tau} = b \frac{\partial^2 \theta}{\partial x^2} + c(T - \theta) \tag{6}$$

where $a = k_{fx}/(Gc_f L)$, $b = k_{sx}/(Gc_f L)$, and $c = h_v L/(Gc_f)$. The finite difference method is based on introducing a discrete grid in time and space: $\tau_n = n\Delta\tau$,

$n = 0, 1, \ldots$ and $x_j = j\Delta x, j = 0, 1, \ldots, J$ where $\Delta x = 1/J$. The values of the nondimensional temperatures are designated by $T_{n,j} = T(\tau_n, x_j)$ and $\theta_{n,j} = \theta(\tau_n, x_j)$. The partial derivatives in Eq. (5) are approximated with difference quotients as follows (forward central for time and space derivatives, respectively):

$$\frac{\partial T}{\partial \tau} \cong \frac{T_{n+1,j} - T_{n,j}}{\Delta \tau},$$

$$\frac{\partial T}{\partial x} \cong \frac{T_{n,j+1} - T_{n,j-1}}{2\Delta x} \text{ or } \frac{\partial T}{\partial x} \cong \frac{T_{n,j} - T_{n,j-1}}{\Delta x}, \text{ and}$$

$$\frac{\partial^2 T}{\partial x^2} \cong \frac{1}{\Delta x}\left\{\left(\frac{T_{n,j+1} - T_{n,j}}{\Delta x}\right) - \left(\frac{T_{n,j} - T_{n,j-1}}{\Delta x}\right)\right\} = \frac{T_{n,j+1} - 2T_{n,j} + T_{n,j-1}}{\Delta x^2}.$$

Similar finite difference approximations for $\partial\theta/\partial\tau$ and $\partial^2\theta/\partial x^2$ will be used in Eq. (6). The spatial difference approximation for $\partial T/\partial x$ that is symmetric about $T_{n,j}$ is called center-spaced. The alternative asymmetric approximation for $\partial T/\partial x$ is called the upwind difference, since the index j increases in the same direction as the flow of the heat transfer fluid. The substitution of these approximations into Eqs. (5), (6) gives explicit formulas for $T_{n+1,j}$ and $\theta_{n+1,j}$ in terms of $T_{n,j}$ and $\theta_{n,j}$. These finite difference schemes have stability requirements, on the size and relationship between $\Delta\tau$ and Δx. The stability requirements are discussed in the next section.

The explicit formulas, to advance the temperature forward in time, are as follows:

$$\lambda\frac{T_{n+1,j}}{\Delta\tau} = \lambda\frac{T_{n,j}}{\Delta\tau} + a\frac{T_{n,j+1} - 2T_{n,j} + T_{n,j-1}}{\Delta x^2} - \frac{T_{n,j+1} - T_{n,j-1}}{2\Delta x} + c\left(\theta_{n,j} - T_{n,j}\right)$$

or

$$\lambda\frac{T_{n+1,j}}{\Delta\tau} = \lambda\frac{T_{n,j}}{\Delta\tau} + a\frac{T_{n,j+1} - 2T_{n,j} + T_{n,j-1}}{\Delta x^2} - \frac{T_{n,j} - T_{n,j-1}}{\Delta x} + c\left(\theta_{n,j} - T_{n,j}\right)$$

and

$$\gamma\frac{\theta_{n+1,j}}{\Delta\tau} = \gamma\frac{\theta_{n,j}}{\Delta\tau} + b\frac{\theta_{n,j+1} - 2\theta_{n,j} + \theta_{n,j-1}}{\Delta x^2} + c\left(T_{n,j} - \theta_{n,j}\right)$$

for $j = 1, J - 1$. For the purpose of implementation, the explicit formula for the fluid temperature is rewritten as

$$T_{n+1,j} = \alpha_f T_{n,j-1} + \delta_f T_{n,j} + \beta_f T_{n,j+1} + \xi_f \theta_{n,j} \tag{7}$$

where, using center-spaced differencing,

$$\alpha_f = \frac{\Delta\tau}{\lambda}\left(\frac{a}{\Delta x^2} + \frac{1}{2\Delta x}\right), \delta_f = \frac{\Delta\tau}{\lambda}\left(\frac{\lambda}{\Delta\tau} - \frac{2a}{\Delta x^2} - c\right), \beta_f = \frac{\Delta\tau}{\lambda}\left(\frac{a}{\Delta x^2} - \frac{1}{2\Delta x}\right),$$

or, using upwind differencing,

$\alpha_f = \frac{\Delta\tau}{\lambda}\left(\frac{a}{\Delta x^2} + \frac{1}{\Delta x}\right), \delta_f = \frac{\Delta\tau}{\lambda}\left(\frac{\lambda}{\Delta\tau} - \frac{2a}{\Delta x^2} - \frac{1}{\Delta x} - c\right), \beta_f = \frac{\Delta\tau a}{\lambda\Delta x^2}$, and $\xi_f = \Delta\tau c/\lambda$, in both cases. The explicit formula for the solid temperature is rewritten as

$$\theta_{n+1,j} = \alpha_s \theta_{n,j-1} + \delta_s \theta_{n,j} + \beta_s \theta_{n,j+1} + \xi_s T_{n,j} \tag{8}$$

where
$\alpha_s = \frac{\Delta\tau b}{\gamma \Delta x^2}, \delta_s = \frac{\Delta\tau}{\gamma}\left(\frac{\gamma}{\Delta\tau} - \frac{2b}{\Delta x^2} - c\right), \beta_s = \frac{\Delta\tau b}{\gamma \Delta x^2}$, and $\xi_s = \frac{\Delta\tau c}{\gamma}$ for $j = 1, J-1$.

The boundary conditions on the fluid temperature, which apply for $j = 0$ and $j = J$, require that the inflow temperature is fixed and that the discrete approximation of the outflow gradient is zero. These conditions take the form $T_{n+1,0} = 1$ and $T_{n+1,J} = T_{n+1,J-1}$. The conditions on the solid temperature are $\theta_{n+1,0} = \theta_{n+1,1}$ and $\theta_{n+1,J} = \theta_{n+1,J-1}$.

Stability Requirements: While implicit finite difference schemes can be unconditionally stable, explicit schemes usually have requirements on the size and relationship between $\Delta\tau$ and Δx to assure stability. For the purpose of this discussion, Eq. (5) is converted to the standard convection-diffusion equation (5). This is accomplished by dividing Eq. (5) by λ and ignoring coupling to obtain

$$\frac{\partial T}{\partial \tau} + b_f \frac{\partial T}{\partial x} = a_f \frac{\partial^2 T}{\partial x^2}$$

where $b_f = 1/\lambda$ and $a_f = a/\lambda$. The stability requirement for forward-time center-spaced differencing is $2a_f \Delta\tau/\Delta x^2 < 1$, which makes the accuracy of the scheme second order in space. The elimination of oscillations [15] in the center-spaced differencing scheme yields the convection-diffusion requirement $\Delta x < 2a_f/b_f$ where a_f/b_f is the ratio of the diffusion coefficient to the convection coefficient. The use of upwind differencing alleviates the convection-diffusion requirement at the expense of producing a scheme that is only first order accurate in space[e]. Upwind differencing, implicitly, adds numerical diffusion so that the convection-diffusion requirement is satisfied. The stability requirement then becomes $(2a_f + b_f \Delta x)\Delta\tau/\Delta x^2 \leq 1$. A simpler stability analysis, without convection, applies to Eq. (6), after ignoring the effect of coupling between Eqs. (5), (6).

3.4 Low-temperature cool thermal storage

Douglas Reindl & Amy Van Asselt

Low-temperature cool thermal energy storage is a strategic technology that can enable building owners to improve site energy efficiency while significantly lowering energy costs associated with building air conditioning system operation. A building equipped with cool thermal energy storage operates energy-intensive refrigeration equipment at nighttime during low electric energy cost

e. F. Brauer, J A. Nohel, Ordinary Differential Equations: A First Course, second ed., W. A. Benjamin, Inc. Menlo Park, California, 1973.

periods to store thermal energy for later use in the building's air conditioning system during daytime when energy costs are higher. This simple concept of a "thermal battery" translates into energy and operating cost savings for the building owner but it also yields benefits for utilities when widely applied. One of the more well-known utility benefits of cool thermal energy storage is its ability to decrease the aggregate electric demand from customers during high-use on-peak periods. This mutual benefit is unique because it represents a single technology that provides benefits to both building owners (end-users) who deploy it and electric utilities serving those application end-users. Beyond building applications, cool thermal energy storage has also been successfully implemented to augment utility power generation and process cooling applications. In utility power generation, cool thermal energy storage can be used to cool inlet air to combustion turbines, thereby increasing their power generating capacity and efficiency. In process cooling applications, cool thermal storage can level the thermal loads in food, chemical, and other industrial processes. More recently, cool thermal storage has found success in providing backup cooling for mission critical data center facilities.

3.4.1 Overview of cool thermal storage applications

Over the past three decades, cool thermal energy storage systems have continued in technology refinement. Cool thermal storage systems have found use in a range of applications that include:

3.4.1.1 Building air conditioning

- Commercial office buildings (large and small)
- Residential dwellings
- Schools
- Healthcare facilities
- Convention centers
- Retail facilities
- Laboratories
- Transportation terminals
- District cooling
- Government facilities (court houses, prisons, etc.)
- Operating rooms (surgery suites)
- Industrial plant air conditioning
- Ground-based aircraft cabin air preconditioning
- Combustion Turbine Inlet Air Cooling (CTIAC)
- Process cooling
- Industrial process cooling
- Data centers/mission-critical facilities

In the sections that follow, a high-level overview is provided for the application of cool thermal storage for building space conditioning, combustion turbine inlet air cooling, and process cooling.

3.4.1.2 Building space conditioning

The air conditioning loads for a given building will vary diurnally with the highest demand for cooling typically occurring during the mid- to late-afternoon periods and lowest cooling demand at nighttime. Fig. 39 illustrates a cooling load profile for a commercial office building with a lightweight wall construction. In this case, the building's aggregate cooling load peaks during the occupied hours at 4:00 pm which is shortly after the highest ambient temperature during the design day. Compared to daytime, the space cooling loads at night are substantially lower. Different building types will have characteristic cooling load profiles that vary from that shown in Fig. 39. For example, a data center will tend to a have a flatter cooling load profile since the majority of the thermal load is due to internal heat gains from computer equipment operating continuously. A 24-h operation healthcare facility will have a time-varying component to the cooling load but the building's cooling load factor (ratio of average cooling load to the peak cooling load) will be relatively high. Cool thermal storage systems tend to be more attractive in applications where there is a time variability or diversity in the cooling load or when there high cooling reliability is required.

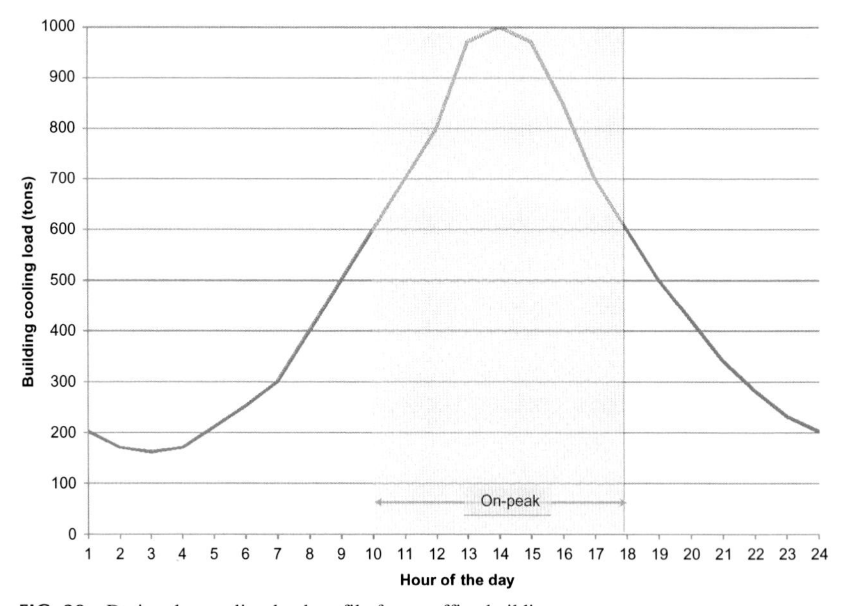

FIG. 39 Design day cooling load profile for an office building.

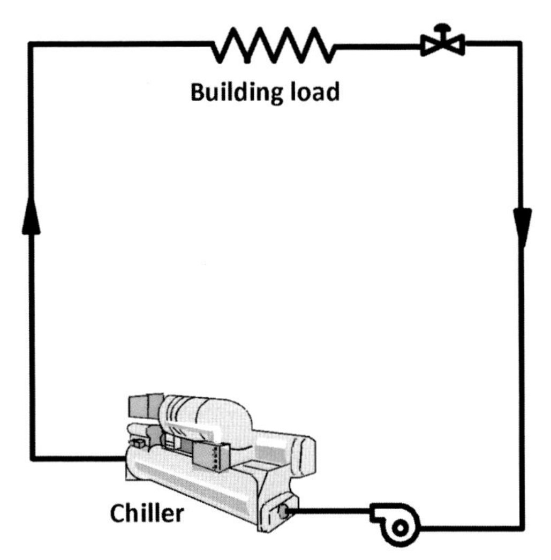

FIG. 40 Space conditioning system providing on-demand cooling to a building (no storage).

Space conditioning for the vast majority of buildings is achieved by the operation of a mechanical refrigeration system, such as the chiller shown in Fig. 40 that converts electricity to cooling energy capable of meeting a building's space cooling load instantaneously as it varies over time. What results is a building electrical demand profile that tends to mimic the building's cooling load profile—high electric demand during the day and low electric demand at night. Thermal energy storage is a strategic technology that enables end-users to decouple the electric demand associated with a building's air conditioning system from its production. At a basic level, a thermal storage system allows the building owner to shift energy usage from periods of high energy cost (on-peak) to periods of low energy cost (off-peak). Since the cost of generating and supplying electricity is generally higher during the day than at night, end-users with cool thermal energy storage systems benefit by shifting their energy usage to these lower cost periods as shown in Fig. 41. Utilities benefit by improved asset utilization when a large number of end-users install cool thermal energy storage systems.

The basic concept of thermal energy storage is to enable the operation of energy-intensive chilling/refrigeration equipment during periods of time when utility costs and cooling loads are low in order to charge a storage system (a "thermal battery"). The stored thermal energy can then be utilized during the following day to meet cooling loads. Thus the storage system provides mechanical designers and building operations staff with a significant degree of freedom or operational flexibility. They can either completely forego the operation of chilling equipment (*full storage)* or allow the operation of much smaller refrigeration equipment during periods of high energy costs by discharging the stored

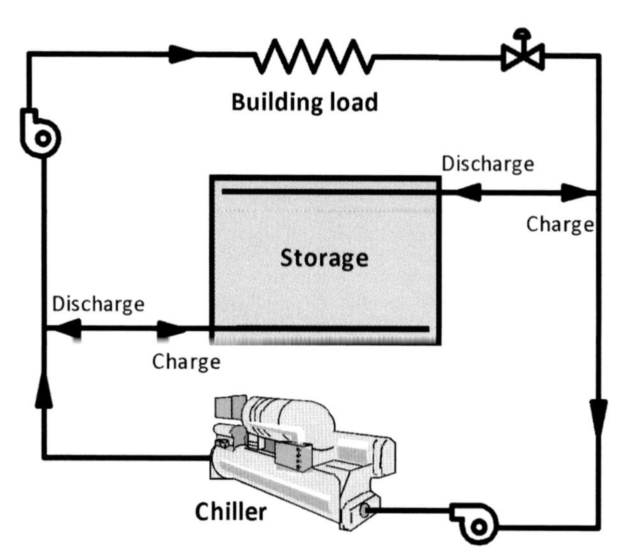

FIG. 41 Space conditioning system providing cooling to a building with thermal energy storage.

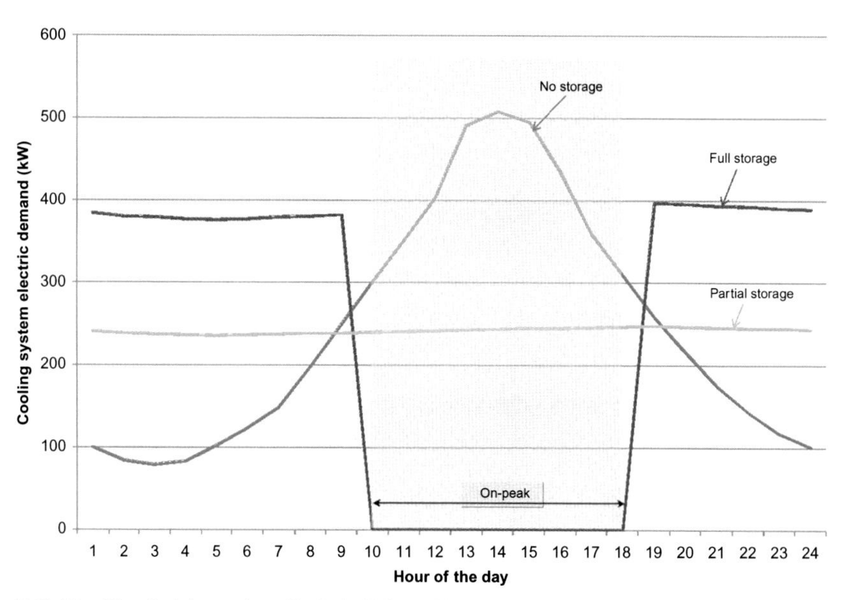

FIG. 42 Electrical demand profile for building without thermal energy storage, with the *full storage* option, and the *partial storage* option.

thermal energy to meet building cooling loads (*partial storage*). Fig. 41 illustrates how a cool thermal energy storage (CTES) system can be integrated into a building's space conditioning system.

Fig. 42 shows three different electricity demand profiles for a building's chilling system: *no storage, full storage, and partial storage*. Similar to

Fig. 39, the *no storage* demand profile shown in Fig. 41 results from the operation of the chilling system directly coupled to meet cooling loads as they occur throughout the day. A building with a *full storage* system operates the chilling system during off-peak periods to build an inventory of "cold" that can be fully utilized to meet the building cooling loads during the entire on-peak period without coincident chiller operation. The *full storage* option achieves the greatest demand reduction during the on-peak period but requires the largest storage system size. In this case, the building's chilling system and storage system are sized to ensure the total integrated building cooling load during the on-peak period is capable of being charged into storage. An alternative approach is the *partial storage* system option. The *partial storage* option significantly reduces but does not eliminate the on-peak demand as it levels the cooling load and corresponding chiller demand over the entire day, thereby requiring a significantly smaller storage system. In this case, the storage and chilling systems are sized such that the combination of chiller operation supplemented with discharging the storage system is sufficient to meet the instantaneous and integrated load during the on-peak period. The *partial storage* option can approach the behavior of the *full storage* option on days outside of the design day whereas a system designed for the *full storage* option will be oversized on these days.

The impact of multiple buildings operating without CTES for space conditioning leads to a very high aggregate utility electrical demand profile during the daytime with a significantly lower aggregate electrical demand at night. Even though peak utility electric demands may only persist for a few hours in a given year, the presence of this single peak requires electric utilities to install sufficient generation, transmission, and distribution capacity in order to satisfy this single instantaneous peak demand as well as additional electrical generation and distribution capacity as a reserve margin. The added electrical generation, transmission, and distribution infrastructure increases the overall costs for delivering electricity to end-users.

The CTES device is an alternative to electrical energy storage (EES) in any application that involves the use of electricity to meet cooling-related thermal processes. At the present time, EES is not a viable technology for storing the magnitude of electrical energy that would be needed for achieving meaningful shaping of a utilities' aggregate electrical demand profile, i.e., MWh or GWh. On the other hand, CTES is an effective and mature technology that can efficiently achieve the equivalent effects of electric energy storage by first undergoing the conversion of electric to thermal energy followed by storage of energy thermally. Charging a storage system using low cost off-peak energy allows owners to forego the operation of energy-intensive refrigeration systems during on-peak operation, thereby offering the potential for substantial operating cost savings. With a significant number of storage systems operating, electric utilities are able to avoid building additional generation and transmission infrastructure "just to meet the coincident customer peak load." CTES systems are able to provide benefits to both owners and utilities at lower cost, with increased end-use chilling net system efficiency, and longer life compared to an equivalent system capable of storing the energy in the form of electricity.

Thermal energy storage systems have found use in a wide array of applications including space conditioning of office buildings, laboratories, convention centers, theaters, retail stores, health care facilities, and other institutional buildings. In addition, practitioners have creatively applied CTES systems in other applications that include power production (turbine inlet air cooling for capacity and efficiency enhancement), aircraft cabin preconditioning, district cooling, industrial process cooling, and backup cooling for thermal energy-intensive datacenters. Although applications are seemingly only limited by the creativity of system designers, there are some key requirements that must generally be met in order for a CTES to be a "good fit."

- *Energy costs:* the greater the time-varying component of the utility rates, the more likely CTES will be economically advantageous
- *Demand costs:* high demand costs incent end-users to shift electrical demand from on-peak to off-peak periods
- *Peak period duration:* short on-peak periods usually result in smaller more cost-effective storage system installations
- *Cooling load diversity:* the best applications for CTES are for loads that have short and relatively high peak periods (i.e., low cooling demands at night to allow charging of the storage system)
- *Load leveling:* a CTES system can be sized to handle the thermal load with smaller compressors than a conventional system if a *partial storage* design is acceptable
- *Reliability:* because of its simplicity, CTES is a technology that can increase the reliability of the building's cooling system
- *Physical Space:* locally, there must be adequate space to site a thermal energy storage system

3.4.1.3 Turbine inlet air cooling

Natural gas-fired combustion turbines are commonly used to meet peak electricity demands due to their ability to come online quickly compared to power generation methods that use steam as the working fluid. Both the power output and fuel efficiency of combustion turbines are negatively correlated with increasing inlet ambient air temperature. Because the density of ambient air decreases as the temperature increases, the turbine experiences a proportionally lower mass flow rate of air through the system which decreases its power output. The rated performance of combustion turbines is defined at ISO conditions of 59°F (15°C) and sea level pressure [29]. Because utility peak electricity demand is driven by building air conditioning loads, the timing of peak demand is often coincident with elevated ambient temperatures. With combustion turbines being used to meet peak demand, their performance is at their worst when the power is needed the most. For example, when inlet air temperatures reach

90–100°F (32–38°C), the combustion turbine power output drops to approximately 80%–85% of its rated capacity. If the inlet air temperatures to a combustion turbine could be decreased to 40°F (4°C), the turbine's power output can be increased to approximately 105% of the ISO rated capacity. In other words, actively cooling inlet air from 100°F (38°C) to 40°F (4°C) can yield an increased power output of approximately 25%. While cooling the inlet air to a combustion turbine to temperatures below this point is possible and can result in further performance gains, it is often not pursued for both economic and technical reasons. One concern with delivering lower inlet air temperatures is the potential for water droplets in the air stream freezing and entering the compressor stage.

There are several different methods available for achieving this inlet air temperature drop including evaporative cooling, absorption cooling, mechanical cooling, and mechanical cooling in combination with a CTES system [30]. Of these technologies, evaporative cooling has been widely deployed for CTIAC purposes due to its low capital cost and simplicity. Unfortunately, evaporative type inlet air cooling systems are bounded by the ambient wet-bulb temperature which is only significantly lower than the dry-bulb temperature in arid climates. In these climates, the water required for this process presents a resource problem that is getting progressively more pronounced. Instantaneous absorption chilling systems can utilize waste heat from the combustion turbine exhaust to fire the chiller which results in reduced parasitic electricity consumption. The disadvantages of CTIAC using absorption chilling are a lower inlet air temperature limit of approximately 48°F (9°C) and high capital costs for the absorption refrigeration equipment as well as the heat recovery heat exchangers required. Mechanical refrigeration systems operating directly to chill inlet air can easily achieve low temperatures but use of this technology without a storage component results in high upfront capital costs to meet the instantaneous inlet air cooling load. Additionally, the parasitic electricity consumption for chiller operation diminishes the turbine's capacity enhancement.

As is the case for building space conditioning, there are sensible and latent energy change technologies used for thermal storage in CTIAC applications. The technical details of these technologies are discussed further later but they have common benefits with respect to their use in CTIAC systems. A CTES system serves to decouple the electricity consumption associated with a mechanical refrigeration system from the combustion turbine inlet air cooling loads experienced during peak demand periods. This means that the cool inlet air temperatures needed to boost the turbine power output are achieved using stored cold liquid or ice. Since the peak power output of a combustion turbine may be required for only a few hours during weekday afternoons, the capacity if the storage system can be sized to meet the integrated cooling load during a relatively short operating period. A significant advantage of CTES over instantaneous mechanical refrigeration comes from the ability to spread out the cooling load over the majority of the day when CTIAC is not required. Although a

CTES system requires a storage tank, the capacity of the refrigeration equipment can be downsized which helps to lower capital cost for the overall system.

When additional combustion turbine capacity is required, addition of a CTES system for CTIAC is less costly than installing new combustion turbines. For a power plant located in Riyadh, Saudi Arabia, the addition of CTES reduced the inlet air temperature from 122°F (50°C) to 55°F (13°C) and resulted in a 30% increase in the net on-peak power output of the plant [30]. This system was installed at less than half the capital cost of the equivalent new capacity of a combustion turbine. The capital costs associated with a CTIAC system will depend on a number of local factors as well as the specific thermal storage technologies (e.g., sensible or latent); most systems have utilized stratified chilled water as the storage media (sensible) rather than ice (latent). Ice-based CTES systems cost 60%–80% as much for incremental power as added turbine capacity while chilled water storage systems cost only 40%–60% as much [29]. With increasingly variable power generation sources, CTIAC using CTES provides a cost-effective method for increasing combustion turbine capacity without installing new generation equipment.

3.4.2 Cool thermal storage technologies

Broadly, CTES technologies are classified into two categories based on the form of stored energy: sensible energy change and latent energy change technologies. Sensible energy technologies rely on the change in temperature of a working fluid as a means of storing thermal energy while latent energy technologies rely on the energy absorption/release associated with the solid-liquid phase change (e.g., melting ice) of a material to accomplish energy storage in a thermal state.

3.4.2.1 *Sensible energy change*

There are two basic types of sensible energy change technologies—chilled water and water solutions utilizing a low-temperature fluid additive. At a basic level, sensible energy change storage systems accomplish the storage of thermal energy by using the heat capacity of a working fluid and causing it to undergo a *temperature change*. With water as the working fluid, 8.34 Btu (8.80 kJ) of thermal energy can be stored in one gallon for 1°F (0.56°C) of temperature change. Chilled water storage systems are typically designed for at least a 20°F (11.1°C) temperature differential which yields a storage density of approximately 0.1 ton-h/ft^3 (12.4 kWh/m^3).

Over the last thirty years, a number of alternative storage tank configurations and designs have been built and tested including spare/empty tanks, series plug flow tanks, baffled tanks, membrane tanks, and stratified tanks. The stratified chilled water tank design has emerged as the most effective and reliable. Interestingly, it is also the simplest. A stratified tank design utilizes the general behavior of liquid density decreasing with increasing temperature. By allowing gravity to naturally separate the more buoyant warmer liquid to the top of the tank and the cooler more dense liquid at the bottom, a stratified sensible energy change storage tank can accomplish its intended purpose of storing thermal energy by naturally separating the warm from the cold fluid.

During charging, warm fluid is drawn from the top part of the storage tank and decreased using a mechanical refrigeration plant (chiller) and subsequently delivered back to the bottom of the storage tank (Fig. 41). The cool fluid within the storage device is now available to meet instantaneous building cooling loads as they occur during a later time. In this case, cool fluid is drawn from the bottom of the tank and supplied to the building where it meets instantaneous space cooling loads resulting in an increase in fluid temperature where it is returned back to the top of the storage tank to dwell before recooling again during the off-peak period using the mechanical chillers.

The storage density for sensible energy change systems is directly proportional to the temperature difference that can be achieved across the storage device. Higher operating temperature differences are desirable because they translate directly into increased capacity of stored energy for a given tank volume. For air conditioning applications, a practical upper limit in return temperature is on the order of 60°F (15.6°C). When water is used as the storage fluid, a low limit in temperature is 40°F (4.4°C) due to the inversion of its density at lower temperatures. This means that a 20°F (11.1°C) temperature differential is readily achievable. During operation, it is important to maintain the high fluid temperature differential in order to maintain stratification within the tank.

If higher storage density is desired or needed, additives can be included in the water to shift its point of maximum density to a lower temperature, thereby permitting cooler supply fluid temperatures. This approach is referred to as "low temperature fluid" (LTF) storage. At present, the only commercially available working fluid specifically designed to extend the operating temperature range of stratified chilled water systems is SoCool®[f]. In retrofit applications, the conversion of a stratified chilled water storage system to a LTF system can yield an increase in the storage density by 50%.

A "small" stratified chilled water storage tank will have 5000 ton-h[g] (17,600 kWh) of thermal capacity while a "large" chilled water storage tank may exceed 150,000 ton-h (530,000 kWh) of capacity. An "average" chilled water storage system has 20,000 ton-h (70,000 kWh) of storage capacity which requires a tank volume on the order 1.75 million gallons (6.44 million liters). A

---

f. SoCool™ is a sodium nitrite solution for use in stratified cool storage systems (U.S. patents 5655377 and 5465585). SoCool™ is a registered trademark of the Chicago Bridge & Iron Company and marketed by The *Cool Solutions Company*.

g. Similar to the concept of a kWh being the energy consumed by a one kW of electricity demanded over a 1-h interval, a "ton-h" is equivalent to meeting a one ton cooling load over a 1-h period. As a quantity of thermal energy, it is equal to 12,000 Btu.

characteristic of stratified chilled water storage systems is that they become increasingly cost advantageous as the total storage capacity requirement increases. A challenge in applying the chilled water storage system technology is allocating a suitable footprint to site the tank. For example, the tank for a 10,700 ton-h (37,600 kWh) chilled water storage tank installed in DuPage County, IL is 73 ft (22 m) in diameter and 40 ft (12 m) tall. A somewhat larger storage system also located in the Chicago area is the Trigen plant with 123,000 ton-h (433,000 kWh) of LTF storage in a single storage tank measuring 127 ft (39 m) in diameter by 90 ft (27 m) tall [29].

Chilled water and LTF storage systems have been applied in a range of applications that include: district systems, large commercial office building space cooling, convention center space cooling, healthcare facility space cooling, and manufacturing/process cooling. In addition to serving as a thermal reservoir for space conditioning, chilled water storage tanks have been leveraged for other uses such as a water source for building fire protection systems. In the latter case, the owners generally receive an additional financial benefit through lower insurance rate premiums.

3.4.2.2 Latent energy change

Latent energy change technologies accomplish the storage of thermal energy through extracting and absorbing heat into a storage medium that undergoes a liquid-solid phase change. The most widely used latent energy change storage medium is ice but other substances, such as eutectic salts, have been developed and applied as well. Fig. 43 shows a simple schematic of an ice storage system operating in a primary/secondary arrangement with building cooling loads. During a charge cycle, the chiller cools a fluid (e.g., ethylene glycol-water solution) to a temperature well below the freezing point of water ca. 24°F (−4.4°C). The cold glycol circulates through a heat exchanger incorporated into the thermal storage device where it absorbs heat from water within the tank causing it to freeze. During a melt period, warm glycol returning from the load can either be prechilled by the mechanical chiller or cooled directly by circulating through the heat exchanger located within the ice storage unit(s). The warm glycol is cooled as it gives up its heat to the accumulated ice within the tank causing it to melt back to a liquid.

The phase change temperature (freezing point) for water is 32°F (0°C) and each pound of ice is capable of absorbing 144 Btu (152 kJ) of heat as it undergoes a phase change from solid to liquid at a freezing constant temperature. This characteristic translates into a significantly higher storage density compared to chilled water. Ice storage systems are capable of yielding storage densities in the range from 0.29 to 0.45 ton-h/ft^3 (36 to 56 kWh/m^3). Compared to chilled water storage, ice storage systems are more compact for a given storage requirement.

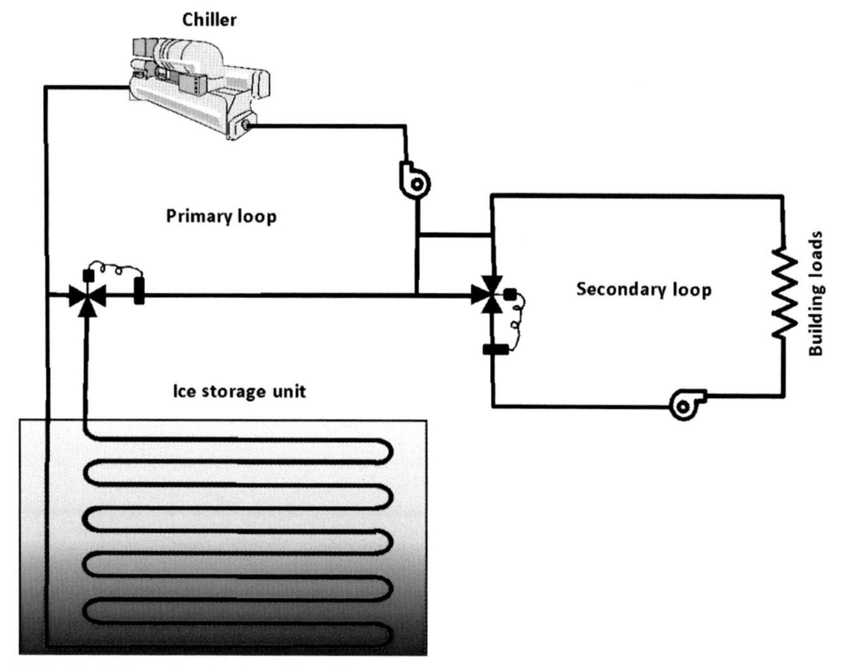

FIG. 43 Schematic of a glycol-based ice storage system.

The use of ice as a thermal storage medium requires a refrigeration system that operates at lower temperatures to enable charging when compared with chilled water storage systems. Lower operating temperatures translate into lower chiller operating efficiencies during the production of ice. To mitigate this operating penalty, alternative storage media such as eutectics[h] were designed such that they undergo a phase change at temperature higher than 32°F (0°C). One of the first eutectics introduced into the marketplace had a phase change transition temperature of 47°F (8.4°C) which, unfortunately, proved to be too high an operating temperature for practical air conditioning applications.

In contrast to sensible energy change storage systems that may rely on a single technology such as a stratified storage tank, a number of technology alternatives exist in latent energy change storage systems. Each latent energy change technology is introduced as follows along with their basic principle of operation.

Static ice internal-melt

A static ice internal-melt thermal storage device is characterized by the manner in which the device discharges its cooling capacity. The vast majority of static

h. A *eutectic* is a mixture of two or more substances to achieve a desired phase-change temperature.

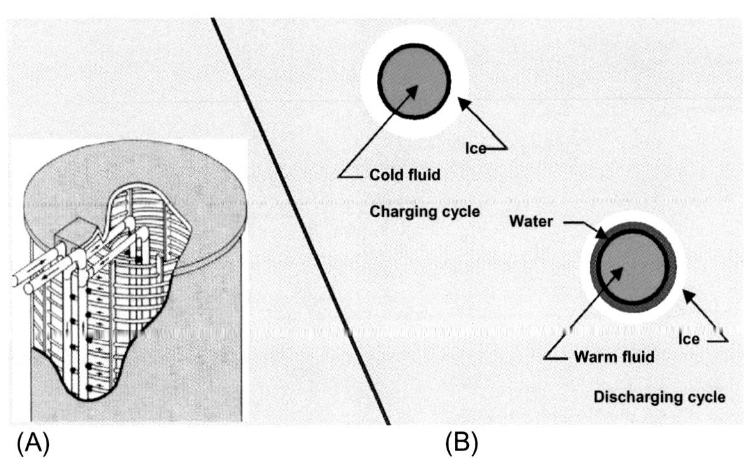

FIG. 44 Rendering of (A) Calmac internal-melt ice-on-coil tank and (B) the processes involved with charging and discharging.

ice internal-melt systems utilize an ethylene- or propylene-glycol secondary working fluid to charge and discharge the storage device. The storage device itself consists of a polyethylene tube heat exchanger inserted into a water-filled container. One configuration of this design uses a fully welded spiral wound heat exchanger inserted into a cylindrical tank as shown in Fig. 44.

During a *charge cycle*, cold secondary fluid circulates through the heat exchanger immersed in the water-filled tank initiating a change in phase of the water surrounding the tube from liquid to ice as shown in Fig. 44B. As the layer of ice on the outside surface of the heat exchanger grows, the resistance to heat transfer increases and continuation of the freezing process requires a greater difference in temperature between the secondary fluid and the freezing point of water to sustain the freezing rate. The consequence of this operation is two-fold. First, with a fixed entering fluid temperature (e.g., 24°F/−4.4°C), the rate of charging will decrease as the storage device state of charge increases. Second, to sustain a fixed charging rate, the entering fluid temperature must decrease as the state of charge increases.

During a *discharge* cycle, warm secondary fluid circulates through the heat exchanger causing the accumulated ice on the tubes to melt from the inside outward (hence the nomenclature "internal melt" system). As the discharge period progresses, an annular layer of water between the heat exchange surface grows. A benefit of the internal-melt design is that a recharge of the storage tank after a partial melt does not impose an energy penalty as is the case with the external-melt design.

Static ice external-melt

A static ice external-melt thermal storage device builds ice on the outside surface of a heat exchanger immersed in a water-filled tank as shown in Fig. 45.

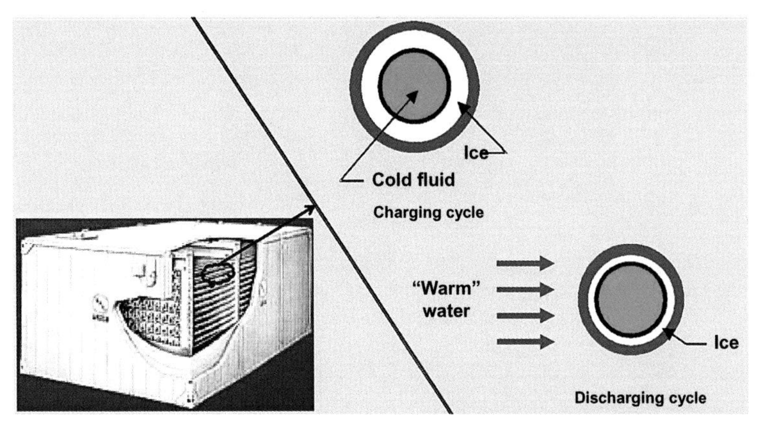

FIG. 45 Rendering of a BAC external-melt ice-on-coil tank (A) and the processes involved with charging and discharging (B).

The external-melt heat exchangers are a serpentine pattern typically constructed of hot-dipped galvanized carbon steel. Although early systems used direct refrigerant within the heat exchanger, most external-melt storage systems today use a secondary heat transfer fluid such as ethylene glycol-water to minimize the refrigerant charge of equipment on-site.

During a *charge* cycle, cold glycol circulates through the heat exchanger coil causing ice to form on the outside surfaces of tubes. To sustain a fixed charging rate, the entering fluid temperature must decrease as the state of charge increases. The layer of ice is allowed to continue growing but unlike the internal-melt system, the ice thickness is limited to avoid bridging between adjacent tubes. Bridging between tubes is undesirable due to the need for external fluid flow paths during the discharge cycle.

During *discharge* the system can be designed to use either cold glycol (as in the internal-melt system) or the near-freezing water residing in the storage tank. The most common approach is to pump the cold water from the storage tank directly to the load and return the warm water back to the storage tank causing the accumulated ice to melt from the outside surface inward (hence the term "external-melt"). Although this approach is convenient for retrofitting buildings equipped with chilled water (since it avoids sending glycol to the building), there is a an operating efficiency penalty associated with recharging the storage device after a partial melt since the residual ice on each tube acts as an insulator to the further freezing of surrounding water.

3.4.2.3 Encapsulated ice and PCMs

An encapsulated ice system is comprised of a multiplicity of water-filled containers immersed in a tank filled with glycol as shown in Fig. 46. Typically, the containers are spherical shaped and available in either a 3 in. (77 mm) or 4 in.

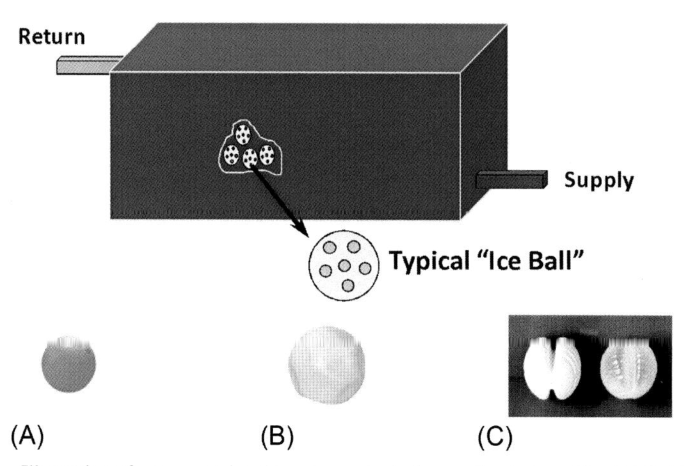

FIG. 46 Illustration of an encapsulated ice storage tank along with photos of the (A) Cristopia ice nodule, (B) Cyrogel Ice Ball, and (C) EnE Systems "Ice Bon" storage nodule.

(98 mm) nominal outside diameter size. Some container designs feature smooth surfaces while others include large "dimples" or a split hemisphere to enhance heat transfer and accommodate water's expansion upon freezing. The most common fluid included in the encapsulated sphere is water with additives to encourage ice formation; however, alternative fluids such as eutectics can be specified to achieve a storage temperature consistent with application requirements.

During a charge cycle, the low-temperature secondary fluid is pumped through the storage tank and past the storage capsules occupying the tank. Water within the individual containers cools and begins to solidify. Nucleating agents added to the water promote the initiation of the phase change. Ice continues to form inside each individual container—building from the outside inward until the capsules are completely frozen. During a discharge cycle, warm secondary fluid is pumped through the vessel with the storage capsules—causing ice to melt from the outside of individual containers inward.

3.4.3 Unitary air conditioning systems

One of the most widely used air conditioning systems in the world is the "unitary system." A *unitary air conditioning system* is a self-contained factory-built packaged configuration that integrates a cooling plant, air moving equipment (fans), and controls. For commercial buildings, the most commonly applied unitary system is the roof-top air conditioner. The cooling capacity for a typical roof-top package ranges from 5 to 20 tons (18 to 70 kW); however, larger packages with capacities on the order of 200 tons (700 kW) are available.

The cooling plant in a unitary package consists of a condensing unit (compressor and condenser), evaporator, and associated expansion devices and refrigerant piping. Generally, unitary systems will feature multiple compressors

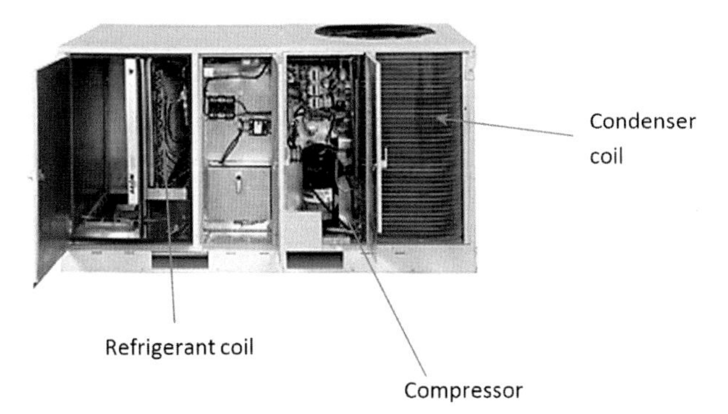

FIG. 47 Cutaway showing the key components of a 5-ton roof-top package (Aaon).

and evaporators in order to achieve the desired overall unit capacity. A fan integrated into the unit draws air across the cooling coil(s) and delivers conditioned air to the interior spaces to meet coincident loads. A separate fan on the other side of the unit draws ambient air across the condenser coils to reject heat from the refrigerant. Fig. 47 shows the primary components that comprise a roof-top package unit.

An operational disadvantage of the traditional unitary system lies in the fact that the refrigeration plant (condensing unit) must operate to meet instantaneous space cooling loads as they occur. This means that the greatest electrical demand from the unit directly coincides with the peak building cooling loads. Unfortunately, it is at this time that the unit's operating efficiency is also at its lowest due to warm outdoor air conditions. The electrical demand for a typical unitary system will generally be greater than 1 kW/ton during peak operating periods and less than 1 kW/ton during off-peak periods.

An emerging technology that aims to achieve energy and energy cost savings by adding thermal storage capability to these systems is known generically as a *unitary ice storage system*. One example of a unitary ice thermal energy storage system is shown in Fig. 48. This particular unit has 40 ton-h

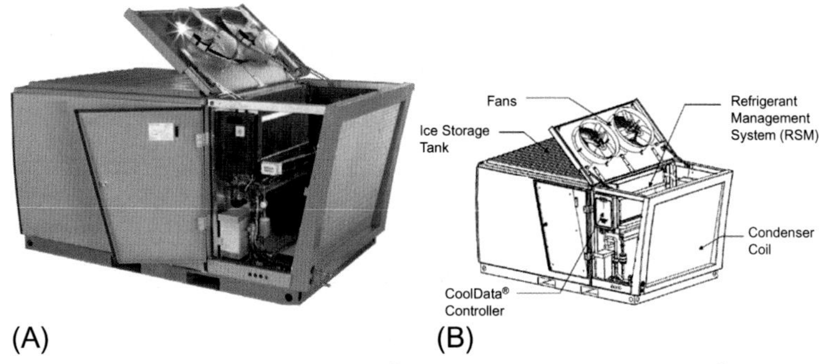

FIG. 48 The Thule Energy Storage Ice Bear® showing (A) and (B) the Ice Bear® 40 condensing unit. *(Courtesy of Thule Energy Storage.)*

(141 kWh) of capacity integrated with a dedicated refrigerant condensing unit (R-410a) and controls.

During a *charge* mode, the self-contained condensing unit provides high pressure liquid refrigerant to the bottom of a series of vertical helical coiled heat exchangers (Fig. 49) immersed in a water tank. Ice forms on the outside surfaces of the helical coils as the refrigerant evaporates inside the coils before returning to the compressor suction. During *discharge*, cold liquid refrigerant is circulated to the evaporator coil dedicated to unit. Here the cold liquid refrigerant absorbs heat from the air stream being cooled and dehumidified and evaporates. The refrigerant vapor is returned back to the top of the helical coil where it heat exchanges with stored ice causing the refrigerant to condense back to a liquid, thereby melting the ice. An onboard refrigerant management system and control properly sequence the unit's operation during both the charging and discharging modes.

The time required to complete a full charge cycle (40 ton-h) is approximately 10.5 h with an outdoor air temperature of 75°F (24°C). During the discharge cycle, the refrigeration system associated with the normal unitary package remains idle—drawing on the stored thermal energy within the Ice Bear®. Thule Energy Storage states that the unit is capable of meeting a 10-ton load continuously for a 4-h period during discharging. Using the ice to avoid the operation of a conventional refrigeration system during on-peak periods, an electric load shift of approximately 10 kW can be achieved.

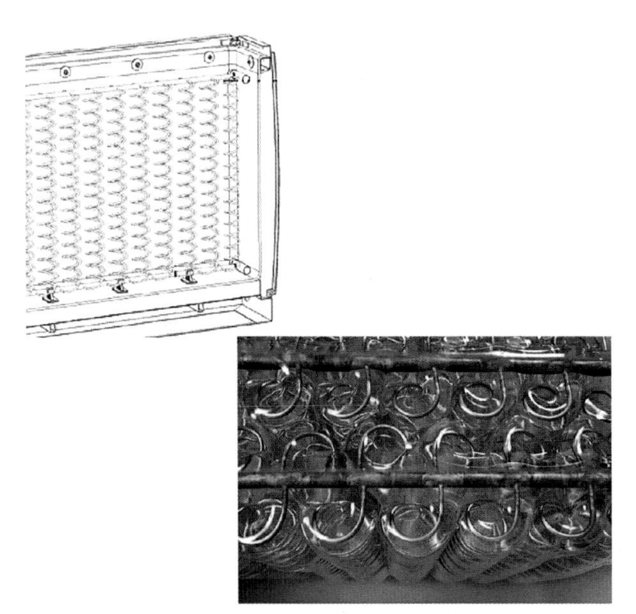

FIG. 49 An illustration of the helical coil arrangement in storage module of the Ice Bear® unit *(left)* and the actual coil arrangement *(right)*. *(Courtesy of Thule Energy Storage.)*

3.4.4 Dynamic ice storage

The previous ice storage technologies can be classified as *static ice* systems. That is, the stored thermal energy, effectively, stays in position during charging and discharging. Another class of ice storage systems is *dynamic ice* storage. In this approach, the production of ice is separated from its location during storage. Dynamic ice systems consist of ice harvesters (plate-type, tube-type) and ice slurry machines.

The key components of a typical ice harvesting system are shown in Fig. 50. The operation of dynamic ice or ice harvesting systems can be characterized by three distinct modes: ice building, ice melting, and chilling. Ice building and chilling require the operation of the refrigeration plant while the ice melt mode only requires operation of chilled water circulating pumps. The ice building and melting operating modes are discussed as follows.

Charging a dynamic ice system involves making or building a sufficient quantity of ice for retention in the storage tank to meet future planned loads. The fundamental principle of operation during ice building is as follows. Water from the storage tank is pumped to a distribution header for distribution over the outside of a series of refrigerated plates or tubes (the evaporator surface). The water falls by gravity over the refrigerated surface while a portion of it freezes to the surface while the remainder falls back into the storage tank. The build

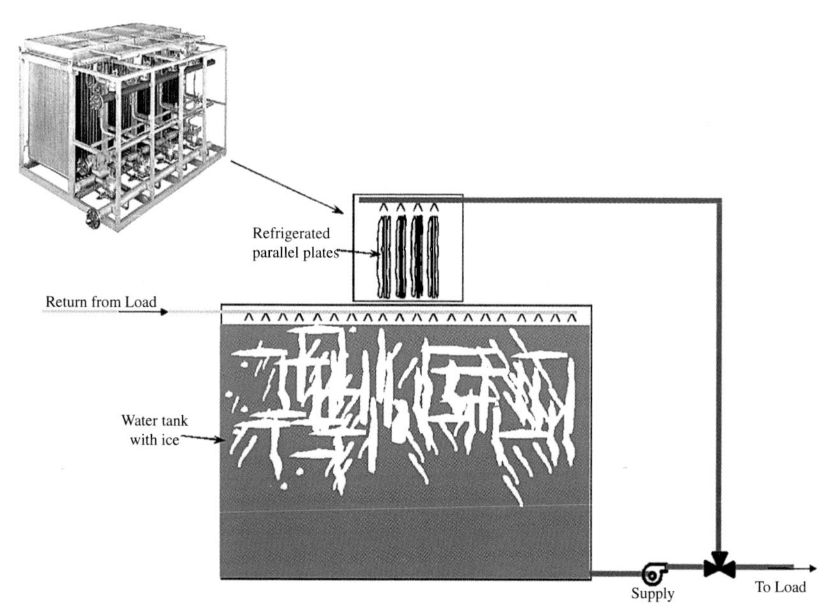

FIG. 50 A Paul Mueller Company falling film ice harvester *(upper left)* and a typical arrangement of an ice harvesting storage system.

process continues until a suitable amount of ice accumulates on the evaporator surface ca. 3/8 in. (9.5 mm) for plate ice and ca. 1" (25.4 mm) diameter for tube ice configurations. Periodically, the accumulated ice is removed or "harvested" from the refrigerated surfaces by initiating a defrost or harvest cycle. During a harvest cycle, hot refrigerant gas from the compressor discharge is directed to one or more of the refrigerated plates (or tubes). The hot gas warms the heat transfer surface slightly, thereby breaking the bond adhering ice to the surfaces, thereby causing the ice to release and discharge by gravity to a tank positioned below the system. This process repeats until a suitable inventory of ice accumulates in the storage tank.

The typical *discharge* of a dynamic ice system is accomplished by pumping water from the bottom of the storage tank to the load and uniformly distributing the warmer return water over the accumulated ice using a spray distribution system located at the top of the storage tank. The ice behaves as a large porous plug heat exchanger rapidly cooling the warm return water as it absorbs heat, melting the ice during its downward, gravity-assisted flow. This storage system design is able to yield extremely high discharge rates while sustaining low supply water temperatures (near 32°F (0°C)). This is due to the large surface area of the ice available for direct contact with the return water.

The energy efficiency of dynamic ice systems strongly depends on the performance of the ice generating refrigeration system which is influenced by the relative duration of ice build and defrost periods. Interesting trade-offs arise because during ice building, the continued growth of ice on the refrigerated surfaces causes decreased thermal performance. This is attributable to the ice acting as a fouling agent; therefore longer build cycles to accumulate thicker ice will result in progressively lower efficiency. Additionally, the defrost cycle itself represents a parasitic load on the system since heat is being reintroduced for the purpose of removing ice from the refrigerated surfaces. Defrost cycles that are too frequent will also erode performance so an optimum ice build/defrost period exists. Field experience has shown that appropriate build periods last about 20–30 min with corresponding harvest periods lasting approximately 45 s. Determining more precise optimal build and harvest periods will depend on the refrigeration system's low-side design approach temperature (nominally, the difference between water freezing point and the saturated evaporator temperature) and the saturated discharge temperature.

Although harvesting types of dynamic ice storage systems offer high thermal storage discharge rate capabilities, their overall operating energy efficiency is relatively low and capital cost high. To address these issues, the development of another type of storage system in the dynamic ice category was pursued—ice slurry systems. Simply defined, an ice slurry is a mixture of ice crystals suspended in a liquid. The liquid used is typically water with a small amount of additive (e.g., glycol, ethanol, calcium carbonate, and others) that slightly lowers its freezing point. Depending on the specific ice slurry technology, the *initial* concentration of the additive may vary from 2% to over 10% by weight.

During a *charging* process, ice crystals are formed by freezing the water portion of the liquid solution. The liquid solution serves as a carrier to suspend and transport the ice crystals of the slurry to a separate storage tank. As more ice is formed, the liquid solution becomes further concentrated necessitating successively lower evaporator temperatures to continue the charging process. While the slurry resides in the storage tank, mechanical agitation (a source of parasitic energy) is required to maintain the ice crystals in suspension.

During a *discharging* process, warm fluid returning from the load is uniformly distributed (sprayed) across the top of the slurry-filled tank to achieve an even melt-out. Uneven flow distribution contributes to thermal "channeling" and inconsistent discharge of the stored slurry.

The most widely used approach for generating ice in slurry form is by the use of variants of a scraped surface heat exchanger. In a scraped surface heat exchanger, water (with additives) circulates through a tube jacketed with refrigerant. The mechanical action by either a rotating rod or dasher blades continually agitates the boundary layer of water near the inside of the tube's surface, thereby preventing ice crystals from accumulating on the inside of the tube's surface. The ice crystals are swept from the tube surface into the free stream of flowing water and ejected out of the slurry generator and directed to a storage tank or load.

Although slurry systems have been applied for the primary purpose of storing thermal energy, they offer the potential for significantly reducing the piping size requirements for meeting thermal loads when used as a secondary heat transfer fluid. The size reduction lies in the fact that circulating a small ice fraction (e.g., 10 wt%) can increase the thermal capacity of a given pipe size by 140%! Although this benefit has been recognized for some time, the ice slurry technology faces a number of significant challenges—particularly in air conditioning system applications. One challenge lies in maintaining hydronic system operability by avoiding the accumulation of ice crystals with subsequent plugging. Other challenges involve maintaining high system energy performance given that the efficiency of the charging process decreases with increasing state of charge for storage.

Ice slurries have found successful use in food processing and specialty storage applications. Fig. 51 shows one such application where ice slurry is being applied to preserve freshly caught fish.

FIG. 51 Ice slurry used for preservation of fresh fish. *(Source: Sunwell Technologies.)*

References

[1] International Renewable Energy Agency, Concentrating solar power, in: Renewable Energy Technologies: Cost Analysis Series Volume 1: Power Section Issue 2/5, 2012. https://www.irena.org/-/media/Files/IRENA/Agency/Publication/2012/RE_Technologies_Cost_Analysis-CSP.pdf.

[2] Andasol-1 in SolarPaces Project Database, 2017: https://solarpaces.nrel.gov/andasol-1

[3] Giuseppe Casubolo, Senior Commercial Director Industrial Nitrates SQM International N.V. Antwerp—Belgium. Personal communication 2019.

[4] T. Bauer, et al., Overview of Molten Salt Storage Systems and Material Development for Solar Thermal Power Plants, DLR Koeln. https://ases.conference-services.net/resources/252/2859/pdf/solar2012_0122_full%20paper.pdf, 2012.

[5] Bilfinger Industrial Services GmbH, References for Heat Accumulators, https://bis-austria.bilfinger.com/en/references/energy-and-distribution/heat-accumulators/2-zone-accumulators/atmospheric-2-zone-accumulator-duisburg/, 2020.

[6] K. Frick, Modeling and Design of a Sensible Heat Thermal Energy Storage System for Small Modular Reactors, PhD Dissertation, North Carolina State University, Department of Nuclear Engineering, 2018. https://repository.lib.ncsu.edu/bitstream/handle/1840.20/34975/etd.pdf.

[7] M. Hämmerle, M. Haider, R. Willinger, K. Schwaiger, R. Eisl, K. Schenzel, Saline cavern adiabatic compressed air energy storage using sand as heat storage material. J. Sustain. Dev. Energy Water Environ. Syst. 5 (1) (2017) 32–45, https://doi.org/10.13044/j.sdewes.d5.0131.

[8] ADELE-ING Engineering-Vorhaben für die Errichtung der ersten Demonstrationsanlage zur Adiabaten Druckluftspeichertechnik Öffentlicher Schlussbericht, DLR, 2018.

[9] B. Xu, P. Li, C.L. Chan, Extending the validity of lumped capacitance method for large Biot number in thermal storage application, Sol. Energy 86 (2012) 1709–1724.

[10] D. Laing, D. Lehmann, M. Fiss, C. Bahl, Test results of concrete thermal energy storage for parabolic trough power plants. J. Sol. Energy Eng. 131 (4) (2009). https://doi.org/10.1115/1.3197844.

[11] N. Hoivik, et al., Demonstration of Energy Nest thermal energy storage (TES) technology. SOLARPACES 2016, AIP Conf. Proc. 1850 (1) (2017) 080011.

[12] E. John, M. Hale, P. Selvam, Concrete as a thermal energy storage medium for thermocline solar energy storage systems, Sol. Energy 96 (2013) 194–204.

[13] L. Nielsen, R. Leithner, Dynamic simulation of an innovative compressed air energy storage plant—detailed modelling of the storage cavern, WSEAS Trans. Power Syst. 4 (2009) 253–263.

[14] M. Medrano, A. Gil, I. Martorell, X. Potau, L.F. Cabeza, State of the art on high-temperature thermal energy storage for power generation. Part 2—case studies, Renew. Sust. Energ. Rev. 14 (1) (2010) 56–72.

[15] V. Kodur, F. Asce, W. Khaliq, Effect of temperature on thermal properties of different types of high-strength concrete, J. Mater. Civ. Eng. 23 (June) (2011) 793–801.

[16] D. Laing, C. Bahl, T. Bauer, M. Fiss, N. Breidenbach, M. Hempel, High-temperature solid-media thermal energy storage for solar thermal power plants, Proc. IEEE 100 (2) (2012).

[17] D. Laing, W.-D. Steinmann, R. Tamme, A. Wörner, S. Zunft, Advances in thermal energy storage development at the German Aerospace Center (DLR), Energy Storage Sci. Technol. (2012) 13–25. ISSN 2095-423.

[18] https://www.globenewswire.com/news-release/2019/07/01/1876757/0/en/EPRI-to-Lead-5-Million-DOE-NETL-Project-to-Test-Concrete-Thermal-Energy-Storage-Technology.html.

[19] GuD-Druckluftspeicherkraftwerke-hocheffiziente Übergangs- und Dauerlösung, VGB PowerTech 1/2,2018.

[20] G. Zanganeh, R. Khannaa, C. Walser, A. Pedrettib, A. Haselbachera, A. Steinfeld, Experimental and numerical investigation of combined sensible–latent heat for thermal energy storage at 575°C and above, Sol. Energy 114 (2015) 77–90.

[21] S. Zavattoni, M. Barbato, A. Pedretti, G. Zanganeh, CFD Simulations of a Pebble Bed Thermal Energy Storage System Accounting for Porosity Variations Effects, SolarPACES, 2011.

[22] D.M. Crandall, E.F. Thatcher, Segmented thermal storage. Sol. Energy 77 (4) (2004) 435–440, https://doi.org/10.1016/j.solener.2003.08.011.

[23] A.J. White, J.D. McTigue, C.N. Markides, Analysis and optimisation of packed-bed thermal reservoirs for electricity storage applications. Sage Eng. (2016). https://doi.org/10.1177/0957650916668447.

[24] J. McTigue, Analysis and Optimisation of Thermal Energy Storage, Department of Engineering, University of Cambridge, 2016.

[25] J.D. McTigue, A.J. White, A comparison of radial-flow and axial-flow packed beds for thermal energy storage, Appl. Energy 227 (2018) 533–541.

[26] J.E. Pacheco, S.K. Showalter, W.J. Kolb, Development of a molten-salt thermocline thermal storage system for parabolic trough plants. J. Solar Energy Eng. 124 (2002)https://doi.org/10.1115/1.1464123#.

[27] A.J. Willmott, Simulation of a thermal solid TES under conditions of variable mass flow, Int. J. Heat Mass Transf. 11 (7) (1968).

[28] A. Stückle, D. Laing, H. Müller-Steinhagen, Numerical simulation and experimental analysis of a modular storage system for direct steam generation. Heat Transfer Eng. 35 (9) (2014) 812–821, https://doi.org/10.1080/01457632.2013.828556.

[29] J.S. Andrepont, Combustion turbine inlet air cooling (CTIAC): benefits, technology options, and applications for district energy, Energy Eng. 98 (3) (2001) 52–69.

[30] K.M. Liebendorfer, J.S. Andrepont, Cooling the hot desert wind: turbine inlet cooling with thermal energy storage (TES) increases net power output 30%, ASHRAE Trans. 111 (2) (2005) 545–550.

Chapter 4

Mechanical energy storage

Aaron Rimpel[a],*, Klaus Krueger[b], Zhiyang Wang[c], Xiaojun Li[d], Alan Palazzolo[e], Jamshid Kavosi[e], Mohamad Naraghi[e], Terry Creasy[e], Bahar Anvari[f], Eric Severson[g], and Eugene Broerman[a]

[a]Southwest Research Institute, San Antonio, TX, United States, [b]Voith, Heidenheim, Germany, [c]Vycon Energy, Cerritos, CA, United States, [d]Gotion Inc., Fremont, CA, United States, [e]Texas A&M University, College Station, TX, United States, [f]ABB, Greenville, SC, United States, [g]University of Wisconsin, Madison, WI, United States

Chapter outline

4.1 Pumped hydroelectric storage

Klaus Krueger

4.1.1 Overview and basics design parameters

This chapter explains the basics of pumped storage systems starting with the objectives of pumped hydro. Other topics include its historical development, the basic types of pumped hydroplants, and technical aspects. The chapter concludes with application objectives and business opportunities for pumped storage.

Pumped storage has remained the most proven large-scale power storage solution for over 100 years. The technology is very durable with 80–100 years

* Chapter lead/organizer.

Thermal, Mechanical, and Hybrid Chemical Energy Storage Systems
https://doi.org/10.1016/B978-0-12-819892-6.00004-6

of lifetime and more than 50,000 storage cycles. It is further characterized by round trip efficiencies between 78% and 82% for modern plants and very low-energy storage costs for bulk energy in the GWh-class. This is the reason why pumped storage predominates the electricity bulk storage market. The total installed power of pumped storage worldwide was around 153 GW by the end of 2017 [1], a share of 97% of all front-of-meter storage applications in the electricity market.

4.1.1.1 Working principle and basic design parameters

Pumped storage stores electricity in the form of potential energy. The basic principle of energy conversion is shown in Fig. 1. In pump mode (charging), electrical energy is taken from the electrical grid to feed a motor that mechanically drives a pump. The water is pumped from the lower basin into the upper basin. In the turbine mode (discharging), the water simply flows downhill via gravity and drives a turbine. The turbine then drives mechanically the generator rotor and the electrical energy returns to the electrical grid. Fig. 2 shows the application range in terms of head and flow for single-stage reversible pump turbines and multistage pumps for high heads.

The following main parameters have a significant influence on the technical solution. The electrical power P is proportional to the head H and to the flow Q. The flow Q influences directly the size of the power units, the power house size, and also the water ways. In contrast, the head H drives the plant type and the type of hydraulic machine. Fig. 3 indicates application ranges for different types of turbines and pumps with respect to head and power for pumped hydroapplications.

It should be pointed out that for low head there are also solutions in operation like Deriaz pump turbines and reversible bulb units.

Besides gross head H and flow Q as indicated Fig. 1, one more parameter is necessary in order to preselect a suitable power unit configuration. It is the gross head variation factor H_{max}/H_{min}, obtained from the division of the maximum gross head (upper reservoir completely full, lower reservoir at minimum level) by the minimum gross head (upper reservoir at minimum level, lower reservoir at maximum level). In practical cases the factor H_{max}/H_{min} varies between 1.05 (very flat reservoirs) and 1.45 (very narrow and deep reservoirs or low head applications). It has to be noted that gross heads can be used for a first approach but that the total delivery heads for the net heads have to be used for a final consideration.

These three parameters H, Q, and H_{max}/H_{min} are decisive criteria for the power unit type preselection. By executing a quick design of the power unit, the speed of unit, the main dimensions of the power unit and also the necessary submergence can be found. The necessary submergence is the head difference between the lowest level of the lower reservoir and the pump setting elevation (see example in Fig. 4 for a single-stage reversible pump turbine). In the case of

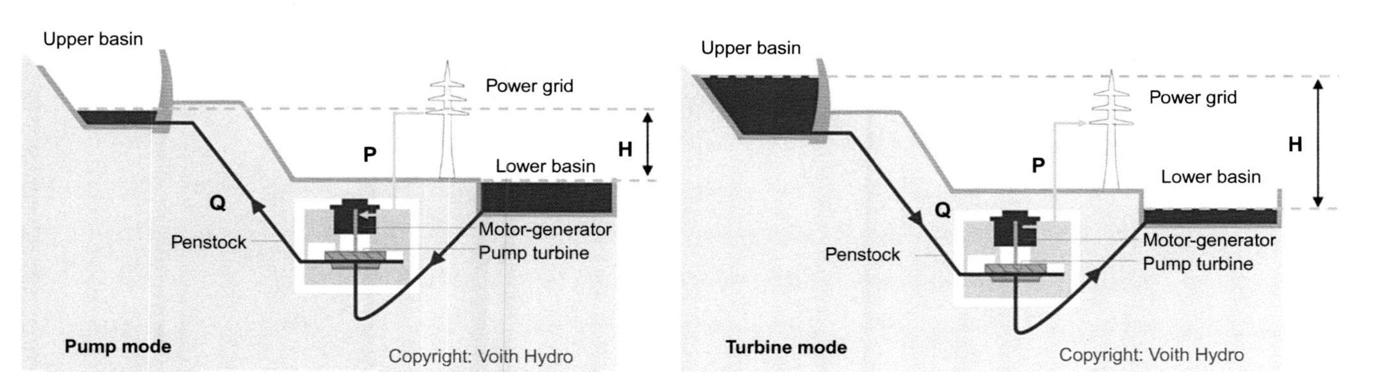

FIG. 1 Energy conversion principle from electrical power to potential energy and vice versa.

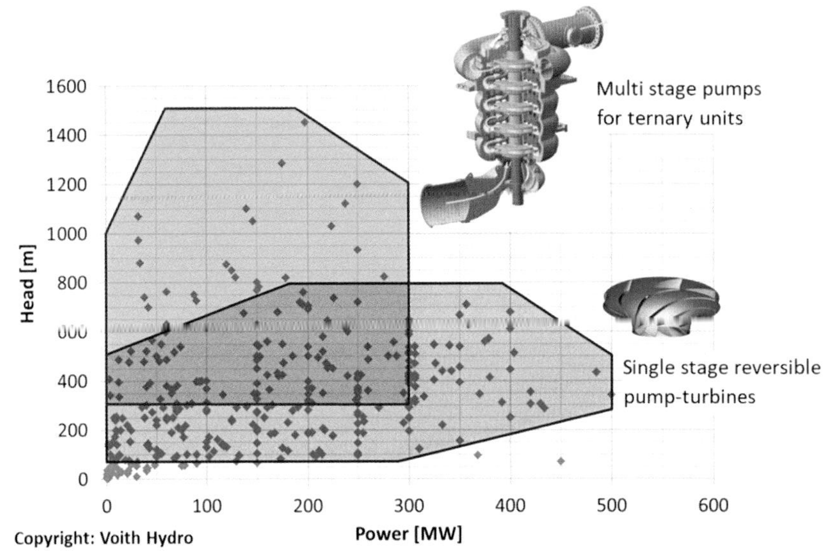

FIG. 2 Application range for single-stage reversible pump turbines (charging and discharging) and multistage pumps for high heads (charging).

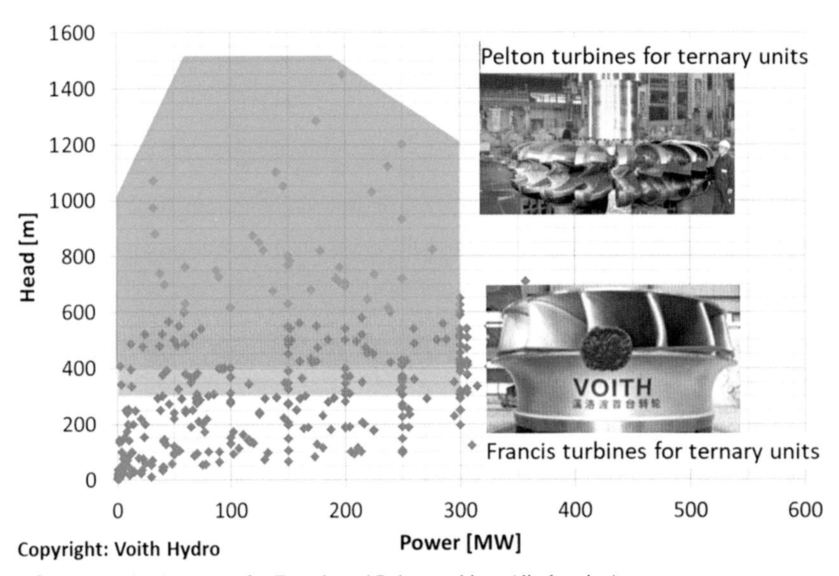

FIG. 3 Application range for Francis and Pelton turbines (discharging).

multistage pumps the elevation of the deepest stage is used. Submergence must be satisfied in order to prevent cavitation in operation, which would damage the runner material. Since the necessary submergence has a huge impact on the civil cost (power house type and setting, length of access tunnels and waterways, etc.), the minimum submergence is calculated at a very early stage of the project development, already.

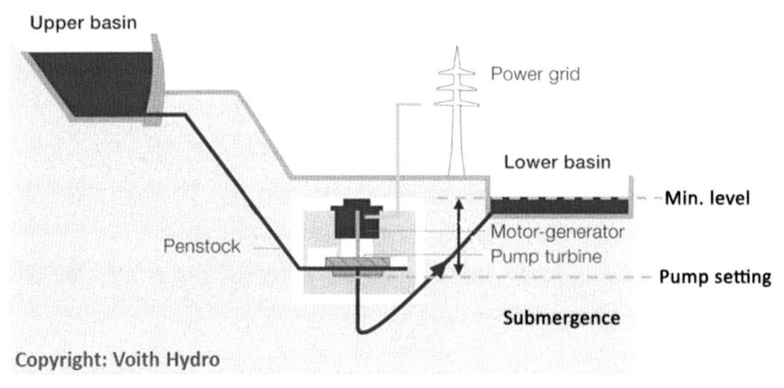

FIG. 4 The submergence of the reversible pump-turbine runner is indicated by the red arrow.

As a rule of thumb, the necessary submergence for single-stage reversible PT can be between 10% and 20% of net head! The net head is the gross head H indicated in Fig. 1 minus the head losses, which are proportional to Q^2 (power of 2 of the flow). Multistage pumps need less submergence than single-stage pump turbines.

4.1.1.2 Types of pumped storage plants

The hydrological conditions at a given site divide pumped storage power plant into two main types: closed- and open-loop configuration. A significant share of pumped storage plants has one upper and one lower reservoir that are operated as a closed loop, that is, there is no natural inflow or outflow in the upper reservoir available. In the most cases the upper reservoir is artificially built (example see Fig. 5). The maximum water volume, which can be circulated between the lower and upper reservoir, remains constant and is determined by the smaller reservoir. The pump and turbine power of each power unit are in the same power range and the gross delivery head and gross head are also similar. Typical "energy to power" ratios for these plants vary between 4 and 10 Wh/W, which corresponds to the maximum full load power duration in turbine mode. One reason for the range of ratios is the compromise between storage duration and related civil costs for the reservoirs and a second reason is that in the past these plants were primarily built for peak shaving and daily cycling operation in order to ensure base load operation of thermal plants (fossil and nukes). One important characteristic metric for closed-loop pumped storage is the storage energy content per load cycle (GWh/LC). The load cycle considers the complete recirculation of the entire water volume from the lower basin to the upper basin and back to the lower basin for its energetic use [2]. In moderate climate zones the evaporation losses in both reservoirs are often compensated by rain and snow fall. In cases where this is not the case a water-refilling system is necessary to compensate the losses.

FIG. 5 Example of a closed-loop pumped storage in Goldisthal, Germany, with an artificial upper reservoir. The powerhouse is equipped with 4 × 250 MW units and the energy to power ratio 8 Wh/W.

Open-loop pumped storage systems like those in central Europe have been in continuous operation for decades due to their economical and operational benefits. In these cases, the upper reservoir and/or intermediate reservoirs have significant natural inflows. Such a system is then a combination of a hydrostorage or even run-of-the-river hydrofacility with pumped storage capabilities. For example in the Black Forest there is a three-stage pumped storage cascade since 1951 in operation with a total power of 470 MW in turbine mode and 308 MW in pump mode (see Fig. 6). The upper reservoir called Schluchsee was in former times a natural lake whose volume was increased significantly by building a new dam. The maximum storage capacity of this reservoir is 133 GWh, but only 62 GWh can be utilized practically due to water level regulations. In 2017, this open-loop configuration has generated 183 GWh of electricity only by the three natural inflows indicated in Fig. 6 without any pump activities. In contrast to daily cycling operation of conventional pumped storage installations (closed loop) such open-loop applications can provide seasonal bulk energy storage for weeks, months or even a year due to their large reservoirs and can also be operated as a run-of-the-river arrangement in parallel. Depending on the natural inflow quantities, the necessary pump power is smaller than the turbine power for such open-loop configurations.

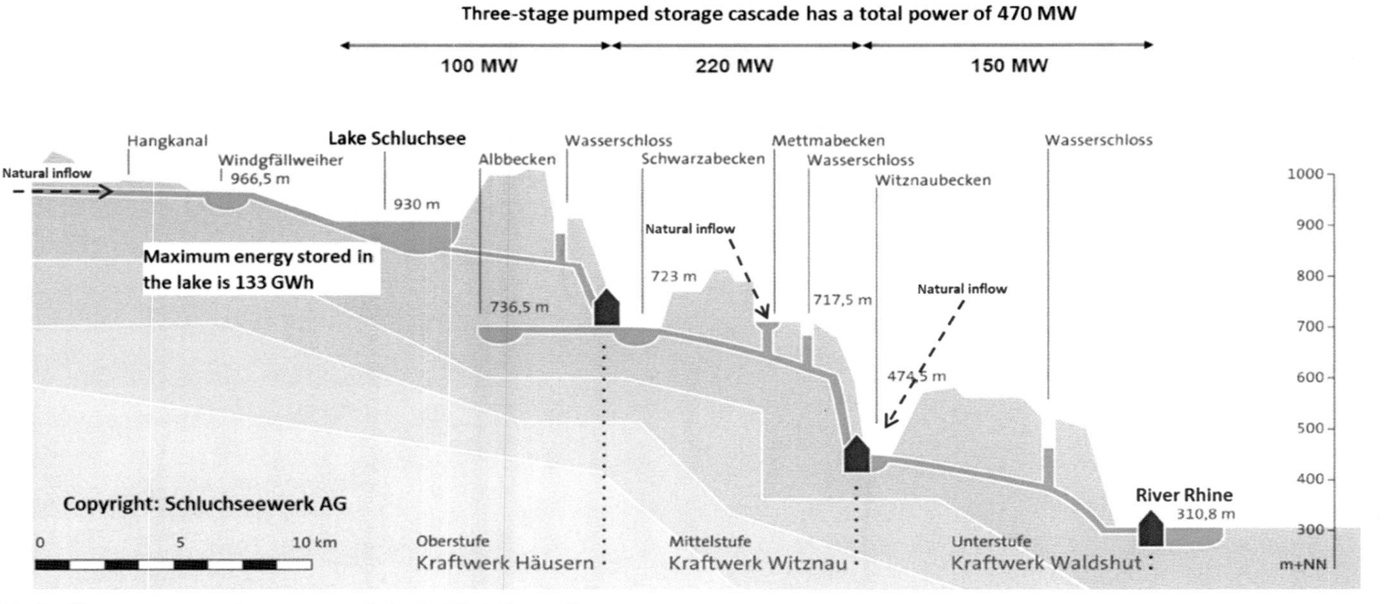

FIG. 6 Three-stage pumped storage cascade in the Black Forest, Germany.

Pumped storage power plants are commonly categorized by the type of construction of their power houses:

- Surface power house (see Fig. 7),
- Deep-excavated or shaft (silo type) power house with access via stairs or elevator (see Fig. 8),
- Cavern-type power house with dedicated large access tunnels (see Fig. 9).

The decision of which type of power house should be applied is a result of the overall design of the hydraulic machinery (e.g., necessary submergence) and waterways, as well as calculations of the key operational parameters like

FIG. 7 Salina pumped storage with surface power house located in Oklahoma, United States.

FIG. 8 Waldeck I deep-excavated power house in Germany.

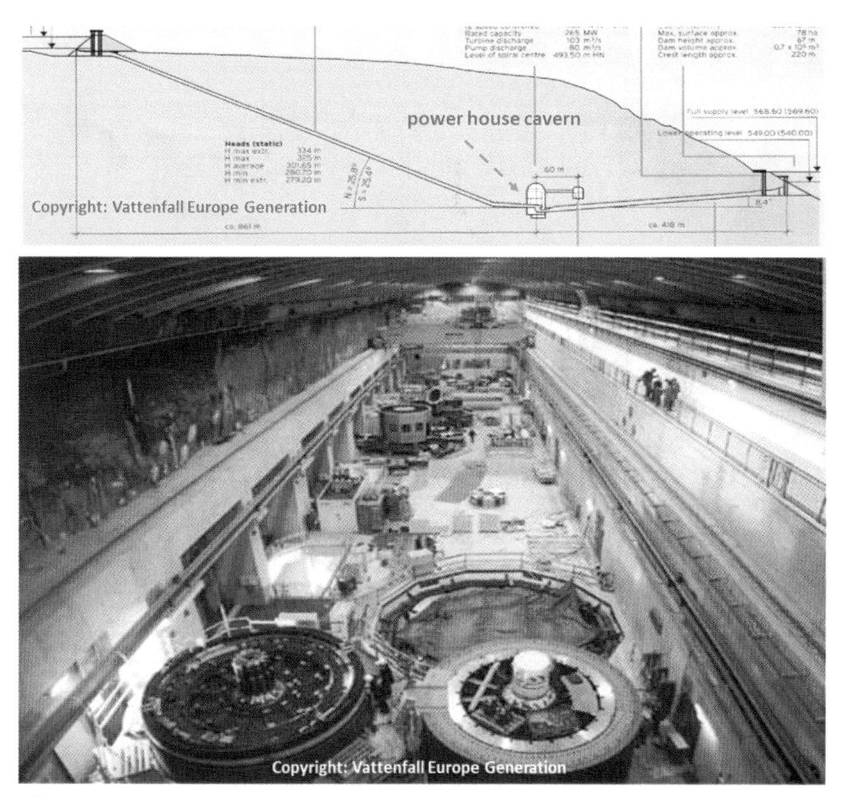

FIG. 9 Waterways (top) and the cavern of the Goldisthal pumped storage in Germany during construction (bottom)

pressure levels, etc., including during transient conditions (e.g., caused by start or stop of pumps and turbines, by closing of valves, etc.). This information is obtained by a detailed transient calculation study, including water hammer calculations. The outputs of such a study may influence the power house arrangement, the waterways design, the necessity of surge tanks or surge chambers, etc.

4.1.1.3 Historical development and types of pumped storage units

The idea of hydraulic energy storage by means of pumps and turbines was born at the end of the 19th century in Switzerland and in Germany. The first pumped storage plant was built in Zurich in 1891 at the Limmat river followed by a second installation 1894 at lake Maggiore and a third one in 1899 at the Aare river [2]. The principle of pumped storage was first realized in Germany in 1891, where a steam machine was driving a centrifugal pump for dewatering the Rosenhof ore mine in the Upper Harz mountain by filling an upper reservoir, which was serving a separate water wheel [2]. As indicated in Fig. 10, the first pumped storage applications were four machine solutions. The pump (*P*) had an

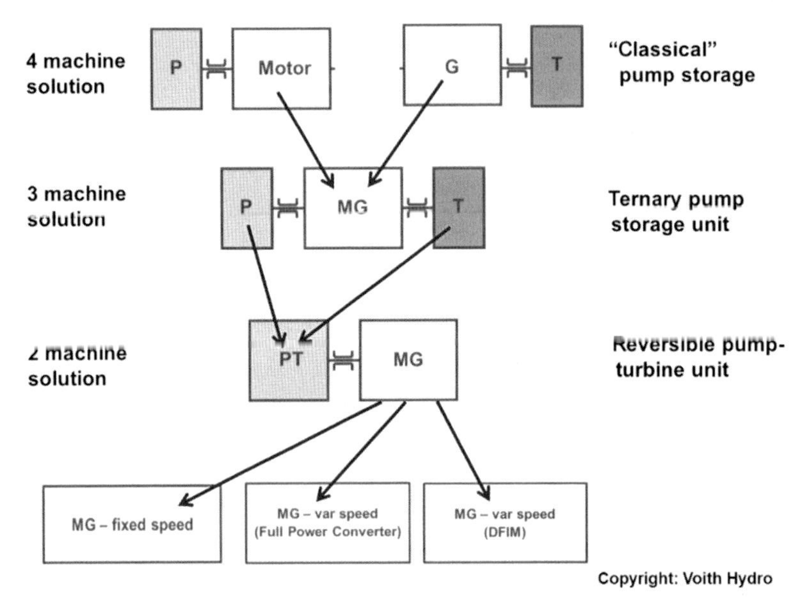

FIG. 10 Sequence of historical development of power unit arrangements.

individual motor (*M*) and the turbine (*T*) had an individual generator (*G*). Such classical configurations can also be found in current applications, where an existing hydrostorage power plant was converted to a pumped storage by construction of a new lower reservoir and a separate pump station (e.g., pumped storage on islands like El Hierro, Madeira, Ikaria). The location of the pump house and turbine power house can be at separate locations and elevations, that is, the gross heads for pumps and turbines can be different—even using separate water ways.

In the next step of historical development the four-machine solution merged to a three-machine solution with one common shaft. Such a configuration is called ternary unit and it is characterized by one rotational direction in pump and in turbine mode. Between the hydraulic machines and the motor-generator (MG) couplings and/or hydraulic torque converters may be foreseen in order to decouple or couple the hydraulic machines. Ternary sets can be arranged both vertically and horizontally. Vertical arrangement has advantages during erection works and operation (e.g., easier shaft alignment, bearing forces can be easier transmitted the civil structures). In the 1930s the pump and turbine were then merged to one hydraulic machine called a reversible pump turbine (*PT*), which is arranged vertically in general. In this configuration the rotational direction needs to be changed between pump and in turbine mode, which takes more, so called changeover time compared to a ternary unit arrangement. All mentioned configurations before were operated with a fixed speed, when connected to the electrical grid.

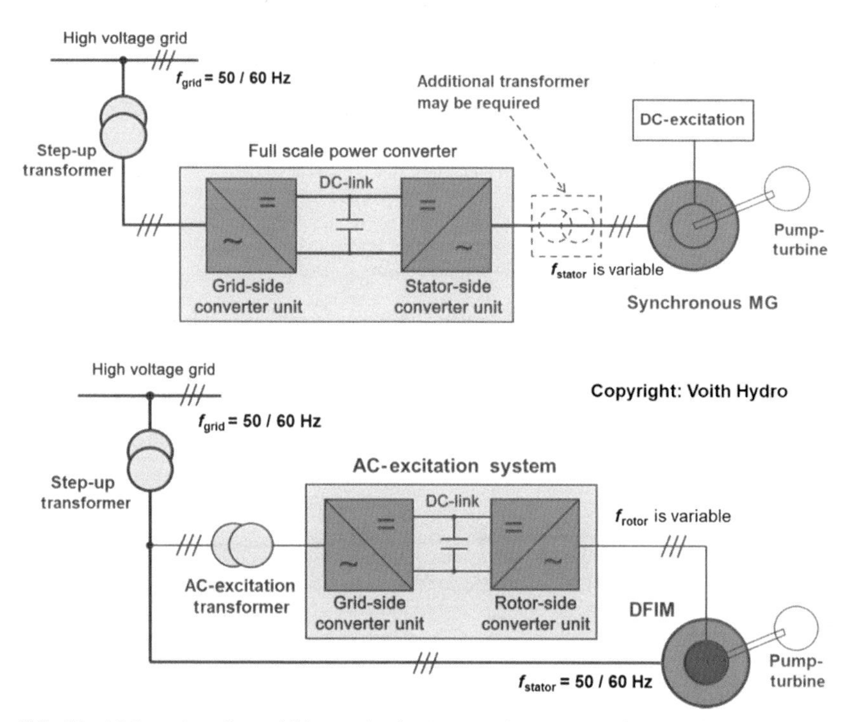

FIG. 11 Main options for variable speed units (top: synchronous machine with full-scale power converter; bottom: double-fed induction machine with an AC excitation).

At the end of the 1980s, the first electrical machines were equipped with converters utilizing power semiconductors, allowing to run hydraulic machines with variable speed. By this, optimized operation, especially flexible charging (pump mode) of pumped hydroplants, became possible without the need of so-called hydraulic short-circuit operation. There are two main design variants common for variable speed (see Fig. 11) machines:

- Synchronous motor generator with a full-scale power converter located between step-up transformer and stator and
- Double-fed induction machine with an AC excitation system for the rotor and the stator is connected directly to the grid via the step-up transformer. The rotor of the DFIM has, like the stator, a three-phase winding instead of the poles used in a synchronous motor generator.

The following three types of power units have been built in the last decades, mainly:

- Conventional reversible pump-turbines units with fixed speed ($PT + MG$), which represent the majority of the installations,
- Variable speed reversible units ($PT + MG$),

- Ternary unit arrangements with fixed speed ($P + MG + T$).

Although the decision which type of power unit has to be implemented depends on given hydrological and geological aspects like gross head H, reservoir size, cross-section, etc., a significant factor for power unit selection is the aforementioned gross head variation factor H_{max}/H_{min}. The following rule of thumb may help for a basic orientation:

- Reversible pump-turbine units with fixed speed can manage $H_{max}/H_{min} < 1.2$,
- Ternary units with fixed speed can manage $H_{max}/H_{min} < 1.2$,
- Reversible pump-turbine units with variable speed using a DFIM can manage
  $H_{max}/H_{min} < 1.5, \ldots, 2$,[a]
- Reversible pump-turbine units with variable speed using a full-size power converter can manage $H_{max}/H_{min} < 2, \ldots, 4$[a]

It should be emphasized that in some executed projects the decision to implement the more expensive ternary units or reversible pump-turbines with variable speed was also driven by the operational flexibility requirements for the power units and not only from the gross head variation factor H_{max}/H_{min}.

4.1.1.4 Power unit concepts and their main operation modes

Reversible power units

The most common power unit concept is a vertical arrangement consisting of a reversible pump turbine and a motor generator with different rotation directions for pump and turbine mode (see Fig. 12). Such power units have up to 12 transition modes as indicated in Fig. 13. Besides standstill, pump mode, and turbine mode, very often these units are capable to operate in the so-called synchronous condenser mode (SC) in both rotational directions. In the SC mode, the runner is rotating in air and the motor generator is supplying reactive power according to the setpoint of the grid dispatcher. In this SC mode, the power unit consumes some active power from the grid in order to compensate mainly the bearing and ventilation losses. For this SC mode a so-called air blow-down system with pressure ranges up to 80 bar is required, which can inject in standstill pressurized air via the turbine cover or via the draft tube in order to push the draft tube water level below the runner.

The synchronous condenser mode has two different states, due to the two rotation directions of reversible pump turbines. The mode changes from turbine rotation to pump rotation direction and vice versa always need a disconnection from the electrical grid using the circuit breaker between the motor-generator

---

a. The upper value for the range depends on the electrical system design, on the required load control bands at individual heads, and on the available submergence.

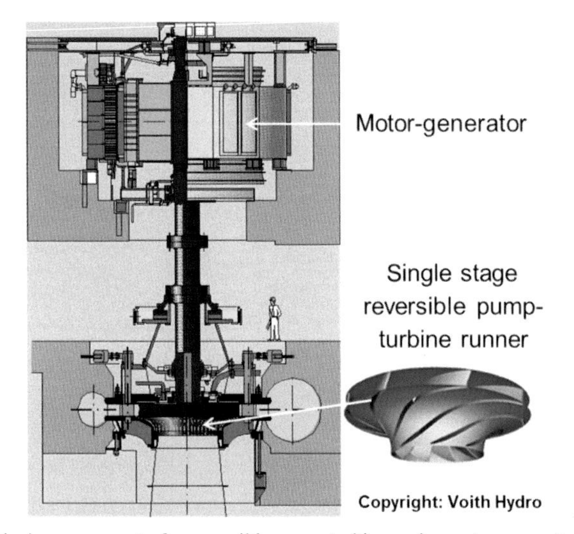

FIG. 12 Vertical arrangement of a reversible pump turbine and a motor generator.

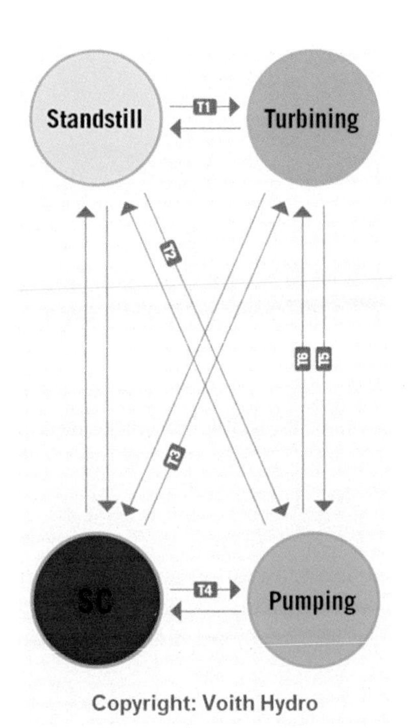

FIG. 13 Operation modes and possible transitions (SC: synchronous condenser mode).

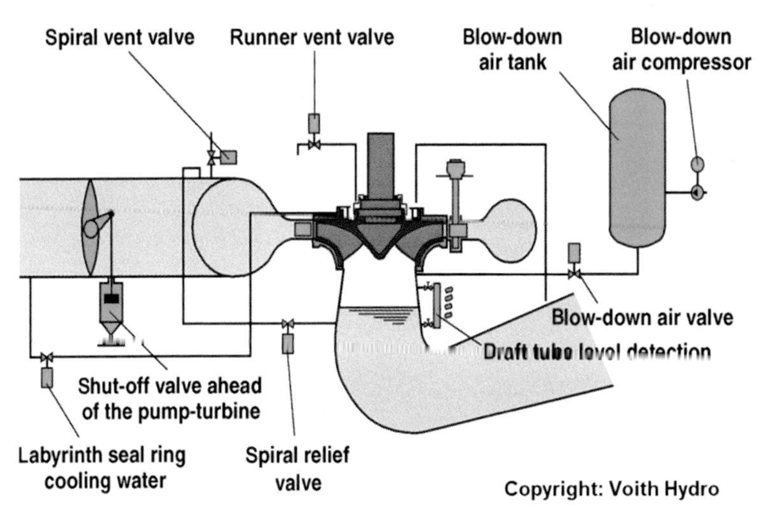

FIG. 14 Overview of involved components and systems necessary in synchronous condenser mode.

terminals and step-up transformer. After the electrical disconnection, a phase reversal switch in series with the circuit breaker is also needed to transpose two phases in order to reverse the rotation direction.

Fig. 14 shows an overview of necessary additional components and system, which are necessary to run a reversible power unit in synchronous condenser mode:

- Blow-down system (compressor, blow-down air tank, blow-down air valve, etc.),
- Draft tube water-level instrumentation,
- Deaeration system (here runner air vent and spiral vent valve),
- Dedicated fresh water supply from the penstock for continuous labyrinth cooling, and a
- Spiral relief valve in a pipe allowing water to flow to the draft tube in this mode because during the synchronous condenser mode the spiral case water gets heated up due to energy dissipation in the labyrinth seals. The rotating runner "feeds" the spiral case through the closed distributor with warm water, since a closed distributor is not perfectly water tight.

The start in turbine mode is quite simple since gravity helps by forcing water through the turbine. In contrast, to start in pump mode requires either dedicated start-up equipment or a start-up procedure with significant power until synchronization with the electrical grid. One can choose between several options to start in pump mode:

- Starting and synchronization via pony motor. The runner rotates in air, which means the power unit must be equipped with a blow-down system (see on the left of Fig. 15).

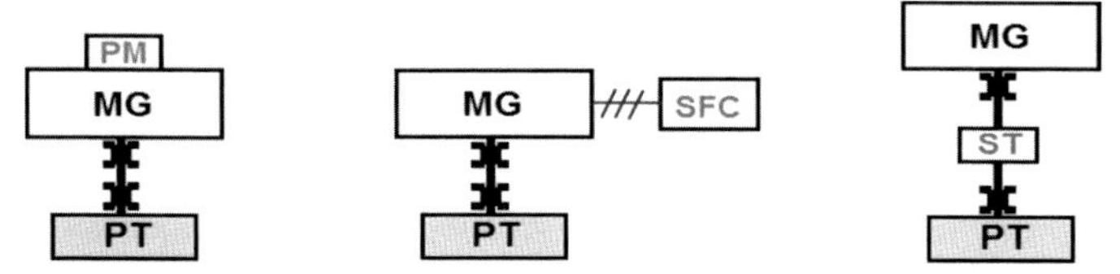

FIG. 15 Examples of start-up equipment for reversible pump turbines.

- Starting and synchronization via a static frequency converter (SFC, see in the middle of Fig. 15.) During start the runner rotates in air, which means the power unit has to be equipped with a blow-down system.
- Starting and synchronization via start-up turbine. The runner of pump turbine can remain filled with water (see on the right of Fig. 15). The start-up turbine needs to be emptied afterwards. Advantages: reduces the start-up time in pump mode and it is suitable for multistage pump turbines (3+ stages) also. Disadvantages are additional investments in a separate starting turbine; longer shaft arrangement; additional ventilation and friction losses for the start-up turbine in dewatered conditions.
- Back-to-back start in case of multiple units. In this case the unit started in pump mode can remain filled, that is, no blow-down system is required. The starting energy comes from an adjacent power unit in turbine mode. This mode requires dedicated electrical connections between all motor generators since the frequency during the start-up is different from the grid frequency.
- Direct start of a power unit by connecting the stator with the grid (in some cases, some current-limiting devices might be required to limit the maximum currents). In such cases the rotor of the motor-generator needs a robust mechanical and thermal design since the damper winding acts like a squirrel cage rotor during the acceleration of the unit.

Reversible power units with variable speed

The following advantages can be achieved by applying variable speed technology as shown in Fig. 11:

- Possibility to control the consumed motor power in pump mode (see Fig. 16). This feature also increases the total number of pump utilization hours in a year (see Fig. 17).
- Larger head range variations are possible, that is, better utilization of the reservoir volumes (important for low head applications or deep and narrow reservoirs). Please refer to the previous subchapter explaining the gross head variation factor H_{max}/H_{min}.
- Larger control band in turbine mode can be achieved due to lower part load (down to 25%) and with higher efficiency (see Fig. 16).

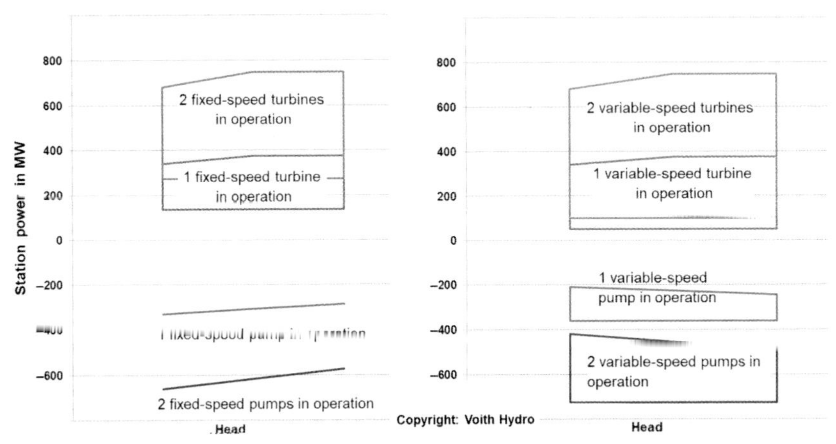

FIG. 16 Illustrative example of power regulation chart: (left) two units with fixed speed and (right) two units with variable speed.

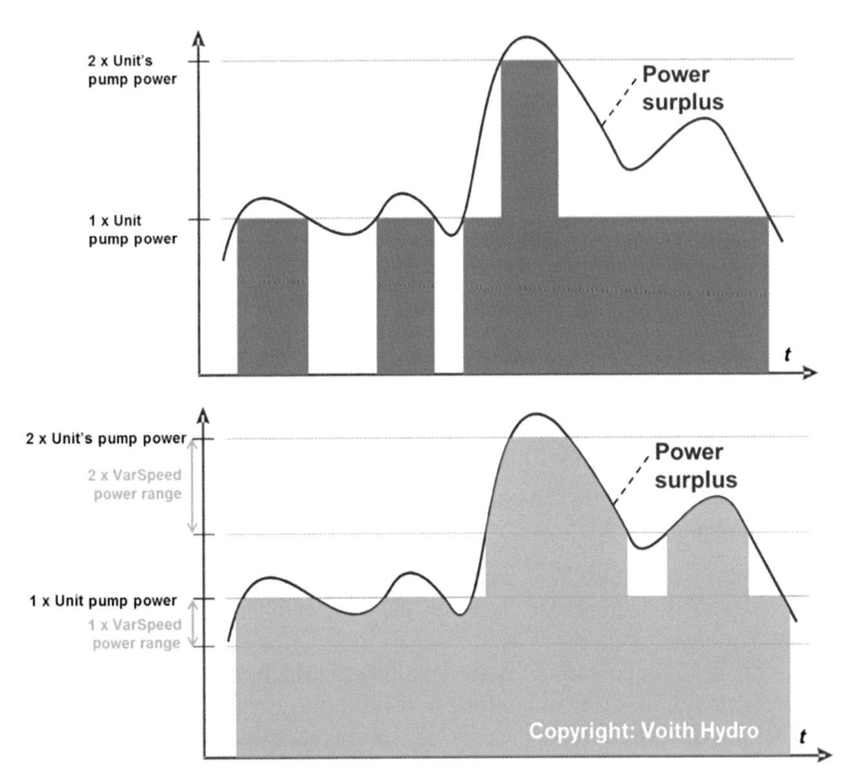

FIG. 17 Illustrative comparison of flexibility for integrating surplus power from the grid: (top) two units with fixed speed and (bottom) two units with variable speed

- Faster load ramping in pump and in turbine mode utilizing the AC excitation system for active and reactive power control. This is useful when offering important ancillary services to the grid such as primary and secondary frequency control reserves in both modes.
- Potential to improve the grid stability in case of grid faults (e.g., low-voltage ride through LVRT) by injecting fast active and reactive power in both pump and turbine mode.

Ternary power units with fixed speed

Fig. 18 shows as an example of the vertical arrangement of one of the three ternary units of Kopswerk II commissioned 2008 (3×175 MW in turbine mode, 3×160 MW in pump mode) with a head range between 721 and 818 m. The Pelton turbine cannot be disconnected mechanically from the motor-generator shaft, that is, the turbine is always coupled with the motor-generator shaft. The ventilation losses of the Pelton runner rotating in air with closed nozzles in pumped mode or in the SC mode are low. The three-stage pump can be connected and disconnected by the hydraulic torque converter.

The second example is a horizontal arrangement for the ternary units in Obervermuntwerk II located also in Austria (see Fig. 19). Both units were commissioned in 2019 and the power is 2×180 MW both in turbine and pump mode with a head range between 243 and 311 m. Different from Kopswerk II the pump has a single stage and there is a claw coupling arranged between motor generator and Francis turbine. In pumped mode or in SC mode the Francis turbine is mechanically decoupled in order to avoid losses.

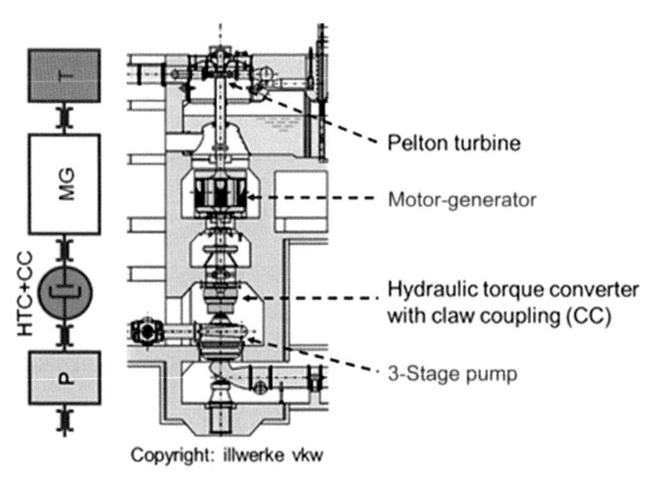

FIG. 18 Vertical ternary unit configuration of Kopswerk II, *Austria HTC*, hydraulic torque converter; *CC*, claw coupling.

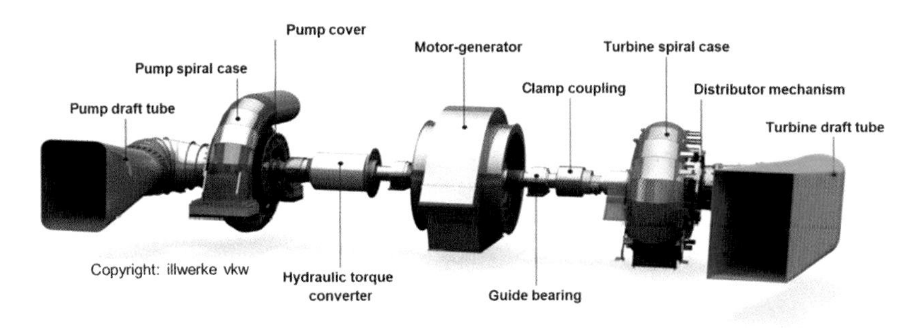

FIG. 19 Horizontal ternary unit configuration of Obervermuntwerk II, Austria

There are some important differences between reversible and ternary power units:

(a) Ternary units for medium heads can be arranged horizontally and vertically. For very high heads the ternary units are also arranged vertically like reversible units.

(b) Unlike reversible power units, ternary units rotate in the same direction for both pump and turbine modes, that is, a change of rotation direction is not necessary to switch between pumping and generating. This leads to significantly faster change over times and the motor generator is a standard generator solution (see Fig. 23).

(c) Very many ternary units are equipped with a hydraulic torque converter (HTC) between pump and motor generator (see Fig. 19). This HTC is designed to start the pump in watered condition without stopping the spinning motor generator, which saves significant mode change up time. The acceleration energy of the pump until synchronous speed via this HTC comes either from the mechanically coupled turbine or from the electrical grid via the motor generator. In some cases the pump start can be managed in less than 60 s.

(d) Unlike reversible power units, ternary units have a fifth mode of operation, which is shown in Figs. 20 and 21 called hydraulic short circuit. In this mode both hydraulic machines are operated in parallel, which allows to control the overall consumed load of the pumped storage plant by regulating the turbine power generation. The resulting continuous load control range is illustrated in the upper part of Fig. 24. The hydraulic short circuit is the only solution to vary the pump power with units operated at fixed speed. In case of reversible pump turbines, a hydraulic short circuit is not possible within one unit. The operator has to utilize two units in parallel: one in turbine mode in one in pump mode.

The number of different mode transitions of ternary units shown in Fig. 21 is only 9, whereas a reversible pump turbine may have up to 12 transition modes (see Fig. 13). Each transition is programmed in a dedicated step sequence as part

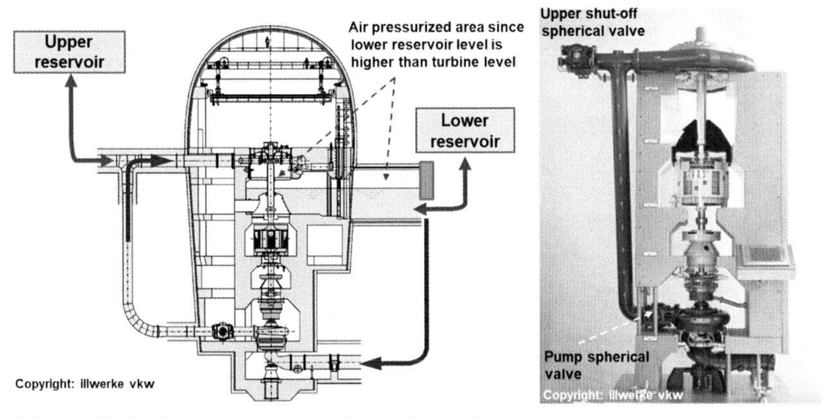

FIG. 20 Hydraulic short-circuit operation in Kopswerk II, Austria.

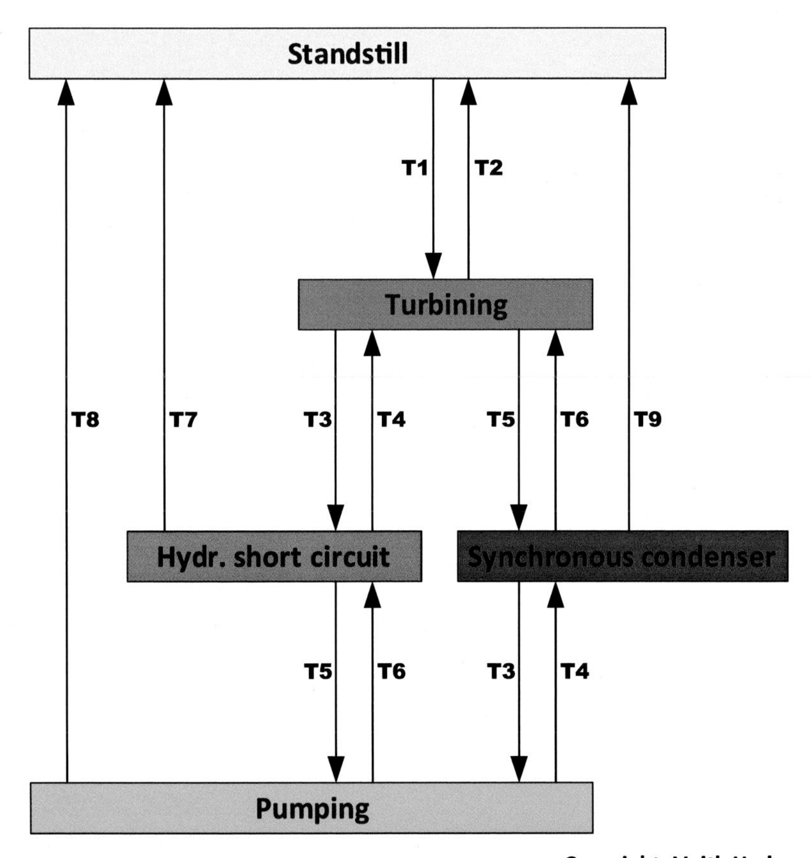

FIG. 21 Operation modes and possible transitions for a ternary unit.

of the control system. In this particular ternary unit the transition sequences T3, T4, T5, and T6 are used two times. For instance T3 is responsible to start the pump in order to change from turbine mode to hydraulic short-circuit mode or in order to change from SC mode to pump operation. In these four transition sequences (T3, …, T6) either the pump or the turbine is switched on or off, whereas the other hydraulic machine remains in its current state. In order to achieve this reduction of step sequence programs, some auxiliary systems—like the oil supply system for the pump and turbine—are common systems, which run continuously.

Comparisons of reversible with ternary power units

Fig. 22 gives a first orientation regarding application head ranges and main utilization purposes for reversible and ternary power units. For the last row in the table shown in Fig. 22, it is assumed that the break-even point where the costs of DFIM, including an AC excitation, start to be more attractive compared with a synchronous motor generator with a full-scale power converter is in the range of 100 MW. Future decreasing costs of the converter technology based on power electronics can shift this "break even area" to higher unit power output values than 100 MW.

An important difference between ternary units and reversible units is the time required to switch between operation modes. The transition times shown in Fig. 23 are examples from executed projects, with only 6 transitions listed.

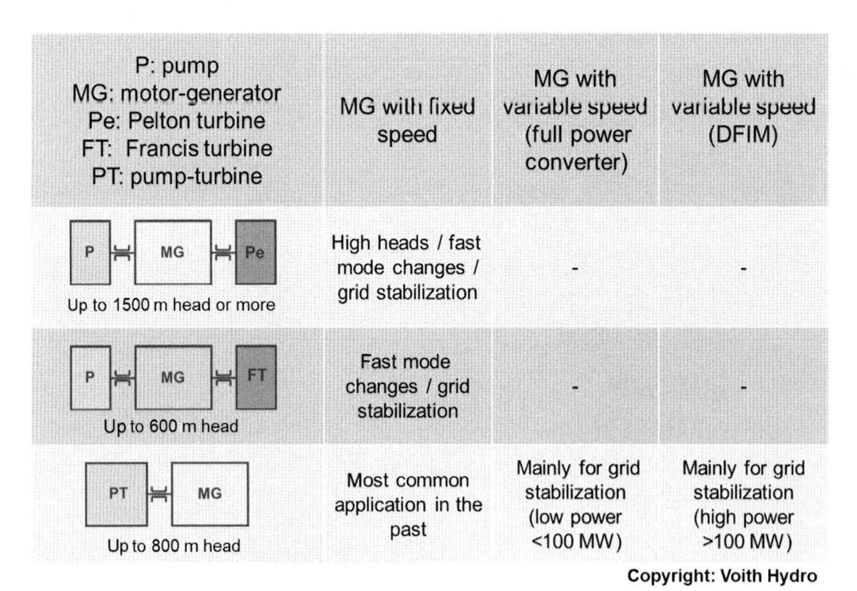

P: pump MG: motor-generator Pe: Pelton turbine FT: Francis turbine PT: pump-turbine	MG with fixed speed	MG with variable speed (full power converter)	MG with variable speed (DFIM)
P – MG – Pe Up to 1500 m head or more	High heads / fast mode changes / grid stabilization	-	-
P – MG – FT Up to 600 m head	Fast mode changes / grid stabilization	-	-
PT – MG Up to 800 m head	Most common application in the past	Mainly for grid stabilization (low power <100 MW)	Mainly for grid stabilization (high power >100 MW)

FIG. 22 Overview table with application range of ternary and reversible power units with respect to head and main utilization purpose.

T	Pump-turbine Mode change			Time [s] A	B	C	D	E
1	Standstill	→	Turbining	90	75	90	90	65
2	Standstill	→	Pumping	340	160	230	85	80
3	SC-mode	→	Turbining	70	20	60	40	20
4	SC-mode	→	Pumping	70	50	70	30	25
5	Turbining	→	Pumping	420	300	470	45	25
6	Pumping	→	Turbining	190	90	280	60	25

FIG. 23 Examples of mode change times of reversible power units (columns A–C) and ternary units (columns D and E).

The first three columns are project examples from reversible units with fixed speed (A and B) and variable speed using a DFIM (C). The longest transition times for reversible units occur during the start in pump mode, or reversing from turbine to pump mode, and vice versa. The start of the pump is mainly determined by the time for blow down until the runner can start to rotate in air and by the time to speed-up the unit (using a static frequency converter (SFC)). The long transition times T5 and T6 are result of the necessary change of rotation direction. This requires the mechanical stop of the unit and also the electrical disconnection of the MG from the grid in order to transpose two phases via the already-mentioned phase reversal switch. Power plant B features significantly reduced transition times in comparison with the classical reversible unit in column A. They are achieved by a large implemented package of measures, resulting in, for example, faster opening and closing of spherical valves, faster blow-down operation, or faster acceleration of the pump by an oversized SFC.

Columns D and E indicate transition times for two projects with ternary units. Pumped hydroplant D is a horizontal and E is a vertical unit arrangement. In general transition times of the ternary units are smaller compared with reversible units, especially the transition times T4 to T6 are remarkably shorter. It must be mentioned that Fig. 23 is intended for general comparison purposes. Transition times for individual projects will differ, since not only the power unit concepts determine transition mode times, but also the inertia of the water ways and the design of the auxiliary systems.

The additional flexibility regarding load control flexibilities for the different power unit concepts is illustrated in Fig. 24 both for the pump and turbine mode. The largest gain in terms of control band is achieved by ternary units in hydraulic short-circuit operation. Variable speed units can also offer some control band in pump mode, although a much narrower band than ternary units. Reversible pump turbines with fixed speed cannot regulate the pump load for a given head at all.

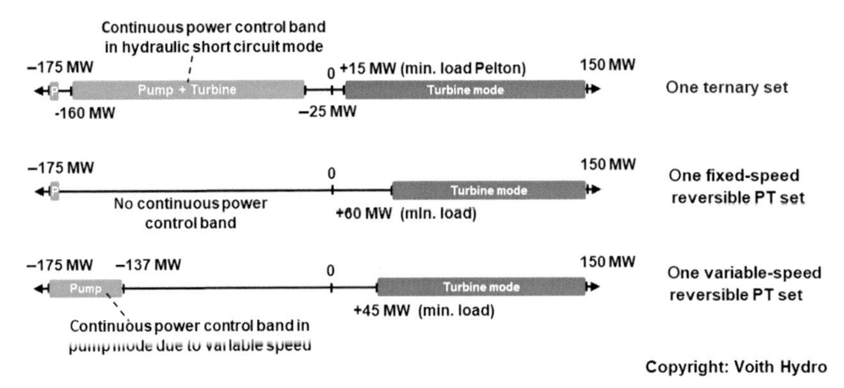

FIG. 24 Comparison of the load control band in turbine and pump mode for a ternary unit (top), fixed speed unit (middle), and variable speed unit using a DFIM as an electrical machine (bottom).

The table in Fig. 25 helps for orientation in order to identify the main cost drivers for the different power unit concepts with the classical reversible pump-turbine unit at fixed speed for reference. The main cost drivers for variable speed units utilizing a DFIM are the electrical machine (MG) itself and the AC excitation illustrated in Fig. 11. In the third variant—variable speed with full-scale power converter—the converter is the dominant cost driver. Both variants with variable speed have also an impact on the necessary power house volume, since the converters and their transformers need additional rooms or caverns. The ternary variant displayed in the last row of the table does not need any electrical power converters and also no SFC, since the pump start is done via a hydraulic torque converter (HTC) in watered conditions. The cost drivers for this last variant are the cost for the two hydraulic machines (pump and turbine), including two spherical valves, and the power house size—since a ternary shaft arrangement is significantly longer (see Figs. 18 and 19).

Fig. 26 gives a final overview of the comparative advantages and disadvantages of ternary units versus reversible pump-turbine units.

4.1.1.5 Application objectives and business opportunities for pumped storage

In a deregulated electricity market transmission and distribution systems operators (TSO, DSO) consider energy storage as one of several flexibility options in order to control the residual load. The term "residual load" was introduced 2009 in a German study from Frauenhofer Institute [3] and a common definition for it is shown in Fig. 27. Residual load refers to the instantaneous difference between supplied and consumed electricity. The residual load can be positive but also negative and is actively controlled by flexible thermal power plants, by energy storage systems, by electricity import and export options, by curtailments of renewables or by adapting the load demand (demand-side

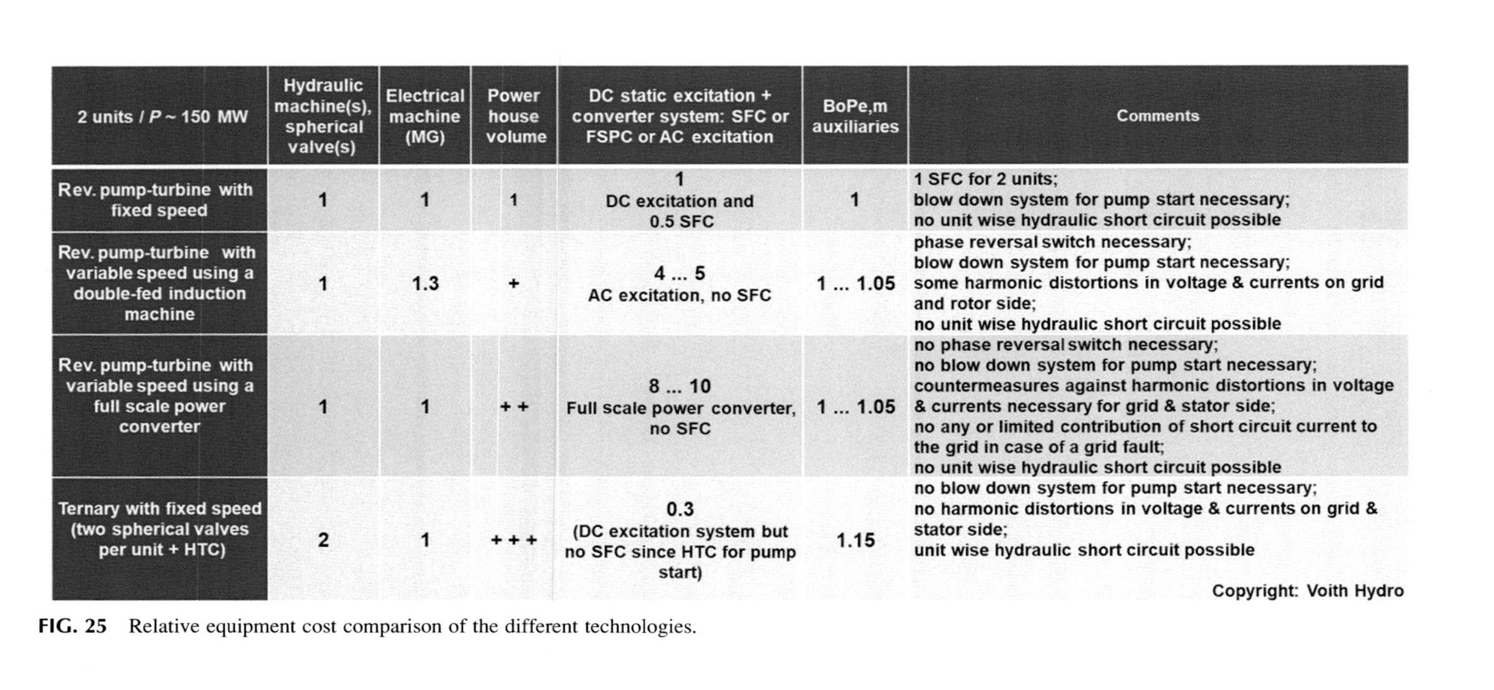

2 units / $P \sim 150$ MW	Hydraulic machine(s), spherical valve(s)	Electrical machine (MG)	Power house volume	DC static excitation + converter system: SFC or FSPC or AC excitation	BoPe,m auxiliaries	Comments
Rev. pump-turbine with fixed speed	1	1	1	1 DC excitation and 0.5 SFC	1	1 SFC for 2 units; blow down system for pump start necessary; no unit wise hydraulic short circuit possible
Rev. pump-turbine with variable speed using a double-fed induction machine	1	1.3	+	4 ... 5 AC excitation, no SFC	1 ... 1.05	phase reversal switch necessary; blow down system for pump start necessary; some harmonic distortions in voltage & currents on grid and rotor side; no unit wise hydraulic short circuit possible
Rev. pump-turbine with variable speed using a full scale power converter	1	1	+ +	8 ... 10 Full scale power converter, no SFC	1 ... 1.05	no phase reversal switch necessary; no blow down system for pump start necessary; countermeasures against harmonic distortions in voltage & currents necessary for grid & stator side; no any or limited contribution of short circuit current to the grid in case of a grid fault; no unit wise hydraulic short circuit possible
Ternary with fixed speed (two spherical valves per unit + HTC)	2	1	+ + +	0.3 (DC excitation system but no SFC since HTC for pump start)	1.15	no blow down system for pump start necessary; no harmonic distortions in voltage & currents on grid & stator side; unit wise hydraulic short circuit possible

FIG. 25 Relative equipment cost comparison of the different technologies.

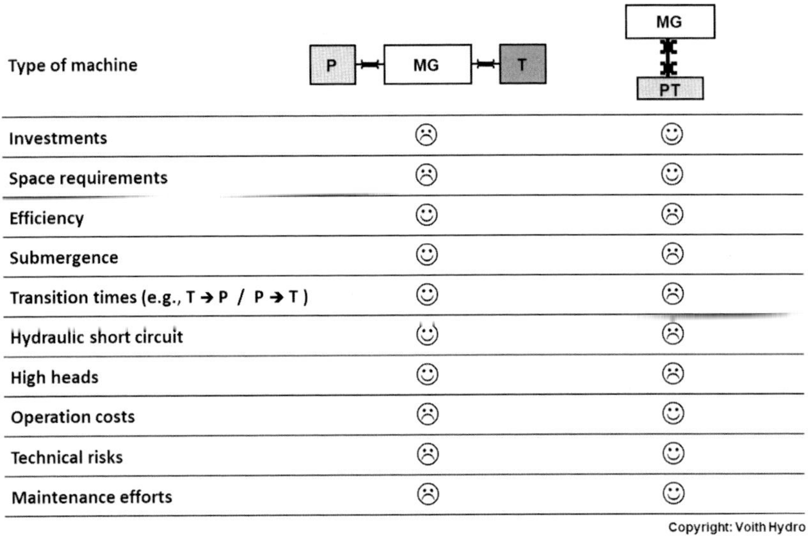

Type of machine	P–MG–T	MG / PT
Investments	☹	☺
Space requirements	☹	☺
Efficiency	☺	☹
Submergence	☺	☹
Transition times (e.g., T → P / P → T)	☺	☹
Hydraulic short circuit	☺	☹
High heads	☺	☹
Operation costs	☹	☺
Technical risks	☹	☺
Maintenance efforts	☹	☺

FIG. 26 Advantages and disadvantages of ternary units in comparison with reversible pump turbines.

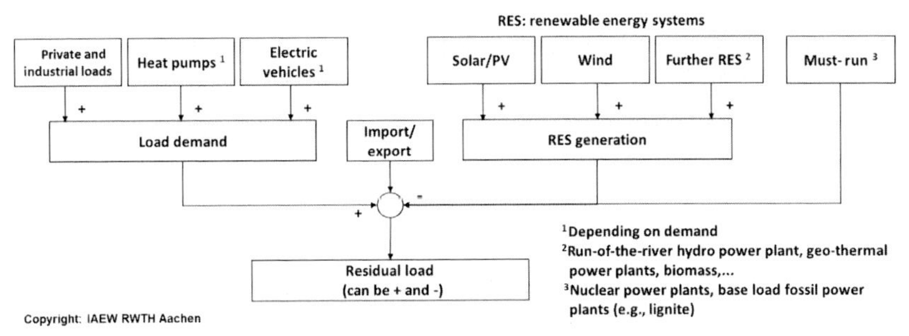

FIG. 27 Definition of residual load.

management). These flexibility options along with new grid extension or grid reinforcements are in competition with new storage expansions.

Fig. 28 illustrates the future regional distribution of the residual load superimposed on the map of Germany for one particular time in a year with the same weather conditions regarding sun irradiation and wind conditions (*Note: the two scales differ by factor of 5*). Such simulations were done for several years using past representative weather years with hourly time resolution and in accordance with the German Power Network Development Plan. The distribution of the residual load is very heterogeneous over time and space and has to be compensated by the aforementioned flexibility options.

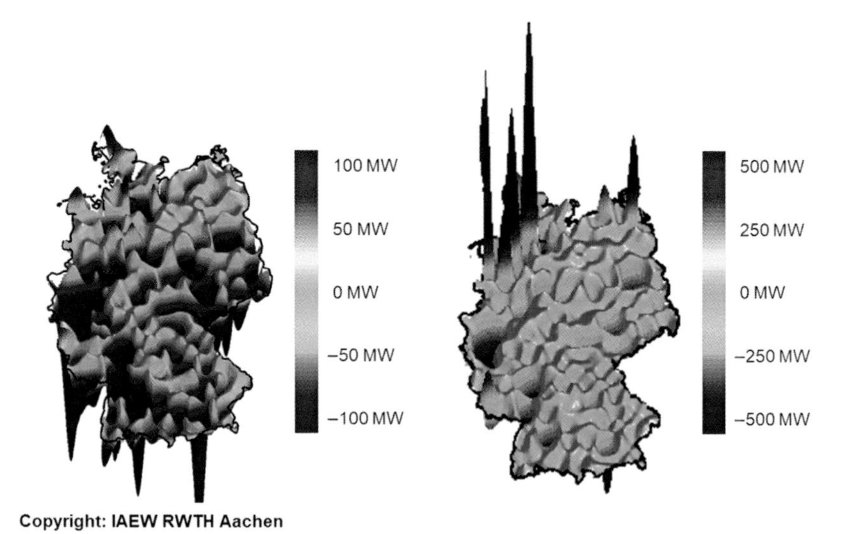

FIG. 28 Regional distribution of residual load 2024 and 2035 in Germany; left: scenario for 2024 with 45% share of renewables, right: scenario for 2035 with 80% share of renewables.

In former nonregulated electricity systems, energy storage systems did not have any important functionality, because a sufficiently high and permanently available reserve capacity was provided by coal and nuclear plants and because the residual load was never negative. Energy storage was previously performed by nature in the primary resources of coal, gas, uranium, or oil, while electricity was generated according to demand, meaning that energy storage took place before production. The generation of renewable energy from run-of-river hydropower power plants, wind, and photovoltaics (PV) is detached from demand. As a consequence, in systems especially with a high share of production from volatile renewable sources, it becomes necessary to store electrical energy instead of fossil energy, especially when the residual load gets negative. This changes the sequence of storage and electricity generation.

In this new context, pumped storage plants gain importance due to their grid-stabilizing properties like:

- Provision of reliable available capacity and fast balancing/regulating power, which helps to flexibly integrate renewable energies and to smoothen the residual loads,
- Redispatch and other ancillary services like reactive power control,
- Grid network reconstruction capability after a black out, etc.

"Redispatch" refers to necessary grid stabilization measures in order to compensate forecast wind and PV errors due to unexpected weather conditions or caused by hard transmission limitations due to congestion of high-voltage lines. Redispatch measures could be the start or stop of fast reserves, for example,

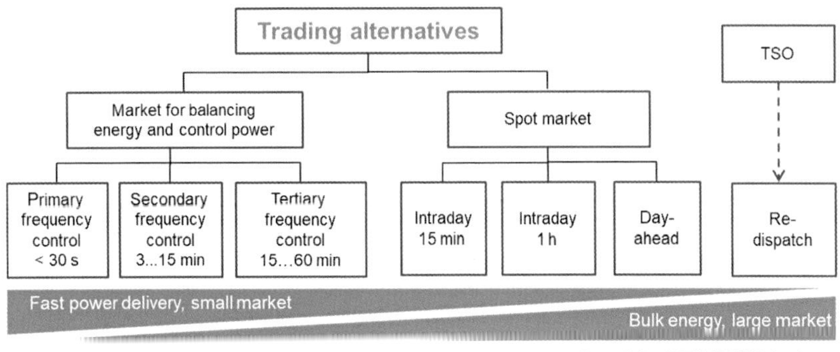

FIG. 29 Electricity trading opportunities central Europe (TSO, transmission system operator).

conventional generation like gas turbines but also storage facilities. For these reasons and because of their black start capability, PSP are an indispensable part in many electricity supply systems worldwide.

Fig. 29 illustrates the portfolio of different trading opportunities for generation and storage power plants in central Europe. The rough "equivalents" in the United States to the European energy balancing and control power products are as follows:

Europe	United States
Primary frequency control (PFC)	Governor response
Secondary frequency control (SFC)	Regulating reserve + spinning reserve
Tertiary frequency control (TFC)	Nonspinning reserve + supplemental reserve

One main revenue stream of pumped storage plants was mainly the arbitrage business between peak and off-peak electricity prices, that is, high revenues from the supply of electricity in turbine mode and relatively low cost during pump operation. In many countries of the world the revenue situation changed (see Fig. 30). In the year 2000 there were sufficient price spreads in Germany around 50 € per MWh and no grid utilization tariffs for pumping electricity were in place. After 2014 the remaining average price spread got reduced to 15 €/MWh and even lower in the following years. One reason for the spread decrease was the introduction of transmission grid utilization fees in 2009, including new surcharges and the second one was the merit order effect in an energy-only market with a significant expansion and subsidization of wind and solar parks, which have zero marginal costs compared with thermal power plants.

In 2019 Europe introduced new definitions for ancillary services for grid frequency support:

- FCR: frequency control reserve, which can be activated within seconds and
- FRR: frequency restoration reserves, which can be activated within minutes (separate positive and negative lots).

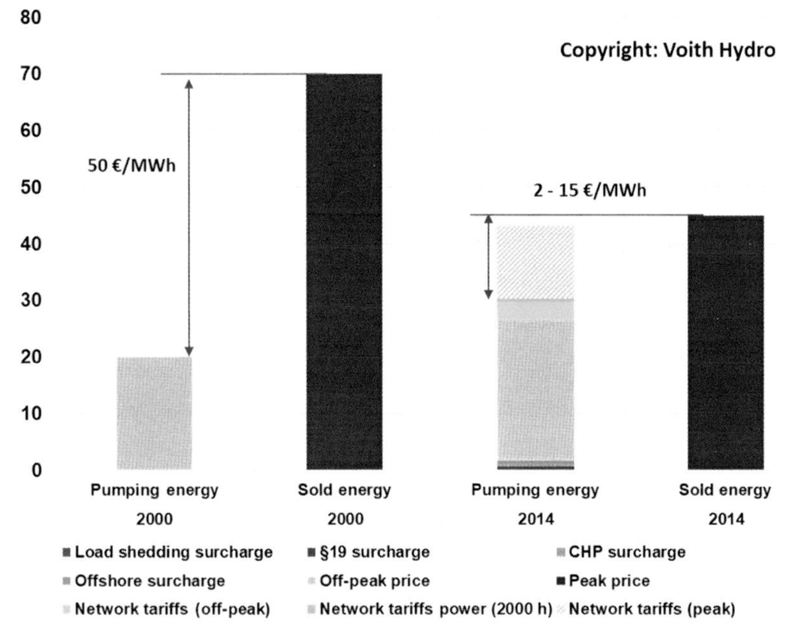

FIG. 30 Development of electricity price spreads for pumped storage in Germany.

The revenues from these ancillary services are showing a sharp price decline over time despite the increase in installed renewable energy output. The average performance prices for FRR have fallen from 2014 to 2017 by 70% to 95% due to the highly competitive situation and limited trading market volume (see Fig. 31).

Several studies emphasize the necessity of additional bulk energy storage systems, if the share of renewables goes beyond 60% of the annual electricity generation or if the absolute short-term weather forecast error for wind and PV reaches critical absolute values, which cannot be compensated by other existing flexibility options connected to the grid [2].

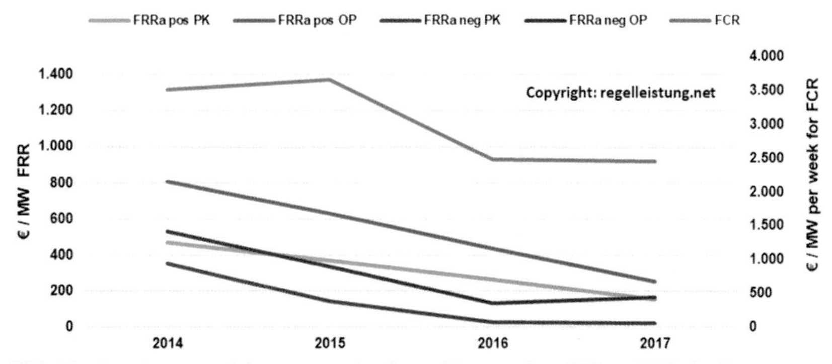

FIG. 31 Development of the revenues for the ancillary services FFR und FCR in Germany.

FIG. 32 Commercial flywheel system (courtesy: Vycon, Inc.).

Missing revenue streams in some deregulated electricity markets like United States or Germany is not the only reason few new pumped storage projects are developed. Another impediment can be the long permitting phases. This phase can be shorter, for example, if countries favor new pumped storage power plants in order to avoid curtailments of wind and solar or to support the base load operation for new nuclear power plants. Currently, China has the largest increase for new pumped storage power plants worldwide. From 2016, China is intending to build 28 new pumped storage power plants with capacity of 30 GW in total. By the end of 2019, 17 pumped storage plants have already been awarded with a total capacity of 22.5 GW.

4.2 Flywheel energy storage

4.2.1 Background

Zhiyang Wang, Xiaojun Li, and Alan Palazzolo

4.2.1.1 Application areas

Flywheels store kinetic energy, which is transformed from surplus electrical power and transformed back into electric power on demand. The kinetic energy is in the form of a high-speed rotating inertia. The electrical-mechanical-electrical power transformation is accomplished with electrical motor generators that are an integral part of the flywheel unit. This energy conversion technology (motor/generator) provides the flywheel energy storage system with a high and easily adjustable/scalable power density. A commercially available flywheel system is shown in Fig. 32. The basic components include a vacuum/containment vessel, spinning wheel (inertia), motor generator, and power and interface electronics.

Due to their advantages in power densities, flywheel systems are widely implemented in energy storage applications where fast charging and discharging is required, during which time the energy transferred is considerably less than the flywheel's capacity. These applications include uninterruptible power supply (UPS), regenerative energy recovery systems, and electrical grid quality improvement facilities, etc. For UPS application in modern data centers and casinos, the electrical power systems are almost always equipped with fast-start generators as backups for a blackout condition, and that can be online in a matter of seconds. The energy storage components in these UPSs will only need to bridge the time between grid loss and available backup sources. The total energy needed during these transient events is much less than the flywheel's capacity. For example, supplying 10 s of backup power at 1 MW, which is a common case, only requires 2.8 kW-h of stored energy.

Consider a comparison of this with electrochemical batteries. The charge/discharge speed can be evaluated with the C-rating for batteries. A rating of 1C means the battery can be fully charged in 1 h at its rated current. For lead acid batteries, the rating is normally under 0.2C, which implies that at full rated current the battery will be fully charged in 5 h. For Li-ion energy cells, the most advanced research studies achieved charge ratings higher than 10C; however, the commonly adopted limit is under 1C [4]. Li-ion batteries can be charged with high C-rates during fast charge, at the risk of a shortened battery life. However, they cannot sustain a high discharge C-rate for an extended period of time. In addition, the charge/discharge C-rate normally is a function of the state of charge (SOC), which means the power output or state of power (SOP) of li-ion batteries decreases as the SOC decreases. This implies that a 10-MW-h Li-ion energy storage system must be implemented if the required power output is 10 MW, and the C rating is 1. Similarly a 50-MW-h lead-acid battery system with a rate of 0.2C is needed to meet a 10-MW requirement. The energy actually used in these systems might be just a very small percentage of the total energy stored. The large majority of the loads on the same 10 MW system, may only require 100 kW-h, for example. An excessively scaled energy storage system could increase the total system installation cost, increase complexity for routine maintenance/supervisory control, and reduce reliability. Flywheel energy storage systems are more suited for these applications, with equivalent C-ratings of commonly over 100C and sometimes 1000C or more.

Flywheel implementation can be categorized as on-board or on-ground for regenerative energy recovery (transportation) applications. The on-board scenarios could include electrical and hybrid cars/buses, large-scale mining vehicles, and harbor cranes. Flywheel application on vehicles can be found as early as the 1950s, when a system called Gyrobus was developed in Switzerland [5]. Emergence of electric cars brought attempts to install small flywheels in cars, and relatively large ones in buses. However, due to competition with batteries and the relatively lower power/energy ratio in these applications, little progress has been made in commercialization. A related potential flywheel application is to utilize it to power the transient peaks of the charge/discharge events, which smooths the charge/discharge demand on the onboard battery system. This will both improve the life of the battery system and increase the energy recovery rate for the regenerative braking system. Flywheel systems received commercial success for large-system applications like harbor cranes and large-scale mining vehicles. In these applications, the flywheel system provides power for transient power peaks during heavy equipment operation. This reduces the transient loading on the diesel engines and leads to smaller-sized engines running under more optimal conditions. For these applications, the usage of flywheels will save fuel cost, reduce emissions, improve engine reliability, and reduce the overall system costs because much smaller diesel engines will be needed to cover nonpeak power loading.

Flywheels have received commercial success installed at power stations for rails and trams for on-ground regenerative energy recovery (transportation) applications. A metro system installation can charge and discharge during a train stop/start within 20 s or less with a very high braking power. This application is very similar to the UPS except for the higher frequency of operation for the former. Ref. [6,7] report on the implementation of Vycon flywheels in the LA Metro and Swiss Tram systems. The LA Metro system was reported to be able to recover up to 66% of the total braking power.

Another area that flywheel energy storage systems are currently being tested in is electrical grid power quality improvement applications. Variations of loads on the grid cause frequency and voltage fluctuations as generating capacity attempts to follow the changing demand. An energy storage system is needed to stabilize the grid and keep it from collapsing once the load fluctuation amplitude exceeds capacity. Flywheel systems can provide a means for stabilizing the transient fluctuations when high power is needed for short durations. For example, in the semiconductor industry, the power quality variations, such as intermittent voltage drop, can create serious yield issues and quality problems for the end products. The instability caused by these transients can be in the scales of seconds or even milliseconds at MW levels. High power with relatively low net energy transfer is needed during these upset events. Traditionally, for these applications, rotary UPS[b] units have dominated the market. In recent years, static UPS units using high-speed flywheels are also being implemented.

There is a promising variation to the aforementioned grid application. For grid stabilization projects, where a longer time and larger energy is needed, attempts are being made to combine the advantages of both flywheels and chemical batteries. The Irish electrical grid company implemented a demonstration project combining both flywheels and lead acid batteries [8]. The flywheel

b. Also known as dynamic UPS: a large, low-speed flywheel coupled with a diesel engine and synchronous generator, for example, Piller UNIBLOCK.

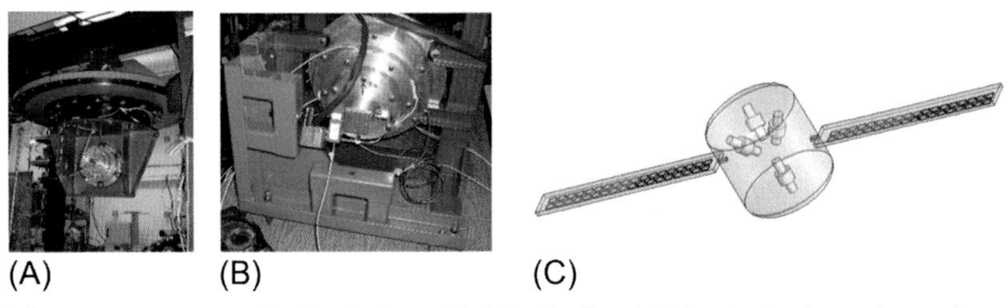

FIG. 33 Prototype satellite flywheel tested in (A) spin pit and (B) bearing load capacity test fixture at Texas A&M, and (C) satellite concept with four flywheels for combined energy storage and attitude control.

subsystem provides for the transient peak and surge of power demand. At the same time, the chemical batteries, which respond at a slower rate, provide for the longer-term power demands.

In addition, flywheels can also be used in areas where there is mismatch between grid power input capacity and equipment consumption level. In other words, converting a low-power input to a high-power output. For example, in many third world countries with fragile grids, the power rating might not be enough for modern medical equipment like MRIs, which consume ten or even a hundred times the power required in the standby mode. In these instances, the operating time might just be seconds or milliseconds every 10–20 min. Apparently, increasing the grid capacity is not an option due to the cost or even the city's infrastructures, so using a flywheel to charge at the rest period and discharge when needed is a very good option in this scenario.

Another example is for oil rig drill string-lifting equipment, such as draw works. Normally the oil rigs are at remote areas where the grids are rated at lower power levels enough for rig operations only, and long heavy pipes are lifted periodically. These periodic operations normally require a power input much higher than what the local power outlet can provide. Flywheel systems fits perfectly in these applications by providing for power bursts when needed, and they can re-charge during other drilling periods.

Another interesting application of flywheels is for the satellite industry. Palazzolo et al. [9] assisted NASA in developing flywheel systems on satellites for combined attitude control and energy storage (Fig. 33). A four-flywheel architecture was introduced so that satellite attitude is changed by individual flywheel inertia torques during spinning up/down. At the same time, the flywheels can be charged/discharged without affecting the satellite positioning.

Flywheel energy storage systems (FESS) have been used as integral parts of renewable energy resources such as solar farms and wind turbines. Due to the fact that the power output of a wind turbine is unpredictable, FESS have been proposed to smooth the output and increase the efficiency of a wind turbine [10–13]. Solar power is unavailable at night and varies during the daytime, so an energy storage system is necessary for photovoltaic (PV) solar systems. A high-temperature superconducting FESS was tested with solar cells experimentally at different spinning speeds to verify its functionality in Ref. [14].

Battery-powered electric vehicle technology has made tremendous strides, but there is room for further improvement. The driving range of a fully charged electric vehicle (EV) is considerably lower than that of an internal combustion engine (ICE) powered vehicle. The SOC estimation of an EV is not as accurate as an ICE vehicle, which only needs a fuel gage, which leads to a concern about the remaining drive range, a.k.a., the range anxiety. Charging an EV requires high-power facilities and considerably longer time compared to gasoline refills. Current EVs are more suitable for short-range trips in the urban environment, where the driver makes frequent stop-and-go actions. Thus, it is desirable to instead use extra high-power energy storage and conversion components so that the battery is prevented from deterioration out of high power cycles [15]. The life cycles of electromechanical batteries are limited, and replacing and recycling batteries is expensive and may be environmentally harmful. High power-to-mass ratio and high life cycles are key characteristics that may enable FESS to be used as a power source/supplement for EV applications. FESS have less environmental impact than other technologies, such as supercapacitors or batteries. Production Li-ion battery have a high specific energy of 100–150 W-h/kg. However, the use of composite materials in FESS can also result in high specific energy. Normally, the filament-wound composite has very high tangential yield stress and relatively low radial stress level. Flywheels, therefore, tend to fail by radial stress. Shrink fit of composite rims can cause an external pressure exerted either on the inner or outer radius of the rim. The pressure causes compression in the radial direction, thus counteracting some of the radial tensile stress caused by the centrifugal load. Some designs have reached specific energy in the range 40–80 W-h/kg [16,17].

4.2.1.2 Comparison to other energy storage technologies

The recent gain in importance of energy storage systems is driven by the increased usage of renewable energy sources. There are various kinds of energy storage technologies, including pumped hydroelectric storage, compressed air, and thermal energy storage using molten salts. Compared to flywheel technology, these technologies are generally less portable, less scalable, require larger infrastructure investments, and have different geographic or environmental limitations. Therefore the major competitors for flywheel energy storage systems, in similar application fields, are electrochemical batteries and ultracapacitors.

Depending on the information source, the numerical ranges for power and energy densities of each energy storage technology vary, as all are evolving. However, there does exist performance boundaries that, with certain overlaps, define appropriate application limitations for each technology. A qualitative evaluation and comparison between batteries, flywheels, and ultracapacitors can be done to provide a rough baseline and idea on which technology to look at when planning or sizing a certain project for each individual scenario. As examples, charts from three different technical sources are listed here to serve for reference purposes only.

A common misconception when discussing flywheel technology is to only consider composite flywheels when calculating volumetric energy and power densities. Although steel flywheels have relatively less energy density in terms of weight, compared to composite flywheels, steel flywheels have comparable energy density in terms of volume due to the higher density of the steel. Steel flywheels generally have much higher volumetric power density. They could also have comparable power density in terms of weight, which comes from the higher motor power density. This in turn results from higher bending modal frequencies due to the extra stiffness that the steel hub contributes to the rotordynamics.

Some general observations may be made concerning the strengths and weaknesses of chemical batteries, flywheels, and ultracapacitors. Chemical batteries, especially the Li-ion types, are generally better for applications where a large supply of reserve energy is needed. Their energy densities in terms of weight are generally much better than flywheels or ultracapacitors. However, battery systems are limited when the power level required is high and the discharge time is short. In other words, if the power/energy ratio is high, a certain system will require fewer supercapacitor or flywheel units than batteries. Financially, this will render the battery system impractical compared to ultracapacitors or flywheels. For high-power and short-duration applications, flywheel and ultracapacitors will have significant merits. Considering volumetric energy and power densities, flywheels are advantageous, especially for steel flywheels. Ultracapacitors could be the weakest among all three, making them relatively hard to implement in limited space applications.

However, flywheels have disadvantages in self-discharge. Batteries can be stored for years and keep most of their energy. Supercapacitors can also keep their charges for a relatively long period. However, the energy in flywheel will be lost in terms of days or hours, depending on the design. The nature of the technology determines that there will be losses coming from a combination of windage, motor losses, and bearing losses. These losses make the flywheel unfit for energy storage applications that will require the system to keep its charge for months.

Another characteristic to consider is the cell uniformity. To construct large energy storage projects that are of MW or GW level, many subsystems/individual cells will need to be combined and work together. The uniformity of each cell is very important as this will affect the system performance both statically and dynamically. Any difference between cells could potentially render the cells to charge and discharge between themselves and create an internal resonance. For batteries, this uniformity is even more important as it is safety related. Unbalanced cells for Li-ion batteries could generate local heat concentration and potentially lead to fire. Flywheels have natural advantages as they are essentially just motors/generators. Most modern flywheels are actively controlled by power electronics and their output can be precisely controlled.

Maintenance is another factor to consider. Both batteries and ultracapacitors have relatively shorter cycle lives compared with flywheels. Even though the lives of individual cells have been extended to higher values as technology evolves, the systemic lives when many cells are assembled together remain an important technical challenge. The greater longevity of flywheels will yield a significant reduction in replacement costs as compared to the other two technologies.

Figs. 34 and 35 provide graphical comparisons of various energy storage technologies.

4.2.1.3 Technology projections

Higher energy and power densities will always be the ultimate goals for any energy storage systems, and the flywheel is no exception. In terms of size, the flywheel energy storage system might progress toward two extremes. The smaller flywheels will be used in portable applications, for example, electrical cars and satellites. The large flywheels that can store more than 50 kW-h per unit will be increasingly more popular for many grid applications that were traditionally the domain of chemical batteries. These applications would be where a large quantity of electrical energy needs to be stored, but batteries might not be optimal due to their limited life cycles and frequent cycling.

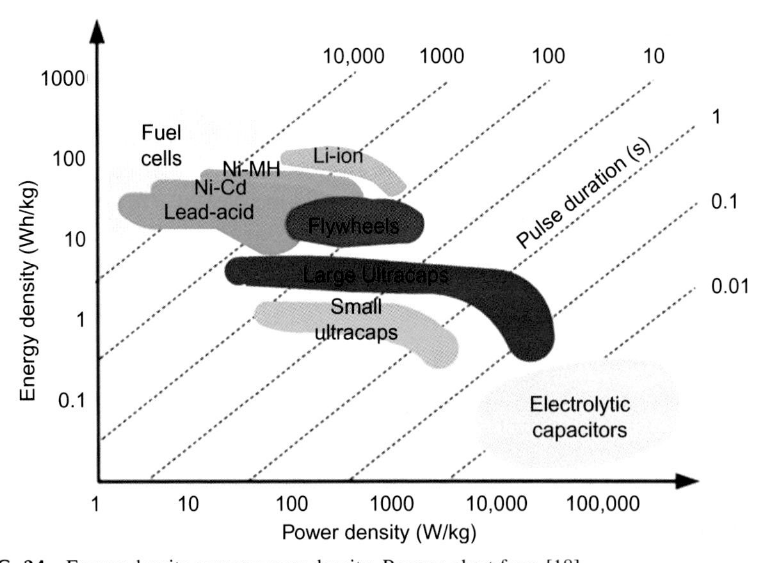

FIG. 34 Energy density versus power density, Ragone chart from [18].

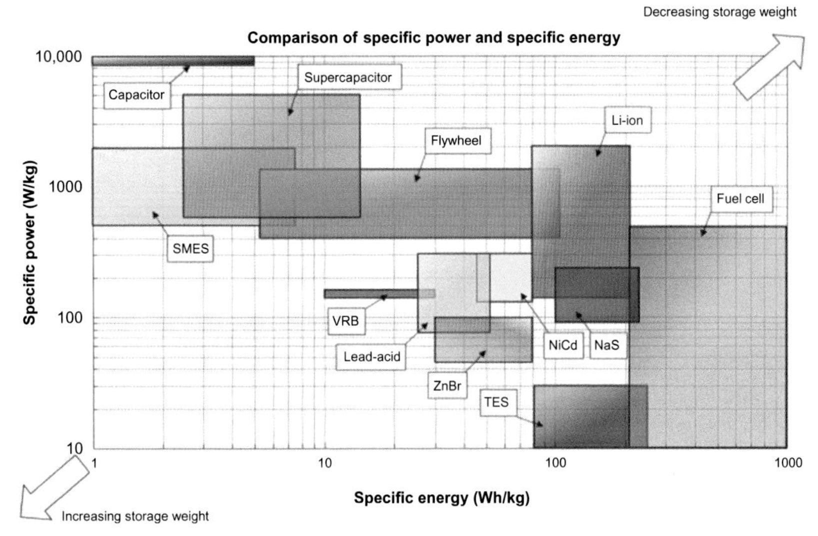

FIG. 35 Energy and power density in terms of weight [19]

Steel flywheels will become more competitive with their composite counterparts. Composite wheels most likely will always have better energy density in terms of weight; however, steel flywheels will most likely maintain their advantage in terms of volumetric energy density. Also, due to the maturity of metallurgy for steels, the steel flywheel's mechanical integrity is more uniformly controllable and inspectable, and their fatigue lives can be easily and accurately estimated. In addition, the steel flywheels can be easily recycled up to more than 95% with little penalty in cost or quality. This makes them much more environmentally friendly than composite flywheels, chemical batteries, or ultracapacitors. These facts, together with a much lower cost, give steel flywheels a commercial advantage over their composite counterparts.

Amber Kinetics built a prototype system capable of 32 kW-h and 8 kW, with continuous operation for 4 h [20]. The system does not need active cooling, and the self-discharge rate is only 65 W. The FESS includes a 10,000-lb, high-strength steel flywheel, which is mainly targeted at renewable energy applications, especially for PV systems. A novel shaftless flywheel design and prototype by researchers at Texas A&M University have the potential to double the stored energy for steel flywheels [21,22]. It is designed to provide 100-kW-h and 100-kW capabilities and weighs over 12,000 lb. The flywheel is magnetically levitated and is driven by an integral motor/generator. The prototype was built and tested with full magnetic levitation and low-speed tests as shown in Fig. 36.

As for future developments, goals may include:

(1) Spin the shaft as fast as possible
(2) Support a hub as heavy as possible

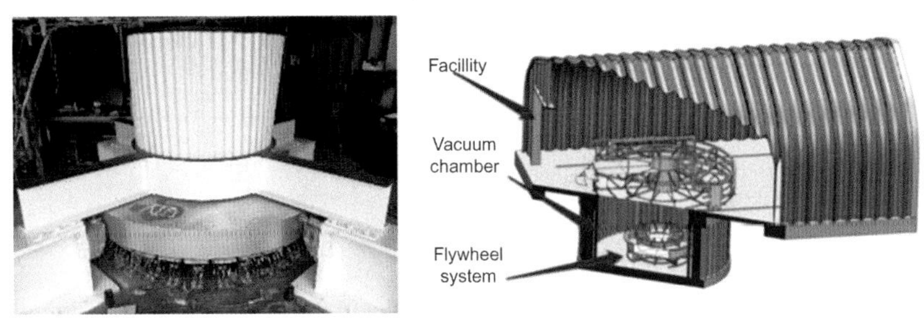

FIG. 36 Prototype 6 Ton 100 kW-h designed flywheel at Texas A&M.

(3) Reduce losses to as little as possible
(4) Increase longevity
(5) Develop more robust, easier to maintain, and foolproof magnetic-bearing support systems
(6) Reduce system cost
(7) Improve motor efficiency during charging/discharging
(8) Reduce parasitic losses during idling
(9) Increase the power density of the motor/generator
(10) Improve sealing and thermal management

A large percentage of the flywheels are hermetically sealed and vacuumed to reduce the windage loss during spinning. Improving the seal and developing standard methods/recommendations on the sealing strategy will reduce the leakage and reduce the operating time for the vacuum pumps. The last objective will provide less accessory power consumption and will help improve the standby time of flywheel energy storage systems. Of course, making the vacuum better may lead to a greater thermal management challenge, especially if the motor is charging/discharging frequently. So, technology developments to improve the thermal management for the system, especially technologies that will improve the heat exchange between rotors and stators and between stators to the outer environment within vacuum, will greatly benefit the flywheel technology in the future.

One of the emerging fields in energy storage systems (ESS) research is hybrid energy storage systems. Some ESS need to meet different requirements for different applications, including specific energy, specific power, energy density, power density, response time, life cycle, cost, environmental impact, environmental sensitivities (e.g., sensitivities to temperature and humidity), predictability (how accurate is the SOC and SOP estimation?), and many others. It is clear that no single technology can provide a perfect solution for all requirements. Since each ESS has its unique characteristics, it is natural to combine several technologies together. Due to the high-power density, quick response and high life cycles, FESS are ideal to be bundled together with Li-ion batteries, which has high energy density, low power density, slower response, and lower life cycles. A layered architecture with the FESS acting as a power buffer and the battery as the main energy sources could give hybrid systems better performance, economy, and robustness. An interesting analogy to this approach is the memory architecture of a computer. It is a more sophisticated layered architecture with registry, cache, random access memory, and hard disk space.

4.2.2 Mechanical design

4.2.2.1 Steel flywheels

Zhiyang Wang, Xiaojun Li and Alan Palazzolo

Geometry and construction

The central objective of flywheel rotor design is to increase the stored kinetic energy by increasing the maximum operating speed. This speed is limited by the centrifugally induced stress exceeding the material strength, leading to some form of rotor burst failure. The strength is determined by the selected rotor materials and fatigue cycle and safety factor requirements. The stress may be reduced by the shaping and construction of the wheel and of the hub that attaches the wheel to the shaft (for flywheels with shafts).

There have been many efforts to optimize the flywheel geometry to achieve this goal of maximizing the top kinetic energy within a given stress limitation. Pena-Alzola et al. [23] provide a good summary for some of these efforts. Assuming the maximum stress limitation (safety factor times the yield stress) that a candidate material can endure is σ_{max}, the maximum energy density in terms of mass and volume can then be expressed as:

$$\frac{E}{m} = K\frac{\sigma_{max}}{\rho} \tag{1}$$

$$\frac{E}{V} = K\sigma_{max} \tag{2}$$

Here, K is a shape factor that relates the flywheel's maximum energy density to its geometry, as shown in Fig. 37.

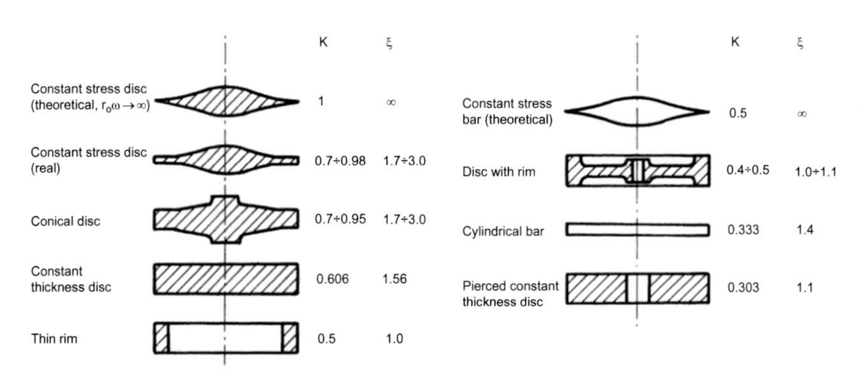

FIG. 37 Shape factor *K* for various flywheel geometries [23].

Material properties

When choosing materials for flywheels to meet desired operating points, the most important are yield strength and the fatigue strength, which are well defined for steel.

Yield strength The maximum stress generated during flywheel rotation must be lower than the material yield strength for the wheel to survive. A correctly setup FEA model will always generate results much closer to reality than handbook formulas, as the model will consider detailed features and stress concentrations of the design. However, theoretical mathematical models are also available for discs and wheels, which could be very useful during the initial planning and sizing of flywheel products. Assuming rotation is the dominant source of induced stresses, and the disk is perfectly axisymmetric, the stress will be nonzero only in the tensile circumferential (hoop) and tensile radial directions.

For *solid discs*, the radial stress and hoop stress can be expressed as [24]:

$$\sigma_r(r) = \frac{\rho\omega^2}{8}(3+\gamma)\left(R_o^2 - r^2\right) \tag{3}$$

$$\sigma_\theta(r) = \frac{\rho\omega^2}{8}\left[(3+\gamma)R_o^2 - (1+3\gamma)r^2\right] \tag{4}$$

Here, σ_r is the radial stress, σ_θ is the hoop stress, ρ is the material density, γ is the Poisson ratio, R_o is the outer diameter of the disk, and r is the radial location where the stress is calculated. The maximum values will appear at center of the disk for both radial and hoop stresses:

$$(\sigma_r)_{max} = (\sigma_\theta)_{max} = (3+\gamma)\left(\frac{\rho\omega^2 R_o^2}{8}\right) \tag{5}$$

The maximum Von-Mises stress is defined as:

$$\sigma_v = \sqrt{\frac{(\sigma_1-\sigma_2)^2 + (\sigma_2-\sigma_3)^2 + (\sigma_1-\sigma_3)^2}{2}} = \sqrt{\sigma_r^2 + \sigma_\theta^2 - \sigma_r\sigma_\theta} \tag{6}$$

Thus, the maximum value for rotational Von-Mises for solid discs will be:

$$(\sigma_v)_{max} = \sqrt{(\sigma_r)_{max}^2 + (\sigma_\theta)_{max}^2 - (\sigma_r)_{max}(\sigma_\theta)_{max}} = (3+\gamma)\left(\frac{\rho\omega^2 R_o^2}{8}\right) \tag{7}$$

For annular discs with an inner radius of R_i, the radial stress induced by rotation is:

$$\sigma_r(r) = \frac{\rho\omega^2}{8}(3+\gamma)\left(R_o^2 + R_i^2 - \frac{R_i^2 R_o^2}{r} - r^2\right) \tag{8}$$

The hoop stress for the annular disk will be:

$$\sigma_\theta(r) = \frac{\rho\omega^2}{8}\left[(3+\gamma)\left(R_o^2 + R_i^2 + \frac{R_i^2 R_o^2}{r}\right) - (1+3\gamma)r^2\right] \tag{9}$$

The maximum value for radial stress will occur at $= \sqrt{R_i R_o}$:

$$(\sigma_r)_{max} = (3+\gamma)\left(\frac{\rho\omega^2 (R_o - R_i)^2}{8}\right) \tag{10}$$

The maximum value for hoop stress will occur at $r = R_i$:

$$(\sigma_\theta)_{max} = \left[(1-\gamma)R_i^2 + (3+\gamma)R_o^2\right]\left(\frac{\rho\omega^2}{4}\right) \tag{11}$$

The maximum radial and hoop stresses appear at different locations. In order to determine the position and value of the maximum Von Mises stress, first consider the radial stress and hoop stress difference:

$$\sigma_\theta(r) - \sigma_r(r) = \frac{\rho\omega^2}{8}\left[(3+\gamma)\frac{2R_o^2 R_i^2}{r^2} + 2(1-\gamma)r^2\right] \geq 0 \tag{12}$$

Then, since both the radial and hoop stresses caused by rotation will be tensile, it results that

$$\sigma_\theta(r) \geq \sigma_r(r) \geq 0 \tag{13}$$

According to Eq. (6),

$$\sigma_v = \sqrt{\sigma_r^2 + \sigma_\theta^2 - \sigma_r\sigma_\theta} \leq \sqrt{\sigma_r^2 + \sigma_\theta^2 - \sigma_r\sigma_r} = \sigma_\theta \tag{14}$$

Since $\sigma_r = 0$ and $\sigma_\theta = (\sigma_\theta)_{max}$ at $r = R_i$, the Von-Mises stress for the annular disk reaches its maximum point:

$$(\sigma_v)_{max} = \sigma_v(R_i) = (\sigma_\theta)_{max} = \left[(1-\gamma)R_i^2 + (3+\gamma)R_o^2\right]\left(\frac{\rho\omega^2}{4}\right) \tag{15}$$

It is interesting to note the limiting case as $R_i \to 0$. The maximum Von-Mises stresses for an annular disc is shown to be twice that of a solid disc for the same outer radius. The maximum stress, as determined by the finite element method FEM or a theoretical model, is checked against the yield strength to determine if the safety factor is sufficient.

Annular flywheel versus shaftless flywheel Typically, an annular flywheel rotor is supported by a shaft, with most of the moment of inertia provided by the flywheel rotor, while the shaft acts as an interface to the magnetic bearings and the motor/generator. An annular flywheel is subject to both rotational and shrink-fit induced stress, and will usually fail at the inner radius of the rotor, where the maximum hoop stress occurs. Thus, the maximum spin speed is

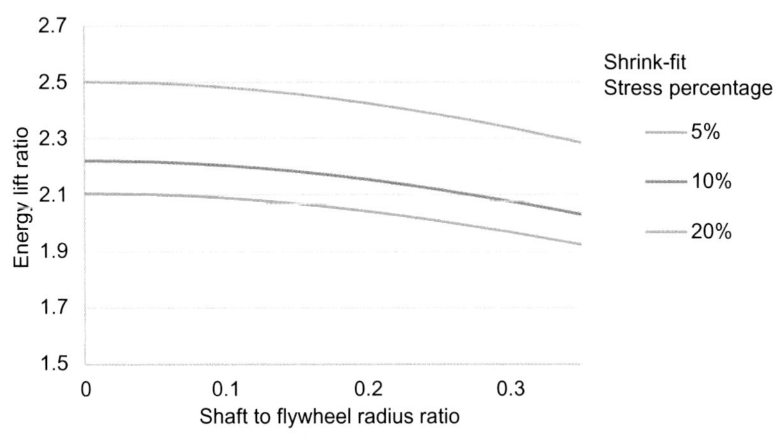

FIG. 38 Ratio of specific energy of shaftless versus annular flywheel [22].

confined by the center hole hoop stress, which in turn lowers the specific energy. In addition, the stress caused by an interference fit is also detrimental to the flywheel. Depending on the inner radius of the flywheel and the shrink-fit stress, it can be concluded that the specific energy of a solid shaftless flywheel can be 100% higher than that of an annular flywheel [22,25]. In Fig. 38, the horizontal axis is the ratio of the shaft radius to the flywheel outer radius and the vertical axis is the ratio of the specific energy of the shaftless flywheel to that of the annular flywheel. The different curves correspond to different shrink-fit allowances. In general, the shaftless flywheel will have twice the energy density. The other notable feature is that the energy density of a flywheel with a shaft hole will always be about 50% of a shaftless flywheel, no matter how small the shaft diameter is.

Fatigue strength Designing to ensure that the peak Von Mises stress is less than the yield strength is a good first step, but it should be followed by considering possible low-cycle fatigue failure resulting from charge and discharge speed variations. A low-cycle fatigue study is useful for estimating the fatigue life of the flywheel. The fatigue strength of a material defines how many stress cycles that it can endure before failing catastrophically. It marks the capability of certain material to endure cycling between its maximum and minimum operating stress. If designed incorrectly, a flywheel can eventually burst after a few repeated cycles, even if the maximum stress level is far below the material yield strength.

There are two methods commonly used in combination for checking fatigue life. The first one is to use published *S-N* curves. The *S-N* life depends on the maximum stress (σ_{max}) and minimum stress (σ_{min}) level experienced by a structure. The stress ratio R_s is defined as:

$$R_s = \frac{\sigma_{min}}{\sigma_{max}} \tag{16}$$

This ratio R_s also defines the flywheels depth of discharge (DoD) as the stress level is in a linear relationship with the stored energy level for any fixed flywheel design, by Eqs. (1), (2). According to Eq. (13), the stress will remain tensile during flywheel operation. This means σ_{max} and σ_{min} will never change sign and the ratio R_s will remain positive. The DoD can then be expressed as:

$$DoD = (1 - R_s) * 100\% \tag{17}$$

Once R_s for the design is fixed, the *S-N* curve for the steel under investigation could be retrieved to check the fatigue life. It must be noted that an *S-N* curve is for a perfect specimen, and the curve will vary depending on the condition and chemical composition of the steel. For example, MIL-HDBK-5H [26] provides *S-N* curves for un-notched AISI 4340 alloy steel bar and billet samples that were hand polished to an RMS 10-microinch surface finish and with ultimate tensile stress of 260 ksi. Forty-one samples were tested and the results yielded the following curve fit equation:

$$\log\left(N_f\right) = 11.62 - 3.75\log\left(S_{eq} - 80.0\right) \tag{18}$$

With

$$S_{eq} = S_{max}(1 - R_s)^{0.44} \tag{19}$$

Here, all stresses S are in ksi, and N_f is the number of cycles to failure. Eqs. (18), (19) provide the number of cycles to failure for a steel flywheel undergoing repeated, identical charge/discharge cycles, based on the calculated stress ratio R_s. The argument of the log function becoming negative in (2.18) indicates that S_{eq} is below the endurance limit, so theoretically the flywheel will never experience fatigue failure. The peak allowable stress increases as R_s increases, which indicates that the flywheel life will decrease as the depth of discharge increases. The *S-N* curves may change due to a number of factors such as actual surface finish, method of loading surface treatments, temperature, etc. These may be accounted for by using Marin factor corrections applied to S_{max}.

The *S-N* life calculation described here assumes that the flywheel surface is crack-free in the highly stressed zones. A more advanced estimate of flywheel fatigue life will also take into consideration the possibility of cracks, based on crack growth theory. Small imperfections in a part may be viewed as small cracks that generate local stress concentrations. A small crack in the flywheel rotor may continue to grow due to the cyclic centrifugal loading and attain a threshold size above which sudden enlargement of the crack and ensuing catastrophic failure may occur. The cycle count before the crack grows to the failure threshold, given certain safety margins, is called the fatigue life of the wheel (in cycles). The fatigue life value relies on four factors: the material properties, stress cycling amplitude (or stress ratio R_s as defined here), the initial crack size, and how the stresses are concentrated around the cracks.

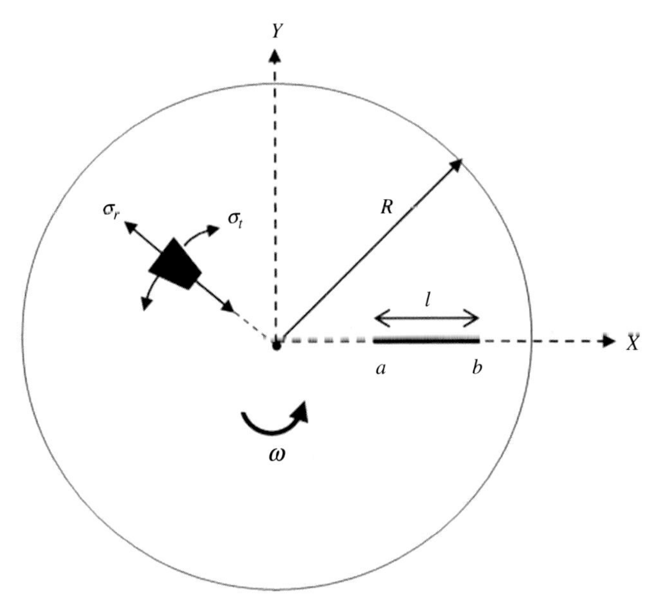

FIG. 39 Geometry and coordinates of a rotating solid disk with radial crack.

A stress intensity factor (SIF) was defined in [27] to calculate the fatigue crack growth rate in rotating discs. It was found that radially oriented, through-the-thickness cracks are the most critical due to the circumferential (hoop) stress, and are of the mode I type (opening mode), acting along the crack. The Stress Intensity Factors (SIF) reach a constant maximum value at the center part of the disc as illustrated in Fig. 39.

Equations for the SIF K_{Ia} and K_{Ib}, at the radial extremes $r = a$ and $r = b$, of the crack are given in Eqs. (20), (21), respectively. Ref. [27] shows that $K_{Ia} \approx K_{Ib}$ at $|r| \leq 0.2R$, and that the worst location of the crack is at the radial midpoint of the disc. Given the crack size is relatively small comparing to the disk radius, this worse-case SIF can be expressed as Eq. (22), for a through-the-thickness crack.

$$K_{Ia} = \frac{\sqrt{2}}{8\sqrt{\pi(b-a)}}\rho\omega^2 \int_a^b \left\{ \left[(3+\nu)R^2 - (1+3\gamma)r^2\right]\sqrt{\frac{b-r}{r-a}} \right\} dr \tag{20}$$

$$K_{Ib} = \frac{\sqrt{2}}{8\sqrt{\pi(b-a)}}\rho\omega^2 \int_a^b \left\{ \left[(3+\nu)R^2 - (1+3\gamma)r^2\right]\sqrt{\frac{r-a}{b-r}} \right\} dr \tag{21}$$

$$K_{Ia} \approx K_{Ib} = \frac{3+\gamma}{8}R^2\rho\omega^2\sqrt{\pi\frac{l}{2}} \tag{22}$$

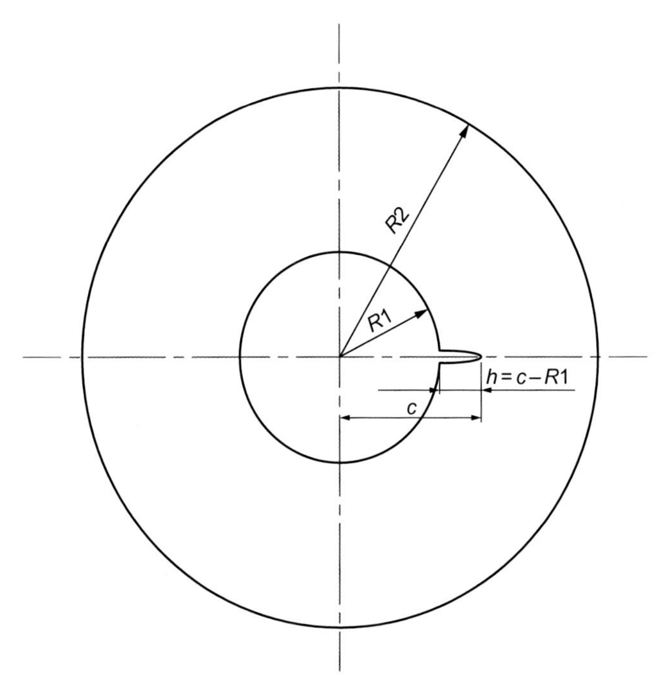

FIG. 40 Annular disk with a radial oriented crack at the inner wall.

Ref. [28] examined the case of an annular rotating disk with a radial crack located at the inner wall as shown in Fig. 40. The authors provided multiple formulas for SIFs depending on the crack sizes. The crack size for flywheel life estimation will be very small compared to the radius of the disk, since cracks in the delivered articles will be ones present after QA screening (e.g., ultrasonic screening). For our purpose of designing a steel flywheel, the SIF can be taken as:

$$K_I = 1.13\frac{3+\gamma}{4}\rho\omega^2 R_2^2\sqrt{h} \tag{23}$$

where γ is the Poisson ratio, ρ is the material density, ω is the angular velocity, R_2 is the outer radius, and h is the crack size.

Ref. [29] also provided a method for stress intensity factor calculation for cases where the cracks appear at the outer radius of a spinning annular disk. The formulas are given in Eqs. (24)–(27) and were derived utilizing approximate methods developed in [30]. The result for the stress intensity factor below holds for both the one crack case and for the case with two cracks, when they are symmetric about the center of the annular disk. The mode I SIF is obtained from:

$$K_I = \sigma'\sqrt{\pi(c-R_1)} \tag{24}$$

where

$$\sigma' = \frac{3\sigma_m + \sigma_n}{4} \tag{25}$$

$$\sigma_n = \frac{1}{R_2 - C}\int_{R_1}^{R_2} \sigma dr \tag{26}$$

$$\sigma_m = \frac{1}{C - R_1}\int_{R_1}^{C} \sigma dr \tag{27}$$

$$\sigma = \sigma_o\left(\frac{R_1^2}{R_2^2} + 1 + \frac{R_1^2}{r^2} - \frac{1+3\gamma}{3+\gamma}\frac{r^2}{R_2^2}\right) \tag{28}$$

$$\sigma_o = (3+\gamma)\frac{\rho\omega^2 R_2^2}{8} \tag{29}$$

Combining terms yields:

$$\begin{aligned}\sigma_n &= \frac{\sigma_o}{R_2 - C}\int_{R_1}^{R_2}\left(\frac{R_1^2}{R_2^2} + 1 + \frac{R_1^2}{r^2} - \frac{1+3\gamma}{3+\gamma}\frac{r^2}{R_2^2}\right)dr \\ &= \frac{\sigma_o}{R_2 - C}\left[\left(\frac{R_1^2}{R_2^2} + 1\right)r - \frac{R_1^2}{r} - \frac{1+3\gamma}{3+\gamma}\frac{r^3}{3R_2^2}\right]\Bigg|_{R_1}^{R_2}\end{aligned} \tag{30}$$

$$\begin{aligned}\sigma_m &= \frac{\sigma_o}{C - R_1}\int_{R_1}^{C}\left(\frac{R_1^2}{R_2^2} + 1 + \frac{R_1^2}{r^2} - \frac{1+3\gamma}{3+\gamma}\frac{r^2}{R_2^2}\right)dr \\ &= \frac{\sigma_o}{C - R_1}\left[\left(\frac{R_1^2}{R_2^2} + 1\right)r - \frac{R_1^2}{r} - \frac{1+3\gamma}{3+\gamma}\frac{r^3}{3R_2^2}\right]\Bigg|_{R_1}^{C}\end{aligned} \tag{31}$$

Ref. [31] notes that there are three types of behavior for fatigue crack propagation, depending on the variation of the stress intensity factor inside the steel flywheel. These three types are called region I, II, and III. When the SIF is below a material-defined characteristic called the threshold stress intensity range (ΔK_{th}), the crack behavior falls into region I. the flaw crack size will stay unchanged while operating in region I. there is also a second material related threshold referred to as the transition value ΔK_T. The crack size will grow according to the Paris law, as defined by Eq. (32), when the SIF is above ΔK_{th} but below ΔK_T. For a loading with $R_s = 0$, namely, the stress fluctuates between zero and maximum value, the transition value of the SIF can be expressed as in Eq. (33).

$$\frac{dh}{dN} = C_I(\Delta K_I)^{m_I} \tag{32}$$

$$K_T = 0.04\sqrt{E\sigma_g} \tag{33}$$

Here, dh/dN is the crack growth rate in terms of stress cycles, C_I and m_I are the material properties that define the crack growth rate, ΔK_I is the SIF change during charge/discharge cycles, E is Young's Modulus and σ_g is the mean value of tensile strength and yield strength. The flywheel operates inside region II until the SIF increases beyond ΔK_T. The material then enters region III, where the crack size grows at a much faster rate than as defined by the Paris law. Designing a flywheel to operate in region I is a conservative approach. However, this may lead to excessive size, weight, and cost. An alternative approach is to design for region II operation and provide a defined, clearly visible, finite life to the flywheel user. This will provide commercial viability along with safety and reliability. Design for operation in region III is highly discouraged since it involves a high risk that a crack will propagate rapidly leading to a potentially catastrophic failure during routine operation.

Therefore, as described, Paris's law can normally be used for the fatigue life estimation of a steel flywheel operating in region II. The fatigue life can be calculated in terms of remaining cycles, using the calculated stress intensity factor updated each charge/discharge cycle (max/min stress level). The flywheel manufacturer Vycon Inc. utilizes variations of the general procedure shown now to obtain desired flywheel life.

(1) Use the maximum crack size that can be detected by the ultrasonic test (UT) (normally C-scan to QA a AAA grade steel) as a starting point for the crack size. This guarantees that the true initial crack sizes existing in the manufactured products will be much shorter than the size used.

(2) Assume the crack happens at the worst SIF location and calculate the SIF variation for the application scenario. This, combined with procedure 1, and assuming full charge/discharge cycles, guarantees the most conservative estimation of the product's fatigue life cycles.

(3) Based on the SIF variation, use the experimentally tested/published fatigue data for the steel with the same chemical composition and heat treatments, and calculate the cycle lives for the product based on the target design and operating conditions.

Conclusion During the design phase of steel flywheel products, it is suggested that the maximum stress of the material needs to be checked to make sure that the flywheel is strong enough, including sufficient safety margins, to handle its maximum speed. The material yield strength will define the maximum energy that a flywheel can store. After the yield strength is checked, a fatigue calculation will need to be carried out with the most conservative assumptions based on material properties and manufacture and inspection capabilities to guarantee the safety of the products within its calculated life.

Actual operation conditions will be at depths of discharge different than the design (full charge/discharge) value, since different energy charge/discharge start points may occur in each cycle. Thus the cycle lives in reality will be higher than the predicted value by the fatigue calculation, as the SIF variation for each operation cycle will be less than full charge/discharge. Many commercial companies, for example, Vycon Inc., will implement counters to calculate how many equivalent full charge/discharge cycles that the product has experienced. There are many methods of doing this kind of equivalent cycle counting. The rainflow method [32] is one of the widely used algorithms for this purpose.

Steel versus composite flywheels

Composite flywheels possess a major advantage over their steel counterparts in terms of the energy density per unit of weight. Carbon fibers are commonly used to build these wheels due to the material's superior characteristics, even though there are also wheels made by stone fibers or glass fibers. Theoretically, the high yield strength of the carbon fiber can lead to flywheels with comparable energy density as chemical batteries, including the Li-ion types, in terms of both weight and volume. However, due to manufacturing limitations, the current technology can only generate composite flywheels with a much lower energy density as compared to their theoretical limits.

Composite flywheels generally spin much faster than steel ones due to their higher tensile strength and the need to compensate for their lower inertia. This requires a much higher 1st bending modal frequency for the composite flywheels, as the operating characteristics of frequently charging/discharging makes the supercritical rotordynamic design less favorable, so the flywheel's operating speed has to stay below the first flexible mode. However, the carbon-fiber-wound hubs generally contribute very little stiffness to the overall bending stiffness of the rotor. This physical limitation reduces the bending frequencies for a given rotor design, as compared to the steel hubs, since the steel hub will contribute considerably to the bending stiffness, and have a higher 1st bending frequency. This rotordynamic limitation further brings the performances for composite and steel flywheels closer.

In addition, if power density is the priority, then steel flywheels have a natural advantage. As stated previously, a higher rotor stiffness will help motors with the same EM design (motor dimensions) to spin faster. For any given EM design, a faster rotor speed will boost the power density of the steel flywheel. This is especially true for surface mount permanent magnet motors as their torque coefficients are almost independent of the speed, and power capability is proportional to the speed. In addition, due to the higher mass densities of steels, the steel flywheel will also find its place for applications where the volumetric energy density is the priority.

Other factors might also play a role in giving steel flywheels some advantages in certain applications. One of them will be temperature behavior. The carbon fibers and their bonding material will be degraded at about 300°C or possibly lower, which is much lower than for steel. These degradations for carbon fibers are normally not reversible. There were incidents where a composite flywheel burst after an unintentional overheat. At least one case was caused by running the flywheel with a faulty vacuum sensor and leaking housing. The air friction overheated the fibers, compromised their strength, and created the accident. Another factor is the manufacturability of the flywheel. Fiber wound composite wheels cost much more than the steel ones. Peeling of layers, deformation, and bonding material degradation can potentially cause manufacturing issues. Quality checks with the fiber wound hub take more effort than ultrasonic test on steels. Finally, the fatigue life calculation for composite wheels is less mature than for steel ones, even though research has been conducted [33,34]. Fatigue models are still under development, and, most importantly, are missing corresponding material data. On the other hand, fatigue theories are well developed for steels and supported by many decades of experimental data in the form of *S-N* curves and crack growth coefficients.

4.2.2.2 Composite flywheels

Jamshid Kavosi, Mohammad Naraghi, Terry Creasy, and Alan Palazzolo

Background

Flywheel energy storage systems (FESS) are a clean and efficient method to level supply and demand in energy grids. The energy is stored in a flywheel as kinetic energy by spinning the flywheel and released on demand; therefore, FESS are "mechanical batteries" [35]. The FESS have been around for over 100 years and were used in early industrial systems. Similar to other energy storage systems, such as electrochemical batteries, energy density (energy stored in the battery per its unit mass) is one of the main parameters, which defines the battery's performance. The achievable energy density of a FESS, such as a circumferentially wound ring or cylinder, is proportional to the material's specific strength. This proportionality favors using materials with high specific strength such as fiber composites to enhance the energy density of the flywheel.

In this section, we present the recent trends in developing composite flywheels in which fiber choice, architecture, and volume fraction are designed to achieve enhanced energy density. First, we present a comparison between various types of energy storage systems, mainly the electrochemical devices such as lithium ion batteries and FESS. This comparison will highlight the benefits of FESS in terms of reliability, safe temperature range of operation, energy, and power density. Next, many manufacturable designs are presented and discussed in some detail. While continuous fiber-reinforced composites have very high strength to density ratio, especially in the direction of the fibers, they are highly anisotropic materials. As such, the performance of the composite flywheel is often limited because of tensile stresses that develop in the wheel in the direction transverse to fibers due to centrifugal forces. Additional tensile stresses may also evolve during the curing process primarily due to anisotropic thermal expansion in individual plies. Therefore, different designs suggested to manipulate the radial stress are also presented here.

Theoretically optimal designs of FESS are also presented. In theory, the optimized design of a FESS would be the rotor with continuously varying stiffness and density along the radial direction. However, manufacturing challenges associated with continuously varying the flywheel properties can make these designs highly uneconomical. Partly to accomplish the radial variation in properties and partly to induce residual compressive stresses during processing, which can balance tensile stresses caused by centrifugal forces multiring designs have also been discussed. The multiring rotor is preferred from a manufacturing standpoint, and there are many options for discrete changes such as density, prefit stresses, material selections, elastomeric interlayers, and fiber selection and orientation.

Lastly, potential future research directions targeting enhanced energy storage in FESS is presented. In particular, two potential future directions are presented: the use of nanotechnology and additive manufacturing to make flywheels out of materials with enhanced strength and more uniform stress, respectively.

Comparison with various energy storage systems

In our modern society, energy storage devices (ESD) benefit two main commercial power divisions: transportation and electrical power utilities [36]. Given the seasonal, weekly, daily, and even hourly changes in these sectors' energy demands, various ESDs can be used in these applications to level demand and supply. Flywheel energy storage systems, chemical batteries, and ultracapacitors are the main contributors to energy storage technologies. Flywheel energy storage devices have certain advantages in both sectors compared to the chemical batteries and supercapacitors because flywheels balance power, energy density, environmental friendliness, lifespan, temperature sensitivity, and safety, as discussed later in detail.

Two main parameters often used to evaluate the performance of an energy storage device are energy density and power density of the total energy stored and the maximum power retrieved per unit mass, respectively. Some current industry technology product data provide a rough comparison of both performance metrics in a Ragone plot, Fig. 41. An ideal ESD should appear in the Ragone plot's top-right portion, which means an ESD can store a large amount of energy and release it quickly [38].

Electrochemical batteries, such as Li-ion batteries (LIB), are well suited for the technologies that demand high energy density. Specific to the chemical batteries, LIBs showed the promising potential to be used on a large scale. LIBs—because they have high specific energy densities and stable cycling performance—have been widely used in various applications, including the electric vehicles (EVs), portable devices, and grid energy storage [42,43]. LIBs showed

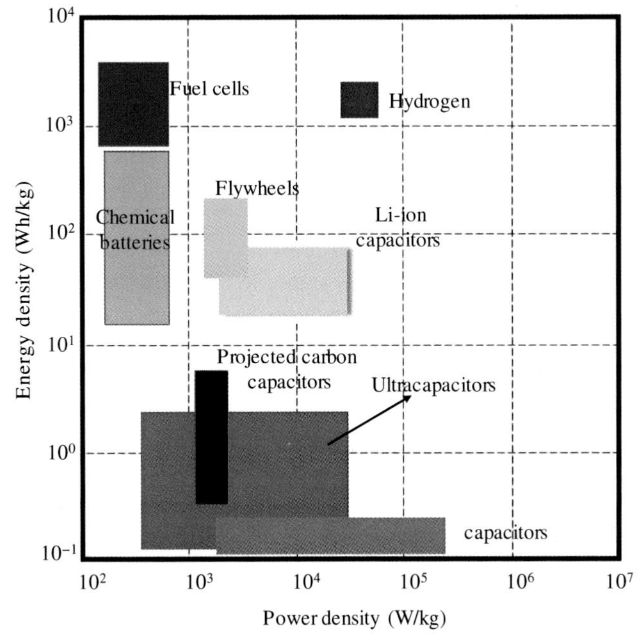

FIG. 41 Ragnone plot showing the energy density versus power density of several energy storage devices—various sources, including [37–41].

promising energy densities—as high as ~250 W-h/kg—that can be further enhanced to ~650 W-h/kg and ~950 W-h/kg with using the Li-S and Li-air system, respectively. Nevertheless, LIBs appear on the Ragone plot's top left, presenting a tradeoff between energy storage and power. This limitation is inherent to ion diffusion [44]. In addition, electrochemical batteries become less efficient the faster they are charged and can be damaged if charged too quickly. Li-ion batteries also have life cycle limitations caused by battery aging [45].

The LIBs also suffer from fire hazards and poor low-temperature performance; both pose major challenges in battery thermal management [42,46–49]. For instance, if LIBs operate improperly, they can release energy abruptly and that can lead to fire or explosion. The main failure mode in the LIBs system is thermal runaway, which happens when the exothermic reaction between Li and air goes out of control. Battery performance depends on operating temperature because high or low temperature affects ion transport. As the battery temperature exceeds ~80°C, the exothermic chemical reaction rate inside the batteries increases and further heats the cell. The continuously rising temperature may result in a fire or an explosion. Also, if the battery is damaged and a short-circuit results, the battery may ignite. Accidents related to the LIBs occur frequently in cellphones, laptops, and electric vehicles. These life-threatening accidents have highlighted the safety issues to be resolved for future high-energy battery systems [50].

Ultracapacitors are also attractive energy storage technologies because they, too, have high power density. Ultracapacitors, also known as supercapacitors, have a structure similar to conventional capacitors, but they store energy using an electrolyte solution between two high surface area electrodes [51,52]. Electric double-layer capacitors, pseudo-capacitors, and hybrid capacitors are three supercapacitor types. High life cycles—as much as 105 cycles for around 40 years—and high power density are distinctive characteristics of supercapacitors [53]. Supercapacitors are more effective at rapid and regenerative energy storage than chemical batteries. However, they also present a tradeoff between energy and power densities, and they appear on the opposite corner of the Ragone plot than LIBs because supercapacitors provide high power but low energy density. Because the ion transport occurs in a short distance without an electrochemical reaction, these devices are highly efficient with low resistive losses. As a result, supercapacitors have significantly reduced cooling requirements. Nonetheless, there are many disadvantages with supercapacitors. The main issue with supercapacitors is high self-discharge rate that keeps designers from using the supercapacitor as the only energy source in vehicles [54]. Also, supercapacitor system needs containment in case there is a dielectric breakdown that makes the supercapacitors explode. Another supercapacitor issue: limited temperature range. The electrolyte is a liquid and at low temperatures the electrolyte solidifies. Even before solidification, ion transport is lowered by cold temperatures, and this reduction increased internal resistance in the supercapacitors [55]. Finally, if a voltage difference is maintained for an extended time, dielectric absorption can occur, preventing the supercapacitors from fully discharging.

Unlike electrochemical energy storage devices, FESS offer a good balance between energy density and power density. The flywheel's energy density scales with the mechanical strength required to withstand high centrifugal forces, and recent efforts put the flywheels energy density as high as ~80–100 W-h/kg, which is similar to that of LIBs [56]. In addition, operating systems can extract the kinetic energy stored in the flywheel controllably in times as short as seconds, with power densities similar to supercapacitors [57]. For instance, in the transportation industry, given wide environmental conditions, there is a need for high power in a compact, reliable, and lightweight package. Moreover, flywheels have operational life that can match the vehicle life. Also, unlike LIBs and supercapacitors, active temperature control might not be required for some flywheel systems; these factors significantly reduce weight, space, and cost requirements [56]. FESS are also useful in the power utility industry that needs a long-lasting, low-maintenance energy storage technology that is less temperature sensitive than chemical batteries. As an example, one utility equipment provider, Beacon Power Corporation, produced commercial flywheels that Beacon claims require maintenance only once every twenty years. Given all the benefits of FESS regarding low maintenance, low temperature sensitivity, and a good balance between energy density and power density, they are also considered as an enabling technology for many other applications, including space satellites in low earth orbits, pulse power transfer for hybrid electric vehicles, and many stationary applications [58,59].

Material selection and geometrical design of FESS

The kinetic energy stored in the flywheel, Ek, is proportional to the second power of the flywheel angular velocity, ω, as

$$E_k = \frac{1}{2} I \omega^2 \tag{34}$$

where I is the moment of inertia. The faster the angular velocity of the flywheel, the higher the energy stored in it. An upper limit on the angular velocity can be determined to prevent material failure subjected to centrifugal forces (burst speed). This upper limit is often expressed in terms of energy density of the flywheel, or the energy stored in the flywheel per unit mass. The mass should be the mass of the whole energy storage system, and the energy should be the energy supplied in the device. Simply put, the energy density of the rotor at the burst speed is proportional to the specific strength of the material of the flywheel, that is, the ratio of the material strength to density (σ_u and ρ). The proportionality constant is the so-called shape factor K, which depends on the geometrical configuration and the failure criterion of the material as shown in Eq. (2). The value of shape factor K often in the range of 0.3–1, as shown in Fig. 37.

While FESS made out of steel have a long history, because the energy density is proportional to the specific strength, the choice of the advanced composite materials is more favorable. Composite materials compared to steels have higher specific strength, and flywheels made from them can store two to three times as much energy per kilogram as steel flywheels. Other attractive characteristics of the composite flywheels are their relatively uncomplicated containment. The failure of composite rotors is not as catastrophic as the metallic rotors. In most of the fiber composite flywheels, fibers are wound circumferentially, and the composite fails because the radial stress produced during the spinning. Hence, failed composite fragments exert contact pressure on their housing, which are lower than steel, nevertheless, dangerous [28]. Despite all the benefits, composites are highly anisotropic. Especially if orthotropic materials such as carbon, glass, or aramid fibers are used with poor transverse strength and modulus properties, then there should be specific consideration on the geometric design as well.

One of the earliest studies that demonstrated the applicability of composite materials with significantly large specific strength for flywheel energy storage applications was by Rabenhorst et al. [60]. DeTeresa et al. [25] examined the performance of commercial high-performance reinforcement fibers for the application in the flywheel energy storage devices. Multiple factors, including inherent strength of the fibers and stress-rupture lifetime, were considered.

Their study showed that carbon fibers are preferred for high-performance energy storage application, owing to the high specific strength of carbon fibers and E-glass for the lower cost.

In addition to the choice of the materials, flywheel geometry also has a major impact on the stress distribution within a flywheel for a given angular velocity and thus the energy density of the flywheel. This dependence is captured with the shape factor. Fig. 37 shows different flywheel geometries and their corresponding shape factor. Accordingly, the optimal shape that exhibits the failure at all points simultaneously and maximizes the energy density is shown in the top left corner of the figure. Other profiles are also related to the shape factor in this figure [39]. Despite very promising shape factors, in the case of fiber-reinforced composites, manufacture of the structure shape factor exceeding 0.5 is impractical and uneconomical.

In 2008, Arslan [61] investigated the effect of various shapes of the flywheel on its energy density via finite element analysis. Although the focus of the studies was mainly on steel, the results can shed light on the significance of shape in distributing stress in composite FESS. He studied the six common geometries, a constant thickness disk, constant thickness ring, parabolic tapering, linear tapering, truncated isostress exponential decay tapering, and a modified isostress case (Fig. 42), and ranked them according to their energy storage performance. As expected, the constant thickness profile was the worst case in terms of the energy storage and performance, while the isostress profile with the specific energy of 8.977 W-h/kg showed the highest energy storage, 48.7% higher than the constant thickness disk case.

Optimal composite flywheel design for enhanced energy density

In addition to various geometries, the proper arrangement of fibers and/or inducing certain residual stress state would also result in improved performance

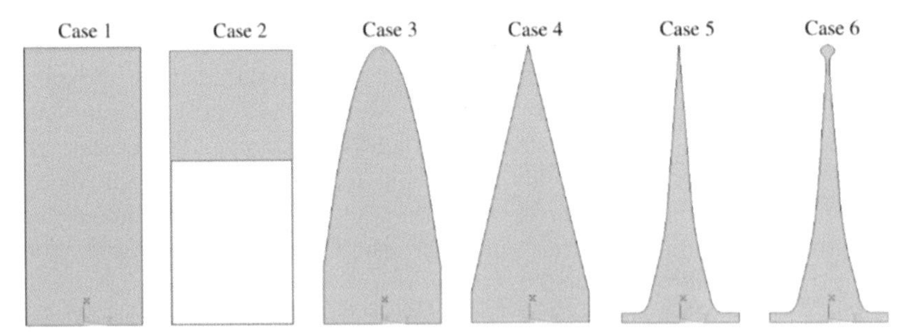

FIG. 42 Effect of geometry on the maximum energy density of a flywheel. Case 1: a constant thickness disk, Case 2: constant thickness ring, Case 3: parabolic tapering, Case 4: linear tapering, Case 5: truncated isostress exponential decay tapering, and Case 6: modified isostress case. *(From J. Park, A. Palazzolo, Magnetically suspended VSCMGs for simultaneous attitude control and power transfer IPAC service, ASME J. Dyn. Syst. Meas. Control 132 (5) (2010) 051001.)*

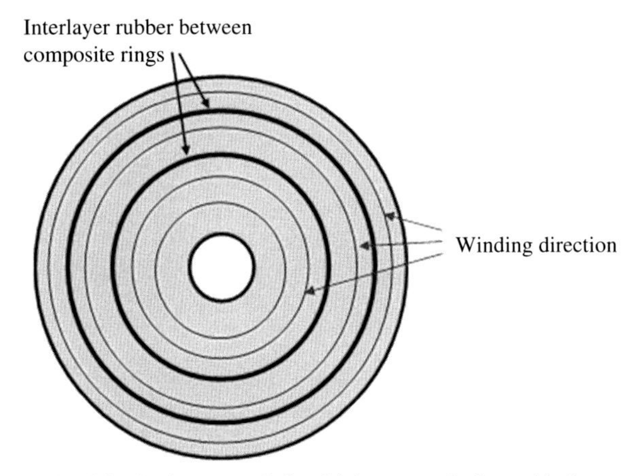

FIG. 43 Schematic of flywheel composed of multiple concentric rings with elastomer interlayer in between.

of the flywheel. The cylindrical rotor proposed by Danfelt et al. [62] in 1977 is one example where the stress field in a multiring flywheel subjected to centrifugal forces was calculated through stress analysis. The authors calculate the radial and hoop stresses in a composite flywheel where soft rings were placed in between fiber-reinforced stiffer rings to manipulate the stress field. The schematic of the multiring composite flywheel is shown in Fig. 43. The rubber rings were placed in between higher stiffness rings to lower the radial stress. Their proposed design is composed of the six concentric rings of Kevlar/epoxy composite system of same radial thickness separated by five thin layers of hyperelastic rubber rings in between in order to reduce the radial stress distribution in the rotor. The authors considered three different scenarios. In one case, the rubber rings were placed in between composite rings, while the density and stiffness of the composite remained the same among all composite rings (benchmark). In the second and third cases, they systematically varied the density and stiffness of the composite rings to manage the stress field at a given angular velocity. The stress distribution analysis of the flywheel with rubber rings and constant composite density and stiffness is shown in Fig. 44A. As shown in the figure, the peak radial stress in the rings varies from ~20 MPa to over 70 MPa. Following this result, Danfelt et al. attempted to find the stress field where the maximum stress in the rings at any given angular velocity are comparable to each other, so that all the rings fail around the same angular velocity.

In their initial analysis, authors decreased the density of the composite rings along the radial direction (from 0.05 to 0.45 lb./in^3) to achieve the constant radial stress, but results showed that this could not be achieved only by changes in the density of the composite rings. In some cases, this approach led to failure at lower stresses, as the maximum stress in the inner rings even went above the value of the bench mark experiment. Followed by this analysis, they varied the

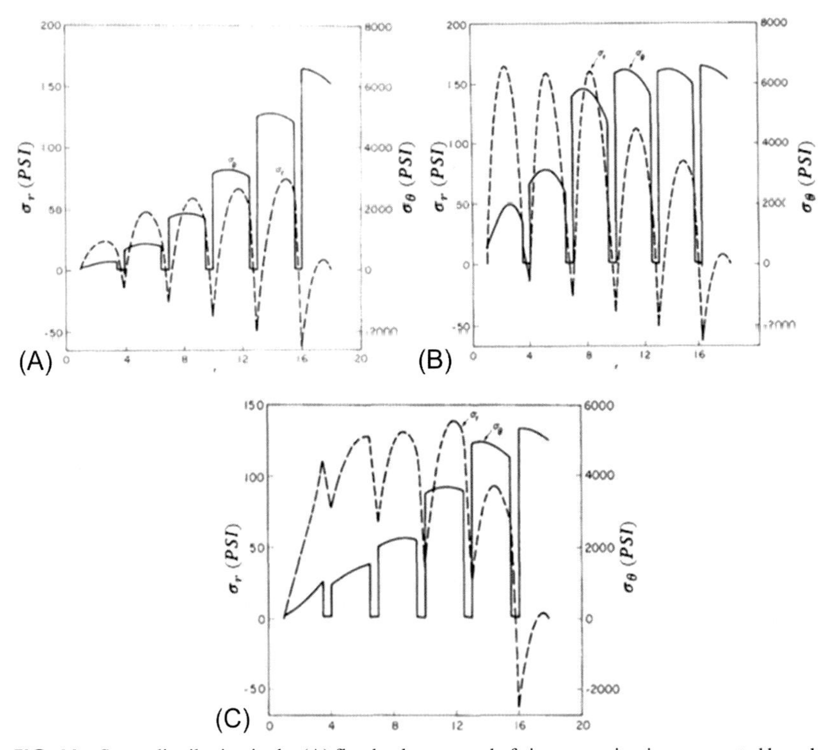

FIG. 44 Stress distribution in the (A) flywheel composed of six composite rings separated by rubber interlayer, (b) flywheel with varying density of the composite rings (low density in the inner ring to high density in the outer ring), (c) flywheel with varying young moduli of the interlayer elastomer ring (from high stiffness of the elastomeric layer to low stiffness of the elastomeric interlayer) *(From R. Pena-Alzola, et al., Review of flywheel based energy storage systems, in: 2011 International Conference on Power Engineering, Energy and Electrical Drives, IEEE, 2011 with permission.)*

stiffness of the inter-ring elastomer layer. The stiffness of the rings varied from very high stiffness (composite stiffness) in the inner rings to the very low stiffness at the outer ring. The obtained stress distributions showed smoother radial stress distribution compared to the first case. According to this analysis multiring flywheels with varying composite density and rubber rings compared to single-ring flywheels are promising in terms of achievable energy storage and efficiency. However, further analysis is required to compare the data in terms of energy density.

Among the early studies to explore the optimal design of the flywheel to maximize the specific energy, Lawrence Livermore National Laboratories (LLNL) carried out a comprehensive study under the department of energy sponsorship. In particular, LLNL sponsored a competition between industry leaders to design and develop flywheel designs for testing. The designs, shown

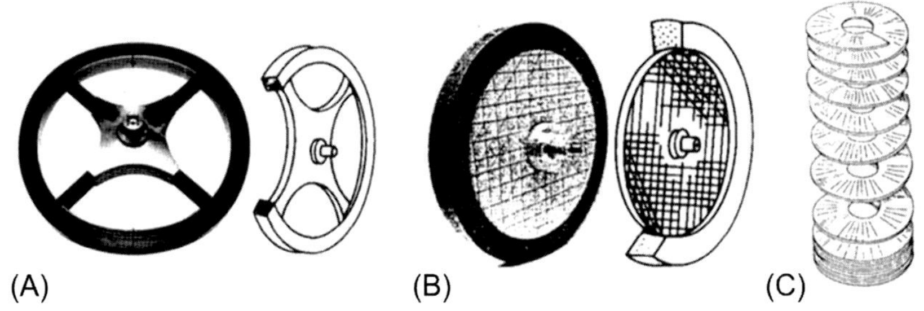

FIG. 45 Flywheel design by (A) Garrett AiResearch, (B) General Electric, (C) AVCO. *(From Pena-Alzola, et al., A review of flywheel based energy storage systems, in: Proceedings of the 2011 International Conference on Power Engineering, Energy and Electrical Drives, 3rd IEEE International Conference on Power Engineering, Energy and Electrical Drives, Power-Eng2011—Torremolinos, Malaga, Spain, 11 May–13 May 2011, 2011,with permission.)*

in Fig. 45, were selected for their performance and practicality. These designs were from Garrett AiResearch, General Electric Company, and the AVCO Corporation.

The Garrett flywheel was a wound rim of glass and Kevlar filament with a spoked hub composed of graphite/epoxy struts bonded to aluminum sheets (Fig. 45A). The Garrett flywheel performed the best with the highest recorded burst energy density of 72.8 W.h/kg. The General Electric (GE) design consisted of an outer ring of filament-wound carbon/epoxy composite attached to the disk hub structure made of the crossply glass filaments with the maximum energy density of 68.0 W.h/kg (Fig. 45B). The AVCO's "constant stress" flywheel comprised a ring that contained a bidirectional weave of the Kevlar or glass fibers were placed in the hoop and radial directions. The goal of this weaving pattern was to produce a constant stress profile where radial fibers would interact with the hoop fibers to transmit the outer hoop stresses to the inner hoop fibers. The AVCO design posed significant manufacturing challenges where they were only able to produce a single flywheel for testing (Fig. 45C). The measure for the goodness of the flywheel was maximum energy density and stored energy in the flywheel rotor.

In 1984, Genta [60] proposed the plane stress calculation that includes the density and stiffness tuning along the radial direction of the flywheel to mitigate the radial stress produced during spinning. In order to optimize the energy stored in the flywheel, Genta assumed zero radial stress and constant radial-hoop Poisson's ratio plus constant hoop stress in the flywheel. Accordingly, this optimized design requires the hoop stiffness and density to be functions of radial distance. He suggested that the optimum design could be obtained in practice by continuously varying the fiber content as a function of radius to tune hoop stiffness. He also suggested that density of the composite along the radial direction can be tuned up by incorporating the ballasting materials such as lead particles in the polymer matrix.

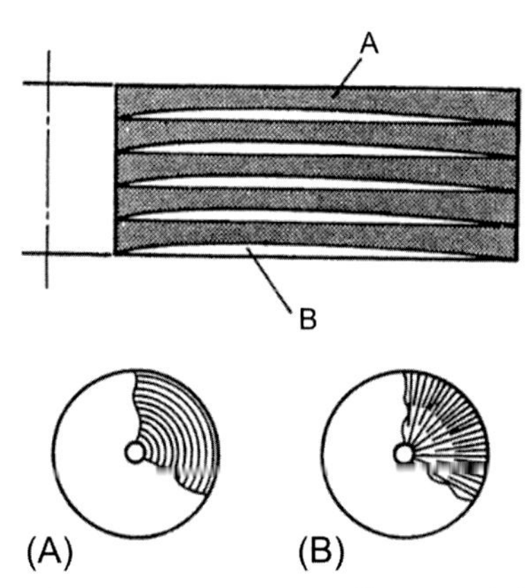

FIG. 46 The optimized design suggested by Miyata; fibers are reinforced both in the hoop and circumferential direction. *(From J.G. Williams, D.P. Isherwood, Calculation of the strain-energy release rates of cracked plates by an approximate method, J. Strain Anal. Eng. Design 3 (1) (1968) 17–22 with permission.)*

In 1985, Miyata [25] proposed the optimal design for the high energy storage flywheel, using a plane stress analysis that includes the axial thickness as a parameter. Miyata's solution suggests using the hoop fiber plies of varying thickness on top of the radially arranged fibers of complementary thickness to tune up the radial and circumferential elastic moduli to more uniformly distribute stress and postpone failure. His suggested model is shown in the Fig. 46. However, this optimized flywheel is yet to be manufactured, and its fabrication is posed with major challenges related to unconventional fiber placement.

In 1999, Ha et al. [61] did a comprehensive study to optimize the stored energy of the flywheel that is held with a permanent magnet along the inner radius of the flywheel. This magnetic material places a compressive stress condition along the inner radius of the multiring rotor. In this analysis, the inner and outer radii of the flywheel are kept constant and the total stored energy of the flywheel is optimized. The authors varied the sequence of the composite rings along the composite rings. The five composite materials selected to be used were Glass, Aramid, and three grades of the carbon fiber as the choice for the rings. After investigating the possible permutations, the authors concluded that having materials that increase the stiffness/density ratio as the radius was increased, provided the optimal sequence. This result would indicate that the more rings the flywheel contains, with the proper order, the higher the total stored energy can be.

In 2012, Ha et al. [63] conducted a study to reduce the radial stress in the experiment by stepwise tuning of the stiffness and density of the fibers in the radial direction. The experimental design is in line with the guidelines presented in the analytical work of Danfelt et al. [62] that manipulated the density and stiffness of the composite to delay failure. In the flywheel in [63], the stiffness and density of the composite ring were varied by comingling carbon and glass fibers. The flywheel was made first as separate rings, each with a different carbon to glass fiber ratio. The inner and outer radii of the rings were designed to be press fit inside one another, resulting in a solid FESS. The press fitting during fabrication was mainly carried out to induce compressive stresses, which will offset the tensile radial stresses that develop in the wheel at high angular velocities. In addition, fiber comingling generates the prestress that mitigates the radial stress build-up in the rotating flywheel. The authors studied three different cases, A, B, and C, of the four-rim flywheel as depicted in Fig. 47. There were two reinforcement materials: carbon fiber and glass fiber, and the reinforcement volume fraction in the matrix was different in each rim. The innermost material rings of all these cases were mainly glass fibers (lower stiffness), transitioned to the mostly carbon fiber content in the outermost ring (higher stiffness). In case A, all the rims were wound simultaneously and cured by continuous filament winding. In case B, all the rims were wound and cured separately and press fitted in the flywheel. In case C, the rims were wounds in two sets. Rim set 1 is composed of rims 1 and 2, and the rim set 2 is composed of the rims 3 and 4, followed by the press fitting of the set 1 to the set 2. The three radii in between the inner and outer radii plus the interference (the difference between the inner radius of an outer ring and the outer radius of an inner ring) for the cases that include the press fitting were the variables of an optimization search as follows. Their stress analysis first calculated the residual stresses for the cured rings. If there was press fitting, then the rings were analyzed at each point in the press fitting to ensure that can be removed without assistance. Once the rotor was fully assembled while surviving the stationary residual stress associated with manufacturing, then stress analysis was carried out at high angular velocities. The parameters to be optimized were the strength ratios, $R_r = \sigma_r/Y_T$ and $R_\theta = \sigma_\theta/Y_\theta$, at a given angular velocity of $\omega_{max} = 15$ krpm. Their analysis and experiment show that case B with the press-fit interference was the

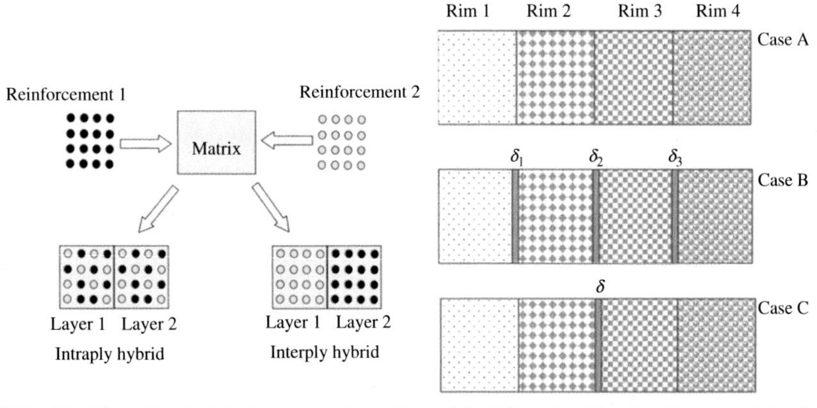

FIG. 47 Three flywheel design approaches. *(From A.B. Palazzolo, Vibration Theory and Applications with Finite Elements and Active Vibration Control, John Wiley, 2016 with permission.)*

optimum design. The radial strength ratio in case B is 0.19 while the strength ratio in cases A and C reaches 0.39 and 0.75, respectively. Case A has the advantage over cases B and C, in terms of manufacturing cost and simplicity.

Referring to the works mentioned here, there are many options for enhancing the performance of flywheels. Theoretically optimized solutions for rotors with continuously varying material or geometric properties can serve as benchmarks against new flywheel design ideas. However, manufacturing difficulties linked with continuously varying the flywheel properties can make these designs highly uneconomical. Given all of these possibilities, there is a clear need to narrow down the research range to the most relevant analyses. Discrete variation (multiring rotors) is preferred from a manufacturing standpoint, but there are many options for discrete changes such as density, press-fit stresses, material selections, elastomeric interlayers, and fiber selection and orientation [52].

Flywheel stress analysis One of the earliest studies to analyze the elastic stresses in the symmetrical disk was carried out by Manson [64]. His analysis was based on a finite-difference solution of the equilibrium and compatibility equations for elastic stresses in an axisymmetric disk. The advent of the composite materials pushed the researches toward new designs to mitigate the radial stress during spinning of the composite flywheel. Following by Manson, Genta [55,60] performed extensive studies to modify the Manson's solution for orthotropic materials. The introduction of the multiring design to manipulate the radial stress in the flywheel attracted a number of researchers to conduct a study to further improve the energy storage capability of the flywheel. In this regard, the literature is abundant with a number of studies using the multiring design approach, attempting fabrication of an optimized design.

This section is dedicated to the stress analysis of multirings flywheel, where the stress analysis of single-ring flywheels is presented as an introduction. Much of the works on the single rings was carried out by Genta [55,60], Lekhnitskii [65], and Garbys and Bakis [66], and the stress and failure formulations presented for the multiple-ring rotor are based on the Arnold formulations [67,68].

Flywheel stress analysis—Single-ring flywheels Stress distribution in the axial orthotropic disk with the constant thickness, under the assumption of the axial symmetry, plane stress state, and linear orthotropic behavior of the materials can be solved following the equilibrium, displacement compatibility, and the constitutive equations for axially orthotropic materials using these relationships:

$$\frac{d\sigma_r}{dr} + \frac{\sigma_r - \sigma_\theta}{r} + \rho\omega^2 r^2 = 0 \quad \text{(Equilibrium)} \tag{35}$$

$$\varepsilon_r = \frac{d}{dr}(r\varepsilon_\theta) \quad \text{(Strain compatibility)} \tag{36}$$

$$\begin{bmatrix} \varepsilon_r \\ \varepsilon_\theta \end{bmatrix} = \begin{bmatrix} \frac{1}{E_T} & \frac{-\nu_L}{E_L} \\ \frac{-\nu_L}{E_L} & \frac{1}{E_T} \end{bmatrix} \begin{bmatrix} \sigma_r \\ \sigma_\theta \end{bmatrix} \text{ (Constitutive equation)} \quad (37)$$

where r is the radial distance, and the centrifugal force is included as a body force term ($\rho\omega^2 r^2$) in which ρ is the density and ω is the angular velocity of the rotating mass. In the equations, σ_r and σ_θ are the radial and circumferential stresses, and ε_r and ε_θ are the radial and circumferential strains, respectively. Also, E_T, E_L, and υ_L are the transverse elastic modulus, longitudinal elastic modulus, and longitudinal Poisson's ratio. Eqs. (35)–(37) are the basic terms used in the elasticity solution. The solution to the set of PDEs can be obtained based on various types of boundary conditions (BCs), and the results are typically presented in terms of nondimensional quantities such as $\mu = \sqrt{\frac{E_L}{E_T}}$, $\beta = \frac{r_i}{r_0}$, and $\chi = \frac{r}{r_0}$, which simplifies the stress equations for the flywheel. A commonly assumed set of BCs is zero radial pressure on the inner and outer radii (traction free surfaces). The particular solution of the set of (2.35)–(2.37) PDEs becomes

$$\sigma_r = \rho\omega^2 r_0^2 \left(\frac{3+\nu}{8}\right)\left(1+\beta^2 - \frac{\beta^2}{\chi^2} - \chi^2\right) \quad (38)$$

$$\sigma_c = \rho\omega^2 r_0^2 \left(\frac{3+\nu}{8}\right)\left(1+\beta^2 + \frac{\beta^2}{\chi^2} - \frac{1+3\nu}{3+\nu}\chi^2\right) \quad (39)$$

Another interesting set of BCs can also be applied to describe the assembly of the rings at zero angular velocity, where constant inner pressure, P, and outer pressure, Q, is applied ($\omega = 0$):

$$\sigma_r = \left(\frac{P-Q}{1-\beta^2}\right)\beta^2\chi^{-2} + \left(\frac{Q-P\beta^2}{1-\beta^2}\right) \quad (40)$$

$$\sigma_h = \left(\frac{P\beta^2-Q}{1-\beta^2}\right) + \left(\frac{Q-P\beta^2}{1-\beta^2}\right)\beta^2\chi^{-2} \quad (41)$$

Genta also suggested the solution for the stress distribution in the flywheel subjected to the temperature gradient. The solution to this problem is dependent on the temperature profile. If the temperature profile follows the polynomial form:

$$T = \sum_{i=0}^{n} b_i \chi^i \quad (42)$$

where the coefficients b_i have dimension of temperature. For each term in these equations, stress distribution can be calculated separately. The stress field coming from the ith term of the temperature profile is

$$\sigma_r = \frac{Eb_i\alpha}{i+2}\left\{\frac{1}{1-\beta^2}\left[1-\beta^{2+i} - \left(\beta^2 - \beta^{2+i}\right)\frac{1}{\chi^2}\right] - \chi^i\right\} \quad (43)$$

$$\sigma_h = \frac{Eb_i\alpha}{i+2}\left\{\frac{1}{1-\beta^2}\left[1-\beta^{2+i}+\left(\beta^2-\beta^{2+i}\right)\frac{1}{\chi^2}\right]-(i+1)\chi^i\right\} \tag{44}$$

In principle, the stress field in a flywheel under the combined effect of centrifugal forces, prescribed inner and outer pressure and change in temperature can be obtained via superposition.

Genta [23,29] also further studied the more general case of the hyperbolic disc using an orthotropic material in different loading conditions, similar to the aforementioned study for the isotropic disk. For simplicity, a general solution form to the constant thickness ring is provided with this equation:

$$\begin{aligned}\sigma_c = {} & \rho\omega^2 r_0{}^2\left(\frac{3+\nu_{hr}}{9-\mu^2}\right)\left[\mu L\chi^{(\mu-1)}+\mu(L-1)\chi^{(-\mu-1)}-\chi^2\left(\frac{\mu^2+3\nu_{hr}}{3+\nu_{hr}}\right)\right] \\ & +\left(\frac{P\beta^{\mu+1}-Q}{1-\beta^{2\mu}}\right)\mu\chi^{\mu-1}+\left(\frac{P-Q\beta^{\mu-1}}{1-\beta^{2\mu}}\right)\mu\beta^{\mu+1}\chi^{-\mu-1}\end{aligned} \tag{45}$$

$$\begin{aligned}\sigma_r = {} & \rho\omega^2 r_0{}^2\left(\frac{3+\nu_{hr}}{9-\mu^2}\right)\left[L\chi^{(\mu-1)}-(L-1)\chi^{(-\mu-1)}-\chi^2\right]+\left(\frac{P\beta^{\mu+1}-Q}{1-\beta^{2\mu}}\right)\chi^{\mu-1} \\ & -\left(\frac{P-Q\beta^{\mu-1}}{1-\beta^{2\mu}}\right)\beta^{\mu+1}\chi^{-\mu-1}\end{aligned} \tag{46}$$

where $L=\left(\frac{\beta^{-\mu-1}-\beta^2}{\beta^{-\mu-1}-\beta^{\mu-1}}\right)$. Genta's solution comprehensively considers different loading conditions, including temperature variation on the wheel. In his analysis the temperature gradient is considered to be in the polynomial form. Other solutions have also been proposed by Arnold et al. [36] with comparable results, where the rotating disk experiences thermal loads (linear temperature gradient), centrifugal forces, and internal and external pressure.

Flywheel stress analysis—Multiring flywheels Stress distribution in the flywheel composed of multiple concentric rings can be solved with an approach similar to the single-ring method by employing appropriate boundary conditions, mainly the continuity of radial displacement and radial stress. Here we are referring to the general approach of Arnold et al. [31] and Ha et al. [32–34] approach. Arnold et al. [31,35] take into account the interference fit between each consecutive rings as well. Radial stress continuity requires the following boundary conditions:

$$\begin{aligned}&\sigma_{r_1}(r=r_1)=P_1\\&\sigma_{r_1}(r=r_2)=\sigma_{r_2}(r=r_2)=P_2\\&\sigma_{r_2}(r=r_3)=\sigma_{r_3}(r=r_3)=P_3\\&\sigma_{r_{n-1}}(r=r_n)=\sigma_{r_n}(r=r_n)=P_n\\&\sigma_{r_n}(r=r_{n+1})=P_{n+1},\end{aligned} \tag{47}$$

where $\sigma_1, \sigma_2, \ldots, \sigma_n$ are the radial stresses, and P_1 and P_{n+1} are the internal and external pressure applied to the flywheel, respectively. $P_2, P_3, \ldots, P_n$ are the interfacial pressures, which are unknowns. Also, $r_1, r_2, \ldots, r_n$ are the radius of the rings. Similarly, the same approach regarding kinematic constraint at the interface (interference fit δ) can be written as

$$\begin{aligned}
&u_{r_2}(r=r_2)-u_{r_1}(r=r_2)=\delta_1\\
&u_{r_3}(r=r_3)-u_{r_2}(r=r_3)=\delta_2\\
&u_{r_4}(r=r_4)-u_{r_3}(r=r_2)=\delta_3\\
&\vdots\\
&u_{r_n}(r=r_n)-u_{r_{n-1}}(r=r_n)=\delta_{n-1}
\end{aligned} \tag{48}$$

where $\delta_1, \delta_2, \delta_3$, are the interference fits between the consecutive rings that produce internal pressure P_2, P_3, P_4, P_n. The continuity equation leads to determination of the remaining interfacial pressures. Eqs. (47), (48), together with force-displacement equation, provide a system of $2n$ equations with $2n$ unknown constants (where n is the number of the rings in the flywheel):

$$[K]\{C\}=\{R\}, \tag{49}$$

where $\{C\}$ is the vector composed of the unknown constants, $[K]$ is the geometry and material matrix, and $[R]$ is a vector of (interference fits δ) or applied forces.

Future trends for composite flywheels Given the various benefits of flywheels as energy storage devices, such as their wide range of operating temperature, low maintenance and long life, there is a desire to further push them toward the upper right corner of the Ragone plot (Fig. 41) to make FESS with remarkably high power and energy density. In this regard, it seems that the bottleneck is improving the energy density of the flywheel. The increase in energy density of a flywheel can be observed from two angles: structural design and materials design. The former includes a host of techniques where the goal was to reduce the maximum stress of a flywheel at a given angular velocity, so that higher angular velocities can be achieved. Several examples were presented in this chapter, such as employing rubber rings to lower radial stress. The second and complementary approach is a materials design approach, where novel materials are to be developed with higher strength and utilized in the flywheel to further increase the safe angular velocity of the wheel. The use of carbon fiber-reinforced composites is an example of that. However, the advent of novel materials calls for a revisit of this approach.

Nanomaterials Nanomaterials such as nanotubes, platelets, nanofibers, etc. with exceptional mechanical properties are promising agents in reinforcing the composites [36–39]. Materials scientists predicted that composites reinforced with nanoscale fillers have the potential to possess exceptional properties ranging from mechanical property through tailoring tensile modulus, strength [40–42], fracture toughness [43], or wear resistance [44], as well as electrical [41] and optical properties [45]. However, based on the results acquired so far, it is still not evident whether such exceptional properties of nanoreinforcement agents can be translated to bulk composite. The reasons include the poor dispersion of nanoscale reinforcements especially at high volume fractions (more than 5%–10%). On the other hand, nanomaterials have a much higher surface area per unit volume compared to the mesosize particles and fibers (such as carbon fibers). Hence, they can bond much more efficiently with a polymer matrix. Hence, compared to microfibers, significantly lower level of nanomaterial is required to alter the properties of a composite dramatically. The number of studies has demonstrated that levels below 5% of nanofillers can significantly alter the properties of the matrix [46]. Another limiting factor is the high cost of nanomaterials. However, the latter is a problem that can be remedied likely with more research on processing and more demand.

Epoxy composites with mesosized particle fillers are attractive for component manufacture by techniques, including injection molding and fiber impregnation because they have relatively low viscosity of the uncured composition. The same manufacturing methods can be applied to resin reinforced with the nanomaterials. However, given the large surface area of these particles, measures need to be taken to avoid agglomeration. Given the dispersion challenges and high costs of nanomaterials at the moment, it seems that FESS with hybrid fillers (microfibers and nanofillers) is a promising path forward that strikes a balance between the benefits of nanomaterials in terms of increasing strength measures and cost. Alternatively, cheap nanomaterials may also be employed.

Boyle et al. [47] investigated the effect of TiO_2 nanowires on the mechanical properties of the composite flywheels. TiO_2 are relatively inexpensive, easy to prepare, and easy to surface function (a process required to achieve good bonding between filler and matrix, and to enhance dispersion). The composite flywheel was fabricated through winding of the carbon fiber in resin bath infused with TiO_2 nanowires functionalized with silane. The results from three-point bending test of the composite flywheel showed that the incorporation of TiO_2 nanowires can increase the strength of the composite by 30%. The authors noted that improvement in the mechanical properties of the composite flywheel has a large economic impact by reduction of about 20%–30% in the flywheel energy storage cost. This improvement can lead to increase in the power capacity from 20 to 26 MW and energy capacity from 6.5 to 7.5 MW-service hours, which effectively decreases the average energy storage cost by 20%–25%.

Lin et al. [48] proposed that incorporation of carbon nanotubes into the epoxy matrix could increase the performance of flywheels. The authors performed finite element analysis to calculate the stress field inside the centrifugally loaded flywheel ring. The authors suggested a design that involves the gradual changes of the nanotube concentration in the rotor along the radial direction. The nanotube concentration is higher in the inner part of the rotor to provide a higher strength for the composite. The nanotube concentration lowered while moving along the radial direction.

Despite the good preliminary work, it appears that a lot more needs to be carried out in terms of application of nanomaterials to flywheels. The most intriguing path forward seems to be addressing the problems that are common between flywheel design and nanocomposite processing in general, such as nanofiller dispersion and alignment to better transfer load. Another intriguing direction is with respect to the specific application of nanomaterials to flywheels, including the variations of filler alignment and concentration as a function of radius to modulate strength and stiffness. This approach should be intended to both reduce stress fields (by modulating stiffness) and strength, allowing for higher angular velocities.

Additive manufacturing Additive manufacturing (AM) facilitates the fabrication of complicated geometries, eliminating the need for expensive tooling and multistep processing [49]. Recently, direct deposition of metals facilitated manufacturing of the functionally graded materials (FGM) to provide exciting new alternatives for the design elements. Morvan et al. [50] fabricated the graded flywheel composed of two materials: Cobalt (light/strong) and Tungsten Carbide (heavy/weak) using the AM method. The authors performed the optimization to find the optimum gradient of materials properties and geometry along the radial direction of flywheel. Following that they 3D-printed a functionally graded composite on the LENS apparatus.

The additive manufacturing, advanced filament-winding processes, can also be used to design FESS with optimal fiber arrangements that are nearly impossible to fabricate using other methods. An example is the development of the AVCO's "constant stress" flywheel that contains a bidirectional weave of fibers were placed in the hoop and radial directions (Fig. 45C).

4.2.2.3 Bearings and rotordynamics

Zhiyang Wang, Xiaojun Li and Alan Palazzolo

Bearings

The flywheel hub design, whether made from steel or composites, is the foundation for characteristics and technical specifications of flywheel energy storage systems. This defines the weight, speed, energy levels, and predicted cycle lives that the system will have. These remain only objectives until the feasibility of the design can be physically proven and implemented. Among the many design factors, the rotor-bearing system, or rotordynamic design, is one of the most important. No matter how good a flywheel hub design looks on paper, it must be physically realized along with its shaft and support system.

Contact-type bearings

Contact-type bearings are an essential part of many flywheel systems. Lots of variations were developed for contact bearings, including angular-contact ball bearings, deep groove ball bearings, tapered roller bearings, cylindrical roller bearings, spherical bearings, etc. Any type of these contact bearings has strengths and weaknesses, and can possibly be selected in the final system design. There are lots of publications on the merits of each type of bearing, which can be used as a guide during the selection process. This chapter will focus on several major technical points during ball-bearing selection for the following reason: ball bearings are the most commonly used type in many flywheel systems; most of these technical highlights could be useful and applicable with small modifications for other type of contact bearings.

For ball-bearing selection, two of the most important technical parameters that will affect the operation and life of the bearings are load and operating speed. It is strongly suggested to prepare the simulated load level, load duty cycle, maximum speed, and nominal operational speed range for the bearing manufacturer to verify their calculated life of the bearing after the initial bearing selection and rotordynamic layout.

Generally speaking, the heavier the load capability is, the slower the bearing can run. Given the same bearing load, the ball operating speed (ball center linear speed and rolling speed) will indeed affect the life of the bearing whether it is made of steel or ceramic. As the center speed of the ball is hard to evaluate, a good approximation for the bearing speed criteria (with some deviation from the true nature) will be to use the inner race diameter (in millimeter) times the rotational speed (in RPM) to obtain a parameter called the DN number. For a medium-loaded ball bearing to work with reasonable life, a good starting point is to keep that number under 0.3 million-DN (experience-based). For flywheel systems, as the load could get much heavier depending on the design, this number could be adjusted to as low as 0.1 million-DN. It must be noted that this number is not a sufficient condition, meaning a bearing with these levels of DN numbers will not necessarily work. However, if an application is designed to be pairing with a bearing at this DN number value, there is a good chance that the required ball bearings can be found or designed by some commercial vendors.

The German bearing company GMN Bearing's catalog [51] provides a good method for first-cut estimation of bearing life. In their methods, the predicted loads were separated into axial (F_a) and radial (F_r) components and converted into an equivalent load (P) based on coefficients derived from bearing types and arrangements (X and Y coefficients):

$$P = X \cdot F_r + Y \cdot F_a \tag{50}$$

Then an equivalent load rating is calculated if multiple bearings(number i) are used to make a bearing assembly:

$$C = i^{0.7} \cdot C_{single} \tag{51}$$

where C_{single} is the load capacity from the catalog.

A nominal service life can then be calculated in hours as:

$$L_{10} = \frac{10^6}{60 \cdot n} \left(\frac{C}{P} \right)^3 \tag{52}$$

In addition to the load and speed rating of the bearing, there are many other parameters that need to be thoroughly considered and discussed with the contact-type bearing vendors. These parameters include ball and race materials, cage types, heat treatment of the races, preload for the bearing, lubricant types, relubrication methods, etc. After everything is finalized, it is strongly suggested to work closely with the bearing vendor to make sure everything is correct and will work as intended.

Besides bearing selection, there are many techniques widely used during flywheel rotor-bearing architecture design for contact-type bearing systems. The most commonly used method is to support the flywheel vertically, and use a set of permanent magnet rings on the housing to generate magnetic attraction forces on upper surface or the rotor or flywheel hub. This method is designed to reduce a major part of the gravity load for the flywheel, from the contact-bearing surface. It can be easily seen from Eq. (52) and by instinct that, by reducing the load on bearings, the bearing lives can be greatly improved.

Another technique is to use the outer race of the bearings to support the rotor and inner race for the stationary support. This will equivalently reduce the DN number for the bearing and help alleviate the technical challenge when the shaft diameter is high, but a higher running speed is also required.

One final factor that needs to be addressed is bearing damping. It could be easily forgotten that contact-type bearings normally have very high stiffness but very low damping. This could mean vibration problems and reliability issues when the actual physical prototype is built and tested. Even though this vibration concern might not happen at all, but once it happens, it could cause devastating issues, and is worth developing a well-thought out plan during the design phase.

Magnetic bearings

There are two major type of contactless bearings used for high-speed rotating machineries: gas bearings and magnetic bearings. Going contactless has many benefits. First, there will be less friction and friction-related loss. Second, the line speed at the bearing surface can be much higher than contact-type bearings. This means the flywheel can theoretically spin much faster than on contact-type bearings. Third, the load capacity of the contactless bearings, especially magnetic bearings, are nearly independent of journal speed. Unlike contact bearings, the journal speed normally does not need to be derated for very high-speed flywheels. Fourth, the contactless bearings normally require much less maintenance and has longer life than their contact counterparts. Finally, the contactless bearings normally run much quieter than the contact ones.

For flywheel application, the usage of gas bearings is very limited, as most of the flywheels are operating in vacuum and gas bearings needs "gas" to operate. Thus magnetic bearings provide the major option if contactless bearing solutions are needed for a flywheel system. For magnetic bearings, many types exist and three types are commonly used in flywheel industry and academic research fields: passive magnetic bearings, active magnetic bearings, and superconductive magnetic bearings. The superconductive bearing requires superconducting material, which has a premium cost and has stringent requirements on its running conditions. This makes the superconductive magnetic bearing less suitable for commercial use. The remainder of this section will focus on passive magnetic bearings and active magnetic bearings.

Passive magnetic bearings use passive magnetic forces to support the journal without any active controls. Due to the instability of attractive magnetic force (the attraction force increases as the two attracting objects move closer), passive magnetic bearings normally utilize repulsive forces to generate the bearing force. The major advantages of the passive magnetic bearings are their simplicities. There are no active controls involved and the bearing itself is naturally stable. However, without the control system, the bearing characteristics are fixed by design, as compared to the flexibilities and possibilities in bearing characteristics that can be provided by adjusting the controls for the active magnetic bearing counterpart.

Another drawback of a passive magnetic bearing PMB is its relatively low load capacity. Almost all flywheels use passive magnetic bearings to reduce the gravity-bearing loads. The low load capacity means that the PMB journal needs to be very long for a heavy flywheel. The extra bearing length will reduce the "effective" working length of the rotor and reduce the rotordynamic margin. In addition, passive bearing-supported flywheels generally have very little mobility, due to the combination of a weak bearing and a heavy rotor. Extra procedures are needed during transportation to protect the bearings and rotor.

Another aspect to consider for passive magnetic bearings is rotordynamic characteristics. PMBs have almost zero damping, which could be a problem if there are external disturbances since there is almost no way to dissipate his energy. Finally, the passive bearings cannot be used to stabilize all degree of freedoms for the rotor. As pointed out by Larsonneur in [52], p. 28, Chapter 2, based on Earnshaw's theorem, when purely supported by permanent magnets, there is always at least one unstable degree of freedom. Flywheel systems supported by PMBs always have one degree of freedom restrained by another type of bearings, it could be an active axial magnetic bearing, or needle axial bearing with very small, round tips contacting a fixed flat surface, or by other mechanical methods.

On the contrary, active magnetic bearings use attractive magnetic force to generate their bearing reactions. A simple illustration to explain the working principles is shown in Fig. 48. Based on the controller outputs, the bearing control coils will create unbalanced electromagnetic fields on each side of the rotor

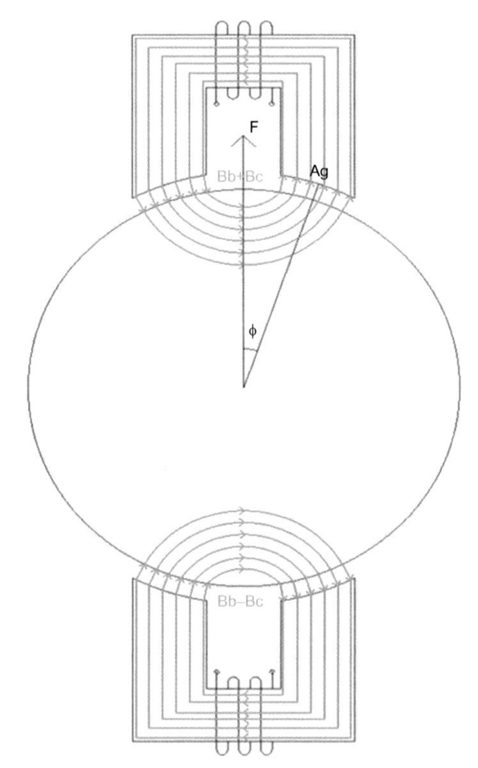

FIG. 48 Illustration for active magnetic bearing working principles.

surface. The unbalanced field will generate different attraction forces at opposite ends of the rotor. Assuming the bearing pole width is relatively small and the air gap nearly constant, the approximate net control force can be written as in Eq. (53).

$$F = \frac{4B_b B_c}{\mu_o} A_g \cos(\varnothing) = \frac{4B_b A_g \cos(\varnothing)}{\mu_o} C_{Bc} N I_c = K_I N I_c \tag{53}$$

where F is the bearing force created by the control coil current I_c, B_b is the bias flux density, B_c is the control flux density, A_g is the pole gap, $\varnothing$ is the pole angle, μ_o is the permeability in free space, C_{Bc} is the magnetic circuit constant function relating the control coil current I_c to the control flux density B_c, and K_I is the effective bearing current stiffness.

Multiple observations can be drawn from Eq. (53): First, the magnetic bearing can be linearized after introducing the bias flux density B_b. Second, the bearing current stiffness is proportional to B_b for any given bearing design. Third, as the maximum flux density is limited by saturation on certain cross sections, there is a constant maximum value for the total air gap flux density: $max(B_b + B_c)$. Considering the maximum bearing force F is proportional to max($B_b B_c$), theoretically there exist an optimum value of B_b for the bearing to generate maximum load capability. This optimum value will be $B_{sat}/2$, where B_{sat} is the maximum flux density in air for a given design, which happens when any section on the magnetic circuit path gets saturated. It must be noted here that, for permanent magnet biased magnetic bearings, the Bb value will be set to a value higher than $B_{sat}/2$, to avoid demagnetization for the permanent magnets, especially if the operating temperature is high for the bearings.

To obtain a rough estimate on how much load a magnetic bearing can support, imagine a flat object under a bearing pole where the magnetic field on the top is saturated and on the bottom is zero. In this scenario, the magnetic force generated on the object will be the theoretical maximum for a fixed bearing pole area. The maximum force per pole area can be expressed as

$$\frac{F_{max}}{A_p} = \frac{B_{sat}^2}{2\mu_o} \tag{54}$$

Commonly used M-19 material has a saturation flux density of approximately 2 T. Hence the theoretical load capacity limitation with M-19 will be 1.57 MPa (228 psi). The best magnetic bearing laminations can be made from Iron Cobalt Alloys, like Hiperco 50 with 2.4-T saturation. For these bearings the theoretical load capacity limitation can be as high as 2.29 MPa (332 psi). It must be noted that the bearing load density will be quite lower than these approximate theoretical values.

The frequency dependent force of an active magnetic bearing pole is:

$$f = \frac{1}{2\mu_0} \oint \left(B_{pm} + B_i\right)^2 dA \tag{55}$$

where B_i is the frequency-dependent control flux. The frequency-dependent component of the force is

$$\begin{aligned} f_i(\omega) &\approx \frac{B_{pm}}{\mu_0} \oint B_i(\omega) dA \\ \frac{f_i(\omega)}{f(0)} &\approx \frac{\phi(\omega)}{\phi(0)} \end{aligned} \tag{56}$$

where ϕ is the magnetic flux flowing across the air gap. The frequency response of the force to current can therefore be approximated by the frequency response of the control flux ($\phi_i(\omega)$) to the excitation current or the average control flux density ($\overline{B}_i(\omega)$) to the excitation current. Thus the dynamic component of force will rolloff with frequency since the flux density will rolloff with frequency, due to eddy currents. The dynamic force also experiences an increasing phase lag for the same reason at high frequencies [57]. The induced eddy currents on the magnetic bearing create an opposing magnetic field to the excitation. The force magnitude reduction and phase lag are dependent on the materials, heat treatment, lamination thickness, etc.

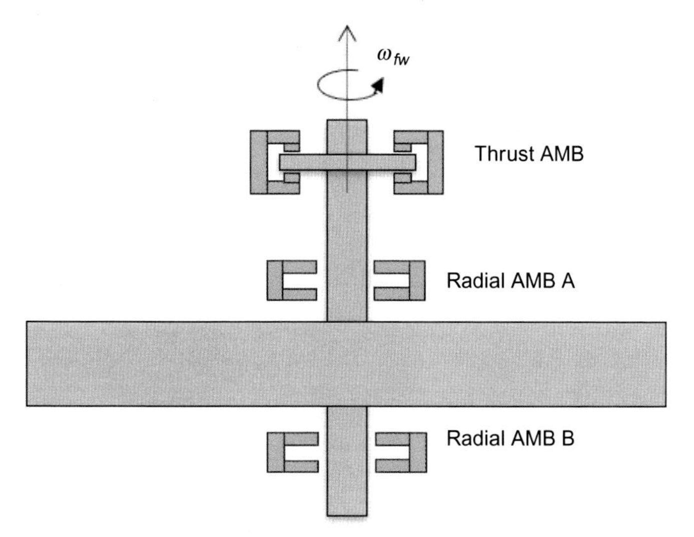

FIG. 49 Typical layout for active magnetic bearing-supported flywheels.

A typical layout of a flywheel supported by active magnetic bearings is illustrated in Fig. 49. Normally a complete system requires two radial bearings at each side of the rotor and at least one axial bearing handles the thrust loads. For some flywheel vendors, for example, Vycon Inc., there are also cases where a radial bearing and a thrust bearing are combined into one integral combo bearing. The topology of the magnetic bearing can be either electrically biased or permanent magnet biased, depending on the bias magnetic flux generation method. The electrically biased magnetic bearings are normally heteropolar, for which the bias magnetic flux polarity will change along the rotor circumference. The PM biased magnetic bearings are normally homopolar, for which the bias magnetic flux polarity will be the same along the rotor circumference. Homopolar designs will generate less eddy currents and associated heating of the rotor. Considering the fact that the bias flux is generated without bias coil currents in permanent magnetic bias bearings, the PM-biased magnetic bearing will generally run much cooler and energy efficient than their electrically biased counterparts.

Mechanical bearings are seldom used to support a flywheel solely. However, some mechanical bearings, such as ball bearings, are used as a part of a hybrid bearing system together with magnetic bearings [54]. Another typical hybrid bearing system is a combination of passive and active magnetic bearing. Ref. [55] presents a combination of a passive magnetic bearing and an active radial magnetic bearing to support a flywheel in the axial direction and to provide stiffness for tilting motion.

Rotordynamics

Flywheel systems store kinetic energy by spinning the flywheel, supported by bearings, to high speeds. This requires careful rotordynamic analysis to avoid resonances and instabilities, as well as to maintain control. There are various details specifically for flywheel systems that need to be addressed when performing rotordynamics studies. Fig. 50 is illustrates a mode shape of a representative rotordynamic model for an energy storage flywheel.

The entire rotor needs to be modeled with high detail, accurate material properties, and correct dimensions. Stress analysis should be performed, especially at stress concentrations near abrupt diameter changes. As the flywheel systems are normally designed to be axisymmetric, a 2D rotordynamic FEA model should be sufficient to study the rotor behaviors, even though 3D FEA tools such as NASTRAN or ANSYS could also be used.

After the rotordynamic model is setup and checked, the first step will be checking the free-free modes of the rotor. Due to the nature of flywheel applications, the spin speed will regularly vary during charge and discharge cycles. Due to this operating behavior most flywheel systems are designed to operate under their 1st forward bending frequency. The maximum operating speed is

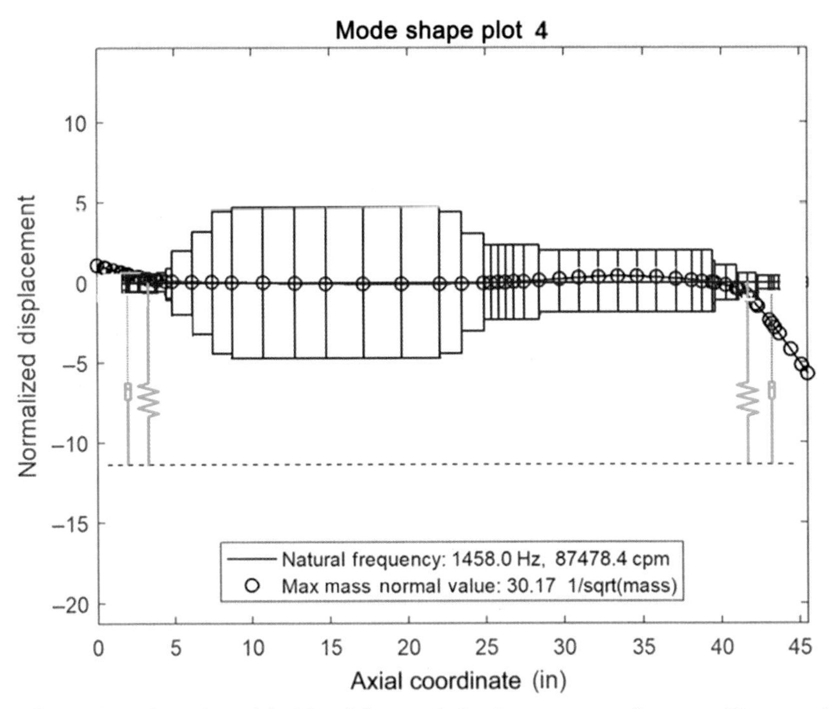

FIG. 50 Rotordynamic model with radial magnetic bearing actuators and sensors. *(Courtesy of Vycon Inc.)*

much stay lower than the first forward bending modes of the rotor with sufficient safety margin. An undamped critical speed map aids the engineer to decide the optimum bearing stiffness and guide the selection of bearing configurations.

After the rotor design is fixed and bearing is selected, it is strongly suggested to run unbalance response simulations based on the bearing characteristics. The ISO1940 standard will regulate how much unbalance weight to use in simulations for various balancing grades, which also needs to be selected by the designer. Of course, a better balancing grade will always benefit the vibrational behavior of the final product. However, commercial costs and manufacturability will limit how well the rotor assembly can be balanced. A balancing grade of G2.5 or better is normally suggested. API 617 will provide a good reference on where to put the unbalance weights and also give guidelines on how to carry out the unbalance response analysis. After the unbalance response simulation is finished, it is always a good idea to provide the loads to the bearing vendor, especially in system using contact-type bearings, to have a detailed analysis on their impact on bearings.

4.2.3 Electrical design

4.2.3.1 Conventional motor/generator design

Bahar Anvari

Motor types and function

An electrical motor/generator M/G is required for charging and discharging the stored kinetic energy in the flywheel. Induction, permanent magnet, and synchronous variable reluctance machines are commonly used for flywheel energy storage applications [56]. Induction machines are usually cheaper and simpler for manufacturing and they can be constructed from high-strength materials. Since they do not use any permanent magnets, there is no concern for demagnetization and loss when the flywheel is spinning in vacuum. However due to lower power density, limited speed, and high rotor temperature they are not the most favorable candidate for extra high-speed applications. They also have poor power factor when there is a light load, high magnetizing inrush current, and the speed control of an induction machine is more difficult compared to permanent magnet and synchronous variable reluctance machines.

Synchronous variable reluctance machines have low loss at starting, they have high efficiency, high power density, and dissipate heat relatively easy. They do not use permanent magnets, so there is no concern for demagnetization and loss when the flywheel is spinning in a vacuum. The rotors can be manufactured with high-strength and low-cost materials. However, they have complex structural features, are difficult to manufacture, have low power factor, have torque ripple and vibration and noise, and it is difficult to regulate their speed. They also have higher manufacturing cost compared to induction machines.

Permanent magnet synchronous machines have higher power density, efficiency, torque density, power factor with range of load and speed, load density, higher speed capability. However, they have high cost due to using magnets, risk of demagnetization, and challenging thermal management issues.

Some unconventional machines designs are also have been investigated for flywheel energy storage applications [57]. Bearing-less homopolar ac machines, axial flux permanent magnet machines, multiphase electric machine, Halbach array permanent magnet array machines [57]. A coreless permanent magnet machine for magnetically levitated shaftless flywheel energy storage application is presented in [58]. This paper presents the use of a coreless, or ironless, stator and a magnetic bearing levitated flywheel. Thus, the winding can be placed on a stationary structure. The proposed design is identical compared to the existing coreless machine for operation in vacuum and has the holding structure on the ground. Fixing the windings onto the base through holding structures located on the ground has high strength, reliability, and simple mechanical design in terms of manufacturing and maintenance.

The power, speed, and torque requirements for flywheel energy storage application is illustrated in Fig. 51. The rated speed is ω_1, the rated torque is T_{rated}, and the rated power is P_{rated}. The required time for the flywheel to be charged up to the rated speed in the motoring mode is t_1. During this period, the flywheel input power and speed are increased to the rated power and rated speed. During t_1 to an arbitrary time t_2, the flywheel spins with the minimum power and torque, indicated by P_1 and T_1, respectively, in the vacuum at the constant rated speed. This power depends on the motor type, loss, and the effectiveness of the pump that is required for creating the vacuum.

The flywheel operates in the generating mode of operation during t_2 to t_3 and is discharged with constant power while its speed is decreasing from rated value down to ω_2 rpm. Time intervals t_3–t_2 can be easily found from the rotational kinetic energy ΔE and the relation between power and energy as follows:

$$\Delta E = \frac{1}{2} J \left({\omega_1}^2 - {\omega_2}^2 \right) \tag{57}$$

$$P_{average} = \frac{\Delta E}{(t_3 - t_2)} \tag{58}$$

where J is the moment of inertia, ω_1 and ω_2 represent the initial and final speeds of the flywheel, respectively. In an ideal case, the required time for the flywheel to get to ω_2, with the average power of $P_{average}$ is t_1. The generator should supply constant power of P_{rated} during t_2 and t_3. The electric motor/generator should generate P_{rated} constant power while the speed is decreased from ω_1 and ω_2 rpm. Time t_2 can be any arbitrary time according to energy storage requirement. Additionally, there might be some limitations on t_2 due to the operation of the flywheel in vacuum condition.

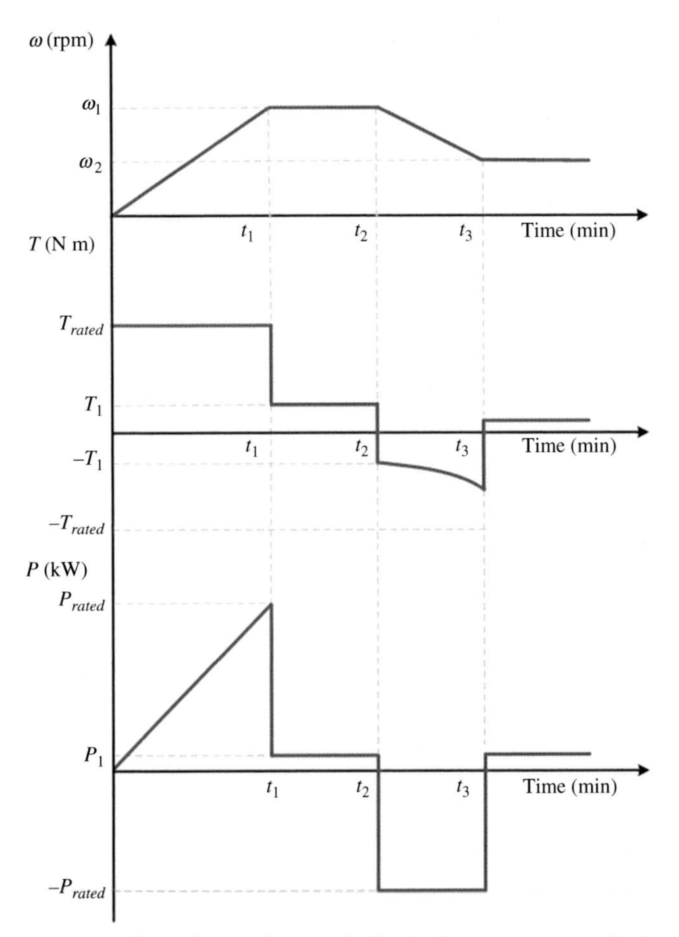

FIG. 51 Electric machine design requirements for flywheel energy storage application.

Motor/generator control

A bidirectional power converter is required given any electric machine design for energy conversion in a flywheel energy storage system. The power electronic converter topologies are DC-AC, AC-AC, and AC-DC-AC. The most common power electronics configuration is the back-to-back or AC-DC-AC configuration, connected to a DC link capacitor, which includes three-phase bridged semiconductor switches that are often controlled by the pulse-width modulation technique. Field-oriented control and direct torque control are two common control strategies used for induction machines. Both techniques control the machine flux and torque to track command values regardless of any disturbances. For synchronous machines, field-oriented control using space

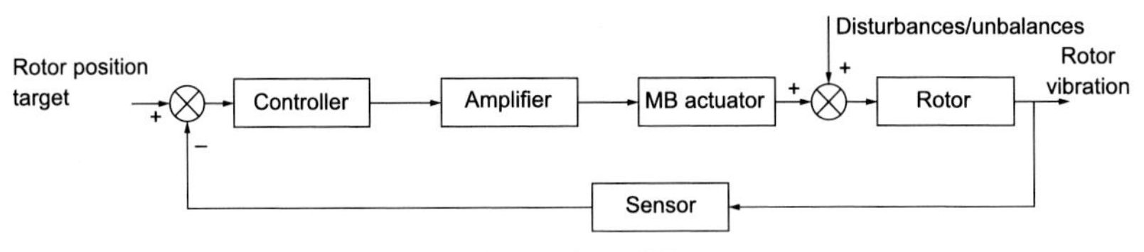

FIG. 52 Closed-loop control for magnetic bearings [59].

vector pulse width modulation is used for motoring modes of operation. Field-oriented control enables controlling of the current and hence the torque and flux of the machine. In the discharge mode of operation, a voltage controller is required to adjust the voltage drop on the DC bus and control the current.

Magnetic bearing control

Zhiyang Wang, Xiaojun Li and Alan Palazzolo

As discussed in previous sections, active magnetic bearings rely on attractive magnetic forces to generate bearing forces. As the attractive magnetic force will increase as two objects get closer, this mechanism is unstable by nature. A closed-loop control system needs to be in place to stabilize the active magnetic bearing support system. A basic architecture for the closed-loop control system is shown in Fig. 52. The sensors measure rotor displacements at the bearing locations. The displacement information is then compared to the target rotor position settings (normally 0). Then the error is sent to the controller to process. The power amplifier will convert the controller outputs into corresponding current excitation to energize the actuator and generate bearing forces on the rotor. The controller can normally be Single Input Single Output (SISO), which means each magnetic bearing channel will act independently. However, for dynamically complicated systems gyroscopic and other effects need to be compensated. In these instances a Multiple Input Multiple Output (MIMO) control system may be required, where multiple magnetic bearing axes become coupled.

A simplified equation of motion (EOM) for a flywheel modeled as a rigid body is given by

$$\boldsymbol{M}_f \ddot{\boldsymbol{q}}_f + \boldsymbol{G}_f(\omega)\dot{\boldsymbol{q}}_f = \boldsymbol{f}_{em} + \boldsymbol{f}_{im} \tag{59}$$

in which $\boldsymbol{q}_f = [\theta_{x,f}, \theta_{y,f}, x_f, y_f, z_f]^T$ denotes the flywheel position vector. The mass imbalance force is defined as $\boldsymbol{f}_{im}$, and $\boldsymbol{f}_{em}$ is the magnetic control force. The term $\boldsymbol{G}_f(\omega)$ denotes the speed-dependent gyroscopic matrix:

$$\boldsymbol{G}_f = \begin{bmatrix} 0 & I_p\omega & 0 & 0 & 0 \\ -I_p\omega & 0 & 0 & 0 & 0 \\ 0 & 0 & 0 & 0 & 0 \\ 0 & 0 & 0 & 0 & 0 \\ 0 & 0 & 0 & 0 & 0 \end{bmatrix} \tag{60}$$

The gyroscopic effect generates pairs of forward and backward conical modes. As the spin speed increases, the rigid body forward mode will converge linearly with the spin speed ($\omega I_p/I_t$), as the backward mode converges to zero. This effect is more evident for flywheels with high I_p/I_t ratio. Gyroscopic effect alone positions the poles on the imaginary axis, which cause the system to be marginally stable. However, phase lag from the AMB, amplifiers, or other parts of the system destabilizes the modes. While a single-input- single-output (SISO) control algorithm will levitate the flywheel, it may not be able to stabilize the system when the flywheel is rotating. Multiple-input-multiple-output (MIMO) algorithms can be used more effectively for this purpose. For instance a MIMO PD controller may utilize proportional and derivative gain matrices of the form

$$\boldsymbol{P} = \begin{bmatrix} P_x & & & & \\ & P_y & & & \\ & & P_{\theta x} & P_{\theta c} & \\ & & -P_{\theta c} & P_{\theta y} & \\ & & & & P_z \end{bmatrix} \quad (61)$$

$$\boldsymbol{D} = \begin{bmatrix} D_x & & & & \\ & D_y & & & \\ & & D_{\theta x} & -D_{\theta c} & \\ & & D_{\theta c} & D_{\theta y} & \\ & & & & D_z \end{bmatrix} \quad (62)$$

The crosscoupled feedback applies torque in one plane per the angular position and velocity in its quadrature plane. This strategy can reduce the speed dependency of system dynamics [62]. The following simulation results illustrates how MIMO control is useful for suppressing vibration of a high I_p/I_t ratio flywheel. The with MIMO control results in Fig. 53 shows an at–minimum fourfold reduction in rotor vibration by including a MIMO controller.

The control laws most commonly used in commercialized flywheel systems are traditional PIDs, even though there are efforts in improving the control robustness and performance by applying modern control theories like H-infinity. For PID control, the proportional and integral term will determine the bearing stiffness and the derivative term will affect the bearing damping. Lead-lag compensators are used to improve the gain and phase margin for stability of closed-loop system, and to boost bearing stiffness and damping in specific frequency ranges.

Due to the large overall weight of flywheel rotors, even the best balancing instruments and efforts may leave relatively large unbalance residues on the flywheel rotors.

These unbalances, together with the flywheel's high spinning speed, could impose large bearing loads on contact bearings and magnetic bearings. The large bearing loads could reduce the bearing life for contact bearings and

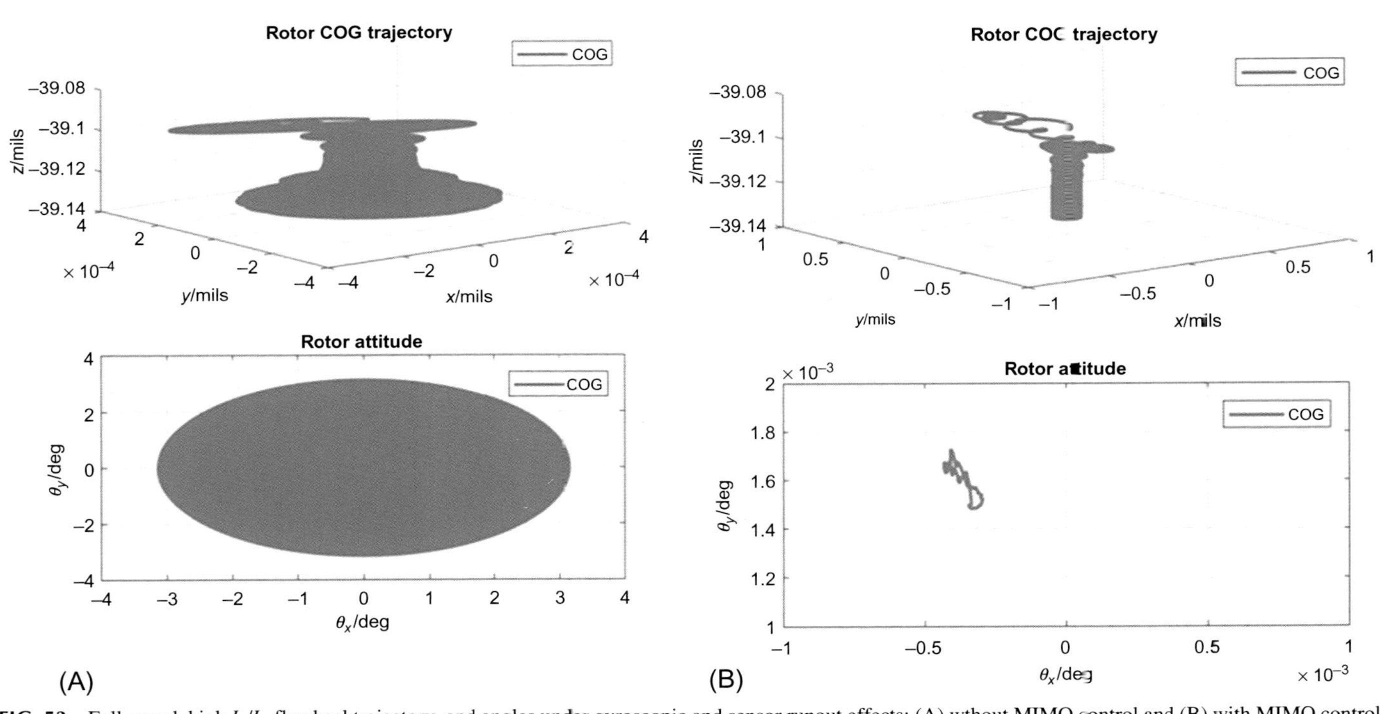

FIG. 53 Full-speed, high I_p/I_t, flywheel trajectory, and angles under gyroscopic and sensor runout effects: (A) wthout MIMO control and (B) with MIMO control.

increase the loss for magnetic bearings. Further, the loads transmitted through the bearings to the housing could create large vibrations and noise in the surrounding environment. This may limit the application scopes of the system and reduce the reliability for the supporting structures. To solve this problem, most magnetic bearing providers/designers will include synchronous cancellation in their bearing control algorithm. By implementing synchronous cancellation, the magnetic bearing controller will demodulate the synchronous displacement information for the rotor and reduce the bearing control efforts at the synchronous frequency. This will effectively remove the magnetic bearing reaction to the unbalance load of the flywheel. As discussed in previous sections, most flywheels run below their first forward bending modal frequency. This means, by integrating synchronous cancellation and removing bearing reaction to the unbalance load, the rotor spinning center will tend to shift from its geometry center to the mass center and become rotordynamically more stable. In addition, as there is no bearing loads coming from rotor unbalance, the flywheels can run much quieter and introduce very little support system vibrations. Therefore, this feature is also one of the most important advantages of magnetic bearing over traditional contact bearings, besides their long lives and low loss.

To carry out synchronous cancellation, rotor synchronous vibration data first need to be demodulated. There are many demodulation methods to retrieve the amplitude and phase information. Frequency domain analysis, or FFT, could be one of them and the simplest to implement. However, for FFT analysis, the frequency resolution will equal to the sampling frequency divided by the FFT size. It means that, to have a 0.1-Hz (6 RPM) resolution, 10 s of data will be needed. This might not work for many applications due to the accuracy, resolution, and calculation speed requirements for synchronous cancellation.

Another very useful method will be adopting a commonly used signal demodulation theory in communication industry, which is based on the heterodyne principle as illustrated by

$$\cos(\omega_1 t) * \cos(\omega_2 t) = \frac{1}{2}(\cos((\omega_1 + \omega_2)t) + \cos((\omega_1 - \omega_2)t) \tag{63}$$

Since the unbalance response stays constant, we treat the synchronous vibration as a complex constant, namely, unbalance response vector, modulated by rotational frequency. Therefore, this constant can be recovered by multiplying a signal with exactly the same carrier frequency (rotating speed) in the time domain. The results, after low-pass filtering, will be the constant with its amplitude reduced by one-half [60]. The synchronous vibration X_{sync} can be written as

$$X_{sync} = \left|X_{sync}\right| e^{i\left(\Omega_{sync} t + \varphi\right)} \tag{64}$$

and if the orbit is circular, and the demodulation is applied to one orthogonal axis, the measured value X_{sync_meas} becomes

$$X_{sync\_meas} = |X_{sync}|\left(\cos(\varphi)\cos(\Omega_{sync}t) - \sin(\varphi)\sin(\Omega_{sync}t)\right), \tag{65}$$

where Ω_{sync} is the rotor synchronous frequency, $|X_{sync}|$ is the amplitude for the synchronous vibration, and ϕ is the phase angle for the synchronous vibration, referenced to the key phasor of the system. Demodulating with signals at same synchronous frequency Ω_{sync}, yields

$$X_{sync\_meas}\cos(\Omega_{sync}t) = \frac{|X_{sync}|}{2}\left(\cos(\varphi)\cos(2\Omega_{sync}t) - \sin(\varphi)\sin(2\Omega_{sync}t)\right) + \frac{|X_{sync}|}{2}\cos(\varphi) \tag{66}$$

$$X_{sync\_meas}\sin(\Omega_{sync}t) = \frac{|X_{sync}|}{2}\left(\cos(\varphi)\sin(2\Omega_{sync}t) + \sin(\varphi)\cos(2\Omega_{sync}t)\right) - \frac{|X_{sync}|}{2}\sin(\varphi) \tag{67}$$

The following process is a standard method for demodulation defined for communication signals: By filtering out the $2\Omega_{sync}$ component (twice of modulation frequency), the original signal amplitude and phase are recovered:

$$|X_{sync}|\cos(\varphi) = LowPassFilter\{2|X_{sync}|\cos(\Omega_{sync}t)\} \tag{68}$$

$$|X_{sync}|\sin(\varphi) = LowPassFilter\{2|X_{sync}|\sin(\Omega_{sync}t)\} \tag{69}$$

After the synchronous vibration information is demodulated, the magnetic bearing controller injects the cancellation signal into the control loop. The essence of synchronous cancellation is generating zero responses to the synchronous or $1\times$ components of rotor displacement. There are multiple ways of achieving this goal. However, one of the easiest and most straightforward methods is to minimize the control current at the synchronous frequency. Without control currents at the synchronous frequency, there will be effectively no bearing response to the unbalances of the rotor.

To achieve this goal, a signal is injected at the controller's input to block out the position error at the synchronous frequency, or at the controller's output to block the input to the amplifier at the synchronous frequency. This is illustrated in Fig. 54.

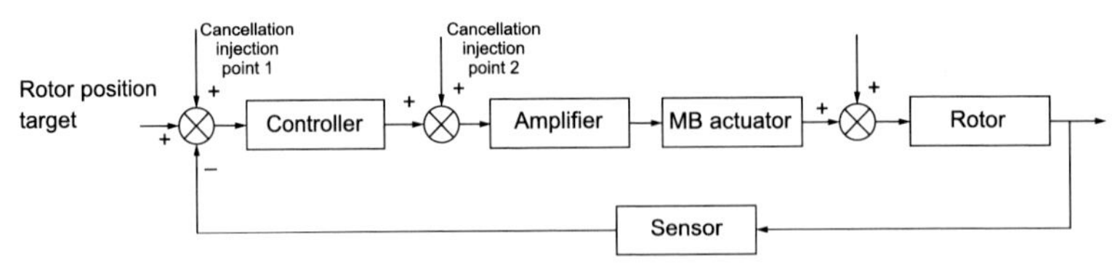

FIG. 54 Synchronous cancellation injection points.

The first method will reduce the control error generated between rotor position control target (normally 0) and the synchronous motion of the motor. The second method will minimize the synchronous control current by removing the corresponding commands. To avoid instability by introducing the synchronous cancellation, the cancellation effort will be introduced gradually using

$$u_{sc}(t_{n+1}) = u_{sc}(t_{n+1}) + \alpha T_{sc} X_{sync} \tag{70}$$

during actual implementation, where u_{sc} is the synchronous cancellation signal, T_{sc} is the affection matrix, α is a factor determining the rate of the algorithm converging. A higher α value will speed up the converging process of synchronous cancellation but might introduce an instability. Its value normally stays below 1.

For method 1 (injection at point 1), the T_{sc} is taken as

$$T_{sc} = \left(\frac{1}{1 + G_{OL}}\right)^{-1} = (G_{sen})^{-1} \tag{71}$$

For method 2 (injection at point 2), T_{sc} is taken as

$$T_{sc} = (G_{sen})^{-1} G_{ctrl}, \tag{72}$$

where G_{OL} is the open-loop transfer function, G_{sen} is the closed-loop sensitivity transfer function, and G_{ctrl} is the controller transfer function.

Bearingless motors for flywheel energy storage

Eric Severson

Introduction Bearingless motors combine the functionality of an active magnetic bearing AMB and an electric motor into a single electric machine. Electric motors inherently feature a large normal force density acting on the surface of their rotor. This normal force density is nearly an order of magnitude larger than the tangential force density used to create torque. However, in a standard electric motor, the machine designer goes to great length to ensure that the design is symmetric in a manner that prevents a net force from acting on the rotor. The premise of a bearingless motor is to disrupt this symmetry and create controllable forces on the rotor. These forces allow the electric machine to be simultaneously operated as a magnetic bearing and as an electric motor.

Integrating magnetic bearing capability into an electric motor offers several critical benefits that are especially relevant for flywheel energy storage. These benefits come from reductions in raw material and overall electric machinery size, which have the potential to significantly reduce the cost of flywheel storage modules and significantly increase the module energy density. From the magnetic bearing perspective, a bearingless motor can be understood as replacing the magnetic bearing's bias field with the magnetizing field of the motor. Compared to magnetic bearings that use current to create the bias field, the bearingless motor further offers a possibility of an efficiency increase. In applications that require a significant power rating, which is typically true of flywheel storage modules, the dimensions of the bearingless motor are driven by the motor design, which means that the magnetic suspension design is inherently overdimensioned. This results in a large safety factor and enables bearingless motor concepts, which are capable of creating large suspension forces with an insignificant penalty to the motor performance.

Various concepts of bearingless motors are described throughout the research literature. These variations can be categorized, based on the degrees of freedom that the bearingless motor is able to control (axial, radial, and/or tilting) and whether forces are actively actuated [25], rely on static passive forces in certain degrees of freedom [61], or use electrodynamic principles [63]. The most common form of a bearingless motor is able to produce radial $\hat{x}$ and $\hat{y}$ forces on the rotor but relies on external bearing support for the remaining degrees of freedom. An illustration of such a system is shown in Fig. 55. Further technology divisions can be distinguished based on the dominant mechanism of creating forces: Maxwell (or reluctance) versus Lorentz forces.

Nearly any type of electric motor can be transformed into a bearingless motor that is capable of producing controllable radial forces by adding a suspension winding to the stator. These radial force producing motors can be categorized as either $p_s = p \pm 1$ or $p_s = 1$, where p is the number of pole pairs of the electric motor and p_s is the number of pole pairs in the suspension winding. Bearingless motors that fall into the $p_s = p \pm 1$ category include bearingless versions of conventional synchronous machines and asynchronous machines. This category of machine requires accurate knowledge of the rotor's angular position within the suspension controller to produce radial force vectors. The $p_s = 1$ category includes the bearingless ac homopolar and consequent pole motors. This category has the distinct advantage of being able to produce radial force vectors without knowledge of the rotor's angular position. This benefit must be weighed against restrictions on the motor design: the motor must have at least eight poles and the types of motors that can be used are typically lower performance. These trades are presented in Table 1.

Historically, the stator winding of radial force producing bearingless motors has been implemented as two separate windings: (1) the traditional motor winding used to produce torque and (2) a new winding used to create the magnetic

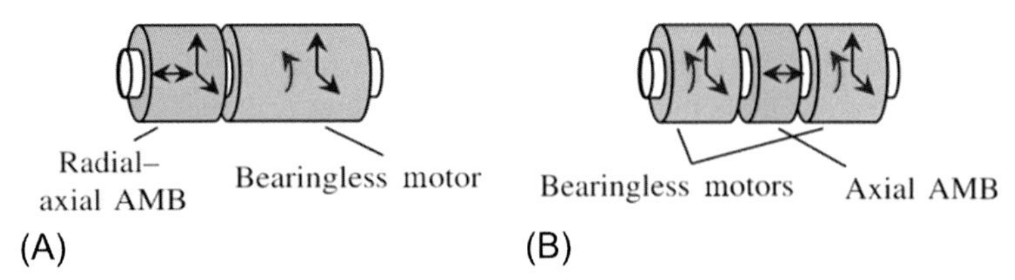

FIG. 55 Example bearingless motor system topologies.

TABLE 1 Comparison of radial force producing bearingless motor categories.

	Category 1	Category 2
Types of motors	Any synchronous or asynchronous motor	AC homopolar or consequent pole motor
Motor pole pairs p	No restriction	$p \geq 4$
Suspension pole pairs p_s	$p_s = p \pm 1$	$p_s = 1$
Rotor angle needed for suspension forces?	Yes	No
Force rating scaling law	$\sigma_s V_R \left(\frac{1}{p_s \delta_{ef}} \pm \frac{1}{r}\right)$	AC homopolar motor: $V_R\left(2\left[\delta_1 + \frac{1}{r}\right]\sigma_{s0} + \delta_2 \sigma_S\right)$ Consequent pole motor: $\sigma_S V_R \delta_{ef2}$
ϕ_k, see (ES17)	$\theta_m + \phi_0$	ϕ_0

suspension field. The coils of these windings reside in the same stator slots, which means that they compete for slot space. Any slot space that is allocated to the coils of the suspension winding is no longer available for use by the coils of the motor winding. The suspension winding needs to be dimensioned with a safety factor relative to a worst-case force scenario (such as passing through a critical speed or a large transient disturbance, i.e., an earthquake). In typical bearingless motors, 25%–40% of the slot space is allocated to the suspension coils. This has significant consequences for the motor performance, reducing torque density and increasing losses. Under normal operating conditions, bearingless motors require only a small fraction of the rated suspension current, meaning that the bearingless motor's slot space is significantly underutilized. This has led to recent research in developing "combined" windings, where the same motor coils are used for creating suspension forces and torque. In these combined windings, the bearingless motor drive is able to dynamically allocate slot space to either the motor operation or the magnetic suspension operation at runtime based on the real-time requirements of the bearingless motor. This approach is particularly effective in high inertia applications, such as flywheel energy storage, where a brief reduction in the torque capability of the motor (to react to a transient force disturbance) has minimal impact on the shaft speed.

Several studies have compared the performance implications of using a combined winding to using separate motor and suspension windings. The general conclusions are that the combined winding structure improves the motor

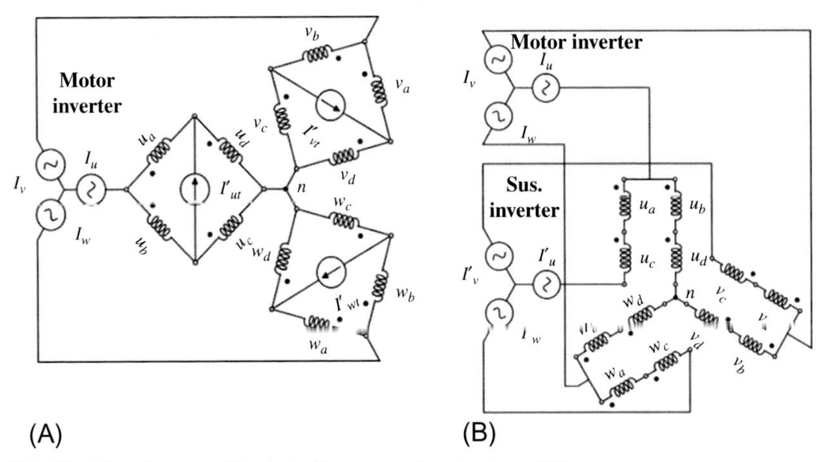

FIG. 56 No voltage combined winding connection structures [69].

performance (e.g., Ref. [64] shows that combined windings can decrease copper loss by 30%–40%), but that they create additional challenges for the power electronics in the form of higher device VA ratings and unconventional converter topologies [65,66]. There are several approaches to implementing combined windings, each with their own tradeoffs in terms of motor performance and power electronic requirements. The approaches include using multiphase windings, where five or six motor phases are connected to a multiphase inverter, for example [67,68]; no voltage combined windings [65,69], which can be further divided into bridge [70] and parallel [71] connection structures that utilize distinct motor and suspension inverters connected to separate terminals as shown in Fig. 56; a middle point injection approach [72]; and using several three-phase inverters connected to distinct sets of pole pairs distributed circumferentially around the stator [73].

Flywheel energy storage concepts based around bearingless motor technology include [74–77]. In [74], an outer-rotor ac homopolar concept is proposed,

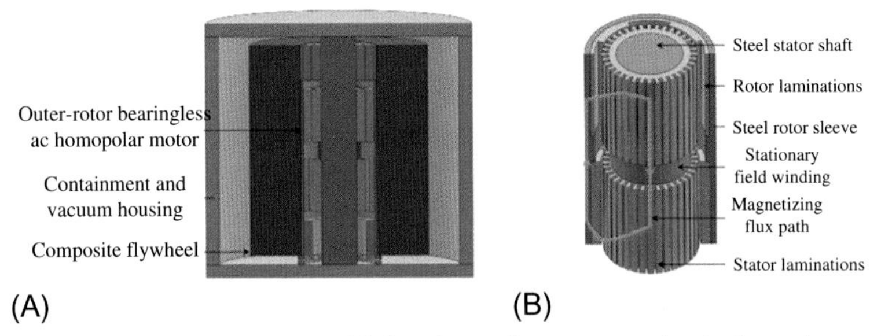

FIG. 57 Flywheel energy storage module based around an outer-rotor bearingless ac homopolar motor [74].

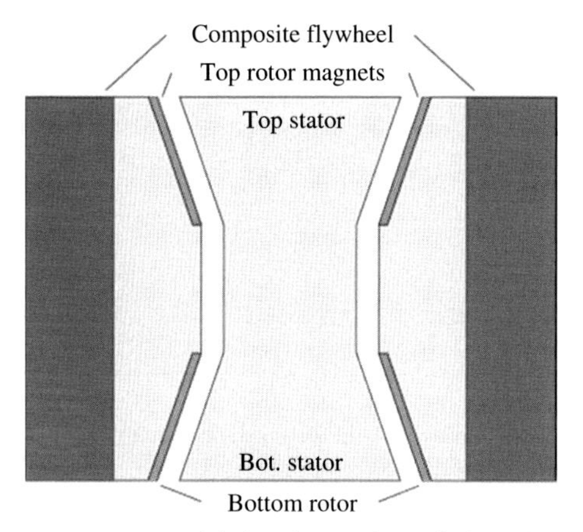

FIG. 58 Flywheel energy storage module based around a conical bearingless motor.

shown in Fig. 57, which is able to actively actuate four degrees of freedom (radial movement and tilting along two axes) and relies on a passive magnetic bearing to support the axial direction.

The ac homopolar motor relies on a field winding, which means that the magnetizing field can be greatly reduced when the flywheel module is not transferring power (idling or free-wheeling intervals) to nearly eliminate self-discharge. In [75], a flywheel system is proposed and tested, which relies on an outer-rotor permanent magnet motor with two cone shaped airgaps, depicted in Fig. 58. This machine is able to actively actuate all five degrees of freedom. The radial and tiling directions are actuated by controlling normal forces in the two airgaps independently and the axial direction is regulated by leveraging the cone-shaped airgap profile and controlling the magnetization state of each airgap through field weakening and field strengthening.

Radial force producing bearingless motors (of both the $p_s = p \pm 1$ and $p_s = 1$ type) are the primary technology category considered for flywheel energy storage. This section will now present the fundamental theory of operation, design sizing laws, and operating approach used for these types of bearingless motors.

Principle of operation The basic principle of creating radial forces is illustrated in Fig. 59, where a two-pole suspension winding has been added to a four-pole motor. The two-pole suspension field enhances the rotor's magnetic field on the right side of the airgap and weakens it on the left side. The Maxwell Stress Tensor can be used to show that the normal force density is proportional to the square of the airgap magnetic field. The normal force density is plotted in Fig. 59B with and without the suspension winding by assuming sinusoidal fields. The force density can be integrated to calculate the total force vector acting on the shaft. In this illustration, the two-pole suspension field has disrupted

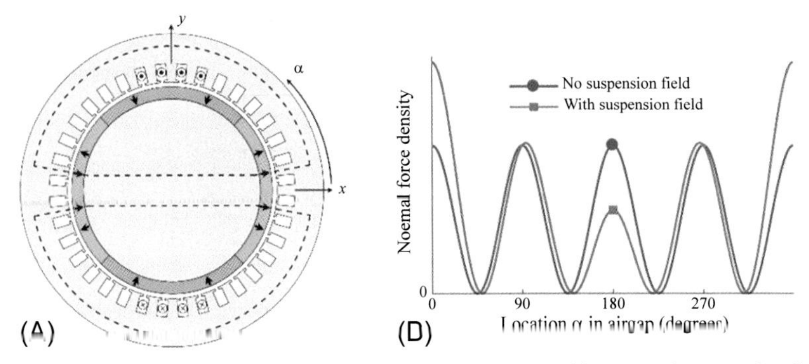

FIG. 59 Example four-pole bearingless permanent magnet motor with a two-pole suspension field ($p_s = p \pm 1$ category motor).

the symmetry of the four-pole magnetizing field to create a net force in the $\hat{x}$ direction, but no net force in the $\hat{y}$ direction.

Force expressions can be derived by using the Fourier Series to decompose the magnetic fields in the airgap of the machine into their harmonic components. The Maxwell Stress Tensor can then be used to determine the force density that is acting on the rotor as (73), where σ_{Fn} and σ_{Ftan} are the normal and tangential force densities, which in cylindrical coordinates act in the radial $\hat{r}$ and circumferential $\hat{\phi}$ directions; B_n and B_{tan} are the normal and tangential airgap fields, again in the radial and circumferential directions.

$$\vec{\sigma} = \begin{bmatrix} \sigma_{Fn} \\ \\ \sigma_{Ftan} \end{bmatrix} = \begin{bmatrix} \frac{1}{2\mu_0}\left(B_n^2 - B_{tan}^2\right) \\ \frac{1}{\mu_0} B_n B_{tan} \end{bmatrix} \tag{73}$$

The Maxwell Stress Tensor can be used to calculate the torque and radial $\hat{x}$ and $\hat{y}$ force components acting on the rotor by integrating (73) over the surface of the rotor, as (74) and (75).

$$T = \int_A \vec{r} \times \vec{\sigma}\, dA \tag{74}$$

$$\begin{bmatrix} F_x \\ F_y \end{bmatrix} = \begin{bmatrix} \int_A \vec{\sigma} \cdot \hat{x} dA \\ \int_A \vec{\sigma} \cdot \hat{y} dA \end{bmatrix} \tag{75}$$

It can be shown that only the airgap fields, which differ in harmonic index (number of pole-pairs) by one, will interact to produce magnetic forces [25].

Bearingless motor sizing laws

Sinusoidal airgap fields can be assumed as well as an equivalent linear current density on the inner bore of the stator to derive the machine sizing laws, as is done in [78], the results of which will now be summarized. This is an extension of the conventional approach used to derive the motor torque-sizing law. For motor-sizing analysis, the equivalent linear current density of (76) is assumed, where $\hat{A}$ is referred to as the electrical loading of the machine and is constrained by thermal considerations. For air-cooled machines, this is typically in the range of 30–80 kA/m, while higher values can be obtained for liquid-cooled machines [79]. The angle ψ of the equivalent current is determined by the phase angle of the stator currents. When (76) is referring to the motor winding, $\hat{A}$ is replaced by $\hat{A}_T$ and $n = p$. When (76) is referring to the suspension winding, $\hat{A}$ is replaced by $\hat{A}_S$ and $n = p_s$. This winding creates both normal and tangential airgap field components. In addition to this, the motor is assumed to have a magnetizing airgap field given by (77) in the normal direction, where $\hat{B}_\delta$ is referred to as the magnetic loading of the machine and is constrained by the saturation field of steel. The angle θ_m is the direct axis of the rotor, in electrical radians.

$$A(\alpha) = \hat{A}\sin(n\alpha - \psi) \tag{76}$$

$$B_\delta(\alpha) = \hat{B}_\delta\cos(p\alpha - \theta_m) \tag{77}$$

Using (74), the well-known machine sizing law (78) can be derived, where $\langle\sigma_{Ftan}\rangle$ is the average tangential force density on the surface of the rotor and V_R is the rotor's volume in an inner-rotor design (outer-rotor designs use the volume of the stator). This expression is useful in motor design because it poses the torque rating in terms of fundamental quantities of the electric machine: the volume of the rotor, the magnetic loading, and the electric loading.

$$T = 2V_R\langle\sigma_{Ftan}\rangle \tag{78}$$

$$\langle\sigma_{Ftan}\rangle = \frac{\hat{B}_\delta\hat{A}_T}{2} \tag{79}$$

$$V_R = \pi r^2 l \tag{80}$$

The bearingless motor $\hat{x}$ and $\hat{y}$ direction forces are calculated from (75) in this same manner for a $p_s = p \pm 1$ bearingless motor with a fixed effective airgap length of δ_{ef} as (81) when $\psi = \theta_m + \phi_s$ and r is the airgap radius. The effective airgap length δ_{ef} includes the physical airgap as well as the thickness of any magnets on the rotor surface. The choice of ψ indicates that the suspension currents have the same frequency as the motor currents. The angle ϕ_s is a phase angle of the currents and determines the direction of the forces produced. In (81), the $\pm$ terms are positive for $p_s = p + 1$ and negative for $p_s = p - 1$, indicating that the inherent force rating is higher for bearingless motors with $p_s = p + 1$. This is due to the Lorentz force terms contributing positively to the

Maxwell force terms in the $p_s = p + 1$ motor, as explained in [78]. The term σ_S does not have physical meaning in the same manner as $\langle \sigma_{Ftan} \rangle$, but it is defined in terms of the magnetic and electric loading in the same form in (82) to facilitate the design sizing analysis.

$$\begin{bmatrix} F_x \\ F_y \end{bmatrix} = \sigma_s V_R \left(\frac{1}{p_s \delta_{ef}} \pm \frac{1}{r} \right) \begin{bmatrix} \cos\phi_s \\ \pm\sin\phi_s \end{bmatrix} \tag{81}$$

$$\sigma_s = \frac{\hat{B}_\delta \hat{A}_s}{2} \tag{82}$$

Eq. (81) indicates that the force rating of the $p_s = p \pm 1$ bearingless motor scales in approximately the same manner as the torque rating. This means that changes to the volume of the rotor or the magnetic loading will scale both the torque and force ratings equivalently.

The sizing equations are now summarized for the $p_s = 1$ bearingless motors. This category of machine corresponds to the consequent pole motor and the ac homopolar motors with at least eight poles. In both of these machines, an effective airgap profile of (83) is used and radial forces are produced via a two-pole suspension field with $\psi = \phi_s$. This angle is independent of θ_m, which is one of the key advantages of this category of bearingless motors. The force expressions for the consequent pole motor are shown in (84) and the expressions for the ac homopolar motor are shown in (85). The ac homopolar motor contains a homopolar bias field, which is denoted as $\hat{B}_{\delta 0}$, which necessitates the definition of a second equivalent stress value as σ_{s0} in (86).

$$\delta_{ef}(\alpha) = \frac{1}{\delta_1 + \delta_2 \cos(p\alpha - \theta_m)} \tag{83}$$

$$\begin{bmatrix} F_x \\ F_y \end{bmatrix} = \sigma_s V_R \delta_2 \begin{bmatrix} \cos\phi_s \\ \sin\phi_s \end{bmatrix} \tag{84}$$

$$\begin{bmatrix} F_x \\ F_y \end{bmatrix} = V_R \left(2 \left[\delta_1 + \frac{1}{r} \right] \sigma_{s0} + \delta_2 \sigma_s \right) \begin{bmatrix} \cos\phi_s \\ \sin\phi_s \end{bmatrix} \tag{85}$$

$$\sigma_{s0} = \frac{\hat{B}_{\delta 0} \hat{A}_s}{2} \tag{86}$$

The results of the sizing analysis are summarized in Table 1. All types of bearingless motors scale with the volume of the rotor and the magnetizing field in approximately the same manner. These equations show that additional nuances exist between the motor types in terms of how the number of suspension poles, airgap length, and airgap radius impact the force ratings.

Operation of bearingless motors Details on the radial force regulator structure for different categories of bearingless motors can be found in [25]. This section will present a brief overview for three phase bearingless motors of both

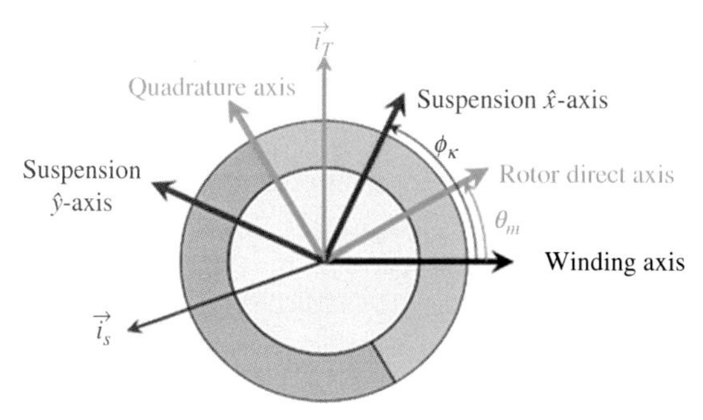

FIG. 60 Motor and suspension reference frames and current space vectors.

the $p_s = p \pm 1$ category and the $p_s = 1$ category based on space vector analysis. The motor current space vector $\vec{i}_T$ and suspension current space vector $\vec{i}_s$ for a three-phase motor system are defined as (87) and (88). The bearingless motor is assumed to have a separate motor winding (with phase currents of i_u, i_v, and i_w) and a separate suspension winding (with phase currents of i_{su}, i_{sv}, and i_{sw}). Through a reference frame change, the motor current space vector can be viewed as consisting of both direct and quadrature current components as $\vec{i}_T = (i_d + ji_q)e^{j\theta_m}$. These space vector quantities are depicted in Fig. 60 and form the basis of field-oriented motor control.

$$\vec{i}_T = i_u + i_v e^{j\frac{2\pi}{3}} + i_w e^{-j\frac{2\pi}{3}} \tag{87}$$

$$\vec{i}_s = i_{su} + i_{sv} e^{j\frac{2\pi}{3}} + i_{sw} e^{-j\frac{2\pi}{3}} \tag{88}$$

Similar to field-oriented motor control, a reference frame transformation can be used to construct equivalent $\hat{x}$ and $\hat{y}$ suspension current components. This transformation can be derived from (89), which maps the suspension current space vector to a radial force vector on the shaft through a force-current gain k_i and an angle offset ϕ_k, depicted in Fig. 60.

The value of the angle ϕ_k depends on the type of bearingless motor as described in (90). For $p_s = p \pm 1$ bearingless motors, ϕ_k rotates with the rotor's direct axis angle θ_m and ϕ_0 is a constant offset value representing an angular offset between the suspension winding and the motor winding. The dependency upon θ_m indicates that to produce a specified force vector, the force regulator needs to have an accurate measurement of θ_m. This is in contrast to $p_s = 1$ bearingless motors. In these motors, ϕ_k is stationary, depending only on ϕ_0, and therefore measurement of θ_m is not required for the magnetic suspension operation.

$$\vec{i}_s = \frac{\vec{F}}{k_i} e^{j\phi_k} \tag{89}$$

$$\phi_k = \begin{cases} \theta_m + \phi_0, & \text{for } p_s = p \pm 1 \\ \phi_0, & \text{for } p_s = 1 \end{cases} \tag{90}$$

$$\vec{i}_s = \left(i_x + j i_y\right) e^{j\phi_k} \tag{91}$$

The force vector can be decomposed into $\hat{x}$ and $\hat{y}$ force terms as $\vec{F} = F_x + jF_y$, leading to the definition of equivalent $\hat{x}$ and $\hat{y}$ suspension current components: $F_x = k_i i_x$ and $F_y = k_i i_y$. The phase currents required to create a desired force vector can be obtained by solving (91) and (88). The transformation of phase currents to a reference frame with $\hat{x}$ and $\hat{y}$ suspension currents enables field-oriented control techniques to be used to regulate the $\hat{x}$ and $\hat{y}$ forces on the shaft. For combined windings, which do not consist of distinct suspension and motor windings, an equivalent suspension winding model can be constructed.

4.2.4 Auxiliary components

Zhiyang Wang and Alan Palazzolo

4.2.4.1 Vacuum systems

Flywheels store rotational kinetic energy, so all types of drag torques will decrease the stored energy. This means intrinsic losses result from windage loss and motor standby loss. To reduce the windage loss, most commercial flywheel systems operate in vacuum. As a rule of thumb, the vacuum level needs to be less than 10 Pa and can be as low as 0.1 Pa for higher-speed flywheels.

Vacuum system layout requires a systematic design process, including a well-sealed housing, air-tight connectors, and long-lasting, high-quality pipes between the housing and vacuum pump, and most importantly, an appropriately selected vacuum pump. There are many types of pumps like rotary vane pumps, diffusion pumps, turbomolecular pumps, etc. Any type of pump can possibly be used for flywheel systems, if it can satisfy the performance requirement with enough efficiency.

Many factors need to be considered when choosing a vacuum pump. The first criterion is to make sure the vacuum pump can meet the target vacuum level. The second factor is the flow rate of the pump. This specification will determine how fast the vacuum level for a flywheel system can be drawn from ambient to working level. This time could determine the lead time for system start in the field during commissioning. In addition, many components for the flywheel, for example, motor windings and laminations, will outgas in a vacuum environment. This parameter effects the ability of the pumps to keep the vacuum level at target value in a stable fashion. It must be noted that most manufacturers mark their working flow rate at their optimum working condition, normally close to one atmosphere pressure. However, for flywheel systems, it is the flow rate at the vacuum level that is of the most importance. Two pumps from different manufacturers with exactly the same specifications on working flow rate can have dramatically different performances at vacuum. This means that various pump models will need to be tested during the implementation of a flywheel system. The third highlight point for pump selection process will be the power consumption. As noted earlier, the vacuum system is implemented to reduce the power loss of the flywheel. Therefore, the power consumption for the vacuum pump selected is very important and could even take priority over all other performances in some cases. Another important factor to consider is the noise level. If the operating noise is high, the application scenario could be limited for a certain design due to environmental regulations. Finally, as the vacuum pump will be working intermittently and only when the sensor detects low vacuum level, the suck back behavior of the vacuum system needs to be verified, to prevent leakage and FOD (Foreign Object Debris) like vacuum oil entering the system via the vacuum pump itself.

To maintain the vacuum rating for the flywheel housing, seals need to be implemented at all assembly interfaces, electrical plugs and any other interfaces to the external environment. There are two types of seals commonly used for this purpose: elastomer seals and metal seals. An elastomer seal is a relative economic option that can reach satisfactory results in most applications. The advantages of the elastomers are their flexibility and reusability. However, the relative soft characteristics render elastomers unideal for applications where very high level of vacuum is needed. Elastomers also become soft at elevated temperature. For applications with high environmental temperature or with elevated housing temperature in operation due to motor loss, elastomer seals might develop leakages. Temporary or transient seal leaks are hard to troubleshoot as the leakage point might change from run to run. To select a correct elastomer seal design, the deciding factors are the elastomer material, cross-section shape and size, elongation rate for the rings and compression rate for the cross section.

On the other hand, metal seals will be a better choice if higher vacuum level or higher temperature is needed. However, most metal seals are single-use or semi-single-use items. This could be an issue for maintenance and repair in some cases. There are many types of metal seals like C-Ring, E-Ring, O-Ring, etc. For flywheel application, the seals will usually be of an internal pressure type. Brochures from metal manufactures will provide detailed guidelines on installation cavity dimensions, seal dimensions, and tolerance requirements, based on the target surface dimensions to seal and pressure differential levels cross the seal.

Support structure

Typically, flywheels are installed underground or inside containment including a support structure. The most important functionality of the support structure is securing the flywheel module in place. To serve this purpose, the structure must be strong enough to hold the flywheel weight and loads generated from various external disturbances. For systems that will be commissioned in California, an

OSHPD test might be needed to certify that the system can survive earthquake shock, up to certain acceleration levels. The structural modes will also need to be thoroughly calculated and examined to make sure the flywheel won't resonate near those frequencies during operation. In addition, the support structure needs to be designed so that the vibration of a single flywheel, even during the occurrence of rotor faults, should not affect nearby equipment, especially other flywheels, to prevent a chain reaction that can cause a potentially devastating event.

For certain flywheels used in environments susceptible to external vibrational disturbances, the support structure might also serve the purpose of vibration isolation. For angular motions, a gimbal type of structure might be used as illustrated in Fig. 61.

Vibration isolators can be implemented for reducing linear vibrations resulting from ground motion. It should be noted that additional damping in the mount will raise the transmissibility ratio of flywheel housing motion to ground motion. This is illustrated in Fig. 62. As can be seen in the plot, the smaller the damping ratio is, the better the isolation results are. Conclusions can also be drawn from the same chart that the isolation is increasingly effective as the disturbance input frequency ω is further above the natural frequency ω_n. From this it can be concluded that (a) the isolation mount stiffness needs to be as small as possible to reduce the mount's natural frequency. Secondly, the isolators are only effective for high-frequency disturbances. For low-frequency disturbances

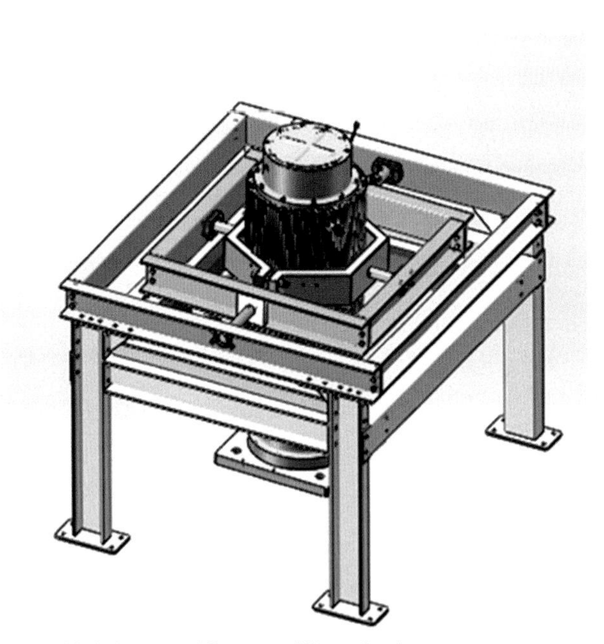

FIG. 61 Flywheel Gimbal system. *(Courtesy of Vycon Inc.)*

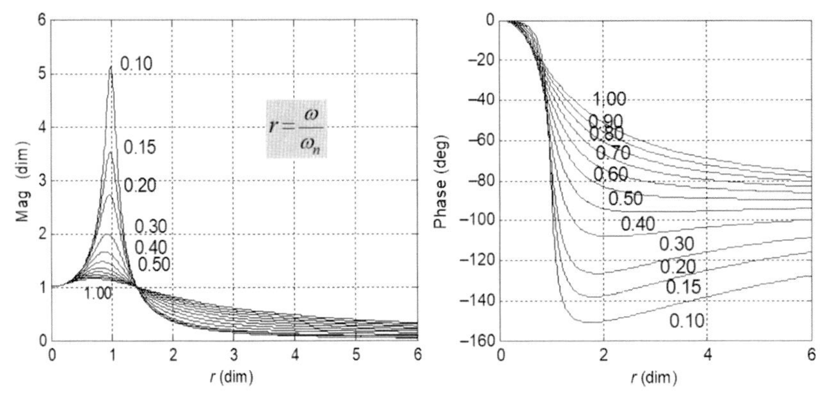

FIG. 62 Transmissibility ratio amplitude and phase angle of flywheel housing motion to ground motion for various damping values.

transferred through the ground, the rotor-bearing system must be designed with enough margin to handle these external forces by itself.

Containment

Rotor burst, or structural failure of flywheels can generate projectiles containing extremely high kinetic energies. These energies can pose significant risks for operators and valuable equipment surrounding the flywheel energy storage system. Therefore, containment design for flywheels is an important considerations that should be addressed at the beginning phase of any project. Virtual containment describes a practice of derating the flywheel rpm to a point that the safety factor is sufficiently large to relax some structural containment requirements. The latter is described later.

The containment needs to survive the first wave of radial impacts with extremely high, impulsive load. Some debris will then develop axial velocities and travel upward or downward, toward the top and bottom of the containment. This second wave of impact can potentially cause severe damage as the top (lid) could be the weakest point of the containment structure. The lid provides an entry port since the flywheel will have to be inserted into the containment, which is made in the form of a pressure vessel or built as a bunker if buried underground. The top lid is normally an individual part separate from the remaining containment structure, and if not properly designed could experience material and/or connection failure and lose functionality. This type of cover failure was reported at Quantum Energy, a flywheel energy storage company in San Diego, California [80].

Colozza [81] provided a method of designing a containment vessel basing on a triburst failure assumption, where the flywheels bursts into three equally sized pieces when its structural integrity fails catastrophically. In his proposed method, the impact energy for each of the three projectile is assumed to be 1/3 of the total flywheel stored energy and converted into the projectile's speed upon impact:

$$V_{imp} = \sqrt{\frac{2E_{fw}}{3M_p}} \quad (92)$$

where E_{fw} is the total flywheel energy, M_p is the projectile mass, and V_{imp} is the velocity upon impact. Assuming the momentum of the projectile is reduced to zero upon impact within a duration of t_{imp} (75uSec in Colozza's article), the impact force F_{imp} will be:

$$F_{imp} = \frac{M_p V_{imp}}{t_{imp}} \quad (93)$$

Based on the calculated impact force value (93) and the projectile impact area A_{imp}, the impact pressure P_{imp}

$$P_{imp} = \frac{F_{imp}}{A_{imp}} \quad (94)$$

is used to calculate the stress σ_{imp} acting on the containment material. This is then used as a guideline to design the containment vessel.

Colozza indicated that the containment design achieved with the method would be very conservative. In fact, if the containment is indeed designed to be able to hold the burst energy with 100% certainty, the volume, weight, and accompanying cost could render the flywheel energy storage system commercially impractical. Therefore, many commercialized systems use other methods and preventions to guarantee product safety, such as underground installation, large safety margin during stress and fatigue verification, strengthened QA criteria, reinforced bearing system to prevent failure due to rotordynamics, and measures to prevent overspeed in case of a malfunction.

Auxiliary bearings

The flywheel will still possess a large amount of energy and angular momentum after a magnetic bearing fails. This may cause severe structural damage if an auxiliary bearing system ABS is not installed. The ABS act only for a brief period, permitting the flywheel to stop, typically under braking action by the motor. Ref. [82] describes a detailed ball-bearing model utilized in auxiliary bearing service. Ref. [83] expands on this model by including temperature calculation and thermal growth effects. The preceding work focused on prediction of the occurrence of severe vibrations after a drop event onto the ABS. These vibrations are typically of a transient nature but may lock into a severe backward of forward harmonic whirl. Fatigue damage and race life expectancy from repeated rotor drops onto the ABS, may be estimated using a rainflow fatigue cycle counting approach as illustrated in [84]. Ref. [85] compares predicted vibrations and ABS loads during drop events and compares the results to predictions. In addition the reference provides a theoretical explanation of forward whirl that has been experimentally observed in flywheel drop events.

4.2.5 Loss mechanisms

Zhiyang Wang, Xiaojun Li and Alan Palazzolo

4.2.5.1 Windage loss

Spin of the flywheel rotor in air provides extra load on the flywheel motor, generates heat on the rotor, and drains stored kinetic energy. This occurs even in a vacuum chamber with a less than perfect vacuum, and may result in an unacceptable drop in operating speed. Windage may be predicted with highest accuracy by utilizing a CFD model, which, however requires much skilled modeling time and effort, and computational time. Vkzncik [86] provided an approximate, analytical/empirical approach for estimating windage losses for a cylindrical rotor spinning in a hollow cylindrical housing. In his article, the theoretical skin friction coefficient C_d for turbulent flow can be determined from Eq. (95), where *Re* is the Reynolds number as defined by Eq. (96).

$$\frac{1}{\sqrt{C_d}} = 2.04 + 1.768\ \ln\left(\mathrm{Re}\sqrt{C_d}\right) \tag{95}$$

$$\mathrm{Re} = \frac{R_{fw} t_{gap} \omega_{fw}}{v_{gap}}, \tag{96}$$

where R_{fw} is the flywheel radius, t_{gap} is the gap between the flywheel and the housing, ω_{fw} is the flywheel spin speed, v_{gap} is the dynamic viscosity or absolute viscosity of the fluid between the flywheel and housing. The windage loss power P_{w_loss} can then be expressed as:

$$P_{w\_loss} = \pi C_d \rho_{gap} R_{fw}^4 \omega_{fw}^3 L_{fw}, \tag{97}$$

where ρ_{gap} is the fluid density and L_{fw} is the length of the flywheel. Depending on the gap between the flywheel and housing, the gas in between can be laminar, turbulent, or mixed. The states of gas will have a huge impact on the windage loss and affect the accuracy for the calculation. Therefore, it is strongly suggested to use a CFD tool to simulate the actual gas states and numerically retrieve the windage loss power basing on the CFD results.

4.2.5.2 Bearing losses

Rotational friction is the mechanism for bearing loss in contact-type bearings. For the reasons given in the contact-bearing section, our discussion will focus on ball bearings. They are the most commonly used in flywheel systems, and similar conclusions can be reached for other types of contact bearings.

There are many factors affecting the friction coefficient for ball bearings. One of them is the ball materials. In general, ceramic balls in hybrid bearings should have lower friction than their steel counterparts. Cage material is also important. Polyamide or PEEK may have advantages in friction performance, compared with the commonly used brass. Besides those factors, the bearing design itself, bearing preload, lubricant types and lubricant fill will also affect the friction coefficients and must be carefully evaluated. As the friction force is decided by the friction coefficient and bearing load condition together, the final loss will also depend on speed and working load of the bearings. To obtain a more accurate estimate of contact-type bearing loss, it is strongly recommended that the designer should work closely with the bearing vendor to identify accurate estimates, based on the actual bearing configuration and the simulated loading conditions.

For passive magnetic bearings the only loss is eddy current generation on the magnets themselves, which is very minor. For active magnetic bearings, the major loss components include copper loss, iron loss, and amplifier loss. The bearing windage loss is normally neglected for flywheel applications as they will be operating in a vacuum, they have relatively slow surface speed compared to the flywheel rim, and they have relatively small surface area. These loss components are very similar to those of motors, but there are some fundamental differences. The most important difference between magnetic bearing and the motor/generator in term of losses comes from their modes of operation. For the synchronous motors used in flywheels, the motor winding will create a rotating magnetic field that forces the rotor to follow the rotating field. The principal harmonics of the voltage, current and magnetic field are $1\times$, $2\times$, or $N\times$ of the rotating frequency, depending on the number of poles (2 poles, 4 poles, or $2N$ poles, correspondingly). Most of the electrical disturbances and noises are multiples of the principal harmonics, and are determined by the motor design and control hardware/software. In contrast, magnetic bearings must counter vibration inputs from a wide frequency spectrum. Although one might expect the synchronous ($1\times$) component due to unbalance response to be the dominant component of bearing control currents, they are quite often near zero. Many magnetic bearing controllers use so-called open-loop cancellation algorithms to intentionally generate close to zero response for the synchronous vibration. This will theoretically allow the rotor to self-balance and spin about its mass center. This also results in reduced losses in the magnetic bearing as a spinoff benefit. Therefore, the significant current frequency components generally cover a wide frequency range for active magnetic bearings. The broad spectrum of the control currents generated by external disturbances makes the loss prediction relatively more difficult for active magnetic bearings. Therefore, a maximum loss is typically estimated during the design phase of the magnetic bearings. The copper loss will consist of two parts: the loss from the bias current and the loss from the control current. Both losses can be calculated by considering the simulated/calculated coil resistance and corresponding maximum currents. Permanent magnet PM-biased active magnetic bearings normally do not have bias currents and their accompanying copper loss. Therefore, the stator of PM-biased active magnetic bearing normally runs much cooler than the EM-biased bearings.

The iron loss also consists of two parts: hysteresis loss and eddy current loss. Traxler provided a theoretical formula for the hysteresis loss P_{MB_hys} in *Section 5: Losses in Magnetic Bearings*, from Ref. [52]. This formula is applicable if the bearing is made from iron and the flux density is between 0.2 and 1.5 T, and is given as

$$P_{MB\_hys} = k_h f_h B_m^{1.6} V_{iron}, \tag{98}$$

where k_h is an experimentally determined constant, f_h is the BH looping frequency, B_m is the amplitude of flux density fluctuation, and V_{iron} is the volume of the iron.

The magnetic bearing's eddy current losses result from both the rotating motion of the rotor and the high-frequency control flux. As the magnetic bearings are normally constructed with thin lamination sheets, the eddy losses are normally very small and can be neglected in most cases. For reference, Traxler also provided a formula for calculation of the eddy current loss at a given frequency:

$$P_{MB\_eddy} = \frac{1}{6\rho_{lam}} \pi^2 e^2 f_r^2 B_m^2 V_{iron} \tag{99}$$

where P_{MB_eddy} is the eddy current loss, ρ_{lam} is the specific electrical resistance of the lamination, e is the lamination thickness, and f_r is the frequency of flux fluctuation.

Eq. (99) provides some interesting implications for comparing EM and PM-bias magnetic bearings. For a PM-biased magnetic bearing, there is no polarity change around the rotor circumference, which explains its name, homopolar. This makes the B_m term much smaller than for EM-biased bearings, since the EM-biased bearings are normally of a heteropolar type, and their magnetic flux changes polarity multiple times around the rotor circumference. In addition, for rotational eddy loss generation in PM-biased bearings, the source frequency f_r will equal to the rotational frequency. However, for EM-biased bearings, the f_r will be $2\times$ the rotational frequency for NSSNNSSN arrangement and $4\times$ for NSNSNSNS arrangements. Both larger f_r and larger B_m values make the eddy current loss higher for EM-biased magnetic bearings than for the PM-biased type. Considering the additional lower copper loss for PM-biased magnetic bearings and the difficulty of heat transfer inside a vacuum environment, these factors give the PM-biased magnetic bearings many advantages over the EM-biased ones for flywheel applications.

Magnetic-bearing power amplifier losses mainly come from cable resistive losses, amplifier switching losses, and controller power consumption. All of these will depend on the design of magnetic-bearing controller and can be obtained from the vendor or power electronics designer.

4.2.5.3 Motor/generator losses

Most flywheels use VFD (Variable Frequency Drive)-driven AC motors as the driving components since their speed and torques need to be accurately controlled and frequently modified at precise power levels. Among the AC motors, surface mount permanent magnet motors are the most widely used due to their high efficiencies, constant torque capability, continuously magnetized circuit, and readiness for high-speed applications.

The majority of the losses in motors/generators come from copper loss, iron loss, power electronics loss, and windage loss. Among these, the windage losses are nearly zero for motors operating in vacuum, and the power electronics losses are generally well defined by the VFD providers. During the design phase, major attention should be given to the copper loss and iron loss. The copper loss comes from the resistive losses of the winding when the motor/generators are energized with currents, to generate torques and power. It is relatively easy to calculate the DC resistance from design parameters such as wire materials, total winding length, and wire diameters. However, during actual operation, it is always found that the copper loss could be much higher than calculated from the DC resistance. Unlike magnetic bearing windings, which uses small diameter wires, motor/generators use cables with much larger diameters due to their high current carrying requirements, which could be hundreds or thousands of amps. There are two effects of the currents inside the motor winding, due to the magnetic fields generated by the high-frequency AC currents. The first effect is the eddy current effect, which will create currents converging at metal surfaces and creating additional losses. The second effect is that the magnetic field generated by parallel cables with current running in the same direction will have a tendency of drawing the flowing electrons to the neighboring surface. This effect of convergence reduces the effective cross section of the electrical wire and equivalently increase the resistance. These AC effects are the reasons why many motor/generators use small wires to construct large cables in each turn of their windings. For copper loss calculation, a coefficient called the AC factor should be multiplied by the calculated DC resistance for the purpose of getting a more accurate estimate of the copper loss. The AC factor is normally simulated through FEA simulation and should be verified on final machines.

For surface-mount PM motors, most iron losses come from stator laminations, rotor magnets, and metal sleeves (if there is any). The iron loss in stator laminations is mainly hysteresis loss and can be estimated, based on the stator lamination weight and manufacturer-provided test data, adjusted by the frequency and amplitude of magnetic field fluctuations. The losses in rotor magnets and metal sleeves are mainly eddy current losses and can be estimated via finite element analysis FEA simulation tools. Even though metal sleeves have huge benefits in rotordynamics, the losses on them render them much inferior to composite sleeves for the sake of thermal management. Both factors should be considered and evaluated during the planning phase of any flywheel projects.

4.3 Gravity and buoyancy-based energy storage systems

Aaron Rimpel, Eugene Broerman

In the category of mechanical energy storage, pumped hydro and flywheels make up the overwhelming majority of commercially implemented energy storage technology. The only other mechanical energy storage concepts[c] are at the developmental stages, and they primarily include the storage of gravity-based potential energy and buoyancy-based potential energy. In essence, gravity energy storage (GES) and buoyance energy storage (BES) are similar topics since they rely on the relative positioning of a static load in a potential energy field. The basic difference between GES and BES is shown in Fig. 63. For GES, an available elevation difference is implemented to be able to raise (during charge mode) and lower (during discharge mode) a massive weight. The energy storage potential for the system is proportional to the magnitude of the weight and the distance it can travel between its maximum and minimum elevations—that is, potential energy $= Wh = mgh$. The gravity force is transferred through cable tension to a winding drum or spool, which is connected to the motor/generator. For BES, the buoyancy force on a buoy acts on a cable in the same way as the weight in GES, except the physical direction is reversed. Potential energy for BES is proportional to the net buoyancy force and the distance the buoy can travel between its maximum and minimum depths in the body of water. The net buoyancy force is equal to the weight of the displaced water (i.e., buoy volume

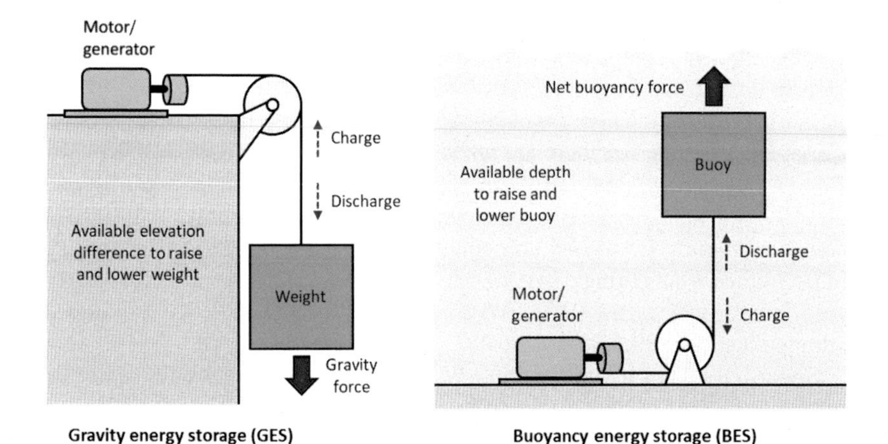

FIG. 63 Basic differences between gravity potential energy storage (left) and buoyancy potential energy storage (right)

---

c. As described in the introduction, CAES is not considered a truly "mechanical" energy storage technology in the scope of this chapter. Its basic operating principle is more consistent with heat engines, so CAES is presented in Chapter 6.

times density of water) minus the gravitational weight of the buoy itself. Fig. 63 shows very generic configurations of GES and BES, though numerous variants have been thought of, of which some are discussed in the subsequent sections.

4.3.1 Gravity energy storage (GES)

There are several GES concepts under development that will have relevant scale demonstrations in the near future. In general, these concepts offer long life and cost advantages over other energy storage technologies.

Cava et al. [87] describe one concept called ARES (Advanced Rail Energy Storage), which incorporates the transfer of weighted rail cars between low and high elevation topographies. It is claimed that ARES can compete with pumped hydro in terms of rated power (thousands of megawatts) and total storage (hundreds of thousands of megawatt-hours) and can provide round-trip efficiencies up to 80%. Advantages over pumped hydro include minimal environmental impact, simpler site requirements and permitting, and lower capital cost (e.g., roughly half on a per-kilowatt capacity basis). Pilot tests for ARES were completed in 2013 at a site in California, and a 50-MW, 12.5 MW-h commercial project is underway in Nevada. The Nevada project will operate rail cars (8600 metric ton combined weight) on over 9 km of track covering 610 m of elevation differential.

More recently, another concept being developed is called Energy Vault, and it consists of a crane that stacks concrete blocks during charge mode and lowers the blocks during discharge mode [88–93]. There are six different crane arms roughly 120 m off the ground, and each block weighs 35 metric tons. The cranes stack the blocks in charge mode, then they lower the blocks in discharge mode. Nominal energy capacity is 35 MW-h with 4-MW peak power and up to 90% round-trip efficiency. There are plans to construct such a storage tower for commercial use in India, but there are no additional data at the time of this publication. Advantages of this concept are relatively low cost since there is really no new technology that needs to be developed, and they can be highly modular. It is anticipated that the 35 MW-h unit could be built for $200–230/kW-h, though higher volume would bring costs down to $150/kW-h, and levelized cost of storage is anticipated to be $0.05/kW-h compared to $0.17/kW-h for pumped hydroapplications [89,91].

Gravitricity plans to build a 250-kW demonstration and a 4-MW full-scale prototype of a GES system that utilizes unused mine shafts as space to raise and lower weights. Fig. 64 shows a rendering of the concept that implements four winding drums that extend and retract cables to move the weight. These systems offer fast response times competitive with lithium-ion battery storage, but with significantly lower cost and long life, and flexible range of power delivery. They cite possible storage depths up to 1500 m and weighs up to 3000 metric tons, which equates to 44.1 GJ of energy storage. Round-trip efficiencies up to 90% are also claimed. At the time of this publication, the demonstration has not

FIG. 64 Rendering of gravitricity concept suspending weight in unused mine shaft. *(From Gravitricity, www.gravitricity.com.)*

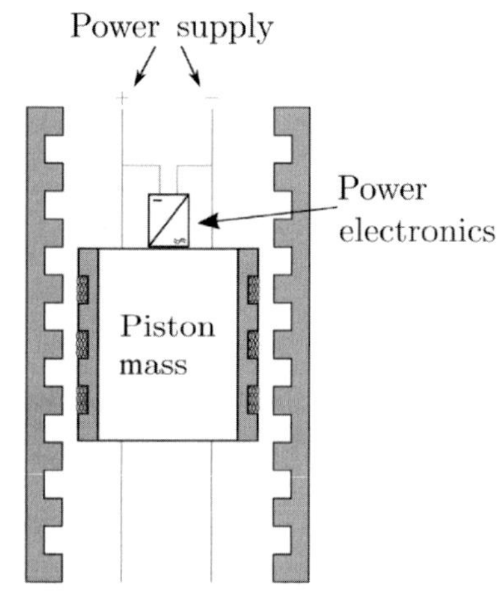

FIG. 65 GES concept utilizing linear electric machine technology. *(From C. Botha, M. Kamper, Capability study of dry gravity energy storage, J. Energy Storage 23 (2019) 159–174.)*

yet been conducted, and the full-scale prototype site has not been selected [94]. Instead of using conventional rotor motor generators, a similar concept shown in Fig. 65 poses to implement linear electric machine technology to raise a weight with electromagnetic force and extract power as the weight falls, though this concept is mostly theoretical at present [95].

Another GES concept is actually a hybrid with pumped hydrostorage. The concept, shown in Fig. 66, is a closed hydraulic system that incorporates a massive piston. The piston is raised using a pump to pressurize the fluid below

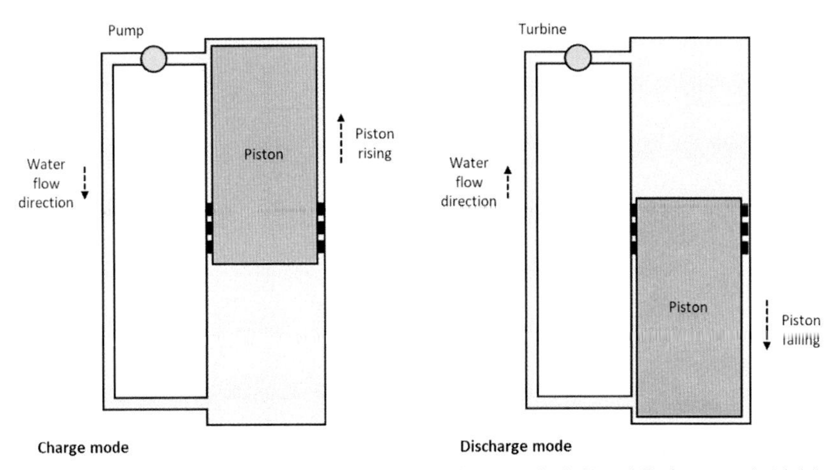

FIG. 66 Hybrid gravity and pumped hydroconcept: charge mode (left) and discharge mode (right).

the piston, storing gravitational potential energy. When the piston is allowed to fall, water flows through a turbine and energy is recovered. There have been several theoretical studies that have modeled this system for different applications [96–98], though there do not appear to be any relevant test data demonstrating performance to date. Commercial concepts up to 10 GW-h have been proposed by different companies—for example, Heindl Energy[d] and Gravity Power[e]—and a megawatt-scale demonstration is anticipated in the near future.[f] The economy of scale is inherent in the design, as energy storage scales with size to the fourth power, while cost only scales with size to the second power. A technical challenge of the concept is the seal between the piston and the wall, which must be able to withstand the hydraulic pressure under the piston (up to 60 bar). A concept is shown in Fig. 67.

4.3.2 Buoyancy energy storage (BES)

There are no commercial-scale BES systems that have been developed to date since the first concepts were proposed about a decade ago [99]. The work to date is mostly theoretical and with small lab-scale experiments. While Fig. 63 depicts a single buoy with a submerged motor/generator, the motor/generator could easily be surface-based, either on a barge or on land with the appropriate cable and pulley arrangement (Fig. 68). The land-based option would allow larger machinery and simplify connection to the grid. Depending on the design conditions, the buoy would be subject to very large compressive stresses at depths below the surface of the water, which would decrease the volume of

d. https://heindl-energy.com.
e. www.gravitypower.net.
f. www.gravitypower.net/megawatt-demo-plant-under-construction.

FIG. 67 Rolling membrane seal concept that allows piston movement. *(From Heindl Energy, http://heindl-energy.com/technical-concept/engineering-challenges/.)*

the buoy if made from a very flexible structure. Internal pressurization and/or more significant structural design would reduce the volume-reduction effect at the cost of greater weight (which offsets the buoyancy force) and more expensive components.

A numerical example [99] of the energy storage capacity of a BES system with a single spherical buoy is as follows: Consider a 10-m diameter buoy pressurized to 10 bar above atmospheric pressure at the surface. The net buoyancy force to submerge the buoy would be 5 MN, so storage at 100 m depth would be 500 MJ or 139 kW-h. For comparison, the average household energy usage is about 30 kW-h per day.[g] Assuming a discharge time of 3 h—that is, 46 kW of power, not including efficiency losses—the average velocity of the rising buoy would be under 10 mm/s, which should be low enough to ignore losses from drag forces. For an actual implementation, losses from motor and generator and friction in the pulleys would need to be considered to determine overall round-trip efficiency. Theoretical estimates suggest losses from motor generators would be less than 10% and losses from pulley friction would be on the order of 1%, making round-trip efficiency targets of 80% reasonable.

Small-scale tests with BES have been accomplished by a few researchers to demonstrate the operating principle. Alami et al. [100–102] tested with an array of conical-shaped buoys that were allowed to rotate (Fig. 69). The buoys also were treated with a helical groove pattern to promote a certain spin rate as the buoy array ascended. The reasoning for this arrangement was to reduce drag during fast ascents ~1.5 m/s. Bassett et al. [103–105] tested using spherical-shaped buoys at similar velocities. They mention that the large hydrodynamic

g. Citing 2018 annual electricity consumption data for United States residential utility customers. Source: U.S. Energy Information Administration, https://www.eia.gov/tools/faqs/faq.php?id=97&;t=3.

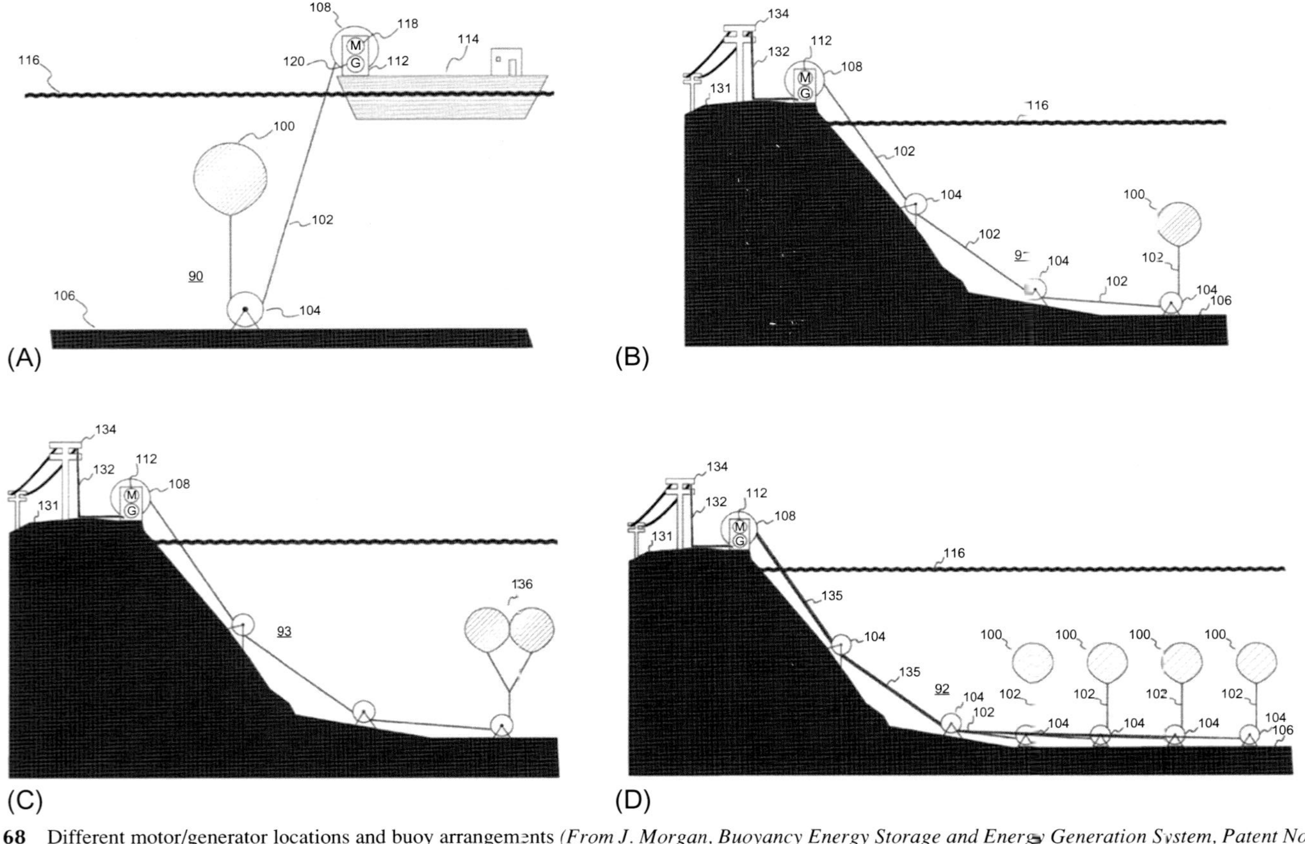

FIG. 68 Different motor/generator locations and buoy arrangements *(From J. Morgan, Buoyancy Energy Storage and Energy Generation System, Patent No. US 0107627 A1, 2010.)*

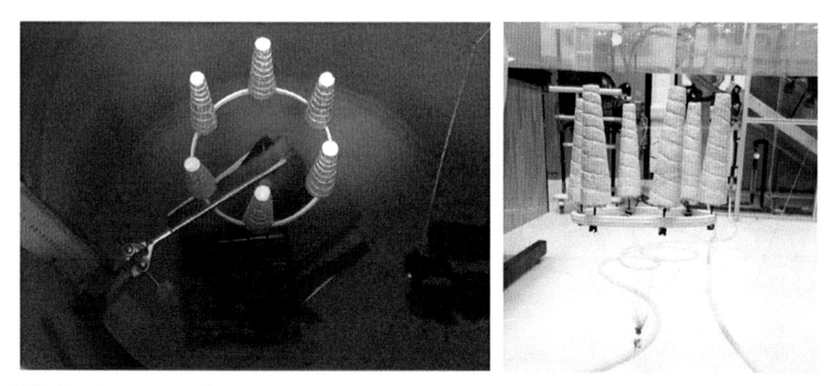

FIG. 69 Buoy array with free-rotating conical-shaped buoys tested by Alami et al. *(From A. Alami, Analytical and experimental evaluation of energy storage using work of buoyancy force, J. Renew. Sustain. Energy 6 (2014) 013137; A. Alami, Experimental assessment of compressed air energy storage (CAES) system and buoyancy work energy storage (BWES) as cellular wind energy storage options, J. Energy Storage 1 (2016) 38–43.)*

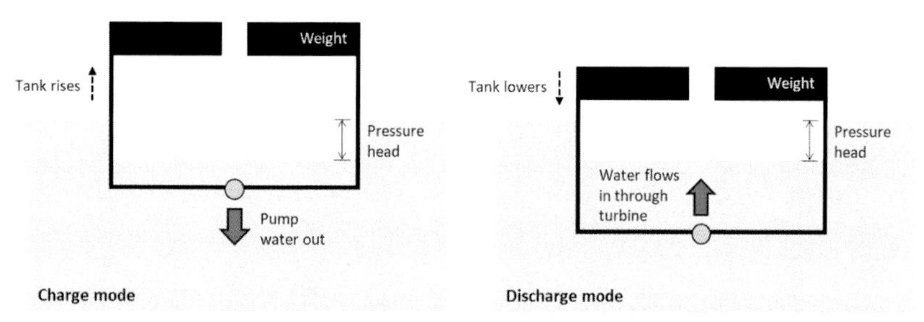

FIG. 70 Hybrid buoyancy and pumped hydroconcept: charge mode (left) and discharge mode (right).

drag for their test conditions account for 20% of the input energy compared to ~1% kinematic losses, so decreasing velocity would be a requirement for more realistic efficiencies.

It is logical that BES at a relevant scale would need to be located near deep bodies of water. However, BES for high altitude has also been proposed [99,106]. In these embodiments, balloons or structures filled with lighter-than-air gases—for example, helium or hydrogen—are positioned in the atmosphere and raised and lowered to release and store energy, respectively. Compared to underwater BES, atmospheric BES would require significantly larger "buoy" structures and considerably more cable travel due to the much lower density of air versus water (over 800 times lower).

Another concept of BES is actually a hybrid with pumped hydroenergy storage. Fig. 70 depicts a concept similar to Klar et al. [107,108]. A weighted vessel or tank displaces a volume of water and floats on the surface. By pumping water from the inside of the tank to the surrounding body, the tank rises and increases its gravitational potential energy (i.e., charge mode). To extract the stored energy, the outside water is allowed to enter the vessel through a turbine, causing the vessel to lower. There are other variants of the same concept that include more buoyant tanks and tanks anchored to the sea floor via submerged buoys and pulleys [107].

Acknowledgments

The authors acknowledge the support from Qatar National Science Funds, under the award number 8-2048-2-804 for the support to prepare Section 4.3.2.2: Composite Flywheels. Technical support for the energy storage flywheels and test facilities at Texas A&M University is provided by Erwin "Tom" Thomas and Randall Tucker (OTBOG Energy LLC).

References

[1] International Energy Association, World Energy Outlook, 2018.
[2] J. Giesecke, S. Heimerl, E. Mosonyi, Wasserkraftanlagen, Planung, Bau Und Betrieb (Hydropower Plants, Planning, Construction and Operation), 6th ed., 2014. Springer Vieweg.
[3] Fraunhofer Institut (IWES) Report, https://www.bee-ev.de/fileadmin/Publikationen/Studien/100119_BEE_IWES-Simulation_Stromversorgung2020_Endbericht.pdf, 2009.
[4] What is C-Rate—Battery University. https://batteryuniversity.com/learn/article/what_is_the_c_rate.
[5] Gyrobus: Flywheel Powered Transportation. https://www.amusingplanet.com/2019/02/gyrobus-flywheel-powered-public.html.
[6] O. Solis, et al., Saving money every day: LA Metro subway wayside energy storage substation, in: ASME/IEEE Joint Rail Conference, April 2–4, Colorado Springs, Colorado, 2014
[7] O. Solis, et al., LA metro red line wayside energy storage substation revenue service regenerative energy saving results, in: ASME/IEEE Joint Rail Conference, April 2–4, Colorado Springs, Colorado, USA, 2014.
[8] Ireland Pilots Hybrid Flywheel Battery System. http://www.bestmag.co.uk/content/ireland-pilots-hybrid-flywheel-battery-system.
[9] J. Park, A. Palazzolo, Magnetically suspended VSCMGs for simultaneous attitude control and power transfer IPAC service, ASME J. Dyn. Syst. Meas. Control 132 (5) (2010) 051001.
[10] R. Sebastián, R. Peña Alzola, Flywheel energy storage systems: review and simulation for an isolated wind power system, Renew. Sust. Energ. Rev. 16 (9) (2012) 6803–6813.
[11] S.M. Mousavi, F. Faraji, A. Majazi, K. Al-Haddad, A comprehensive review of flywheel energy storage system technology, Renew. Sust. Energ. Rev. 67 (2017) 477–490.
[12] H.H. Abdeltawab, Y.A.R.I. Mohamed, Robust energy management of a hybrid wind and flywheel energy storage system considering flywheel power losses minimization and grid-code constraints, IEEE Trans. Ind. Electron. 63 (7) (2016) 4242–4254.
[13] J.M. Carrasco, L.G. Franquelo, J.T. Bialasiewicz, S. Member, E. Galván, R.C.P. Guisado, S. Member, M. Ángeles, M. Prats, J.I. León, N. Moreno-Alfonso, Power-electronic systems for the grid integration of renewable energy sources : a survey, IEEE Trans. Ind. Electron. 53 (4) (2006) 1002–1016.

[14] I. Vajda, Z. Kohari, L. Benko, V. Meerovich, W. Gawalek, Investigation of joint operation of a superconducting kinetic energy storage (flywheel) and solar cells, IEEE Trans. Appl. Supercond. 13 (2) (2003) 2169–2172.
[15] J. Hansen, An Assessment of Flywheel High Power Energy Storage Technology for Hybrid Vehicles, Oak Ridge National Lab Technical Memo ORNL/TM-2010/280, 2011, December.
[16] A.C. Arvin, C.E. Bakis, Optimal design of press-fitted filament wound composite flywheel rotors, Compos. Struct. 72 (1) (2006) 47–57.
[17] X. Li, L. Hu, A. Palazzolo, A lightweight, high-speed kinetic battery for hybrid and electric vehicles, in: 2019 IEEE Transportation Electrification Conference and Expo (ITEC), Detroit, MI, USA, 2019, pp. 1–6, https://doi.org/10.1109/ITEC.2019.8790504.
[18] ECE2795 for U of Pittsburg. http://www.pitt.edu/~akwasins/ECE2795uGridSpr17.html.
[19] X. Luo, et al., Overview of current development in electrical energy storage technologies and the application potential in power system operation, Appl. Energy 137 (2015) 511–536.
[20] S. Sanders, M. Senesky, M. He, L. Hope, Ten, Low-Cost Flywheel Energy Storage Demonstration, Final Report CEC-500-2015-08, California Energy Commission, https://ww2.energy.ca.gov/2015publications/CEC-500-2015-089/CEC-500-2015-089.pdf, 2015.
[21] X. Li, A. Palazzolo, D. Tingey, H. Xu, P. McMullen, Z. Wang, Shaft-less energy storage flywheel, in: ASME 2015 9th International Conference on Energy Sustainability, San Diego, California, USA, June 28–July 2, 2015. ISBN :978-0-7918-5685-7.
[22] X. Li, B. Anvari, A. Palazzolo, Z. Wang, H. Toliyat, A utility-scale flywheel energy storage system with a shaftless, hubless, high-strength steel rotor, IEEE Trans. Ind. Electr. 65 (8) (2018).
[23] Pena-Alzola, et al., A review of flywheel based energy storage systems, in: Proceedings of the 2011 International Conference on Power Engineering, Energy and Electrical Drives, 3rd IEEE International Conference on Power Engineering, Energy and Electrical Drives, PowerEng2011—Torremolinos, Malaga, Spain, 11 May–13 May 2011, 2011.
[24] Z. Wang, A Novel Flywheel and Operation Approach for Energy Recovery and Storage, PhD dissertation, Texas A&M University, 2011.
[25] B. Anvari, X. Li, H.A. Toliyat, A. Palazzolo, A coreless permanent-magnet machine for a magnetically levitated shaft-less flywheel, IEEE Trans. Ind. Appl. 54 (5) (2018) 4288–4296, https://doi.org/10.1109/TIA.2018.2839747.
[26] MIL-HDBK-5H, Metallic Materials and Developments for Aerospace Vehicle Structures, Military Handbook, (1998, 1 December).
[27] J.G. Blauel, J. Beinert, M. Wenk, Fracture-mechanics investigations of cracks in rotating-disks, Exp. Mech. 17 (3) (1977) 106–112.
[28] H.F. Bueckner, I. Giaever, Stress concentration of a notched rotor subjected to centrifugal forces, J. Appl. Math. Mech. 46 (5) (1966) 265–273.
[29] D.R.J. Owen, J.R. Griffiths, Stress intensity factors for cracks in a plate containing a hole and in a spinning disk, Int. J. Fract. 9 (4) (1973) 471–476.
[30] J.G. Williams, D.P. Isherwood, Calculation of the strain-energy release rates of cracked plates by an approximate method, J. Strain Anal. Eng. Design 3 (1) (1968) 17–22.
[31] W.D. Pilkey, Formulas for Stress, Strain, and Structural Matrices, John Wiley & Sons, New York, 1994.
[32] A.B. Palazzolo, Vibration Theory and Applications with Finite Elements and Active Vibration Control, John Wiley, 2016.
[33] J.T. Tzeng, P. Moy, Composite energy storage flywheel design for fatigue crack resistance, in: 2008 14th Symposium on Electromagnetic Launch Technology, Canada, 2008, June.

[34] S.M. Arnold, A.F. Saleeb, N.R. Al-Zoubi, Deformation and Life Analysis of Composite Flywheel Disk and Multi-Disk Systems, NASA/TM-2001-210578, 2001.

[35] S.M. Arnold, A.F. Saleeb, N.R. Al-Zoubi, Deformation and life analysis of composite flywheel disk systems (in English), Compos. Part B-Eng. 33 (6) (2002) 433–459, https://doi.org/10.1016/S1359-8368(02)00032-X.

[36] S.K. Ha, H.I. Yang, D.J. Kim, Optimal design of a hybrid composite flywheel with a permanent magnet rotor (in English), J. Compos. Mater. 33 (16) (1999) 1544–1575, https://doi.org/10.1177/002199839903301605.

[37] S.K. Ha, S.J. Kim, S.U. Nasir, S.C. Han, Design optimization and fabrication of a hybrid composite flywheel rotor (in English), Compos. Struct. 94 (11) (2012) 3290–3299, https://doi.org/10.1016/j.compstruct.2012.04.015.

[38] S.K. Ha, M.H. Kim, S.C. Han, T.H. Sung, Design and spin test of a hybrid composite flywheel rotor with a split type hub (in English), J. Compos. Mater. 40 (23) (2006) 2113–2130, https://doi.org/10.1177/0021998306061324.

[39] S. Arnold, A.F. Saleeb, N. AlZoubi, Deformation and Life Analysis of Composite Flywheel Disk and Multi-Disk Systems, 2001.

[40] S. Chawla, J. Cai, M. Naraghi, Mechanical tests on individual carbon nanofibers reveals the strong effect of graphitic alignment achieved via precursor hot-drawing, Carbon 117 (2017) 208–219, https://doi.org/10.1016/j.carbon.2017.02.095.

[41] J. Cai, M. Naraghi, Non-intertwined graphitic domains leads to super strong and tough continuous 1D nanostructures, Carbon 137 (2018) 242–251, https://doi.org/10.1016/j.carbon.2018.05.030.

[42] M.F. Yu, O. Lourie, M.J. Dyer, K. Moloni, T.F. Kelly, R.S. Ruoff, Strength and breaking mechanism of multiwalled carbon nanotubes under tensile load (in English), Science 287 (5453) (2000) 637–640. (Online). Available: Go to ISI://000084989400043.

[43] C. Lee, X. Wei, J.W. Kysar, J. Hone, Measurement of the elastic properties and intrinsic strength of monolayer graphene, 321 (5887) (2008) 385–388, https://doi.org/10.1126/science.1157996.

[44] S. Peeterbroeck, et al., Polymer-layered silicate-carbon nanotube nanocomposites: unique nanofiller synergistic effect (in English), Compos. Sci. Technol. 64 (15) (2004) 2317–2323, https://doi.org/10.1016/j.compscitech.2004.01.020.

[45] F. Gardea, D.C. Lagoudas, Characterization of electrical and thermal properties of carbon nanotube/epoxy composites (in English), Compos. Part B-Eng. 56 (2014) 611–620, https://doi.org/10.1016/j.compositesb.2013.08.032.

[46] D. Sun, C.C. Chu, H.J. Sue, Simple approach for preparation of epoxy hybrid nanocomposites based on carbon nanotubes and a model clay (in English). Chem. Mater. 22 (12) (2010) 3773–3778, https://doi.org/10.1021/cm1009306.

[47] S.Q. Li, F. Wang, Y. Wang, J.W. Wang, J. Ma, J. Xiao, Effect of acid and TETA modification on mechanical properties of MWCNTs/epoxy composites (in English). J. Mater. Sci. 43 (8) (2008) 2653–2658, https://doi.org/10.1007/s10853-008-2489-1.

[48] E. Vassileva, K. Friedrich, Epoxy/alumina nanoparticle composites. II. Influence of silane coupling agent treatment on mechanical performance and wear resistance (in English). J. Appl. Polym. Sci. 101 (6) (2006) 4410–4417, https://doi.org/10.1002/app.23297.

[49] M. Nyman, L.E. Shea-Rohwer, J.E. Martin, P. Provencio, Nano-YAG:Ce mechanisms of growth and epoxy-encapsulation, (in English). Chem. Mater. 21 (8) (2009) 1536–1542, https://doi.org/10.1021/cm803137h.

[50] Z.H. Guo, T. Pereira, O. Choi, Y. Wang, H.T. Hahn, Surface functionalized alumina nanoparticle filled polymeric nanocomposites with enhanced mechanical properties (in English). J. Mater. Chem. 16 (27) (2006) 2800–2808, https://doi.org/10.1039/b603020c.

[51] T.J. Boyle, N. Bell, M. Ehlen, B. Anderson, W. Miller, Improved Flywheel Materials: Characterization of Nanofiber Modified Flywheel Test Specimen, 2013.
[52] K.-C. Lin, J. Gou, C. Ham, S. Helkin, Y.H. Joo, Flywheel energy storage system with functionally gradient nanocomposite rotor. in: 2010 5th IEEE Conference on Industrial Electronics and Applications, IEEE, 2010, pp. 611–613, https://doi.org/10.1109/ICIEA.2010.5517040.
[53] L. Costa, R. Vilar, Laser powder deposition (in English). Rapid Prototyp. J. 15 (4) (2009) 264–279, https://doi.org/10.1108/13552540910979785.
[54] S. Morvan, G.M. Fadel, J. Love, D. Keicher, Manufacturing of a Heterogeneous Flywheel on a LENS Apparatus, 2001.
[55] GMN Bearing, https://www.gmnbt.com/pdf/catalog-BallBearings.pdf.
[56] G. Schweitzer, E. Maslen, Magnetic Bearings, Springer, 2010.
[57] X. Li, Design and Development of a Next Generation Energy Storage Flywheel, (PhD thesis), Texas A&M University, 2018, May.
[58] C. Zhang, K.J. Tseng, A novel flywheel energy storage system with partially self-bearing flywheel-rotor, IEEE Trans. Energy Convers. 22 (2) (2007).
[59] B. Han, S. Zheng, Y. Le, S. Xu, Modeling and analysis of coupling performance between passive magnetic bearing and hybrid magnetic radial bearing for magnetically suspended flywheel. IEEE Trans. Magn. 49 (10) (2013) 5356–5370, https://doi.org/10.1109/TMAG.2013.2263284.
[60] M.A. Awadallah, B. Venkatesh, Energy storage in flywheels: an overview, Can. J. Electr. Comput. Eng. 38 (2) (2015) 183–193.
[61] X. Li, A. Palazzolo, Multi-input–multi-output control of a utility-scale, shaftless energy storage flywheel with a five-degrees-of-freedom combination magnetic bearing, ASME. J. Dyn. Sys. Meas. Control. 140 (10) (2018).
[62] R. Pena-Alzola, et al., Review of flywheel based energy storage systems, in: 2011 International Conference on Power Engineering, Energy and Electrical Drives, IEEE, 2011.
[63] J. Park, A. Palazzolo, R. Beach, MIMO active vibration control of magnetically suspended flywheels for satellite IPAC service, J. Dyn. Syst. Meas. Control 130 (4) (2008).
[64] MIT Modulation & De-Modulation, http://web.mit.edu/6.02/www/s2012/handouts/14.pdf.
[65] A. Chiba, T. Fukao, O. Ichikawa, M. Oshima, M. Takemoto, D.G. Dorrell, Magnetic Bearings and Bearingless Drives, Elsevier, 2005.
[66] J. Asama, Y. Hamasaki, T. Oiwa, A. Chiba, Proposal and analysis of a novel single-drive bearingless motor, IEEE Trans. Ind. Electron. 60 (1) (2013) 129–138.
[67] J. Van Verdeghem, V. Kluyskens, B. Dehez, Experimental investigations on passively levitated electrodynamic thrust self-bearing motors, IEEE Trans. Ind. Appl. 55 (5) (2019) 4743–4753.
[68] K. Raggl, T. Nussbaumer, J.W. Kolar, A comparison of separated and combined winding concepts for bearingless centrifugal pumps, J. Power Electron. 9 (2) (2009) 243–258.
[69] E.L. Severson, S. Gandikota, N. Mohan, Practical implementation of dual-purpose no-voltage drives for bearingless motors, IEEE Trans. Ind. Appl. 52 (2) (2016) 1509–1518.
[70] E.L. Severson, Reduced hardware parallel drive for no voltage bearingless motors, in: 2018 International Power Electronics Conference (IPEC-Niigata 2018-ECCE Asia), Niigatae, 2018, pp. 4020–4027.
[71] W. Gruber, Bearingless Slice Motor Systems Without Permanent Magnetic Rotors, TRAUNER Verlag and Buchservice, 2017.
[72] G. Sala, et al., Space vectors and pseudoinverse matrix methods for the radial force control in bearingless multisector permanent magnet machines, IEEE Trans. Ind. Electron. 65 (9) (2018) 6912–6922.

[73] E.L. Severson, R. Nilssen, T. Undeland, N. Mohan, Design of dual purpose no-voltage combined windings for bearingless motors, IEEE Trans. Ind. Appl. 53 (5) (2017) 4368–4379.

[74] S.W.K. Khoo, R.L. Fittro, S.D. Garvey, An AC self-bearing rotating machine with a single set of windings, in: 2002 International Conference on Power Electronics, Machines and Drives, Sante Fe, NM, USA, 2002, pp. 292–297.

[75] R. Oishi, S. Horima, H. Sugimoto, A. Chiba, A novel parallel motor winding structure for bearingless motors, IEEE Trans. Magn. 49 (5) (2013) 2287–2290.

[76] A. Chiba, K. Sotome, Y. Iiyama, M.A. Rahman, A novel middle-point-current-injection-type bearingless PM synchronous motor for vibration suppression, IEEE Trans. Ind. Appl. 47 (4) (2011) 1700–1706.

[77] P. Kascak, R. Jansen, T. Dever, A. Nagorny, D.K. Loparo, Bearingless five-axis rotor levitation with two pole pair separated conical motors, in: 2009 IEEE Industry Applications Society Annual Meeting, Houston, TX, 2009, pp. 1–9.

[78] E.L. Severson, Bearingless AC Homopolar Machine Design and Control for Distributed Flywheel Energy Storage, (PhD dissertation), University of Minnesota, 2015.

[79] P. Kascak, Fully Levitated Rotor Magnetically Suspended by Two Pole-Pair Separated Conical Motors, (PhD dissertation), Case Western Reserve University, 2010.

[80] M. Ooshima, S. Kobayashi, H. Tanaka, Magnetic Suspension Performance of a Bearingless Motor/Generator for Flywheel Energy Storage Systems, IEEE PES General Meeting, Providence, RI, 2010, pp. 1–4.

[81] J. Abrahamsson, J. Ögren, M. Hedlund, A fully levitated cone-shaped Lorentz-type self-bearing machine with skewed windings, IEEE Trans. Magn. 50 (9) (2014) 1–9.

[82] E.L. Severson, Bearingless motor technology for industrial and transportation applications, in: IEEE Transportation Electrification Conference and Expo (ITEC), Long Beach, CA, 2018, pp. 266–273.

[83] J. Pyrhonen, T. Jokinen, V. Hrabovcova, Design of Rotating Electrical Machines, John Wiley & Sons, 2013.

[84] Blast Injures 5 at Poway Business, https://www.sandiegouniontribune.com/sdut-explosion-blast-quantum-energy-storage-poway-2015jun10 story.html. (10 June 2015).

[85] A.J. Colozza, High Energy Flywheel Containment Evaluation, NASA/CR–2000-210508, (2000, October).

[86] G. Sun, A. Palazzolo, A. Provenza, G. Montague, Detailed ball bearing model for magnetic suspension auxiliary service, J. Sound Vibrat. 269 (3–5) (2004) 933–963.

[87] F. Cava, J. Kelly, W. Peitzke, M. Brown, S. Sullivan, Advanced rail energy storage: green energy storage for green energy, in: T. Letcher (Ed.), Storing Energy, Elsevier, 2016.

[88] A. Colthorpe, Tata Power in India Agrees to Buy 35 MWh of Gravity-Based Energy Storage from Energy Vault, Energy Storage News, 2018, November.

[89] T. Husseini, Tower of Power: Gravity-Based Storage Evolves Beyond Pumped Hydro, Power Technology, 2019, March.

[90] M. Marani, The Gravity-Powered Battery Could Be the Future of Energy Storage, The Architect's Newspaper, 2018, November.

[91] A. Rathi, Stacking Concrete Blocks Is a Surprisingly Efficient Way to Store Energy, Quartz, 2018, August.

[92] J. Runyon, Gravity-Based Energy Storage Hits the Market, Renewable Energy World, 2018, November.

[93] J. Spector, Can Newcomer Energy Vault Break the Curse of Mechanical Grid Storage? Green Tech Media, 2018, November.

[94] A. Fawthrop, How Gravitricity Aims to Balance Renewables on the Grid Through Gravity-Based Energy Storage, NS Energy, 2019, November.

[95] C. Botha, M. Kamper, Capability study of dry gravity energy storage, J. Energy Storage 23 (2019) 159–174.
[96] A. Berrada, K. Loudiyi, R. Garde, Dynamic modeling of gravity energy storage coupled with a PV energy plant, Energy 134 (2017) 323–335.
[97] A. Berrada, K. Loudiyi, I. Zorkani, System design and economic performance of gravity energy storage, J. Clean. Prod. 156 (2017) 317–326.
[98] A. Berrada, K. Loudiyi, I. Zorkani, Dynamic modeling and design considerations for gravity energy storage, J. Clean. Prod. 159 (2017) 336–345.
[99] J. Morgan, Buoyancy Energy Storage and Energy Generation System, 2010, Patent No. US 0107627 A1.
[100] A. Alami, Analytical and experimental evaluation of energy storage using work of buoyancy force, J. Renew. Sustain. Energy 6 (2014) 013137.
[101] A. Alami, Experimental assessment of compressed air energy storage (CAES) system and buoyancy work energy storage (BWES) as cellular wind energy storage options, J. Energy Storage 1 (2016) 38–43.
[102] A. Alami, H. Bilal, Experimental evaluation of a buoyancy driven energy storage device, Adv. Mater. Res. 816–817 (2013) 887–891.
[103] K. Bassett, R. Carriveau, D. Ting, Experimental analysis of buoyancy battery energy storage system, IET Renew. Power Generat. 10 (10) (2016) 1523–1528.
[104] K. Bassett, R. Carriveau, D. Ting, Underwater energy storage through application of Archimedes principle, J. Energy Storage 8 (2016) 185–192.
[105] K. Bassett, R. Carriveau, D. Ting, Integration of buoyancy-based energy storage with utility scale wind energy generation, J. Energy Storage 14 (2017) 256–263.
[106] E. Kelly, R. Arnold, High Altitude Gravity Energy Storage, 2017. Patent No. US 9701387 B2.
[107] R. Klar, M. Aufleger, M. Thene, Buoyancy Energy—Decentralized Offshore Energy Storage in the European Power Plant Park, University of Innsbruck, Unit of Hydraulic Engineering, 2012. Published online, http://www.buoyant-energy.com/files/buoyant_energy_at_a_glance.pdf.
[108] R. Klar, B. Steidl, T. Sant, M. Aufleger, R. Farrugia, Buoyant energy-balancing wind power and other renewables in Europe's oceans, Journal of Energy Storage 14 (2017) 246–255.

Further reading

X. Kang, A. Palazzolo, Dynamic and thermal analysis of rotor drop on sleeve type catcher bearings in magnetic bearing systems, ASME J. Eng. Gas Turbines Power 140 (2) (2018).
K. Krüger, A. Maaz, N. Rotering, A. Moser, Forecasting for pumped storage in Germany, Hydro Rev. 24 (6) (2016), https://www.hydroreview.com/2016/12/01/forecasting-for-pumped-storage-in-germany/#gref. (01 December 2016).
J.G. Lee, A. Palazzolo, Catcher bearing life prediction using a rainflow counting approach, ASME J. Tribol. 134 (3) (2012) 031101 (15 p.).
G. Sun, A.B. Palazzolo, Rotor drop and following thermal growth simulations using detailed auxiliary bearing and damper models, J. Sound Vibrat. 289 (2006) 334–359 (Note: Corrigendum for authors was published in J. Sound Vibrat. 306 (2007) 975).
J.E. Vkzncik, Prediction of Windage Power Loss in Alternators, Nasa Technical Note, NASA TN D-4849, 1968.
Z. Wang, R. Carriveau, D. Ting, W. Xiong, Z. Wang, A review of marine renewable energy storage, Int. J. Energy Res. 43 (2018) 6108–6150.

Chapter 5

Chemical energy storage

Michael A. Miller[a], Joerg Petrasch[b], Kelvin Randhir[b], Nima Rahmatian[b], and James Klausner[b]
[a]*Department of Materials Engineering, Southwest Research Institute, San Antonio, TX, United States,* [b]*Department of Mechanical Engineering, Michigan State University, East Lansing, MI, United States*

Chapter outline

5.1 Introduction

This chapter describes the current state of the art in chemical energy storage, which we broadly define as the utilization of chemical species or materials from which useful energy can be extracted immediately or latently through the process of physical sorption, chemical sorption, intercalation, electrochemical, or chemical transformation. Here, we further define "sorption" as any of the gas-solid uptake and desorption mechanisms that may involve reversible physisorption (physical ad/absorption and desorption) through nonbonded van der Waals dispersion interactions, or chemisorption (chemical absorption) when the strength of interaction forces between a material or surface and a gas constituent is of the same order as chemical bonding (or both).

Chemical energy storage aligns well with the great challenge of transitioning from fossil fuels to renewable forms of energy production, such as wind and solar, by balancing the intermittency, variability, and distributed generation of these sources of energy production with geographic demands for consumption. Indeed, geographic regions best suited for renewable energy production are too often at a remote location from greatest demand. The most prevalent forms of chemical energy storage in use today are liquid hydrocarbons, electrochemical, such as reversible batteries, biomass, and gas (e.g., hydrogen and methane). Currently, storing electricity directly in batteries or capacitors from wind and solar at scale is challenging because even the most advanced electrochemical or charge storage devices, such as lithium ion batteries or ultracapacitors, have

Thermal, Mechanical, and Hybrid Chemical Energy Storage Systems
https://doi.org/10.1016/B978-0-12-819892-6.00005-8

relatively low volumetric energy densities compared with liquid fuels such as diesel, gasoline, or liquid methane. A sobering representation of such comparisons is illustrated in Fig. 1, which is a plot of the volumetric versus gravimetric energy densities for various forms of energy storage, including the liquid energy carriers derived from fossil fuels [1].

It is important to make a distinction between chemical energy storage and energy carriers. Only renewable energy sources with intermittent generation require energy storage for their base operation, whereas primary energy resources must utilize an energy carrier to provide energy storage for later use, transport of that energy to meet temporal and geographic demands, and utilization in vehicles for transportation. Remarkably, there are only four types of energy carriers available to us in our everyday experience: electricity, hydrogen, biomass, and hydrocarbon fuels. Ubiquitous of the hydrocarbon fuels are those derived from petroleum, like the gasoline and diesel fuels upon which much of our transportation sector relies.

A notable subclass of hydrocarbon fuels are those synthesized from petroleum products and used as liquid carriers for later reformation into gaseous hydrogen. This approach—known as Liquid Organic Hydrogen Carriers (LOHCs)—has inspired in recent years comprehensive studies on its viability, as an alternative to compressed or liquid hydrogen, for delivering hydrogen at scale from large central production facilities to the forecourt of a hydrogen fueling station. A crucial requirement of LOHC candidates is that they must reversibly hydrogenate and dehydrogenate under relatively mild thermocatalytic conditions. For example,

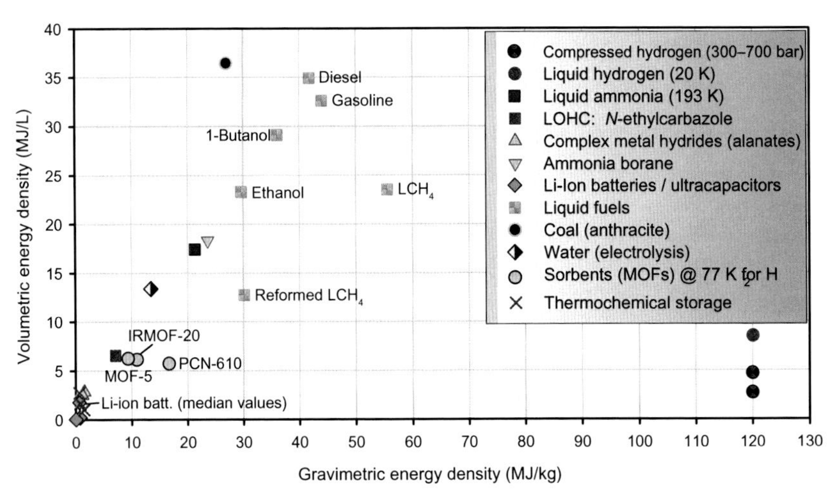

FIG. 1 Relationship between gravimetric and volumetric energy densities mapped out for various hydrogen storage modalities (compressed gas, hydrides, chemical hydrogen, and sorbents), compared with the energy content in liquid fuels or carriers, electrical storage, and thermochemical storage. Energy content for liquid ammonia, reformed liquid methane, *N*-ethylcarbazole, ammonia borane, and water electrolysis is that of the contained H_2.

N-ethylcarbazole (NEC), in its hydrogenated form H_{12}-NEC, has a volumetric energy density of 6.5 MJ/L (hydrogen energy content, Fig. 1), which is 40% greater than compressed hydrogen at 700 bar (300 K, LHV). As illustrated in Fig. 2, its dehydrogenation reaction yielding six hydrogen molecules can be done at 180°C and ambient pressure using a supported Pd catalyst (e.g., Pd/alumina, 5 wt.%). Hydrogenation of NEC occurs at a similar temperature using alumina supported Ru as the catalyst.

In the transportation sector, storage of hydrogen in materials on board vehicles powered by fuel cells has been the central motivation for an intensive, two-decade-long multidisciplinary effort aimed at advancing fundamental understanding of materials-based hydrogen storage under various modalities of gas uptake and release. Power generation systems based on fuel cells can play a central role in a hydrogen-based transportation infrastructure. In the United States and abroad, major research and development initiatives toward establishing a hydrogen-based transportation infrastructure have been undertaken, encompassing key technological challenges in hydrogen production and delivery, fuel cells, and hydrogen storage. However, the principal obstacle to the implementation of a safe, low-pressure hydrogen fueling system for fuel-cell

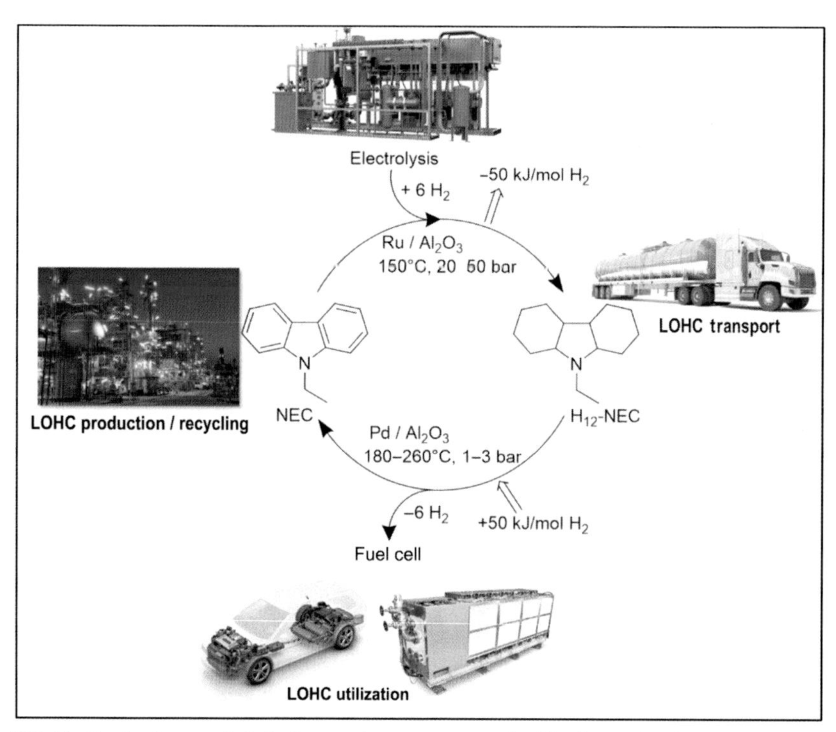

FIG. 2 Production, catalytic hydrogenation, transport, and utilization cycle of *N*-ethylcarbazole (NEC) as a liquid organic hydrogen carrier (LOHC).

powered vehicles remains storage under conditions of near-ambient temperature and moderate pressure [2].

As indicated in Fig. 1, ultra-high surface area sorbents with nano-engineered crystalline structures can exceed the volumetric energy density of liquid hydrogen through reversible physisorption, though such densities are realized at 77 K and ~80 bar of hydrogen pressure. Remarkable though this performance may be when compared with material technologies available less than a decade ago, the gravimetric capacity still falls short of the professed targets for onboard storage of hydrogen in vehicle applications because the mass of the storage system (i.e., storage vessel, tabulation, valves, etc.) must also be taken into consideration. Fortunately, the outlook for achieving higher than present storage capacities under moderate conditions is promising as new material innovations continue to emerge at a remarkable pace.

While most efforts in chemical energy storage focus on electrochemistry [3], a significant portion of the world's primary energy consumption is for thermal use such as space heating and industrial process heat [4]. Thermal storage technologies currently in use, mainly rely on physical energy storage via sensible and latent heat, e.g. hot water, molten salt, and phase change materials such as paraffins. Thermochemical storage constitutes a high energy density, high exergy alternative.

Thermochemical storage relies on high reaction enthalpy, reversible chemical reactions that proceed at elevated temperatures. Important reaction types for thermochemical storage are reduction-oxidation (redox), hydrogenation-dehydrogenation, and carbonation-decarbonation reactions. Furthermore, a range of multistep processes have been considered.

Historically nuclear and concentrating solar applications have been the main drivers behind thermochemical storage. More recently, the increasing share of intermittent renewable electricity generation has spurred interest in thermal buffering technologies summarized as power to heat that may benefit greatly from high energy density thermochemical storage.

5.1.1 Hydrogen storage

Hydrogen as an energy carrier is arguably one alternative to replacing petroleum products for transportation and stationary applications, if it can be produced in large quantities by clean and renewable means. This notion is premised on the fact that molecular hydrogen possesses the largest amount of chemical energy per chemical bond (142 MJ/mol/kg), which is three times higher than hydrocarbons (47 MJ/mol/kg) [5]. Therefore significant benefits can be gained when the energy capacity of hydrogen is combined with more efficient forms of energy consumption for power generation. For example, the maximum efficiency of the internal combustion engine is limited by the Carnot cycle efficiency of approximately 40%, though in practical terms conversion from chemical to mechanical via thermal energy is only 25% at best. Alternatively, the direct process of electron transfer from hydrogen to oxygen in the electrochemical reaction of a fuel cell is not limited by the Carnot cycle, and such devices can attain efficiencies ranging 50%–60% with an enthalpy and free energy of combustion equal to −286 and −237 kJ/(mol H_2), respectively [5].

Considering the thermodynamic arguments given earlier, it is evident that power generation systems based on fuel cells can play a central role in a hydrogen-based infrastructure for both stationary energy production and transportation. However, efficient modalities of hydrogen storage are crucial to both of these applications. Current approaches include compressed hydrogen gas, cryogenic and liquid hydrogen, chemical hydrogen storage, and hydrogen sorption in a solid-state material. The focus of this section of the chapter is on solid-state materials for hydrogen storage because it is applicable to both stationary and the very challenging vehicle applications.

Many transition elements form metallic (interstitial) hydrides, in which hydrogen atoms occupy the interstices of the metal crystal structure to within the covalent radii of hydrogen molecules. Thus such hydrides are not compounds but gas-solid solutions. Early development of low-pressure hydrogen storage systems for fuel-cell applications included metallic (interstitial) hydrides as the storage package for molecular hydrogen. While it is true that metals, such as palladium and niobium, and many alloys, such as $LaNi_5$, are capable of storing significant quantities of hydrogen (2–5 wt.%), problems of a practical nature occur with these storage materials. The most difficult problems to overcome are that the most efficient metallic hydrides are expensive, heavy, and bind hydrogen so tightly that the temperature required to desorb it is impractical for most applications. Accordingly, reversible chemistries must be distinguished from those that are not reversible (and impractical) and would require reprocessing of the spent material (i.e., dehydrided) at a separate processing plant. For example, the $Mg(BH_4)_2$ and $LiBH_4$ have theoretical hydrogen capacities of 14.9 and 18.5 wt.%, respectively, but such salts exhibit dehydriding enthalpies of the order 53 kJ/mol (endothermic), and would require an energy-intensive process to regenerate the spent products (Mg or Li, and B) [6].

5.1.1.1 Reversible solid-state hydrogen storage materials

Solid-state materials that have a propensity to absorb hydrogen can be categorized according to the energy of binding interactions between hydrogen molecules or atoms and the sorbent. While we tend to draw definitive lines between nondissociative physisorption of molecular hydrogen and dissociative chemisorption of atomic hydrogen to form chemical bonds, there is in principle a continuum of binding interactions possible between these two modalities, which are represented in Fig. 3 [7].

The weakest binding interaction is physisorption (i.e., physical adsorption and desorption), which occurs due to nonbonded van der Waals dispersion interactions (typically <0.5 eV). Physisorption typically requires low temperature and/or high pressure of the system, and is generally reversible

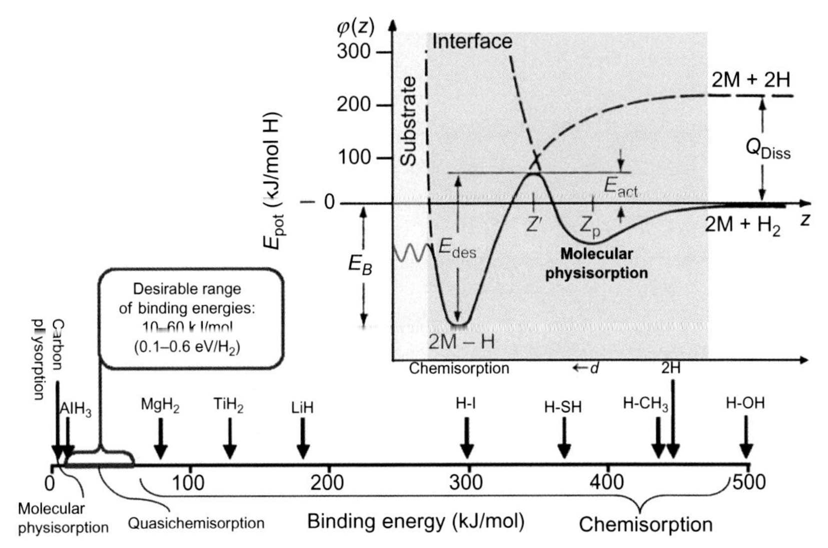

FIG. 3 Continuum of binding interactions between physisorption and chemisorption modalities. Region between 0.1 and 0.6 eV/H_2 is most desirable for room temperature hydrogen storage. Inset—Potential energy diagram for molecular physisorption and chemisorption binding interactions with a surface.

(though hysteresis can occur) by increasing the temperature and/or lowering the pressure. The process is nonactivated, reversible, and offers fast kinetics of gas uptake and release. To maximize gas uptake, materials with the highest possible specific surface area (SSA) must be considered for physisorptive storage because surface area is qualitatively proportional to gas uptake (formally known as the "Chahine rule" [8]).

In light of this rule, molecular-scale engineering of crystalline frameworks, such as metal-organic frameworks (MOFs), has been rigorously employed to develop novel storage materials from a vast selection of readily assembled building units (e.g., [9–12]). Notably, highly porous structures such as MOF-177 ($SSA_{BET} = 4750\ m^2/g$) and the PCN-6X series of isoreticular MOFs ($SSA_{BET} = 5109\ m^2/g$), Fig. 4, are ideal storage materials to use as benchmarks for physisorption owing to their high gravimetric (7.6 and 7.3 wt.%, respectively) and volumetric (48 and 28 g/L, respectively) uptake capacity for hydrogen at 77 K [13, 14]. Along parallel developments, molecular building units engineered to form three-dimensional porous polymer networks (PPNs, $SSA_{BET} = 6470\ m^2/g$), which are completely amorphous, have led to the establishment of a new benchmark in gravimetric uptake: 8.5 wt.% at 77 K and 60 bar [15]. An important advantage that PPN materials have over MOFs lies in their remarkable thermal and chemical stability, which can be attributed to their entirely covalent bonding network.

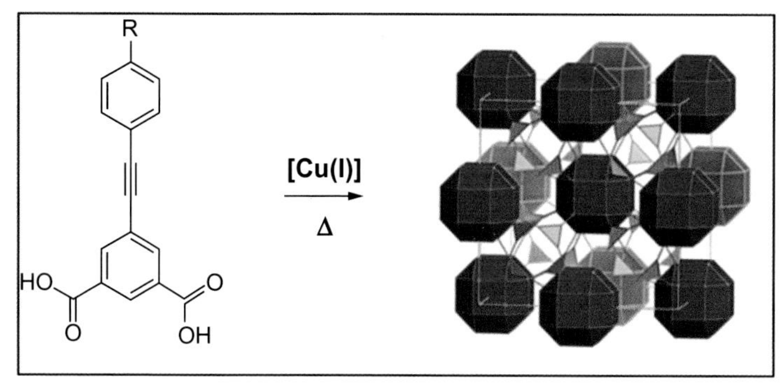

FIG. 4 Building units for synthesis of isoreticular MOF PCN-6X storage material and connected network topology of its crystalline structure and void space for accommodating hydrogen.

MOF chemistry has enabled a successful strategy for developing highly porous physisorption materials of extraordinary surface areas with well defined, ordered, pore spaces. They have been shown to selectively accommodate a high capacity of hydrogen molecules reversibly, though so far only at low temperature (77 K). In the case of hydrogen physisorption, it is now evident that two strategies—optimization of binding interactions and engineering of optimized structural motifs—are both necessary to achieve the volumetric and gravimetric capacities necessary for onboard storage of hydrogen in a vehicle application.

Fig. 5 shows example capacities of three different promising classes of hydrogen storage materials versus the isothermal conditions at which they are active [1]. It is apparent that binding interactions must fall within an optimum range in order to maximize reversible storage [16]. They can be neither too high nor too low, and the optimum range defines the thermodynamic constraints of the sorbent system. At or near room temperature, the desirable range for the binding energy of hydrogen, regardless of modality, is between 0.1 and 0.6 eV/H_2 (10 and 60 kJ/mol). This range was determined from an analysis based on entropic arguments, where the reference entropy values for H_2 are between 0.1 and 10 MPa, and temperature ranging 253–358 K. At 253 K the enthalpy of adsorption ($\Delta H = T\Delta S$) ranges ~21–32 kJ/mol H_2 and at 358 K the enthalpy increases to 51–71 kJ/mol H_2. On this basis, one can determine that the ideal binding affinity falls in the range between ~20 and 70 kJ/mol H_2, leading to a conservative upper range of 58 kJ/mol H_2 (0.6 eV/H_2).

Chemisorption (chemical adsorption) is stronger than physisorption and involves interaction energies on the same order as chemical bonding (>0.5 eV). Chemisorption spans a large range of interaction energies, from weakly interacting or reversible interactions up through strongly interacting and irreversible reactions. The challenge for a storage sorption material is to optimize the strength of the interactions for high uptake under temperature and pressure conditions that are practically accessible by the system and environment, while maintaining high recovery of the gas species under such conditions.

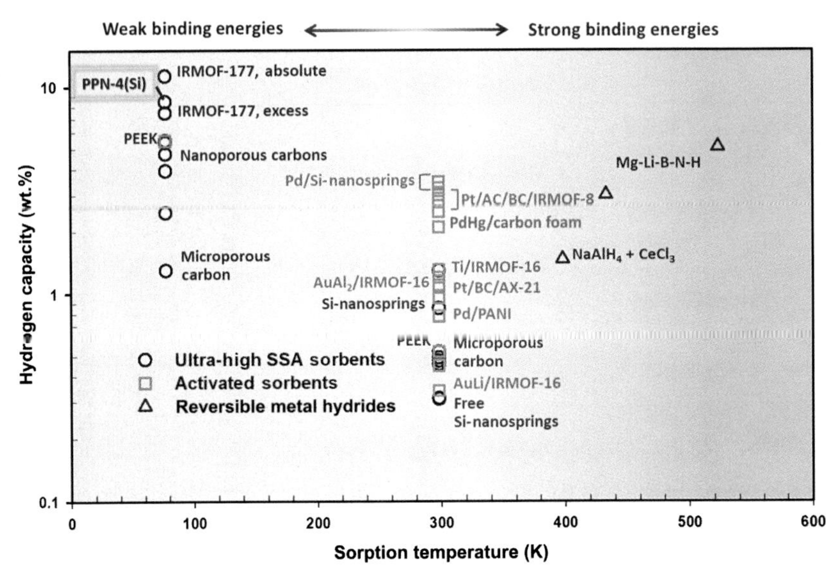

FIG. 5 Range of storage capacities and isothermal conditions for uptake or release of hydrogen measured for three promising classes of storage materials. (Notes: PEEK = polyether ether ketone; Pt/AC/BC/IRMOF-8 = Pt supported on activated carbon and combined with a bridging compound (BC), which is then combined with the "receptor" IRMOF-8 (an isoreticular metal organic framework); X-IRMOF-16 = IRMOF-16 in which the void space of the MOF framework is intercalated with X; Mg-Li-B-N-H = A complex, ternary hydride consisting of $MgH_2/LiBH_4/LiNH_2$).

Advances in chemisorption materials

Significant strides toward improving storage capacity and kinetics have occurred along parallel fronts with the development of new materials and the catalyzation of existing ones. Addition of a catalyst as a homogenous chemical constituent of, or heterogeneous dopant in, the storage material's matrix is a common denominator across various forms of candidate material technologies, most notably for chemisorption materials. For example, the seminal investigation by Bogdanovic and Schwickardi [17] led to the identification of a new class of alkali metal hydride that, when doped with a catalyst such as β-$TiCl_3$, significantly enhances the kinetics of molecular hydrogen desorption and renders the dehydriding process reversible under moderate conditions (Eq. 1). These catalytically enhanced hydride materials, based on aluminum hydride (AlH_4^-) and any Group 1 or 2 metal (Na, Li, Mg), are lightweight, store upwards of 5% H_2 by mass [18], and release it below 200°C, thereby overcoming the limitations of conventional metallic (interstitial) hydrides. The thermodynamic regime within which a few of these materials perform is suitable for both vehicle and stationary applications. However, the kinetics for the absorption and desorption reactions are still slow compared with metallic hydrides.

$$3NaAlH_4 \overset{\beta-TiCl_3}{\Longleftrightarrow} Na_3AlH_6 + 2Al + 3H_2 \overset{\beta-TiCl_3}{\Longleftrightarrow} 3NaH + Al + 3/2H_2 \quad (1)$$

The thermodynamics of the hydrogen-metal (or complex metal hydride) systems can be best understood through the use of phase diagrams. Unlike alloy (solid-liquid) phase diagrams, the metal-hydrogen phase composition is dictated by the temperature and pressure of the surrounding hydrogen gas. Equilibrium metal-hydrogen phase diagrams are often constructed from pressure concentration temperature (PCT) measurements. These diagrams consist of isothermal measurements of the equilibrium hydrogen concentration in a metal as a function of the surrounding hydrogen gas pressure [19].

To develop a basic understanding of the principles behind these thermodynamic measurements, it is useful to start with an idealized representation of a PCT measurement (Fig. 6A). Moving along one isotherm, hydrogen begins to dissolve into the host-metal lattice at low concentrations as the surrounding gaseous hydrogen pressure is increased. This region represents a solid solution of hydrogen in the metal which is denoted as the β phase. Hydrogen continues to be absorbed with increasing pressure until H-H interactions become important. At this point (1 in Fig. 6A), a hydride (denoted β phase) is formed locally by the occupation of particular interstitial lattice sites. The nucleation and growth of the hydride phase may occur at free surfaces, at intergrain boundaries, or throughout the bulk of the metal, depending on nucleation and diffusion mechanisms. Under idealized equilibrium conditions, the hydrogen gas pressure remains constant as hydrogen is absorbed and the α phase is transformed into the β phase. This behavior is observed as a plateau in the PCT diagram. Thus the existence of an equilibrium plateau

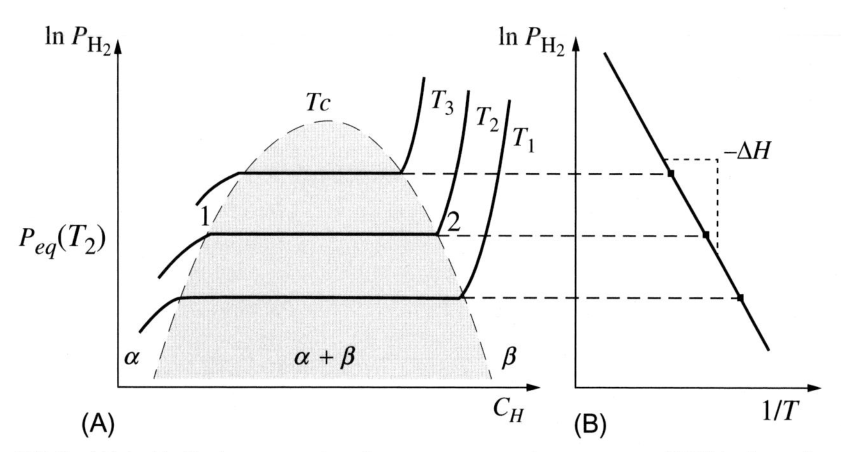

FIG. 6 (A) An idealized representation of pressure concentration temperature (PCT) isotherms for the α solid solution phase and β hydride phase. (B) The enthalpy of hydride formation ΔH is obtained from the slope of a Van't Hoff plot of ln(P_{eq}) as a function of $1/T$ [ln $P_{eq} = (\Delta H/RT) - \Delta S/R$].

signals the coexistence of the two phases. As the hydride phase grows the total hydrogen content of the sample increases. Eventually all of the α phase is transformed into the β hydride phase (point 2 in Fig. 6A). The pressure again rises and the overall hydrogen concentration continues to increase as hydrogen is dissolved as a solid solution in the β hydride phase.

The slope and length of the equilibrium plateau is of particular importance for hydrogen storage applications. A flat plateau enables the reversible absorption and desorption of hydrogen from a metal simply by raising or lowering the surrounding hydrogen pressure above or below the plateau pressure. In raising the pressure, hydrogen is absorbed while forming the β hydride phase. Hydrogen is desorbed by lowering the pressure, transforming the hydride back into the α phase. This change in pressure can be minor (1–2 bar for the classic alloy $LaNi_5$ at 22°C) compared to the pressures needed to store a significant amount of hydrogen by pressurization (200 bar). The length of the plateau determines how much hydrogen can be reversibly stored in a metal hydride.

The ease with which hydrogen can be reversibly absorbed and desorbed depends on the relative thermodynamic stability of hydrogen in the host metal. Hydrogen absorption to form a solid solution and the formation of a hydride phase can be either an exothermic or an endothermic process. The enthalpy of formation defines a phase's relative stability. This can be determined directly from a series of equilibrium PCT measurements at different temperatures as shown in Fig. 6B. The plateau pressures as a function of temperature are represented as a Van't Hoff plot. Those plots determine the enthalpy ΔH and the entropy ΔS of hydride formation.

A real-world example of PCT profiles is shown in Fig. 7 at three different isothermal conditions for the reaction of $NaAlH_4$ when doped with a catalyst (β-$TiCl_3$). Hydrogen uptake occurs during the transformation between two distinct phases: low pressure uptake for the Na_3AlH_6 phase (1.6 wt.%), followed by additional uptake at higher equilibrium pressures for the $NaAlH_4$ phase (2.3 wt. %), yielding a total reversible capacity of ~3.9 wt.% [20]. This absorption capacity is completely reversible during desorption at all three isothermal conditions. The enthalpies of formation derived from analysis of the Van't Hoff plot are −37 and −47 kJ/mol H_2 for the $NaAlH_4$ and Na_3AlH_6 phases, respectively.

Advances in physisorption materials

Statistical thermodynamics and molecular simulations of physisorption are cast in terms of the absolute thermodynamic variables (internal energy, U; Helmholtz free energy, F; grand potential, Ω; enthalpy, H; and Gibbs free energy, G). These absolute variables are not, however, accessible to experimental measurements of adsorption. The accessible quantities in the experimental measurements of adsorption are the *excess* thermodynamic variables. In considering the mass balance of the two-phase system, the excess adsorption measured by experiment is the total amount of gas present in the gas-solid system minus the amount

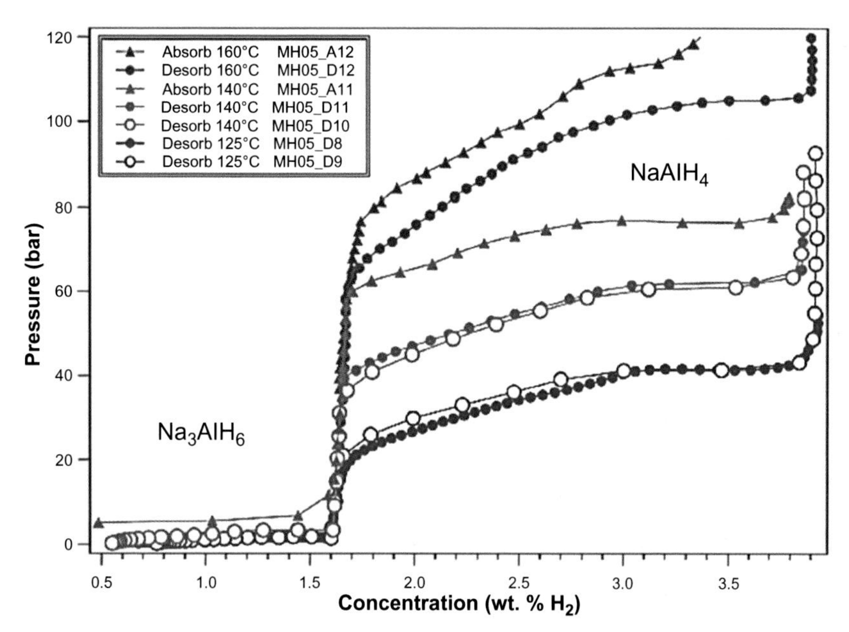

FIG. 7 PCT profile for chemisorption in catalytically doped sodium alanate ($NaAlH_4$) at three different isothermal conditions. Each plateau corresponds to a different phase as noted. Both absorption and desorption directions are shown.

present in the gas phase. According to the Gibbs definition, the excess adsorption is determined by the difference between the densities of the bulk and adsorbed "fluids" in the adsorbed phase [21]:

$$\Gamma = V_a(\rho_a - \rho_b) \tag{2}$$

where V_a is the volume of the adsorbed phase, ρ_a is the density of the adsorbed phase, and ρ_b is the density of the bulk phase. In supercritical fluids adsorbed at high pressures in the range 10–1000 bars, for example, the absolute amount adsorbed approaches a saturation value, whereas the excess amount adsorbed, Eq. (2), reaches a maximum then decreases with pressure. This experimentally accessible measurement is illustrated in Fig. 8 for a variety of nanoporous carbon materials, including single wall carbon nanotubes (SWNTs) at room temperature and 77 K. Further discussion of the significance of excess and absolute adsorption will be discussed later in the context of establishing a benchmark material for physisorption.

The simplest approach to deriving thermodynamic properties in adsorption is the Langmuir model [22]:

$$n = n_0\left(\frac{BP}{1 + BP}\right) \tag{3}$$

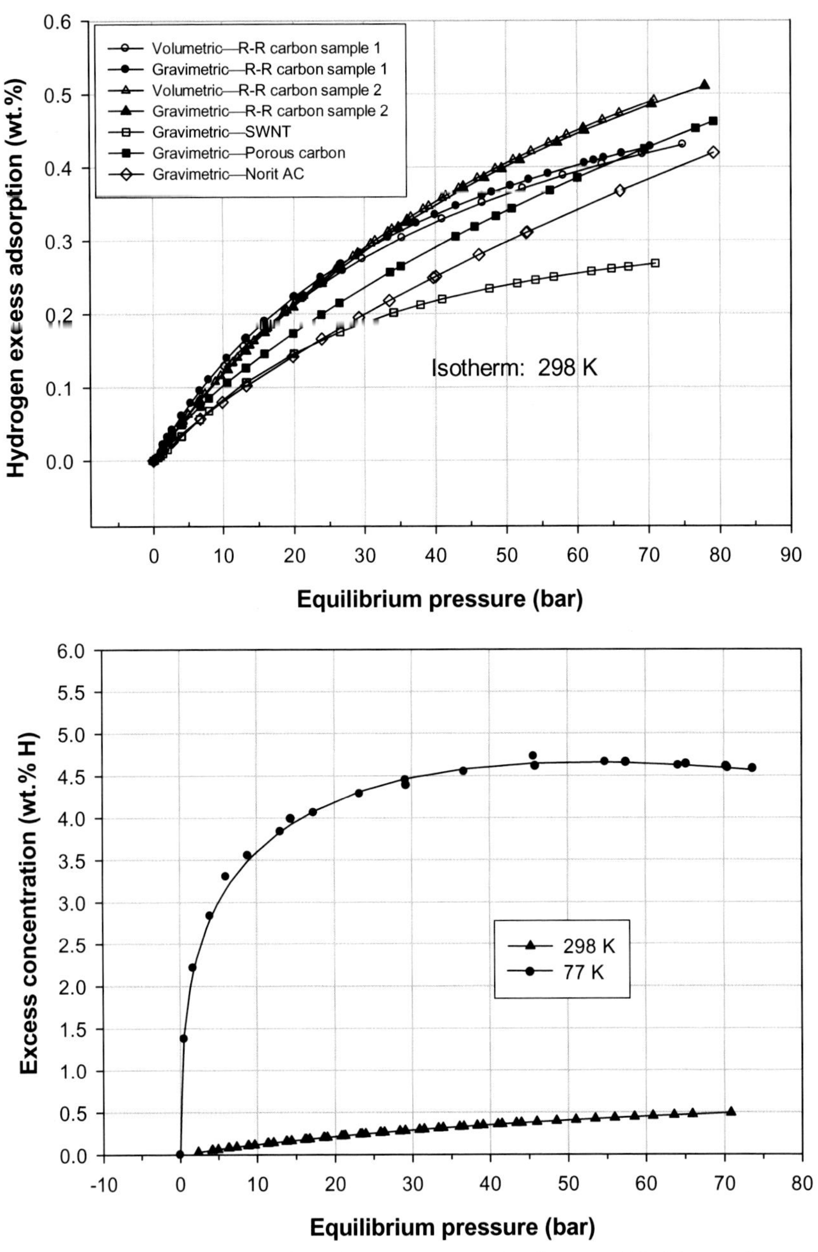

FIG. 8 Hydrogen physisorption in various nanostructured carbon materials. Top panel—room temperature physisorption of SWNT and porous carbon. Bottom panel—comparison between low-temperature (77 K) and room temperature physisorption. Note that the Gibbs excess concentration reaches a peak value once the density of the bulk gas is equal to that of the adsorbed layer on the sorbent—this is the only directly accessible measurement of gas uptake.

where P is the bulk-gas pressure, n is the surface concentration in terms of the number of moles of gas adsorbed per mass of sorbent, and n_0 is the saturation concentration. This model takes into account only the interactions between adsorbate and adsorbent, neglecting the interactions between adsorbate species. Under isothermal conditions, the Langmuir equation increases monotonically as a function of pressure. The saturation concentration and the coefficient B can be obtained from a linear fit of Eq. (3) to the measured data. However, this equation is applicable only to homogeneous surfaces and, consequently, does not adequately describe isotherms obtained on heterogeneous surfaces [23, 24]. In heterogeneous systems, such as porous sorbents, an additional parameter, m, is added to the Langmuir equation as an exponent of the pressure. This extended model is referred to as the Langmuir-Freundlich (L-F) equation.

However, neither the Langmuir nor the L-F models can describe the characteristic maximum in excess adsorption isotherms that occurs in the supercritical region. In the very low-pressure, subcritical and subatmospheric region, these models are accurate and can be used to derive thermodynamic quantities. In the supercritical region, a virial-type expression consisting of the temperature-independent parameters a_i and b_i must be applied to model the maximum in excess adsorption and estimate the isosteric heats of adsorption [25–27]:

$$\ln P = \ln N + \frac{1}{T}\sum_{i=0}^{m} a_i N^i + \sum_{i=0}^{n} b_i N^i \quad (4)$$

where P is the pressure, N is the adsorbed amount, T is the temperature, and m and n represent the number of coefficients required to adequately describe the isotherms. Concentration-dependent isosteric heats of adsorption are calculated from the best-fit parameters using:

$$Q_{st} = -R\sum_{i=0}^{m} a_i N^i \quad (5)$$

where R is the universal gas constant. It should be emphasized that the accuracy of this method relies on the number of isotherm curves considered in the fitting routine. Isotherm curves spaced apart by small temperature intervals would of course be most desirable, providing a more complete description of the thermodynamic surface tying together the isotherms and enabling accurate derivation of the heats of adsorption.

Illustrative calculations of the virial method are shown in Fig. 9. In this case, Eq. (4) was fitted to two hydrogen adsorption isotherms (77 and 298 K) for an ultra-microporous carbon (Takeda 4A). The fitted parameters were then used in Eq. (5) to arrive at the isosteric heats of adsorption shown in Fig. 10. Notice that these computed heats of adsorption quickly fall below the desirable range indicated in Fig. 3 at room temperature.

This method has been found to be valid for a great many different types of porous materials with very small pore volumes. It is apparent from Fig. 9 that

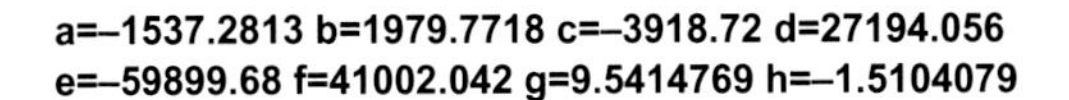

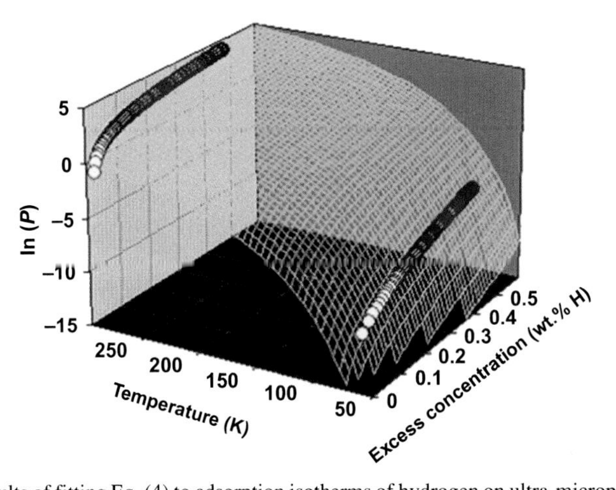

FIG. 9 Results of fitting Eq. (4) to adsorption isotherms of hydrogen on ultra-microporous carbon at 77 and 298 K.

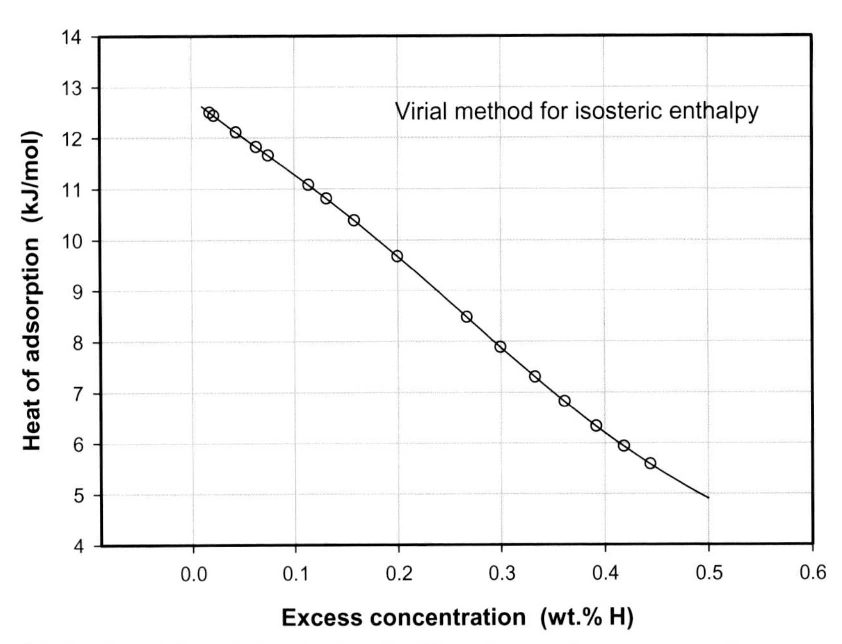

FIG. 10 Isosteric heat of adsorption from Eq. (5) as a function of excess concentration.

the virial model reproduces the qualitative features of the adsorption isotherms spanning sub- and supercritical pressure regimes.

Under practical conditions, such as actual gas cylinders, it is important to realize that the discharge pressure of a cylinder needs to be at least 1.5 bar [25]. A large or even moderate isosteric heat of adsorption may reduce the amount of H_2 available for use. Accordingly, the isosteric heat of adsorption must be carefully managed even when using weakly interacting materials for physisorption.

Benchmarks for low-density, ultra-high surface area storage materials

In contrast to the stagnant levels of performance of pristine carbon materials, synthetic framework materials like MOFs have evolved in measured performance much more rapidly than zeolites or even nanoporous carbon materials principally due to the relative ease with which topology and other properties can be engineered into their structural motifs, starting from a large selection of structural building units and employing systematic methods for synthesis. Both the volumetric and gravimetric energy densities of a cross section of different framework materials have steadily increased as a consequence of a concerted effort to design structural motifs that result in higher specific surface areas and stronger van der Waals binding interactions between hydrogen molecules and the building units of the framework than the materials that preceded them.

MOF-177 (Fig. 11) has been deemed an ideal material to use as a benchmark for hydrogen physisorption for the following reasons: (1) it has high gravimetric and volumetric uptake capacity, (2) its synthesis is simple and highly reproducible, and (3) it has a crystalline structure that is well characterized in atomic connectivity and chemical composition.

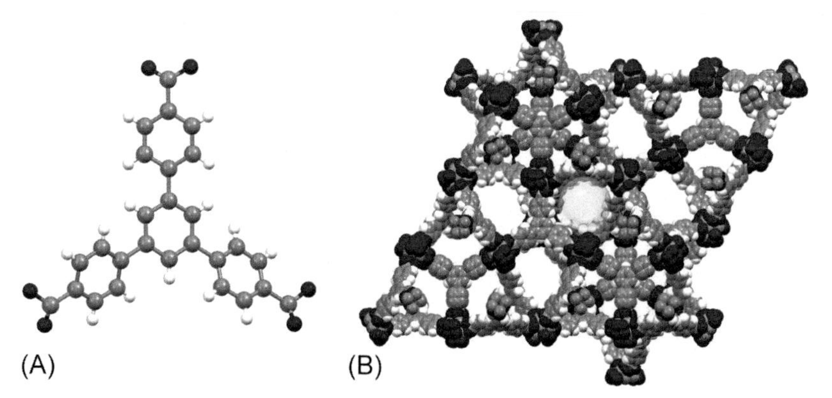

FIG. 11 (A) Molecular structure of 4,4′,4″-benzene-1,3,5-triyl-tribenzoic acid (BTB) linker and (B) crystal structure of MOF-177. An approximately spherical pore is shown as large sphere of 17 Å in diameter. *((Reproduced from H. Furukawa, M.A. Miller, O.M. Yaghi, Independent verification of the saturation hydrogen uptake in MOF-177 and establishment of a benchmark for hydrogen adsorption in metal-organic frameworks, J. Mater. Chem. 17 (2007) 3197.))*

Although the surface excess concentration is a useful concept, from the viewpoint of hydrogen storage, the total amount that a material can store has been found to be more relevant to use for hydrogen as a fuel. However, at high temperatures and pressures (i.e., above the critical temperature and pressure of hydrogen), the density profile of the adsorbed phase becomes more diffuse and, therefore it is not possible to distinguish between the adsorbed and bulk phases. In this situation the surface excess is the only experimentally accessible quantity, and therefore is not a reliable method to estimate the absolute adsorbed amount with high accuracy, although many efforts have been devoted to resolve this issue [28].

To estimate the absolute amount of gas adsorbed, the thickness or volume of the adsorbed layer must be known. However, this variable, and its spatial profile, cannot be measured experimentally. Instead, the absolute adsorption can only be estimated theoretically using, for example, Monte Carlo simulations [29]. Under these circumstances, the absolute amount of hydrogen adsorbed is estimated using a simple equation [30]:

$$N_{abs} = N_{ex} + \rho_{bulk} V_{pore} \tag{6}$$

where N_{abs} is the absolute adsorbed amount, N_{ex} is the surface excess amount, ρ_{bulk} is the bulk density of H_2, and V_{pore} is pore volume for MOF-177. For crystalline materials such as MOFs, the hydrogen-accessible pore volume can be estimated from classical void-filling computations of the unit cell using the kinetic diameter of hydrogen (2.6 Å).

The validity of MOF-177 as a benchmark material was shown based on the complementarity between volumetric (7.5 ± 0.1 wt%) and gravimetric (7.3 ± 0.1 wt%) H_2 adsorption measurements at 77 K (Table 1). Experimental details can be found in [13].

As noted earlier, PPN storage materials exhibit exceptionally high surface areas ($SSA_{BET} = 6470\ m^2/g$) with a correspondingly high gravimetric excess capacity: 8.5 wt.% at 77 K and 60 bar. The sorption profiles for PPN-4 and MOF-177 are compared in Fig. 12 at 77K. In the case of PPN-4, the usable hydrogen stored in this material is represented by the absolute gravimetric profile, which is slightly above 11 wt.% at 80 bar.

In the search for novel forms of materials for physisorption storage, crystalline materials of well-defined pores and structural motifs have emerged with remarkably high surface areas and gas storage capacities. Indeed, MOFs (including PCNs) and the amorphous PPN-type materials have set a new standard in hydrogen storage capacity via low temperature physisorption, and they can be readily synthesized at commodity scales. While nanoporous carbons also exhibit promising characteristics, the overall comparisons suggest that the development of MOF-based and PPN-type storage materials is a more promising venue to meeting the much sought-after storage targets as long as the engineering requirements for densification of the sorbent in a real-world storage tank can be achieved.

TABLE 1 Summary of high-pressure H_2 adsorption measurements of MOF-177 (all values ±1% SD).

		Surface excess amount (Pressure)			Absolute adsorbed amount (Pressure)			
Analysis	Surface area[a] (m^2/g)	(wt.%)	(g/L)	(bar)	(wt.%)	(g/L)	(bar)	Method
Sample 1	5640 (4750)	7.6	32	(66)	11.2	48	(72)	Volumetric
Sample 2	5640 (4750)	7.4	32	(57)	11.5	49	(72)	Volumetric
Sample 3	5250 (4630)	7.3	31	(52)	11.1	48	(75)	Gravimetric

[a]*Langmuir model (BET model).*

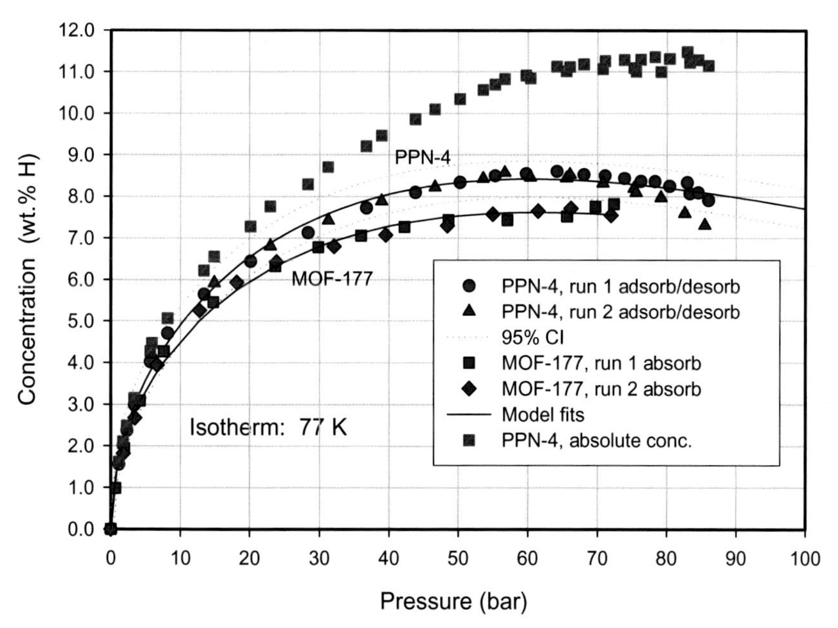

FIG. 12 Gibbs excess hydrogen concentration versus pressure profiles for MOF-177 and PPN-4 porous sorbents at 77 K. The estimated absolute concentration is also shown for PPN-4.

5.1.1.2 Thermochemical energy storage concepts by reaction type

Redox reactions

A reduction-oxidation (redox) thermochemical energy storage (TCES) system consists of solid metal oxide material in porous or particulate form reversibly releasing or consuming oxygen for storing or releasing energy. This reversible reaction is generally represented by

$$MO_x \leftrightarrow MO_{x-y} + \frac{y}{2}O_2 \tag{7}$$

Here MO_x is the oxidized form of the metal oxide, which is thermally reduced to MO_{x-y}. In most redox TCES systems, the reduction (release of oxygen) occurs when the oxidized form of the metal oxide is heated to a temperature higher than its equilibrium transition temperature (T_{eq}). T_{eq} is the temperature, at which the Gibbs free energy change (ΔG) of the redox reaction (Eq. 7) is zero. However, metal oxides are generally heated to 100–200°C higher than T_{eq} for achieving faster reaction rates. This reduction temperature is referred to as T_{red}. The reduced form of the metal oxide may be cooled in the absence of oxygen and stored for long-term storage or the hot reduced material can be stored in well-insulated containers in low oxygen pressure (P_{O_2}) environments for short-term storage. The reduced material is oxidized exothermically using air when stored energy is required for a specific application such as power generation, heating, etc. The energy release step (also referred to as oxidation) is performed at a temperature lower than T_{eq} and will be referred to as T_{ox}. The major advantage of metal oxide redox thermochemical system lies in the use of air as the gaseous reactant: gas storage is not required, which reduces costs. An additional advantage is production of high purity oxygen as a potential source of revenue during thermal reduction.

Pure metal oxides redox systems

Several candidates for pure metal oxides redox systems have been extensively investigated for TCES. The primary objective of these studies was to ascertain the viability of the metal oxide for TCES application. Screening criteria were recyclability, target temperature, energy density, and cost.

BaO₂/BaO Simmons [31] was the first to present a theoretical discussion on the advantages of energy storage device that uses the heat of reaction associated with the reversible oxidation of metal oxides, particularly those oxides, which can be reacted with air. He presented theoretical calculations showing that BaO/BaO_2 has considerable potential as a storage system. The chemical reaction associated with this metal oxide is given by

$$BaO_2 \leftrightarrow BaO + \frac{1}{2}O_2 + 72.5\,\text{kJ/mol} \tag{8}$$

T_{eq} for the reaction was reported to vary between 740°C and 880°C with the gravimetric energy density varying between 390 and 526 kJ/kg. These variations are due to the presence of impurities such as $Ba(OH)_2$ and $BaCO_3$ leading to side reactions [32, 33]. Various studies Fahim [34], Andre [35], and Carrillo [32] showed that T_{eq} decreases with decreasing P_{O_2}. This is in agreement with the thermodynamic model provided by Till [36] who had combined the available thermochemical data for decomposition of BaO_2 and presented the equilibrium P_{O_2} as a function of T_{eq}. This is given by

$$\ln(\mathrm{P_{O_2}}/\mathrm{atm}) = -4170/\mathrm{T_{eq}} + 3.74. \tag{9}$$

The first experimental demonstration for TCES via BaO/BaO_2 was done by Bowrey [37]. Air can be used as the heat transfer medium and as the source of oxygen during the oxidizing step. It can also be used as the heat transfer fluid and as a purging fluid for removing the oxygen produced during the reduction step. Mass transfer presents no difficulties provided the maximum temperature and the maximum rate of temperature rise are carefully selected. The most recent study on BaO/BaO_2 TCES was performed by Carrillo et al. [32]. They demonstrated that the oxidation conversion can be enhanced by thermal pretreatment of the sample at high temperatures, improving the overall performance due to elimination of carbonate and hydroxide impurities in the sample.

CuO/Cu₂O The chemical reaction associated with CuO/Cu_2O TCES is as follows:

$$\mathrm{CuO} \leftrightarrow \mathrm{Cu_2O} + \frac{1}{2}\mathrm{O_2} \tag{10}$$

Chadda et al. [38] were the first to propose the use of CuO/Cu_2O for TCES in 1989. They studied the cyclic decomposition of cupric oxide followed by the oxidation of cuprous oxide in air and found the reactivity essentially unchanged for up to 20 cycles. Recyclability of CuO/Cu_2O was attributed to the swelling of the CuO particles and the development of a highly porous structure on repeated cycling. Renewed interest in the CuO/Cu_2O TCES system led to the work of Hänchen et al. [39], and Alonso et al. [40]. Alonso et al. demonstrated its suitability for energy storage experimentally using a solar-heated rotary kiln setup. T_{eq} for the reaction was reported to vary between 740 and 1030°C with P_{O_2} of 0.21 atm. The relation between equilibrium oxygen partial pressure and transition temperature according to Chadda et al. [38] is as follows

$$\ln(\mathrm{P_{O_2}}/\mathrm{atm}) = -30431/\mathrm{T_{eq}} + 21.8 \tag{11}$$

The energy storage density for CuO is 811 kJ/kg. One of the major challenges is the melting of Cu_2O at 1235°C close to the reduction temperature. This results in sintering and significant grain growth during thermal cycling as observed by Deutsch et al. [41] and Hänchen et al. [39]. Photographs of their experimental samples are shown in Figs. 13 and 14, respectively.

FIG. 13 Sintering of the granulate material in the fixed-bed reactor: (A and B) top and side views after 1 cycle, (C) top view after 20 cycles, and (D) inside view of the material after 20 cycles. *(Reprinted with permission from M. Deutsch, F. Horvath, C. Knoll, D. Lager, C. Gierl-Mayer, P. Weinberger, F. Winter, Hightemperature energy storage: kinetic investigations of the CuO/Cu2O reaction cycle, Energy Fuel 31 (3) (2017) 2324–2334.)*

Hänchen et al. [39] also found that the O_2 uptake capacity decreases with the number of redox cycles due to material degradation. Similarly, Deutsch et al. [41] compared CuO/Cu_2O redox behavior in TGA and in packed bed. Copper oxide samples after 20 cycles are shown in Fig. 13. It was observed that the material maintained its granular structure internally, while the surface appeared to have sintered. This indicated that only the surface of the material had participated in the redox reaction resulting in poor cyclability. However, the small sample tested in the TGA showed high reversibility over 20 cycles even after strong sintering. This was attributed to the small sample size.

FIG. 14 Bulk morphology of (A) the starting copper wool, (B) the Cu wool oxidized in air at 1300 K, (C) the material from (B) reduced in Ar at 1223 K, and (D) the material from (B) after four oxidation-reduction cycles [39].

Alonso et al. [40] applied the concept of a rotary reactor to counter the sintering problem in Cu_2O. The basic idea to avoid sintering was to keep the copper oxide particles in continuous motion by rotating and mixing the reactor. Their rotary kiln reactor is shown in Fig. 15.

They demonstrated almost 80% conversion to CuO in argon atmosphere at reduction temperatures below 900°C. They observed that rotary movement led to the particles' coagulation in small spheres. These favorable outcomes were not observed when the experiments were performed under air. It was found that oxygen release was two orders of magnitude lower and the chemical conversion

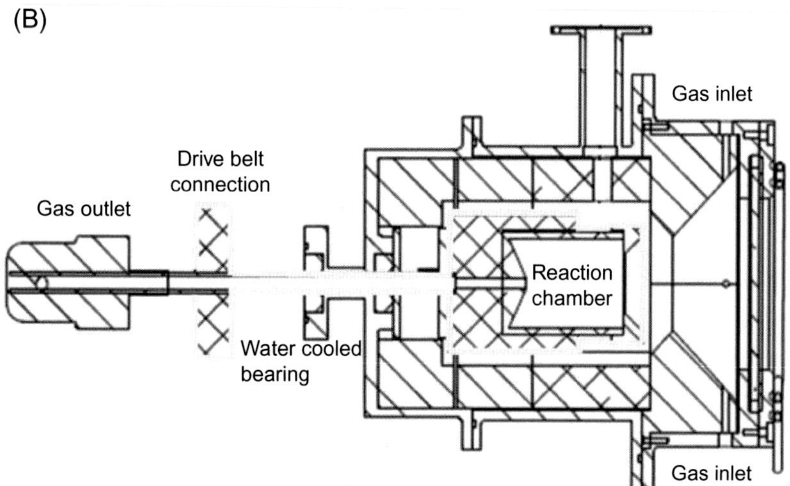

FIG. 15 (A) Photograph of the rotary solar reactor installed in the focal zone of the solar furnace (window cooling circuit removed). (B) Sketch of the solar reactor with rotary parts marked in *gray* [40].

was significantly lower than under Argon. Overall, CuO/Cu_2O has advantages such as high energy density and resource abundance, yet efforts are still needed to address the incomplete oxidation of CuO to Cu_2O and the sintering/shrinkage issue. Moreover, the reduction of copper oxides was found to be generally faster than the oxidation process, and both reactions depend strongly on the O_2 concentration of the carrier gas.

Fe_2O_3/Fe_3O_4 Iron oxide is the most abundant transition metal oxide in the earth's crust. This makes the Fe_2O_3/Fe_3O_4 redox couple very attractive for low-cost TCES. This redox reaction reads:

$$3Fe_2O_3 \leftrightarrow 2Fe_3O_4 + \frac{1}{2}O_2 \tag{12}$$

The theoretical energy density is 505 kJ/kg. André et al. [35] have found that Fe_2O_3 was reduced to Fe_3O_4 at a temperature greater than 1361°C in 20% O_2, The theoretical T_{eq} is 1290°C. As expected, the onset of thermal reduction decreased to 1145°C under pure Ar. André et al. mentioned the possibility of using iron oxide as TCES material as the conversion rate of Fe_3O_4 to Fe_2O_3 can reach 92%. Monazam et al. [42] have performed investigations on the same redox couple for chemical looping. The material was chemically reduced using carbon monoxide at 900°C and subsequently oxidized at the same temperature using air. Their experimental results suggested that the oxidation of magnetite occurs in two stages: (1) the conversion corresponding to the initial oxidation reached ~80% at 900°C, (2) as the reaction progressed beyond the surface, diffusion through the porous oxide layer became the rate-controlling step. The conversion during the second, diffusion-controlled step decreased as temperature was increased possibly due to the increased thickness of the oxidized layer. No data on cyclic stability for TCES has been reported so far.

Mn_2O_3/Mn_3O_4 Manganese oxide is one of the most versatile transition metal oxides because of its ability to attain +4 to +2 oxidation states. Pure manganese can form MnO_2, Mn_2O_3 and Mn_3O_4, and MnO depending on the temperature and oxygen partial pressure. The phase diagram and the CALPHAD model by Grundy et al. [43] shown in Fig. 16 show the equilibrium phases in Mn-O system at different temperatures and oxygen partial pressures.

The most widely investigated reaction for pure manganese oxide is the Mn_2O_3/Mn_3O_4 redox couple given by

$$3Mn_2O_3 + \text{heat}\,(96\,\text{kJ/mol}) \leftrightarrow 2Mn_3O_4 + \frac{1}{2}O_2 \tag{13}$$

Practical operating regime for this reaction is marked by a dotted red ellipse in Fig. 16. Theoretical energy density of the redox couple is 202 kJ/kg. The reduction and oxidation temperatures range from 550°C to 1000°C. Higher

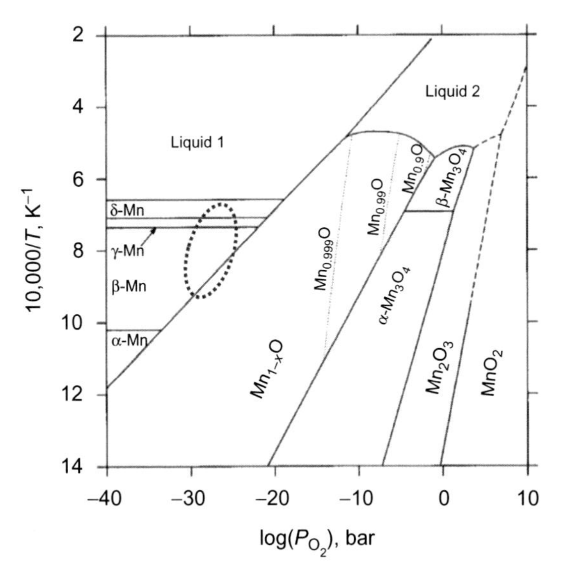

FIG. 16 Calculated potential phase diagram of the Mn-O system [43].

reduction temperatures are thermodynamically more favorable for reduction at high oxygen partial pressure of 0.21 atm. However, severe sintering and deactivation of the material as observed by Carrillo et al. [44] limits the maximum reduction temperature to below 1000°C. Another major drawback of this redox couple is the slow reoxidation kinetics of Mn_3O_4 to Mn_2O_3. This has been confirmed by many researchers [45, 46]. Carrillo [32] and Agrafiotis et al. [46] have shown that low cooling rate can enhance oxidation. Oxidation may be improved by addition of impurities such as iron as suggested by Fetisov et al. [47]. Karagiannakis et al. [45] have performed reduction in the temperature range of 920°C and 1000°C and oxidation in the range of 500–850°C in air and obtained an energy density of 110 kJ/kg.

Co_3O_4/CoO Cobalt is an expensive transition metal with concentrations on the order of 1 ppm in the earth crust. This makes Co_3O_4/CoO redox couple expensive for use in large-scale TCES. Furthermore, CoO is considered to be toxic and carcinogenic. However, Co_3O_4/CoO is still the most widely investigated pure metal oxide redox couple for TCES due to fast reaction rates. The chemical reaction of this redox pair is given by

$$Co_3O_4 + \text{heat}\,(203\,\text{kJ/mol}) \leftrightarrow 3CoO + \frac{1}{2}O_2 \tag{14}$$

The equilibrium transition temperature for the reaction in air is 895–935°C. Theoretical energy density of Co_3O_4/CoO redox pair is 844 kJ/kg. Fig. 17 shows the equilibrium phases in Co-O system at different temperatures and

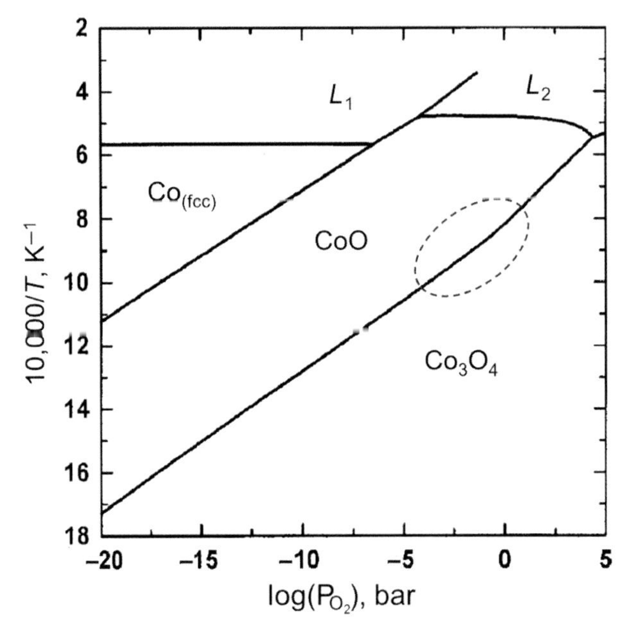

FIG. 17 Calculated oxygen potential phase diagram of the Co-O system [48].

oxygen partial pressures [48]. The red dotted ellipse in Fig. 17 indicates the practical operating regime for the reaction.

Energy densities of 400 and 525 kJ/kg were obtained experimentally by Neises et al. [49] and Karagiannakis et al. [45] in a rotary kiln and packed bed reactor, respectively. Several cyclic stability studies have been performed on the Co_3O_4/CoO redox system. All studies indicate that the material is cyclically stable with far better reaction kinetics than other pure metal oxide systems. Neises et al. [49] reported severe sintering of the powdered form of the metal oxide that resulted in less than 50% of theoretically achievable energy density. Coating of the material on ceramic foams and forming porous structures have resulted in conversions of more than 90% with exceptional stability [46, 50].

Mixed metal oxides redox systems

Doping Co_3O_4/CoO redox couple Attempts have been made to improve the performance of cobalt oxide TCES by adding varying quantities of other metal oxides. Mixing Co_3O_4 with Mn_2O_3 (% Mn $\leq$ 1.8) resulted in narrowing of the operating redox temperatures, but no significant improvement in energy density was observed [44]. Furthermore, Block and Schmücker [51] report that more than 10 mole % of Mn in Co_3O_4 resulted in lowering of energy density compared to pure Co_3O_4. Oxidation rates increase with addition of more than 0.5–5.6 wt.% of Cr_2O_3 in Co_3O_4. The percentage conversion of Co_3O_4 to CoO decreased to 84% at more than 5.6 wt.% of Cr_2O_3 in Co_3O_4.Similar effects were observed when Fe_2O_3 was added to Co_3O_4. Block and Schmücker [51], and Pagkoura et al. [52] confirm higher reduction rates than pure metal oxide. Wong suggested that with a 23.2 wt.% addition of Fe_2O_3 has the highest oxidation rate while maintaining a high energy storage density. Higher iron content resulted in lowering of the reaction enthalpy. Mixing Al_2O_3 with Co_3O_4 resulted in the formation of $CoAl_2O_4$ spinel in which Co cation exists in reduced state. Therefore materials with Co/Al molar ratio greater than ½ appeared to have stable reactive cyclability but much lower energy density compared to pure Cobalt oxide. Other dopants that were investigated by researchers were Cu, Ni, Mg, and Na. None of these had any beneficial effect of the performance of Co_3O_4/CoO TCES system.

Doping Mn_2O_3/Mn_3O_4 redox couple Mixing small quantities of cobalt oxide and copper oxide did not have any improvement in energy density or reaction rate of Mn_2O_3/Mn_3O_4 redox couple [32, 44, 51]. Doping of transition metal ceramic oxides such as TiO_2, ZrO_2, and Al_2O_3 improved the oxidation rates but no significant benefit was observed with respect to energy density. Fe_2O_3, on the other hand, appeared to improve the reaction rates and energy density. The oxidation fraction was found to increase from 36% to a maximum of 100% when iron oxide was added for an equilibrium transition temperature of around 950°C. Carrillo et al. [32] suggest incorporation reduction of the bond energies between divalent Mn cations and oxygen anions due to the presence of Fe cations in the lattice which results in enhanced mobility of oxygen anions during the oxidation. 20.8 mole% of Fe_2O_3 in Mn_2O_3 was reported to produce the best performance with energy storage density of 267 kJ/kg, which is superior to 202 kJ/kg of the pure Mn_2O_3. Similar improvements in energy density were reported by Block and Schmücker [51] and Andre [35] who investigated 0–33 and 15–50 mole% Fe_2O_3 in Mn_2O_3.

Varsano et al. [53] identified the following reaction upon mixing Li_2O_2 with Mn_2O_3

$$Li_2MnO_3 + LiMn_2O_4 + \text{heat} \leftrightarrow 3LiMnO_2 + \frac{1}{2}O_2 \tag{15}$$

They report the oxidation and reduction to be fast and highly reversible with no side reactions. Reaction temperature range was reported to be between 500°C and 1000°C with energy storage density of 200 kJ/kg.

Perovskites The class of inorganic compounds having a general formula of ABO_3 is called perovskites.

Several variations of such compounds are possible and for suitable combinations of cations high amounts of oxygen can be evolved during thermal reduction without causing any changes to the molecular structure. Oxygen mobility in a perovskite lattice occurs via vacancy diffusion mechanism and has been observed to be extremely fast at high temperatures. The general nonstoichiometric reaction of a perovskite for TCES is given by

$$ABO_{3-\delta_1} + \text{heat} \leftrightarrow ABO_{3-\delta_2} + \frac{\delta_1 - \delta_2}{2} O_2 \quad (16)$$

Here δ_i is the oxygen vacancy concentration per mole of the perovskite. The enthalpy of the reaction is the function of δ_i and is usually extracted from experimental data via Van't Hoff's analysis.

Babiniec et al. [54] were the first to report the use of lanthanum-based perovskites for use in TCES. $La_xSr_{1-x}Co_yMn_{1-y}O_{3-\delta}$ (LSCM) and $La_xSr_{1-x}Co_yFe_{1-y}O_{3-\delta}$ (LSCF) with varying values of x and y were investigated. They concluded that LSCM were more suitable for TCES due to higher oxygen exchange capacities (higher change in δ_i). LSCF perovskite showed higher changes in δ_i with a maximum of 0.49. However, the overall mass specific enthalpy change was observed to be higher for LSCM perovskite with the maximum of 250 kJ/kg attained by $La_{0.7}Sr_{0.3}Co_{0.9}Mn_{0.1}O_{3-\delta}$. One of the drawbacks of using lanthanum-based perovskites is high cost of lanthanum oxide compared to abundant transition metal and alkaline metal oxide. Thermal reduction in most cases has to be carried out in low oxygen partial pressure resulting in parasitic energy loss.

After having investigated lanthanum-based perovskites, Babiniec et al. [55] investigated another two types of calcium-based perovskites. These were $CaAl_{0.2}Mn_{0.8}O_{3-\delta}$ (CAM) and $CaTi_{0.2}Mn_{0.8}O_{3-\delta}$ (CTM). They reported energy storage densities to be 390 and 370 kJ/kg, respectively. This showed significant improvement compared to $La_{0.7}Sr_{0.3}Co_{0.9}Mn_{0.1}O_{3-\delta}$. Thermal reduction experiments in air up to 1250°C indicated the onset of oxygen release at 900°C and 750°C for CAM and CTM, respectively.

Another class of perovskites investigated for TCES is $Ba_xSr_{1-x}Co_yFe_{1-y}O_{3-\delta}$ (BSCF) by Zhang et al. [56]. For perovskites with no Barium, they conclude that $SrCoO_{3-\delta}$ shows the largest O_2 exchange capacity and $SrFeO_{3-\delta}$ shows the best reversible performance. They concluded that $SrFeO_{3-\delta}$ and $SrCoO_{3-\delta}$ may be suitable for TCES while $SrMnO_{3-\delta}$ is not suitable for temperatures below 1050°C. Ba containing systems ($BaCoO_3$, $BaFeO_3$, and $Ba_{0.5}Sr_{0.5}CoO_3$) showed the largest oxygen exchange capacity when reduced under inert ($P_{O_2} = 10^{-6}$ atm) and oxidized at $P_{O_2} = 0.2$ atm up to 1050°C. Furthermore, $BaCoO_3$ was found to fully reoxidize back to its original mass at 600°C at $P_{O_2} = 0.2$ atm. $BaCoO_3$ also showed the largest gravimetric energy density of 292 kJ/kg. This value is, however, lower than the energy density of $CaTi_{0.2}Mn_{0.8}O_{3-\delta}$ reported by Babiniec et al. [55].

Spinels/monoxide Spinels are being considered for TCES due to their reactive stability and high operating redox temperatures. One of the possible candidates that was investigated for TCES was the Hercynite cycle [33]. The chemical reaction for Hercynite-based redox cycle is given by

$$(CoNi)Fe_2O_4 + 3Al_2O_3 + heat \leftrightarrow (CoNi)Al_2O_4 + FeAl_2O_4 + \frac{1}{2}O_2 \tag{17}$$

Theoretical enthalpy of reaction 131 kJ/kg was considered to be relatively low compared to most aforementioned redox couples. It has been inferred from TGA experiments the = reactions were incomplete. Furthermore, XRD analysis of reduced and oxidized samples indicated side reactions might exist. Due to the low energy storage density, this redox material was not found more advantageous than other available candidates for TCES.

Randhir et al. [57] investigated magnesium-manganese oxides for high-temperature thermochemical energy storage. However, the actual reaction is highly complex and nonstoichiometric in the spinel and monoxide phases. A simplified stoichiometric representing the chemical reaction may be given by

$$MgMn_2O_4 + heat \leftrightarrow MgO + 2MnO + \frac{1}{2}O_2 \tag{18}$$

Three variations of material with molar ratios of manganese to magnesium of 2/3, 1/1, and 2/1 were cycled 1000°C and 1500°C under an oxygen partial pressure of 0.2 atm. The material showed excellent reactive stability at high temperatures. The energy density of the material between 1000°C and 1500°C was determined through a combination of acid-solution calorimetry and drop calorimetry. The total gravimetric energy densities (chemical and sensible) obtained for Mg-Mn oxides are in excess of 1000 kJ/kg out of which 55%–60% is chemical. The chemical gravimetric energy densities are compared in Table 2 for the prominent redox TCES materials presently discussed.

Hydration reactions The hydration of metal oxides is an exothermic chemical reaction in which the metal oxides, preferably from alkali and alkaline earth metals, combine with water to produce metal hydroxides [58]. This reaction is followed by a thermal decomposition reaction (dehydration) to provide a complete cycle for thermochemical energy storage. The hydration reaction normally occurs at intermediate temperatures of 250–800°C [59]. The reversible hydration/decomposition reaction for alkaline earth metal oxides can generally be written as follows:

$$MO + H_2O(g) \leftrightarrow M(OH)_2 + heat \tag{19}$$

where MO represents the metal oxide. The most promising candidates for M are Ca, Mg, Sr, and Ba. Li(OH) may also be effective for heat storage [58, 60, 61]. Since the hydration reaction requires steam, the latent heat for generation accounts for a substantial part of the heat of reaction, which means, in order to achieve high efficiencies, the heat of condensation must be recuperated. For instance, the maximum theoretical efficiency for $CaO/Ca(OH)_2$ without recovering heat of condensation is only 46%.

TABLE 2 Chemical energy stored by some prominent redox TCES materials.

		Gravimetric energy density		
		Theoretical	*Max. Experimental*	**Temp.**
	Redox material	*kJ/kg*	*kJ/kg*	°C
1	BaO_2/BaO	390	331	400–1100
2	CuO/Cu_2O	811	NA	750–1030
3	Fe_2O_3/Fe_3O_4	505	NA	1100–1400
4	Mn_2O_3/Mn_3O_4	202	110	550–1000
5	Co_3O_4/CoO	844	525	795–950
6	10 wt.% Fe_2O_3 in Co_3O_4	NA	464	NA
7	33 wt.% Fe_2O_3 in Mn_2O_3	233	NA	NA
8	33 wt.% Mn_2O_3 in Fe_2O_3	188	NA	NA
9	$LiMnO_2$	200	NA	500–1000
10	$LaxSr_{1-x}Co_yMn_{1-y}O_{3-\delta}$	NA	250	400–1250
11	$Ca_xSr_{1-x}Co_yMn_{1-y}O_{3-\delta}$	NA	555	500–1000
12	$CaTi_{0.2}Mn_{0.8}O_{3-\delta}$	NA	370	400–1250
13	$CaAl_{0.2}Mn_{0.8}O_{3-\delta}$	NA	390	400–1250
14	$BaCoO_3$	NA	292	600–1000
15	$(CoNi)Fe_2O_4 + 3Al_2O_3$	131	NA	1000–1400
16	Mg-Mn-O	NA	550	1000–1500

$CaO/Ca(OH)_2$ is the most widely studied hydration-based energy storage system. $Ca(OH)_2$ is cheap and has good cyclical performance. It has a theoretical energy storage of 2214 MJ/m^3 while practical energy densities up to 882 MJ/m^3 have been reported [62]. Ervin performed cycling test on a small sample of $Ca(OH)_2$ powder in a tube furnace by operating at constant temperature of about 525°C and switching the steam pressure between 0.03 atm for dehydration and 2 atm for hydration [63]. The capacity of $Ca(OH)_2$ to lose and absorb water remained unchanged after 211 cycles. However, reaction rates gradually decreased.

Schaube et al. studied $Ca(OH)_2$ dehydration and rehydration reactions at high partial pressures of H_2O. They found the following equilibrium for $Ca(OH)_2 \rightleftharpoons CaO + H_2O$:

$$\ln\left(\frac{P_{eq}}{10^5\ \text{Pa}}\right) = \frac{-12845\ \text{K}}{T_{eq}} + 16.508 \quad (20)$$

Therefore at H_2O partial pressure of 1 bar the equilibrium temperature is 505°C. The reaction enthalpy for hydration/decomposition of $Ca(OH)_2$ is reported to lie in the range of 94.6–111.8 kJ/mol [64–67]. Detailed analysis of the reaction mechanism for hydration and dehydration and their corresponding apparent activation energy, temperature, and pressure dependencies are gathered in Schaube et al. [67]. The activation energies of 162 and 89 kJ/mol are suggested for decomposition and hydration reactions, respectively.

An alternative system is $MgO/Mg(OH)_2$. Here, hydration does not occur above 230°C, hence the efficiency of thermal power cycles used in conjunction with Mg-hydrate systems is very limited [63]. Cycling tests indicated that while the reaction rates stay unchanged, there is a significant drop in the extent of the hydration/decomposition reactions within the first 40 cycles—from 95% to 60%–70%. However, that lower capacity was maintained for the next 500 cycles [63].

Experimental results for both the $Ca(OH)_2$ and $Mg(OH)_2$ cycles suggest that the reactions are usually heat transfer limited. Low thermal conductivities of the reactive material, especially in powder form or porous structures, enforce the use of thin layers in the case of external heating or the use of high temperature gas to heat the bed of reactive material to achieve fast kinetics [63, 68].

Hartman et al. [69] studied the decomposition kinetics for $Mg(OH)_2$ and $Ca(OH)_2$ particles using thermogravimetry at temperatures from 290°C to 430°C. The onset of the decomposition is found to be at 290°C and 325°C for $Mg(OH)_2$ and $Ca(OH)_2$ respectively. They also investigated the sintering of the dehydrated oxides in a nitrogen atmosphere at temperature between 400°C and 700°C. MgO lost 5–6.5 times less surface area than CaO. However, the heat of the reaction for $Mg(OH)_2$ decomposition decreases significantly with the increase of the temperature while $Ca(OH)_2$ shows a relatively steady heat of reaction over temperatures from 0°C to 500°C [69]. Therefore $Ca(OH)_2$ can deliver heat at higher temperatures with larger energy density.

Fujii et al. studied the importance of additives and pellet size on the reactivity and strength of the $CaO/Ca(OH)_2$ pellets [70]. They obtained an expression for estimating pellet shrinkage, which showed good agreement with the experimental findings. They found that aluminum-doped pellets retain their shape on hydration. Other additives for stabilizing the material and improving the reaction kinetics are $Ni(OH)_2$, $Zn(OH)_2$, and $Al(OH)_2$ [71]. The mechanism for the decomposition is suggested to be heterogeneous nucleation and the enhanced rate of decomposition observed, when minute amounts of Ni, Zn, Al, etc. are added, is due to the increase in number of nuclei from which the interface propagates. These additives also reduce the decomposition temperature significantly [71].

Other potentially interesting hydration-based energy storage systems are Ba $(OH)_2$ and $Sr(OH)_2$ [72, 73]. Thermogravimetric results indicated that for Ba $(OH)_2$, decomposition occurs at temperatures above 650°C while the melting point is only 408°C. When adding MgO, the sample did not melt and the kinetics were faster, however, strong sintering was still observed.

Similarly, $Sr(OH)_2$ requires stabilization with MgO to avoid melting at reaction temperatures [72]. $Sr(OH)_2$ decomposition temperatures from TGA are higher than the melting point and material residues were found on TG parts proving the melting of the sample.

The energy density of different metal hydroxide systems is summarized in Table 3.

The main benefits of hydroxide systems are high theoretical energy density, low cost, and abundance of the reactive materials. The main drawback is sintering and loss of reactivity at high temperatures.

The best candidate for hydration thermal energy storage seems to be the $CaO/Ca(OH)_2$ system. Its performance has been demonstrated experimentally in several lab-scale storage modules [74, 75].

Carbonation reactions

Metal carbonate thermochemical energy storage systems (TCES) can theoretically achieve very high energy densities (>1000 kJ/kg) while releasing heat at high temperatures, leading to efficient conversion to work and high overall

TABLE 3 Energy density of metal oxide/metal hydroxide pairs.

Reaction	*T* (K)	Energy density by mass (MJ/kg)	Energy density by volume (MJ/m³)	Reference
$Ca(OH)_2 \rightleftharpoons CaO + H_2O$	713-783	1.352	882-1200 (practical) 2214 (theoretical)	[63, 72]
$Mg(OH)_2 \rightleftharpoons MgO + H_2O$	598	0.710	1674 (theoretical)	[63]
$Ba(OH)_2 \rightleftharpoons BaO + H_2O$	848–933	0.545	2757 (theoretical)	[72]
$Sr(OH)_2 \rightleftharpoons SrO + H_2O$	848–858	0.728 (with 39 wt.% MgO)	–	[72]

system efficiency. The carbonation reaction, i.e. the uptake of CO_2 by a metal oxide, is exothermal, while the reverse decarbonation reaction is endothermal

$$MCO_3 \leftrightarrow MO + CO_2 \tag{21}$$

where M is the metal cation (Mg, Ca, Sr, Ba, or Pb). Metal carbonate systems with potential for TCES are $CaCO_3$, $PbCO_3$, $SrCO_3$, and $BaCO_3$. Pure $MgCO_3$ has less potential due to the low reaction temperature (<360°C) and due to slow carbonation rates. However, $MgCO_3$ may be mixed with other carbonates for TCES. Major advantages of metal carbonate TCES systems are (1) no catalyst is required, (2) gas-solid separation is easy, and (3) raw material costs are low. Major drawbacks are (1) slow energy discharge rates, (2) poor recyclability, and (3) need for CO_2 storage.

Alkaline earth carbonates ($CaCO_3$, $SrCO_3$, $BaCO_3$)

The reversibility of $CaCO_3$ decomposition/carbonation was first examined by Barker et al. [76]. Decomposition of $CaCO_3$ to CaO and CO_2 has a reaction enthalpy of 178 kJ/mol which corresponds to 1780 kJ/kg (Table 4). The equilibrium transition temperature for a carbon dioxide partial pressure of 1 atm is 760°C. It is unanimously observed by different researchers that the decomposition $CaCO_3$ to CaO is always close to complete but the reaction rate of the oxide so formed with carbon dioxide slows down quickly after a rapid initial reaction. The surface area substantially increases during the reaction from non-porous calcium carbonate to the porous oxide. Carbonation of CaO to $CaCO_3$ has two separate reaction regimes. A fast surface reaction is followed by a slow diffusion limited step [77]. The achievable reversible extent of reaction reduces with the number of cycles, rapidly at first and then more slowly. This loss in

TABLE 4 Energy density of some select carbonate base TCES systems.

	Enthalpy	Gravimetric energy density		Temp.
		Theoretical	*Practical*	
	kJ/mol	*kJ/kg*	*kJ/kg*	°C
$CaCO_3$	178	1780		650–850
$SrCO_3$	234	1591		900–1200
$BaCO_3$	272	1380		1000–1500
$Sr_{0.4}Ba_{0.6}CO_3$	257	1448	790	
$CaCO_3 + Ca_2SiO_3$	–	1186		400–900
$SrCO_3$ + 40 wt.% $SrZrO_3$	–	2360	572	1150–1235

capacity is attributed to (1) a loss of pore volume in the oxide and (2) to sintering of the carbonate [78]. Aihara et al. [78], Zhou et al. [79] and Zhao et al. [80] attempted improving the cyclability by inhibiting sintering via addition of inert $CaTiO_3$, $Ca_3Al_2O_6$, and $CaZrO_3$ particles, respectively. Cyclical stability improved; however, conversions were only up to 60%, i.e., 40% of CaO and the inert spacers contributed to parasitic sensible heating. The highest extent of conversion with good cyclability was shown by Zhao et al. [80] using Ca_2SiO_4 as an inert polymorphic spacer. Polymorphic expansion and contraction while transitioning from α- Ca_2SiO_4 to β-Ca_2SiO_4 to γ-Ca_2SiO_4 during decomposition and carbonation of $CaCO_3$/CaO induces cracks in the sintered reactant and results in surface regeneration. This leads to improved recyclability.

Pure $SrCO_3$ was investigated for TCES because of potentially higher operation temperatures and hence higher thermal efficiencies. The equilibrium transition temperature for the decomposition of $SrCO_3$ to SrO at carbon dioxide partial pressure of 1 atm is 1175°C. The enthalpy of reaction is 234 kJ/mol which corresponds to 1591 kJ/kg. Pure $SrCO_3$ showed poor recyclability for reasons similar to those in the $CaCO_3$ case. Attempts to stabilize the pore structure via the sacrificial pore formation method [81] did not show any improvements. Rhodes et al. [82] did achieve promising stability using 40 wt.% $SrZrO_3$ as material spacers. However, after 45 cycles moderate deactivation of the material was observed.

Pure $BaCO_3$ was first investigated for TCES in 2016. This system exhibits a very high equilibrium transition temperature of 1540°C at a carbon dioxide partial pressure of 1 atm. The enthalpy of decomposition of $BaCO_3$ to BaO is 272 kJ/mole. This corresponds to 1380 kJ/kg of $BaCO_3$. A phase transition from orthorhombic to hexagonal at 800°C and hexagonal to cubic at 980°C has been suggested [83, 84]. These phase transformations may boost the energy density of the material.

Mixing of $SrCO_3$ and $BaCO_3$ in the molar ratio of 2/3 was suggested as possible route to enhance the stability of $SrCO_3$ as well as for lowering the decomposition temperature of $BaCO_3$ [81]. According to Weinbruch et al. [85], the hexagonal phase is stable at Sr to Ba molar ratios greater than 2/3. This molar ratio resulted in good cyclic stability. More than 50% of the reactive material was converted to carbonate/oxide in each cycle and the estimated energy density was 790 kJ/kg. Similarly, Andre et al. [86] used MgO as a stabilizer to study the thermochemical recyclability of Ca, Sr, and Ba carbonates. They concluded that MgO did not change the reaction kinetics but clearly enhanced the cyclical stability of the carbonates. In the case of $BaCO_3$ and $CaCO_3$, the materials gradually kept losing carbonation capacity over cycles. However, in case of $SrCO_3$, addition of MgO resulted in exceptional thermochemical recyclability.

Alkali metal carbonates (Li_2CO_3, Na_2CO_3, K_2CO_3) were also considered for TCES. However, these metal carbonates melt before decomposing to their respective oxides and form clinkers that need pulverization for recarbonation. $PbCO_3$ was only considered temporarily for TCES due to toxicity.

Other reactions

Closed-loop reversible reactions Reversible gas phase reactions with large reaction enthalpies may be used to facilitate energy transport over long distance [87]. They can also be used for thermochemical energy storage. However, due to the gaseous nature of reactants and products volumetric energy density would be relatively low. Nevertheless, closed-loop energy distribution networks have the potential to substitute today's natural gas networks in a future carbon-neutral energy economy (see Fig. 18). The endothermic reaction may be driven by concentrated solar radiation, by nuclear reactions, or by direct electrical heating during times of excess renewable generation. Reaction products are then transported to consumers via pipelines. Heat is generated on site via the reverse, exothermic, reaction.

A closed-loop process based on methane steam reforming (see Table 5, #1) was first proposed for use with heat from nuclear reactors [88–90] using Ni [91] and Ru [92] catalysts. It is advantageous due to its high reaction enthalpy. The effectiveness of steam reforming is reduced by the exothermic water gas shift reaction:

$$CO + H_2O \leftrightarrow CO_2 + H_2$$

Similarly, dry reforming of methane has been considered for a closed-loop solar thermochemical energy system [93–101]. This process is also referred to as Solar Chemical Heat Pipe (SCP).

The main drawback of dry reforming is catalyst deactivation by coking due to the reverse Boudouard reaction:

$$2CO \leftrightarrow C + CO_2, \Delta H_{T=296\,K} = 172\,kJ/mol$$

Ammonia synthesis (Table 5, #3) and dissociation has been studied at the Australian National University for several decades [102–111]. Ammonia-based processes compare favorably to reforming reactions because there are no

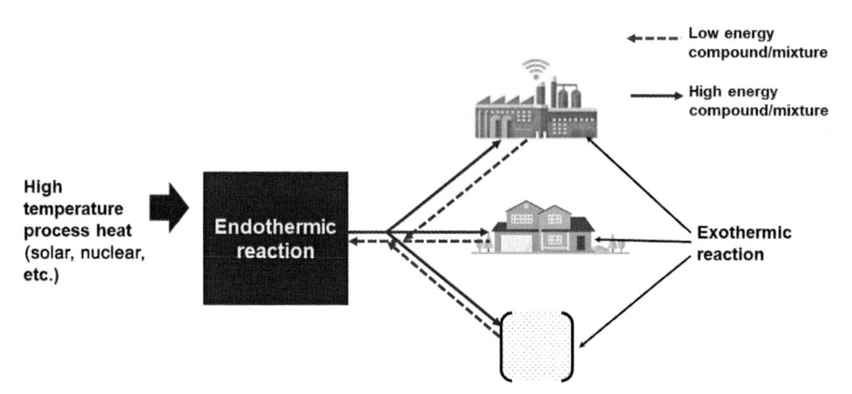

FIG. 18 Schematic depiction of closed-loop reversible thermal energy storage systems.

TABLE 5 Other reactions summary.

#	Name	Reaction	Temperature (°C)	Enthalpy of reaction (kJ/mol)
1	Steam reforming of methane	$CH_4 + H_2O \leftrightarrow CO + 3H_2$	750–1100 (forward), 400 (backward)	206
2	Dry reforming of methane	$CH_4 + CO_2 \leftrightarrow 2CO + 2H_2$	800–1100 (forward), 400 (backward)	261
3	Ammonia synthesis	$NH_3 \leftrightarrow \frac{1}{2}N_2 + \frac{3}{2}H_2$	250–600	66.5
4	Sulfur oxide dissociation	$SO_3 \leftrightarrow SO_2 + \frac{1}{2}O_2$	500–1000	98
5	Phosgene dissociation	$COCl_2 \leftrightarrow CO + Cl_2$	300–600	108
6	Nitrogen trifluoride dissociation	$NF_3 \leftrightarrow \frac{1}{2}N_2 + \frac{3}{2}F_2$	400–750	129
7	Sulfur-iodine cycle	I: $H_2SO_4 \rightarrow H_2O + SO_2 + \frac{1}{2}O_2$ II: $2H_2O + I_2 + SO_2 \rightarrow H_2SO_4 + 2HI$ III: $2HI \rightarrow H_2 + I_2$	I: 800 II: 20–100 III: 400– 700	–
8	NIS	I: $SO_2 + I_2 + 2H_2O \rightarrow 2HI + H_2SO_4$ II: $HI + H_2SO_4 + 2Ni \rightarrow NiI_2 + NiSO_4 + 2H_2$ III: $NiI_2 \rightarrow Ni + I_2$ IV: $NiSO_4 \rightarrow NiO + SO_3$ V: $SO_3 \rightarrow SO_2 + 1/2O_2$ VI: $NiO + H_2 \rightarrow Ni + H_2O$	I: Ambient, aqueous solution II: Aqueous/gas III: 650 IV: 850 V: 850 VI: 600	–

9	UT-3: Calcium-Iron-Bromine	I: $CaBr_2 + H_2O \rightarrow CaO + 2HBr$ II: $CaO + Br_2 \rightarrow CaBr_2 + \frac{1}{2}O_2$ III: $Fe_3O_4 + 8HBr \rightarrow 3FeBr_2 + 4H_2O + Br_2$ IV: $3FeBr_2 + 4H_2O \rightarrow Fe_3O_4 + 6HBr + H_2$	I: 700–750 II: 500–600 III: 200–300 IV: 550– 600	–
10	Ispra Mark 1	I: $CaBr_2 + 2H_2O \rightarrow Ca(OH)_2 + HBr$ II: $2HBr + Hg \rightarrow HgBr_2 + H_2$ III: $HgBr_2 + Ca(OH)_2 \rightarrow CaBr_2 + HgO + H_2O$ IV: $HgO \rightarrow Hg + 1/2O_2$	I: 777 II: 177 III: 177 IV: 627	–
11	Ispra Mark 9: Iron-chlorine	I: $6FeCl_2 + 8H_2O \rightarrow 2Fe_3O_4 + 12HCl + 2H_2$ II: $2Fe_3O_4 + 12HCl + 3Cl_2 \rightarrow 6FeCl_3 + 6H_2O + O_2$ III: $6FeCl_3 \rightarrow 6FeCl_2 + 3Cl_2$	I: 577–677 II: 127–177 III: 277–377	–
12	Ispra Mark 15	I: $3FeCl_2 + 4H_2O \rightarrow Fe_3O_4 + 6HCl + H_2$ II: $Fe_3O_4 + 8HCl \rightarrow FeCl_2 + 2FeCl_3 + 4H_2O$ III: $2FeCl_3 \rightarrow Fe_2Cl_6 \rightarrow 2FeCl_2 + Cl_2$ IV: $Cl_2 + H_2O \rightarrow 2HCl + ½O_2$		–
13	Calcium-iodine	I: $6CaO + 6I_2 \rightarrow Ca(IO_3)_2 + 5CaI_2$ II: $Ca(IO_3)_2 \rightarrow CaO + I_2 + 5/2O_2$ III: $5CaI_2 + 5H_2O \rightarrow 5CaO + 10HI$ III′: $5CaI_2 + 10H_2O \rightarrow 5Ca(OH)_2 + 10HI$ IV: $10HI \rightarrow 5H_2 + 5I_2$	I: 100 II: 550–800 III: 500–700 III′: 300–500 IV: 300–700	–

Continued

TABLE 5 Other reactions summary—cont'd

#	Name	Reaction	Temperature (°C)	Enthalpy of reaction (kJ/mol)
14	Magnesium-iodine	I: $6MgO + 6I_2 \rightarrow Mg(IO_3)_2 + 5MgI_2$	I: 100–150	–
		II: $Mg(IO_3)_2 \rightarrow MgO + I_2 + 5/2O_2$	II: 600	
		III: $5(MgI_2 \cdot H_2O) \rightarrow 5MgO + 25H_2O + 10HI$	III: 400	
		IV: $10HI \rightarrow 5H_2 + 5I_2$	IV: <400	
15	Ispra Mark 13: Sulfur-bromine cycle	I: $HBr \rightarrow H_2 + Br_2$(electrochemical)	I: Ambient	–
		II: $SO_2 + Br_2 + 2H_2O \rightarrow 2HBr + H_2SO_4$	II: Ambient	
		III: $H_2SO_4 \rightarrow H_2O + SO_2 + \frac{1}{2}O_2$	III: 800	
16	Copper-chloride	I: $2Cu + 2HCl \rightarrow 2CuCl + H_2$	I: 430–475	–
		I: $4CuCl \rightarrow 2CuCl_2 + 2Cu$ **(electrochemical)**	II: Electrochemical, 25–75	
		III: $2CuCl_2 + H_2O \rightarrow CuO{*}CuCl_2 + 2HCl$	III: 350–400	
		IV: $CuO{*}CuCl_2 \rightarrow 2CuCl + ½O_2$	IV: 530	
17	Magnesium-chloride	I: $MgCl_2 + H_2O \rightarrow 2HCl + MgO$	I: 450	–
		II: $MgO + Cl_2 \rightarrow MgCl_2 + ½O_2$	II: 500	
		III: $2HCl \rightarrow H_2 + Cl_2$ **(electrochemical)**	II: 80	

relevant side reactions, and the process is routinely used to superheat steam in chemical plants. However, the reaction enthalpy is comparably low and the exothermic synthesis reaction requires very high pressures (300 bar), while the usable temperature is relatively low (~300°C).

Reversible processes that have been studied to a lesser degree [112, 113] include Sulfur trioxide dissociation, phosgene dissociation, and nitrogen trifluoride dissociation (Table 5, #4,5,6)

One major obstacle to industrial-scale application of closed-loop solar chemical cycles is catalyst deactivation. Affordable base metal catalysts, e.g., Fe, Co, or Ni can degrade within hours. In contrast, noble metal catalysts, particularly Ru and Rh, remain active much longer, but impose significant costs when they need to be replaced.

Multistep reactions for hydrogen production

A wide range of three- and four-step thermochemical cycles were originally proposed for hydrogen production using heat generated in nuclear reactors in the 1970s and 1980s. Later, some were adapted for concentrated sunlight as the heat source. In these processes, energy would eventually be stored and transported in the form of hydrogen.

The Sulfur-iodine (S-I) cycle, proposed by General Atomics (see Table 5, #7, [114]), has been studied in great detail: The decomposition of sulfuric acid (I) is endothermic and requires temperatures exceeding 1073 K. It has been demonstrated experimentally, using Pd-Ag and ferric oxide catalysts [115]. A solar volumetric absorber reactor with Pt catalyst has been built and tested at the German Aerospace Center, DLR [116]. The exothermic Bunsen reaction (II) involves gaseous SO_2 absorbed at temperatures between 293 and 373 K. The slightly endothermic HI decomposition reaction (III) proceeds between 673 and 973 K [115]. Due to the moderate temperature requirements it is considered a possible near-term solution for solar hydrogen [117]. However, separation of H_2SO_4 and HI as well as gas phase separation of H_2 and I_2 have not been solved technologically [117, 118]. The Nickel-sulfur-iodine (NIS, Table 5, #8) cycle [119] is a variant of the S-I cycle that does not require acid separation and concentration.

The UT-3 cycle (Table 5, #9) is a four-step thermochemical process developed at the University of Tokyo [120–125]. It requires a maximum reaction temperature of 1023 K. According to Teo et al. [126] the upper efficiency limit of the UT-3 cycle is only about 12%. An assessment of research on thermochemical cycles carried out in the Ispra program is presented in Beghi [127]. Cycles studied in this program include the four-step Mark 1 process (see Table 5, #10) with a maximum reaction temperature of 1050 K—which was abandoned because it requires Mercury, the Mark 1C (Cu) cycle, the Mark 2 (Mn) cycle, the Mark 3 (V) cycle, and the Mark 6 cycles were investigated to avoid the use of mercury—with mixed results. Cycles of the Fe-Cl family were subsequently seen as superior because the elements involved are cheap, abundant, and relatively benign. The point of departure was the Ispra Mark 9 (Table 5, #11) and research cumulated in the Mark 15 (Table 5, #12); however, all FeCl cycles suffered from low thermodynamic efficiency and corrosion. They were eventually abandoned and later research focused on Sulfur-based cycles such as the hybrid Ispra Mark 13 and S-I cycles. Other thermochemical hydrogen production cycles include the Calcium Iodine [128] and the Magnesium Iodine cycle [129] (Table 5, #13 and # 14, respectively).

All of these cycles are complex; most suffer from low thermal efficiency and many have unfavorable thermodynamics or kinetics. With the advent of cheap, intermittent electricity from renewable sources, more suitable for electrochemical hydrogen production, industrial-scale deployment of these cycles is becoming increasingly unlikely.

Hybrid processes

Hybrid processes are a departure from purely thermochemical hydrogen production cycles. They substitute reactions that are hard to drive thermochemically or are otherwise unfavorable with electrochemical processes. One proponent is the Sulfur-Bromine (Ispra Mark 13, Table 5, #15) cycle, where the dissociation of HBr is driven electrochemically. Further examples are the Copper-Chloride and Magnesium-Chloride Cycles ([130], Table 5, #16 and #17, respectively). These cycles may have certain merits where large amounts of heat are produced in conjunction with electricity, i.e., in thermal (particularly nuclear) power plants. These cycles are not well suited for the most likely future scenarios based on cheap electricity from renewable, distributed, and intermittent sources. However, if nuclear fusion should eventually become viable or a new generation of nuclear fission plants is successful, these processes may see renewed interest.

References

[1] M.A. Miller, R.A. Page, Standardized Testing Program for Solid-State Hydrogen Storage Technologies, Final Report for the Department of Energy hydrogen and fuel cell program, Contract No. DEFC3602AL67619, July 30, Southwest Research Institute, San Antonio, TX, 2012.

[2] S. Satyapal, J. Petrovic, C. Read, G. Thomas, G. Ordaz, Catal. Today 120 (2007) 246–256.

[3] M. Armand, J.M. Tarascon, Building better batteries, Nature 451 (7179) (2008) 652.

[4] I. Gur, K. Sawyer, R. Prasher, Searching for a better thermal battery, Science 335 (6075) (2012) 1454–1455.

[5] A. Léon, A. Léon (Ed.), Hydrogen Technology, Springer-Verlag Berlin Heidelberg, 2008.

[6] L. Klebanoff, Hydrogen Storage Technology Materials and Applications, CRC Press, Taylor & Francis Group, Boca Raton, FL, 2013.

[7] R.C. Lochan, M. Head-Gordon, Phys. Chem. Chem. Phys. 8 (2006) 1357–1370.

[8] P. Bénard, R. Chahine, Storage of hydrogen by physisorption on carbon and nanostructured materials, Scr. Mater. 56 (10) (2007) 803–808.

[9] (a) H. Li, M. Eddaoudi, M. O'Keeffe, O.M. Yaghi, Nature 402 (1999) 276. (b) O.M. Yaghi, M. O'Keeffe, N.W. Ockwig, H.K. Chae, M. Eddaoudi, J. Kim, Nature 423 (2003) 705.
[10] (a) N.L. Rosi, J. Eckert, M. Eddaoudi, D.T. Vodak, J. Kim, M. O'Keeffe, O.M. Yaghi, Science 300 (2003) 1127. (b) J.L.C. Rowsell, J. Eckert, O.M. Yaghi, J. Am. Chem. Soc. 127 (2005) 14904.
[11] J.L.C. Rowsell, O.M. Yaghi, J. Am. Chem. Soc. 128 (2006) 1304.
[12] A.G. Wong-Foy, A.J. Matzger, O.M. Yaghi, J. Am. Chem. Soc. 128 (2006) 3494.
[13] H. Furukawa, M.A. Miller, O.M. Yaghi, Independent verification of the saturation hydrogen uptake in MOF-177 and establishment of a benchmark for hydrogen adsorption in metal-organic frameworks, J. Mater. Chem. 17 (2007) 3197.
[14] D. Yuan, D. Zhao, D. Sun, H.C. Zhou, Angew. Chem. Int. Ed. 49 (2010) 5357–5361.
[15] D. Yuan, W. Lu, D. Zhao, H.-C. Zhou, Highly stable porous polymer networks with exceptionally high gas-uptake capacities, Adv. Mater. 23 (32) (2011) 3723–3725.
[16] K.S. Chan, M.A. Miller, X. Peng, First-principles computational study of hydrogen storage in silicon clathrates, Mater. Res. Lett. 6 (1) (2017) 72–78.
[17] B. Bogdanovic, M. Schwickardi, J. Alloys Compd. 253–254 (1997) 1–9.
[18] G. Sandrock, K. Gross, et al., J. Alloys Compd. 330–332 (2002) 696–701.
[19] P. Dantzer, Metal-hydride technology: a critical review, in: H. Wipf (Ed.), Hydrogen in Metal III, Topics in Applied Pysics, vol. 73, Springer-Verlag, 1997, p. 279.
[20] K.J. Gross, G.J. Thomas, C. Jensen, J. Alloy. Compd. 330 (2002) 683–690.
[21] A.M. Puziy, A. Herbst, J.G. Poddubnaya, P. Harting, Langmuir 19 (2003) 314–320.
[22] P. Bénard, R. Chahine, Langmuir 17 (2001) 1950–1955.
[23] L. Zhou, J. Zhang, Y. Zhou, Langmuir 17 (2001) 5503–5507.
[24] R. Sips, J. Chem. Phys. 18 (1950) 1024.
[25] M. Dincă, A. Dailly, Y. Liu, C.M. Brown, D.A. Neumann, J.R. Long, J. Am. Chem. Soc. 128 (2006) 16876.
[26] L. Czepirski, J. Jagiełło, Chem. Eng. Sci. 44 (1989) 797.
[27] J. Jagiello, T.J. Bandosz, K. Putyera, J.A. Schwarz, J. Chem. Eng. Data 40 (1995) 1288.
[28] K. Murata, M. El-Merraoui, K. Kaneko, J. Chem. Phys. 114 (2001) 4196.
[29] H. Frost, T. Düren, R. Snurr, J. Phys. Chem. B 110 (2006) 9565.
[30] A.V. Neimark, P.I. Ravikovitch, Langmuir 13 (1997) 5148.
[31] J.A. Simmons, Reversible oxidation of metal oxides for thermal energy storage, in: Sharing the Sun: Solar Technology in the Seventies, vol. 8, 1976, pp. 219–225.
[32] A.J. Carrillo, D. Sastre, D.P. Serrano, P. Pizarro, J.M. Coronado, Revisiting the BaO_2/BaO redox cycle for solar thermochemical energy storage, Phys. Chem. Chem. Phys. 18 (11) (2016) 8039–8048.
[33] S. Wu, C. Zhou, E. Doroodchi, R. Nellore, B. Moghtaderi, A review on high-temperature thermochemical energy storage based on metal oxides redox cycle, Energy Convers. Manag. 168 (2018) 421–453.
[34] M.A. Fahim, J.D. Ford, Energy storage using the BaO_2/BaO reaction cycle, Chem. Eng. J. 27 (1) (1983) 21–28.
[35] L. Andre, S. Abanades, G. Flamant, Screening of thermochemical systems based on solid-gas reversible reactions for high temperature solar thermal energy storage, Renew. Sust. Energ. Rev. 64 (2016) 703–715.
[36] L. Till, Thermochemical data of barium peroxide from thermogravimetric measurements, J. Therm. Anal. Calorim. 3 (2) (1971) 177–180.
[37] R.G. Bowrey, J. Jutsen, Energy storage using the reversible oxidation of barium oxide, Sol. Energy 21 (1978) 523–525.

[38] D. Chadda, J.D. Ford, M.A. Fahim, Chemical energy storage by the reaction cycle CuO/Cu_2O, Int. J. Energy Res. 13 (1) (1989) 63–73.
[39] M. Hanchen, A. Stiel, Z.R. Jovanovic, A. Steinfeld, Thermally driven copper oxide redox cycle for the separation of oxygen from gases, Ind. Eng. Chem. Res. 51 (20) (2012) 7013–7021.
[40] E. Alonso, C. Pérez-Rábago, J. Licurgo, E. Fuentealba, C.A. Estrada, First experimental studies of solar redox reactions of copper oxides for thermochemical energy storage, Sol. Energy 115 (2015) 297–305.
[41] M. Deutsch, F. Horvath, C. Knoll, D. Lager, C. Gierl-Mayer, P. Weinberger, F. Winter, High-temperature energy storage: kinetic investigations of the CuO/Cu_2O reaction cycle, Energy Fuel 31 (3) (2017) 2324–2334.
[42] E.R. Monazam, R.W. Breault, R. Siriwardane, Kinetics of magnetite (Fe_3O_4) oxidation to hematite (Fe_2O_3) in air for chemical looping combustion, Ind. Eng. Chem. Res. 53 (34) (2014) 13320–13328.
[43] A.N. Grundy, B. Hallstedt, L.J. Gauckler, Assessment of the Mn-O system, JPE 24 (2003) 21–39. https://doi.org/10.1007/s11669-003-0004-6.
[44] A.J. Carrillo, D.P. Serrano, P. Pizarroa, J.M. Coronado, Thermochemical heat storage based on the Mn_2O_3/Mn_3O_4 redox couple: influence of the initial particle size on the morphological evolution and cyclability, J. Mater. Chem. A. 2 (2014) 19435.
[45] G. Karagiannakis, C. Pagkoura, A. Zygogianni, S. Lorentzou, A.G. Konstandopoulos, Monolithic ceramic redox materials for thermochemical heat storage applications in CSP plants, Energy Procedia 49 (2014) 820–829.
[46] C. Agrafiotis, M. Roeb, C. Sattler, A review on solar thermal syngas production via redox pair-based water/carbon dioxide splitting thermochemical cycles, Renew. Sust. Energ. Rev. 42 (2015) 254–285.
[47] V.B. Fetisov, N.V. Korchemkina, G.A. Kozhina, S.A. Petrova, R.G. Zakharov, E.A. Pastukhov, A.V. Fetisov, A.N. Ermakov, K.S. Mitrofanov, Kinetics of redox processes in manganese oxides, Inorg. Mater. 42 (4) (2006) 374–376.
[48] M. Chen, B. Hallstedt, L.J. Gauckler, Thermodynamic assessment of the Co-O system, J. Phase Equilibria 24 (3) (2003) 212–227.
[49] M. Neises, S. Tescari, L. de Oliveira, M. Roeb, C. Sattler, B. Wong, Solar-heated rotary kiln for thermochemical energy storage, Sol. Energy 86 (10) (2012) 3040–3048.
[50] C. Agrafiotis, A. Becker, M. Roeb, C. Sattler, Exploitation of thermochemical cycles based on solid oxide redox systems for thermochemical storage of solar heat. Part 5: testing of porous ceramic honeycomb and foam cascades based on cobalt and manganese oxides for hybrid sensible/thermochemical heat storage, Sol. Energy 139 (2016) 676–694.
[51] T. Block, M. Schmücker, Metal oxides for thermochemical energy storage: a comparison of several metal oxide systems, Sol. Energy 126 (2016) 195–207.
[52] C. Pagkoura, G. Karagiannakis, A. Zygogianni, S. Lorentzou, M. Kostoglou, A.G. Konstandopoulos, M. Rattenburry, J.W. Woodhead, Cobalt oxide based structured bodies as redox thermochemical heat storage medium for future CSP plants, Sol. Energy 108 (2014) 146–163.
[53] F. Varsano, C. Alvani, A. La Barbera, A. Masi, F. Padella, Lithium manganese oxides as high-temperature thermal energy storage system, Thermochim. Acta 640 (2016) 26–35.
[54] S.M. Babiniec, E.N. Coker, J.E. Miller, A. Ambrosini, Investigation of $La_xSr_{1-x}Co_yM_{1-y}O_{3-\delta}$ ($M = Mn, Fe$) perovskite materials as thermochemical energy storage media, Sol. Energy 118 (2015) 451–459.
[55] S.M. Babiniec, E.N. Coker, J.E. Miller, A. Ambrosini, Doped calcium manganites for advanced high-temperature thermochemical energy storage, Int. J. Energy Res. 40 (2) (2016) 280–284.

[56] K. Zhang, C. Zhang, L. Zhao, B. Meng, J. Liu, S. Liu, Enhanced oxygen permeation behavior of $Ba_{0.5}Sr_{0.5}Co_{0.8}Fe_{0.2}O_{3-\delta}$ membranes in a CO_2-containing atmosphere with a $Sm_{0.2}Ce_{0.8}O_{1.9}$ functional shell, Energ. Fuels 30 (3) (2016) 1829–1834.
[57] K. Randhir, K. King, N. Rhodes, L. Li, D. Hahn, R. Mei, N. Au Yeung, J. Klausner, Magnesium-manganese oxides for high temperature thermochemical energy storage, J. Energy Storage 21 (2019) 599–610.
[58] M. Felderhoff, R. Urbanczyk, S. Peil, Thermochemical heat storage for high temperature applications–a review, Green 3 (2) (2013) 113–123.
[59] C. Prieto, P. Cooper, A.I. Fernández, L.F. Cabeza, Review of technology: thermochemical energy storage for concentrated solar power plants, Renew. Sust. Energ. Rev. 60 (2016) 909–929.
[60] S. Li, H. Huang, X. Yang, Y. Bai, J. Li, N. Kobayashi, M. Kubota, Hydrophilic substance assisted low temperature $LiOH{\cdot}H_2O$ based composite thermochemical materials for thermal energy storage, Appl. Therm. Eng. 128 (2018) 706–711.
[61] M. Kubota, N. Horie, H. Togari, H. Matsuda, Improvement of hydration rate of LiOH/LiOH·H2O reaction for low-temperature thermal energy storage, in: *Procceding of the 2013 Annual Meeting of Japan Society of Refrigerating and Air Conditioning Engineers, Tokyo, Japan, September*, 2013, pp. 10–12.
[62] G. Ervin, D.K. Chung, T.H. Springer, A Study of the Use of Inorganic Oxides for Solar Energy Storage for Heating and Cooling of Buidings, (1974). *Final Report, June, 1.*
[63] G. Ervin, Solar heat storage using chemical reactions, J. Solid State Chem. 22 (1) (1977) 51–61. Elsevier.
[64] P.E. Halstead, A.E. Moore, 769. The thermal dissociation of calcium hydroxide, J. Chem. Soc. (1957) 3873–3875.
[65] B.V. L'vov, V.L. Ugolkov, Kinetics and mechanism of free-surface decomposition of Group IIA and IIB hydroxides analyzed thermogravimetrically by the third-law method, Thermochim. Acta 413 (1-2) (2004) 7–15.
[66] J.A.C. Samms, B.E. Evans, Thermal dissociation of $Ca(OH)_2$ at elevated pressures, J. Appl. Chem. 18 (1) (1968) 5–8.
[67] F. Schaube, L. Koch, A. Wörner, H. Müller-Steinhagen, A thermodynamic and kinetic study of the de-and rehydration of $Ca(OH)_2$ at high H_2O partial pressures for thermo-chemical heat storage, Thermochim. Acta 538 (2012) 9–20.
[68] A. Kanzawa, Y. Arai, Thermal energy storage by the chemical reaction augmentation of heat transfer and thermal decomposition in the $CaOCa(OH)_2$ powder, Sol. Energy 27 (4) (1981) 289–294.
[69] M. Hartman, et al., Decomposition kinetics of alkaline-earth hydroxides and surface area of their calcines, Chem. Eng. Sci. 49 (8) (1994) 1209–1216. Elsevier.
[70] I. Fujii, M. Ishino, S. Akiyama, M.S. Murthy, K.S. Rajanandam, Behavior of $Ca(OH)_2/CaO$ pellet under dehydration and hydration, Sol. Energy 53 (4) (1994) 329–341.
[71] M.S. Murthy, P. Raghavendrachar, S.V. Sriram, Thermal decomposition of doped calcium hydroxide for chemical energy storage, Sol. Energy 36 (1) (1986) 53–62.
[72] L. André, S. Abanades, Investigation of metal oxides, mixed oxides, perovskites and alkaline earth carbonates/hydroxides as suitable candidate materials for high-temperature thermochemical energy storage using reversible solid-gas reactions, Mater. Today Energy 10 (2018) 48–61.
[73] W.E. Wentworth, E. Chen, Simple thermal decomposition reactions for storage of solar thermal energy, Sol. Energy 18 (3) (1976) 205–214.
[74] M. Schmidt, C. Szczukowski, C. Roßkopf, M. Linder, A. Wörner, Experimental results of a 10 kW high temperature thermochemical storage reactor based on calcium hydroxide, Appl. Therm. Eng. 62 (2) (2014) 553–559.

[75] M. Schmidt, M. Linder, Power generation based on the $Ca(OH)_2/CaO$ thermochemical storage system–experimental investigation of discharge operation modes in lab scale and corresponding conceptual process design, Appl. Energy 203 (2017) 594–607.
[76] R. Barker, The reversibility of the reaction $CaCO_3 \rightleftarrows CaO + CO_2$, J. Appl. Chem. Biotechnol. 23 (10) (1973) 733–742.
[77] S.K. Bhatia, D.D. Perlmutter, Effect of the product layer on the kinetics of the CO_2-lime reaction, AICHE J. 29 (1) (1983) 79–86.
[78] M. Aihara, T. Nagai, J. Matsushita, Y. Negishi, H. Ohya, Development of porous solid reactant for thermal-energy storage and temperature upgrade using carbonation/decarbonation reaction, Appl. Energy 69 (3) (2001) 225–238.
[79] Z. Zhou, P. Xu, M. Xie, Z. Cheng, W. Yuan, Modeling of the carbonation kinetics of a synthetic CaO-based sorbent, Chem. Eng. Sci. 95 (2013) 283–290.
[80] M. Zhao, J. Shi, X. Zhong, S. Tian, J. Blamey, J. Jiang, P.S. Fennell, A novel calcium looping absorbent incorporated with polymorphic spacers for hydrogen production and CO_2 capture, Energy Environ. Sci. 7 (10) (2014) 3291–3295.
[81] K. Randhir, Porous Reactive Structures for Thermochemical Processes (Doctoral dissertation), University of Florida, 2016.
[82] N.R. Rhodes, A. Barde, K. Randhir, L. Li, D.W. Hahn, R. Mei, … N. AuYeung, Solar thermochemical energy storage through carbonation cycles of $SrCO_3/SrO$ supported on $SrZrO_3$, ChemSusChem 8 (22) (2015) 3793–3798.
[83] J.J. Lander, The Phase System BaO-$BaCO_3$, J. Am. Chem. Soc. 73 (12) (1951) 5893–5894.
[84] E.H. Baker, 137. The barium oxide–carbon dioxide system in the pressure range 0.01—450 atmospheres, J. Chem. Soc. (1964) 699–704.
[85] S. Weinbruch, H. Büttner, M. Rosenhauer, The orthorhombic-hexagonal phase transformation in the system $BaCO_3$-$SrCO_3$ to pressures of 7000 bar, Phys. Chem. Miner. 19 (5) (1992) 289–297.
[86] L. Andre, S. Abanades, Evaluation and performances comparison of calcium, strontium and barium carbonates during calcination/carbonation reactions for solar thermochemical energy storage, J. Energy Storage 13 (2017) 193–205.
[87] Q. Ma, L. Luo, R.Z. Wang, G. Sauce, A review on transportation of heat energy over long distance: exploratory development, Renew. Sust. Energ. Rev. 13 (6-7) (2009) 1532–1540.
[88] K. Kugeler, H.F. Niessen, M. Röth-Kamat, D. Böcker, B. Rüter, K.A. Theis, Transport of nuclear heat by means of chemical energy (nuclear long-distance energy), Nucl. Eng. Des. 34 (1) (1975) 65–72.
[89] H. Fedders, R. Harth, B. Höhlein, Experiments for combining nuclear heat with the methane steam-reforming process, Nucl. Eng. Des. 34 (1) (1975) 119–127.
[90] H. Fedders, B. Höhlein, Operating a pilot plant circuit for energy transport with hydrogen-rich gas, Int. J. Hydrog. Energy 7 (10) (1982) 793–800.
[91] A.M. Gadalla, B. Bower, The role of catalyst support on the activity of nickel for reforming methane with CO_2, Chem. Eng. Sci. 43 (11) (1988) 3049–3062.
[92] A. Berman, R.K. Karn, M. Epstein, Kinetics of steam reforming of methane on Ru/Al_2O_3 catalyst promoted with Mn oxides, Appl. Catal. A Gen. 282 (1–2) (2005) 73–83.
[93] R. Buck, J.F. Muir, R.E. Hogan, Carbon dioxide reforming of methane in a solar volumetric receiver/reactor: the CAESAR project, Solar Energy Mater. 24 (1-4) (1991) 449–463.
[94] A. Wörner, R. Tamme, CO_2 reforming of methane in a solar driven volumetric receiver–reactor, Catal. Today 46 (2-3) (1998) 165–174.
[95] D. Fraenkel, R. Levitan, M. Levy, A solar thermochemical pipe based on the CO_2 CH_4 (1:1) system, Int. J. Hydrog. Energy 11 (4) (1986) 267–277.

[96] R. Levitan, H. Rosin, M. Levy, Chemical reactions in a solar furnace—direct heating of the reactor in a tubular receiver, Sol. Energy 42 (3) (1989) 267–272.
[97] M. Levy, R. Levitan, H. Rosin, R. Rubin, Solar energy storage via a closed-loop chemical heat pipe, Sol. Energy 50 (2) (1993) 179–189.
[98] J.F. Muir, R.E. Hogan Jr., R.D. Skocypec, R. Buck, Solar reforming of methane in a direct absorption catalytic reactor on a parabolic dish: I—test and analysis, Sol. Energy 52 (6) (1994) 467–477.
[99] R.D. Skocypec, R.E. Hogan Jr., J.F. Muir, Solar reforming of methane in a direct absorption catalytic reactor on a parabolic dish: II—modeling and analysis, Sol. Energy 52 (6) (1994) 479–490.
[100] J.H. Edwards, K.T. Do, A.M. Maitra, S. Schuck, W. Fok, W. Stein, The use of solar-based CO_2/CH_4 reforming for reducing greenhouse gas emissions during the generation of electricity and process heat, Energy Convers. Manag. 37 (6-8) (1996) 1339–1344.
[101] R. Levitan, M. Levy, H. Rosin, R. Rubin, Closed-loop operation of a solar chemical heat pipe at the Weizmann Institute solar furnace, Solar Energy Mater. 24 (1–4) (1991) 464–477.
[102] P.O. Carden, Energy corradiation using the reversible ammonia reaction, Sol. Energy 19 (4) (1977) 365–378.
[103] H. Kreetz, K. Lovegrove, Theoretical analysis and experimental results of a 1 kWchem ammonia synthesis reactor for a solar thermochemical energy storage system, Sol. Energy 67 (4-6) (1999) 287–296.
[104] H. Kreetz, K. Lovegrove, Exergy analysis of an ammonia synthesis reactor in a solar thermochemical power system, Sol. Energy 73 (3) (2002) 187–194.
[105] K. Lovegrove, A. Luzzi, Endothermic reactors for an ammonia based thermochemical solar energy storage and transport system, Sol. Energy 56 (4) (1996) 361–371.
[106] K. Lovegrove, A. Luzzi, I. Soldiani, H. Kreetz, Developing ammonia based thermochemical energy storage for dish power plants, Sol. Energy 76 (1-3) (2004) 331–337.
[107] K. Lovegrove, A. Luzzi, H. Kreetz, A solar-driven ammonia-based thermochemical energy storage system, Sol. Energy 67 (4-6) (1999) 309–316.
[108] K. Lovegrove, A. Luzzi, M. McCANN, O. Freitag, Exergy analysis of ammonia-based solar thermochemical power systems, Sol. Energy 66 (2) (1999) 103–116.
[109] A. Luzzi, K. Lovegrove, A solar thermochemical power plant using ammonia as an attractive option for greenhouse-gas abatement, Energy (Oxford) 22 (2–3) (1997) 317–325.
[110] A. Luzzi, K. Lovegrove, E. Filippi, H. Fricker, M. Schmitz-Goeb, M. Chandapillai, Base-load solar thermal power using thermochemical energy storage, J. Phys. 4 9 (1999) 3–105.
[111] A. Luzzi, et al., Techno-economic analysis of a 10 MW solar thermal power, Sol. Energy 66 (2) (1999) 91–101.
[112] A.A. Koutinas, A. Lycourghiotis, P. Yianoulis, Industrial Scale Modelling of the Thermochemical Energy Storage System Based on CO/sub 2/+ 2NH/sub 3/in Equilibrium NH/sub 2/COONH/sub 4/Equilibrium, Name: Energy Convers. Manage, 1983.
[113] T.A. Chubb, Analysis of gas dissociation solar thermal power system, Sol. Energy 17 (1975) 129–136.
[114] D. O'Keefe, et al., Preliminary results from bench-scale testing of a sulfur-iodine thermochemical water-splitting cycle, Int. J. Hydrogen Energ. 7 (1982).
[115] V. Barbarossa, S. Brutti, M. Diamanti, S. Sau, G. De Maria, Catalytic thermal decomposition of sulphuric acid in sulphur-iodine cycle for hydrogen production, Int. J. Hydrog. Energy 31 (7) (2006) 883–890.
[116] A. Noglik, M. Roeb, T. Rzepczyk, J. Hinkley, C. Sattler, R. Pitz-Paal, Solar thermochemical generation of hydrogen: development of a receiver reactor for the decomposition of sulfuric acid, J. Solar Energy Eng. 131 (1) (2009) 011003–011007.

[117] C. Perkins, A.W. Weimer, Likely near-term solar-thermal water splitting technologies, Int. J. Hydrog. Energy 29 (15) (2004) 1587–1599.

[118] S. Abanades, P. Charvin, G. Flamant, P. Neveu, Screening of water-splitting thermochemical cycles potentially attractive for hydrogen production by concentrated solar energy, Energy 31 (14) (2006) 2805–2822.

[119] S. Sato, S. Shimizu, H. Nakajima, Y. Ikezoe, A nickel-iodine-sulfur process for hydrogen production, Int. J. Hydrog. Energy 8 (1) (1983) 15–22.

[120] A. Aochi, T. Tadokoro, K. Yoshida, H. Kameyama, M. Nobue, T. Yamaguchi, Economical and technical evaluation of UT-3 thermochemical hydrogen production process for an industrial scale plant, Int. J. Hydrog. Energy 14 (7) (1989) 421–429.

[121] K. Yoshida, et al., A simulation study of the UT-3 thermochemical hydrogen production process, Int. J. Hydrog. Energy 15 (3) (1990) 171–178.

[122] M. Sakurai, M. Aihara, N. Miyake, A. Tsutsumi, K. Yoshida, Test of one-loop flow scheme for the UT-3 thermochemical hydrogen production process, Int. J. Hydrog. Energy 17 (8) (1992) 587–592.

[123] M. Sakurai, A. Tsutsumi, K. Yoshida, Improvement of Ca-pellet reactivity in UT-3 thermochemical hydrogen production cycle, Int. J. Hydrog. Energy 20 (4) (1995) 297–301.

[124] M. Sakurai, E. Bilgen, A. Tsutsumi, K. Yoshida, Adiabatic UT-3 thermochemical process for hydrogen production, Int. J. Hydrog. Energy 21 (10) (1996) 865–870.

[125] M. Sakurai, N. Miyake, A. Tsutsumi, K. Yoshida, Analysis of a reaction mechanism in the UT-3 thermochemical hydrogen production cycle, Int. J. Hydrog. Energy 21 (10) (1996) 871–875.

[126] E.D. Teo, N.P. Brandon, E. Vos, G.J. Kramer, A critical pathway energy efficiency analysis of the thermochemical UT-3 cycle, Int. J. Hydrog. Energy 30 (5) (2005) 559–564.

[127] G.E. Beghi, A decade of research on thermochemical hydrogen at the Joint Research Centre, Ispra, Int. J. Hydrog. Energy 11 (12) (1986) 761–771.

[128] K. Fujii, W. Kondo, W. Mizuta, T. Kumagai, The calcium-iodine cycle for the thermochemical decomposition of water, Int. J. Hydrog. Energy 2 (4) (1977) 413–421.

[129] Y. Shindo, N. Ito, K. Haraya, T. Hakuta, H. Yoshitome, Kinetics of the catalytic decomposition of hydrogen iodide in the thermochemical hydrogen production, Int. J. Hydrog. Energy 9 (8) (1984) 695–700.

[130] M.C. Petri, B. Yildiz, A.E. Klickman, US work on technical and economic aspects of electrolytic, thermochemical, and hybrid processes for hydrogen production at temperatures below 550 C, Int. J. Nucl. Hydrog. Prod. Appl. 1 (1) (2006) 79–91.

Further reading

K. King, K. Randhir, J. Klausner, Calorimetric method for determining the thermochemical energy storage capacities of redox metal oxides, Thermochim. Acta 673 (2019) 105–118.

Y. Tadokoro, T. Kajiyama, T. Yamaguchi, N. Sakai, H. Kameyama, K. Yoshida, Technical evaluation of UT-3 thermochemical hydrogen production process for an industrial scale plant, Int. J. Hydrog. Energy 22 (1) (1997) 49–56.

M. Zhao, M. Bilton, A.P. Brown, A.M. Cunliffe, E. Dvininov, V. Dupont, ... S.J. Milne, Durability of $CaO–CaZrO_3$ sorbents for high-temperature CO_2 capture prepared by a wet chemical method, Energy Fuel 28 (2) (2014) 1275–1283.

Chapter 6

Heat engine-based storage systems

Jeff Moore[a], Natalie R. Smith[a], Gareth Brett[b], Jason Kerth[c], Rainer Kurz[d], Sebastian Freund[e], Miles Abarr[f], Jeffrey Goldmeer[g], Emmanuel Jacquemoud[h], Christos N. Markides[i], Karl Wygant[j], Michael Simpson[i], Richard Riley[b], Scott Hume[k], and Josh D. McTigue[l]

[a]Southwest Research Institute, San Antonio, TX, United States, [b]Highview Power, London, United Kingdom, [c]Siemens Energy, Houston, TX, United States, [d]Solar Turbines, San Diego, CA, United States, [e]Energiefreund Consulting, Munich, Germany, [f]Carbon America, Arvada, CO, United States, [g]General Electric, Schenectady, NY, United States, [h]MAN Energy, Zurich, Switzerland, [i]Imperial College London, London, United Kingdom, [j]Hanwha Power Systems America, Houston, TX, United States, [k]Electric Power Research Institute, Charlotte, NC, United States, [l]National Renewable Energy Laboratory, Golden, CO, United States

Chapter outline

Thermal, Mechanical, and Hybrid Chemical Energy Storage Systems.
https://doi.org/10.1016/B978-0-12-819892-6.00006-X

6.1 Thermodynamic cycles and systems

Contributors: Natalie Smith, Ph.D., Michael Simpson
Review: Rainer Kurz, Ph.D., Jeff Moore, Ph.D.

6.1.1 Heat engines and heat pumps

Thermodynamic cycles are powerful applications of the laws of thermodynamics, benefitting modern society. By enabling process directions that would not automatically occur, such as increasing pressure, producing work from other forms of energy, and moving heat from cold to warm, thermodynamic cycles are fundamental to the energy sector, and thus energy storage. Generally, thermodynamic cycles accomplish one of two functions and are categorized as such:

(1) **Heat Engine:** converts thermal energy from a high-temperature source to work output from the cycle,
(2) **Heat Pump:** converts work input to the cycle to transfer heat from low- to high-temperature reservoirs.

Both cycles are depicted in Fig. 1. Often heat engines are referred to as power cycles, as producing power from heat is their primary function. Heat pumps are also referred to as refrigeration cycles when the objective is to create a cold environment in the low-temperature reservoir. Additional cycle classifications are made based on working fluid phase (whether the working fluid is always a gas or the cycle crosses the vapor dome), indirect versus direct heating or combustion, open versus closed, etc. (Fig. 2).

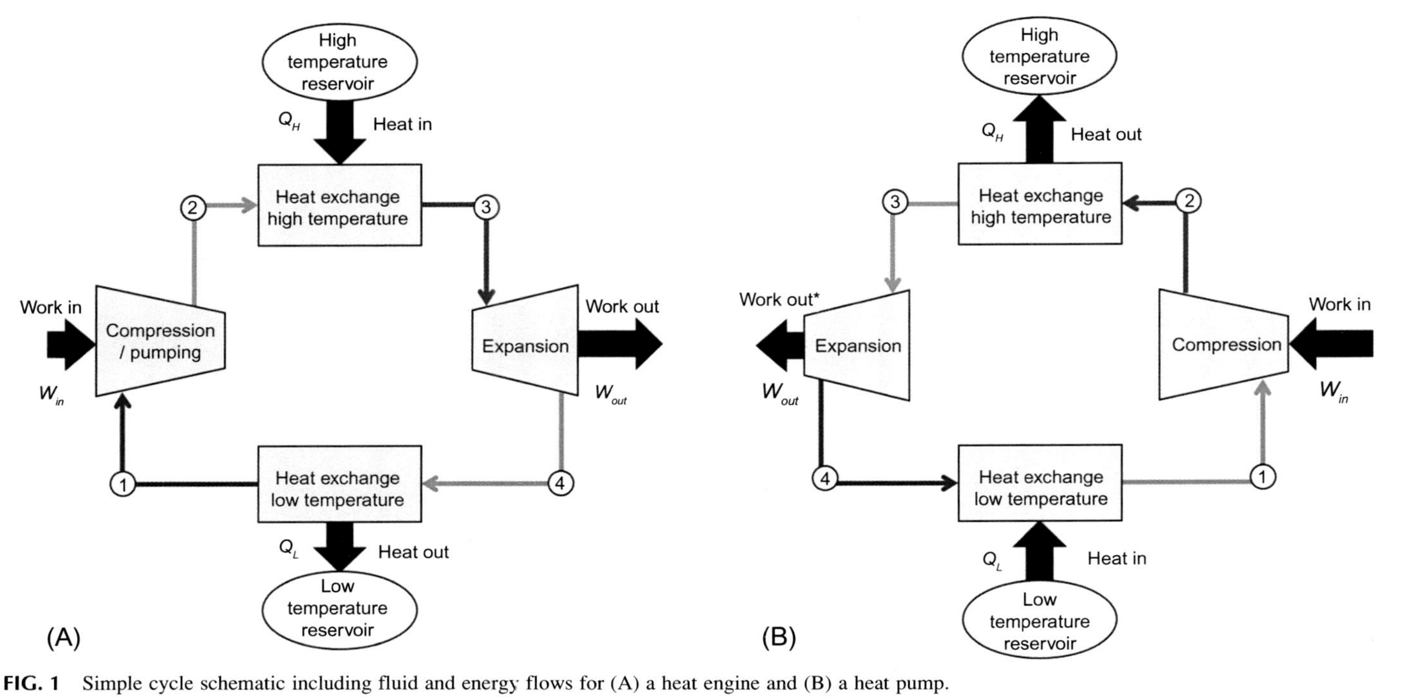

FIG. 1 Simple cycle schematic including fluid and energy flows for (A) a heat engine and (B) a heat pump.

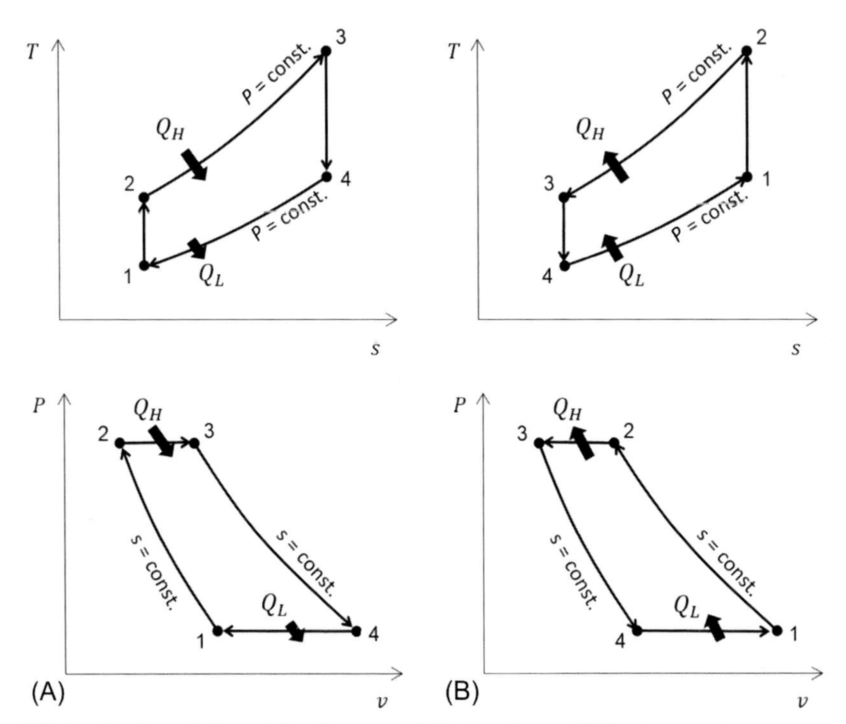

FIG. 2 Simple cycle *T*-*s* and *P*-*v* diagrams. (A) Heat engine and (B) heat pump.

Simple cycles involving four fundamental steady flow processes are depicted for both types in Fig. 1. These processes include compression or pumping, heat addition, expansion, and heat rejection and operate between two thermal reservoirs, high and low temperature. The same processes typically occur in both cycles just in reverse order of one another. For some heat pumps, expansion can be accomplished simply with an expansion valve, and therefore would have no work output. Performance of each cycle is evaluated as a comparison of the energy associated with the desired objective compared to the required energy associated with producing the objective. This results in slightly different parameters for each.

To define cycle performance, heat engines use thermal efficiency, η_{th}, which is the ratio of net work output, $W_{net,out}$, to heat addition, Q_H,

$$\eta_{th} = \frac{W_{net,out}}{Q_H}.$$

To define cycle performance, heat pumps use coefficient of performance, COP_{HP}, which is the ratio of heat exchange at high temperature, Q_H, to the net work input required, $W_{net,in}$,

$$COP_{HP} = \frac{Q_H}{W_{net,in}}.$$

For refrigeration heat pumps where the low-temperature reservoir is the object, the definition of coefficient of performance, COP_R, is altered to be the ratio of heat exchange at low temperature, Q_L, to the net work input required, $W_{net,in}$,

$$COP_R = \frac{Q_L}{W_{net,in}}.$$

Therefore it is important to maintain clear definitions of *COP* for discussions of heat pump and refrigeration cycle performance.

6.1.2 Carnot and reverse Carnot cycle

The ideal case for both of these types of cycles includes four completely reversible processes: adiabatic compression, isothermal heat addition, adiabatic expansion, and isothermal heat rejection, shown in Fig. 3. These cycles are referred to as the Carnot and reverse-Carnot cycles for the heat engine and heat pump, respectively. While not physically possible, these cycles provide a benchmark for the highest possible efficiency achievable between two temperature reservoirs. Combining the definition of thermal efficiency and the energy balance of the cycle ($W_{net,out} = Q_H - Q_L$), thermal efficiency can be rewritten as follows

$$\eta_{th} = \frac{Q_H - Q_L}{Q_H} = 1 - \frac{Q_L}{Q_H}.$$

For the fully reversible heat engine, such as the Carnot cycle, the ratio of heat transfers can be replaced with the ratio of absolute temperatures of the two thermal reservoirs, derived from a second law analysis of the cycle with the entropy flows without irreversibility:

$$\eta_{th,Carnot} = 1 - \frac{T_L}{T_H}.$$

In a similar manner, the coefficient of performance for a heat pump and for a refrigerator can be combined with the energy balance of the cycle ($W_{net,\ in} = Q_H - Q_L$)

$$COP_{HP} = \frac{Q_H}{W_{net,in}} = \frac{1}{1 - {Q_L}/{Q_H}}$$

$$COP_R = \frac{Q_L}{W_{net,in}} = \frac{1}{{Q_H}/{Q_L} - 1}.$$

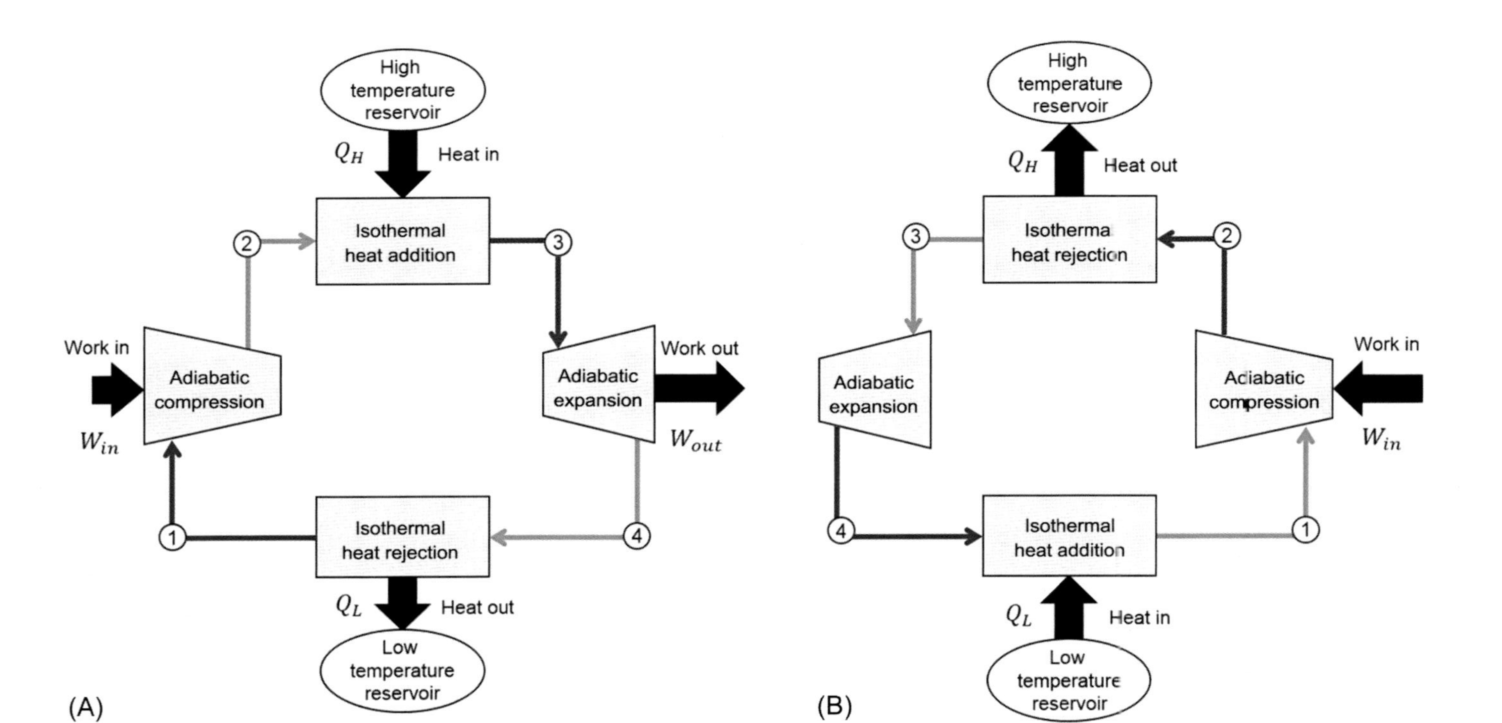

FIG. 3 Carnot cycle schematics. (A) Carnot cycle and (B) reverse Carnot cycle.

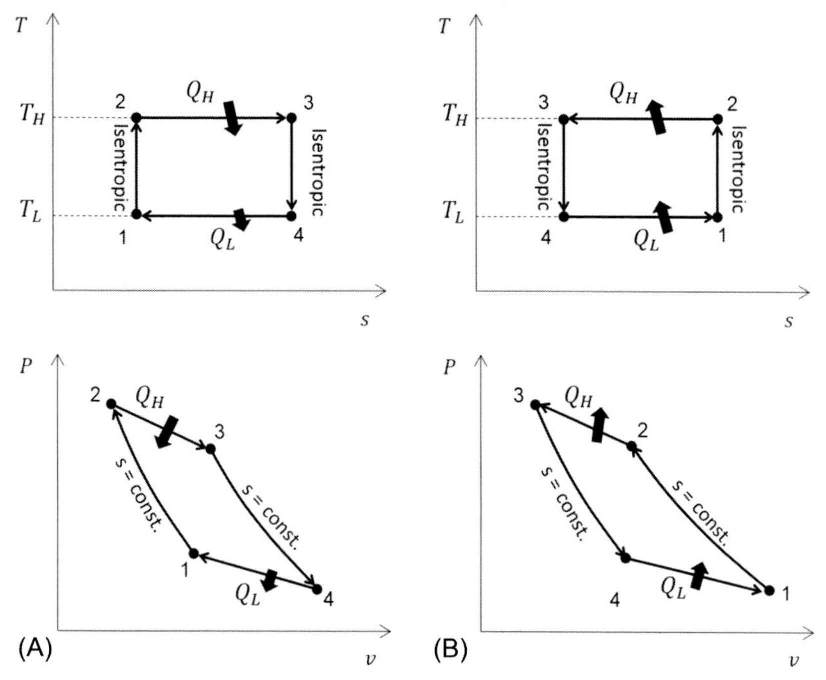

FIG. 4 Carnot cycle *T-s* and *P-v* diagrams. (A) Carnot cycle and (B) reverse Carnot cycle.

For the fully reversible heat pump, such as the reverse Carnot cycle, the ratio of heat transfers can be replaced with the ratio of absolute temperatures of the two thermal reservoirs (Fig. 4),

$$COP_{HP,Carnot} = \frac{1}{1 - {T_L}/{T_H}}$$

$$COP_{R,Carnot} = \frac{1}{{T_H}/{T_L} - 1}.$$

6.1.3 Round-trip efficiency

The energy storage methods discussed in this chapter will utilize versions of heat engines and heat pumps. In some cases, they may not use the complete cycle, and in other cases, they will use more complex versions which incorporate recuperation, various flow splits, intercooling, reheat, etc. Generally, a charge cycle which converting excess power or thermal energy into stored energy will use a heat pump cycle, and a discharge cycle which converts stored energy to work will use a heat engine. Each subsection will define the details of typical cycle arrangements used for that energy storage technology.

To evaluate the total system performance of an energy storage technology, the performance metric of round-trip efficiency, *RTE*, is often defined. It compares the energy input to stored energy in charge mode to the energy output from stored energy in discharge mode,

$$RTE = \frac{W_{net,outin} * Duration}{W_{grossnet,inout} * Duration} = \frac{Power_{net,out} * Duration\_Discharge}{Power_{gross,in} * Duration\_Charge}.$$

The details of this definition may vary depending on the specific energy storage technology.

6.1.4 Exergy

Exergy is the amount of energy that can be used within a thermal system. While energy conservation is tied to the 1st law of thermodynamics, exergy is closely tied to the 2nd law of thermodynamics. Exergy is thus the part of the energy that, within a certain environmental conditions, can be converted into any other type of energy. For the remaining part of the energy (often referred to as Anergy), this is not possible. In a process in a closed system, energy stays constant (1st law), but exergy is reduced in processes, unless the processes are reversible.

The exergy of heat is the portion of heat that can be converted to other forms of energy. This is related to the Carnot efficiency, because the exergy of heat relative to the energy for a reversible process is equal to the Carnot efficiency. The exergetic efficiency is thus a better indicator of the quality of a process, since it compares the actual process with the best possible process. Thermal efficiency, on the other hand, compares the process with an impossible process.

6.1.5 Working fluids

All thermodynamic cycles have at their very core the properties of the working fluid that define their constituent processes. The working fluid, in this context, is the substance which is used to convert between thermal energy and mechanical energy, exploiting property changes during compression and expansion, and/or changes in the phase of the fluid. Therefore a broad range of considerations naturally come into play when choosing a suitable fluid. Some of the key ones are as follows:

- Working temperature range
- Heat transfer properties
- Pressure range
- Safety/environmental impact
- Density (Power density is not a fluid property?)
- Cost

- Degradation and material compatibility
- Phase transition pressure and temperature?

Each of these is further discussed in the following sections.

6.1.6 Working temperature range

Arguably the most obvious criterion for working fluid selection, a brief consideration of the working temperature range of the proposed thermodynamic cycle quickly screens out many candidate fluids. Cycles which are reliant on phase change at given temperatures, such as Rankine cycles, lead to a sharp delineation of viable and nonviable working fluids according to their critical temperatures, while thermal decomposition temperatures can present upper limits on cycle temperatures, particularly for organic fluids. Similarly, the freezing temperature of the fluid must be well below the lowest operating temperature of the cycle. Nevertheless, the exercise often remains nontrivial after this filtering. For example, pumped heat energy storage designs that rely on organic Rankine cycles (e.g., [1]) still have a vast range of fluids to choose from. Approaches such as computer-aided molecular design (CAMD) have been proposed that are highly suitable for fluid selection or design, allowing a systematic search of familiar and as yet unexplored molecular structures for high-performing options [2–4] (Figs. 5 and 6).

For energy storage technologies that store thermal energy at temperatures above ambient, a selection of heat-engine cycles is shown in Figs. 1–3. The plot shows transitions between different working fluids and cycles for different heat source temperatures, which in this context could refer to a high-temperature thermal store. For thermal store temperatures below 400°C and at smaller scales, organic Rankine cycle (ORC) engines offer some of the highest thermal efficiencies [5], while steam Rankine cycles have historically been favored for temperatures between around 400°C and 800°C and larger systems. Supercritical carbon dioxide (sCO_2) cycles have important advantages (e.g., low environmental impact of the working fluid, compact turbines) and promise high efficiencies over a broad range of temperatures, although typically at higher temperature than ORCs, exploiting the abrupt change in the compressibility factor of CO_2 close to its critical point [6]. However, this promise has not yet been proven in practice, at scale, with significant developments continuing to take place this space.

6.1.7 Pressure range

A related consideration is the pressure range across which the fluid achieves the desired temperatures and phase behavior. While fluids at high pressure tend to offer high densities and compact systems, the cost of containment can be high. Supercritical CO_2 cycles, for example, permit high thermal efficiencies within a

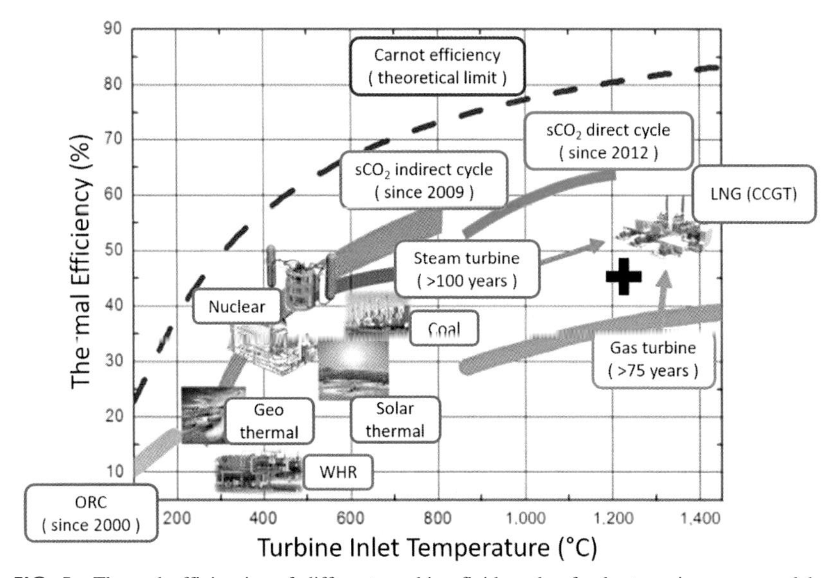

FIG. 5 Thermal efficiencies of different working-fluid cycles for heat engines, arranged by maximum cycle temperature, which takes place at the turbine inlet. *(Figure taken from Y. Ahn, et al., Review of supercritical CO2 power cycle technology and current status of research and development, Nucl. Eng. Technol. 47 (6) (2015), 647–661.)*

small physical footprint, but require components to withstand pressures that are, by definition, in excess of 73 bar [6]. Conversely, the use of fluids at low pressure can lead in some cases to thermodynamic benefits, in increasing power density and efficiencies, but subatmospheric operation and the potential for air ingress can lead to engineering complications and added costs.

The specific heat ratio, γ, of the fluid is relevant to the rate of change of pressure with temperature: monatomic gases with higher γ values require a smaller change in pressure for a given temperature change, relative to diatomic or heavier gases, leading to gases such as argon being considered for pumped thermal energy storage [7]. This thermodynamic effect must be considered alongside other important fluid properties, such as thermal conductivity and viscosity, which influence heat and mass transfer, and therefore the overall performance of relevant systems from cost and power-density perspectives. Gases such as helium and hydrogen, with their low molecular weights, also have high sonic velocities, meaning that turbine designs are less likely to be limited by nozzle exit velocities [8].

6.1.8 Heat transfer properties

As mentioned previously, a number of properties influence the performance of a working fluid with regard to heat transfer. The specific heat capacity and

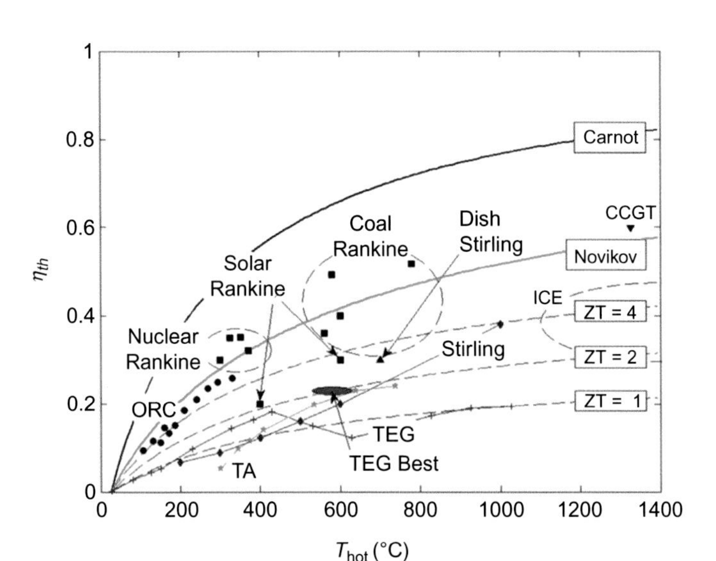

FIG. 6 Thermal efficiencies of different heat engines, arranged by heat source temperature. *CCGT*, combined cycle gas turbine; *ORC*, organic Rankine cycle; *TA*, thermo-acoustic engine; *TEG*, thermo-electric generator. ZT is a parameter used to characterize thermo-electric generators: ZT = 2 is currently achievable, while a value of 4 is well beyond present capabilities. *(Figure taken from C. N. Markides, Low-concentration solar-power systems based on organic rankine cycles for distributed-scale applications: overview and further developments, Front. Energy Res. 3 (December) (2015), 1–16.)*

thermal conductivity at relevant conditions are central, while the latent heat of vaporization is also a driver where phase change is present. A working fluid with a high thermal conductivity is desirable, to enable high heat transfer coefficients and minimize the required area for heat exchange [9]. Accurate matching of heat capacity flows ($\dot{m}c_P$) during heat exchange is also important for high cycle efficiency. Consequently, fluids with low specific heat capacities can cause difficulties by requiring high mass flow rates to absorb the available heat, leading to physically larger components and/or increased pressure drop. The pressure drop is also strongly influenced by the viscosity of the fluid.

6.1.9 Safety/environmental impact

The safety and environmental impact of a working fluid should be considered on a whole-life basis, taking into account the raw material inputs, production of the fluid (if not naturally occurring), transport and handling, manufacture and assembly of the energy storage system, operational considerations, and end of life disposal/recycling.

Among the working fluids described within this chapter, a broad range of safety aspects need to be handled appropriately. Some of the safety hazards include:

- Flammability (e.g., hydrocarbons, hydrogen)
- Oxygen depletion (liquid nitrogen, carbon dioxide)
- Extreme heat (multiple) or cold (cryogenic cycles)
- High pressures (compressed air, hydrogen, carbon dioxide cycles)
- Toxicity (e.g., ammonia)

The environmental impact of the working fluids is equally relevant, particularly in light of the need to build a more sustainable energy system, which has been a major driver of the need for energy storage in the first place. In this context, the global warming potential (GWP) of certain refrigerants becomes a challenge for certain heat pump/engine cycles, though concerted efforts are underway to move to low-GWP fluids in sectors such as refrigeration, which expands the range of options. Similarly, the ozone depletion potential (ODP) of some refrigerants poses an environmental concern, though again, more advanced alternatives are often available to traditional high-ODP choices [10].

More broadly, life cycle analysis is a key tool in helping to assess the overall environmental impact of the chosen working fluid, of which GWP and ODP are just two categories. Others include depletion of natural resources, toxicity to marine life, and effect on air quality.

6.1.10 Power density

A working fluid that gives rise to a higher power density for the energy storage system is desirable in order to make a given system more compact, and to realize potential cost savings from smaller components and land requirements. For working fluids operating in the vapor phase, a high vapor density is valuable, reducing the volumetric flow rates and pressure drops in any heat exchangers [9].

6.1.11 Cost

The choice of fluid drives cost both directly and indirectly, through purchase of the working fluid itself, and through its influence on the hardware components that constitute the heat engine. The latter is typically a much larger effect for commonly used working fluids. Where costing data is not readily available for precise comparisons, preliminary estimates can be obtained using costing approaches such as the module costing technique [11], allowing early stage technologies to be compared ahead of fully costed plant designs [12]. A small sample of purchase prices of possible working fluids is listed in Table 1—it is quickly apparent that the cost per unit mass of working fluids varies over a very wide range.

TABLE 1 Indicative prices for various working fluids, for small-batch purchases (10–100s of kg) in Europe.

Working fluid	Indicative price ($/kg)	Source
Argon	7	[13]
Carbon dioxide	1	[13]
Helium	450	[14]
Propane	3	[13]
R1233zd(e)	26	[15]
R245fa	38	[15]

Prices vary according to location, scale, and distribution channel. Presented in USD (2019).

6.1.12 Degradation and material compatibility

Degradation of working fluids can take place through contamination, thermal and chemical decomposition. Contamination with air, water, and dirt can lead to reduced performance (e.g., due to the presence of noncondensable gases), corrosion, and erosion (e.g., turbine blade damage). Insufficient thermal stability of the working fluid at the required operating conditions can lead to the formation of corrosive or fouling decomposition products [16], leading to reduced performance and additional maintenance costs. Similarly, compatibility issues can lead to chemical decomposition, for example due to reaction with lubricating oils or metals. It is important to test working fluids in the presence of the materials that they will be exposed to [17]: for example the thermal decomposition of refrigerant R-11 is accelerated by a factor of 20 in the presence of 18-8 stainless steel compared to platinum [18], while the presence of air almost doubles the decomposition of cyclopentane at 300°C [19].

Compatibility needs to be considered regarding not only undesirable reactions and corrosive effects, but also for the implications on how the system is operated. On the latter point, refrigerants, for example, exhibit a large variation in their miscibility with and solubility in lubricating oil [20]. While some can be readily used with a range of oils, others require particular lubricants such as polyolester oils, which require additional operating protocols to manage water absorption.

6.2 Cryogenic energy storage

Contributors: Gareth Brett, Scott Hume, Richard Riley, Josh McTigue.

This section examines the use of cryogenic liquefied gas as energy storage media and working fluids for rechargeable energy storage systems.

Although there are several potential gases which can be used in this way, the cycling of very large volumes of working fluid gas has limited the practical application of this storage type to air—an abundant and free process fluid. In general, the processes examined in this section are based on existing power generation and industrial gas liquefaction technologies (and their associated turbomachinery) as a basis for a large-scale storage system which use liquid air as its working fluid and energy storage medium, commonly referred to as a Liquid Air Energy Storage (LAES) system.

6.2.1 Background

Liquid Air Energy Storage (LAES) as a utility-scale concept was first proposed independently by Shepherd in 1974 [21] and Smith in 1977 [22]. This was followed by patents from Mitsubishi, Hitachi, and others in the 1990s. LAES is a potential alternative to pumped hydro or cavern-based compressed air energy storage (CAES), both of which have restrictive geographical requirements which engender significant site survey costs and permitting issues such as the environmental impact assessment of flooding large areas of land required for reservoirs in the case of pumped hydro.

The early LAES systems proposed were essentially variants on first generation CAES which is based on separating (in time) the compression and combustion/expansion sections of the Brayton cycle. In the liquid air case, instead of storing the compressed working fluid in a cavern, the air is refrigerated, liquefied, and stored (at low/close to atmospheric pressure) in a large insulated tank, which can be sited, within reason, anywhere (Fig. 7).

The drawback of this kind of CAES system (diabatic CAES) is the requirement to burn fuel (usually natural gas) to provide heat for the power recovery part of the cycle.

CAES developers have addressed this shortcoming through two alternative designs (both considered as second generation CAES systems):

- "Advanced Adiabatic Compressed Air systems" (an example being the MHPS project in Utah)[a]; and,
- "Isothermal Compressed Air systems" (such as those proposed by the US early stage companies SustainX and LightSail).

Both of these approaches are transferable to LAES systems with both isothermal compression [23] and adiabatic systems [24] proposed in the literature. In developing LAES systems, one of the leading technology developers in the field, Highview Power, has chosen to adopt a hybrid approach involving aspects of the adiabatic and isothermal CAES systems to remove the requirement for a fuel source while maintaining efficiency.

a. https://arstechnica.com/science/2019/05/power-systems-company-to-build-worlds-first-1-gw-energy-storage-project-in-utah/.

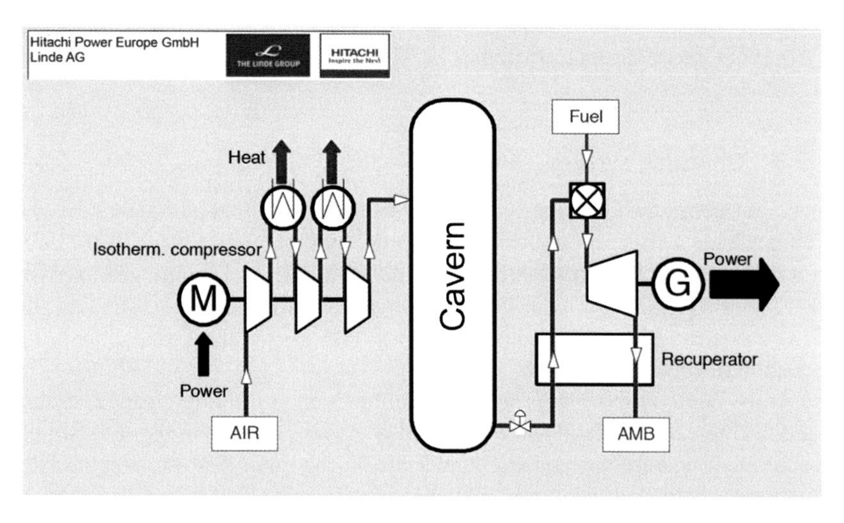

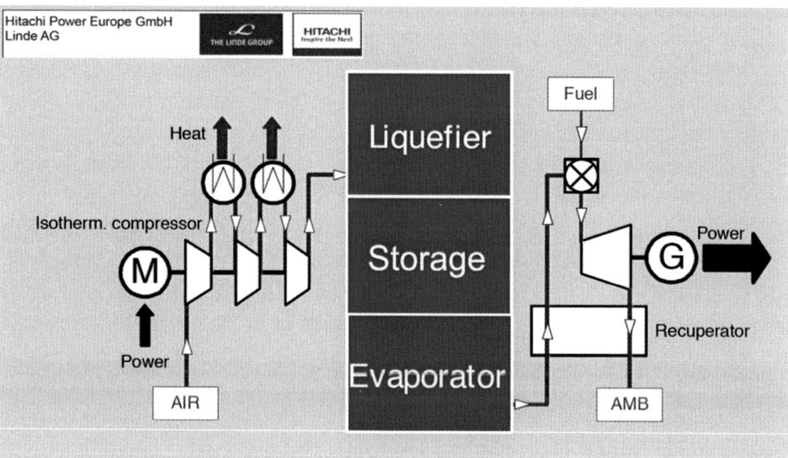

FIG. 7 First generation compressed air and equivalent liquid air energy storage systems. *(From R. Stover, S. Alekseev, Liquid Air Energy Storage: A flexible and Widely Applicable Medium Term Large Scale Energy Storage Concept, 2013 (Linde and Hitachi).)*

6.2.2 LAES system description

- A LAES system can be described as a thermomechanical energy storage system, where electrical energy is stored in the form of thermal energy—the primary form of which is a cryogenic liquid storage system. Hot and cold thermal storage systems may be used in parallel with the cryogenic store to capture and redistribute energy during charging and discharging of LAES, to maximize the storage efficiency. Mechanical equipment (e.g., compressors, expanders, heat exchangers) is used in the charging and discharging processes to convert the electrical energy to and from these thermal energy storage systems.

6.2.3 Charging system

In simple terms, the charging system may be based on either a Linde Cycle or a Claude Cycle refrigeration process and works by first compressing ambient air in a compression train (typically to around 60 barg), cooling the gas to cryogenic temperatures in a main heat exchanger, and then finally expanding the air across a throttling valve, which induces flash evaporation, thus liquefying a portion of the air.

At the beginning of the process, the ambient air is initially cleaned of dust using a filter. Subsequently, it is cleaned of water and carbon dioxide, which would otherwise freeze, and eventually block, the main heat exchanger. This is typically achieved using a standard air purification skid based on pressure and temperature swing adsorbers, with the captured CO_2 and water rejected back into the atmosphere during the regeneration part of the skid cleaning cycle. If economically justified, permanent capture of the atmospheric CO_2 could be implemented using appropriate technology as an alternative to the currently commercially available systems (Fig. 8).

A temperature-entropy diagram for the charging phase of LAES is shown in Fig. 9. Atmospheric air is first compressed to supercritical pressures in two stages of compression, and the heat of compression is stored in hot storage. Prior to expansion, this heat can then be supplied to the working fluid of the discharge system, thereby increasing the system power output and avoiding the need for firing from gas or other fuel source. The number of compression stages and pressure ratio may be varied to achieve a range of hot storage temperatures. The illustrated cycle is a Linde cycle: the supercritical air (5) is cooled further to cryogenic temperatures (6) by a combination of recycled air and cold storage. The cryogenic fluid (6) is expanded in a Joule-Thomson valve creating a liquid-vapor mixture (7). The mixture is separated, with roughly 75% of the fluid stored as liquid (7f) and the remaining vapor (7v) recycled and used to cool the fluid from 5 to 6.

If the liquefaction cycle is based on the Claude Cycle then some of the compressed air is diverted from the main process stream and is expanded to provide a refrigeration source, which is then used in the main heat exchanger to cool the process gas that is to be liquefied. The source of refrigeration for cooling can be supplemented by use of high-grade cold recycle; this is where cold from the discharging process previously captured and stored in a thermal energy storage system is applied to the main process flow.

The liquid product is stored at low pressure in insulated cryogenic liquid vessels.

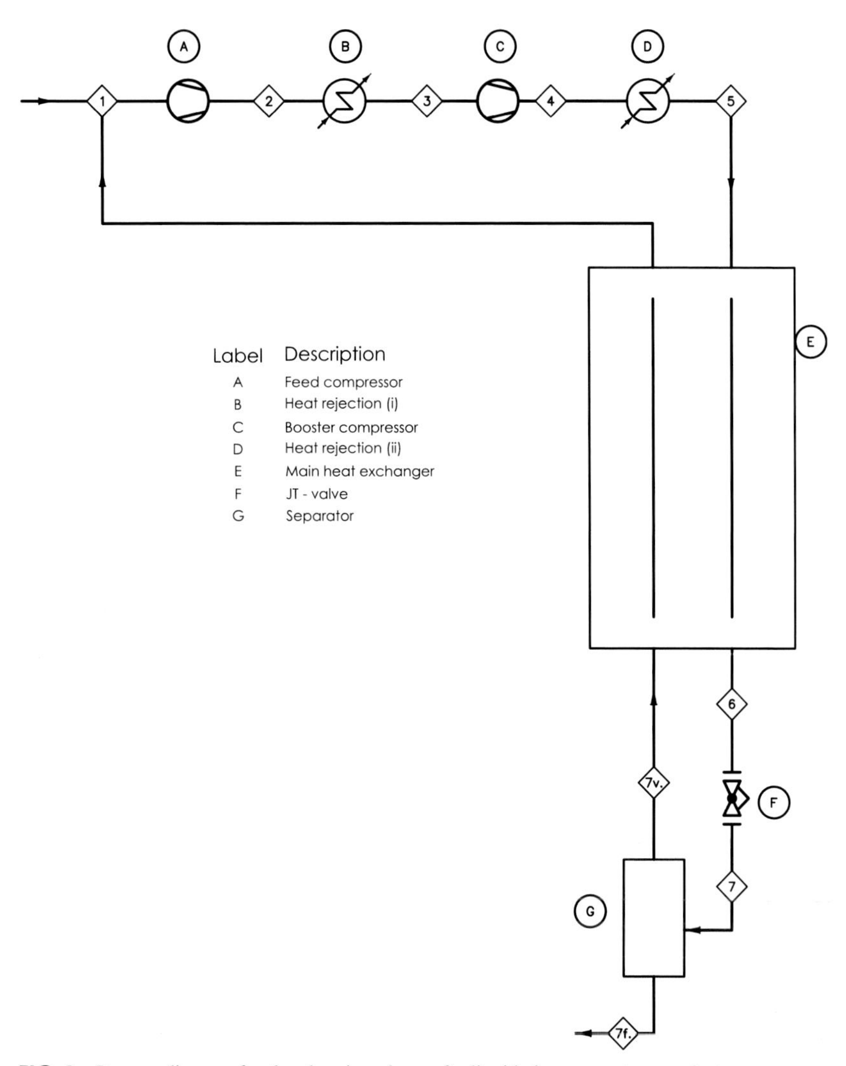

FIG. 8 Process diagram for the charging phase of a liquid air energy storage plant.

6.2.4 Discharging system

The discharging system is akin to a supercritical Rankine cycle. Liquid air is drawn from the cryogenic vessels and pumped to high pressure using a cryogenic pump (typically up to around 150 barg). Heat is then added to the working fluid in two stages: the first stage heats the fluid in a heat exchanger, where the fluid changes phase from a liquid to a gas, exiting the heat exchanger close to ambient temperature; the second stage is where a heat source is used to further increase the temperature of the working

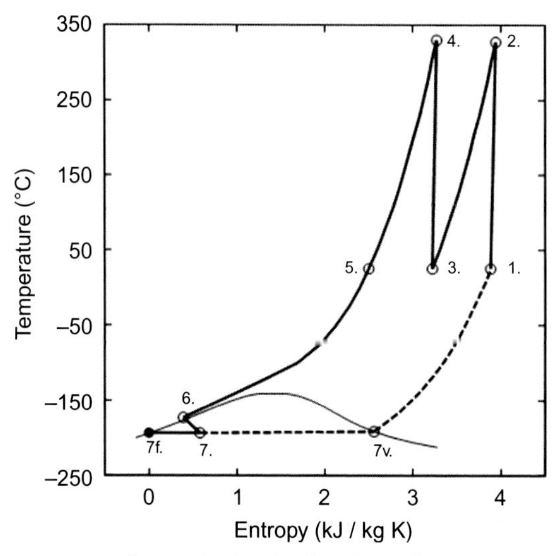

FIG. 9 Temperature-entropy diagram for the charging phase of a liquid air energy storage plant.

fluid.[b] The air is now expanded across a turbine to extract mechanical work from the working fluid. To increase the efficiency of the process, maximizing the work output, multiple expansion stages with reheat stages are often desirable.

During the first stage of heating of the working fluid, valuable cold can be captured from the cryogenic fluid and stored in a cold thermal storage system. This cold can then be used in the charging system to supplement the refrigeration of the process gas, minimizing the required electrical load for liquefaction (Fig. 10).

6.2.5 Pilot plant

Highview has demonstrated its cycle at pilot scale: its 300 kW pilot plant operated from 2011 to 2014. A simplified process flow diagram of Highview's pilot plant is shown in the following diagram. Note that in this system, the heat of compression is rejected through a cooling system; however, the cold captured from evaporation of liquid air is retained in a high-grade cold store and recycled to enhance the liquefier performance during the charging process.

A photograph of the pilot plant and a diagram showing its component parts are also shown so the physical configuration can be seen (Figs. 11–13).

---

b. As the working fluid is at a supercritical pressure when heated, the fluid undergoes what is sometimes termed "pseudoboiling": the heat provides the energy to overcome the fluid's intermolecular forces, but simultaneously increases its sensible energy. Hence, the temperature of the fluid increases throughout this process.

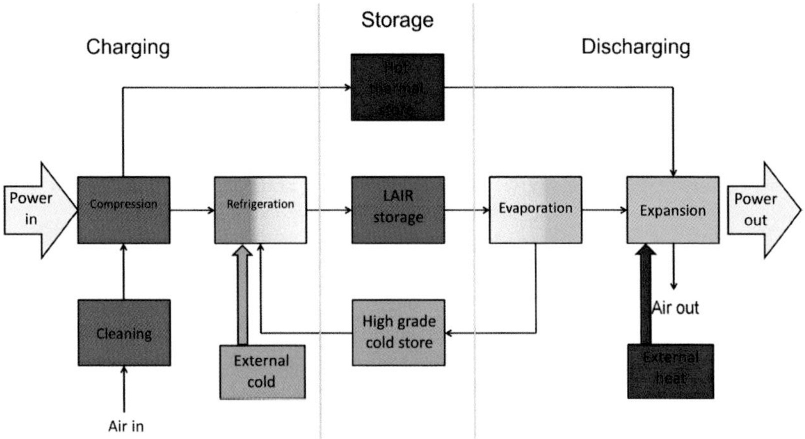

FIG. 10 Conceptual model of Highview LAES system.

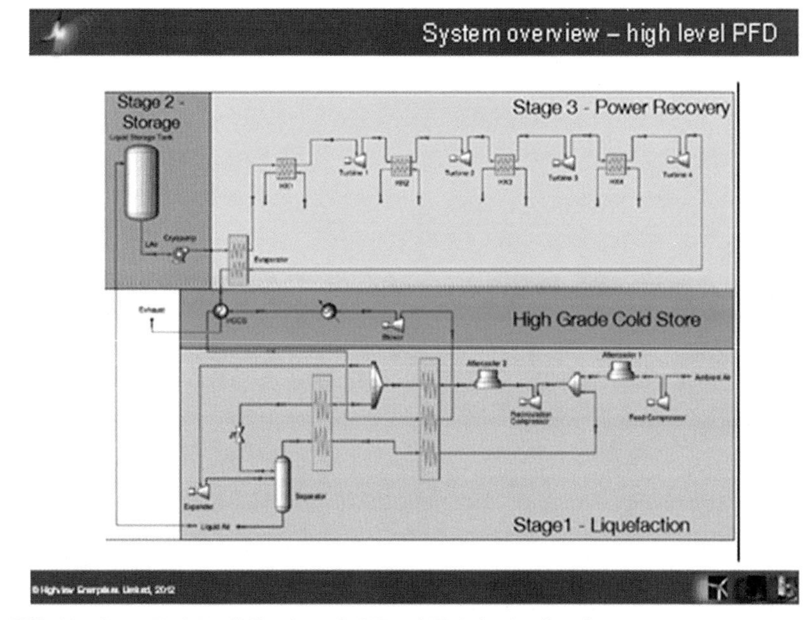

FIG. 11 AspenTech HYSYS schematic PFD of Highview's pilot plant.

FIG. 12 Photograph of Highview's pilot plant at slough heat and power.

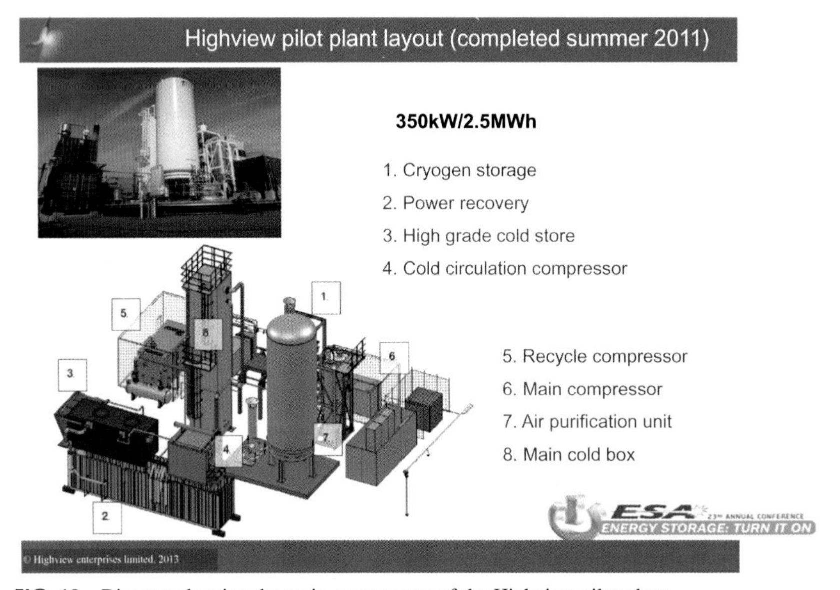

FIG. 13 Diagram showing the main components of the Highview pilot plant.

6.2.6 Performance

In general, LAES performs better at larger scale; turbomachinery is typically more efficient (lower relative leakage) and the standing losses (thermal leakage) from the storage tanks are lower due to the relative change in system volume and surface area. In addition, the specific cost of systems is lower at larger scale, improving the system economics (dealt with separately later).

The key indicators for storage system performance usually relate to energy density, AC to AC round-trip efficiency, flexibility and response time, and standing losses.

- *Energy density*—the electrical energy recoverable from 1 kg of liquid air: approximately 0.15 kWh/kg for a 50 MW scale stand-alone plant (assuming max. reheat temperature of 370°C recovered from liquefier compressors).
- *AC to AC round-trip efficiency*—Discharging net energy delivered/total charging energy consumed. For a stand-alone unit at suitable scale (see later) efficiencies of up to 60% are possible. Adding external waste heat or cold can significantly increase the useful work available from the cycle, which makes colocation with other thermally based processes attractive if the appropriate conditions are available.
- *Response time*—provided that the cryogenic feed pump(s) are kept cold and the turbine oil warm, starting the power turbine to synchronization is usually achievable in around 10 min (Highview's most recent demonstration plant, a 5 MW/15 MWh discharge system located at Pilsworth Quarry near Manchester in the United Kingdom, can fully load in $<$10 min). On the charging side, provided the liquefier is cold (which usually means it has been at operating temperature within 24 h), it takes approximately 5 min to liquid production although the compressors will be loaded shortly after starting. If the liquefier has not been operated recently and is at room temperature, a cooling cycle could take around 2 h or more, depending on the liquefier size.
- *Standing losses*—A liquid air system loses its stored energy through heat leakage into the main cryogenic storage tank which causes boil-off. This is counteracted with insulation and losses of less than 0.2% per day are readily achievable based on established industrial gas industry equipment. This makes LAES comparable to self-discharge rates of battery technologies (Table 2).[c]

6.2.7 Scale

LAES is one of the only storage technologies which can offer large-scale storage without geographical restrictions. However, if preexisting, commercially available components are to be employed there are limits to the scale of systems which are readily deployable as examined as follows on a (major) component by component basis.

- *Storage Tanks*—using Liquid Natural Gas (LNG) storage technology would permit very large quantities of liquid air to be stored. Cryogenic storage of 200,000 m^3 using LNG technology would be capable of storing in

c. Sandia Report "Long vs Short-term Energy Storage Technologies Analysis" http://prod.sandia.gov/techlib/access-control.cgi/2003/032783.pdf.

TABLE 2 Sample self-discharge rates of various storage technologies from Sandia report.[a]

Technology	Electricity requirement % stored per day	Loss mechanism
Lead-acid battery, flooded	0.1	Trickle charge
Lead-acid battery, VRLA	0.1	Trickle charge
Ni/Cd	0	Insignificant
Na/S	0.05	Heating
Zn/Br	0.24	Flow losses
Li-ion	0.24	Trickle charge
V-redox	0.2	Flow losses
High speed flywheel	1.25	Bearing losses, windage

[a] *Sandia Report "Long vs Short-term Energy Storage Technologies Analysis" http://prod.sandia.gov/techlib/access-control.cgi/2003/032783.pdf.*

the >20 GWh range of deliverable energy. As a result, tanks are not seen as a limiting factor in deployable scale.

- *Liquefier*—a typical, readily available large-scale liquefier would have a daily production capacity of about 2500 tonnes per day (tpd) of liquid air. Employing the relevant cold recycle systems integrated into the Highview design, this would have a net electrical demand of around 25 MW. At present, the likely cheapest way to provide higher capacity while maintaining flexibility of power demand would be to utilize multiple 2500 tpd trains.
- *Cryogenic Feed Pump*—the required flow of liquid air for a 50 MW power turbine is in excess of 100 kg/s at approximately 150 barg. The LNG industry is a useful source of such pumps, which are typically used in regasification terminals.
- *Power Turbine*—For a power output of 50 MW, multistage axial machines, typically used in the power generation sector, are ideally suited for this application and are readily available from a number of original equipment manufacturers (OEMs).

Drawing conclusions about an example practical plant discussed earlier, a system which utilized 2 × 2,500 tpd liquefiers (operating only during a 12 h off-peak period) would generate 2500 tonnes of liquid air on a daily basis (representing around 250 MWh of stored power). Allowing for a level of "spare" inventory this would be kept in a 3000 t cryogenic tank and would feed a power recovery system of 50 MW giving a total maximum power output of 50 MW for up to 5 h.

6.2.8 Engineering considerations

In order to develop a practical system, several compromises are inevitably encountered between ultimate performance and cost. The points detailed as follows represent some of the more common points encountered.

Heat storage considerations—for the heat required to drive the discharge cycle in a stand-alone system:

- High temperature improves overall performance (in terms of round-trip efficiency) but increases cost of the storage medium (for example hot water versus thermal oil versus molten salt). The eventual design decision will often be driven by site location and the desire for a common, repeatable design and may involve a combination of more than one medium and temperature.
- High-Grade Cold Store for recycling cold to the liquefier during recharging—due to the very low temperatures involved (−160°C) a packed bed is the preferred solution (low cost). Some alternatives are available in liquid form, but are usually hydrocarbon based, which pose safety issues when used in large volumes due to flammability or GHG and or cost concerns if low-temperature refrigerants are used.

Transient thermal storage performance—the design of the thermal stores (both hot and cold) must pay significant attention to preservation of the quality of the heat or cold stored to prevent a drop off in performance of the system near the end of the cycle. Degradation of heat causes a reduction in the power output during discharge and degradation of cold causes a reduction in the quantity of liquid air produced per kWh of charging power consumed.

Rotating equipment requirements—the variation in process conditions between the charging and discharge sides of a LAES system rules out the potential for dual purpose compressor/expanders. From a cost perspective, the use of "off the shelf" turbomachinery systems is preferred (although a compromise between performance and cost inevitably results) which can readily deliver the temperatures required to achieve a 60% RTE, which limits the requirements for special thermal management of these systems. Turbo Expanders in the 50 MW range are usually based on small axial steam turbines with modest modifications to suit the change in working fluid from steam to air.

Heat exchanger requirements—performance of heat exchangers is very important in LAES systems to preserve overall system performance. Low approach temperatures, low pressure drops are critical and a small footprint helps to minimize the site area. In some parts of the system extreme temperature gradients coupled with high pressures require specialist equipment (e.g., at the exit of the main cryogenic feed pump −170°C to 15°C across one unit at a pressure of 150 barg). Diffusion bonded units have been found to fit the process conditions well in these more challenging parts of the system.

6.2.9 Alternative working fluids

The large volumetric gas flow in a LAES system means it is very difficult to recirculate the working fluid from one cycle to the next—this points toward atmospheric air being an ideal working fluid as it can be used in a "once through" manner. The main downsides of air are as follows:

- a requirement to clean it up (remove CO_2 and water) to prevent fouling and eventual blockage of the liquefier; and
- oxygen concentration of the stored liquid air, which due to preferential evaporation of N_2 in the stored liquid air over storage periods of several months means a requirement to manage the inventory by either using it up, or adding liquid nitrogen to replace the N_2 which has evaporated.

An alternative working fluid which could be used in this type of cycle is nitrogen (the downside is that N_2 is an asphyxiant, so appropriate measures must be taken to ensure O_2 levels are maintained in nearby enclosed spaces in the event of leaks), but it can be used to avoid O_2 concentration if very long duration storage is required (e.g., interseasonal).

6.2.10 Advanced concepts

LAES involves the transfer of electricity and hot and cold energy, which provides numerous opportunities to integrate LAES into other energy or thermal processes. For instance, if external heat is available from a source such as an open cycle gas turbine (OCGT) or an industrial application, then this heat can be integrated into the discharging process to further increase the efficiency and power output from power recovery of the liquid air [25]. This can potentially also remove the need for hot thermal storage, thereby reducing the cost.

If an external cold resource is available, such as that from a LNG regasification terminal, then this can be integrated into the liquefaction plant to provide some, or all, of the required refrigeration to liquefy the air [26–29]. The reduced compressor electrical load leads to an increase in the resultant LAES system efficiency. Systems which combine both waste heat sources and external cold sinks with LAES see additional benefits to cost and efficiency [25]. In practice, LAES typically has significant amounts of unused low-grade cold which is unsuitable for recycling to the liquefier to enhance performance. This presents a potential integration opportunity for the provision of chilled water as part of a polygeneration system.

LAES may be integrated with electricity generation systems. While an advantage of nuclear power and geothermal power is a constant baseload supply of electricity, these systems may be required to operate flexibly if high proportions of variable renewable generation are present on the grid. Operating at part load reduces the plant efficiency and may also reduce the lifetime of components. Therefore LAES may be colocated with baseload power generation, such as nuclear [30] and geothermal [31] to enable the plants to operate continuously at their design point. The LAES plant absorbs excess electricity which is not required by the grid, and then dispatches it at times of high demand. The LAES system may be directly integrated into the thermal flows of these baseload plants in order to boost efficiencies and reduce the required number of components. LAES can also be integrated with variable renewable generation such as solar thermal [32, 33] with the objective of improving dispatchability.

LAES has similarities with both CAES and PTES which facilitates the integration of these systems. For instance, like CAES, LAES compresses a gas to high pressures (and the heat of compression may be stored). However, unlike CAES the high-pressure gas is then throttled to a liquid state to enable high energy densities. A hybrid LAES-CAES concept was envisioned whereby CAES provided short-term energy storage services, and a coupled LAES system provided longer duration storage [34]. By storing compressed air in aboveground pressure vessels, the geographic limitations of CAES could be lifted. The CAES system stores energy for short durations, and if further energy storage is required, some of the compressed air is liquefied and then stored for longer durations in large insulated storage tanks.

LAES has conceptual similarities with PTES as energy is stored using a heat pump (refrigeration) cycle and is subsequently dispatched through a heat engine. Since PTES operates over a hotter temperature range than LAES, these two cycles may be combined with the PTES system acting as a topping cycle to the LAES cycle, as described in [35]. The cold energy generated in the PTES charging system is not stored but is instead used to cool supercritical air to cryogenic temperatures, similar to the LAES-LNG system mentioned previously. The combined cycle has higher efficiencies and energy densities than stand-alone PTES or LAES systems [35].

6.3 Pumped heat

Contributor: Jason Kerth
Review: Miles Abarr, Ph.D., Sebastian Freund, Ph.D., Emmanuel Jacquemoud, Christos Markides, Ph.D., Josh McTigue, Ph.D., and Jeffrey Moore, Ph.D.

6.3.1 Introduction

In the Pumped Heat Energy Storage (PHES) process, electrical energy is used to drive a heat pump, which moves thermal energy from a cold store to a hot store, effectively charging the system. Then, as desired, the energy stored in the thermal stores can be discharged by using it to drive a heat engine, which returns

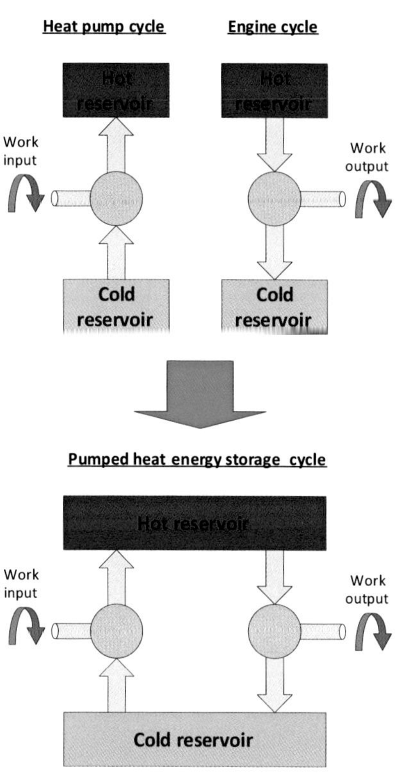

FIG. 14 Generalized illustration of pumped heat energy storage.

electricity to the grid. The basic realization of a PHES system as a combination of heat pump and a heat engine with shared thermal reservoirs is illustrated in Fig. 14. The pioneers of the Pumped Heat Energy Storage Concept [36, 37] promoted charging and discharging cycles based gas-phase machinery and stationary solid-phase thermal stores. However, as is alluded to in [36], any thermodynamic heat pump and engine schemes, in combination with any means of storing thermal energy, can form the basis of a pumped heat energy storage system. Later practitioners have accordingly considered liquid-phase thermal stores [38] and also cycles involving a condensing working fluid [39]. This section reviews variations on cycles, thermal stores, and other implementation details which may be applied toward the goal of realizing practical PHES system.

It is worthwhile at the outset to examine a few concepts and features that are common among all PHES schemes. First, the following definitions of the component cycle efficiencies are useful:

Heat Pump Coefficient of Performance—Ratio of thermal energy provided to the hot reservoir to the mechanical work input

$$COP_{HeatPump} = \frac{Q_H}{W_{input}}$$

Engine Efficiency—Ratio of mechanical work output to thermal energy consumed from the hot reservoir

$$\eta_{Engine} = \frac{W_{output}}{Q_H}$$

This leads to the following understanding of Round-trip efficiency for a PHES system

PHES Round-Trip Efficiency—Ratio of mechanical work out to mechanical work in

$$\eta_{PHES} = \frac{W_{output}}{W_{input}}$$

$$= \frac{Q_H}{W_{input}} \cdot \frac{W_{output}}{Q_H}$$

$$= COP_{HeatPump} \cdot \eta_{Engine}$$

If the reversible Carnot cycle is considered for both the heat pump and the heat engine, this leads to the following:

$$COP_{HeatPump} = \frac{T_H}{T_H - T_C}$$

$$\eta_{Engine} = \frac{T_H - T_C}{T_H}$$

Then given the PHES system round-trip efficiency above

$$\eta_{PHES} = COP_{HeatPump} \cdot \eta_{Engine}$$

$$= \frac{T_H}{T_H - T_C} \cdot \frac{T_H - T_C}{T_H}$$

$$= 1$$

Thus with reversible cycles, the round-trip efficiency of PHES is unity and independent of the temperatures. The efficiency shortcoming of real systems is attributable to the usual loss mechanisms present in all thermodynamic power systems, with predictable directional impacts. These factors include heat exchange temperature approaches, hydraulic losses in piping and equipment, efficiency of gas machinery, mechanical losses in the drivetrain elements, and losses in the electrical machines. The quantitative sensitivity of different cycle embodiments to the different factors will be discussed further as follows.

Another important observation from the earlier development is that for the idealized Carnot cycles, the efficiency is independent of temperature choices, since higher maximum temperature and lower minimum temperature improve the efficiency of the engine, but have the reciprocal effect on the heat pump

COP. This temperature independence of idealized cycles, however, does not perfectly hold for practical embodiments of PHES. For practical systems, factors that favor high engine efficiency (e.g., high temperature) tend to improve the PHES round-trip efficiency, but with less significance than the stand-alone engine on account of the offsetting affect to the heat pump discussed here. A very nice derivation of the theoretical round-trip efficiency of a PHES system (ideal gas recuperated cycle) is provided in [38].

While the temperatures may not have such a significant effect on the round-trip efficiency, it is important to note their effect on the stores. If the engine cycle is more efficient (i.e., operating with a higher maximum temperature and lower minimum temperature), this means that less thermal energy must be stored for a given amount of energy delivery from the system. This generally has a positive impact on the cost and size of the stores. Furthermore, the greater the temperature range of the store, the less thermal mass is required for a given quantity of thermal energy storage. A lower thermal mass requirement of course generally translates into smaller and less costly stores.

6.3.2 The basic ideal gas cycle

The basic ideal gas PHES cycle is illustrated schematically in Fig. 15. The main features of this cycle are described as follows, but it should be apparent that most of these characteristics are in common with any of the ideal gas cycle variants:

6.3.2.1 *Charging cycle*

- Consists of a hot compressor and a cold expander
- $\dot{m}_{compressor} = \dot{m}_{expander}$
- $Pr_{compressor} = Pr_{expander}$
- $W_{compressor} > W_{expander}$
- $W_{in} = W_{compressor} - W_{expander}$

6.3.2.2 *Discharging cycle*

- Consists of a cold compressor and a hot expander
- $\dot{m}_{compressor} = \dot{m}_{expander}$
- $Pr_{compressor} = Pr_{expander}$
- $W_{compressor} < W_{expander}$
- $W_{out} = W_{expander} - W_{compressor}$

The main distinction of the basic cycle is the hot machine experiences the complete temperature range between the limiting cycle peak temperature and the ambient temperature. The temperature range of the cold machines is then established by the isentropic process over the same pressure ratio, having maximum temperature at the ambient. Given the large temperature range of the machine, the required pressure ratio of the basic cycle is correspondingly high.

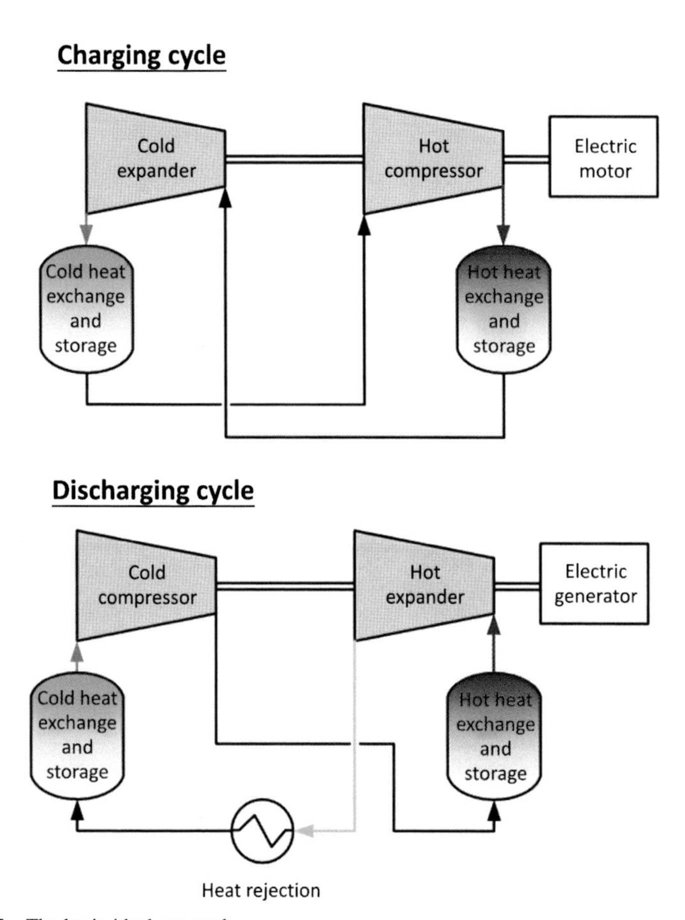

FIG. 15 The basic ideal gas cycle.

This linkage between the pressure ratio and the temperature ratio motivates the use of Argon as a working fluid. This is because monatomic gases exhibit the highest rise in temperature per unit rise in pressure, owing to their high ratio of specific heats (k value). Among the monatomic gases, Argon is the most readily available and economical, so it is the most reasonable choice. Helium can also be considered, but given its low molecular weight, it is disadvantaged for systems using turbomachinery. Low molecular weight gases require more compression head to achieve a given pressure ratio, which in general means more stages and thus more cost. There are however some examples of Helium turbomachines which have been studied and implemented. Some discussion and examples are provided by [40]. If a monatomic gas is selected for a working fluid, this seriously discourages the operation of the system in any such way that involved regular venting and makeup of the working fluid. The motivation for venting is further discussed in a later section.

The problems alluded here with the monatomic gases are avoided in a system that uses reciprocating machinery. Reciprocating machines can achieve higher pressure ratios in a single stage, and operational control of the system can likely be achieved without venting or makeup of the working fluid. Also, a major benefit arising from the high-temperature range of the basic cycle is that the thermal store mass is minimized for a desired quantity of stored energy. Undoubtedly these factors contributed to the selection of the basic cycle with reciprocating machinery by one of the first pioneers of PHES, Isentropic, Ltd.

6.3.3 The ideal gas cycle with recuperation

The cycle with recuperation is shown in Fig. 16. The effect of recuperation is that the temperature range of the hot machine is reduced, having its maximum temperature still at the limiting cycle peak temperature, but with the lower temperature of the hot machines at a rather discretionary point between the ambient and peak temperatures. The effect of this discretionary "warm" temperature on the cycle performance is as follows:

(1) With increasing warm temperature, the pressure ratio of the cycle decreases.
(2) With consistent assumptions regarding pressure drops and machinery efficiencies, the overall round-trip efficiency reaches some maximum value with the optimal selection of the warm temperature.
(3) With increasing warm temperature, the thermal mass (i.e., size) of the thermal stores increases, since roughly the same quantity of thermal energy moves into and out of storage, while the temperature range of the store decreases.
(4) The volume flows for the machinery do not increase with recuperation, assuming a constant limiting high-side pressure. The importance of volume flow is that this parameter more than most others impacts the size and cost of equipment and piping. The cycle mass flow increases with higher warm temperature since, as with the stores, the same amount of energy must be extracted over a narrower temperature range; however, the effect on the volume flow at the low side of the cycle is offset, since low-side pressure is simultaneously increasing. Note that the increase in low-side pressure follows from the decrease in cycle pressure ratio with higher warm temperature, thus if the high-side pressure is held constant, low-side pressure increases with reduced ratio.

Note that in the limiting case where the warm temperature approaches the ambient, the recuperator duty reduces to nil and we return essentially to the basic cycle. The recuperated cycle generally demonstrates an optimal round-trip efficiency value with a warm temperature significantly above the ambient, so it seems that selecting the basic cycle implies sacrificing efficiency for the sake

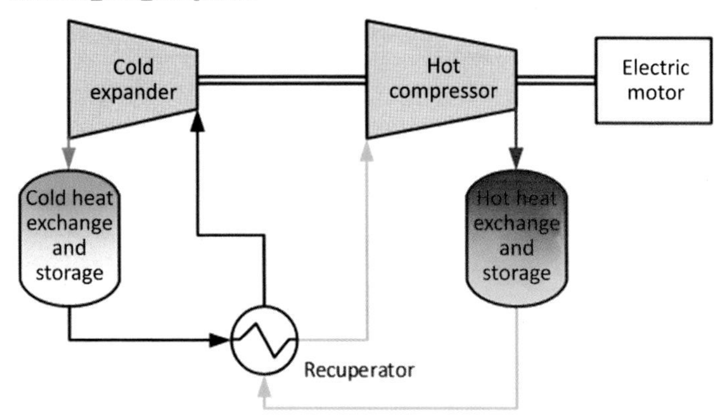

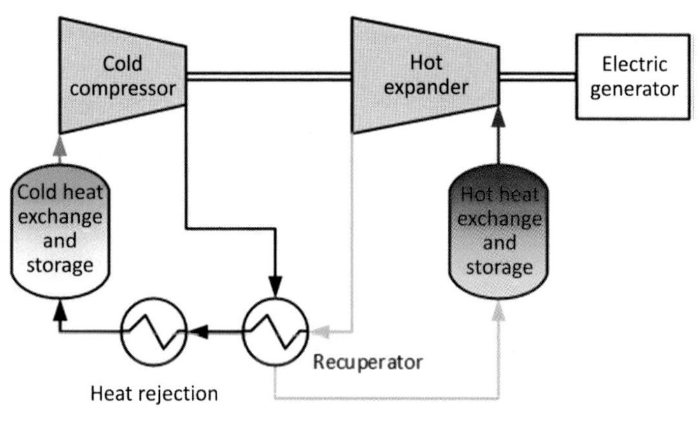

FIG. 16 Recuperated ideal gas PHES cycle.

of reducing thermal store size. In most practical embodiments of PHES, the cost of the energy conversion equipment (machinery, heat exchangers, piping) should be expected to outweigh the cost of the stores. Also, the lower pressure ratio of recuperated embodiments is favorable with regard to charging and discharging rate control via inventory management, which is further discussed as follows. Therefore it appears that the recuperated cycle has several important advantages over the basic cycle variant, especially when considering an embodiment in turbomachinery.

Finally, because the low temperature end of the hot store of the recuperated cycle operated significantly above ambient, this cycle is well suited for use with a molten salt hot media. Molten salt solidifies at temperatures well above ambient. Therefore in order to practically apply molten salt, it must be always maintained at an elevated temperature. This requirement is easily achieved with appropriate selection of parameters using the recuperated cycle. The attraction to achieve a PHES embodiment with liquid-phase media, and in particular with molten salt, which has a proven track record in large-scale thermal energy storage in concentrating solar power applications, motivates the recuperated cycle. A thorough analysis of the cycle is provided by Laughlin [38]. Development and commercialization of a PHES embodiment based on the recuperated cycle with molten salt thermal store is underway at the time of writing. The development has been undertaken by Malta, Inc., which is a spinout from a development project by the same name at Alphabet's moon-shot factory, X [41].

6.3.4 The ideal gas overlap cycle

The recuperated cycle demonstrates that by reducing the temperature range of the machines, the pressure ratio is correspondingly reduced, resulting in numerous benefits to the system. In the recuperated cycle, the temperature range of the machines is reduced by virtue of the recuperator—a physical heat exchanger transferring energy between the high-pressure and low-pressure working fluid. It is also possible to achieve a similar thermodynamic effect of recuperation by operating the stores over a wider temperature range, specifically in such a way that there is an overlap between the temperature range of the hot and cold stores. Fig. 17 provides a cycle schematic of the overlap cycle and Fig. 18 provides a graphical illustration for the temperature ranges of the different cycles. Note that in this schematic, losses are not considered and so the ranges represent the idealized cycle variants. A cycle with losses will need to include heat rejection as discussed previously, which will have some effect on the temperatures. In each of the main cycle variants there are different options about how the heat rejection may be accomplished. The next section offers further discussion on this topic.

The feasibility of the overlapped cycle depends primarily on the choice of the store concept and available thermal media. As illustrated in Fig. 18, the overlap cycle requires a larger operable range for the media, which can be challenging. For example, molten salt, which is a candidate for the hot media, generally has a freezing point above the ambient, so that the overlap cycle cannot be applied with this choice of media. The more likely media choice for the overlap cycle variant is a solid-phase media. The overlap cycle with solid-phase media was the embodiment described by one of the first investigators into PHES [37].

The overlap version might be advantageous with some form of media, since it reduces some working fluid piping associated with connecting the recuperator. However, known practical media choices all struggle to have the necessary attributes. For example, suitable low-cost liquid-phase media seem to lack sufficient temperature range. Solid media, on the other hand, have favorable temperature range, but solid-phase store solutions generally lead to additional cost

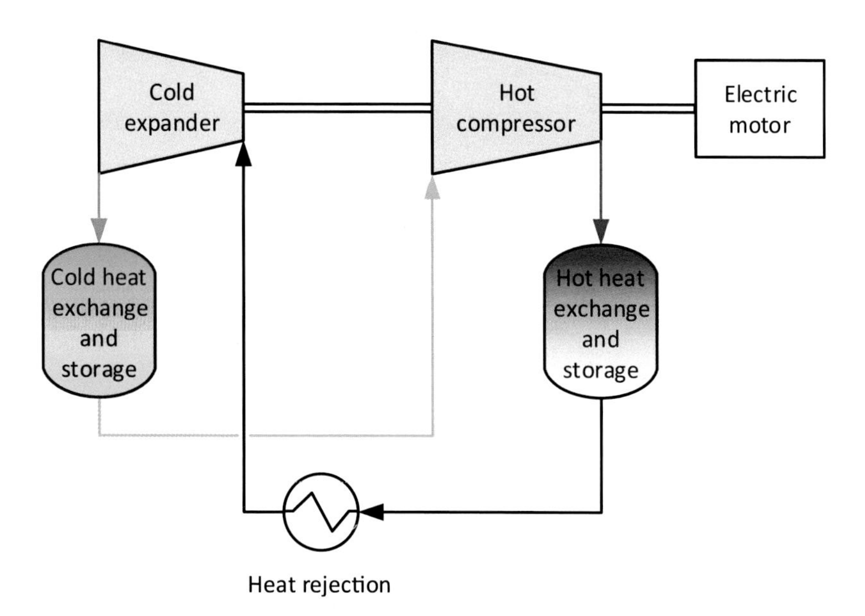

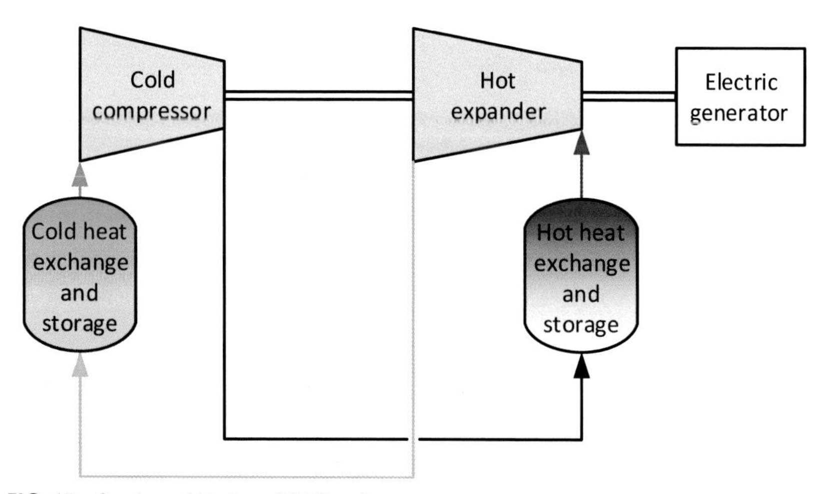

FIG. 17 Overlapped ideal gas PHES cycle.

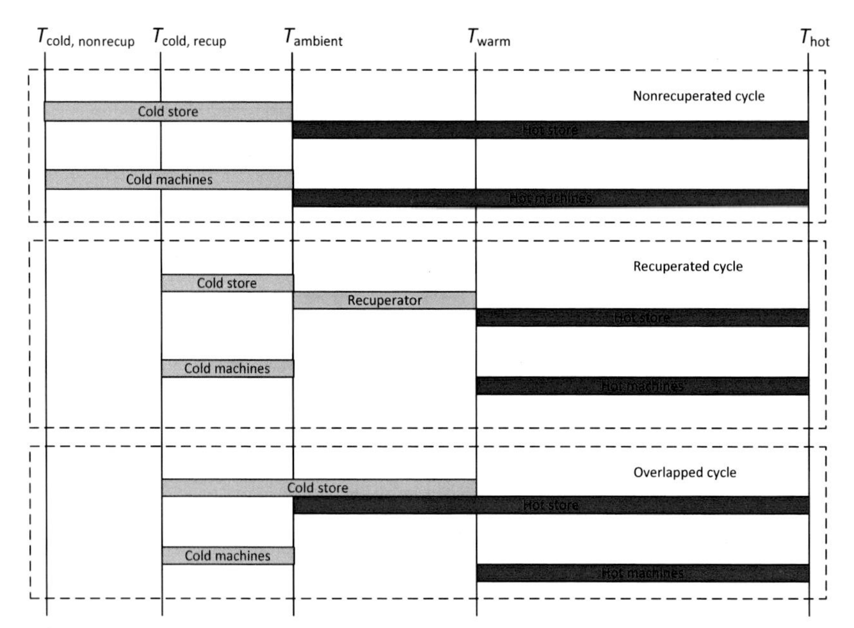

FIG. 18 Temperature ranges of the various cycles.

and/or complexity. Trade-offs in this regard will be discussed further in a later section.

It is noteworthy that even though the overlap cycle changes the operable temperature range of the media, this does not dictate more or less media mass for a specific energy storage capacity as compared to the recuperated cycle.

6.3.5 Influencing factors on ideal gas cycles

In the comparative analysis that follows, a number of key parameter decisions impact many of the solutions in similar ways. The following paragraphs identify such parameters and offer some qualitative remarks regarding the effects that each has on the operable cycles.

6.3.5.1 Influence of pressure

For ideal gases, enthalpy is primarily a function of temperature. Therefore the temperatures at the various state points for an ideal gas-based PHES are driven mainly by pressure ratios of the machinery, and not by absolute pressure. Absolute pressure may be selected almost arbitrarily. The main considerations for pressure selection are as follows:

(1) At higher pressure, machinery volume flows are lower, leading to more compact and generally more cost-effective gas machinery.

(2) At higher pressure, density of the gas is higher, leading to higher blade bending stress in the machines. This concern is most pronounced for the hot compressor, since compressor blading ideally is thinner than expander blading operating at similar conditions. Thin blades, in turn, experience higher stress levels for a given gas load and also have lower-frequency vibratory modes, leading to more modes potentially in the desired operable speed range.
(3) Higher pressure leads to higher wall thicknesses of piping and pressure casings. Note that since volume flow decreases with increasing pressure, the weight of metal per unit length dedicated to pressure containment may not change very much as a function of pressure. However, the length of the machinery flow paths, exchangers, and piping may be expected to decrease at higher pressure yielding some net material savings at higher pressure.
(4) In cases of indirect heat exchange, higher pressure containment requires thicker metal sections separating the working fluid from the thermal media. Thicker sections have higher conductive thermal resistance, leading to higher temperature approaches in heat exchangers. This in turn decreases efficiency, which is discussed further as follows.
(5) At higher pressure, hydraulic losses in pipes and heat exchangers are generally less as a percentage of operating pressure. The reason is that for a given mass flow, volumetric flow decreases at higher pressure. Therefore for a given target velocity, the duct diameter necessary to achieve such velocity also decreases with higher pressure. For an ideal gas, the Reynolds number is proportional to mass flow and inversely proportional to the duct diameter, thus higher pressure flows generally yield higher Reynolds numbers. Higher Reynolds number in turn leads to lower hydraulic friction factor. The loss percentage is in direct proportion to the friction factor and to the velocity squared.
(6) Thicker walls, driven by pressure as described previously, may introduce limitations on thermal ramp rates. If fast starting and stopping are deemed important, thick metal sections should be avoided.
(7) Expansion joints, which may be helpful to achieve sufficient flexibility in piping to accommodate thermal growth of the equipment or otherwise prevent excessive piping loads from being transmitted to equipment, might be limited in size, temperature, and pressure rating capability. Note that piping flexibility can also be achieved by adding expansion loop features, but these features also increase piping hydraulic losses.

6.3.5.2 Influence of temperatures

There are four temperatures that must be considered in a PHES cycle: these are the minimum and maximum temperatures for each of the cold and the hot heat exchange processes. Several factors influence selections that include the following:

(1) At extreme temperature conditions, more costly metallurgy and/or fabrication methods must generally be considered. A full discussion of options for various components is beyond the scope of this section, but many elements of the system are impacted, most notably the machinery rotating gas path components, machinery casing components, heat exchanger metallurgy, materials and construction methods for the thermal stores.
(2) Acceptable temperature range of the thermal media. Again a comprehensive discussion is beyond the scope of this section, but some key considerations include the following:
 a. Vapor pressure of liquid-phase thermal media at desired temperature. In general, the thermal store must be pressure rated to this vapor pressure.
 b. Chemical degradation of the thermal media at high temperature
 c. Freezing of media at low temperature. This issue may pertain to the cold media, but might also pertain to the lower limiting temperature for the hot media, especially or particularly in the case of molten salt hot media. This is discussed in further detail later.
 d. Unreasonably high viscosity at low temperature. High viscosity in liquids results in low Reynolds number, which in turn generally translates to high hydraulic losses and also poor heat transfer characteristics. High viscosity might also necessitate the use of reciprocating pumps rather than centrifugal pumps, the former being in general more costly and less reliable.

6.3.5.3 *Heat rejection*

Since the PHES cycle, like all thermodynamic systems, involves irreversibility, it is therefore necessary to incorporate some heat rejection facility into the cycle design. For example, if a PHES system absorbs 100 MWh of electricity via its charging system and has 70% round-trip efficiency, this means it returns 70 MWh of electricity via its discharging system and therefore must reject 30 MWh of electricity. The absence of heat rejection would imply a perpetual motion machine, which is a well-proven impossibility. Analysis by Laughlin [38] indicates that round-trip efficiency improves with lower temperature for the heat rejection.

It is worth noting that as parts of the PHES cycle may operate at temperatures below ambient, it is not out of the question that a PHES system may absorb some energy from the ambient. Such absorption of energy from the ambient seems to be implied in some of the early work, notably [36], but such absorption will never release the designer from the need for rejection. Absorption of energy from the ambient equal to or exceeding heat rejection to the ambient would again imply a perpetual motion machine, in violation of thermodynamic principles. For example, a system might be realized where 100 MWh of electricity is absorbed via the charging system and 20 MWh of energy is absorbed from the ambient via heat transfer to a part of the cold cycle. However, given similar loss characteristics as the 70% efficient cycle described previously, it should not be expected that the latter cycle with absorption from ambient will still realize 30 MWh of rejection, thereby effectively increasing AC-to-AC efficiency to 90%. More likely, the latter cycle with absorption will find itself rejecting 50 MWh to ambient and returning 70 MWh AC, resulting in comparable net AC-to-AC efficiency and with higher heat exchange duty and thus higher cost. Nevertheless, the opportunity to configure a PHES cycle which incorporates some absorption of heat may be interesting in some cases. For example, it could realize some efficiency advantage resulting from diurnal temperature variation in combination with anticipated use case or it might find some economic merit in providing distinct cooling. This topic may merit further exploration.

6.3.6 Inventory control in ideal gas cycles

The ideal gas PHES with turbomachines offers the opportunity to control the rate of energy charging or discharging via the inventory level in the respective circuits. Increasing the inventory in the circuit increases the pressure and therefore increases mass circulation and therefore rate of energy charging or discharging in direct proportion for a given volume flow of the machines. Holding a constant volume flow rate over the operable range has the favorable effect that efficiency likewise remains at levels near the design point efficiency. In order to make use of this efficient control mechanism, it must be feasible to add or remove working fluid from the loop. The most practical method to achieve this is to use air as a working fluid, or possibly nitrogen which can be fairly easily derived from air. In this manner, working fluid can simply be vented to the atmosphere to reduce the rate of energy charging or discharging, or made up from the atmosphere to increase it. In order to conveniently vent inventory to atmosphere (i.e., without using additional costly mechanical means of evacuation), the pressure at the low side of the cycle should always be greater than atmospheric pressure over the entire operable range. The following equation reasonably characterizes the situation:

$$P_{max} = \frac{(PR)(P_{baro} + MarginForVenting)}{MinRate\%} \tag{1}$$

where:

P_{max} = maximum pressure in the cycle
PR = pressure ratio of the cycle
P_{baro} = barometric pressure
$MarginForVenting$ = DP allowance for a vent valve
$MinRate\%$ = minimum rate of energy charging or discharging relative to cycle design rate.

For example, given a target minimum charge/discharge rate of 20% of design, a barometric pressure of 1.013 bara, a margin for venting of 1 bar, and a cycle pressure ratio of 5, this would indicate a max design pressure requirement of (5)(1.013 + 1)/0.2 = 50.325 bara. Note that the margin for venting factor derives from the desired speed of response while moving to the minimum charging or discharging rate. Given any target response speed, this factor may be controlled to a lower figure by using larger valves (i.e., with higher expense). Also, when increasing the rate of charging or discharging, working fluid must be supplied to the loop via an external compressor. This discharge pressure rating of this compressor less any line and control valve losses must exceed at least the low-side pressure of the loop at the maximum rate of charging or discharging in order to be capable of loading the loop to the design level.

While venting the working fluid to atmosphere provides a simple way to reduce inventory, this fluid must subsequently be made up in order to restore the inventory to its design level. Considering the objective to restore the inventory to design level in as small a time as possible, the compression and conditioning equipment required to achieve the desired speed of inventory management might become large and thus cost prohibitive. Therefore it is most likely favorable not to vent the working fluid, but rather to store it in such a way that it can be quickly reintroduced into the system. A feasible approach using a system of pressurized vessels has been used similarly for the control of closed-cycle gas turbines [40]. The basic operation for the discharging cycle of an ideal gas-based PHES system is described as follows. In this example, an array of three pressure vessels are considered, but the number and sizing of vessels can be optimized depending on turndown goals. Also, the same concept and in fact the same vessels can be applied for the charging cycle.

- Working fluid moves from the high-pressure side of the loop into storage vessels (increasing pressure in the vessel).
- As the pressure in each vessel approaches the pressure in the high side of the loop, the pressure gradient driving this process declines.
- In order to further unload, a subsequent lower-pressure store is required, and so on until the desired minimum output level is achieved.
- The loading process is the reverse of the unloading process. The sequence of operations for unloading, then loading in this case would be:
  - Unloading inventory from cycle: Fill HP Store → Fill MP Store → Fill LP Store
  - Loading inventory to cycle: Empty LP Store → Empty MP Store → Empty LP Store

Fig. 19 illustrates the unloading process and Fig. 20 illustrates the reverse loading process.

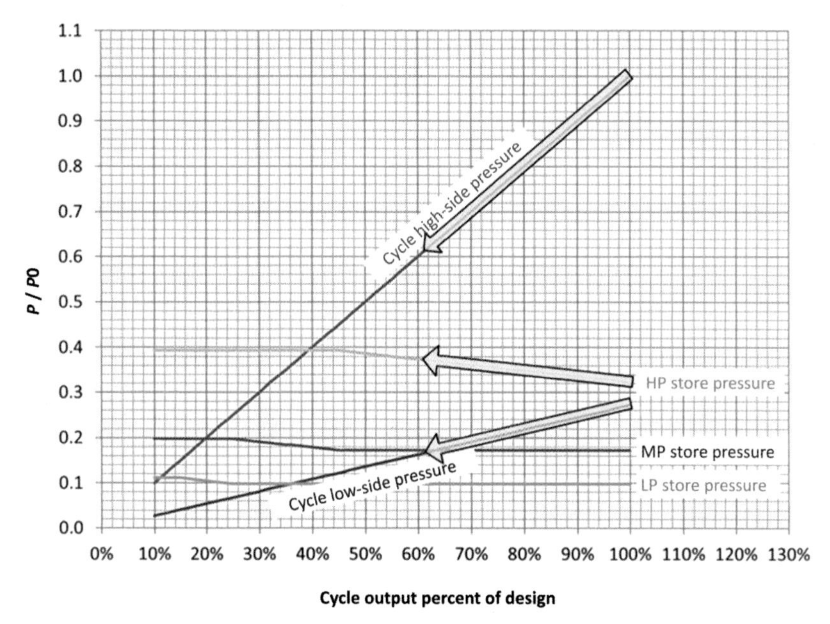

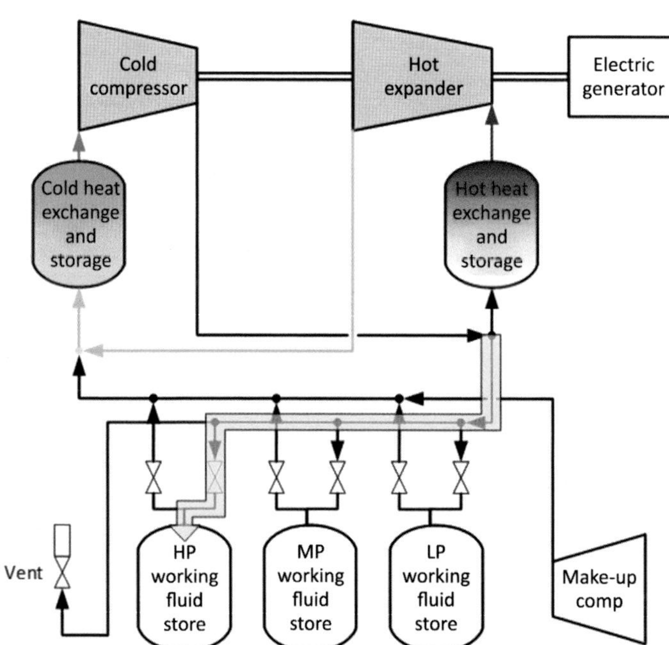

FIG. 19 Pressure vs cycle output percentage and system schematic for inventory control of PHES discharging cycle: Decreasing output from 100% to 60% illustrated.

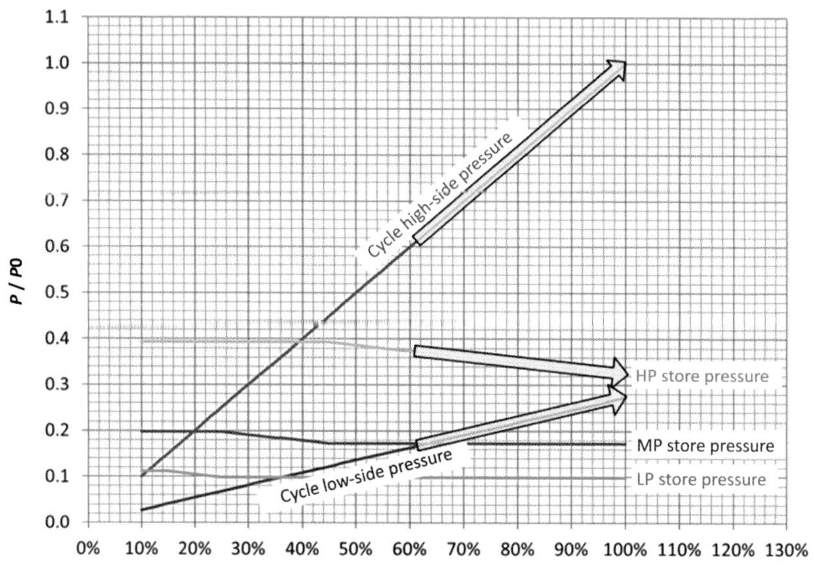

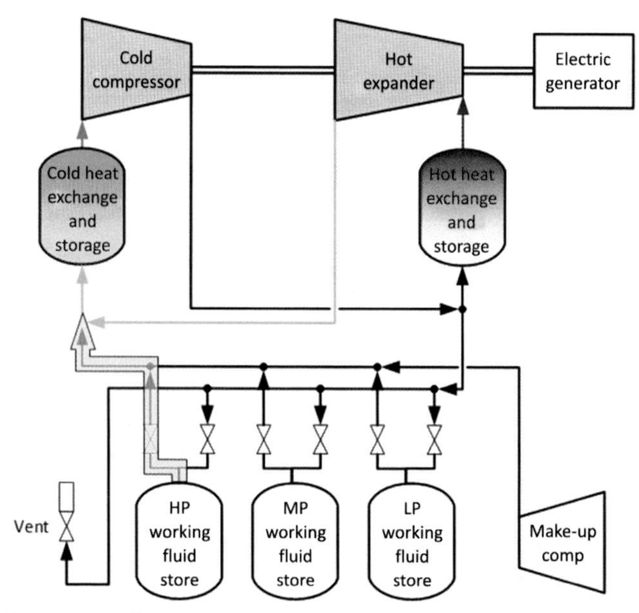

FIG. 20 Pressure vs cycle output percentage and system schematic for inventory control of PHES discharging cycle: Increasing output from 60% to 100% illustrated.

6.3.7 Parametric sensitivity for the recuperated cycle

This section explores the sensitivity of cycle performance attributes to various design decisions and component performance attributes, including design temperature ranges, machinery efficiency, heat exchanger temperature approaches, and pressure losses. In the sensitivity evaluation, the treatment of heat exchangers' loss and approach assumptions merits special consideration. In reality, achievable temperature approaches and pressure losses in heat exchange equipment depend on a rather large number of variables, including fluid properties as well exchanger geometry. A comprehensive treatment of the situation cannot be practically handled within the sensitivity analysis that follows, but it is reasonable to incorporate certain key high-level performance characteristics trends. If a fixed ΔP or ΔP percentage is assumed for a particular heat exchanger over various comparative cases, then this unfairly penalizes cases where the heat exchange duty is lower compared to the base case. Also, if closer heat exchange approach temperatures are considered without additional ΔP, then this fails to capture the reality that lower approach with constant ΔP implies additional surface area "in parallel" which in turn implies reduction in flow velocity and thus heat transfer coefficient. In order to represent effects such as these without incorporating a comprehensive heat exchange analysis program in the sensitivity evaluation, a sort of reference condition for heat exchange can be considered and then the effects can be captured by variation from this arbitrary reference condition according to the following formula:

$$\Delta P = \Delta P_{ref}\left(\frac{\Delta h}{\Delta h_{ref}}\right)\left(\frac{\Delta T_{LM,ref}}{\Delta T_{LM}}\right)\left(\frac{P}{P_{ref}}\right)^{\lambda}$$

The final term in this formula captures in a simplified form the dependence of both heat transfer coefficient and pressure loss (via the friction factor) on the Reynolds Number. In the sensitivity analysis that follows, this parameter should be viewed as a somewhat arbitrarily selected "bias factor" favoring high-pressure heat exchange over low-pressure heat exchange, rather than a value that is somehow mathematically founded in first principles of hydraulics and heat transfer.

It is also helpful to consider the implication of this relationship in the context of the basic heat exchanger equation:

$$Q = U \bullet A \bullet \Delta T_{LM}$$

$$\dot{m}\Delta h = U \bullet A \bullet \Delta T_{LM}$$

$$\frac{\Delta h}{\Delta T_{LM}} = \frac{UA}{\dot{m}}$$

Substituting this back into the earlier expression and simplifying permits a reduction in the number of variables as constants can be combined:

$$\left(\frac{\Delta h_{ref}}{\Delta T_{LM,ref}\Delta P_{ref}}\right)=\left(\frac{UA}{\dot{m}\Delta P}\right)\left(\frac{P}{P_{ref}}\right)^{\lambda}$$

$$\frac{\left(\frac{UA}{\dot{m}\Delta P}\right)}{\left(\frac{UA}{\dot{m}\Delta P}\right)_{ref}}=\left(\frac{P}{P_{ref}}\right)^{-\lambda}$$

In other words, taken together, the identification of a reference condition for enthalpy change, pressure loss, and approach combines to define a reference condition for specific total heat transfer coefficient times area ("UA") per unit pressure loss. This parameter may in turn be considered as a sort of heat exchange performance achievement per unit cost, such that at constant cost, trade-offs can be made to "shift" performance from one parameter to the other, so long as the product remains the same. Likewise, increasing or decreasing the product entails, respectively, increasing or decreasing the cost. The assumptions for the heat exchange performance characteristics are defined in Table 3. The reference approach temperature for the HTHX is arbitrarily doubled compared to what is used for other exchangers, since the system RTE is less sensitive to approach at the HTHX.

Table 4 identifies all other key attributes of the "base case" and provides some rationale for why these values were selected. In the discussion that follows, if parameters are not explicitly varied they remain equal to these base case figures. The parameters provided by the table are sufficient to fully constrain the

TABLE 3 Heat exchange performance characteristic.

Parameter	Symbol	Reference value for HTHX	Reference value for all other HX
Specific enthalpy change across the heat exchanger	Δh	290.7 kJ/kg	290.7 kJ/kg
Log Mean temperature difference	ΔT_{LM}	11.11 K	5.6 K
Pressure loss across the heat exchanger	ΔP	0.52 bar	0.52 bar
Specific total heat transfer coefficient times area per unit pressure loss	$\frac{UA}{\dot{m}\Delta P}$	50.32 kJ/ kg K bar	99.83 kJ/ kg K bar
Heat exchanger inlet pressure	P	34.47 bara	34.47 bara
Pressure Exponent	λ	0.7	0.7

TABLE 4 Base case parameters for cycle sensitivity analysis.

Parameter	Value	Rationale
Ambient Condition		
Ambient Dry Bulb	15°C	ISO Condition for GT Performance
Ambient Relative Humidity	60%	ISO Condition for GT Performance
Ambient Wet Bulb Temperature	10.8°C	ISO Condition for GT Performance
Barometric Pressure	1.013 bara	ISO Condition for GT Performance
Cycle Operating Parameters		
Charging Compressor Discharge Temperature (T1 in the following charts)	565.6°C	Typical maximum for industrial steam turbine metallurgy and manufacturing techniques
Charging Compressor Discharge Pressure	34.57 bara	See Note[a]
High Temperature Store, Lower Temperature (T2 in the following charts)	293.3°C	Optimal value given other design choices
Discharging Compressor Discharge Pressure	34.76 bara	Somewhat arbitrary, see Note[b]
Scaling Parameters		
Discharging Size	1.0 MW	Presented values are normalized to 1 MW discharging size and 1:1 duration ratio
Ratio of Charging to Discharging Duration	1:1	
Cooling System[c]		
Cooling Water Approach to T_wb	5.6 K	A slightly aggressive value
Cooling Water Design Temperature Rise	8.3 K	A typical value
Machinery		
Isentropic Efficiency, All Machines	90%	An aggressive target
Included Parasitic Effects[d]		
Motor and Generator Efficiency	98.5%	A typical value
Gear efficiency, Charging and Discharging	97.7%	A typical value

Continued

TABLE 4 Base case parameters for cycle sensitivity analysis—cont'd

Parameter	Value	Rationale
Working Fluid	Dry Air	

[a]*Based on other assumptions of the base case and a 0.5 bar venting margin (refer to Eq. 1), this maximum pressure affords turndown to 17% of design for charging system and 22% of design for discharging.*
[b]*Discharging compressor discharge pressure is set to equalize pressures for the high-temperature heat exchange for charging and discharging cycles. Using equal pressure ensures equal duty vs temperature profile and thus constant temperature approach across the entire duty.*
[c]*Cooling system is assumed as evaporative cooling tower with circulating water system.*
[d]*Parasitic effects in direct path between AC electric input and AC electric output are considered in the RTE figures, but other parasitic effects, including but not limited to cooling fans, thermal media circulation, facility lighting and HVAC, etc. are not included here.*

heat and material balance (HMB). A complete set of state points for the base case is provided in Table 5.

Fig. 21 shows the effect of the main cycle temperature design variables on the round-trip efficiency (RTE). Round-trip efficiency increases with the maximum cycle temperature, thus the designer is incentivized to the highest practical value. As described earlier in the paper, the limit may relate to choices of metallurgy and construction methods in the turbomachines as well as the capability of the selected thermal media. For a given cycle maximum temperature, there is an optimum of the ratio of the maximum temperature to the hot store lower temperature (T2/T1). Note that this ratio is stated in terms of absolute temperature (K). The optimal ratio appears near to 0.75 over the temperature range considered.

Fig. 22 shows that the duty for both the hot and cold heat exchange services per unit discharging power is reduced as temperature increases, owing most likely to the simultaneous increase in cycle efficiency. The heat exchange duty is minimized by higher T2/T1 ratio, but this effect is not very strong and diminishes at higher temperature.

Fig. 23 shows the effect of temperature selections on charging and discharging pressure ratio. Higher temperature ratio (i.e., lower T2/T1) of course requires higher pressure ratios. Higher maximum cycle temperature also increases the pressure ratio. The pressure ratio of the charging system is less than that of the discharging system in this configuration on account of the location of the heat rejection in the cycle. The ratio of the discharging pressure ratio to that of the charging system increases with lower T2/T1. This has impact to the achievable turndown via inventory control, with equivalent pressure ratios between charging and discharging being more favorable as they lead to equivalent turndown for both systems.

TABLE 5 State points for base case.

Leaving	Entering	Mass flow (kg/s)	Pressure (bara)	Enthalpy (kJ/kg)	Temp (°C)
Charging					
Recuperator LP	Compressor	7.834	9.02	273.4	293.3
Compressor	HTHX	7.834	34.47	567.6	565.6
HTHX	Recuperator HP	7.834	33.97	278.8	298.9
Recuperator HP	Expander	7.834	33.62	−6.1	26.4
Expander	LTHX	7.834	9.27	−87.9	−58.7
LTHX	Recuperator LP	7.834	9.16	−11.5	15.8
Discharging					
HTHX	Expander	7.834	33.97	555.2	554.4
Expander	Recuperator LP	7.834	6.91	275.8	295.6
Recuperator LP	Ref HX	7.834	6.81	75.1	100.5
Ref HX	LTHX	7.834	6.80	−4.6	22.0
LTHX	Compressor	7.834	6.72	−81.1	−53.1
Compressor	Recuperator HP	7.834	34.76	65.7	95.0
Recuperator HP	HTHX	7.834	34.47	266.4	287.2

Fig. 24 shows the minimum temperature of the working fluid in the cycle, which is pertinent to the selection of the low-temperature media. It is a strong function of T2/T1, but a weak function of the cycle maximum temperature.

Dividing the heat exchange duty by the temperature range provides an indication of the thermal mass of the store per unit of energy discharged. This parameter is plotted in Fig. 25. Higher cycle temperature drives smaller stores. Higher T2/T1 drives larger stores. It is interesting that the thermal mass for the high-temperature store and the low-temperature store are not much different,

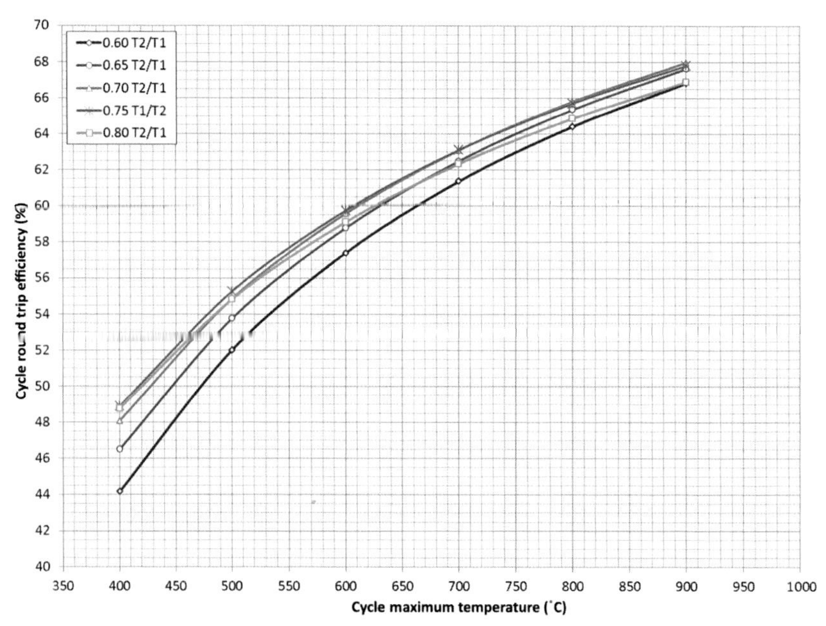

FIG. 21 Cycle temperature effect on round-trip efficiency.

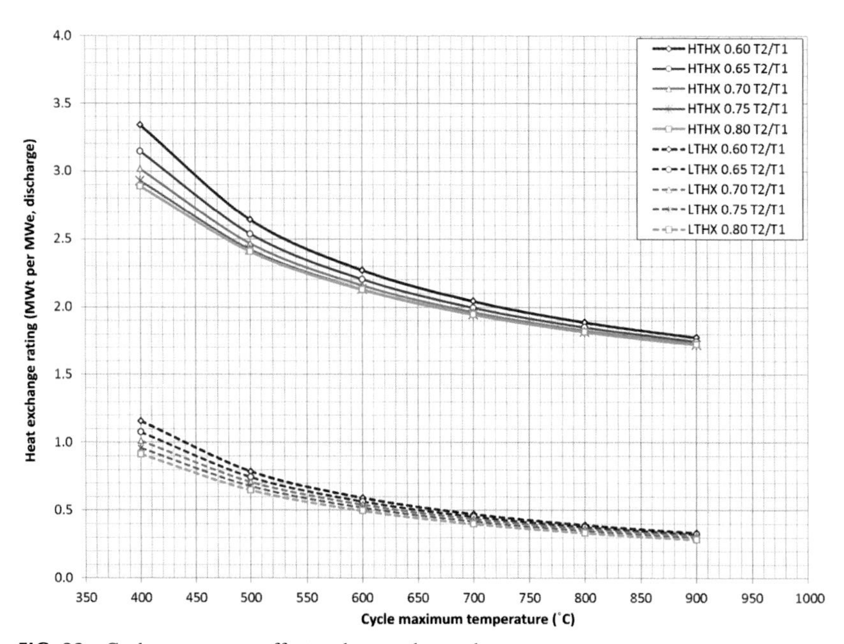

FIG. 22 Cycle temperature effect on heat exchange duty.

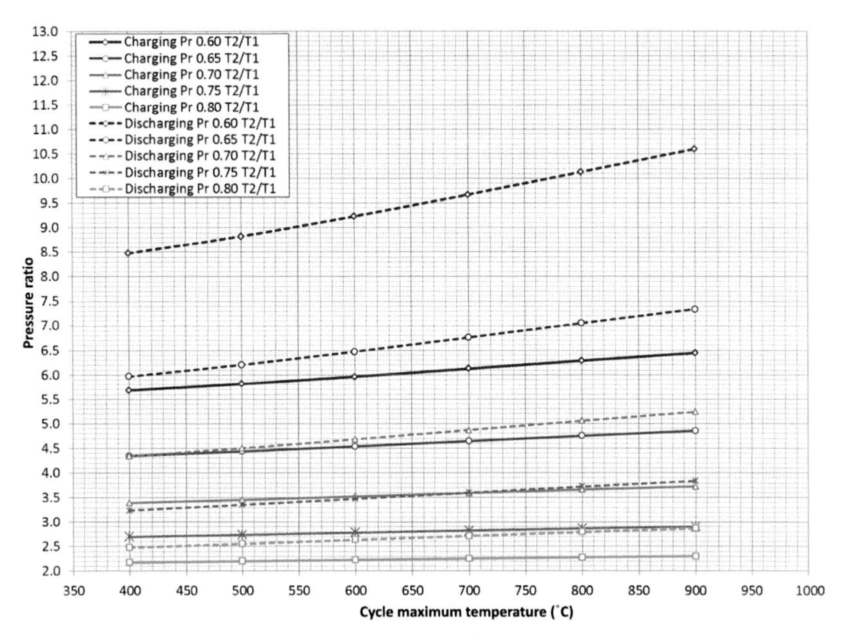

FIG. 23 Cycle temperature effect on pressure ratio.

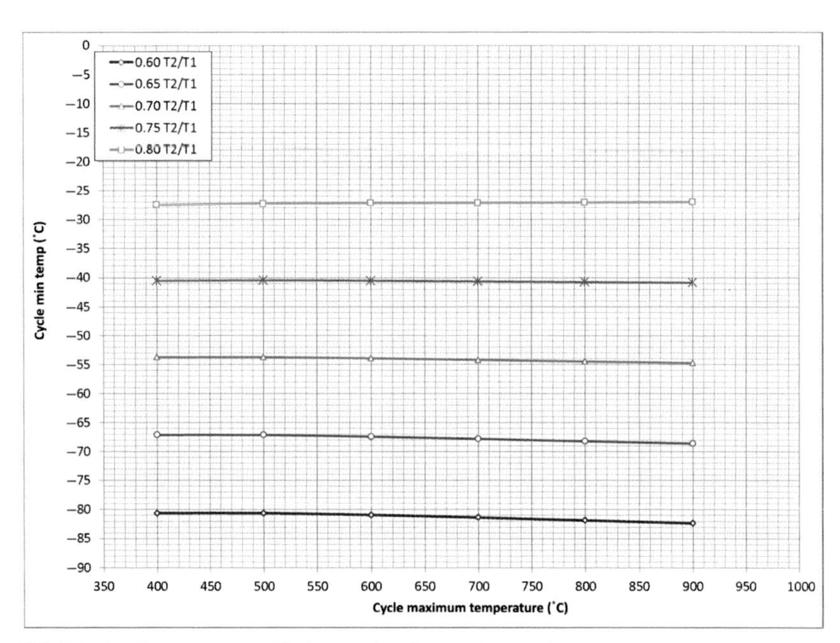

FIG. 24 Cycle temperature effect on cycle minimum temperature.

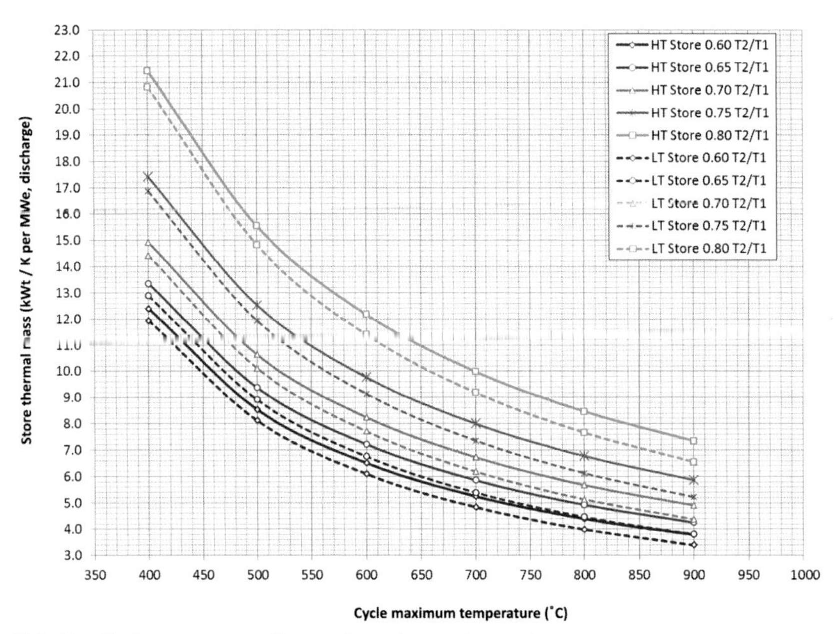

FIG. 25 Cycle temperature effect on thermal store size requirement.

the latter having lower duty but over a smaller temperature range compared to the former, such that the effects roughly offset.

Fig. 26 shows the effect of temperatures on machinery volume flows. As mentioned earlier, volume flow is a main cost driver for turbomachinery. Volume flows in general decrease with higher temperature, owing to the simultaneous increase in the cycle round-trip efficiency (lower mass flow needed to deliver a unit of discharging energy). Also, volume flows reach a minimum value which seems to be near the optimal selection for T2/T1 for the charging machines, and at perhaps slightly higher T2/T1 for the discharging machines. Of course, a much more pronounced effect on volume arises from the choice of the cycle pressure level. This effect is not studied in the sensitivity evaluation, but it should be easily recognized that volume flow varies inversely with the selected pressure level.

Fig. 27 shows that RTE improves with lower heat rejection temperature. The independent variable is the temperature of the working fluid post heat rejection, such that this figure is applicable for a wide array of choices with regard to the cooling system design, e.g., air cooling vs evaporative cooling, different selection of heat exchange approach, etc.

Fig. 28 shows the impact of machinery efficiency on the RTE. The order of importance with regard to machinery efficiency achievement is as follows, from

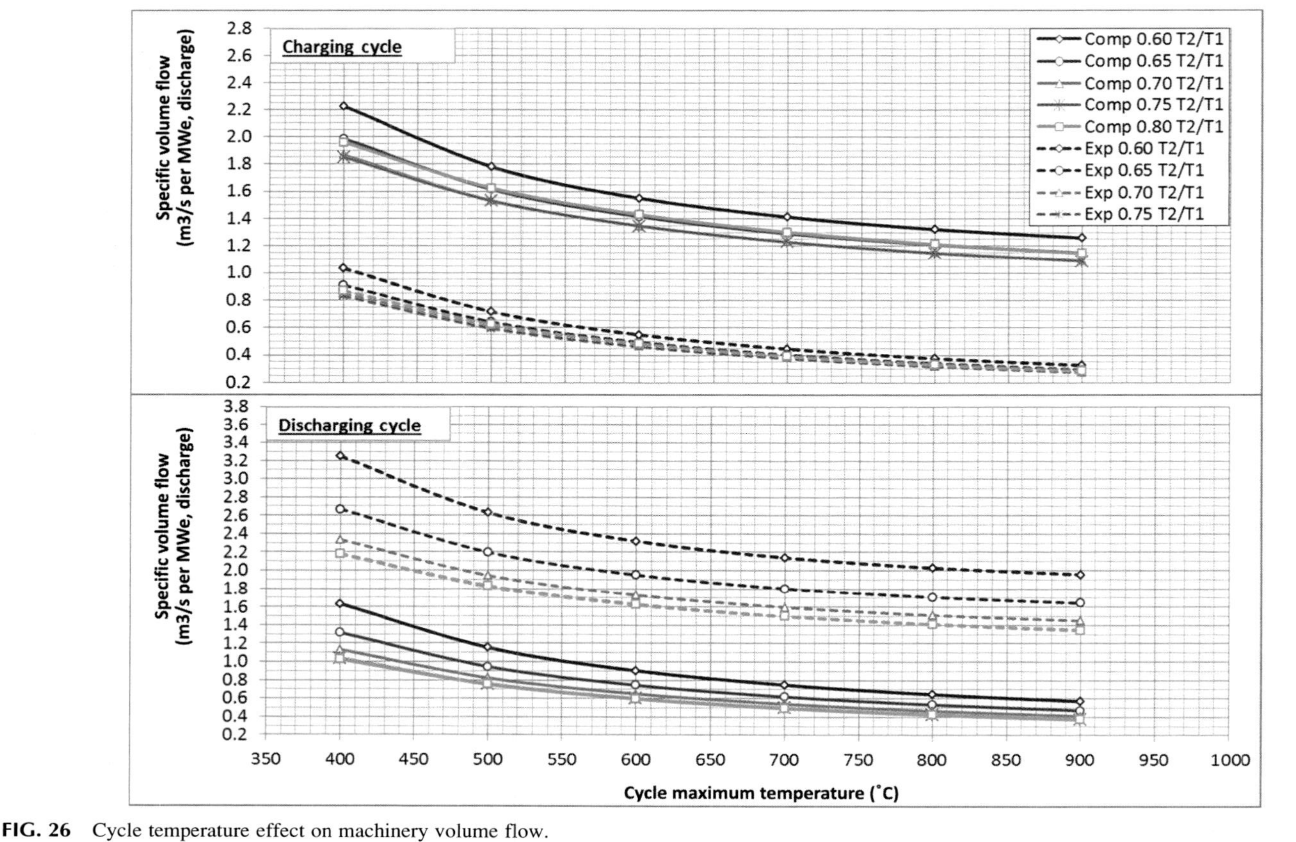

FIG. 26 Cycle temperature effect on machinery volume flow.

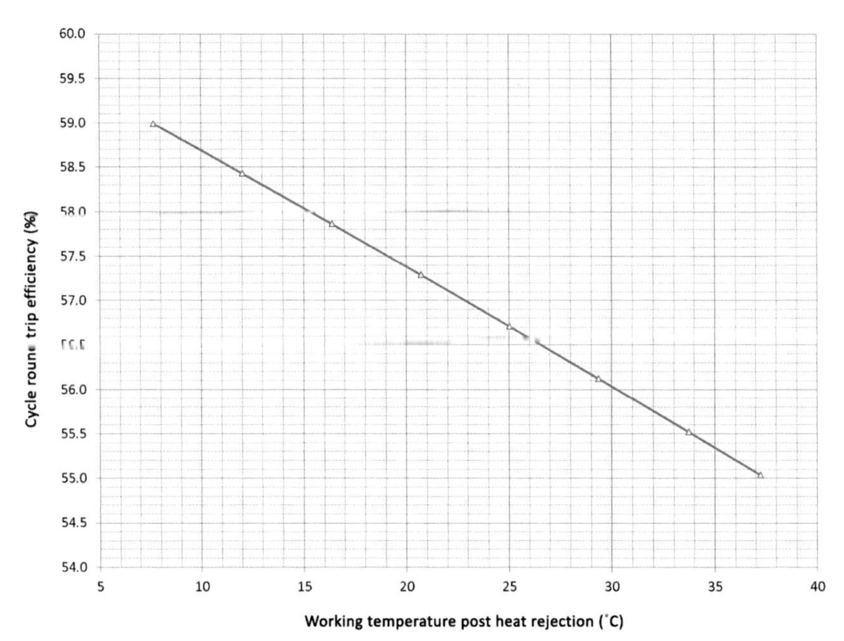

FIG. 27 Heat rejection temperature effect on round-trip efficiency.

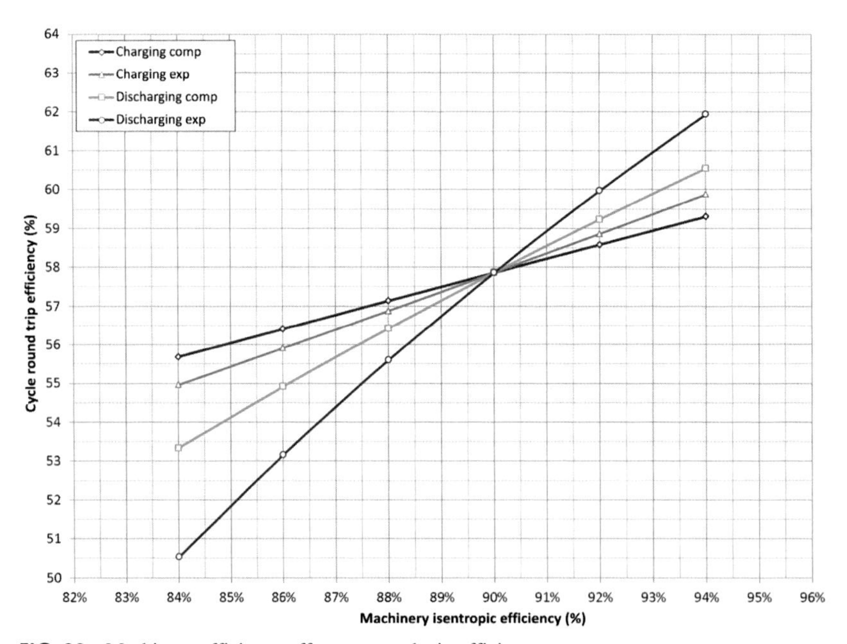

FIG. 28 Machinery efficiency effect on round-trip efficiency.

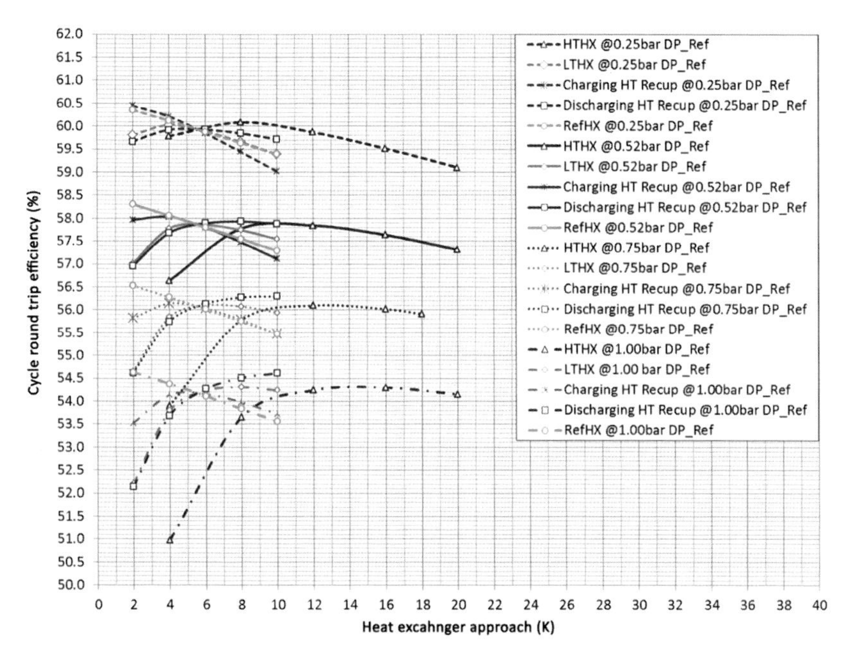

FIG. 29 Heat exchanger design effect on round-trip efficiency.

highest priority to lowest: discharging expander, discharging compressor, charging expander, charging compressor.

Fig. 29 illustrates the effect of heat exchanger performance characteristics on the RTE. An important takeaway is that most of the heat exchangers demonstrate an optimal approach temperature within the practical range where temperature varies slightly for each exchanger. The optimal value occurs where the irreversibility arising from the exchanger pressure loss is optimally balanced with the irreversibility arising from the heat transfer across the finite temperature difference, where according to the methods described previously closer approaches require more pressure loss for a given level of performance achievement. The different families of curves (identified as different DP_ref) represent variations on this level of performance achievement. While each family shows optimal approaches per heat exchanger, it is clear that moving from one family (i.e., identifying and applying an optimal level of performance achievement per unit cost) has much stronger relevance to RTE than selecting the optimal approach for the selected technology.

6.3.8 Options regarding the point of heat rejection

As mentioned previously, all PHES systems must reject some heat to ambient. The designer has some latitude to where in the cycle the heat rejection occurs. Each of the cycle diagrams earlier, Figs. 15–17, shows a heat rejection in the most favorable location for the cycle with respect to round-trip efficiency. In each case, it is also possible to incorporate heat rejection at other points in the cycle, and also it is possible to consider heat rejection from multiple points, with a distribution of the overall heat rejection service across these multiple heat rejection elements. The reasons to consider alternate or additional points for heat rejection may include one or more of the following:

(1) The heat rejection generally entails some auxiliary load for the system, such as the energy required to run coolant circulating pumps, evaporative cooling tower fans, direct air-cooled heat exchange fans, or the like. In many cases, it will be favorable to incur this extra load concurrently with the charging operation, since electrical energy typically has lower value at the time of charging.

(2) As mentioned earlier, lower heat rejection temperature is advantageous to round-trip efficiency. If the charging or discharging cycle tends to operate at night, when the average temperature is generally lower, it may be favorable to shift heat rejection duty toward this part of the cycle. The tendency of a PHES system to charge or discharge at day or night is dictated by the specifics of a particular project, for example some projects may aim to absorb excess solar energy production (charging during the day), while other projects may aim to absorb excess wind energy production or baseload production while demand is lower (charging during the night).

(3) Dividing the heat rejection duty may allow a reduction in the size of the heat rejection equipment where the charging and discharging cycles do not operate simultaneously. Instead of having a cooling tower and water circulating pump sized to deliver 100% of the heat rejection service either during the charging or discharging cycle, these components could be reduced, for example to 50% size, if the heat rejection duty could be evenly divided between the charging and discharging cycles.

(4) Adjusting the distribution of the heat rejection service may change the operable temperature ranges of the machines, which in turn may affect the pressure ratio required to achieve such temperature range. This feature could be used to provide a degree of trimming control in the system. Adjusting the heat rejection toward achieving equal pressure ratios between charging and discharging cycles might also offer some advantage toward the situation of turndown operation, as is informed by Eq. (1).

Figs. 30–32 illustrate heat rejection at multiple points for each of the basic cycle variants illustrated, respectively, in Figs. 15–17.

Another option that may exist with some selections of thermal media is heat rejection from the media rather than from the working fluid. This option is most clearly available for liquid-phase media, but also may be possible with conveyable solid media. Only the stationary solid-phase store effectively precludes this concept. Fig. 33 provides an illustration of this concept and it is also illustrated

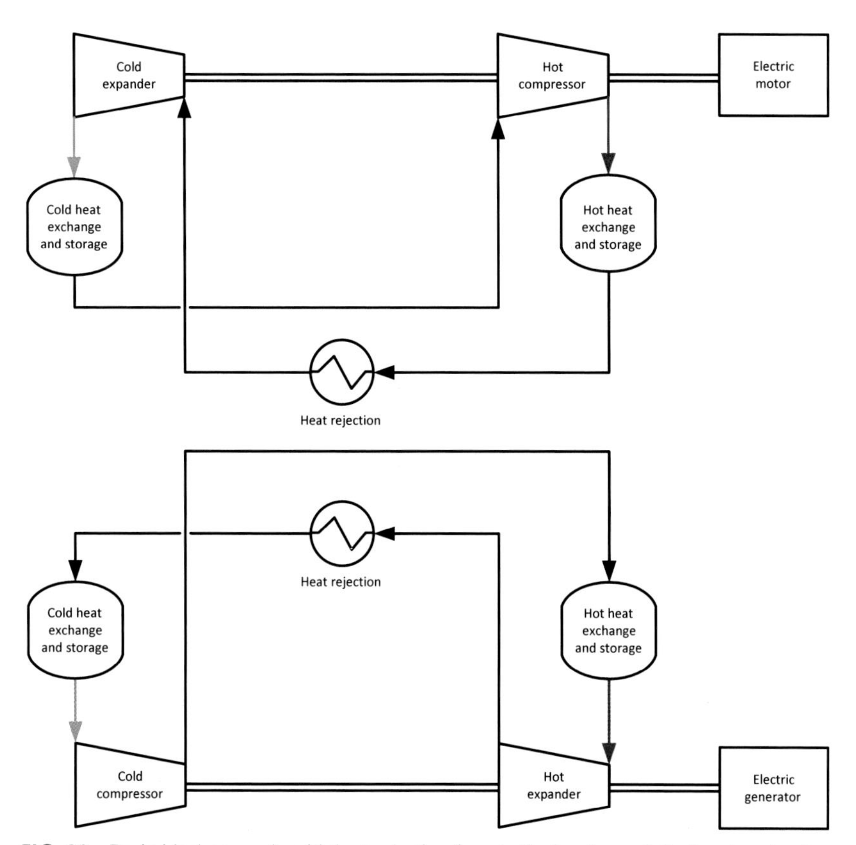

FIG. 30 Basic ideal gas cycle with heat rejection from both charging and discharging circuits.

by Laughlin [38]. Depending on the cycle variant and design goals (e.g., rejection during charging or discharging) the basic premise illustrated in Fig. 33 might be applied with the heat rejection exchanger upstream or downstream of the media-to-working fluid heat exchanger and/or the concept might be applied to the hot media or the cold media.

This option could be used, for example, to shift heat rejection duty from one operating mode to the other (for example from charging to discharging). It might also reduce the pressure drop for the working fluid that would be associated with coolers in the loop as illustrated in earlier schematics; however, it should be noted that heat rejection duty shifted into the media implies additional heat transfer from the cycle to the media, which also incurs pressure drop. It is likely that only the incremental pressure drop from connecting piping and headers is saved. Heat exchanger core pressure drop is merely transferred from the working fluid direct heat rejection cooler to the working fluid-to-media exchanger. Furthermore, circulating energy for the media likely increases in this

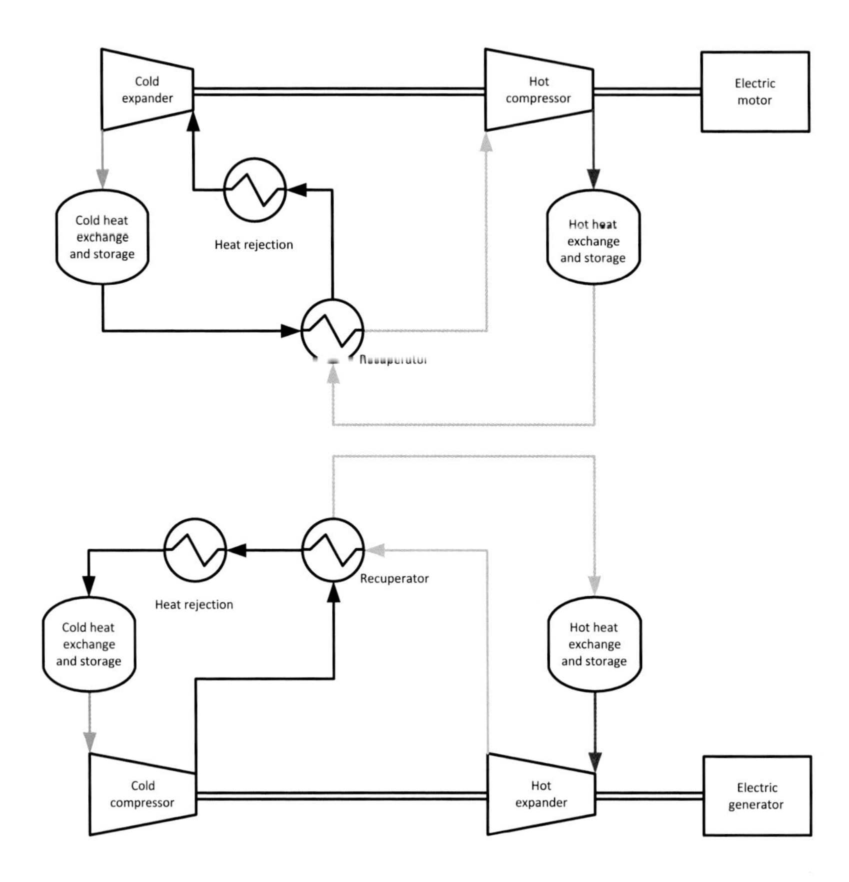

FIG. 31 Recuperated ideal gas cycle with heat rejection from both charging and discharging circuits.

concept since additional losses in the media-to-working-fluid exchanger are incurred and the media cooling heat exchanger is added.

6.3.9 Trans-critical CO_2 cycle

Several PHES practitioners have investigated and/or advocate PHES cycles based on trans-critical CO_2 working fluid. This section provides an exemplary comparative analysis between t-CO_2-based cycles and a recuperated ideal gas cycle. The main motive to consider a trans-critical cycle is that selection of the appropriate low-side pressure for the CO_2 leads to the possibility to utilize water ice for the low-temperature store. This presents a clear advantage in terms of cost and energy density of the low-temperature storage media. However, some challenges are also introduced with a phase-change storage concept.

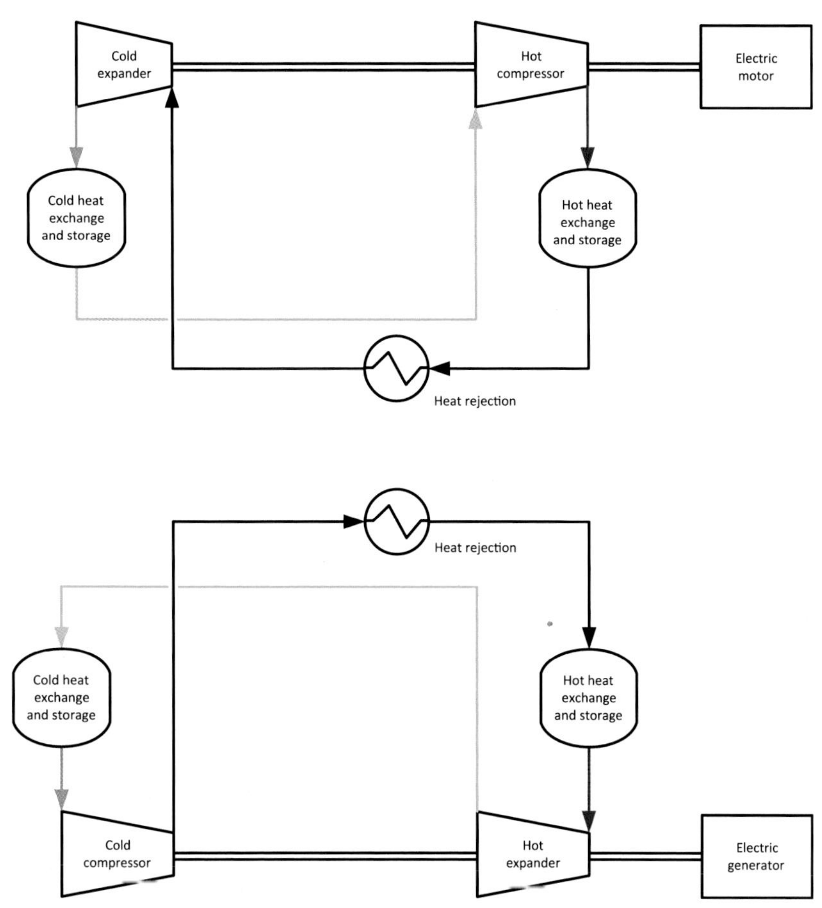

FIG. 32 Overlapped ideal gas cycle with heat rejection from both charging and discharging circuits.

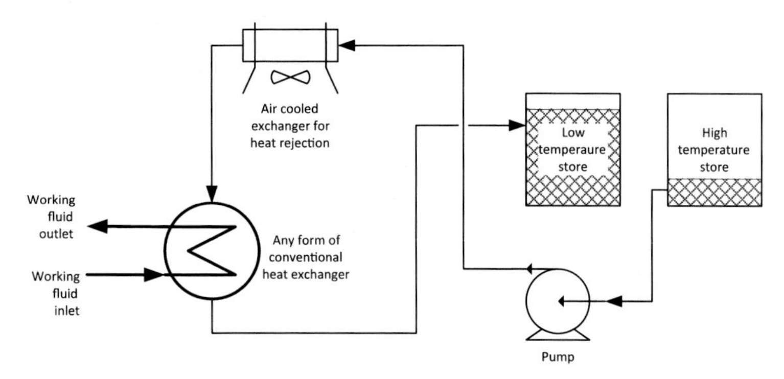

FIG. 33 Heat rejection from circulating media in lieu of directly from working fluid.

In this section, two variants of t-CO_2 cycle are analyzed. In the first case (called Alt1), the low-side pressure is set to utilize water-ice for the cold store, while the maximum cycle pressure is aligned roughly to the capabilities of a high-temperature thermal oil (Therminol66 is considered here). These selections result in a high-side pressure within range of current supercritical steam power cycles. In the second case (called Alt2), the low-side pressure is again tied to the water-ice phase change, but the high-side pressure is aligned toward the capabilities of molten salt. The purpose of this Alt2 evaluation is to discern if the higher temperature level possible with molten salt has a meaningful positive impact on the store size and efficiency situation of the t-CO_2 concept. While these factors are both positively affected, the pressure required to achieve the desired temperature is quite high by power cycle standards. Such pressures, however, have been demonstrated, including with CO_2, in some gas injection projects in the oil and gas industry [42, 43]. Fig. 34 provides a cycle schematic for the ideal gas cycle considered and Table 6 provides the state points. Fig. 35 provides a schematic for the t-CO_2 cycle (which schematic is unchanged for Alt1 and Alt2 embodiments). Table 7 provides the state points for Alt1 and Table 8 provides the state points for Alt2.

For the case of the ideal gas cycle, thermocline stores were considered for both hot and cold, with a molten salt hot fluid and hexane cold fluid. For the t-CO2 cycle, thermocline stores were also considered for the hot cycle. For the cold cycle, an encapsulated PCM concept was considered with water encapsulated in plastic spheres. An appropriate water-glycol heat transfer fluid was used to carry heat between the cycle to the store.

The following points and analysis summarize a comparison between the recuperated ideal gas cycle and a t-CO_2 cycle

(1) The Ideal Gas Cycle had better round-trip efficiency (RTE) at design point compared to the t-CO_2 embodiment with Therminol66 (Alt1), but the t-CO_2 embodiment with molten salt (Alt2) had the best design point efficiency. The high efficiency of Alt2 most likely arises from the low sensitivity of the cycle to machinery efficiency, especially on the cold end of the cycle where the machine powers are so small. The lower efficiency of Alt1 seems to be driven by the low-temperature level as well as the significant heat exchanger pinch issues throughout the cycle which lead to overall degradation of the RTE. It may be possible to mitigate some of these effects by minor modifications of the t-CO_2 cycle compared to what has been considered.

(2) The t-CO_2 Alt1 variant has a higher heat pump coefficient of performance (COP) but lower engine efficiency. Because of this, the thermal stores of this embodiment must store more thermal energy per unit of electrical energy delivery. The stores for t-CO_2 also operate over a narrower temperature range, which increases the requirement for thermal mass (mass x specific heat) per unit of thermal energy storage. This generally leads to larger stores (refer to Table 9).

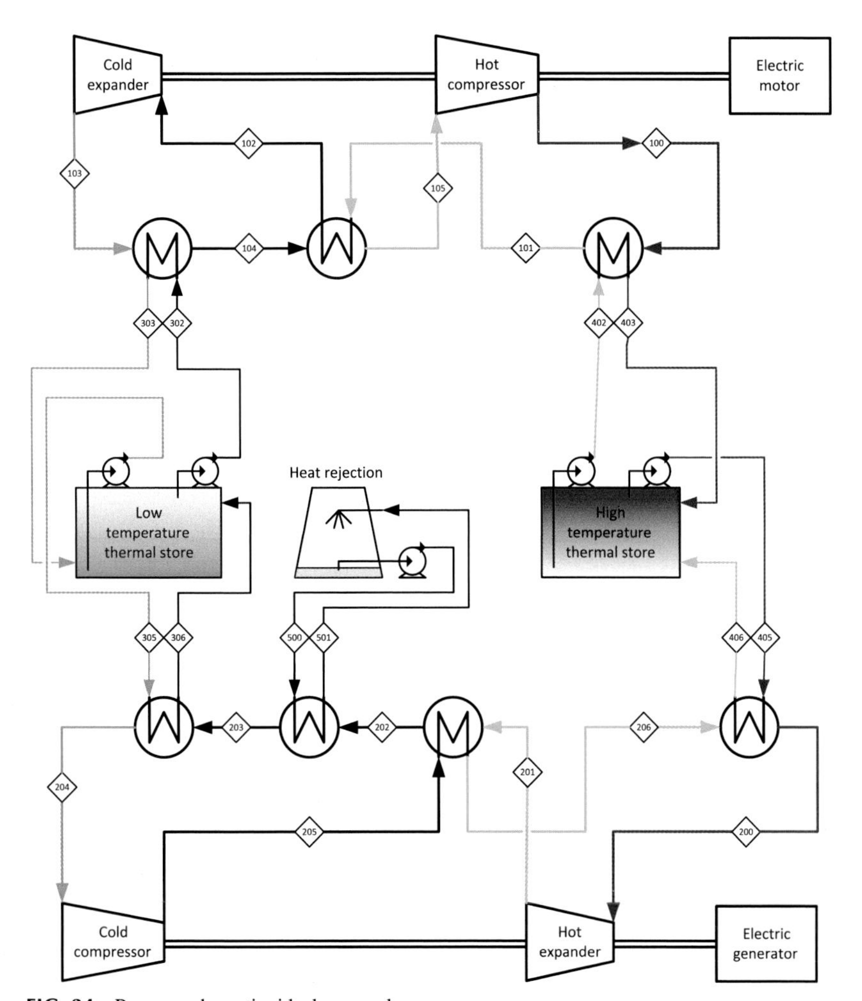

FIG. 34 Process schematic, ideal gas cycle.

(3) Machinery volume flows are significantly smaller for the t-CO_2 scheme leading to smaller and probably less costly turbomachinery. While the smaller size presents a cost and size advantage, the smaller flows and higher pressures likely mandate radial machinery, which is generally less efficient. That said, the t-CO_2 cycle should be less sensitive to machinery efficiency.

(4) The pressure containment for the t-CO_2 cycle would require significantly thicker walls. The mass of piping may be comparable for the cycles. The thick pipes of the t-CO_2 system might be a disadvantage when it comes to achieving piping flexibility (bellows type expansion joints are impossible, so only expansion loops can be considered, i.e., longer pipe runs with more losses). Also, if a fast-reacting system is desired, as is likely in

TABLE 6 Heat and material balance data for ideal gas cycle.

State point	Description	*P* (psia)	*T* (°F)	Mass flow per MW discharging with 1:1 CT/DT (lbm/h)
Charging Cycle Working Fluid				
100	Comp Discharge/ Hot HX Inlet	500.0	1050.0	62,179
101	Hot HX Exit/Recup HP Inlet	492.8	570.0	62,179
102	Recup HP Outlet/ Exp Inlet	487.6	79.0	62,179
103	Exp Outlet/Cold HX Inlet	134.4	−74.0	62,179
104	Cold HX Outlet/ Recup LP Inlet	132.9	60.0	62,179
105	Recup LP Outlet/ Comp Inlet	130.8	560.0	62,179
Discharging Cycle Working Fluid				
200	Hot HX Outlet/Exp Inlet	492.8	1030.0	62,179
201	Exp Outlet/Recup LP Inlet	100.2	564.0	62,179
202	Recupeator LP Outlet/Cooler Inlet	98.8	213.0	62,179
203	Cooler Outlet/ Cold HX Inlet	98.7	72.0	62,179
204	Cold HX Outlet/ Comp Inlet	97.5	−64.0	62,179
205	Comp Outlet/ Recup HP Inlet	504.2	203.0	62,179
206	Recup HP Outlet/ Hot HX Inlet	500.0	549.0	62,179
Cold Media Circulation—Hexane				
301	Amb Tank Outlet/ Pump Inlet	30.0	64.0	30,453
302	Pump Discharge/ Charging HX Inlet	80.0	64.0	30,453

TABLE 6 Heat and material balance data for ideal gas cycle—cont'd

State point	Description	*P* (psia)	*T* (°F)	Mass flow per MW discharging with 1:1 CT/DT (lbm/h)
303	Charging HX Outlet/Cold Tank Inlet	30.0	−70.4	30,453
304	Cold Tank Outlet/ Pump Inlet	30.0	−70.4	30,453
305	Pump Discharge/ Discharging HX Inlet	80.0	−70.4	30,453
306	Discharging HX Outlet Amb Tank Inlet	30.0	64.0	30,453
Hot Media Circulation—HITEC				
401	Warm Tank Outlet/Pump Inlet	30.0	560.7	43,100
402	Pump Discharge/ Charging HX Inlet	80.0	560.7	43,100
403	Charging HX Outlet/Hot Tank Inlet	30.0	1042.0	43,100
404	Hot Tank Outlet/ Pump Inlet	30.0	1042.0	43,100
405	Pump Discharge/ Discharging HX Inlet	80.0	1042.0	43,100
406	Discharging HX Outlet/Warm Tank Inlet	30.0	560.7	43,100
Cooling Water Circulation				
500	Pump Outlet/ Cooling HX Inlet	45.0	63.5	84,966
501	Cooling HX Outlet/Tower Inlet	30.0	88.5	84,966
502	Tower Discharge/ Pump Inlet	15.0	63.5	84,966

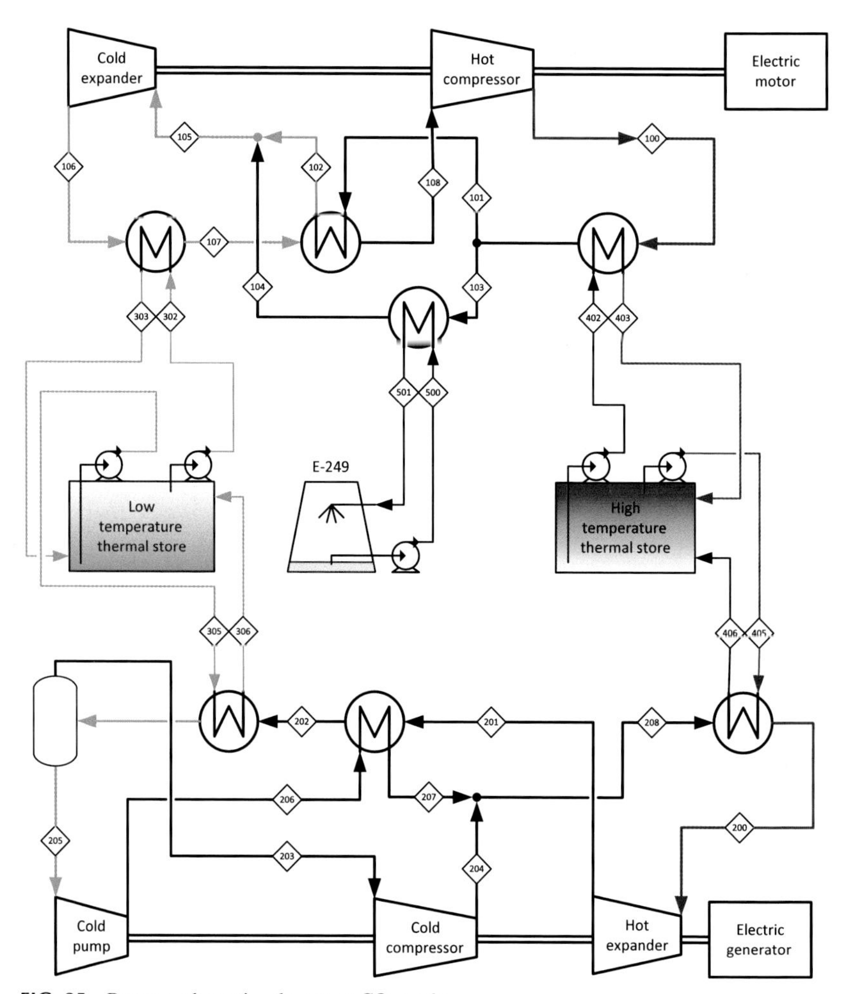

FIG. 35 Process schematic, alternate t-CO_2 cycle.

energy storage applications, thick pipes could lead to rate change limitations if temperatures change with the load point.

(5) At turndown conditions, the ideal gas cycle should have an efficiency advantage because the output can be controlled by loop inventory/pressure with consistent volume flow at any load, leading to very flat efficiency. The t-CO_2 cycle is tied to the CO_2 condensing pressure on the low side, so inventory control does not seem a favorable means of cycle control. Also, flow control or high-side pressure control of t-CO_2 will lead to significant temperature excursions leading to loss of efficiency and potentially getting the high temp and low temp reservoirs "out of balance," e.g., more thermal energy left in the cold reservoir than the

TABLE 7 Heat and material balance for Alt1 t-CO_2 cycle configuration.

State point	Description	*P* (psia)	*T* (°F)	Quality	Mass flow per MW discharging with 1:1 CT/DT (lbm/h)
Charging Cycle Working Fluid					
100	Comp Discharge/ Hot HX Inlet	3301.2	610.0	1.000	87,651
101	Hot HX Exit/ Recup HP Inlet	3269.0	230.4	1.000	53,761
103	Recup HP Outlet/Mixer Inlet 1	3236.4	76.5	1.000	53,761
102	Hot HX Exit/ Cooler Inlet	3269.0	230.4	1.000	33,890
104	Cooler Outlet/Mixer Inlet 2	3236.4	76.5	1.000	33,890
105	Mixer Outlet/ Expander Inlet	3236.4	76.5	1.000	87,651
106	Exp Outlet/ Cold HX Inlet	441.8	23.0	0.141	87,651
107	Cold HX Outlet/Recup LP Inlet	437.3	31.3	1.000	87,651
108	Recup LP Outlet/Comp Inlet	432.9	221.4	1.000	87,651
Discharging Cycle Working Fluid					
200	Hot HX Outlet/Exp Inlet	3827.3	538.0	1.000	87,651
201	Exp Outlet/ Recup LP Inlet	587.5	230.7	1.000	87,651
202	Recup LP Outlet/Cold HX Inlet	581.6	84.4	1.000	87,651
203	Cold HX Vapor Outlet/ Comp Inlet	575.8	41.0	1.000	14,502

Continued

TABLE 7 Heat and material balance for Alt1 t-CO_2 cycle configuration—cont'd

State point	Description	*P* (psia)	*T* (°F)	Quality	Mass flow per MW discharging with 1:1 CT/DT (lbm/h)
205	Comp Outlet/ Mixer Inlet 1	3865.8	328.5	1.000	14,502
204	Cold HX Liquid Outlet/ Pump Inlet	575.8	41.0	0.000	73,149
206	Pump Outlet/ Recup HP Inlet	3905.0	75.4	1.000	73,149
208	Recup HP Outlet/Mixer Inlet 2	3865.8	171.9	1.000	73,149
209	Mixer Outlet/ Hot HX Inlet	3865.8	194.7	1.000	87,651
Cold Media Circulation—SR1 15%					
301	Amb Tank Outlet/Pump Inlet	30.0	42.0	0.000	560,000
302	Pump Discharge/ Charging HX Inlet	80.0	42.0	0.000	560,000
303	Charging HX Outlet/Cold Tank Inlet	30.0	26.0	0.000	560,000
304	Cold Tank Outlet/Pump Inlet	30.0	26.0	0.000	560,000
305	Pump Discharge/ Discharging HX Inlet	80.0	26.0	0.000	560,000
306	Discharging HX Outlet Amb Tank Inlet	30.0	42.0	0.000	560,000

TABLE 7 Heat and material balance for Alt1 t-CO_2 cycle configuration—cont'd

State point	Description	*P* (psia)	*T* (°F)	Quality	Mass flow per MW discharging with 1:1 CT/DT (lbm/h)
Hot Media Circulation—Therminol66					
401	Warm Tank Outlet/Pump Inlet	30.0	200.0	0.000	66,000
402	Pump Discharge/ Charging HX Inlet	80.0	200.0	0.000	66,000
403	Charging HX Outlet/Hot Tank Inlet	30.0	545.5	0.000	66,000
404	Hot Tank Outlet/Pump Inlet	30.0	545.5	0.000	66,000
405	Pump Discharge/ Discharging HX Inlet	80.0	545.5	0.000	66,000
406	Discharging HX Outlet/ Warm Tank Inlet	30.0	200.0	0.000	66,000
Cooling Water Circulation					
500	Pump Outlet/ Cooling HX Inlet	45.0	63.5	0.000	112,962
501	Cooling HX Outlet/Tower Inlet	30.0	88.5	0.000	112,962
502	Tower Discharge/ Pump Inlet	15.0	63.5	0.000	112,962

TABLE 8 Heat and material balance for Alt2 t-CO_2 cycle configuration.

State point	Description	*P* (psia)	*T* (°F)	Quality	Mass flow per MW discharging with 1:1 CT/DT (lbm/h)
Charging Cycle Working Fluid					
100	Comp Discharge/ Hot HX Inlet	6162.5	1049.0	1.000	38,262
101	Hot HX Exit/ Recup HP Inlet	6100.9	441.3	1.000	26,215
103	Recup HP Outlet/Mixer Inlet 1	6039.8	76.5	1.000	26,215
102	Hot HX Exit/ Cooler Inlet	6100.9	441.3	1.000	12,046
104	Cooler Outlet/Mixer Inlet 2	6039.8	76.5	1.000	12,046
105	Mixer Outlet/ Expander Inlet	6039.8	76.5	1.000	38,262
106	Exp Outlet/ Cold HX Inlet	441.8	23.0	0.141	38,262
107	Cold HX Outlet/Recup LP Inlet	437.3	31.3	1.000	38,262
108	Recup LP Outlet/Comp Inlet	432.9	432.3	1.000	38,262
Discharging Cycle Working Fluid					
200	Hot HX Outlet/Exp Inlet	8529.1	1009.4	1.000	38,262
201	Exp Outlet/ Recup LP Inlet	587.5	486.5	1.000	38,262
202	Recup LP Outlet/Cold HX Inlet	581.6	120.6	1.000	38,262

TABLE 8 Heat and material balance for Alt2 t-CO_2 cycle configuration—cont'd

State point	Description	*P* (psia)	*T* (°F)	Quality	Mass flow per MW discharging with 1:1 CT/DT (lbm/h)
203	Cold HX Vapor Outlet/ Comp Inlet	575.8	41.0	1.000	7476
205	Comp Outlet/ Mixer Inlet 1	8615.2	463.3	1.000	7476
204	Cold HX Liquid Outlet/Pump Inlet	575.8	41.0	0.000	30,785
206	Pump Outlet/ Recup HP Inlet	8702.3	111.6	1.000	30,785
208	Recup HP Outlet/Mixer Inlet 2	8615.2	410.2	1.000	30,785
209	Mixer Outlet/ Hot HX Inlet	8615.2	420.4	1.000	38,262
Cold Media Circulation—SR1 15%					
301	Amb Tank Outlet/Pump Inlet	30.0	42.0	0.000	250,000
302	Pump Discharge/ Charging HX Inlet	80.0	42.0	0.000	250,000
303	Charging HX Outlet/Cold Tank Inlet	30.0	26.4	0.000	250,000
304	Cold Tank Outlet/Pump Inlet	30.0	26.4	0.000	250,000
305	Pump Discharge/ Discharging HX Inlet	80.0	26.4	0.000	250,000

Continued

TABLE 8 Heat and material balance for Alt2 t-CO_2 cycle configuration—cont'd

State point	Description	*P* (psia)	*T* (°F)	Quality	Mass flow per MW discharging with 1:1 CT/DT (lbm/h)
306	Discharging HX Outlet Amb Tank Inlet	30.0	42.0	0.000	250,000
Hot Media Circulation—HITEC					
401	Warm Tank Outlet/Pump Inlet	30.0	428.0	0.000	33,700
402	Pump Discharge/ Charging HX Inlet	80.0	428.0	0.000	33,700
403	Charging HX Outlet/Hot Tank Inlet	30.0	1019.5	0.000	33,700
404	Hot Tank Outlet/Pump Inlet	30.0	1019.5	0.000	33,700
405	Pump Discharge/ Discharging HX Inlet	80.0	1019.5	0.000	33,700
406	Discharging HX Outlet/ Warm Tank Inlet	30.0	428.0	0.000	33,700
Cooling Water Circulation					
500	Pump Outlet/ Cooling HX Inlet	45.0	63.5	0.000	112,962
501	Cooling HX Outlet/Tower Inlet	30.0	88.5	0.000	112,962
502	Tower Discharge/ Pump Inlet	15.0	63.5	0.000	112,962

TABLE 9 Recuperated ideal gas vs t-CO_2 parametric comparison.

Parameter	Unit	Ideal gas cycle	Transcritical CO_2 Alt 1	Transcritical CO_2 Alt 2
Round Trip Efficiency	%	57.86	51.54	61.24
Heat Pump COP	–	1.31	1.78	1.34
Engine Efficiency	–	0.44	0.29	0.46
Charging Pr	–	3.82	7.63	14.23
Discharging Pr	–	5.17	6.78	15.11
Cycle Max Temperature	°F	1050.0	610.0	1049.0
Cycle Min Temperature	°F	−73.6	23.0	23.0
Cycle Max Pressure	psia	500.0	3905.0	8702.3
Charging Comp Gas Power	MW	2.31	2.09	1.75
Charging Exp Gas Power	MW	0.64	0.22	0.17
Charging Net Gas Power	MW	1.66	1.87	1.57
Discharging Comp Gas Power	MW	1.15	0.25	0.25
Discharging Comp2 Gas Power	MW		0.17	0.16
Discharging Exp Gas Power	MW	2.19	1.46	1.45
Discharging Exp2 Gas Power	MW			
Discharging Net Gas Power	MW	1.04	1.04	1.04
Charging Comp Head_Is	ft lbf/lbm	88,595	56,829	108,756
Charging Exp Head_Is	ft lbf/lbm	−30,433	−7424	−13,333
Discharging Comp Head_Is	ft lbf/lbm	44,216	8241	19,396
Discharging Comp2 Head_Is	ft lbf/lbm		28,861	50,185

Continued

TABLE 9 Recuperated ideal gas vs t-CO_2 parametric comparison—cont'd

Parameter	Unit	Ideal gas cycle	Transcritical CO_2 Alt 1	Transcritical CO_2 Alt 2
Discharging Exp Head_Is	ft lbf/lbm	−103,876	−49,157	−111,296
Discharging Exp2 Head_Is	ft lbf/lbm			
Charging Comp Inlet Volume Flow	ACFM	3002	524	314
Charging Exp Outlet Volume Flow	ACFM	1085	61	18
Discharging Comp1 Inlet Volume Flow	ACFM	1543	22	9
Discharging Comp2 Inlet Volume Flow	ACFM		33	17
Discharging Exp1 Outlet Volume Flow	ACFM	3933	383	245
Discharging Exp2 Outlet Volume Flow	ACFM			
HT Store Discharging Rate	MW	2.26	3.46	2.18
LT Store Discharging Rate	MW	0.60	1.99	0.83
Heat Rejection Rate	MW	0.62	0.83	0.53
Hot Media Circulation Rate	gpm	51.5	137.3	35.1
Cold Media Circulation Rate	gpm	92.0	1088.9	486.1
HTHX Surface Area	ft2	19,209	3996	1552
LTHX Surface Area	ft2	11,908	2840	1319
Recuperator Surface Area	ft2	29,962	870	1810
Cooler Surface Area	ft2	1364	64	26

TABLE 9 Recuperated ideal gas vs t-CO_2 parametric comparison—cont'd

Parameter	Unit	Ideal gas cycle	Transcritical CO_2 Alt 1	Transcritical CO_2 Alt 2
Hot Store Details		**HITEC**	**Therminol66**	**HITEC**
		Sand	**Sand**	**Sand**
		Quartzite	**Quartzite**	**Quartzite**
Total Mass	klbm	5750	13,901	4592
Total Volume	kft^3	45	123	36
Cold Store Details		**Hexane**	**SR1 15%**	**SR1 15%**
		Sand	**4" Ice Balls**	**4" Ice Balls**
		Quartzite		
Total Mass	klbm	8668	7751	3390
Total Volume	kft^3	79	123	54
Summation of Hot and Cold Stores				
Total Mass	klbm	14,419	21,651	7982
Total Volume	kft^3	124	246	90

corresponding requirement for the hot reservoir. This situation would have to be corrected by some variety of inefficient means. While inventory control presents an option for efficient turndown for the ideal gas variants, the equipment to implement this scheme adds cost to the system. However, if output control proves not to be a relevant value proposition topic, the system can be omitted for the ideal gas cycle, whereas there is no good alternative for t-CO_2 if efficient turndown is.

(6) Considering Therminol66 high-temperature fluid, the specific heat of this fluid increases with temperature, while the specific heat of the t-CO_2 working fluid decreases with temperature over the relevant range. This leads to a pinched high-temperature heat exchange process that reduces the RTE (refer to Fig. 36). This problem is reduced in Alt2 with molten salt, which has a nearly constant specific heat over the range. A graphic is not presented for the molten salt case, but the effect can be discerned in the temperature approaches apparent in the state points (Table 7 for Alt1 and Table 8 for Alt2).

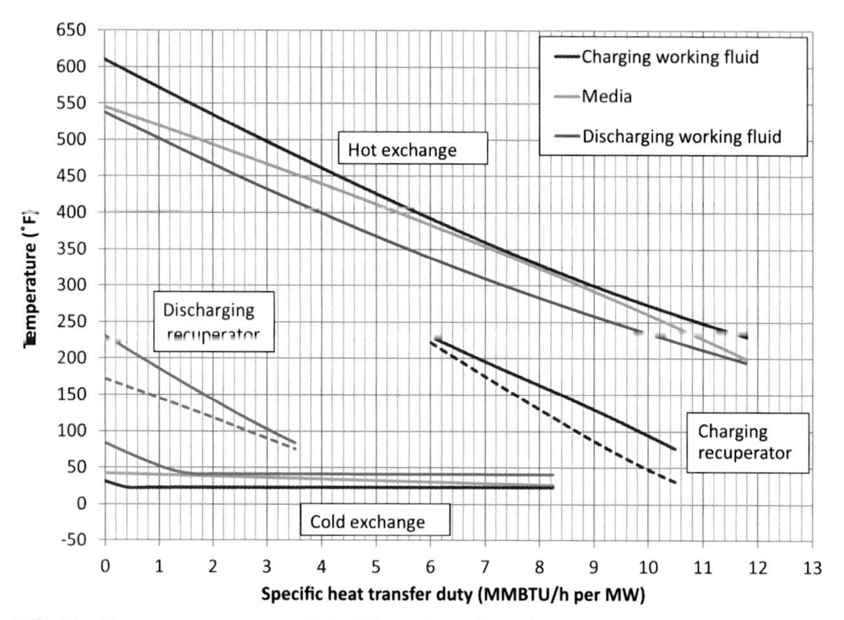

FIG. 36 Recuperator processes, Alt1 t-CO_2 cycle configuration.

(7) In the pressure and temperature ranges of interest for the recuperators in the t-CO_2 cycle, the fluid has considerable higher specific heat at high pressure than at low pressure. In the charging cycle, the high-pressure fluid is on the hot side and the low-pressure fluid is on the cold side. This leads to a pinch at the hot end of the recuperator. Conversely, in the discharging cycle, the high-pressure fluid is on the cold side and the low-pressure fluid is on the hot side. This leads to a pinch at the cold end of the recuperator. These pinches, which are detrimental to RTE, are much more prevalent with CO_2 than with ideal gas (compare Fig. 36 to Fig. 37). For an ideal gas, enthalpy is principally a function of temperature and a very weak function of pressure, such that with the small pressure ratio as exists in the recuperated ideal gas cycle, the specific heats are nearly identical for the fluid on either side of the recuperator, leading to consistent temperature differential across the exchanger. This temperature differential can be made arbitrarily small via application of sufficient heat exchange surface area (of course with cost increasing generally in proportion to area).

(8) For the water/ice low-temperature store of the t-CO_2 cycle, there are two embodiments which are well proven. The first approach is an "ice on coil" approach where water and the cycle working fluid are separated by one heat exchange surface and ice forms and melts on this surface. The advantage to this approach is that lower approach temperatures can be achieved

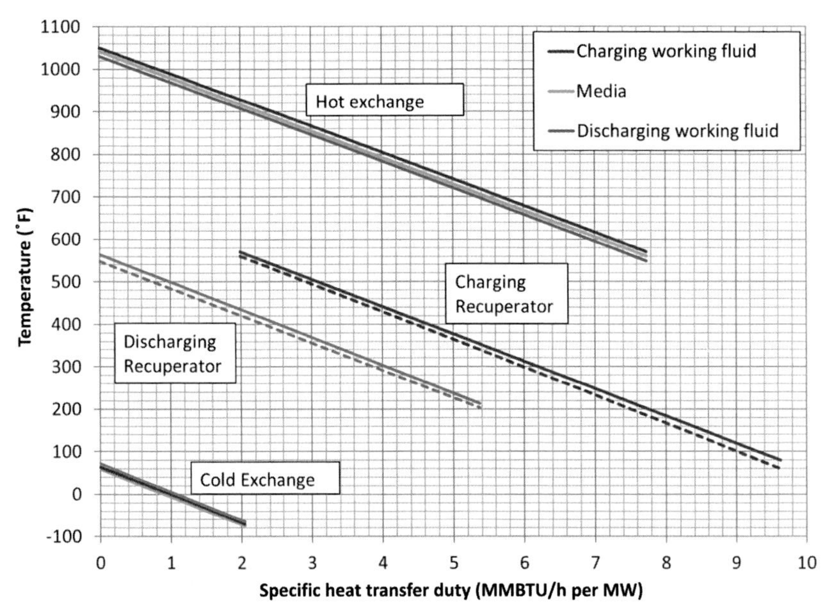

FIG. 37 Heat exchange processes, recuperated ideal gas cycle.

and there is no need to circulate a low-temperature media or heat carrier fluid. The disadvantage is that the thickness of ice that forms on the surface significantly impedes the conduction of heat, such that in order to maintain a close approach temperature, not much thickness of ice can be permitted. This in turn leads to very large heat transfer surface requirement, which would be costly as this surface must also contain the high cycle working pressure. Furthermore, the surface area requirement for the low-temperature exchange process is in this case a function of the store size. The second approach is to circulate a heat carrier fluid and apply an encapsulated phase change material to accomplish the bulk of the thermal energy storage. This is the approach that has been considered here, with an appropriate ethylene glycol mixture applied as the heat carrier. However, since the low-temperature heat exchange processes are at constant temperature (latent heat process) for both charging and discharging while the heat carrier is only undergoing a sensible heating or cooling process, a very high rate of circulation of a heat carrier is required in order to minimize temperature approaches in both the charging and discharging cycles. In this scheme, the total temperature difference between the charging and discharging low-temperature heat exchange processes is comprised of the media approach to the working fluid in charging cycle, the media temperature range, and the media approach to the working fluid on the discharging cycle (compare Fig. 36 to Fig. 37).

(9) Considering the possibility to use solid-phase high-temperature store media, this seems less accessible with the t-CO_2 concept owing to the higher pressure. An indirect HX with solid-phase media would be costly owing to the poor heat transfer characteristics of the media and could present a serious erosion concern. A direct heat exchange would require either pressurizing the store volume or passing the solid media across a pressure boundary to cycle pressure. The former approach is considered infeasible for the high pressure of t-CO_2 along with the large store volume and the latter approach would require solid air-locking solutions that are well beyond proven equipment. Technical requirements for solids air-locking are beyond proven experience for the flow rates and pressure levels that are needed for the ideal gas variant. They are much more so for the t-CO_2 variant, which would require both higher flow of solids and higher pressures. Furthermore, for the ideal gas variant, this scheme is available for both the high- and low-temperature stores if it is advantageous. For the t-CO_2 variant it seems unlikely to find any advantage replacing the water/ice latent heat store for the low-temperature store.

(10) Round-trip efficiency for the t-CO_2 cycle with the configuration considered here seems to have important dependence on the tolerance of vapor evolution in the liquid turbine of the charging cycle. An exit quality of 25% has been assumed in the analysis here.

(11) Hexane is currently considered for the low-temperature fluid for ideal gas cycle, as it is the cheapest and most technically effective fluid for the purpose. It might be possible to find a specialized glycol mixture that can work, perhaps with some cycle compromise, but this will be more expensive if it is viable. While this is a disadvantage for the ideal gas variant, liquid hydrocarbons are stored and managed in many industrial settings. Therminol66 fluid that has been considered here for the high-temperature store is also flammable. It is less hazardous than hexane owing to its lower volatility, but still similar design provisions would be required, e.g., area classification, containment, firefighting in the event of a rupture of the storage tank, etc.

(12) Heat exchangers pose a significant challenge for the ideal gas cycle variant in terms of achieving the desired performance at an acceptable cost point. A preliminary assessment of the exchangers required for each cycle is included based on some idealized assumptions regarding each heat exchanger's channel sizes and velocities. In each case the heat exchange equipment is presumed to be shared between charging and discharging cycle, such that the larger area requirement determines the design (refer to Table 9). While the exchangers for the ideal gas variant admittedly represent a higher degree of technical risk in the commercialization, the printed circuit heat exchangers considered for the t-CO_2 cycle variant are generally also costly, limited in their scalability, and may be limiting their transient characteristics.

6.3.10 Options regarding the thermal stores

There are many possible options to realize a thermal energy store. The compatibility of a given store alternative with one of the PHES cycle alternatives described previously (or application of thermal stores to other cycles or purposes, for that matter) depends on a number of factors. These might include, but are not limited to the following:

- Operable temperature range for the store: The thermal media should not suffer significant physical or chemical degradation while operating within the necessary temperature range.
- Pressure losses imposed toward the cycle should be minimized. This presents a particular challenge in the case of the packed solid bed, since the pressure drop through the bed is related to the desired magnitude of energy storage.
- In store concepts where direct contact between thermal media and working fluid is considered, contamination of either of the substances must be mitigated or managed to acceptable levels.
- Energy consumption and complexity for circulation of thermal media should be minimized

In this section various store configurations are illustrated and discussed. In the illustrations that follow, flow directions and temperature orientations ("Low Temperature" and "High Temperature") characterize one exemplary service (the hot store or the cold store), operating in one exemplary mode (charging or discharging). It should be clear that each of the concepts presented here may be applied to either service (hot store or cold store), and that operation of the system with regard to flow directions to realize either mode of system operation (charging or discharging) should be readily deduced by the reader.

The early investigators into PHES systems focused on solid-phase stationary packed bed thermocline stores, as illustrated by Fig. 38. In this concept, the working fluid passes through a packed bed of porous thermal media, to which or from which the working fluid transfers heat via direct contact with the thermal media. Exemplary operation of a high-temperature store would proceed as follows:

(1) Starting from a state with the entire bed at low temperature, hot working fluid would be fed in at the top of the store vessel and extracted at the bottom. In this context "low" could mean ambient in the case of the basic cycle or it could be the "warm" temperature of the recuperated or overlapped cycle.
(2) As the working fluid moved downward, it would heat up the media, with the upper part of the store heating up first, while the bottom part (and thus the exiting working fluid) remained at ambient temperature.

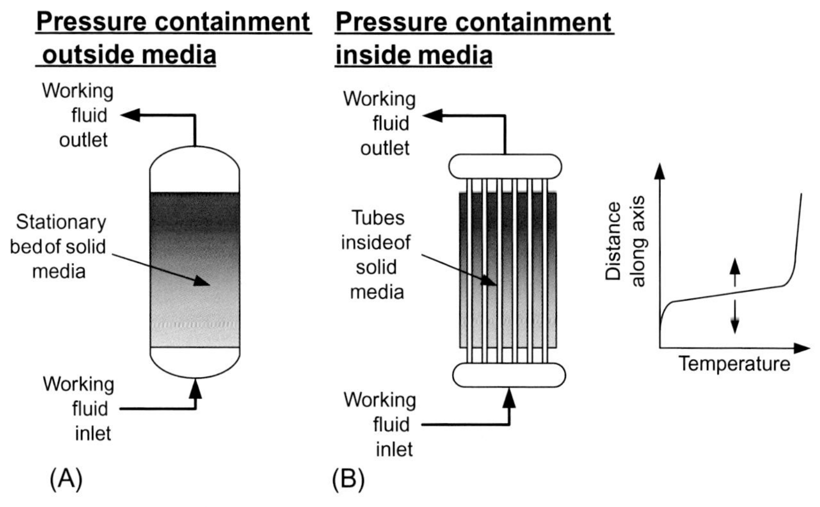

FIG. 38 Direct contact solid stationary thermocline. (A) Pressure outside media (B) pressure inside media.

(3) As operation progressed, a larger and larger fraction of the upper portion of the store would be at a temperature near the hot incoming working fluid.
(4) The store would reach the fully charged state when the media at the bottom of the store began to experience elevated temperatures, and thus the working fluid exit would likewise be above the low temperature of the store.
(5) The discharging would be conducted by reversing the flow of working fluid. Low-temperature working fluid would be introduced at the bottom and would exit the store at the hot temperature near the top.
(6) Discharging could proceed until the media near the top of the store began to experience reduced temperature, at which point the store would be fully discharged and ready for another cycle.

The main advantages of this store concept are as follows:

(1) There is no circulation of the thermal media, so complexity, cost, and energy consumption associated with circulation equipment are avoided.
(2) Solid media is stable over a wide temperature range, so that fairly extreme temperatures are achievable. For example, solid media is a better candidate to realize the nonrecuperated or overlap cycles.
(3) If a natural solid media can be applied, the media cost per unit stored energy is quite low.
(4) The variation of specific heat with temperature of the thermal media does not seem to impose temperature pinch constraint in the packed bed configuration as it does in a counter-flow heat exchange configuration. The shape of the temperature profile within the media "adjusts itself" to balance the

temperature differential which drives heat transfer over the effective length of the temperature gradient

(5) Due to the direct contact and the high availability of heat transfer surface area with low cost, temperature approach can be minimized.

(6) If the pressure vessel is internally insulated, the pressure-containing structure can operate basically at ambient temperature even while the media inside operates at very high temperature. This permits the use of regular carbon steel, which has both high strength at ambient temperature and low cost. Note, however, that if the vessel is externally insulated, then the pressure-containing structure must maintain its strength at the elevated process temperature. In most practical cases this requires the application of alloy steel, stainless steel, or in extreme cases nickel alloys.

The main disadvantages are as follows:

(1) The stores must operate at cycle pressure, giving rise to large (and therefore costly) pressure vessels. At extremely high or low temperatures, either costly metallurgy or internal insulation of the store vessels might also be needed, exacerbating this cost. There also arises a trade-off in this case between store cost and machine cost resulting from cycle pressure selection (and also affecting the turndown capability via inventory control as discussed previously): Increasing the cycle pressure reduces the necessary volume flows, thereby tending to decrease the machinery size and cost, but higher pressure necessitates thicker shell requirements for a pressurized store, increasing the costs for these elements.

(2) Without further corrective measures, there is a difficult coupling between store size, velocity, and pressure drop imposed on the working fluid. For example, if a target flow velocity is selected for the sake of heat transfer, then this dictates a cross-section for the store for a given size of the charging or discharging system. Then as longer store duration is needed, the length of the store must be increased, which in turn increases the pressure drop. A solution to this problem might be realized by subdividing the store into segments and using some configuration of isolation to expose only a fraction of the store to the working fluid at a given time; however, this adds some complexity and cost to the solution. This problem is addressed in a handful of patents by Isentropic, including but not limited to [44].

(3) The length of the thermal gradient represents store volume that cannot be utilized. This wasted volume still must be contained within the pressure vessel, adding to the cost of the store.

(4) The requirement for low pressure drop of the working fluid tends to favor lower density of the store, and thus higher store volume. Store volume in turn is an important cost driver if the store must be pressurized.

(5) If control of the charging/discharging rate by inventory control is considered, the pressurized store represents a very large amount of working fluid that has to be loaded or unloaded in order to affect the control. This means

either the equipment to accomplish the control becomes very large or the control process takes a very long time.

(6) If an internally insulated vessel is used, the insulation then becomes safety critical. If the insulation is breached and hot fluid is permitted to heat the vessel wall above its strength rating temperature, then a vessel rupture could occur. The occurrence of insulation failure might also be difficult to predict or detect since the internal insulation could not be easily inspected and the vessel might not show clear outward signs of the insulation breech. (Some possibility here might exist in the way of infrared imagery of the vessel or exterior coating that changes color with temperature.)

It is possible to avoid the large pressure vessel discussed previously and replace it instead with small diameter tubes routed inside of the solid thermal media. This approach has been developed and deployed by Energy Nest [45]. The main advantages of this concept are as follows:

(1) The thermal store is more modular since any number of tubes with solid media surrounding them can be banked together to form a store of arbitrary size. Furthermore, the size of the store is readily adjusted over the life of the project by adding more modules or taking modules away.
(2) The modularity also leads to a simplified transportation logistic. A store of arbitrary size can easily be assembled from modules with dimensions that permit shipment by standard means, for example ISO container size. Large stores of the aforementioned pressure vessel type will likely involve shipments of oversized loads for the vessels, and then the media would likely have to be assembled into the stores at the project site for systems of appreciable scale.
(3) Minimal risk of rupture, as discussed previously for the case of the internally insulated vessel, and also low consequence of rupture for a single tube.

The main drawbacks to this concept are as follows:

(1) Solids that have a low cost generally also have low thermal conductivity. The distance between every bit of store volume and active heat transfer surface area must accordingly be minimized. Therefore the entire store volume must be essentially filled with pressure-containing heat transfer surface (generally metal). While the media cost is generally low, the cost for this heat transfer surface area is likely to be significant, especially in comparison to other store concepts.
(2) Compared to the concept of media inside an internally insulated pressure vessel as discussed previously, this concept is much more limiting in terms of temperature. In the former concept, the temperature is mainly limited by capability of the media, which can be very high. In the latter concept, the heat transfer surface area must retain sufficient strength for the pressure containment at the maximum store temperature. Therefore at higher

temperatures, more costly metallurgy is required, advancing from plain carbon steel to alloy steel to stainless steel to nickel alloys. This coupled with the high density of the heat transfer surface within the store as mentioned previously likely makes this concept increasingly cost prohibitive at higher temperature.

A thermal store variant that attempts to avoid the problems of the pressurized store discussed previously is the concept of the conveyable solid store. In the embodiment with indirect heat exchange, illustrated in Fig. 39, granular solid media is introduced at the top of a solid-to-working-fluid heat exchanger. Working fluid flows in the counter-current orientation and heat is exchanged through an intermediate metal plate while the working fluid and the media remain segregated.

The main advantages of this store concept are as follows:

(1) The advantages with regard to the temperature range and cost of the solid media discussed previously are still realized.
(2) The store volumes remain at atmospheric pressure, minimizing their cost. Only the relatively small passages of the heat exchanger must contain the cycle pressure.
(3) There is no risk of contamination of the working fluid by the thermal media as the two remain separated.
(4) The stores may realize higher density since there is no need for porosity of the store to facilitate low-resistance passage of the working fluid. Also, no store volume is lost to the thermal gradient

The main disadvantages are as follows:

(1) The system must operate in a "two tank" regime, such that the total tank volume required is $2\times$ the volume of the media.

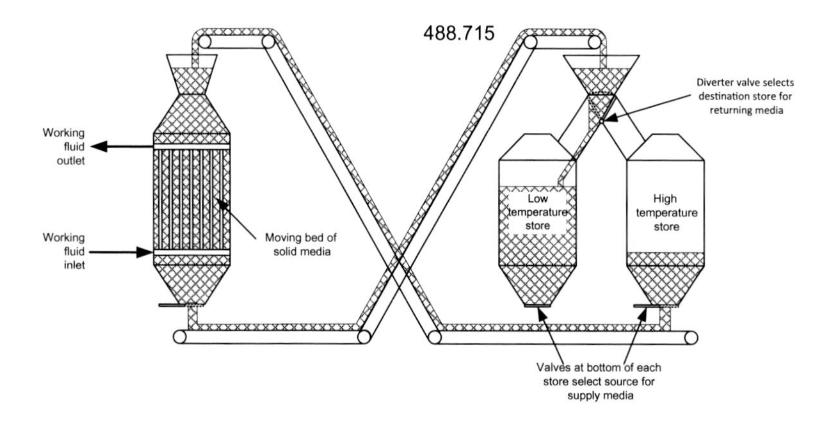

FIG. 39 Indirect contact conveyable solid.

(2) The solid media must be moved from a source store to the exchanger and then returned to a receiving store. Solid material conveyance adds significant complexity and cost compared to no conveyance, as the case of the stationary beds, or fluid conveyance as the case of the liquid-phase store options.
(3) Preliminary investigations into the design of the solid-to-working fluid heat exchanger suggest that this could be an item of prohibitive size and cost for PHES embodiments of desired size and temperature approach.
(4) An array of technical challenges relating to the solid-to-working-fluid heat exchanger may be expected. There are no proven implementations of heat exchangers of this sort having the necessary flows, pressures, temperatures, pressure drops, temperature approaches desired here. Challenges could include, but are not limited to, erosion concerns, novel manufacturing methods to achieve the necessary performance characteristics on the working fluid side, maldistribution and/or nonuniform flow of the solid phase.

This approach is described in a patent by Kerth [46].

Another approach to the conveyable solid concept is the direct contact concept, illustrated in Fig. 40. This method avoids the challenges and cost associated with the solid-to-working-fluid heat exchanger, but in this case, the solid media must enter and exit a vessel that operates at the cycle pressure via a series of airlock feeders. The concept of a "falling bed" or "immersed particle" heat exchanger has been considered for gas turbine recuperator or gas turbine external firing purposes [47, 48]. The adaptation of the falling bed heat exchanger to PHES is the subject of a patent application by Kerth [49]. There is no presently available robust feeding technology that is proven at the necessary pressures and temperatures that are desirable for PHES. The primary candidates include rotary airlocks, double dump valves, and screw feeders. A comprehensive review of candidate concepts for solid feeding is given in [50].

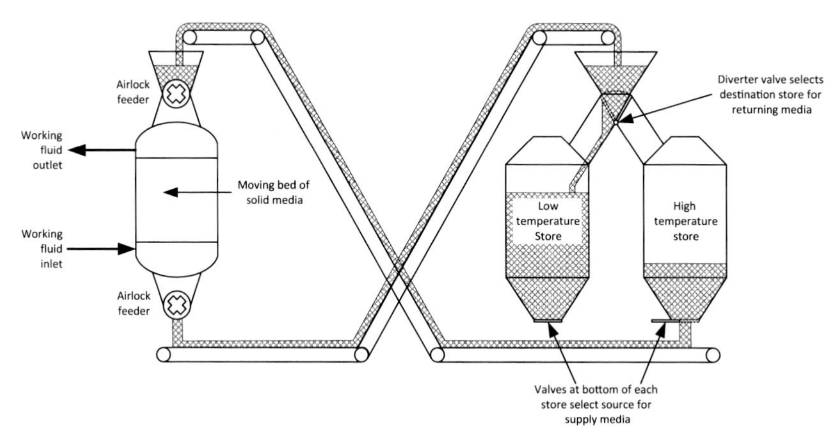

FIG. 40 Direct contact conveyable solid with continuous airlock feeders.

It is also possible to consider a batch process in lieu of a continuous feeding process for the direct contact with solids concept. In the batch scheme, an array of vessels is used. Solid media is loaded and unloaded from one vessel at atmospheric pressure, while another vessel is exposed to the pressurized, flowing working fluid. The vessels are then be switched in their roles. In this scheme, each vessel exhibits a mini-thermocline performance, with the thermal wave moving up during the operating mode as the media is exposed to the working fluid, then moving back down during the loading mode as new media is fed in at the top and used media is extracted from the bottom. Additional vessels might also be employed in this scheme to aid in the pressure transitions between the operating and loading modes. Operation according to this batch scheme could mitigate temperature pinch problems that arise with solid media on account of the significant temperature dependence of their specific heat. The batch concept could, however, result in significant loss and corresponding makeup requirement for working fluid if steps are not taken to manage or mitigate such loss. Note that the loss of working fluid entails not just the loss of the mass but also the loss of energy from the system, such that these losses may reduce the system round-trip efficiency.

The advantages and disadvantages of the various direct contact concepts are similar to those described for the conveyable solid concept with indirect heat transfer, with the exception that technical challenges and costs relating to the heat exchanger are replaced with technical challenges and cost relating to the feeder systems (Fig. 41).

A direct contact arrangement with liquid thermal media, as illustrated in Fig. 42, could also be considered. The main motive for the direct contact liquid concept is to avoid the high cost of indirect heat exchange while also avoiding the challenges of solid conveyance and the airlock feeders. However, the selection of a liquid thermal media with acceptable attributes is quite challenging. The liquid should have low vapor pressure at the maximum cycle temperature. This is necessary to facilitate economical storage of a large volume of the liquid, but also to prevent too much of the liquid from entering the vapor phase in the direct contact process. Any liquid that does enter the vapor phase may be difficult to subsequently separate from the working fluid, which means that the mixture of working fluid and thermal media would travel around the rest of the cycle where it would experience a wider range of pressure and temperature conditions that exist in the exchanger. This in turn could lead to the media recondensing and potentially solidifying elsewhere in the cycle, which could be problematic if not properly managed. In addition, if direct contact is considered for both high-temperature and low-temperature services, the consequences of cross-contamination of the high- and low-temperature thermal media must also be considered. Finally, if inventory control is considered, the consequence of venting the working fluid also needs to be considered. Silicone-based fluids were considered preliminarily for a direct contact embodiment, as they have reasonably high operable temperature range. However, these fluids are likely

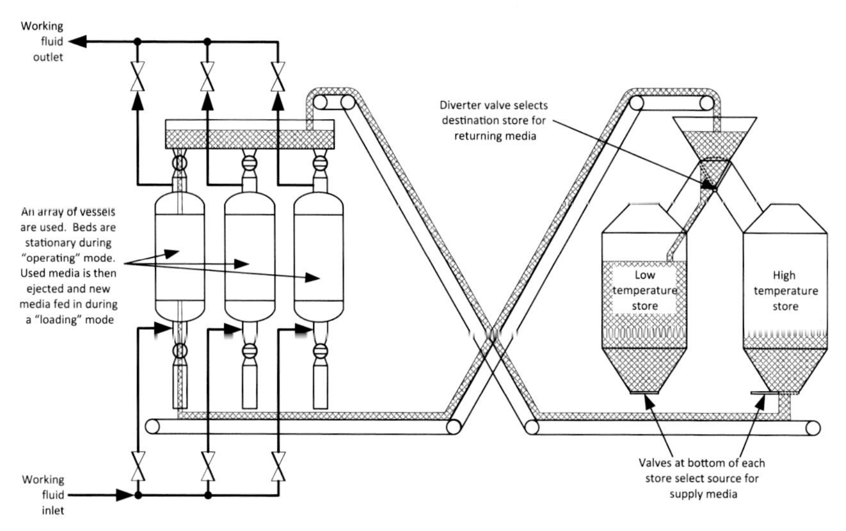

FIG. 41 Direct contact conveyable solid with batch feeding.

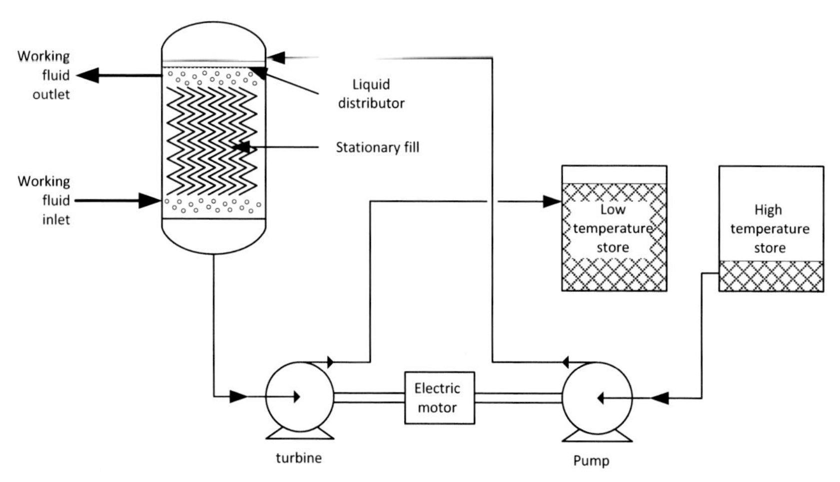

FIG. 42 Direct contact liquid.

to contaminate the working fluid with light hydrocarbons, which presents a problem if the system would be managed via inventory control with working fluid venting to the atmosphere. Also, silicone fluids are rather expensive, so that with large store volumes, the cost savings by eliminating the heat exchanger(s) might be fully offset by the higher thermal media cost.

Returning now to a more "conventional" approach toward the heat exchange and storage, the function can be achieved using liquid-phase stores with a conventional heat exchanger transferring heat between the thermal media and the working fluid. A classical "two tank" embodiment is illustrated in Fig. 43.

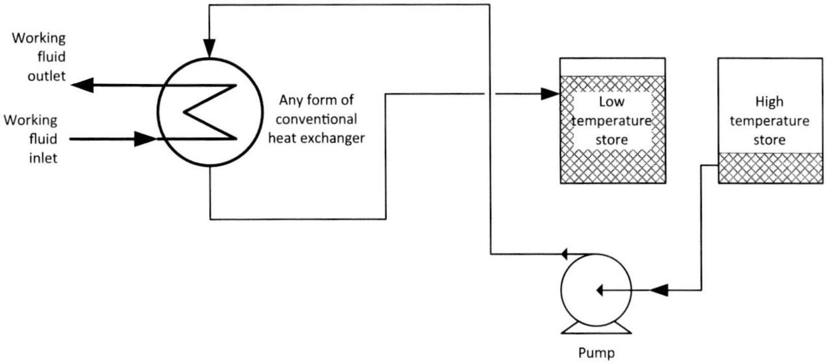

FIG. 43 Indirect contact liquid.

Thermal storage systems based on liquid stores have been extensively demonstrated in concentrated solar power (CSP) applications. Laughlin [38] describes application of liquid-phase stores for the purpose of PHES. The main consideration is the selection of the recuperated cycle, which allows the hot store to operate with a low temperature that is still warmer than the freezing point of the molten salts that are currently applied for CSP services. For the low-temperature store, hexane is considered. Hexane has relatively low viscosity even at minimum cycle pressure, leading to satisfactory heat transfer properties and efficient pumping. Glycol-based fluids might also be considered, but, in addition to being more expensive, have higher viscosity at low temperature leading either to a compromise on the minimum temperature, and/or higher costs in the low-temperature pumps and heat exchangers. Of all the thermal store concepts discussed, the indirect contact with liquid is arguably the most "straightforward" approach. The only deterrent to its universal adoption is the potential for high cost in the heat exchangers. A variety of more traditional heat exchanger concepts have been investigated, including multiple variant types of shell and tube exchangers, brazed plate-fin exchangers, and printed circuit exchangers. All of these appear to have higher-than-desirable cost for that target performance attributes. However, advanced heat exchange designs and manufacturing methods, specifically designs having exceptionally high surface area-to-volume ratios and manufacturing approaches involving a high degree of automation, might still present a viable path for cost-effective indirect heat exchange for PHES.

An alternate store approach that still involves the same indirect contact heat exchange between the thermal media and the working fluid is the replacement of the classical two-tank store with a single thermocline tank, as illustrated in Fig. 44. This approach has been investigated and tested for CSP applications by Pacheco and others [51–53]. The use of the thermocline in lieu of the two-tank approach offers the following significant advantages:

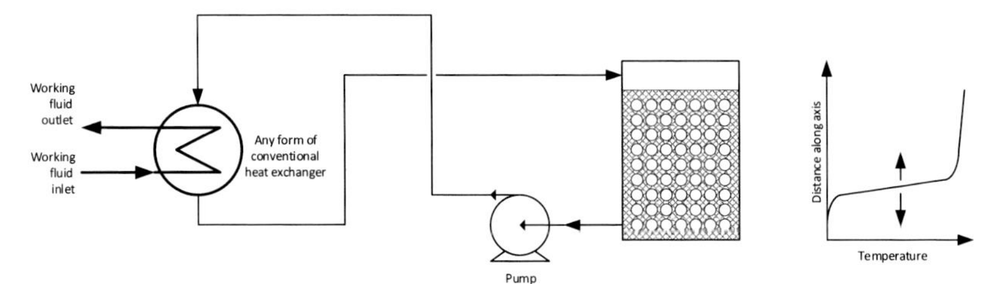

FIG. 44 Intermediate fluid thermocline.

(1) The cost and space requirements for the tank are reduced by almost 50%. As with the stationary packed bed, some of the store length is wasted on account of the length of the thermal gradient.

(2) The tank can be filled with a low-cost filler material, such as natural stone, greatly reducing the amount of thermal fluid required. The thermal fluid generally represents that largest part of the cost of the stores, so reduction of this quantity has a significant impact on the cost.

Despite the potential for cost savings, there has been no real uptake of thermocline tank technology in CSP applications to date. Also, there has been little or no consideration of the thermocline store approach to the low-temperature store that would be needed for PHES or any other similar low-temperature purpose.

In the case of a thermocline with intermediate fluid or with a stationary packed bed, it is possible and potentially advantageous to incorporate encapsulated phase change materials (PCM) into the packed solid bed. To be advantageous, the phase change should occur between the operable low and high temperature of the store. The modeling of this concept is described by Li et al. [54]. Preliminary investigations into application of this concept toward PHES suggest that the approach can significantly reduce the volume of the thermal stores, but it is unlikely to reduce the cost compared to natural stone fill. In almost any practical case, the cost of the encapsulated PCM exceeds the cost to store a comparable magnitude of thermal energy in sensible heat of natural stone.

It is also noteworthy that the concept of a thermocline store with an intermediate fluid has also been considered with a gas-phase intermediate fluid [55].

The advantages of gas-phase intermediate fluid are as follows:

(1) Assuming air is used, the intermediate fluid is essentially without cost, clearly offering a savings compared to any choice of liquid thermal media, all of which are with cost.

(2) Air imposes little constraint with regard to operating temperature limits. It can be used at arbitrarily high temperature and assuming it is adequately dehydrated, it is operable at temperatures well below what is necessary for PHES.

The disadvantages of gas-phase intermediate fluid are as follows:

(1) The heat-carrying capacity of air per unit volume is poor compared to most liquids, resulting in very large volume flows for the intermediate fluid. This in turn leads to large and costly ductwork for the intermediate fluid.
(2) The heat transfer characteristics of atmospheric pressure air are also quite poor compared to liquids, leading to significantly higher heat transfer surface area requirement and generally increased size and cost of the heat transfer equipment for a given heat duty.
(3) The equipment cost and energy consumption needed to circulate a large volume of air is significant. Note that for a hot store, the work consumption is additive to the store energy, so that this energy is not completely lost, albeit mechanical energy is converted to thermal energy with coefficient of performance equal to unity, which is less than any heat pump. For a cold store, the circulation energy is counterproductive to the store's purpose.

6.3.11 Heat exchanger service integration

In the concepts of indirect heat exchange with liquid and especially the recuperated versions, some cost advantage and performance improvement may be realized via integration of multiple heat exchange services. Integration of heat exchange services may eliminate the need for headers and interconnecting piping between some sections of heat exchange. Such heat exchange integration is fairly common, for example in LNG applications where such techniques are applied with both plate-fin and spiral-wound shell-and-tube type heat exchangers. Fig. 45 illustrates an example where the high-temperature exchanger, recuperator, and cycle cooler are integrated into a single physical piece of hardware.

6.3.12 Equipment sharing between charging and discharging cycles

In some cases it may be advantageous to share equipment between charging and discharging cycles. This could be the case, for instance, if heat exchange equipment makes up a substantial part of the cost of the PHES system, such that replication of this equipment for both charging and discharging cycles would substantially and unnecessarily increase the total system cost. There is, however, a trade-off to equipment sharing. While counterintuitive, there are cases where it is economically attractive to operate both systems simultaneously. A specific case is operation of the charging system while energy prices are low (or even negative!) while simultaneously operating the discharging system at low load in order to receive an additional payment for providing ancillary service (reserve or regulation service). Unfortunately, equipment sharing most likely prohibits the use of the PHES system in this manner. In the case of the stationary packed bed store, sharing of the packed bed store is imperative.

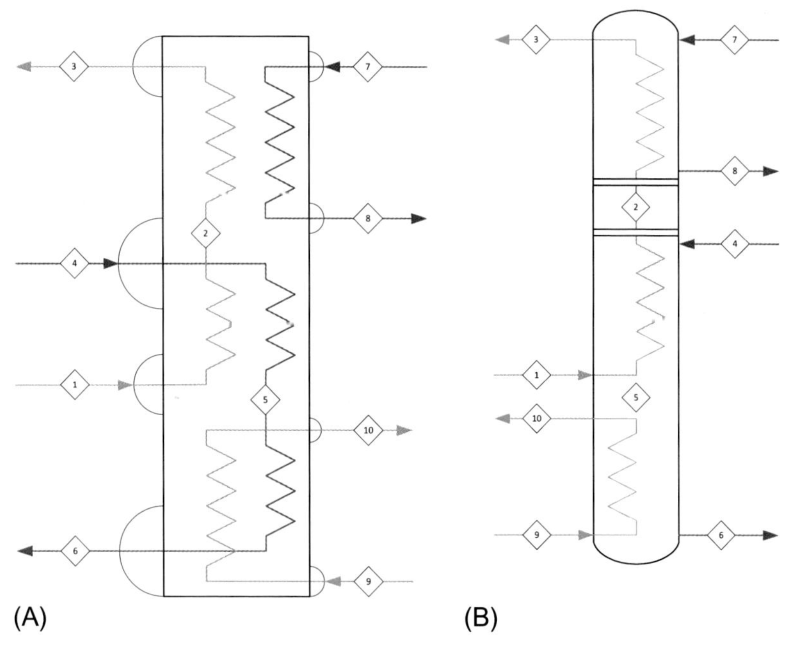

Stream numbers (Common for A and B)

1. Recuperator high pressure (HP) inlet
2. Recuperator HP Outlet / hot exchanger working fluid inlet
3. Hot exchanger working fluid outlet
4. Recuperator low pressure (LP) inlet
5. Recuperator LP outlet / cycle cooler working fluid inlet
6. Cycle cooler working fluid outlet
7. Hot exchanger thermal media inlet
8. Hot exchanger thermal media outlet
9. Cycle cooler coolant inlet
10. Cycle cooler coolant outlet

FIG. 45 Heat exchanger service integration with: (A) Plate-type exchanger (B) shell-and-tube type exchanger.

In order to share process equipment, it is necessary to connect the equipment to the system that is in use, while isolating the equipment from the system that is not in use. This generally is achieved using isolation and/or check valves. Depending on the concept selection for the store and heat exchange, it may be advantageous or necessary to operate the system in such a way as to have a consistent direction of flow through the equipment in either charging or discharging mode. For example, in the direct contact conveyable solid concept or direct contact liquid concept, the working fluid flow must always be oriented upwards while the thermal media flow must always be oriented downwards, whether the system is charging or discharging. Conversely, for the stationary packed bed of solid, the thermal gradient orientation must be preserved whether charging or discharging, while the flow direction is reversed. Either objective is readily achieved by the correct application of valves. Figs. 46–49 illustrate valve usage to achieve the various goals (maintaining thermal gradient or

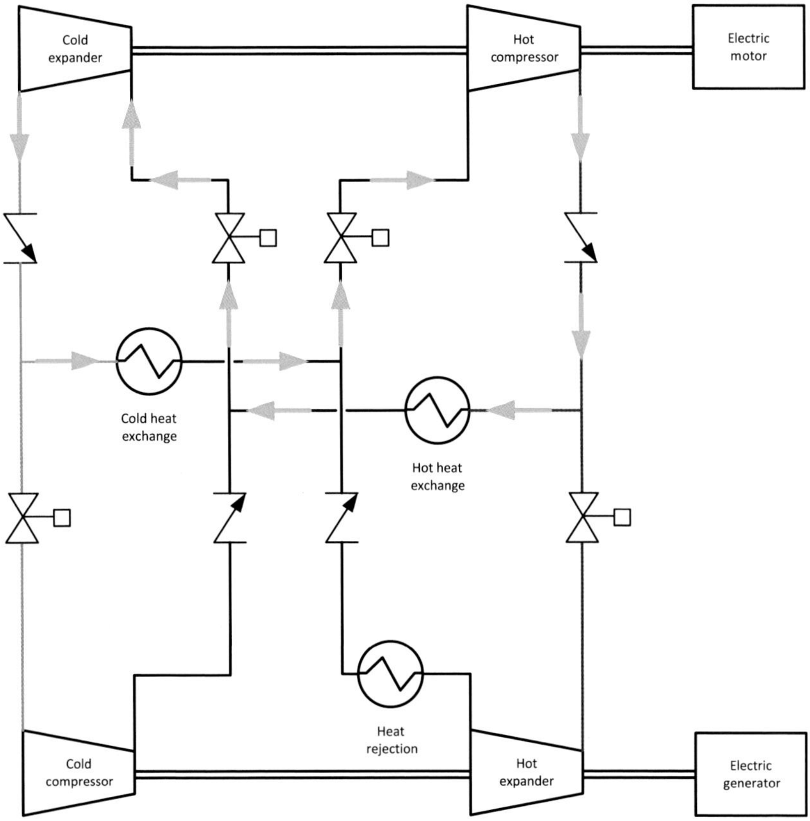

FIG. 46 Shared heat exchanger configuration with preserved thermal gradient of the exchangers—Flow path while charging.

maintaining flow direction) across charging and discharging operating modes. In these figures, the basis of design is that the loop that is not in use (charging or discharging) is deinventoried to atmospheric pressure. By doing this, the isolation can be in general managed using a combination of (passive) check valves and actuated valves. This is advantageous as check valves are significantly less costly. The following illustrations show the minimum requirements for the valves; however, the reader should appreciate that in any case the check valves illustrated as follows could be replaced by isolation valves if so desired, or in fact the combination of check and actuated valve could be replaced by a single 3-way valve.

Another opportunity for equipment sharing in an AC-to-AC energy storage system is the electrical machine. This can result in substantial cost savings, resulting not only from the cost elimination for the second electrical machine

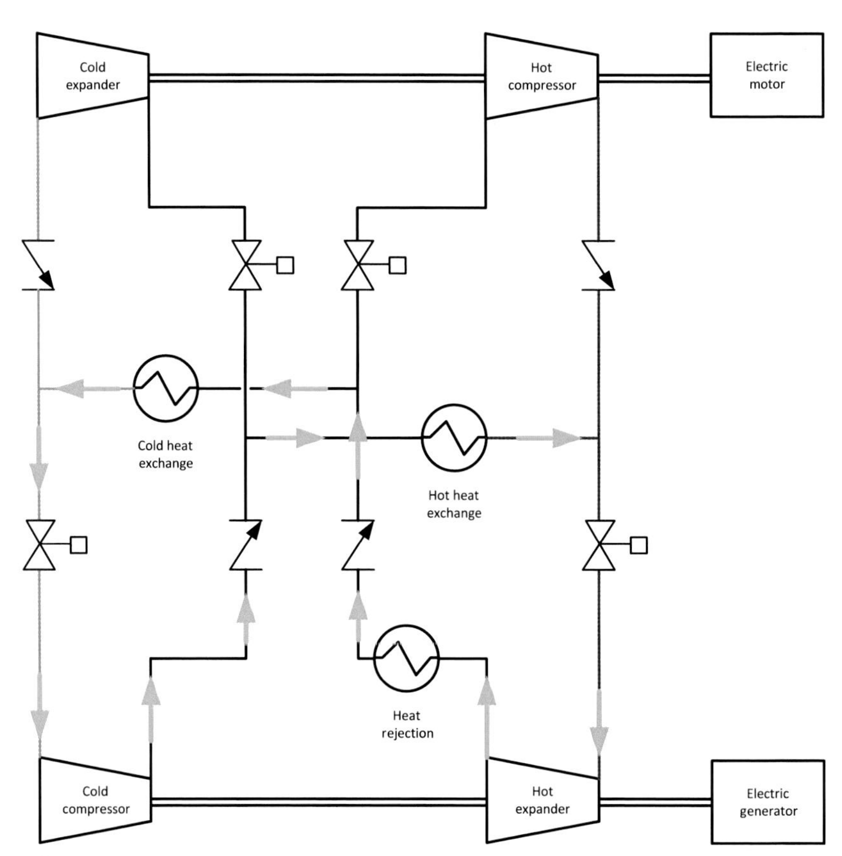

FIG. 47 Shared heat exchanger configuration with preserved thermal gradient of the exchangers—Flow path while discharging.

itself, but also for all the associated auxiliary equipment such as cabling or bus work, circuit breaker, step-up transformer if applicable, and metering and protection, etc. Synchronous and induction electrical machines can be operated selectively as either motors or generators. For synchronous machines, the driver or driven must operate at the grid synchronous rotational speed (1800 rpm, 3600 rpm, etc., as determined by number of poles) and power flow is controlled via the excitation. For induction machines, if the connected machine runs faster than the grid synchronous speed (i.e., the connected machine is a driver), the electrical machine will behave as a generator delivering power to the grid. If the connected machine cannot overtake the grid synchronous speed (i.e., connected machine is a load) then the electrical machine will behave as a motor. Note that in both cases, provisions must be considered for operation prior to synchronization. Also, in the case of the induction machine, additional

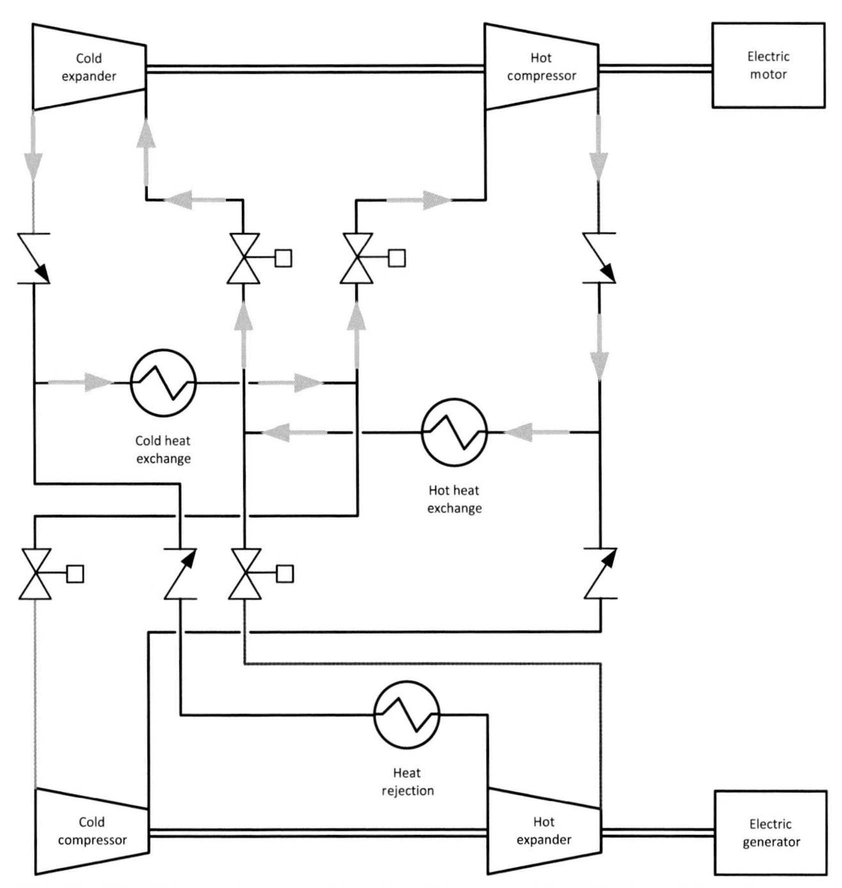

FIG. 48 Shared heat exchanger configuration with preserved flow direction of the exchangers—Flow path while charging.

consideration is needed if power generation is required while disconnected from the grid. An induction generator depends on an external AC supply to develop rotor current, which is needed for generation. If the electric machine is shared, then disconnecting clutches must be applied to determine which system is engaged. Similar to the situation with check valves, a (passive) overrunning clutch can be applied to connect the discharging machine to the electrical machine. To connect the charging machine, a clutch with a controllable disengagement is needed. The shared electrical machine concept is illustrated in Fig. 50.

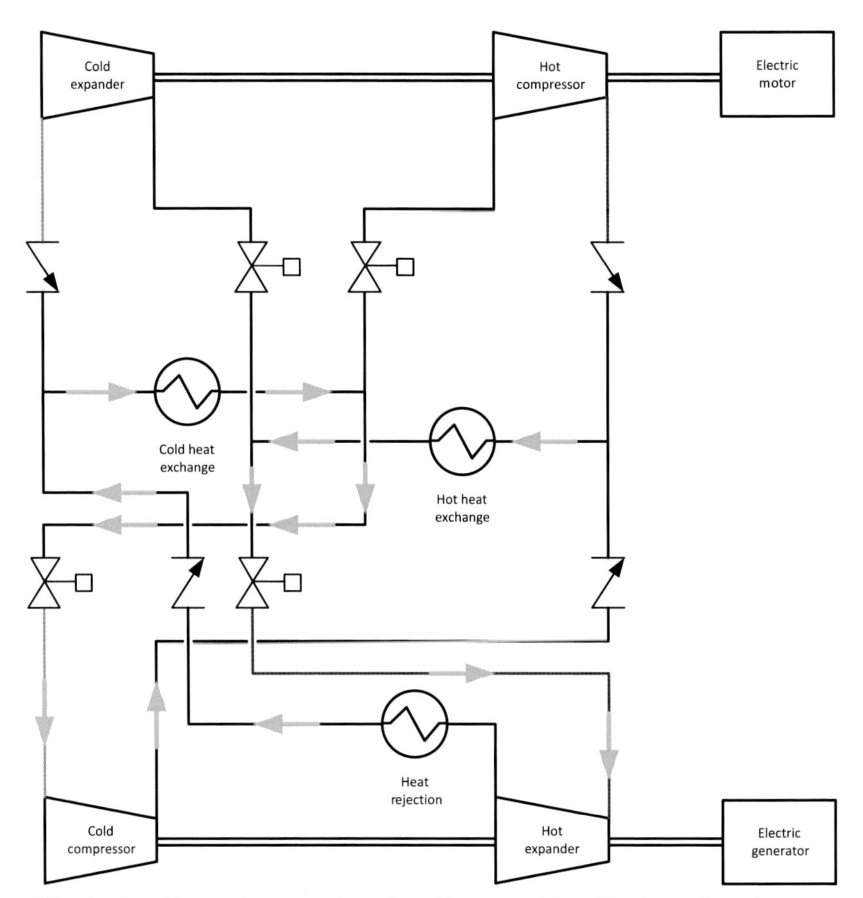

FIG. 49 Shared heat exchanger configuration with preserved flow direction of the exchangers—Flow path while discharging.

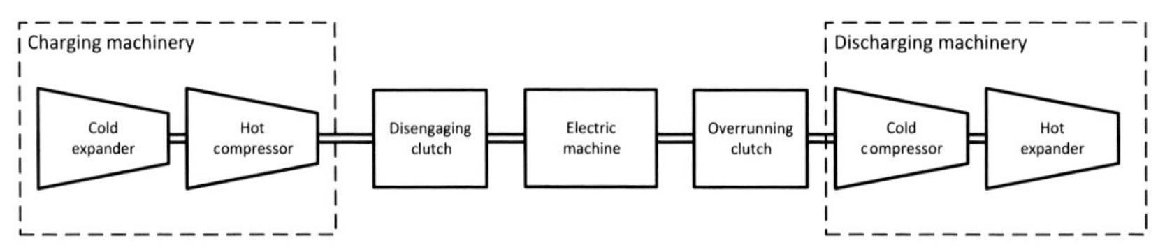

FIG. 50 Shared electrical machine.

Summary of PHES concepts.

	Cycle concept	Nonrecuperated ideal gas PHES	Recuperated ideal gas PHES	Overlapped ideal gas PHES
Thermal Store and Heat Exchange Concept	General Remarks on Thermal Store and Heat Exchange Concept	General Remark on Cycle Concept Leads to high pressure ratio to achieve good round-trip efficiency and stronger preference for argon working fluid. With argon working fluid, system control via inventory management is not practical due to high cost of argon makeup. High pressure ratio requirement means that higher max pressure is needed if the system would be controlled via inventory management	Recuperated cycle mainly developed to allow use of molten salt as the hot storage media, as it permits the lower temperature of the hot store to be above the freezing point of the salt. The cycle also results in lower pressure ratio and better round-trip efficiency. The lower pressure ratio attribute is favorable for system control via inventory management.	Given a thermal storage media with a wide operable temperature range (such as many practical solids), the overlapped cycle offers similar round-trip efficiency and pressure ratio characteristics as the recuperated cycle, but does not require the recuperator and associated piping. Instead, the low-temperature heat exchange process is extended (upward) to a temperature comparable to the recuperator hot-end temperature while the high-temperature heat exchange process is extended (downward) to a temperature comparable to the recuperator low-end temperature. As such the operable temperature ranges of the respective heat exchange process overlap.
Direct Contact Solid Stationary Thermocline	Store must be pressurized to cycle pressure (large, expensive pressure vessel). Managing pressure drop and other inefficiencies in the large vessel has some challenge.	Concept of PHES pioneer Isentropic	Most likely not advantaged compared to Overlapped cycle and not strictly necessary given operable range of store media	Concept of PHES pioneer Saipem

Continued

Summary of PHES concepts—cont'd

	Cycle concept	Nonrecuperated ideal gas PHES	Recuperated ideal gas PHES	Overlapped ideal gas PHES
Direct Contact Conveyable Solid	Significant challenge around effectively transporting the solids into and out from the pressure boundary to facilitate direct contact with the pressurized working fluid. The stores with solids at atmospheric pressure are very cheap, but the requisite solid handling equipment might still offset and make the concept uneconomical. Also, some challenges arise from variation of specific heat of practical solids over the operable temperature range, which leads to reduced efficiency.	Not being considered as no advantage is seen compared to Recuperated or Overlapped Cycle Concepts	The recuperated version would somewhat mitigate the problem arising from the nonconstant specific heat of solid media, however with the higher cost and technical challenges of the recuperator.	Possible, but see General Remarks on Thermal Store and Heat Exchange Concept
Indirect Contact Conveyable Solid	Heat exchangers in this embodiment likely have higher cost compared to Indirect Contact with Liquid Store and not enough savings in the store itself to offset.		Same comment as previously regarding specific heat of media issue. If the solid-to-gas exchangers are expensive compared to gas-gas for same duty, the recuperated version may have cost advantage compared to Overlapped.	
Direct Contact Liquid (1-tank or 2-tank store concept)	Might be possible, but significant challenge to identify a liquid that would not cause problematic contamination of the working fluid. Contamination of the working fluid introduces challenges to manage the system via inventory control since contaminants would be vented to atmosphere. By the time challenges are addressed, the cost of this system likely exceeds the embodiment with Indirect Contact with Liquid Store	Unlikely that acceptable round-trip efficiency could be achieved with operable temperature range of practical liquids for this embodiment	Could be considered if molten salt could be used, but anticipate problems of molten salt carry over. The salt would freeze elsewhere in the cycle leading to significant performance degradation.	Success considered a long shot due to issues discussed in General Remarks for this Thermal Store and heat Exchange Concept.

Indirect Contact Liquid (1-tank or 2-tank store concepts)	Main challenge around heat exchanger design and manufacturing technique to be cost effective.	No advantage is seen compared to Recuperated or Overlapped Cycle Concepts.	Used if a molten salt liquid is selected which has an elevated freezing/melting temperature.	Not being considered since no advantage compared to Recuperated cycle. Specifically, liquids that would support the necessary operable range are much more costly than molten salt. Also, stores would then need to be pressure rated to the vapor pressure of such liquid media at maximum cycle temperature, which could be substantial. Only cost savings would be elimination for the recuperator, which is deemed insufficient to offset these issues.

6.4 Hydrogen storage

Contributors: Rainer Kurz, Jeffrey Goldmeer, Jeffrey Moore.

6.4.1 Introduction

Increasing the use of renewable energy requires new approaches to energy storage and energy transport. An elegant solution for the energy storage problem is the use of the existing natural gas pipeline system, as storage and transport vehicle. In these concepts, surplus electricity from renewables (Wind, Solar) is used to create hydrogen via electrolysis. This hydrogen is then injected into natural gas pipelines [56]. Current European plans call for the capability to add up to 10% hydrogen into the natural gas stream. Similar ideas are discussed in North America [57]. Adding hydrogen to the natural gas requires considerations regarding combustion systems, the impact on compressors, pipeline hydraulics, as well as industrial and residential users.

It must be noted that hydrogen can be generated either from fossil fuels like coal or natural gas, or it can be generated by electrolyzing water. Only if the hydrogen in question is created using electrolysis, and if the electricity for the process comes from renewables, is the hydrogen carbon neutral.

The transport efficiency of the pipeline, safety aspects, and in particular questions about the capability of existing and new infrastructure to use natural gas-hydrogen mixtures as fuel are to be addressed.

Hydrogen decreases the volumetric heating value of the blended fuel, while increasing the reactivity of natural gas fuels, increasing flame velocity, reducing autoignition delay times, and widening the range of flammability. The handling of failed starts, where unburned fuel can be present in the exhaust system and might create the potential for an explosion hazard, has to be addressed. Increasing hydrogen content also increases flame temperature which can lead to higher NO_x emissions; mitigation strategies are discussed. Results from analysis and rig testing of the combustion components with hydrogen and natural gas mixtures will be presented and discussed.

Creating hydrogen is an expensive process, and the transport of pure hydrogen is also energetically very expensive. Mixing lower levels of hydrogen into existing pipelines, rather than transporting pure hydrogen solves some of these problems. A natural gas-hydrogen mixture with moderate levels of hydrogen can be burned in more or less standard gas turbine lean premix combustion systems, the gas compressors that are already installed in the pipelines may be able to remain in service, and the safety issues regarding failed starts, and leaks are manageable.

Blending hydrogen and natural gas also solves another problem: the infrastructure for this kind of storage is already available and does not have to be built up. However, there is no indication that the necessary infrastructure is being built to generate hydrogen from renewables in an amount that would even create a 1% hydrogen content in natural gas pipelines. For example, in 2018 the natural gas distribution system in the United States had a volume of 2.54 trillion cubic feet; 1% of this pipeline volume is 25.2 billion cubic feet [58]. For reference and comparison, in 2019 refineries in the United States generated approximately 3 billion cubic feet of hydrogen per day [59]. This is roughly 0.1% of the US total pipeline volume or about 3% of the US annual natural gas consumption by nominal volume.

Another option being pursued is underground reservoirs suitable for hydrogen storage. These include salt formations, depleted oil and gas fields, and deep aquifers. The man-made salt domes have the lowest risk of reaction with the hydrogen (e.g., rocks, fluids, and microorganisms). There are three facilities in operation in Texas (Moss Bluff, Spindletop, and Clemens Dome) [60]. It has been reported that salt dome H_2 storage can store 1000 GWh of energy (Richard Voorberg talk, Siemens Energy, UTSR meeting, Orlando, FL, Nov. 2019). This approach could provide days and even weeks of electrical energy supply while remaining carbon free. While not covered in this chapter, significant technology development of utility-scale electrolyzers will be required to support this storage solution due to their relatively high price ($850–2000/kWch) [60].

6.4.2 Gas turbine combustion systems

A gas turbine is composed of three main sections: compressor, combustor, and turbine or expansion section as shown in Fig. 51. In normal operation, the compressed air from the compressor section enters the gas turbine combustor. Here, the fuel (natural gas, natural gas mixtures, hydrogen mixtures, diesel,

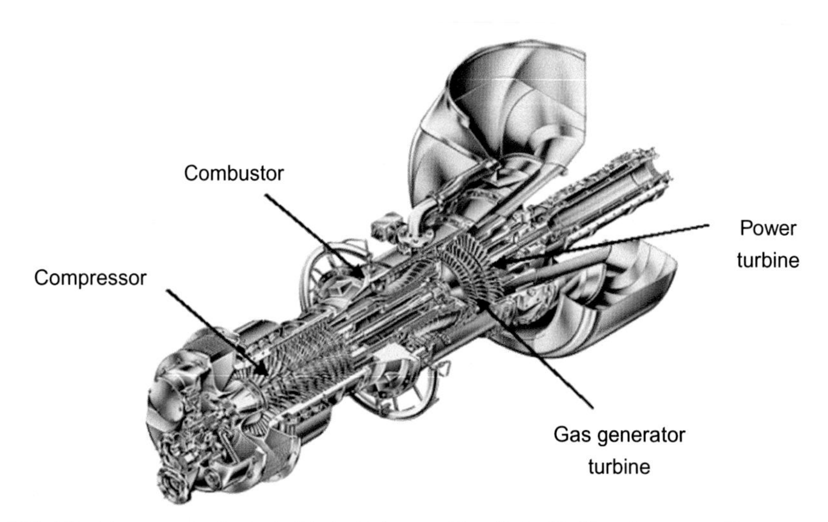

FIG. 51 Components of a typical industrial gas turbine (Solar Turbines Incorporated).

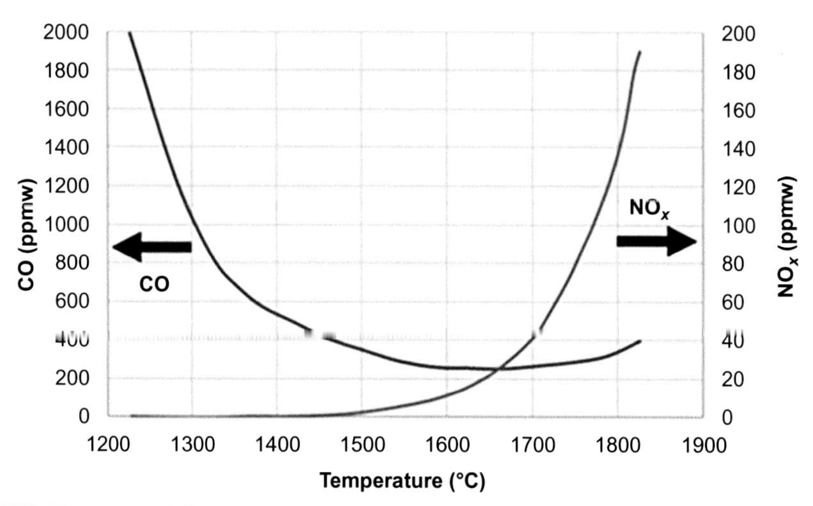

FIG. 52 Impact of flame temperature on gas turbine emissions. *(From S. Gülen, Combustion, in: Gas Turbines for Electric Power Generation, Cambridge University Press, Cambridge, 2019, pp. 308–361. https://doi.org/10.1017/9781108241625.013.)*

kerosene, and many others) is injected into the pressurized air and burns in a continuous flame. The flame temperature is usually so high that any direct contact between the combustor material and the flame has to be avoided, and the combustor has to be cooled using air from the compressor. Additional air from the compressor section is mixed into the combustion products for further cooling (Fig. 52).

Unlike reciprocating engines, gas turbine combustion is continuous. This has the advantage that the combustion process can be made very efficient, with very low levels of products of incomplete combustion like carbon monoxide (CO) or unburned hydrocarbons (UHC). The other major emissions component, oxides of nitrogen (NO_x), is not related to combustion efficiency, but strictly to the temperature levels in the flame (and the amount of nitrogen in the fuel). The production of NO_x is an exponential function of the flame temperature [61]. So, the solution to reduction in NO_x emissions therefore lies in the lowering of the flame temperature. Initially, this was accomplished by injecting massive amounts of steam or water in the flame zone, thus "cooling" the flame. This approach has significant drawbacks, not the least the requirement to provide large amounts (fuel to water ratios are approximately around 1) of extremely clean water. Since the 1990s, combustion technology has focused on systems often referred to as dry low NO_x (DLN), dry low emissions (DLE) combustion systems, or lean-premix combustion. The idea behind these systems is to make sure that the mixture in the flame zone has a surplus of air, rather than allowing the flame to burn under stoichiometric conditions. This lean mixture, assuming the mixing has been done thoroughly, will burn at a lower flame temperature and thus produce less NO_x. One of the key requirements is the thorough mixing of fuel and air before the mixture enters the flame zone. Incomplete mixing will create zones where the mixture is stoichiometric (or at least less lean than intended), thus locally increasing the flame temperature, and as a result create more NO_x. The flame temperature has to be carefully managed in a temperature window that minimizes both NO_x and CO. Lean-premix combustion systems allow emissions of NO_x, CO, and UHC to be kept within prescribed limits for a wide range of loads, usually between full load and minimum load that can range from 30% to 50% of full load output. In order to accomplish this, the air flow into the combustion zone has to be manipulated over the load range.

A key feature of lean premix systems is the fact that a combustible mixture exists in the fuel injectors, and the system has to be configured such that the combustion does not start until this mixture enters the combustor. Therefore a number of characteristics of the fuel have to be evaluated to ensure successful operation of the combustion system. Adding hydrogen to a gas fuel will significantly change the fuel gas properties (Table 10) [62].

TABLE 10 The variation of key gas turbine fuel characteristics with hydrogen additions to pipeline gas [62].

	H_2% with balance pipeline NG					
H_2 blend (% by volume)	*0%*	*5%*	*10%*	*20%*	*30%*	*100%*
Combustion Parameters						
Laminar Flame Speed (cm/s)[a]	124	127	130	139	150	749
Autoignition Delay Time (ms)[b]	124	112	107	104	103	76
Wobbe Index (btu/scf)	1215	1199	1183	1150	1116	1039
Flame Temperature (°F)[c]	4206	4210	4215	4225	4238	4510
Package and Fuel System						
Flammability (% volumetric LEL)	4.88	4.83	4.79	4.71	4.63	4
Maximum Experimental Safe Gap (MESG)	1.10	1.06	1.02	0.94	0.86	0.28
NEC/CSA and IEC Gas Groups	D and IIA	D and IIA	D and IIA	D and IIA	D and IIB	B and IIC

[a]*Calculated for equivalence ratio = 1.0 and mixture temperature and pressure of 600°F and 1 atm.*
[b]*Calculated for equivalence ratio = 0.4 and mixture temperature and pressure of 1200°F and 10 atm.*
[c]*Adiabatic stoichiometric flame temperature calculated for a 20,000 hp gas turbine at full load conditions.*

Heating value is the measure of the energy provided by the fuel, measured on a per volume or per mass basis. Hydrogen has a different heating value than natural gas. On a volumetric basis, hydrogen has a heating value of 274.7 BTU/ft^3 (10.8 MJ/m^3), while 100% methane, the main constituent of natural gas, has a volumetric heating value of 911.6 BTU/ft^3 (35.8 MJ/m^3). From this data it can be noted that hydrogen's volumetric heating value is approximately one-third the heating value of natural gas. This impacts not only the sizing of fuel orifices in the fuel injectors, but also this impacts the fuel accessory systems. Changing the fuel orifice sizing can also impact combustor operability.

6.4.2.1 Flame speed

The speed that a flame will propagate through an air-fuel mixture at a given temperature and pressure is known as flame speed. The laminar flame speed of methane in air gas turbine conditions is approximately 100 cm/s (at stoichiometric conditions), while the laminar flame speed for 100% hydrogen ranges from about 500 to 1000 cm/s at gas turbine conditions [63]. The laminar flame speed of a hydrogen-natural gas blend increases nearly exponentially with hydrogen concentration due to the impact of the flame speed of hydrogen. In the range of 0%–30% hydrogen by volume in pipeline gas, the methane reactions dominate in the combustion process and the increase is relatively modest (Fig. 53).

Each combustion system is designed for select range of flame speed variation. Diffusion flame or conventional systems generally do not have an upper level but do have a lower level where the flame speed becomes too slow and they "blow out." (This happens when the characteristic combustion reaction time is slower than the characteristic flow time.) This is clearly not an issue with

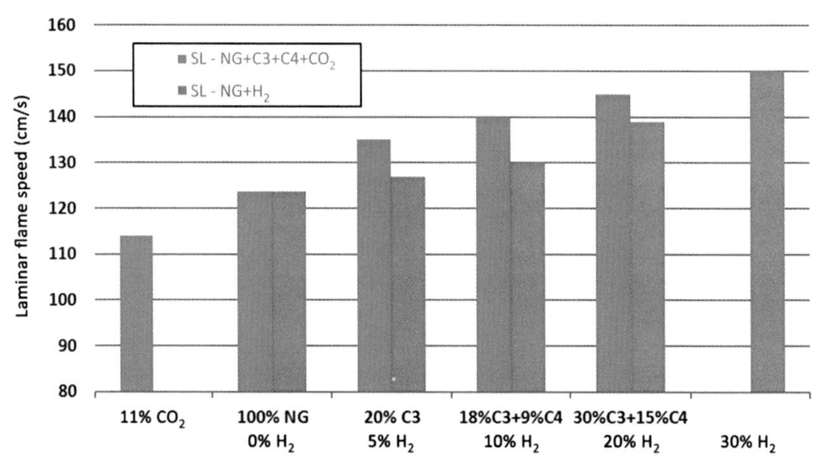

FIG. 53 Flame speed variation calculated for test fuels with varying levels of propane, butane and CO_2 mixed with natural gas compared to mixtures of hydrogen and natural gas.

hydrogen addition. For lean premix combustion systems there is an upper limit as well. The flame speed must be significantly less than the mixture velocity in the injector in order to prevent the flame from pulling into the injector premixer and causing damage. A flame propagating upstream into lean premixed fuel injector is often called "flashback." For lean premixed fuel injectors designed for pipeline natural gas, flashback will occur at very high levels of flame speed. Determining this point for current premixed combustion systems is a key requirement whenever using a fuel different than pipeline gas.

6.4.2.2 *Flame temperature*

The pollutant emissions (NO_x, CO, and UHC) from a gas turbine engine are most directly influenced by flame temperature. The adiabatic flame temperature is the maximum temperature that the products of a given combustion reaction can reach without heat loss. In a gas turbine combustion system, the majority of pollutant emissions will vary proportionally with that of fuels adiabatic flame temperature. In general, fuels with higher adiabatic flame temperature will create more NO_x and less CO and UHC. As can be seen in Table 10, the flame temperature for hydrogen and natural gas mixtures in the range of 0%–30% varies by approximately 30°F which will increase NO_x emissions modestly for a conventional combustion system and very slightly for a lean premixed combustor. The corresponding change in CO or UHC is even less at less than 1 ppm within the typical gas turbine operating range.

6.4.2.3 *Combustion stability*

The stability of a gas turbine combustion system is typically characterized by the presence or lack of significant levels of combustor pressure oscillations (also known as combustion dynamics or combustion acoustics) or combustor rumble. Combustor pressure oscillations occur when the heat release from the flame couples with pressure waves in an acoustic mode of the combustor. Combustor rumble occurs when the combustor or some portion of the combustion volume is operating near the flame extinction point. In either case, if an instability reaches a critical pressure amplitude, damage to the combustor liner or attachments to the turbine section will occur. The formations of these oscillations are highly dependent on the acoustic boundary conditions that exist within the combustion system, and small changes in geometry may trigger larger amplitude oscillations. Therefore extensive analysis, lab testing and, in some cases, engine qualification are required to verify that different fuel compositions do not significantly change the combustion stability characteristics.

6.4.2.4 *Flammability range (lower explosion limit—LEL, upper explosion limit—UEL)*

Hydrogen is highly flammable with a very broad flammability range of 4%–75% in air. It has a slightly lower autoignition temperature and must be treated more carefully than when using natural gas fuels to manage the risk of fire or explosion. This is clearly a concern if there is a gas leak near or in the gas turbine package but is also a concern for failed gas turbine ignition or flameouts when unburned fuel will enter the gas turbine exhaust system. The amount of fuel that can enter the exhaust system between the time the control system detects the failure or flameout and the fuel valve closes could be long enough to completely fill the exhaust ducting. The fuel-air ratio of this mixture in the exhaust is generally below the LEL when burning natural gas so it will not burn. However, with increasing hydrogen this mixture becomes flammable. So, if this mixture were to ignite in the exhaust a fire would occur with some level of pressure rise that may cause damage. Looking at Table 10 it is clear that there is only a modest decrease in LEL for hydrogen and natural gas mixtures of 30% or less. This risk is minimal for most of the power to gas hydrogen mixture scenarios where the hydrogen will be less than 20% (by volume). However, at 20% (by volume) to 30% (by volume) hydrogen there remains the possibility that an exhaust mixture from a failed start or flameout may be flammable and additional study is in progress to completely characterize and mitigate this risk.

6.4.2.5 Gas group and maximum experimental safe gap (MESG)

The gas turbine operator with hydrogen-containing fuels needs to properly assess the gas for the appropriate industry Gas Group. Based on the Gas Group, the hazardous area and the selection of equipment, such as electrical instrumentation and electrical enclosures, should conform to the appropriate industry code. As an indication of the risk with hydrogen and natural gas blends the MESG is included in Table 10. MESG is a standard measurement of how easily a gas flame will pass through a narrow gap bordered by heat-absorbing material. It is a primary factor in determining the Gas Group—for IEC with MESG $\leq$0.5. Table 10 indicates the gas group does not change until the hydrogen mixtures in natural gas increase over 20% (by volume).

6.4.2.6 Hydrogen diffusivity

As the smallest element in nature, hydrogen is very light and very permeable. Common fuel system seals that are leak tight with natural gas fuels may not seal effectively with hydrogen. High hydrogen fuels may require special leak testing of gas systems and/or welding of flange connections. Elastomers, including O-rings and diaphragms, are more susceptible to explosive decompression problems.

6.4.2.7 Hydrogen embrittlement

Absorption of hydrogen into metals can cause a general loss of ductility, which is termed hydrogen embrittlement. High strength martensitic steels are particularly susceptible to embrittlement and should not be used with hydrogen-rich fuels. Per NACE MR0175/ISO 15156 2003 carbide-stabilized grades and the

300 series stainless steels should be used for hydrogen fuels. These requirements are applicable for hydrogen mixtures greater than 4%. In addition, Sandia National Labs has an online technical database relating to hydrogen embrittlement [64].

6.4.3 Application for industrial gas turbines

Industrial gas turbines are used in many applications that support and use pipeline natural gas that will be impacted with the addition of hydrogen. These include gas transmission applications to drive pipeline compressors to transport the gas and for local power generation, often in combined heat and power (CHP) configurations, to generate electricity and steam for end users.

The good news is that there is considerable experience in the industry using low heating value fuels, i.e., syngas, steel mill gases, refinery gases, etc. The value in this experience is that the volumetric heating value of these fuels is similar to the heating value of hydrogen/natural gas blends. Fig. 54 shows a plot of mass versus volumetric heating values for a range of gaseous fuels. In the figure, region 1 encompasses natural gas, LNG, and shale gases, while region 2 includes higher molecular weight hydrocarbons, i.e., ethane and propane. Region 3 fuels are medium BTU gases also known as lean methane. Region 4 includes a variety of process gases, including syngas, steel mill gases, refinery waste gases, etc. These fuels have volumetric heating values in the range of 5000–10,000 kJ/m^3. The importance of this fuel experience is that blends of hydrogen and natural gas have similar volumetric heating values as shown

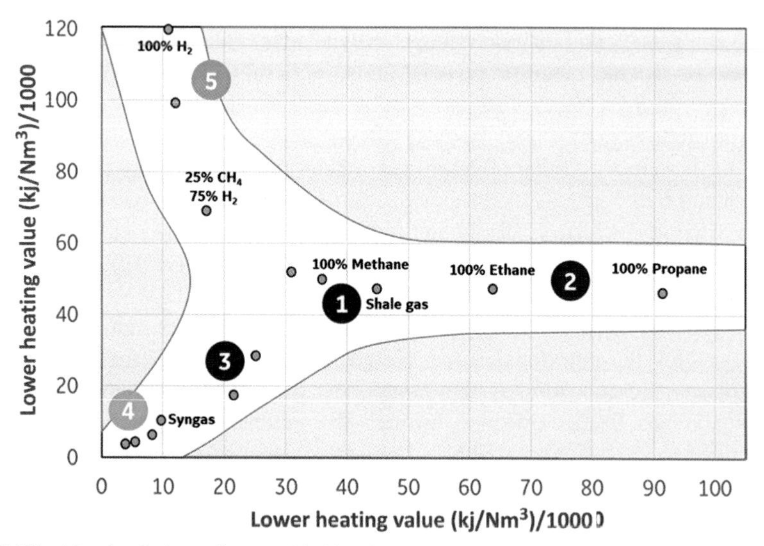

FIG. 54 Mapping fuel experience and fuel heating values. *(Courtesy of General Electric Company.)*

by region 5. Fuels with nearly 100% hydrogen have volumetric heating values that directly overlap the industry experience with these low heating value fuels.

The industry's experience with these types of low heating value fuels is well documented by the various gas turbine OEMs [62, 65–71].

In general, the majority of existing applications use diffusion flame combustion systems. More recently experience is increasing with gas turbines configured with lean premixed combustion system with considerable concentrations of hydrogen. The unique requirements and qualifications along with field experiences for both diffusion and lean premixed combustion systems are discussed in relation to using the expected hydrogen and pipeline natural gas fuel blends.

6.4.3.1 Gas turbines configured with diffusion flame combustors

Gas Turbines with conventional (diffusion) combustion systems have been available for decades. These combustion systems are capable of operating on a very wide variety of gaseous and liquid fuels. This includes the capability to operate on a broad range of hydrogen-rich fuels. Historically for applications the amount of hydrogen in fuel has been over 30% (by volume). Typical hydrogen-rich fuels used in gas turbine applications have been refinery waste gases (~30% by volume hydrogen), steel mill waste gases such as Coke Oven Gas (COG) (~60% by volume hydrogen), and a variety of industrial process gases (30%–100%).

Multiple OEMs have installed a variety of gas turbines across the globe in a variety of industrial applications including steel mill operation on COG and operation on syngas from a variety of solid feedstocks [62, 72–75].

Combustion system

Diffusion flame combustion systems have the advantage today of being able to operate on a wide range of gaseous fuels, including very reactive fuels with hydrogen. Images of a typical can-annular diffusion flame combustion systems are shown in Fig. 55.

The higher reactivity of hydrogen makes the combustion process more robust and flameout less likely. For conventional combustion there are two primary areas of concern: (1) the potential for higher combustor liner wall or injector tip temperatures that may shorten operating life and (2) higher NO_x emissions that result from increased flame temperature.

In general, there are strategies to mitigate the potential for reductions in part life. An OEM has reported that operation with higher hydrogen fuels has not had an impact on combustion system component life. This was predicted analytically and has been confirmed through the extensive operating experience described in [62]. Note that although the flame temperature does increase significantly as hydrogen increases, its effect is muted by gas turbine controls which limit the gas temperature entering the turbine section keeping it nearly constant regardless of the fuel type used. Therefore the effect of hydrogen is

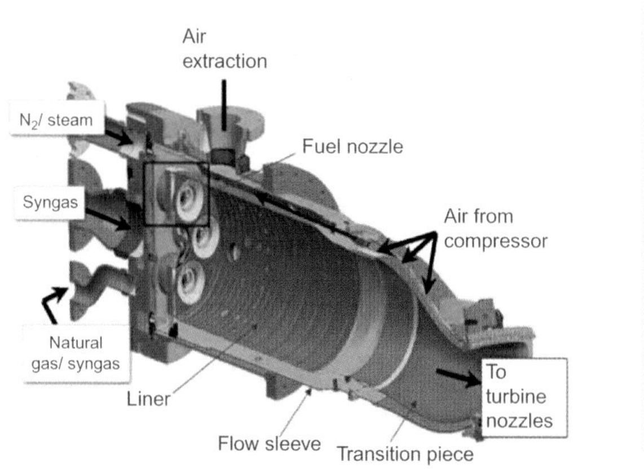

FIG. 55 (L) Multinozzle diffusion combustor schematic; (R) head end and fuel nozzles. *(Courtesy of General Electric Company.)*

very localized creating a more compact and hotter flame front but globally the average temperatures within the zones within the combustor are not substantially different.

The NO_x emissions, however, are increased as depicted in Fig. 56 which compares NO_x produced with hydrogen-rich fuels with NO_x from high methane pipeline natural gas. The NO_x is increased substantially due to the high-temperature flame front. The NO_x emissions with conventional combustion can be reduced by as much as 80% through injection of an inert gas or diluent such as water, steam, or nitrogen. This has two effects: it dilutes percentage of hydrogen and acts as a thermal sink. Both effects reduce NO_x emissions.

Clearly, for hydrogen and natural gas blends of 5%–20% (by volume) the effect on the conventional combustion system will be minor with less than 5% increase in NO_x compared to natural gas [62] alone with limited to no impact on durability.

Package and balance of plant impacts

As the level of hydrogen (and other more reactive gases) increases, additional requirements and limitations are placed on the gas turbine package and the overall balance of plant, see Fig. 57.

For these applications the following list of additional safety requirements is added for gas turbine packages

- Configure and equip packages to meet Gas Group B per Table 10.
- Update gas detection devices as conventional hazardous gas detectors become less effective as the concentration of hydrogen in the fuel increases.
- Update fire detection systems as hydrogen flames are significantly less visible than a hydrocarbon flame.

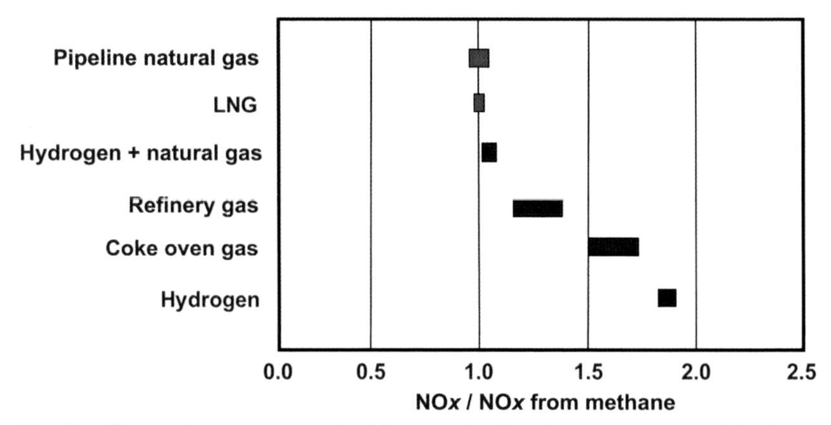

FIG. 56 NO_x emissions increase trends with conventional combustion operating with hydrogen-rich fuels compared to hydrogen in natural gas in the range of 5%–20% (Solar Turbines Incorporated).

FIG. 57 Impact of hydrogen on power plant systems. *(Courtesy of General Electric Company.)*

- For generator packages the risk of flameout is decreased by limiting applications to those that are tied to the power grid.
- Duct firing should not be performed with a fuel containing hydrogen unless the vendor has specifically reviewed the potential risks and provided the appropriate safety systems.
- Ignition and start-up on pipeline quality natural gas or diesel fuel may be required depending on the percentage of hydrogen in the fuel. If required the unit should be able to transfer to the hydrogen-containing fuel at a low load.

- Special exhaust purge sequences are added and used when there is a failed start or after a flameout before a subsequent attempt to restart.
- The fuel system is configured to prevent leakage in the package by using appropriate materials and following industry appropriate guidelines, i.e., NACE compliant materials and appropriate fuel system seals. Some OEMs may require certain fuel connections to be welded instead of using a sealed flange connection.
- In some instances, fuel system piping may be required to undergo an X-ray inspection process to further reduce the risk of leaks.

The power-to-gas renewable energy scenario has caused a reexamination of these package requirements for applications with 5%–20% (by volume) hydrogen in natural gas. Many OEMs have set a hydrogen level at which little to no change is required relative to pipeline quality natural gas. This range can vary from 5% to 10% by volume. However, if the hydrogen is increased from 10% to 20% then many of these requirements are still justified and will need to be implemented for new shipments as well as upgrades to existing gas turbines. A possible exception is the requirement that a standard fuel be used for start-up. This requirement becomes more burdensome for existing gas turbine packages where an alternate fuel to what is provided by the pipeline is not available. The industry has started to examine the broader issues associated with starting a gas turbine with more reactive fuel.

In addition to the package impacts, the addition of hydrogen has other plant-level implications. As mentioned previously, the addition of hydrogen to natural gas can increase NO_x emissions. For existing power plants interested in using hydrogen blending to reduce carbon dioxide emissions, there could be implications regarding increases in NO_x emissions relative to existing regulatory or permitted emission limits. Depending on the level of hydrogen blended in with natural gas, existing power plants may still be able to meet their permitted NO_x emission levels without any aftertreatment. Some gas turbine combined cycle power plants are configured with a NO_x emissions reduction system (i.e., selective catalytic reduction), but this system as initially configured may not be able to remove the additional NO_x associated with combustion of hydrogen, depending on the level of NO_x increase and the margin in the system.

6.4.3.2 Gas turbines configured with lean premixed (DLE and DLN) combustion systems

DLE and DLN combustion systems were specifically developed to reduce NO_x emissions with a primary focus on natural gas as the primary fuel. Over time the development of these combustion systems has demonstrated the capability to operate on gaseous fuels with a much wider range heating values or Wobbe range [76–78].

The ability to utilize hydrogen in gas turbines configured with lean premixed combustion systems is an area of active research and development for most OEMs. The initial assessment of multiple OEMs indicates that existing lean premixed combustion systems with the latest combustion system technology with pipeline gas mixed with 5%–15% hydrogen will not require significant modification [62, 79–81].

In contrast to the conventional combustion, lean premixed combustion operating experience on hydrogen is more limited, although DLE/DLN experience with associated and raw natural gases has become very extensive. Significant amounts of heavier hydrocarbons found in associated gas and raw natural gas lead to flame speeds and temperatures comparable to natural gas hydrogen mixtures. A few examples of lean premixed combustion experience with hydrogen and similar low heating value fuels are as follows:

- A Solar Titan 130S has operated with natural gas mixed with up to 9% (by volume) hydrogen [62]. Qualification and mapping was completed with the unit demonstrating 15 ppm NO_x and no operational issues. The unit is started on 100% natural gas and the package was updated to be compliant with the requirements for applications greater than 4% hydrogen. However, due to customer requirements the operating time accumulated with the 9% hydrogen fuel mix has been brief.
- In 2009 GE collaborated with Compañia Española de Petróleos (CEPSA) to upgrade an existing Frame 6B gas turbine to a DLN combustion system for operation on a fuel with up to 33% (by volume) hydrogen. The gas turbine was operating on a refinery gas that was a by-product of the refining operations at the site. The upgrade from the existing diffusion flame combustor allowed the customer to operate with the refinery gas without needing diluent injection meet the NO_x emissions requirements. In addition to the NO_x emission benefit, the DLN combustion system provided an inspection interval three times longer that available on the previous combustion system [82].
- A set of four GE 7F gas turbines configured with DLN 2.6 combustion system operated on a blend of 5% (by volume) hydrogen and natural gas at a Dow Chemical Company petrochemical plant in the United States. The hydrogen was blended with the natural gas at site upstream of the gas turbines [76].

Units with high and medium Wobbe associated and raw natural gases are much more extensively used and tested, and have few modifications from the standard configurations supporting operation on pipeline gas. The earliest shipments have been in operation for multiple years with many of these shipments reaching the overhaul interval. Operationally, these lean premixed engines run on associated gases in much the same way as they operate on pipeline natural gas. As indicated earlier on the applications with fuels with higher adiabatic flame temperatures the NO_x emissions are higher by 2–5 ppm. As with all gas turbines configured with lean premixed combustion systems, fuel quality with adequate fuel treatment is a prerequisite for trouble-free operation.

It should be noted that the ability of earlier generations of lean premixed combustion systems to use these levels of hydrogen will vary by OEM; some gas turbine models are capable of operating on blends while in other cases the capabilities are still being investigated.

Lean premixed combustion systems

Lean premixed combustion systems are limited by the same fuel and system characteristics that were described earlier for the conventional gas turbines. However, due to nature of the combustion system design several of these characteristics are more restrictive.

As described earlier the NO_x emissions of lean premixed combustion system are controlled by operating the combustion system at fuel lean conditions that are inherently closer to the lean extinction point. In addition, in order to prevent local hot spots, where NO_x formation rates can be considerable, the fuel injector includes a fuel and air premixer section. These design differences present several challenges as natural gas is mixed with hydrogen. First, due to its higher flame speed there is a greater risk for the flame to "flashback" into the injector premixer, which is not designed for high temperature. Second, as with conventional systems the flame temperature changes can impact NO_x emissions. Finally, lean premixed are sensitive to combustor pressure oscillations that have been "tuned out" for natural gas but as hydrogen is added to the fuel the flame shape may change due to variations in flame speed, flame temperature, and fuel density that may cause an increase in pressure oscillation amplitude levels.

These areas of concern for lean premixed combustion system are being actively investigated. Multiple OEMs have been working to qualify their lean premixed gas turbines to allow usage of a broader range of fuels. This activity has included analytical and test assessments of how variations in flame speed, flame temperature, and fuel density impact the combustion characteristics of emissions, combustion stability, and durability (component temperature).

Extensive combustion rig and gas turbine testing has been completed with a range of fuels with variable flame speed and flame temperature as reported in Ref. [78]. In this study flame speed and temperature were changed by adding propane (C3), butane (C4), and CO_2 into natural gas to simulate "associated gases" (raw gas recovered during oil extraction) and raw natural gas.

As outlined, emissions of NO_x and CO are most influenced by flame temperature (Fig. 58). Just as in the case for conventional combustion, the gas turbine controls keep the overall gas temperature entering the turbine constant

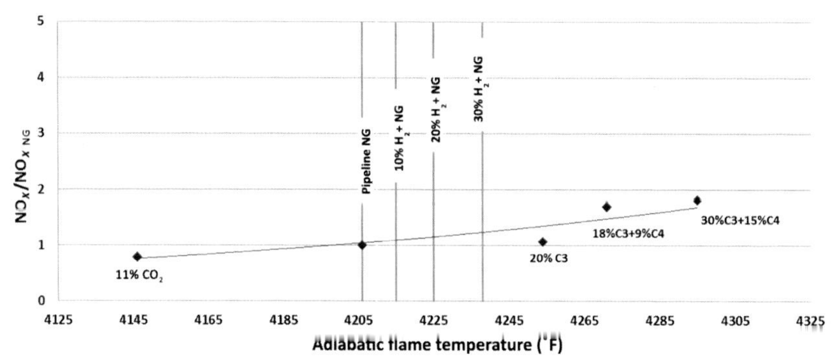

FIG. 58 NO_x emissions variation on a 20,000 hp gas turbine at full load and standard pilot with associated gas test fuels with different values of adiabatic flame temperature [62].

regardless of the fuel being used. However, as the adiabatic flame temperature increases, the NO_x emissions will increase due to the flame becoming more compact and burning hotter locally. Over the range of adiabatic flame temperature typical of these blends of hydrogen, a DLE combustor gas turbine is expected to show a very slight increase in NO_x of 1–2 ppm. Data for CO emissions are not included as these emissions were less than 2 ppm (parts per million), and similarly low levels are expected with hydrogen and natural gas mixtures (Fig. 59).

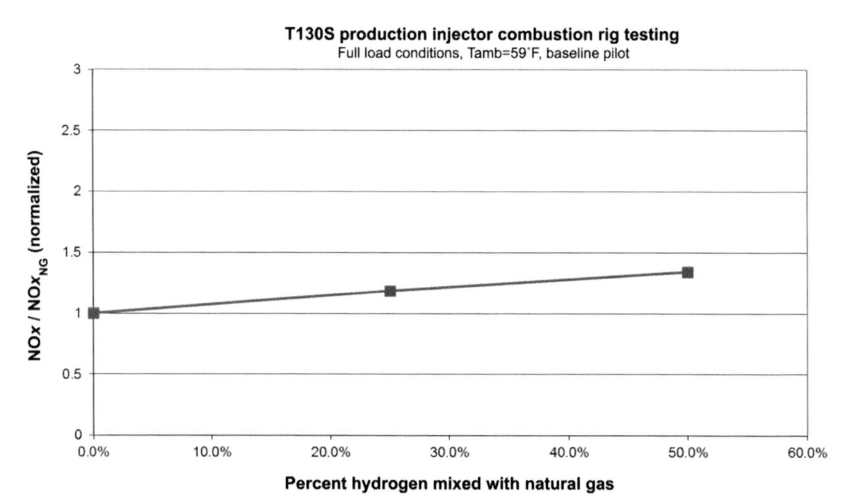

FIG. 59 NO_x emissions variation of a SoLoNO$_x$ injector in combustion rig testing at simulated full load conditions for a 59°F day and constant pilot level with varying fractions of hydrogen mixed with natural gas [62].

However, it should be noted that this DLE configuration uses an added pilot fuel circuit to augment flame stability at low loads and during transients. The pilot control schedule is set experimentally and may need to be adjusted differently with hydrogen mixes as compared to the fuels tested. Due to the enhanced stability generated while burning hydrogen-containing fuels, the analysis indicates that lower levels of pilot gas fraction and usage may be possible.

The testing completed in the fuel variation study also indicated that with the range of fuels tested the gas turbine indicated no change in combustion stability characteristics or the component temperature. Hydrogen in the range of 5%–20% is expected to behave in a similar way. For the component temperature, an assessment against the change in flame temperature compared with the test program is entirely adequate. Similarly, in the range of 5%–10% hydrogen, little to no change in combustion stability characteristics is expected. Engine testing will be conducted for hydrogen concentrations of 20% to confirm the analytical assessments.

Direct testing of hydrogen and natural gas fuel blends is also in progress using combustion rigs with a single fuel injector. Early results were taken on a 15 MW industrial gas turbine with DLE combustion system with hydrogen blended with natural gas. This rig was operated at simulated full load flow conditions at nominal day temperatures (Fig. 60). As expected the NO_x emissions do increase slightly as the adiabatic flame temperature of the fuel gas is increased. However, the magnitude is only 3 ppm. CO and unburned hydrocarbon emissions were throughout the testing. Component temperatures were mapped and little variation was evident. Testing is ongoing to assess the flashback robustness of the DLE injectors at varying levels of hydrogen content. In

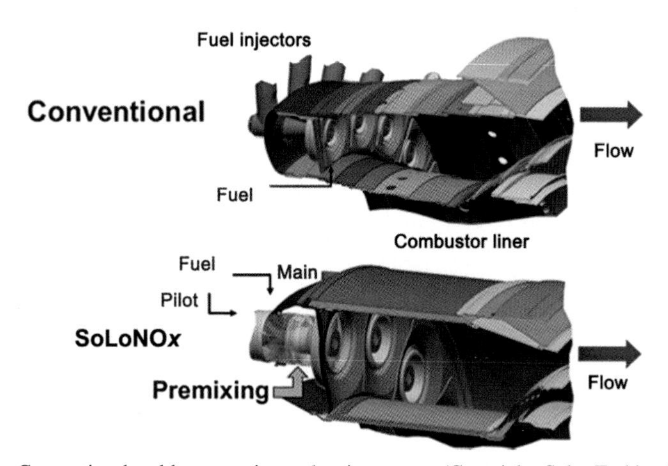

FIG. 60 Conventional and lean premix combustion system (Copyright: Solar Turbines).

test work to date, no flashback events were observed under any test conditions with hydrogen content less than 30%. This work is ongoing to cover other engine models and different DLE legacy configurations.

Various combustion test programs and field operations have been performed in recent years to evaluate the capability of dry low NO_x (DLN) combustion systems to operate on blends of ethane and hydrogen with natural gas. One study was performed on the Advanced Environmental Burner (AEB) that is part of the annular combustor for the GT13E2 gas turbine [83]. This is an annular combustion system configuration with 48 separate burners. This study used a single burner installed in a high-pressure test rig; during the tests, hydrogen was blended into the natural gas supply. No significant temperature excursions were measured at the burner during the tests. One conclusion of the study was that operation on fuels blends of up to 60% (by volume) hydrogen in natural gas is able to achieve NO_x emissions below 15 ppmv without dilution and no indications of flame flashback.

One of the challenges of lean premixed combustion systems is that they are configured for operation on pipeline natural gas, which provides challenges when operating on a fuel like hydrogen that is significantly more reactive. Resolving the issues of operation on a fuel with dramatically different properties may require a new combustor configuration. As part of the US Department of Energy's High Hydrogen Gas Turbine program, GE evaluated a series of new combustion system configurations and developed a combustion system capable of operating on high concentrations of hydrogen [84, 85]. Fig. 61 shows a fuel nozzle scale multitube mixer used for single nozzle combustion testing. As part of this process, a series of combustion evaluation tests were performed in both single nozzle and full chamber combustion tests at full-scale temperature and pressure conditions.

This combustor has become the basis for GE's [86] DLN 2.6e combustion system. This combustion system has been tested not only on natural gas, but on

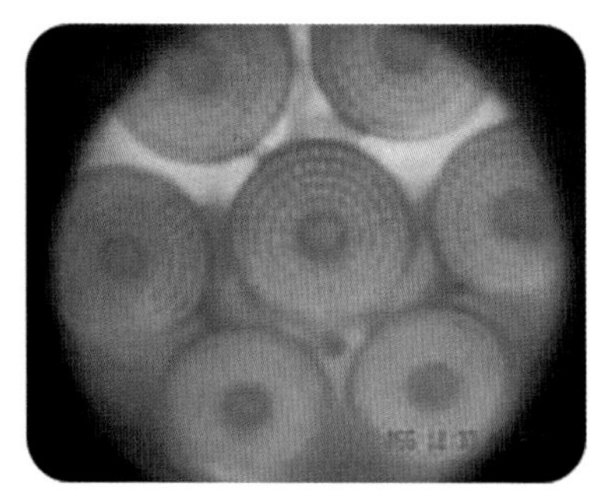

FIG. 61 (L) GE prototype multitube mixer; (R) multitube mixer fuel nozzles in operation in full combustor test. *(Courtesy of General Electric Company.)*

blends of hydrogen and natural gas. The testing demonstrated operation on blends up to 50% (by volume) hydrogen [71].

Similarly, other OEMS have been developing technology for operation on fuels that contain higher percentages of hydrogen. Siemens has evaluated and continues development of the DLE combustion system on their industrial gas turbines [87]. Their SGT-800 is capable of operating on blends up to 50% hydrogen and Siemens has announced testing up to 100% [88]. The MHPS technology, which has been labeled a multicluster combustor, consists of coaxially installed fuel nozzles and air holes. This combustor has been evaluated on syngas and for use with H_2 fuels [89, 90]. MHPS announced in early 2018 that this technology has been successfully tested using a 30% hydrogen mixture [91].

Ansaldo Energia has experience with hydrogen fuels and also continues development of their combustion technology. They have developed a dual zone combustion system that utilizes a trapped vortex for flame stability; see Fig. 62 [93]. In this figure the blue arrows represent cold air and the red arrows indicate hot air past the combustion stabilization zone. Based on combustion testing, this system is noted as being capable of operating on fuel blends with up to 65% (by volume) hydrogen. For the GT36 gas turbine Ansaldo Energai has announced a sequential combustion system (Fig. 63) that is capable of operating on natural gas blends with up to 50% (by volume) hydrogen [92].

6.4.4 Package impacts

The requirements and limitations for the conventional gas turbine package also apply for when using lean premixed combustion systems. Since DLE and DLN combustion systems are operating at leaner conditions the margin with flameout is generally reduced. This increases the risk of a combustible mixture reaching

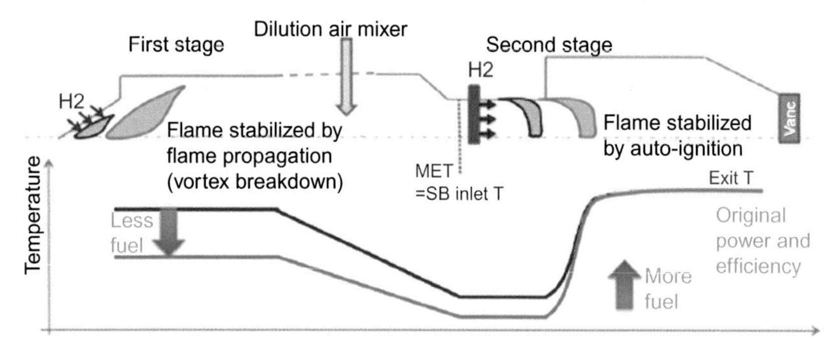

FIG. 62 Schematic of Ansaldo Energia's constant pressure combustion system (Copyright: ASME) [92].

FIG. 63 Hydrogen blending system. *(Courtesy of General Electric Company.)*

the exhaust in the event of a flameout. This risk is always more pronounced during the start sequence due to higher initial fuel flow rates and the potential that combustor light-off is not successful, i.e., fail to light. As stated previously, each OEM has different limits on the allowable percentage of hydrogen allowed for start-up.

Given that initial power to gas scenarios required blending of hydrogen into natural gas, there are two generic options to mix these fuels: deliver a blended fuel to the power plant or supply the fuels independently and blend at the power plant. As an example, Fig. 63 shows a fuel mixing system that was used to blend up to 5% (by volume) hydrogen into natural gas.

6.4.5 CO_2 emissions reduction

From a power to gas view, the fundamental reason behind the use of hydrogen as a power generation fuel is the reduction of carbon emissions. It is clear that operating a gas turbine on 100% hydrogen will result in near zero carbon dioxide emissions (even with 100% hydrogen, the emissions from a gas turbine will include a very small amount of carbon dioxide due to the presence of a small amount of carbon dioxide in the air and in the fuel). However, determining the amount of carbon reduction when using a blend of hydrogen and natural gas is not a simple linear function of the volume flow of the fuels due to the fact that emissions are typically calculated in mass while fuel flows are typically quoted in volume units. Fig. 63 shows the relationship between volumetric flows and heat input for a gas turbine. Using this data, one can determine the impact of carbon emission reduction, which is shown in Fig. 64.

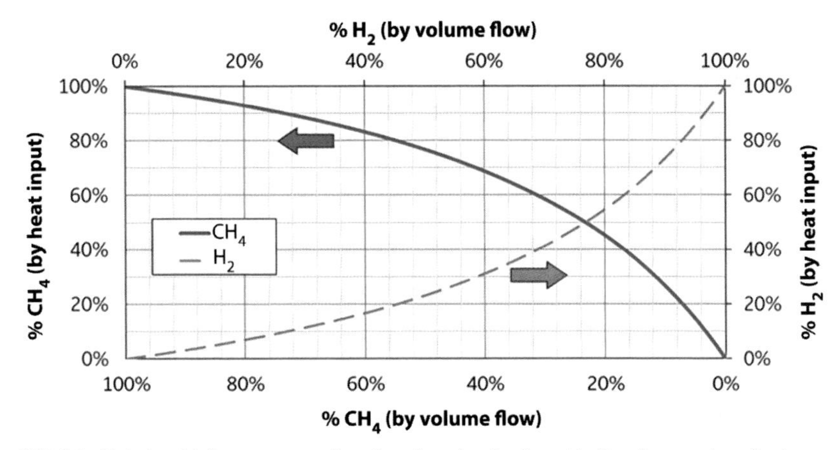

FIG. 64 Relationship between mass flow (heat input) and volumetric flow for a methane/hydrogen fuel mixture. *(Courtesy of General Electric Company.)*

With these figures, we can understand a few key points related to carbon emissions reduction. Current power to gas scenarios are examining the potential to blend small amounts of hydrogen with natural gas, potentially up to 10% or even 20% (by volume) hydrogen. In the case of a 20% (by volume) blend, per Fig. 61, this would result in roughly 8% of the heat input into the gas turbine being provided by hydrogen. Although the hydrogen would be 20% of the volumetric flow, the mass of the hydrogen is still very small compared to the total mass of methane being consumed. Since most of methane's mass derives from carbon, most of what is consumed by the gas turbine is carbon. The net result is that a blend with 20% (by volume) hydrogen only yields a carbon dioxide emissions reduction of approximately 7%. To achieve a 50% reduction in carbon emissions when operating a gas turbine on a mixture of hydrogen and natural gas requires that the blend is approximately 75% (by volume) hydrogen (Fig. 65).

6.4.6 Pipeline transportation

6.4.6.1 Hydrogen gas properties relevant for pipeline transport

Hydrogen gas has a higher mass calorific value than methane gas. Because of this property, molecular hydrogen is appreciated for space shuttle engines. However, it also has a lower mass density than methane gas, leading to a lower volumetric calorific value. Because pipeline losses depend on volumetric flow, the lower volumetric calorific value of hydrogen is important for the transportation efficiency of gas in a pipeline. There are other important differences between methane and hydrogen. In the relevant range of pressures and temperatures, the Joule-Thomson coefficient has a different sign for hydrogen and

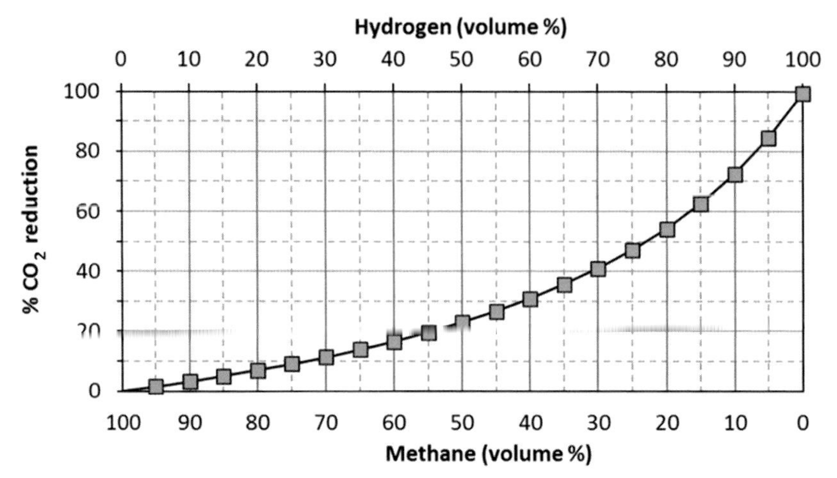

FIG. 65 Relationship between carbon dioxide emissions and percent hydrogen in a hydrogen/methane blended fuel. *(Courtesy of General Electric Company.)*

methane, and the compressibility factor has the opposite trend when the gas is compressed.

The compressibility factor (*Z*) of Hydrogen becomes higher than 1 at higher pressures. Compared to Methane, where the compressibility factor is reduced when the gas is compressed (for typical pipeline applications, say from 40 to 80 bar), the compressibility of Hydrogen increases in the same situation. In other words, the volume reduction is lower for Hydrogen than for Methane [94].

The dynamic viscosity, heat capacity, isentropic exponent, and the thermal conductivity are also different. What are the impacts of these hydrogen characteristics on the transport capacity and its efficiency in the case of blending in a gas transport network?

The knowledge of the dynamic viscosity of the studied gas is necessary to calculate the pressure losses in a pipeline. The pressure losses correlate with the Reynolds number, and the Reynolds number is inversely proportional to the dynamic viscosity at constant velocity and density.

Fig. 66 shows the change in mixture viscosity with the hydrogen concentration. It shows a minimum viscosity at about 65% hydrogen content, and therefore the gas flow will be the least constrained at around 65% hydrogen concentration.

The hydrogen concentration is the primary source of variability in the dynamic viscosity calculation. Pressure and temperature have little impact. The knowledge of the Joule-Thomson coefficient and specific heat capacity at constant pressure (C_p) of the studied gas is necessary to calculate the temperature evolution along the pipeline, and increasing the hydrogen content does lead to a higher temperature along the pipeline after compression, thus increasing the pressure losses as the hydrogen concentration increases [94].

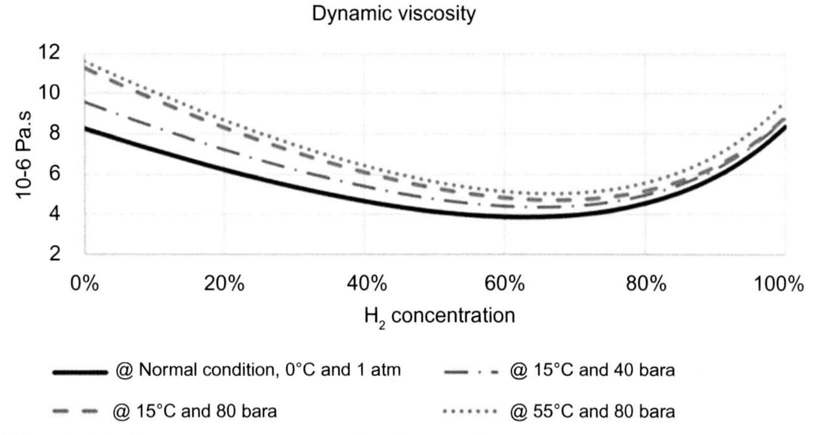

FIG. 66 The figure shows the dynamic viscosity depending on the hydrogen concentration at different pressure and temperature conditions.

Due to the low molecular weight of hydrogen, hydrogen compression requires significantly more work and higher speed or more stages than methane compression. Fig. 67 shows the operating points of a gas compressor, where the inlet pressure and temperature, and the discharge pressure were kept constant, while the hydrogen content was increased. The flow through the compressor was adjusted to keep the energy flow constant. For these conditions, compressing 100% hydrogen gas would increase the work by a factor of 10.

Bainier et al. [94] have studied the impact of hydrogen on the transportation efficiency. Transportation efficiency essentially compared the amount of fuel burned to transport a given amount of energy over a certain distance. Using energy rather than standard flow (or mass flow) allows a direct comparison of the impact of different gas mixtures. A compressor station was modeled assuming the output of two subsequent compressor stations being the same (Fig. 68), which is a good representation of a compressor station in the middle of a multistation pipeline. Fig. 69 shows indeed an increase in power consumption in the compressor station.

The power consumption for a situation where, for different hydrogen concentrations, the same amount of energy is transported is shown in Fig. 69.

The results of the study show fundamental relationships for the discussion on mixing hydrogen into natural gas pipelines:

- At the same pressure conditions and the same suction temperature, the compression work increases with the increase of hydrogen concentration.
- Hydrogen has a negative Joule-Thomson Coefficient, and therefore its temperature increases when the pressure drops. For the gas, flowing into the pipeline downstream of the station cooler, the higher the hydrogen concentration, the harder it is for the gas temperature to decrease along the pipeline. This characteristic has two consequences:

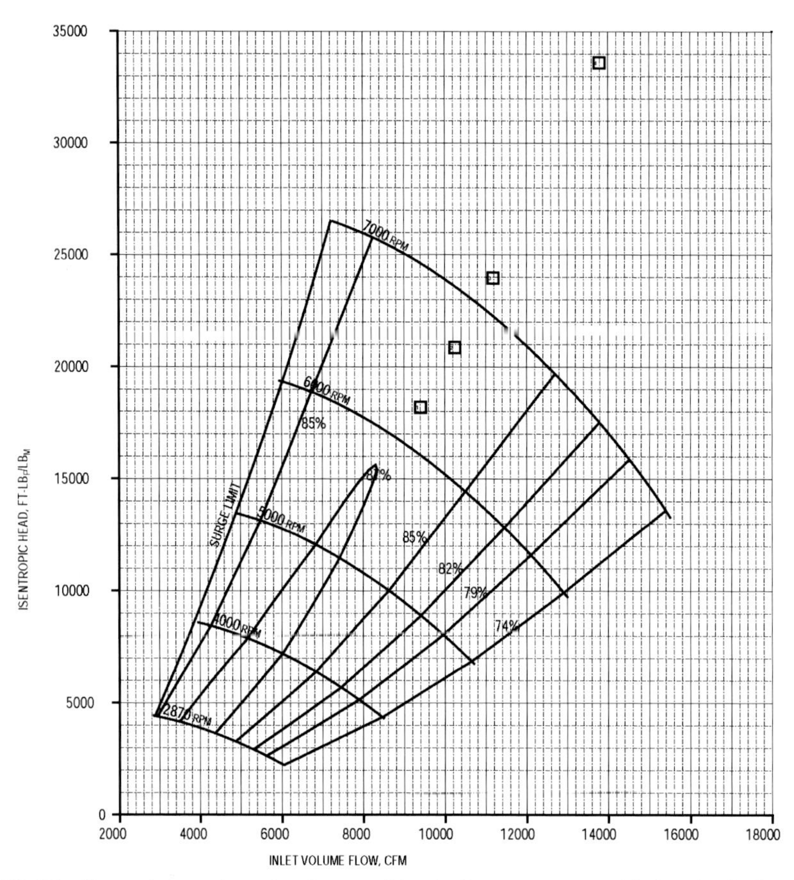

FIG. 67 Change in operating points for one of the pipeline compressors with changing hydrogen content in the pipeline gas. Compressor sized for the pipeline with 0% hydrogen content. The amount of energy transported in the pipeline as well as inlet pressure and temperature and discharge pressure were all kept constant.

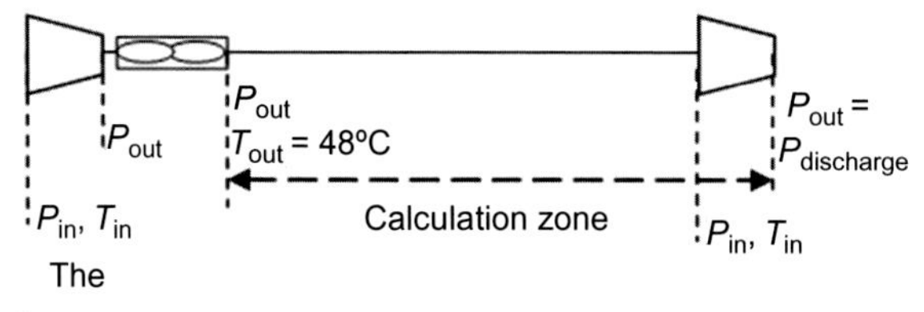

FIG. 68 Pipeline segment studied.

- Pressure losses increase with the H_2 concentration. And the higher the hydrogen concentration, the higher the influence of the soil conductivity.
- If the pipeline is not long enough, the compressor inlet temperature and the required compression power increase with the hydrogen concentration.

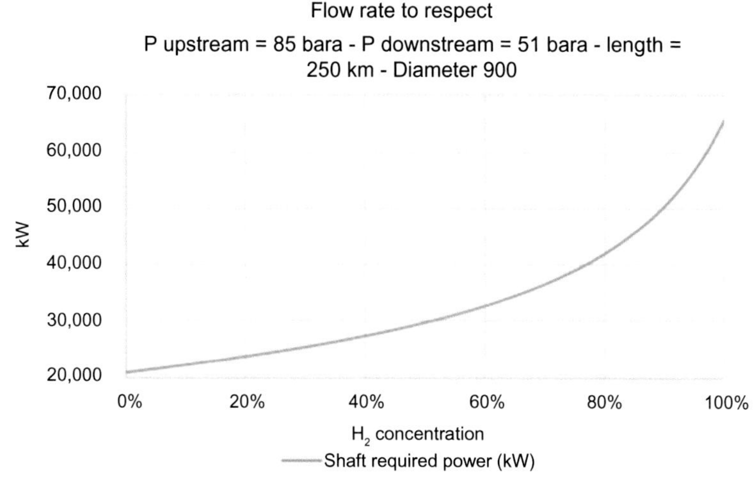

FIG. 69 Power consumption for pipeline compressor station.

- Fig. 70 shows the required increase power to transport the same quantity of energy. For the given parameters, the power increase reflects a reduction in transport efficiency.

Finding a reduction in transportation efficiency when hydrogen is mixed into the pipeline is a serious drawback in the discussion on usage of hydrogen. One has to take into consideration, however, that the yardstick to evaluate the use of hydrogen may not be the transportation efficiency, but rather the fact that pipelines allow for storing hydrogen. In other words, hydrogen injection into pipelines may not have to compete in terms of transportation efficiency, but rather in terms of round-trip efficiency compared to other storage methods, like compressed air storage or batteries. Obviously, in this discussion, the efficiency of the processes that generate hydrogen, using electricity from renewables, has a big impact.

The findings discussed previously indicate that injection of Hydrogen into a natural gas pipelines in moderate rates is manageable with today's technology:

- Conventional combustion systems are proven for hydrogen-natural gas blends up to 30%. Starting on these fuels is the only risk.
- Even for Lean Premix systems, hydrogen-natural gas mixtures of 5%–10% are not a problem today.
- Concerns are related to safety, for example at failed starts. These are manageable with today's technology
- Gas compressors are able to handle hydrogen in natural gas, but they will have to run faster (i.e., restages may be required on existing units), and will consume more power.
- The transportation efficiency of pipelines will be reduced when hydrogen is added.

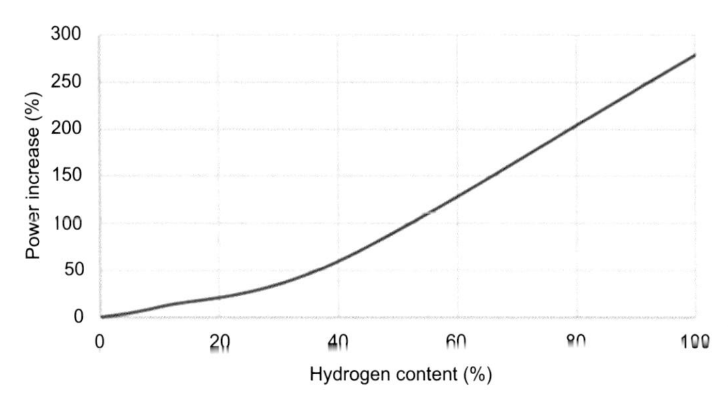

FIG. 70 Power increase to transport the same amount of energy in a pipeline.

6.5 Compressed air energy storage (CAES)

Contributors: Sebastian Freund, Jason Kerth, Karl Wygant.

6.5.1 Introduction

Compressing air while using power, storing the compressed air in a tank, and using it subsequently to produce work when needed is a common concept and industrially in widespread use. Stored compressed air has been used not just for pneumatic actuators and tools but also to start large engines or propel locomotives, small cars, and torpedoes, besides many other purposes. One drawback of compressing and expanding a gas is the change in temperature associated with these processes. A gas typically needs to be cooled during compression to limit the peak temperatures and reduce the power required to reach a certain pressure. For expansion across a significant pressure ratio, a gas needs to be heated to gain power, which would diminish as the temperature drops. In principle to keep temperatures constant, as much heat needs to be removed or added, as energy is used for compression or gained from expansion, respectively. One advantage of using a compressed gas to store energy is that gas can easily be stored without any losses as long as sufficiently sized pressurized volumes are available and that the heat in the compression and expansion processes can be managed. The most common process gas applied for compression storage is air, due to its obvious availability and nonreactive nature. Compressed air storage cycles are not necessarily related to Carnot cycles and the efficiency is not simply linked to process temperatures, although some cycles thermodynamically resemble gas turbine cycles.

In order to increase the power density of the equipment and to maximize the use of the available storage volumes, pressures in Compressed Air Energy Storage (CAES) plants can range up to 200 bar, with multistage compression and expansion.

The self-evident idea of using compressed air to store electricity has been developed into CAES power plants. Those plants use electrically driven compressors to compress air into an air store volume. When electricity is needed, the air is withdrawn, heated via combustion with gas or via a thermal storage system, and run through expansion turbines to drive an electric generator. The compressed air can be stored in large scale in underground volumes, including porous rock formations, depleted gas/oil fields, and caverns in salt or rock formations. To reduce CO_2 emissions and reduce/eliminate fuel use, CAES plant concepts have been developed that use thermal storage systems to capture heat from the compression process and later use that heat during the power generation process. Two CAES plants with natural gas fired heating are in operation, one in Huntorf, Germany, since 1978 and one in McIntosh, Alabama, since 1991, no large-scale adiabatic (ACAES) plant with thermal storage has currently been built. Small demonstrators of water-cooled isothermal CAES plants in the range of about 200 kW to 2 MW have been attempted in the 2010s by SustainX of Seabrook, NH; General Compression of Gaines, TX; and Lightsail of Berkeley, CA. Those efforts were commercially unsuccessful. Alacaes of Airlight Energy SA, Lugano, Switzerland [95, 96] has built an air store including a packed bed regenerative thermal energy storage in a modified road tunnel in a Swiss mountain [95, 96]. Since 2016 they tested this possibly first adiabatic (A-) CAES demonstrator system with compressed air at 7 bar and electrically heated to 550°C. Several 4-h charge (compression) and discharge cycles (through a throttle) were run and the performance of the tunnel and its novel concrete plugs and of the thermal storage (placed inside the pressurized tunnel to obviate the pressure vessel) were measured. Hydrostor of Toronto, Canada has built a 200 kW demonstrator of underwater CAES with two inflatable “balloon” storage tanks and a 3 km pipeline 55 m deep on the ground of lake Ontario. As of 2019 they announced more single MW-scale ACAES projects and published plans to scale their prototypes to potentially 100s of MW, add thermal storage and use man-made underground caverns in combination with a surface lake and water shaft to enable constant pressure in the air reservoir with the water column. This method reduces the required size of the cavern to store a desired mass of air within a pressure range allowable for efficient compressor and expander operation (no “cushion gas” volume needed). No commercial adiabatic CAES plant with thermal storage has been built as of 2019.

Many CAES projects have been entertained since building the McIntosh plant in 1991, several based upon the Dresser-Rand design with two or more trains, but no commercial projects as of 2019 have broken ground. A good overview of the history of CAES since the earliest patents in 1940 to some giant projects in the 2000s like “Norton” (2.7 GW), PG&E’s exploration of depleted gas reservoirs in the Central Valley in California for 300 MW CAES or ACAES plant is given in a comprehensive review paper on CAES by Budt et al. [97]. One of the latest big announcements came in 2014 by Burbank Power & Water to supply wind power from Utah to Los Angeles via the 1.2 GW "Intermountain CAES" based on four trains of recuperated CAES from Dresser-Rand [98].

The efficiency of compressed air cycles is typically given as a "round-trip" efficiency as the ratio of electricity output and energy input for one storage cycle. The energy input includes both electricity to drive the compressors and gas (LHV) to heat the air before the turbines. As electricity and gas are of different value, alternative efficiencies definitions have been proposed that account for the lower value of gas by factoring in, e.g., the efficiency of a reference gas fired power plant to convert the gas LHV into an equivalent electric energy. Only in adiabatic or isothermal CAES plants, where no natural gas fuel is used and electricity is the only input, this efficiency is well defined and comparable with other electricity storage processes like batteries or pumped hydro.

Air storage requires huge pressurized reservoirs. For utility-sized CAES, underground reservoirs such as solution mined gas storage caverns in salt domes are proven. About 0.5 m^3/kWh (delivered) of volume are required for adiabatic CAES at 70 bar pressure, somewhat less when pressures are higher, or about half as much for natural gas fired CAES with higher power generation per unit of air. Gas storage caverns can range beyond 1 MM m^3 in size and the total mining and construction cost are on the order of \$50/m^3 but can be much less for large caverns or when the salt brine can be dealt with at low cost; about 3 years of construction time from start of drilling over completion of the production string and wellhead to first fill can be expected. Compared to the common natural storage, larger and costlier wells, production strings, and well heads are required for CAES as the flow rates are much higher, pressure changes frequent, and corrosion issues more prevalent. Alternatives to salt dome caverns that have been considered for past projects include, e.g., depleted gas fields with porous rock, deep abandoned mine shafts, artificial tunnels and seabed-anchored spheres or balloons, in addition to aboveground pipeline arrays. Such pipeline arrays are common for short-term natural gas storage and maybe economically feasible for small, short-duration CAES.

In summary, CAES remains a promising energy storage technology that is proven by two successful power plants that employ salt caverns and use natural gas to boost power generation. For the future, high storage efficiency and reduced use of natural gas to store "green" electricity without further CO_2 emissions seem desirable characteristics for CAES plants and several concepts have been developed aiming to commercialize utility-scale CAES plants—on the order of 100 MW and multiple hours of storage.

6.5.2 Types of CAES cycles

6.5.2.1 Simple-cycle CAES

The simple-cycle CAES plant is essentially an intercooled gas turbine with reheat, in which the compressor and the turbine shaft are decoupled and the air after compression is stored before being admitted to the combustors and turbines. The flow diagram shows the intercooled compression train, an aftercooler before the air storage (cavern), and a combustor before the expander. The advantage of this cycle is its simplicity and compatibility with commercial compressors and turbines; the first CAES plant developed in the 1970s and built in Huntorf, Germany, is based upon this configuration (Fig. 71).

The efficiency of such a cycle is quite limited and falls below that of a gas turbine when accounting for the gas as input and the net electricity produced as output. This is the main drawback of this cycle. The efficiency in terms of electricity output over electricity plus gas input reaches about 40%. Because of the high pressure ratio between ambient air and cavern pressure, typically about 40–100, multistage intercooling between compression stages is necessary, and reheat combustion is needed to gain maximum power from the turbines. The compressors can advantageously be of axial type (derived from gas turbines or large process compressors, for high flow rate and best efficiency) for the low-pressure unit and of multistage radial type for the medium- (split case process compressor) or high-pressure units (barrel case process compressor). At least two turbines need to be used, e.g., a high-pressure and a low-pressure turbine on one shaft. The high-pressure turbine may be derived from a steam

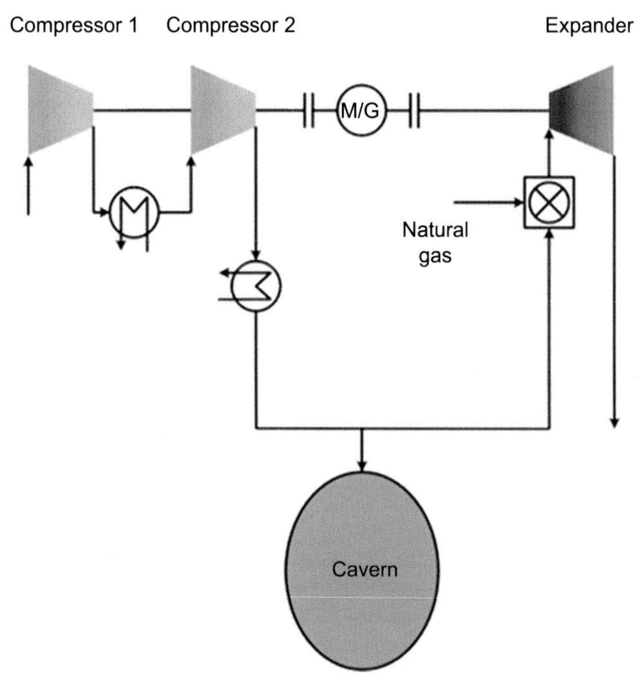

FIG. 71 Process flow diagram of a CAES plant. *(From S. Freund, R. Marquardt, P. Moser, ADELE Adiabatic Compressed Air Energy Storage—Status and Perspectives, 2012, VGB Powertech, 5.2013.)*

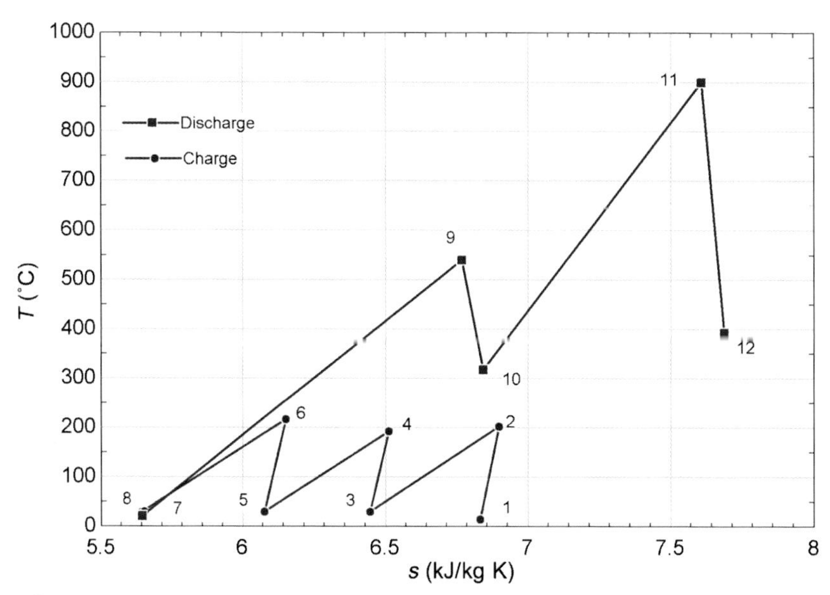

FIG. 72 *T-S* diagram of a CAES cycle with three-stage intercooled compression (1–7) and two-stage reheated expansion (8–12).

turbine with an inlet valve and scroll adapted for reduced pressure loss. The high-pressure combustor may be a silo or can combustor, with ample cooling and dilution air to maintain a turbine inlet temperature compatible with uncooled steam turbines of about 600°C, while keeping the flame stable and unburned hydrocarbon emissions low. The low-pressure turbine can be derived from utility or industrial gas turbines along with a silo or can-annular combustor and suitable adaptation of the transition piece for the high-pressure turbine exhaust at the inlet. Because of the moderate temperatures compared to typical gas turbines and the lean high-pressure combustion, NO_x abatement in CAES was of lesser concern than in modern gas turbine in general. However, NO_x mitigation in as much as needed for regulatory compliance can be achieved by dry low emission (DLE) combustor design, by aftertreatment with, e.g., SCR or NCR scrubbers, and by water injection (Fig. 72).

6.5.2.2 *Recuperated cycle CAES*

For an improvement of the efficiency of the simple-cycle CAES, this type adds a recuperator heat exchanger to the turbine train for preheating the high-pressure air before the combustor and first expansion step, using hot turbine exhaust after the last expansion step. By recovering exhaust heat, fuel use for the same turbine inlet temperature is reduced compared to simple-cycle CAES and the efficiency rises above 50% in terms of electricity output over energy input. Because the cavern air is stored at roughly ambient temperature, nearly all the waste heat in the exhaust can be recovered in this manner, greatly reducing the fuel consumption and yielding a low stack temperature comparable to an efficient combined cycle gas turbine plant. This CAES configuration was implemented in the first plant built in 1991 for PowerSouth in McIntosh, AL, and further developed the cycle concept to improve efficiency and power output [99]. The recuperator for the plant design developed by Dresser-Rand is essentially a fin-tube waste heat recovery unit, heating cold high-pressure air coming from the cavern with hot exhaust air from the low-pressure turbine. It is similar to an HRSG with tube bundles and manifold headers arranged to heat the air in counter-cross flow, suspended on a steel frame and an insulated rectangular sheet metal box with an exhaust diffuser duct at the inlet and a stack at the outlet. For efficiency of compression, intercooling is used in this CAES concept as well. As with simple-cycle CAES, reheated expansion is employed because of the high pressure ratio with a turbine train with steam- and gas turbine-derived machines. The turbomachinery with multistage intercooled compression and reheated expansion is in general the same as for the simple CAES concept (Fig. 73).

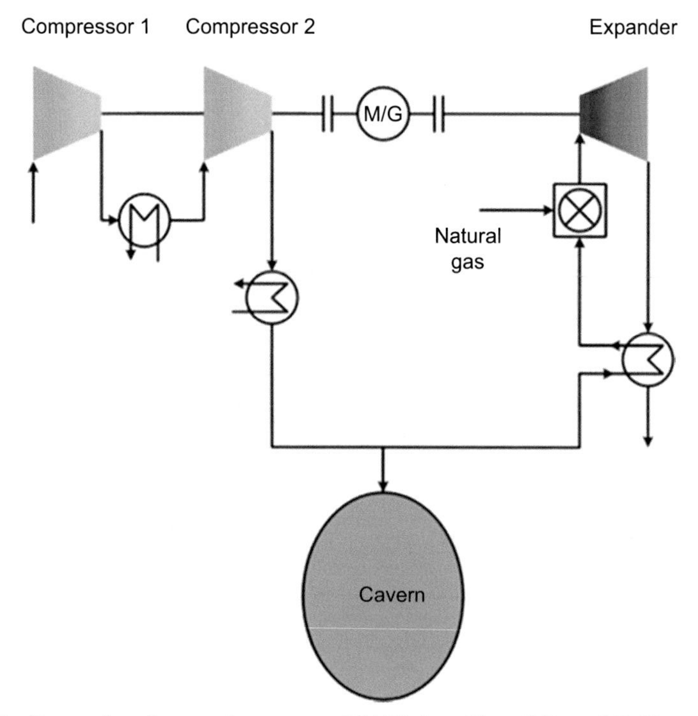

FIG. 73 Process flow diagram of a recuperated CAES plant. *(From S. Freund, R. Marquardt, P. Moser, ADELE Adiabatic Compressed Air Energy Storage—Status and Perspectives, 2012, VGB Powertech, 5.2013.)*

6.5.2.3 Adiabatic CAES

Adiabatic CAES, sometimes also referred to as "advanced" CAES, is differentiated from other CAES by the storage of compression heat for use during expansion. Rather than using as much intercooling as possible during compression like in simple CAES, the compression happens with large pressure ratio to gain significant compressor discharge temperature, followed by a heat transfer and thermal storage process. Rather than using natural gas combustion to heat the air for expansion, the stored heat from compression is recycled. Since about as much heat is produced during compression as power is expended to drive the compressors, the efficiency drastically increases when the internal compression heat rather than external heat from natural gas is used and this heat does not enter the denominator in the efficiency equation. In addition to higher efficiencies, a central advantage of ACAES is that it eliminates fuel combustion and CO_2 emissions, therefore becoming a decarbonized option for pairing with renewable power sources. The disadvantage of ACAES, however, is the cost and complexity associated with heat storage as compared to natural gas combustion (Fig. 74).

The figure shows a process flow diagram of a high-temperature ACAES plant with a solid media heat storage. Limited intercooling after the low-pressure compressor (1) is used to maintain the temperature of the high-pressure compressor (2) discharge within the limits of machines and heat storage with

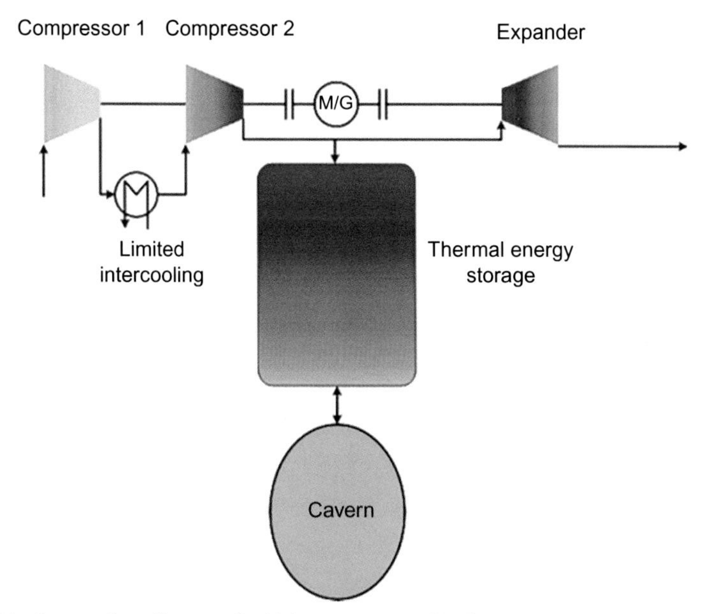

FIG. 74 Process flow diagram of a high-temperature ACAES plant. *(From S. Freund, R. Marquardt, P. Moser, ADELE Adiabatic Compressed Air Energy Storage—Status and Perspectives, 2012, VGB Powertech, 5.2013.)*

variable cavern pressure through intercooler temperature control. Without such intercooling, the temperatures may exceed material limits or the cavern pressure would have to be limited and the thermal storage could not be filled with constant temperature. The *T-S* diagram of this process with the parameters of 600°C storage temperature and 65 bar cavern pressure is shown as follows. Compression of ambient air starts at state point 1, followed by intercooling to state point 3, and high-pressure compression to state point 4. Between 4 and 5, the heat is stored in the thermal energy storage, and the air is stored in the cavern between 5 and 6, losing some remaining heat. Subsequently the air coming from the cavern is heated to state point 7 with a temperature and pressure just below point 4 because of pressure and temperature losses and then expanded for generation of power to state point 8 (Fig. 75).

Thermal storage systems for ACAES plants have been conceptualized mostly as solid regenerators, packed beds of granular material inside a pressure vessel. Regenerators are simple in principle and do not require heat exchangers and fluid handling. However, the pressure vessel becomes challenging for longer duration or higher pressure. Liquid thermal storage with oil for medium temperature or molten salt for higher temperature is one alternative, but large heat exchangers and tanks are required. Another alternative is a regenerator where the pressurized air is kept inside pipes embedded in special concrete to store the heat instead of a packed bed. This regenerator type does not require a pressure vessel, but it may have higher temperature and pressure losses than a packed bed or liquid store. Besides the high-temperature ACAES described

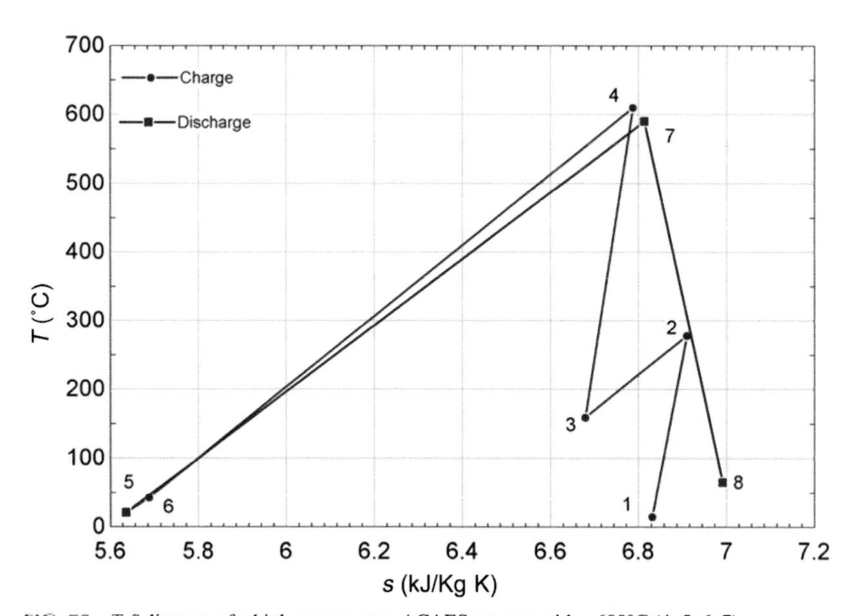

FIG. 75 *T-S* diagram of a high-temperature ACAES process with a 600°C (4–5, 6–7) regenerator and an intercooler (2–3).

previously, plants with thermal storage at lower temperature have been conceptualized as well. Those concepts employ two or more pressure stages at which heat is taken from the compression train, stored, and later given to the expansion train. The thermal energy storage at lower temperature can be accomplished also with pressurized water (~180°C), thermal oil (<300°C...400°C), or molten salt (<565°C). For plants with lower temperature and two-stage storage, the mass of thermal storage material increases compared to single-stage high-temperature storage, but the costs may be lower when a large pressure vessel packed bed regenerator can be replaced with liquid storage and insulated atmospheric pressure tanks. ACAES concepts with two-stage thermal storage at low and high pressure have been described by Schainker et al. [100], De Biasi [101] and engineered in the ADELE project [102, 103]. Both lower storage temperatures and higher cavern pressures can be reached with multistage TES because the pressure ratio between the stages is reduced. If the amount of intercooling required to not exceed temperature limits can be reduced in favor of more heat stored in the TES, the efficiency can be improved as well. It is noteworthy that some form of cooling is required in an ACAES cycle to dissipate the waste heat originating from the inefficiencies (pressure losses) in the system. Without inter- or after cooling during compression, both the lower and upper temperatures in the TES will rise, efficiency will fall, and the cavern inlet temperature as well as the expander exhaust temperature will increase (waste heat discharge) (Fig. 76).

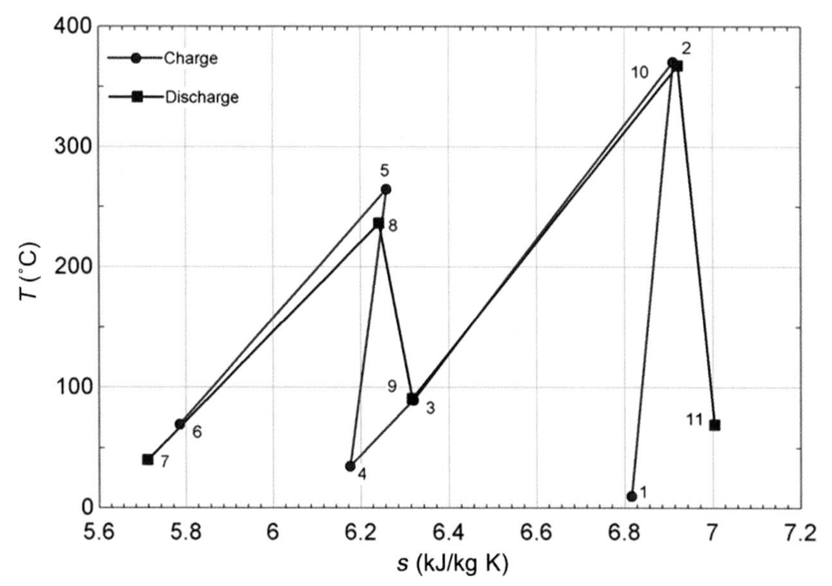

FIG. 76 *T-S* diagram of a low-temperature two-stage ACAES process with a low-pressure 360°C regenerator (2–3, 9–10) and a high-pressure 270°C oil storage (5–6, 7–8) and an intercooler (3–4).

A large engineering effort toward a commercial ACAES plant was the ADELE project [102]. An industry and research agency consortium in Germany developed ACAES technology from 2009 to 2017 with funding both from the corporations involved and from the German Ministry of Economics. The ambitious goal of the project was to design, develop, and build a first-of-its-kind commercial CAES plant showcasing the feasibility and economy of this new storage type to support the energy transition toward a dominant share of renewables on the German grid. The plant specifications were min. 65 MW of generation power per train and min. 4 h of storage with a salt cavern at the designated site in Stassfurt, Sachsen-Anhalt in Germany, and a possible multi-train extension up to 260 MW. Start-up time from standstill to full power of less than 5 min in both charge and generation mode was a key requirement for participating in the reserve and frequency control markets (Fig. 77).

Within the first several years, the research and development focus was on both a high-temperature packed bed regenerator with concrete pressure vessel and on high-temperature radial compressors, as these were the key components that had not been demonstrated for temperatures up to 650°C. Engineering of this high-temperature or "advanced" ACAES concept progressed to a point where cost, efficiency, and technical challenges were well enough understood to be compared to alternatives. In a second phase of the program, preliminary

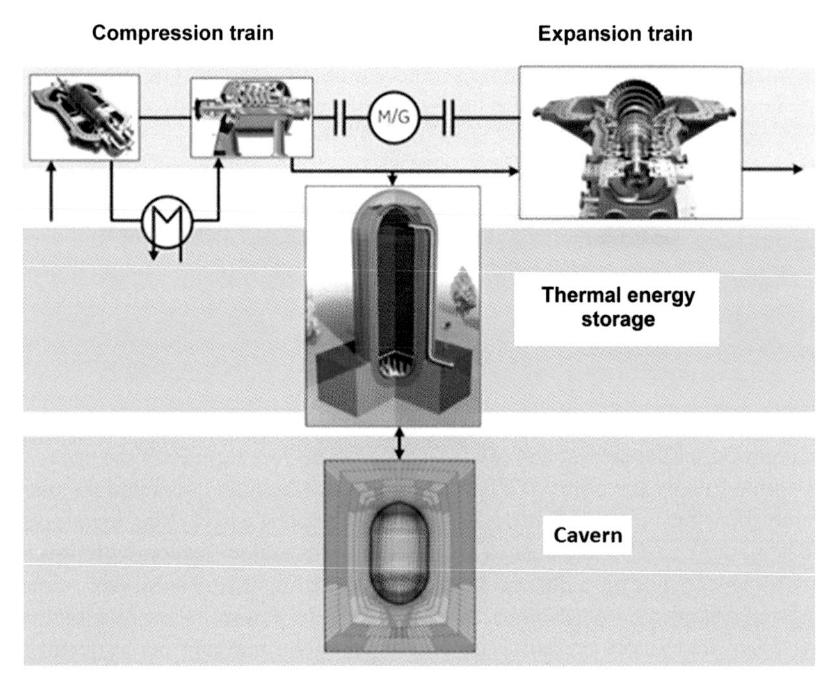

FIG. 77 Adele overview of partner structure. *(VGB-Kongress "Kraftwerke 2012," Mannheim, 10.-12.10.2012.)*

engineering of ACAES plants with four different heat storage configurations was completed, the single-stage, high-temperature, high-pressure solid regenerator TES, and three plant concepts with lower temperature TES [104]. Those concepts are based upon two-stage thermal storage, one at low and one at high pressure, with temperatures between 300°C and 400°C and consequently less costly design of piping, valves, heat storage, and the high-pressure compressor. Unlike for the high-temperature ACAES, technology for turbomachines and heat storage for low-temperature ACAES is commercial and components are available or can be derived from machines on the market with much less effort. The main cost driver in the high-temperature ACAES, larger than the development of a high-temperature compressor and a suitable turbine, was the solid packed bed thermal energy storage and associated piping. The combination of high pressure (65 bar) and high temperature (650°C) with the required vessel volumes (>2000 m^3) and pipe diameters (~1 m) drives material costs up and makes this concept costlier by a factor of two compared to low-temperature ACAES concepts. The key to reduce cost and design challenges is the decoupling of pressure and temperature in the thermal storage. For low-pressure air up to about 16 bar, regenerators with solid inventory housed in insulated steel or concrete pressure vessels can be used cost effectively. For high-pressure air of about 70 bar, using oil or molten salt liquid TES with shell and tube heat exchangers for the air is a viable option. Alternatively, a regenerator thermal storage made of tube bundles for the air that are embedded in heat storing concrete can be used. Such a TES had been developed and tested with steam and thermal oil by DLR and Züblin as a prototype, whereas oil and molten salt had been used for concentrated solar power plants commercially already. Although peak operating temperatures are lower than for high-temperature ACAES, similar round-trip efficiencies of 65%–68% can be expected for low-temperature ACAES plants. Project costs in a range of $1400 ± 200/kW for 6 h of storage can be expected for commercial plants, excluding some initial on-off component engineering.

However, with the changing economics of the electricity market, the arbitrage business case for buying low and selling high on a regular daily schedule fall apart and the future outlook remained uncertain. During the mid-2010s, electricity prices had come under severe pressure from market liberalization and merit order effects when growing renewables squeezed more expensive natural gas out of the market and low cost generation from lignite set the price. In 2015 the FID by the utility RWE who would have built and operated the plant came out negative. Some further development took place to explore downsizing possibilities of ACAES plants to about 10 MW for demonstration with limited investment cost or for industrial behind-the-meter applications. Market studies showed that even in future scenarios with growing renewable share and increasing fossil fuel prices the arbitrage economics alone may not be sufficient to incentivize investment in storage plants, mostly because of a low number of annual full load hours, a problem known from peaking gas turbine plants. Cost optimization of regenerative heat storage through supplemental electric heating to allow size reduction by raising and homogenizing the inlet temperature was also completed before the project was wrapped up and the development put on hold pending renewed market pull [102] (Fig. 78).

A second ACAES endeavor that has actually built prototype and pilot plants is Hydrostor Inc. of Toronto, Canada. This start-up firm has of 2020 developed three small ACAES projects, progressively increasing in power. Their ACAES technology is based on liquid thermal storage with water and oil and compressed air storage in tanks or caverns with displacement of water for constant pressure and smaller storage volumes. The initial prototype can store 1 MW for 3 h and the plant is located on Lake Ontario. This plant stores the compressed air 55–60 m below lake level approximately 3 km offshore from Toronto Island in specially designed balloons. The second plant, sited in Goderich, Ontario, Canada, can store 1.75 MW for 4 h and uses an abandoned mine shaft partially filled with brine for air storage. A third site is planned in Southern Australia with 5 MW of power for multiple hours. Low-temperature, multistage thermal storage has been used for the ACAES cycle. To help minimize costs at small size, integrally geared industrial compressors and expanders derived from standard equipment made by Hanwha are used. The eventual target are grid-scale applications much above 5 MW with the same technology including man-made underground caverns with hydraulic pressure equalization in a plant as shown in Fig. 79.

6.5.2.4 *Isothermal CAES*

The prospect of the Isothermal CAES concept is to deal with the heat in the opposite way of the ACAES—by removing all heat while being created during the compression process, the temperature does not rise and no heat needs to be stored. In practice, however, heat can only be removed across a temperature difference and reducing the rise in temperature means raising the flow rate of cooling medium. This means that isothermal CAES technology relies on ample flow of cooling media and high heat transfer rates during compression and expansion of the air to absorb and provide heat, respectively, along with storage of the cooling medium in addition to the compressed air. In this regard, isothermal (I-) CAES becomes very similar to ACAES. The fundamental challenges of isothermal CAES are first keeping the temperatures near constant to avoid losses and second keeping the equipment small despite the low power density because of low temperatures and the low-speed compression and expansion to enable heat transfer. Furthermore, low-temperature processes tend to lead to low efficiencies because friction and temperature losses create relatively more entropy and destruct more availability than at higher temperature. Operational challenges in I-CAES result from the complexity of fluid handling systems for cooling, storage, and heating of valved multistage compression and expansion trains.

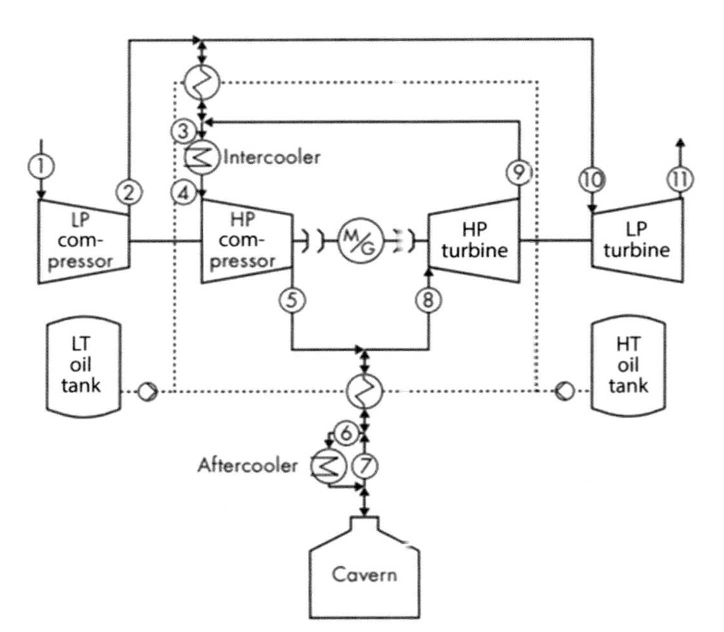

FIG. 78 Comparison of two ACAES concepts from the ADELE program [103].

FIG. 79 Rendering of Hydrostor's grid-scale ACAES concept with plant, heat storage and a pressure balancing pond on the surface above the caverns.

Three companies have worked on I-CAES. Lightsail of Berkeley, CA, has been developing a process using in-cylinder water spray cooling with a modified standard high-speed reciprocating compressor of about 200 kW and carbon fiber pressure tanks. SustainX Inc. of Seaport, NH, developed a compressor based upon a MAN two-stroke marine Diesel engine with three stage compression and expansion with actively controlled valves and water foam for cooling and heating, designed to reach about 1.5 MW with air storage in steel tanks. General Compression of Gaines, TX, developed since about 2010 a 2 MW CAES demonstrator with a large hydraulically driven piston compressor and was partially funded by ConocoPhillips and ARPA-E. The company did not publish much material and in 2015 merged with SustainX. None of these technologies resulted in a commercial system.

6.5.2.5 Gas turbine integrated CAES

CAES plants are like gas turbines with separate timing of compression and expansion and intermediate air storage. Utility gas turbines and CAES plants in generation mode operate principally on a grid at the same time, when electric power is in high demand. The time shift and storage of power is the advantage of CAES, while a utility gas turbine has an advantage of higher electricity to natural gas efficiency. Ideas have been conceptualized to combine both utility gas turbines and CAES for economic synergies. The first such concept described here is a CAES plant combined with a peaker gas turbine. This type of simple-cycle CAES uses waste heat from a gas turbine supplied to compressed air in a waste heat recovery unit instead of combustors. It benefits from the assumption that both peaking gas turbines and CAES plants may always run concurrently when power demand is high, and that the waste heat from peaking gas turbines is typically not used. The compression train uses multistage compression, with two or more intercoolers depending on the desired efficiency and the cavern pressure. The expansion train would have one or more expansion stages, with the air from the cavern heated in a waste heat recovery unit (WHRU) using the gas turbine exhaust heat, similar to an HRSG with fin-tube bundles arranged in counter-cross flow. A reheat step can be introduced for air leaving the high-pressure turbine with an intermediate pressure circuit within the same WHRU. The electric round-trip efficiency of such a waste heat fueled CAES system is estimated to be about 75% when not accounting for the waste heat as an input. The power range is determined on the expansion side by the available waste heat, and on the compression side by the sizing of the compressors and the motor (or motor/generator). For a large peaking plant with, e.g., a 7FA gas turbine (180 MW), the CAES plant could add 240 MW of expansion power. The compression power for same mass flow (about 475 kg/s) and duration as the expansion would be about 300 MW with 75 bar cavern pressure. To reduce the size of the compression train and improve plant economics, a longer charging duration with smaller mass flow and power would be designed for (Fig. 80).

This concept has been described by Robert Schainker et al. in an EPRI study presented in 2007 [100]. Several variations were also presented to integrate the CAES more tightly with the gas turbine for higher efficiency or power. One variation includes single stage expansion without reheat, which at high cavern pressure leads to very cold expander exhaust air that is directed into the gas turbine inlet to boost power and efficiency (similar to inlet cooling).

Another variation to boost gas turbine power is using a split stream of intermediate pressure air after the reheat and inject this air upstream of the combustor into the gas turbine. Air injection in gas turbines increases the power output over-proportional to the amount of extra air, since this air does not absorb any power in the compressor. It may slightly raise efficiency if the firing temperature is kept high because of the increase in pressure ratio. However, the amount of air that can be safely injected is very limited, first because of compressor stall with increasing pressure ratio, and second because of torque and power limits on cold days when the flow rate is high. Utility gas turbines of E, F, G, and H class can be expected to have between 3% and 10% capacity for injecting extra air within the surge and mechanical margins. 10% injected air means approximately 20% extra power (Powerphase Inc. of Jupiter, FL, sells the "Turbo-Phase" air injection system with a gas engine driven compressor for that purpose) (Fig. 81).

The second family of gas turbine integrated CAES concepts grows around the combined-cycle gas turbine plant. In a combined cycle, the gas turbine

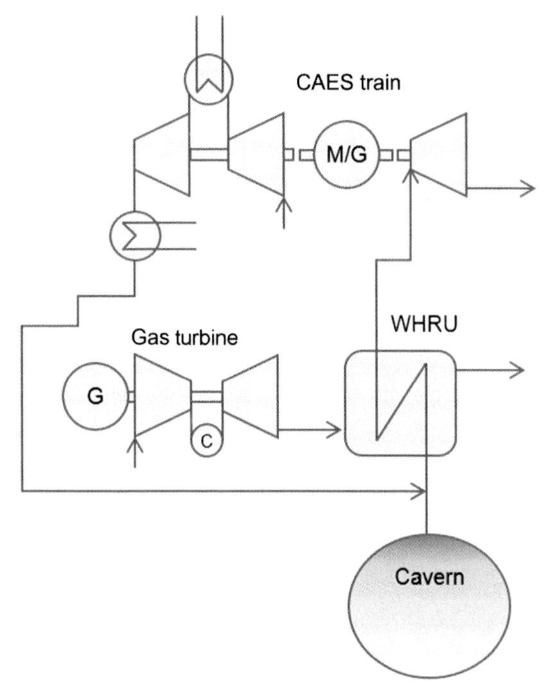

FIG. 80 Gas turbine integrated CAES process using heat from a peaker with a waste heat recovery unit (WHRU).

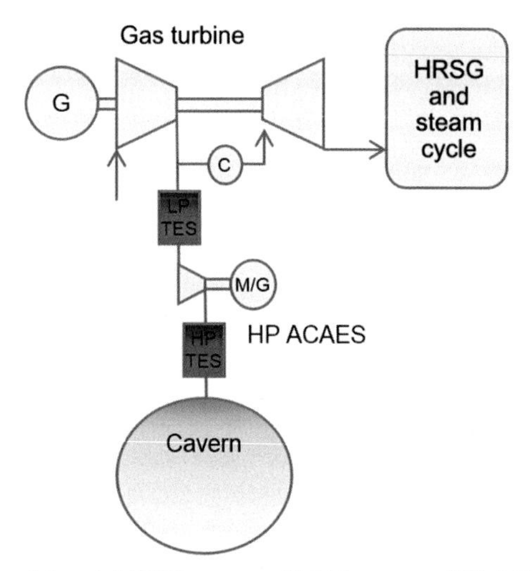

FIG. 81 Gas turbine integrated CAES process with high-pressure (HP) ACAES system and gas turbine injection.

waste heat is used for a steam cycle to bring the power plant LHV efficiency up to 60%. For combining this cycle with CAES, a fraction of the air from the gas turbine compressor discharge is subtracted for storage. This air is sent to a high-pressure (HP) ACAES plant with low- and high-pressure TES and high-pressure compressors, before being cooled and stored in a cavern. Upon discharge, the air is heated in the HP TES and expanded in an HP expander to intermediate pressure in the high-pressure ACAES driving a generator, then injected back into the gas turbine after being heated in the LP TES before the combustors. In the same way as described previously for the waste heat CAES concept with split flow injected into the gas turbine, the injected air raises the pressure ratio and adds overproportional power to the gas turbine during discharge. The amount of air that can be injected into typical gas turbines again is limited to about 3%–10% for stall margin and mechanical limits, depending on the gas turbine make, the ambient and operation conditions. In contrast to the previously described concept of peaker gas turbine CAES, here the air for storage is compressed to an intermediate pressure in the gas turbine compressor before entering the small HP-ACAES system. Using the gas turbine compressor and subtracting flow lowers the gas turbine output power substantially during charge and reduces the pressure ratio, as the turbine work is reduced by the amount of subtracted air at constant compressor mass flow. If equipped with compressor variable inlet and stator vanes, the pressure ratio and power can be somewhat compensated for both the decrease during charge and the rise during discharge of the CAES. This concept offers a high round-trip efficiency, with estimates up to 80%. The reasons for the high efficiency of this CAES cycle are twofold: First, losses attributed to the storage only occur in the smaller HP ACAES system, while the low-pressure section losses are accounted for in the gas turbine. Second, both the HP ACAES and especially the gas turbine combined cycle are very efficient power cycles (Fig. 82).

An example calculation using a cycle simulation programs including gas and steam turbine off-design data has been done for an air-injection CAES integrated with a General Electric 109FA 390 MW/50 Hz combined cycle plant to show the promise of this concept. The 9FA utility gas turbine may handle up to 10% of injected air and 7.5% (48 kg/s) is assumed in this example for the injection and subtraction flow rate going through the high-pressure ACAES. A high-pressure radial compressor of 12 MW brings the pressure of the subtracted air from about 15 bar at the gas turbine compressor discharge up to the storage pressure of 100 bar when charging. A low-pressure and a high-pressure TES with liquid (oil) storage media are used for heat management of the ACAES, and a radial expander to generate 7 MW with the same mass flow during discharge before injection at about 20 bar into the gas turbine combustor section (Table 11).

The round-trip efficiency with a baseline of 57% and 390 MW of the combined cycle is 76%, when accounting for a reduction in efficiency of the combined cycle during part load, the round-trip efficiency of this ACAES concept goes beyond 80%.

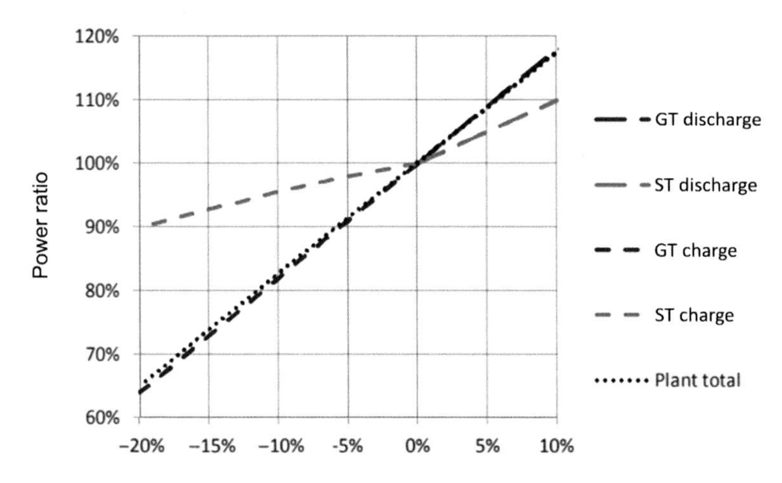

FIG. 82 Power output of a gas turbine combined cycle plant with air-injection high-pressure ACAES.

TABLE 11 Summary of performance data of a 109FA combined cycle with air injection ACAES concept.

Operation mode	GT	ST	HP ACAES	Plant tot.	Plant tot.	NG LHV
	MWel	*MWel*	*MWel*	*MWel*	*rel. MWel*	*MWth*
CCGT only	256	134	0	390	100%	684
CAES Charge	220	130	−11.6	338	87%	653
CAES Discharge	291	142	7	439	113%	724

This CAES concept has been studied in the past at least by the ADELE team in General Electric Global Research in Munich in 2015 and by Isentropic Ltd. in Fareham, United Kingdom, during the same time. A similar although simpler storage concept with injected air has been conceptualized by Turbophase, Inc., makers of the PowerPhase air injection system, for increasing the power output of existing gas turbines. However, no attempts of building a prototype or demonstrator have been published by either group.

The big advantage of such a gas turbine integrated CAES system is the high efficiency and ~2× higher power density compared to stand-alone CAES plants, leading to a smaller size with lower specific cost of the CAES portion.

The size could potentially be small enough and the pressure high enough that pipeline air tanks instead of caverns could be economical, removing the location limitation of CAES. The disadvantage, however, is the reliance of the CAES on the gas turbine operation both for charge and discharge. A further and much bigger gas turbine integrated CAES concept overcomes this disadvantage.

The concept of combining the efficient gas turbine combined cycle with the efficient ACAES into one enormous CCGT-ACAES plant has been investigated by several groups because of its merits: scale on the order of pumped hydro, high efficiency for both storage and generation, reliance on proven power plant technologies. As the process flow diagram shows, the gas turbine, with separated compressor and turbine and a motor/generator, functions as the low-pressure compressor and turbine of an ACAES system added to the combined cycle plant. The ACAES system consists of a low-pressure TES, a high-pressure compressor, a high-pressure TES, and a high-pressure expander. Compressor and expander are connected to a motor and a generator, respectively (or a combined unit). Three operation modes are possible: (1) normal gas turbine operation, without the ACAES active, and with heat recovery and steam turbine generation; (2) charging operation, with only the gas turbine compressor, driven by the motor/generator, and the ACAES high-pressure compressor driven by the motor active, while the cavern is being filled with air and both TES are being heated; (3) discharge operation, with the gas turbine and its combustor running, the steam turbine running and the ACAES providing air to the gas turbine while the cavern and both TES are being depleted. Assuming a generic modern GTCC plant as a basis with 500 MW and 60% LHV efficiency, a conceptual GTCC-ACAES would have the following approximate data for 100 bar cavern pressure with two-stage TES of the ACAES (Table 12).

This concept turns a 500 MW combined cycle plant into a flexible plant that can function as a 900 MW peaker or a 450 MW load on the grid to charge the storage. The round-trip efficiency defined as in CAES using both electricity and natural gas as inputs is approximately 70%. Because of the large air flow rates

TABLE 12 Comparison of CAES plant with CCGT plant.

Operation mode	GT	ST	HP ACAES	Plant tot.	Plant tot.	NG LHV
	MWel	*MWel*	*MWel*	*MWel*	*rel. MWel*	*MWth*
CCGT only	340	160	0	500	100%	833
CAES Charge	−306	0	−140	−446	−89%	0
CAES Discharge	646	160	91	897	179%	833

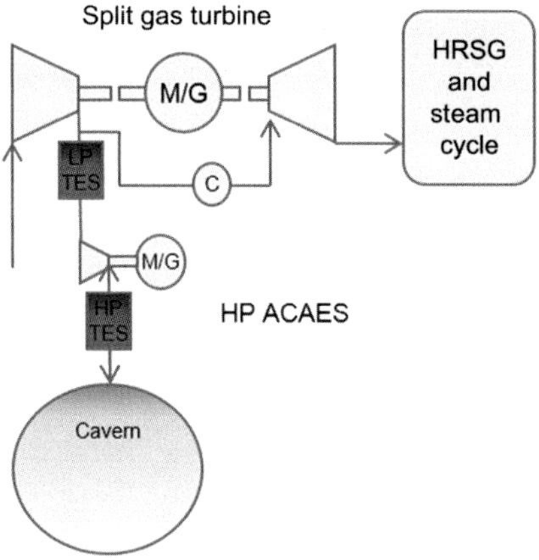

FIG. 83 Gas turbine integrated CAES process with high-pressure (HP) ACAES system and split gas turbine.

involved in utility gas turbines for combined cycles of 500 kg/s or more, the caverns and the TES required for this concept become very large. Several groups have studied such a concept. Among them are Leithner et al. [105] with a TES made of helically arranged ducts filled with pebbles or gravel, Hämmerle et al. [106] in conjunction with a TES-containing fluidized bed heat exchangers using sand stored in silos, and by Isentropic Ltd. of Fareham, United Kingdom, to be used with their layered packed bed thermal store using fine gravel.

The advantages of this concept are the large size, on par with big pumped hydro stations, that can lead to economies of scale more so than any other CAES plant concept, and the flexibility of operating this plant in both CAES charge/discharge mode and as a combined cycle power plant, each with unrivaled efficiency. The disadvantages are the requirement for using natural gas, which runs counter efforts to phase out fossil fuel, and the challenges of equipment design as well as the development of a very large first-of-its-kind power plant project. The required "split turbine" derived from large utility gas turbines is a device that none of the manufacturers has any plans to develop and would take several years and multiple millions USD in engineering (Fig. 83).

6.5.3 Current CAES power plants

Although various concepts for CAES plants have been proposed for many decades and a number of projects have been in an early stage of development, only two commercial CAES plants have been built to date. The operation experience of these two plants, Huntorf and McIntosh, is positive and successful according to earlier reports [107, 108]. However, in the market environment of the past three decades, the economic viability has not been promising enough for building further plants of either simple, recuperated, or any other type (Table 13).

TABLE 13 The two operational CAES plants and their basic specifications.

Type of process	Simple CAES with two-stage combustion turbines, three-stage intercooled compressors	CAES with recuperator and two-stage combustion turbine, intercooled compression
Location	Huntorf, Germany	McIntosh, Alabama
Commissioning	1978	1991
Turbine power	320 MW_{el}	110 MW_{el}
Compressor power	60 MW_{el}	50 MW_{el}
Storage capacity	0.6 GWh_{el}	2.6 GWh_{el}
Thermal efficiency	~42%	~52%
Specific cost	320 DM/kW_{el}	\$591/kW_{el}
Turbine start-up time	9 min	14 min
Image sources: BBC [109] Daly, CAES reduced to practice, ASME 2001; http://www.pennenergy.com		

The plant in Huntorf had several purposes when it was built besides showcasing successful operation of a novel type of storage power plant. Mostly it was to run as a peaking gas turbine, doing peak/off-peak electricity arbitrage, using low-cost electricity of a nearby nuclear power plant (Unterweser), which was owned and operated by the same utility (Preussen Elektra). Because of substantial natural gas fuel cost and moderate electricity prices, peak/off-peak arbitrage has been economically difficult for many years but ancillary services including minute reserve and voltage regulation through VAr injection of the massive synchronous motor/generator as well as provision of backup/blackstart power enabled the plant to stay online for more than 40 years. As the CAES plant is black start capable (only valves need to be opened to start the turbines) and has a direct grid connection via 220 kV (Tennet) to the 1410 MW-sized Unterweser nuclear plant and to a neighboring 700 MW coal plant, both it is meant to enable blackstart of which it can restart the coal plant and the nuclear plant in case of a major outage. For many years this CAES plant with 290 MW was the most powerful "gas turbine" ever built. BBC (later ABB/Alstom) built the turbines based upon a medium-pressure steam turbine fitted with a custom silo combustor for high pressure (42 bar, 560°C) and a GT13 utility gas turbine sans compressor with a silo combustor for low pressure (11 bar, 825°C). The turbine train was upgraded in 2006 after 28 years of operation during a repair overhaul to generate now 321 MW. To accommodate the axial thrust, a balance piston was developed that uses turbine and combustor cooling air. The compression train was made by Sulzer Escher Wyss (now MAN Turbo) and has a power of up to 60 MW. The train consists of a synchronous speed 20-stage axial low-pressure compressor and a 6-stage radial split case compressor connected to the shaft with a gear box. A total of three intercooling and one aftercooling stages with water-cooled shell and tube heat exchangers and a cooling tower are used to manage the temperature rise when compressing the filtered ambient air to an initial well head pressure of 40 bar up to the full cavern pressure of above 70 bar. The driveshaft has a 3000 rpm, 340 MVA synchronous machine centrally located with one shaft flange connected via a self-synchronizing overrun clutch (SSS) to the turbines and the other one via a switchable clutch to the compression train. Start and grid synchronization of the compression train is accomplished using the turbines with the high-pressure combustor only at reduced temperature to accelerate the train and allow full load compression within 6 min. Starting the turbines to full load takes 6 min. For a fast start and 11 min. For a regular start, with synchronization happening after 2 min of acceleration. The site in Huntorf in the north of Lower Saxony near the North Sea coast was chosen because of the salt domes in the area that have been solution mined for gas storage caverns. Two caverns, 650–800 m underground with a volume of about 300,000 m^3 total provide a capacity of 2 h full load plus several hours in partial load when needed (Fig. 84).

The second CAES plant in the world was built at the McIntosh Energy Center in Alabama and commissioned in 1991, it is today operated by the utility

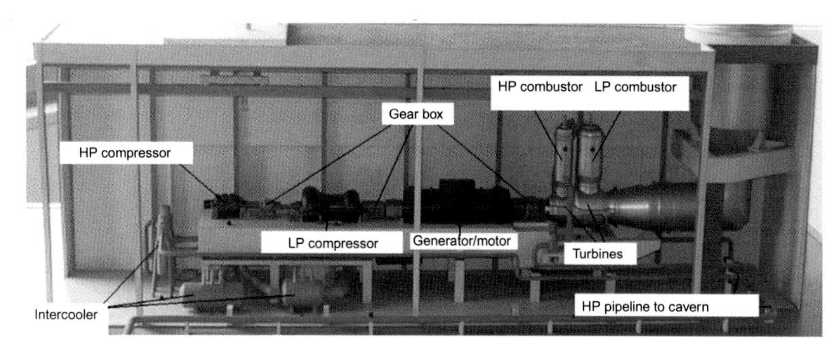

FIG. 84 Model of the Huntorf CAES plant. *(Derived from Wikimedia commons, 2019.)*

PowerSouth. The recuperated CAES cycle employed here adds a heat exchanger to the expander train which uses the waste heat of the exhaust gas to preheat the high-pressure air withdrawn from the cavern prior to combustion and expansion and raises the round-trip efficiency to up to 54%. This includes both natural gas (1.17 kWh) and electricity (0.69 kWh) input for one kWh of generation. The plant's expansion train power output is 110 MW for up to 26 h with air from a salt dome cavern with a volume of 538,000 m^3 and a pressure up to 75 bar. The expansion train comprises a steam-turbine-derived high-pressure expander and a gas-turbine-derivative low-pressure expander. A high-pressure combustor, receiving air preheated by the recuperator at 42 bar, is fired to 538°C, while the low-pressure combustor cans heat the low-pressure turbine inlet air at 15 bar to a temperature of 870°C. The air exits the turbine and enters the recuperator at about 370°C before exiting through the stack. The compression train has a power of only 50 MW and consists of four sections, one axial LP compressor and three centrifugal medium and HP compressors. A shared synchronous electrical machine is employed, that connects to either the compression or expansion train via Synchronous Self-Shifting (SSS) clutches. The start-up of the compression train until synchronization is accomplished using the self-starting expanders with remaining cavern air pressure (Fig. 85).

More details of the McIntosh CAES project can be found in a multivolume EPRI Report [108]. Dresser-Rand from 2008 to 2016 further developed this plant design in response to customer demand for lower cost, higher efficiency, improved emissions, and uprated power; a summary of this evolution provided as follows:

- Expander train output was increased from 110 to 135 MW via minor changes to expander metallurgy and flow-path geometry
- An optional split-train configuration having a separate motor and generator was offered. This configuration offers additional flexibility in that the expander and compressor trains can operate simultaneously and start and stop independently. This principally offers the opportunity to sell ancillary service products using the expander train even during the compression mode.

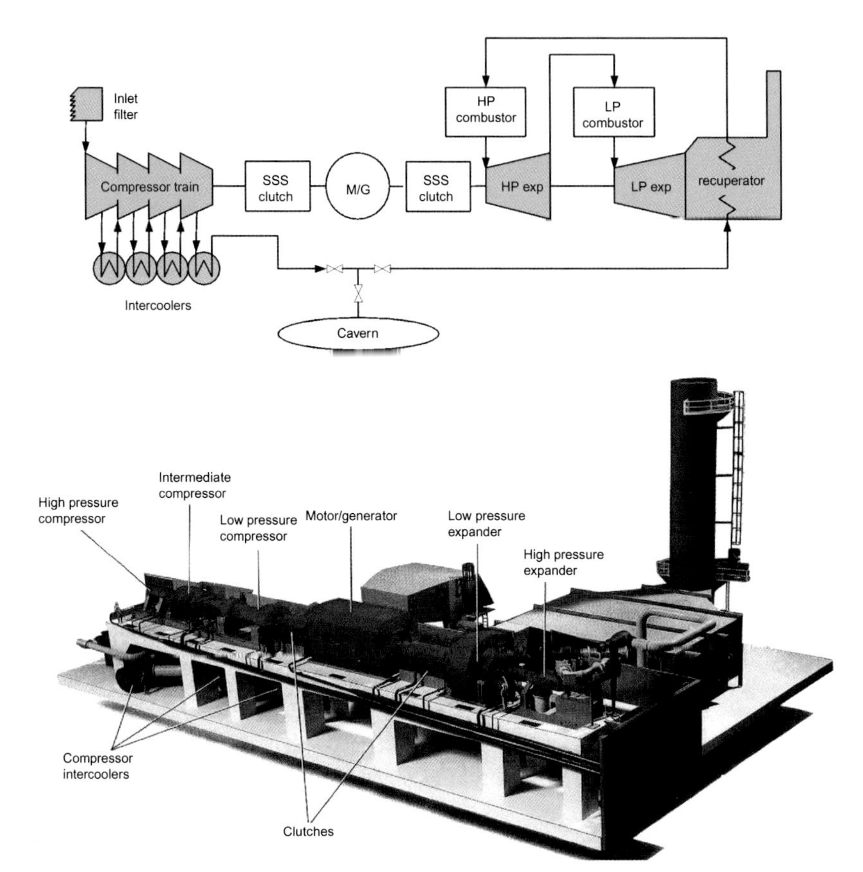

FIG. 85 Flow diagram and layout of the McIntosh CAES plant (Dresser-Rand).

- Recuperator effectiveness was increased from 85% to 90% for lower fuel consumption.
- A higher-pressure option for deeper reservoirs with a third very high-pressure expander section only heated by the recuperator was offered.
- Larger compressor flow rate and train power with almost twice the flow rate of the McIntosh system were offered.
- A Variable Speed Drive system was used to start-up the compressor train instead of using the expansion train. This extends the expander maintenance interval, which is based in part on the number of starts. Recirculation and blow-off valves were applied to reach full speed while consuming only about 25% of normal compression power.
- A combination of combustor water injection and selective catalytic reduction to meet modern air emissions standards was offered. Development of DLE combustion systems was not considered viable because of the low market volume for CAES expanders. There were no other products in Dresser-Rand's portfolio that could share the CAES combustor designs.

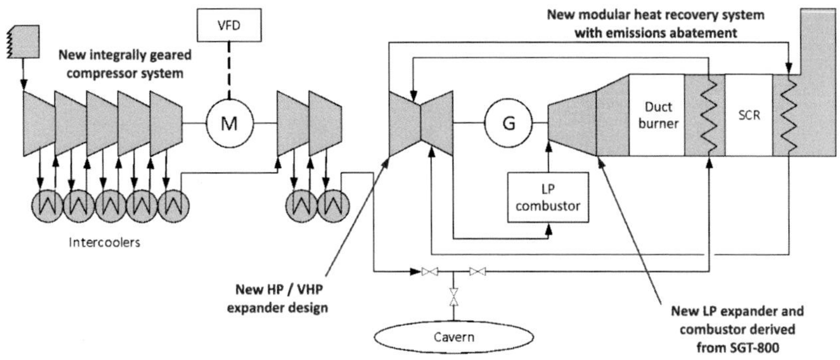

FIG. 86 Flow diagram of the latest Siemens recuperated diabatic CAES concept with separate compression and expansion trains including very high-pressure (VHP) sections.

After Dresser-Rand was acquired by Siemens in 2015, the development continued with opportunities to use equipment and technologies from a wider portfolio. These developments of the recuperated CAES concept include a modular heat recovery system design by NEM (a Siemens business) that offered lower equipment cost and lower site installation labor, and the replacement of the high-pressure combustion system applied in the McIntosh solution by an indirect heating scheme via duct burners located upstream of the heat recovery system. The high-pressure combustion system had limited part load capability and emission limits required water injection, both of which could be overcome by the new solution, in addition to retiring the need for a high-pressure natural gas booster compressor. Also a range of new turbomachinery became available and chiefly a new low-pressure expander and combustor with DLE technology (no water injection for NO_x control). The use of Siemens' large integrally geared type compressors for medium pressure offered ~10% improvement in compressor efficiency from higher per-stage efficiency and more intercooling, with a project-dependent range of 6 to 8 compression sections (for a rendering of the concept see also Fig. 86 in heat exchanger section).

While no utility-scale CAES projects have been successfully completed since the McIntosh project, the increasing need for energy storage has promoted numerous efforts to develop CAES projects. Table 14 shows an overview of recent project developments as of 2019.

On account of Dresser-Rand's CAES experience in the McIntosh project, all of these projects sought equipment supply offers from Dresser-Rand/Siemens as well as associated general project development support.

6.5.4 Rotating equipment requirements

The rotating equipment in CAES plants chiefly includes compressors, expanders, gear boxes, clutches, motors, and generators. Compressors and expanders for CAES power plants are turbomachines because of their flow rate and power density, except for very small scale prototypes, where piston engines

TABLE 14 Overview of recent CAES projects.

Developer	Location	Storage concept
Haddington Ventures	Norton, Ohio, United States	Abandoned Limestone Mine
Iowa Stored Energy Plant Association ISEPA [110]	Iowa, United States	Aquifer
New York State Electric & Gas Corporation (NYSEG) [111][a]	Seneca, NY, United States	Salt (Bedded)
Gaelectric	Larne, Northern Ireland	Salt (Bedded)
APEX CAES	East Texas, United States	Salt (Domal)
Chamisa Energy	West Texas, United States	Salt (Bedded)
Pacific Gas & Electric (PG&E) [112][a]	California, United States	Depleted gas field
Magnum Development[b]	Delta, Utah, United States	Salt (Domal)
Range Energy Storage Systems[b]	Delta, Utah, United States	Salt (Domal)

[a]*Projects received funding under American Reinvestment and Recovery Act of 2009.*
[b]*At time of writing, two developers are effectively competing to develop CAES at this location.*

have been used. As CAES plants are very rare, there are generally no specific turbomachines available for the application and a great deal of either customization or even new development is needed unless process conditions happen to match machinery available on the market for other applications. This fact has long since hampered the development of CAES plants and in many concepts dictated the design flow rates, temperatures, and pressures. Gear boxes, clutches, motors, and generators, however, are typically commercially available for CAES plants in the right size. Efficiency is the key concern for CAES turbomachinery as this factors directly into the round-trip efficiency both for the charge and the discharge side directly into the round-trip efficiency. Turbomachines for CAES plants are also required to have a wider operation range than in many other applications because of load changes to match desired output or input power, pressure changes in the storage, temperature changes from ambient and from heat storage. Together with high efficiency, these operability requirements give rise to specifications that are particularly challenging for compressors (Fig. 87).

A CAES plant compression train is typically arranged with low-, mid-, and high-pressure compressors and gearboxes where needed to increase the speed of

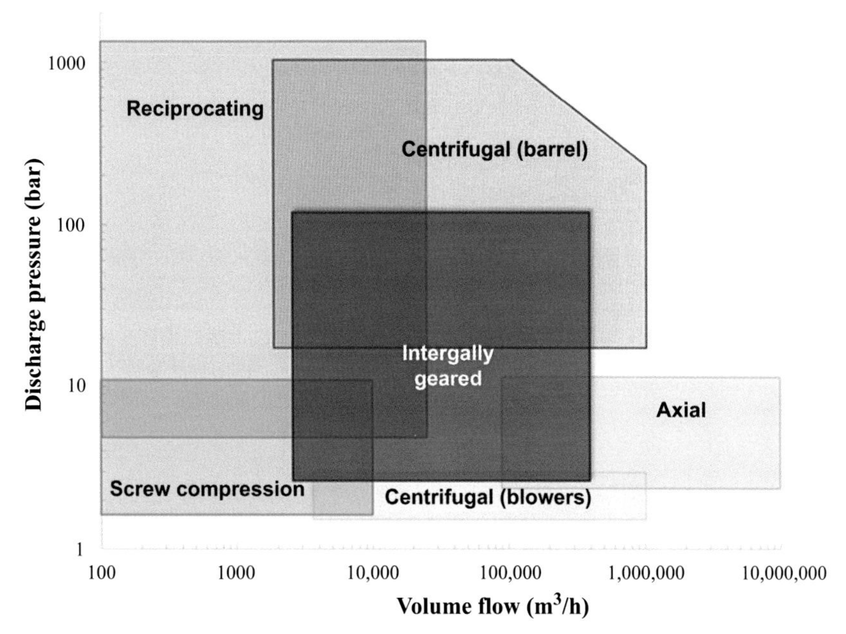

FIG. 87 Overview of compressor types [113].

the motor especially for smaller units or high-pressure compressors. For large plants, axial low-pressure compressors may be used for high volumetric flow rates. These compressors can be commercial process compressors or derived from gas turbines. Process compressors are more limited in size and outlet temperature, but are more readily available/applicable through customization of commercial units. Axial compressors can have polytropic efficiencies above 90%, about 5%...10% higher than radial compressors. For smaller size and for medium and high pressure, radial compressors are used. These have up to eight stages per unit and may have intermediate inlets and outlets for intercooling. For medium pressure, horizontally split casings are typical, and for high-pressure duty, barrel-type casings are used. Since radial compressors are predominantly used in the oil and gas industry, they are manufactured to standard API 617 with few exceptions (Fig. 88).

For ACAES processes, a discharge temperature in excess of 270°C is often desired, which is beyond the standard specification of typical compressors and requires a departure from the API standard, which manufactures in the industry may be reluctant to do. Significant engineering effort especially for temperatures above 450°C is required to deal with, e.g., reduced yield strength and creep of materials, differential thermal expansion, and bearing temperatures. During the ADELE program over the course of a multiyear engineering project, barrel-type radial high-temperature compressors have been designed by GE O&G (now Baker-Hughes). Temperatures up to 650°C could be reached with cooled last stage impellers, casing and bearing cooling, but temperatures limits

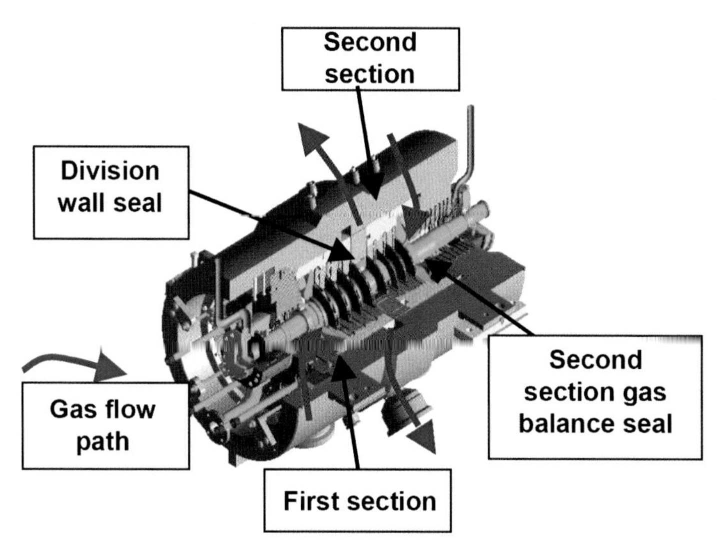

FIG. 88 Cutaway of a barrel-type high-pressure centrifugal compressor. *(From J.J. Moore, T.S. Soulas, Damper seal comparison in a high-pressure re-injection centrifugal compressor during full-load, full-pressure factory testing using direct rotordynamic stability measurement, in: Proceedings of the DETC '03 ASME 2003 Design Engineering Technical Conference, Chicago, IL, September 2–6, 2003.)*

between 560°C and 610°C would lead to higher efficiencies and lower cost solutions. A commercial high-temperature compressor remains unavailable as of the year 2020. For smaller and very small compression trains, integrally geared multistage radial compressors can be used within their flow rate and temperature limits (Fig. 89).

The integrally geared compressor (IGC) has centrifugal compressor stages that are over-hung from pinion ends. The overhung stage arrangement has the advantage of allowing easy interstage cooling to improve efficiency of compression. The IGC is common for many small- and midsize industrial applications and has several distinct advantages with wide range and good efficiency because of optimum speed for each stage. Some of the challenges in application to CAES are power and pressure casing limitations, IGCs are therefore more suited toward smaller size CAES plants (Fig. 90).

When starting or shutting down a CAES compression train, the compressors would be forced to pump against the cavern pressure with reduced speed and reduced mass flow rate. This would lead first to stall and then to surge-violent back flow that can destroy the machines. Antisurge valves and recirculation loops that route air from the compressor discharge to the inlet are used to prevent surge by allowing the pressure ratio to decrease and the mass flow to increase during transients with lower speed, while a check valve to the cavern prevents the high-pressure air from flowing backwards when the compressors are not spinning fast enough. Recirculation loops may need coolers or be

FIG. 89 Integrally geared compressor with intercoolers.

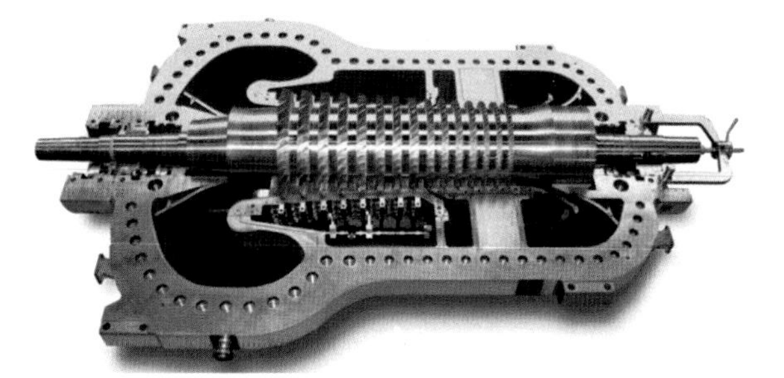

FIG. 90 Axial process compressor, rotor in open casing (AN series). *(Courtesy of General Electric Company.)*

combined with intercoolers. For reducing the mass flow at constant pressure, compressors can be equipped with variable inlet vanes and axial compressor even with variable stator vanes between the stages. The flow rate can be adjusted by up to 30%. However, part load operation with a turndown to, e.g., 25% as in other power plants is generally not possible in CAES plants during compression. Bypass systems or use of the turbines at the same time as the compressors to reduce compression power are possible in principle but reduce efficiency of compression dramatically. Ambient temperature affects the compression train power and mass flow. At lower temperature, more air can be compressed at constant speed and the maximum pressure ratio before stalling is higher. With increasing ambient temperature, the stall margin becomes smaller, less mass of air is compressed, overall the compression power is reduced, while the round-trip efficiency falls slightly as compressor discharge temperatures and intercooling waste heat increase. Inlet guide vanes can help mitigate the effects of ambient temperature and improve the operability of the plant (Fig. 91).

The expansion train in CAES plants includes several expansion stages to drive the generator. Whereas for the compression train commercial compressors are available within flow and temperature limits in principle, designing CAES expansion trains requires creative engineering to make use of customized commercial equipment without developing new turbines from scratch, which can be cost prohibitive. However, as fewer individual expansion stages are needed than in compression and since the aerodynamic design of turbines is fundamentally simpler easier than that of compressors, manufacturers may more readily offer new solutions for expansion trains than for compression trains. In general, CAES expanders can be derived from high- and medium-pressure steam turbines, industrial and utility gas turbines, aeroderivative gas turbines, and even radial expanders.

For high pressures and low flow rates, radial expanders from the O&G industry can be used. These, however, are in general not designed for high temperatures as required for most ACAES concepts and for fired CAES. Small CAES systems have been conceptualized with such expanders. The advantage of radial expanders is their high expansion ratio per stage, simplicity, similarity with common compressors, low cost, and ruggedness compared to axial machines; the disadvantage is an efficiency at least 5% lower than that of axial machines, and a limited size range (Fig. 92).

High-pressure expanders based on steam turbine technology have been used in both commercial CAES plants. Steam turbines readily tolerate temperatures

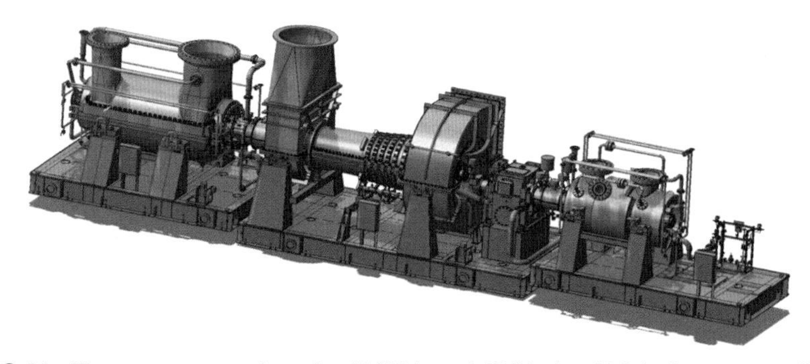

FIG. 91 Three-stage compression train with HP *(green)*, IP *(blue)*, and LP *(red)* compressors [99].

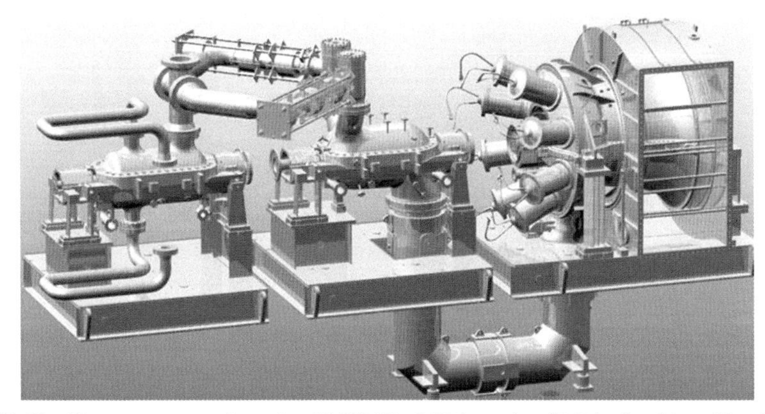

FIG. 92 Three-stage expansion train with HP *(blue)*, IP *(green)*, and LP *(red)* turbines, IP and LP can combustors [99].

of 540°C or above and are available in a wide size range. Large steam turbines run at synchronous speed such that gear box cost and losses can be avoided. To improve the efficiency when used in CAES rather than in a steam cycle with different working fluid conditions, redesign of nozzles, stator vanes, and even buckets (turbine blades) may be required along with an adaptation of piping and inlet valve system for lower pressure losses.

For high flow rates, high efficiencies and medium or low pressures, CAES expanders can be derived from gas turbines. Temperature limits are much higher by default than for any other turbomachines even without blade cooling and efficiencies of gas turbine stages can reach up to 94%. In principle flow rates up to 800 kg/s could be achieved with expanders derived from utility gas turbines, down to about 50 kg/s for industrial turbines. To use parts of a gas turbine as a CAES expander, the aerodynamic flow conditions must match and the speed needs to be lowered when the inlet temperature is reduced; a gear box may be required to couple the expander to a synchronous generator shaft. In simplified terms, aero matching can be approached when the flow function (mass flow/pressure × square root temperature) and the speed divided by square root temperature for the CAES application are similar to the original design conditions of the gas turbine. A detailed investigation stage by stage with meanline analysis or even CFD is required to estimate performance and adjust nozzle, vane (stator), and blade (bucket) profiles. Mechanically, to turn a gas turbine into a stand-alone expander, a new shaft and bearings need to be designed, especially a new thrust bearing and balance piston, as the thrust is not balanced as in the gas turbine with a compressor on the same shaft. Also a new housing and inlet needs to be designed to integrate the rotor and stator assembly. It seems plausible to derive axial CAES expanders gas turbines if available and matching aerodynamically. Such gas turbines are used in the

FIG. 93 Gas turbine rotor in open casing (GE 5002E). *(Courtesy of General Electric Company.)*

oil and gas and power generation industries. The efficiency of the expansion train factors proportionally into the CAES round-trip efficiency (Fig. 93).

Combustion systems in CAES expansion trains have been derived from utility gas turbines, both silo combustors (Huntorf) and can-annular combustors (McIntosh) are used. For new plants, more stringent NO_x and UHC emission controls would have to be implemented but a relatively low temperature ("E-class" of gas turbines) together with progress in "DLE" (Dry Low Emissions) premixed combustors will allow CAES plants to meet regulations more easily than gas turbines even without aftertreatment. For low-pressure turbines derived from utility gas turbines, the original combustor section may be used in a CAES expander to avoid costly new combustor development. Dresser-Rand for many years was about the only manufacturer engaged in CAES turbomachinery sales and marketing based upon the compressors, turbines, and combustors built and tested at the McIntosh plant and improved version of those. In 2015 Siemens acquired Dresser-Rand and brought new opportunities to apply Siemens technologies to further improve the CAES offering. The CAES low-pressure expander can be derived from gas turbines such as Siemens SGT-800, along with the proven DLE combustion system, yielding higher power turbine inlet temperatures, efficiencies, and lower emissions, using less air for higher round-trip efficiency, shorter compression time, or smaller cavern compared to McIntosh-type machinery. In higher-pressure applications where three expander sections are applied, the high-pressure and very high expander can be combined into one unit derived from Siemens SST industrial reheat steam turbines. On the compression side, Siemens integrally geared radial compressors can be used with optimum speed and several intercooling stages for medium pressure compression, along with a back-to-back arrangement of barrel-type

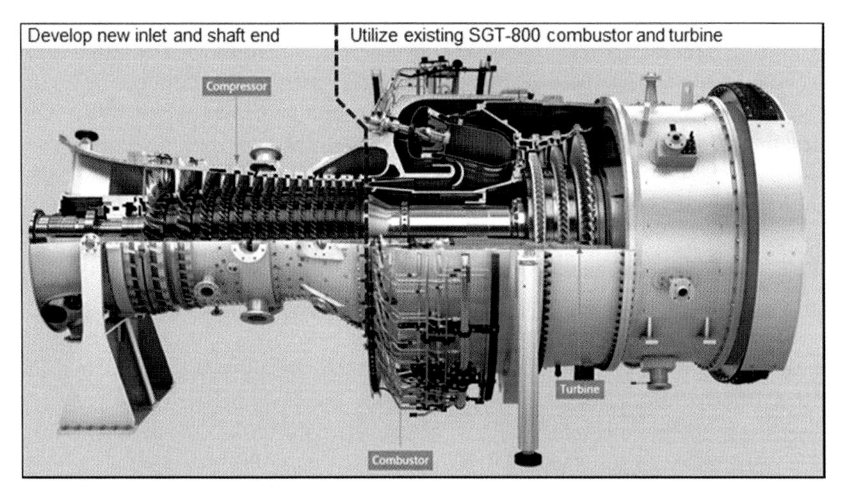

FIG. 94 A Siemens SGT-800 gas turbine and modifications for use as a low-pressure CAES expander. *(Siemens.)*

high-pressure compressors in the train. Due to the companies' size and broad scope of supply, products and services, better economy, faster project development, commissioning, and more comprehensive aftermarket service and maintenance may become available for CAES despite its small niche (Fig. 94).

A motor and a generator are used in CAES plants in the compression and expansion trains. Those are at multi-MW size typically "synchronous" electric machines with two or four poles that spin at full or half grid frequency, respectively, and are cooled with air (or hydrogen) on the primary side of the rotor gap and stator channels and sometimes cooling water on the secondary side. Especially for larger plants it has been economical to use a single synchronous machine with two clutches on either shaft end, one for driving the compression train during charge, and one for taking the power from the expander during discharge. Both current plants, Huntorf and McIntosh, use such a combined electric machine instead of two individual ones. There is no difference between a synchronous generator and a synchronous motor.

The clutch between the expanders and the shaft of the electric machine is a self-synchronizing overrun type to allow the generator to spin independently of the turbines when they slow down. The turbines can connect automatically to a synchronously spinning generator when starting by engaging just when the turbine speed reaches that of the generator. Similarly, the clutch disengages when the turbine valve closes and the torque becomes negative, i.e., the generator can continue to spin on the grid while the turbines disconnect and ramp down. The compressors, on the other hand, cannot start on their own and synchronize to the motor, the compressor clutch needs to be engaged to the motor during standstill before starting and does not need to be disengaged for shutdown of the compression train. The electric machine spins in the same direction in both operation modes because of the stator field's three-phase grid connection. Starting of the expansion train and generator can be done using the turbines with throttled pressurized air to spin up the train to synchronous speed before connecting to the grid. Alternatively, a large transistor-based variable frequency drive (bidirectional), a thyristor-based Load Commutated Inverter drive, or other large motor starting means including transformers or pony drives can be used. The same applies to starting the compression train. The expansion train can also be used to start the compression train and disengage the turbines after synchronization. More power is needed to start the compression train compared to the principally self-starting expansion train. Unloading the compressors to reduce power can be done by antisurge valve opening, blow-off, and recycling loops. Because any kind of starting means is typically sized much smaller than the motor power, only after grid synchronization the full amount of power from the electric machine is available and air can be compressed to sufficient pressure to fill the cavern. A check valve of course can provide protection against backflow during start-up before enough pressure is built up.

Turbomachines can in principle be scaled to any desired plant size. A large turbine or compressor may be a direct scale and aerodynamically similar to a machine of half the diameter with 25% of the mass flow and twice the speed. Below a certain diameter of axial machines, radial machines are used. This is a discontinuity in technology but still allows scaling down the plant to smaller size, from 100s of MW down to several MW. Mass of the machines scales directly with the cube of the diameter but cost for the machines does not scale directly and depends much on availability, customization and engineering, production volume, and last but not least the market environment. For synchronous electric machines, basically two-pole configurations (3000 rpm/3600 rpm) and four-pole (1500 rpm/1800 rpm) are available. Larger machines for 100 MW and above tend to be two-pole, smaller ones are often four-pole. Electric machines are scaled by the length of rotor and stator. They are adapted to the speed of the compressors and turbines by gear boxes if necessary or by electric frequency converters.

6.5.5 Heat exchanger requirements

Heat exchangers are used in CAES plants mainly for intercooling in the compression train. In ACAES plants, heat exchangers are also used for heating of the air in the expansion train. The requirements are to meet cooling/heating duties of a total approaching that of the total electric compression or expansion power while having small temperature and pressure losses. Temperature losses in the ACAES heat exchangers of the thermal storage system affect the roundtrip efficiency by about 2.5% per 10 K of loss in each charge and discharge. Besides aiming for low-temperature losses, the minimization of pressure losses in heat exchangers, piping, heat exchanges, and the inlet and exit sections of

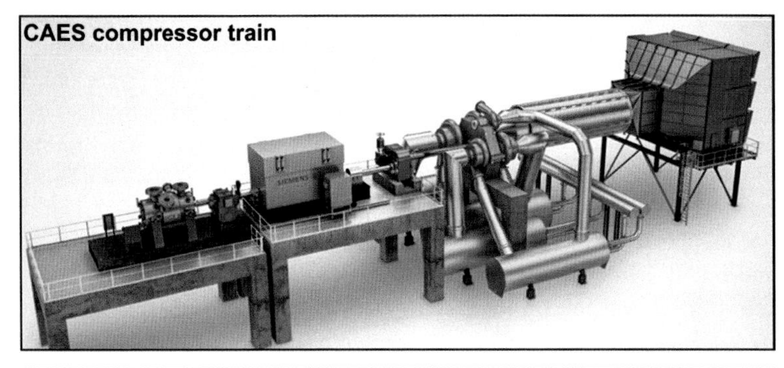

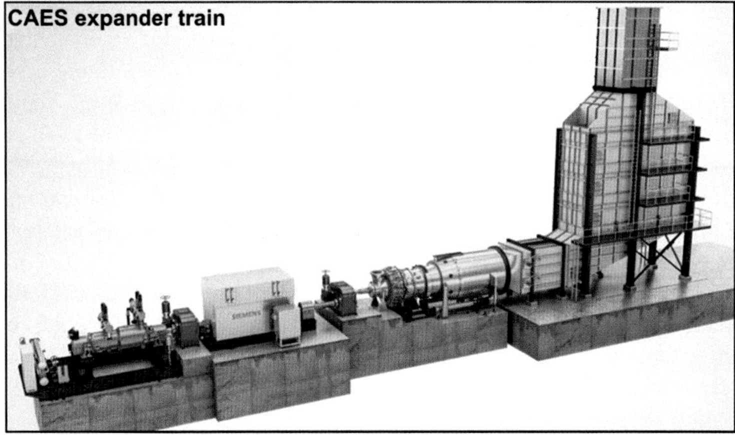

FIG. 95 CAD drawings of a Siemens three-stage recuperated diabatic CAES concept with separate compression and expansion trains. Four Shell and Tube intercoolers are under the low- and high-pressure compressors. A two-stage, high- and very-high pressure finned-tube recuperator is below the stack, duct burners are mounted in the transition duct before the recuperator inlet diffuser.

expanders (diffuser) is important as the CAES round-trip efficiency is reduced by approximately 0.1%-point per each 1% of pressure loss (Fig. 95).

Heat exchangers used for CAES intercoolers in past plants and concepts were either water-cooled shell and tube units or air-cooled finned-tube process gas coolers with forced or induced draft fans. Short tube lengths and many tubes in parallel are needed to reduce pressure losses; enhanced inner surfaces or twisted tape inserts can improve the poor air-to-tube heat transfer coefficient and save on surface area. For ACAES plants with liquid media heat storage, shell and tube heat exchangers have been conceptualized just as for intercoolers. Since the approach temperatures must be small for high efficiency, several shells in series may be required. This requirement, in addition to high volumetric air flow rates, results in large and costly heat exchangers. Alternative to shell and tube heat exchangers, plate-fin heat exchangers have been considered. Those are much more compact and have smaller volumes because of smaller channels and high heat transfer coefficients. However, the shape of fin-tube heat exchangers and size limitations of the stacked plate pack may make inlet and outlet headers and manifolds more difficult to design than for larger but fewer shell and tube units. Also, pressure limits may be lower in plate-fin than in shell and tube heat exchangers. Plate-fin heat exchangers still hold much promise for reducing the cost and footprint of ACAES heat exchangers and warrant more research and development.

For recuperators in CAES plants, with exhaust gas from the low-pressure expander at ambient pressure on one side and high-pressure air from the cavern or first stage expansion on the other side, fin tube units very similar to heat recovery steam generators (HRSG) are suitable. These are made from large bundles (or "harps") of parallel tubes, arranged in counter-cross flow within an exhaust air duct, upstream of the stack. Finned tubes are used to enlarge the heat transfer area, the same kind used in HRSG made from temperature resistant steel. Although the tubes run in cross flow relative the exhaust air, the high-pressure air undergoes heating in multiple passes with many rows of tubes. The rows are oriented such that subsequent passes of the high-pressure air are in counterflow relative to the exhaust air, and the recuperator can reach a high effectiveness (~90%) and small approach temperatures on the hot side (~20°C) for high efficiency. CAES recuperators have been built by subcontractors to Dresser-Rand in 1991 and more recently designed by NEM (Siemens). Generally those can be designed and built by many power plant boiler manufacturers.

Heat exchangers can usually be scaled to any desired plant size by increasing the frontal area, i.e., the tube count, and leaving pressure and temperature loss constant. To keep aspect ratios reasonable, larger heat exchangers tend to have longer tubes with larger diameters. For very small CAES plants, tubular heat exchangers may be replaced by plate or plate-fin heat exchangers.

References

[1] H. Jockenhöfer, W.-D. Steinmann, D. Bauer, Detailed numerical investigation of a pumped thermal energy storage with low temperature heat integration, Energy 145 (2018) 665–676.

[2] A.I. Papadopoulos, M. Stijepovic, P. Linke, On the systematic design and selection of optimal working fluids for organic Rankine cycles, Appl. Therm. Eng. 30 (6–7) (2010) 760–769.

[3] J. Schilling, D. Tillmanns, M. Lampe, M. Hopp, J. Gross, A. Bardow, From molecules to dollars: integrating molecular design into thermo-economic process design using consistent thermodynamic modeling, Mol. Syst. Des. Eng. 2 (3) (2017) 301–320.

[4] M.T. White, O.A. Oyewunmi, A.J. Haslam, C.N. Markides, Industrial waste-heat recovery through integrated computer-aided working-fluid and ORC system optimisation using SAFT-γ Mie, Energy Convers. Manag. 150 (2017) 851–869.

[5] C.N. Markides, Low-concentration solar-power systems based on organic Rankine cycles for distributed-scale applications: overview and further developments, Front. Energy Res. 3 (December) (2015) 1–16.

[6] Y. Ahn, et al., Review of supercritical CO2 power cycle technology and current status of research and development, Nucl. Eng. Technol. 47 (6) (2015) 647–661.

[7] A. White, G. Parks, C.N. Markides, Thermodynamic analysis of pumped thermal electricity storage, Appl. Therm. Eng. 53 (2) (2013) 291–298.
[8] J.C. Lee, J. Campbell, D.E. Wright, Closed-cycle gas turbine working fluids, J. Eng. Gas Turbines Power 103 (1) (1981) 220–228.
[9] J. Bao, L. Zhao, A review of working fluid and expander selections for organic Rankine cycle, Renew. Sust. Energ. Rev. 24 (2013) 325–342.
[10] H. Chen, D.Y. Goswami, E.K. Stefanakos, A review of thermodynamic cycles and working fluids for the conversion of low-grade heat, Renew. Sust. Energ. Rev. 14 (9) (2010) 3059–3067.
[11] R. Turton, R. Bailie, W. Whiting, J. Shaeiwitz, D. Bhattacharyya, Analysis, Synthesis, and Design of Chemical Processes, fourth ed., Prentice Hall, Upper Saddle River, NJ, 2013.
[12] S. Georgiou, N. Shah, C.N. Markides, A thermo-economic analysis and comparison of pumped-thermal and liquid-air electricity storage systems, Appl. Energy 226 (2018) 1119–1133.
[13] BOC, Industrial Gases Price List 2019, Available from: https://www.boconline.ie/en/images/industrial-gas-price-list-ie_tcm674-512090.pdf, 2019.
[14] BOC, Helium, A Grade, 300bar Cylinder, Available from: https://www.boconline.co.uk/shop/en/uk/helium–a-grade–300bar-cylinder-169800, 2019.
[15] S. Eyerer, C. Wieland, A. Vandersickel, H. Spliethoff, Experimental study of an ORC (Organic Rankine Cycle) and analysis of R1233zd-E as a drop-in replacement for R245fa for low temperature heat utilization, Energy 103 (2016) 660–671.
[16] W.C. Andersen, T.J. Bruno, Rapid screening of fluids for chemical stability in organic Rankine cycle applications, Ind. Eng. Chem. Res. 44 (15) (2005) 5560–5566.
[17] C.M. Invernizzi, D. Bonalumi, Thermal stability of organic fluids for Organic Rankine Cycle systems, Org. Rank. Cycle Power Syst. (2017) 121–151.
[18] E.W. Saaski, P.C. Owzarski, Two-Phase Working Fluids for the Temperature Range 50° to 350°C, Richland, Washington, 1977.
[19] D.M. Ginosar, L.M. Petkovic, D.P. Guillen, Thermal stability of cyclopentane as an organic Rankine cycle working fluid, Energy Fuels 25 (9) (2011) 4138–4144.
[20] H.L. von Cube, F. Steimle, Heat Pump Technology, Butterworths, 1981.
[21] D.G. Shepherd, A low-pollution on-site, energy storage system for peak-power supply, in: ASME 1974 Int. Gas Turbine Conf. Prod. Show, vol. 1B, 1974.
[22] E.M. Smith, Storage of electrical energy using supercritical liquid air, Arch. Proc. Inst. Mech. Eng. 1847-1982 (Vols 1-196) 191 (1977) 289–298, https://doi.org/10.1243/PIME_PROC_1977_191_035_02.
[23] R. Morgan, S. Nelmes, E. Gibson, G. Brett, Liquid air energy storage—analysis and first results from a pilot scale demonstration plant, Appl. Energy 137 (2015) 845–853, https://doi.org/10.1016/j.apenergy.2014.07.109.
[24] B. Ameel, C. T'Joen, K. De Kerpel, P. De Jaeger, H. Huisseune, M. Van Belleghem, M. De Paepe, Thermodynamic analysis of energy storage with a liquid air Rankine cycle, Appl. Therm. Eng. 52 (2013) 130–140, https://doi.org/10.1016/j.applthermaleng.2012.11.037.
[25] S. Hamdy, T. Morosuk, G. Tsatsaronis, Exergetic and economic assessment of integrated cryogenic energy storage systems, Cryogenics (Guildf) 99 (2019) 39–50, https://doi.org/10.1016/j.cryogenics.2019.02.009.
[26] X. Peng, X. She, C. Li, Y. Luo, T. Zhang, Y. Li, Y. Ding, Liquid air energy storage flexibly coupled with LNG regasification for improving air liquefaction, Appl. Energy 250 (2019) 1190–1201, https://doi.org/10.1016/j.apenergy.2019.05.040.
[27] I. Lee, F. You, Systems design and analysis of liquid air energy storage from liquefied natural gas cold energy, Appl. Energy 242 (2019) 168–180, https://doi.org/10.1016/j.apenergy.2019.03.087.

[28] X. She, T. Zhang, L. Cong, X. Peng, C. Li, Y. Luo, Y. Ding, Flexible integration of liquid air energy storage with liquefied natural gas regasification for power generation enhancement, Appl. Energy 251 (2019) 113355, https://doi.org/10.1016/j.apenergy.2019.113355.

[29] T. Zhang, L. Chen, X. Zhang, S. Mei, X. Xue, Y. Zhou, Thermodynamic analysis of a novel hybrid liquid air energy storage system based on the utilization of LNG cold energy, Energy 155 (2018) 641–650, https://doi.org/10.1016/j.energy.2018.05.041.

[30] Y. Li, H. Cao, S. Wang, Y. Jin, D. Li, X. Wang, Y. Ding, Load shifting of nuclear power plants using cryogenic energy storage technology, Appl. Energy 113 (2014) 1710–1716, https://doi.org/10.1016/j.apenergy.2013.08.077.

[31] T.H. Cetin, M. Kanoglu, N. Yanikomer, Cryogenic energy storage powered by geothermal energy, Geothermics 77 (2019) 34–40, https://doi.org/10.1016/j.geothermics.2018.08.005.

[32] S. Derakhshan, M. Khosravian, Exergy optimization of a novel combination of a liquid air Energy storage system and a parabolic trough solar collector power plant, J. Energy Resour. Technol. 141 (2019) 081901, https://doi.org/10.1115/1.4042415.

[33] Y. Li, X. Wang, Y. Jin, Y. Ding, An integrated solar-cryogen hybrid power system, Renew. Energy 37 (2012) 76–81, https://doi.org/10.1016/j.renene.2011.05.038.

[34] B. Kantharaj, S. Garvey, A. Pimm, Thermodynamic analysis of a hybrid energy storage system based on compressed air and liquid air, Sustain. Energy Technol. Assess. 11 (2015) 159–164, https://doi.org/10.1016/j.seta.2014.11.002.

[35] P. Farres-antunez, H. Xue, A.J. White, Thermodynamic analysis and optimisation of a combined liquid air and pumped thermal energy storage cycle, J. Energy Storage 18 (2018) 90–102, https://doi.org/10.1016/j.est.2018.04.016.

[36] J. Howes, Concept and development of a pumped heat electricity storage device, Proc. IEEE 100 (2012), 0195-SIP-2011-PIEEE.

[37] T. Desrues, J. Ruer, P. Marty, J.F. Fourmigué, A thermal energy storage process for large scale electric applications, J. Appl. Therm. Eng. 30 (2010) 425–432.

[38] R. Laughlin, Pumped thermal grid storage with heat exchange, J. Renew. Sustain. Energy 9 (2017) 044103.

[39] M. Mercangoez, J. Hemrle, L. Kaufmann, F. Buchter, C. Ohler, Thermoelectric energy storage with transcritical CO2 cycles, in: Proceedings of the 24th International Conference on Efficiency, Cost, Optimization, Simulation, and Environmental Impact of Energy Systems; 2011 Jul 4–7; Novi Sad, Serbia, 2011.

[40] H.U. Frutchi, Closed Cycle Gas Turbines: Operating Experience and Future Potential, ASME Press, 2005.

[41] Malta, Inc. website. https://www.maltainc.com/.

[42] G. Colby, M. Gupta, S. Nove, T. Griffin, H. Miller, N. Sehlstedt, High pressure CO2 compressor testing for Tupi 1, Tupi 2, and Tupi 3, in: Proceedings of ASME Turbo Expo 2012 GT2012-70137 Copenhagen, Denmark, 2012.

[43] B. Hopper, L. Baldassarre, I. Detiveaux, J. Fulton, P. Rasmussen, A. Tesei, J. Demetriou, S. Mishael, World's first 10,000 psi sour gas injection compressor, in: Proceedings of the Thirty-Seventh Turbomachinery Symposium, 2008.

[44] J. Howes, J. Macnaghten, R. Hunt, Thermal Storage System, 2017, US Patent 9,709,347 B2.

[45] Energy Nest website. https://energy-nest.com/.

[46] J. Kerth, J. Williams, Pumped Heat Energy Storage System Using a Conveyable Solid Thermal Storage Media, 2017, US Patent Application US 2017/0350658 A1.

[47] R. Amirante, P. Tamburrano, High temperature gas-to-gas heat exchange based on a solid intermediate medium, Adv. Mech. Eng. 2014 (2014), Article ID 353586.

[48] L. Catalano, F. De Bellis, R. Amirante, W. Rignanese, An immersed particle heat exchanger for externally fired and heat recovery gas turbines, J. Eng. Gas Turbines Power (2011).
[49] J. Kerth, Pumped Heat Energy Stroage System with Conveyable Solid Thermal Storage Media Directly Thermally Coupled to Working Fluid, 2019, US Patent Application US 2019/0162482 A1.
[50] J.M. Craven, Energy Efficient Solids Feed System for High Pressure Process, Department of Chemical and Biological Engineering, The University of Sheffield, 2014 (PhD thesis).
[51] J. Pacheco, S. Showalter, W. Kolb, Development of a molten salt thermocline thermal storage system for parabolic trough plants, in: Proceedings of Solar Forum 2001, April 21-25 Washington, DC, 2001.
[52] D. Brosseau, P. Hlava, M. Kelly, Testing Thermocline Filler Materials and Molten Salt Heat Transfer Fluids for Thermal Energy Storage Systems Used in Parabolic Trough Solar Power Plants, (2004), Sandia Report SAND2004-3207.
[53] S.M. Flueckiger, Z. Yang, S.V. Garimella, Design of Molten-Salt Thermocline Tanks for Solar Thermal Energy Storage, Cooling Technologies Research Center Publications, 2013 Paper 191.
[54] P.W. Li, J. Van Lew, W. Karaki, C.L. Chan, J. Stephens, J.E. O'Brien, Transient heat transfer and energy transport in packed bed thermal storage systems, in: M.A.D.S. Bernardes (Ed.), Developments in Heat Transfer, InTech, 2011. ISBN: 978-953-307-569-3, Available from: http://www.intechopen.com/books/developments-in-heat-transfer/transient-heat-transfer-and-energytransport-in-packed-bed-thermal-storage-systems.
[55] T. Barmeier, Thermal Energy Storage Plant, 2018, US Patent Application US 2018/0106165 A1.
[56] C.J. Quarton, S. Samsatli, Power-to-gas for injection into the gas grid: what can we learn from real-life projects, economic assessments and systems modelling, Renew. Sust. Energ. Rev. 98 (2018) 302–316.
[57] J. Adolf, M. Fischedick, Shell Hydrogen Study-Energy of the Future, Shell Deutschland Oil GmbH, Hamburg, Germany, 2017.
[58] US Department of Energy, Natural gas pipeline volumes, https://www.eia.gov/dnav/ng/ng_move_poe2_a_EPG0_ENP_Mmcf_a.htm.
[59] US Energy Information Administration, US Refinery H2 production capacity, https://www.eia.gov/dnav/pet/hist/LeafHandler.ashx?n=PET&s=8_NA_8PH_NUS_6&f=A.
[60] Hydrogen-based energy conversion, A.T. Kerney Energy Institute presentation, 2014.
[61] I. Glassman, Combustion, second ed., Academic Press, London, 1987.
[62] L.H. Cowell, T. Tarver, R. Kurz, A. Singh, Combustion systems for natural gas hydrogen mixtures, in: APGA Conference, Australia, 2019.
[63] M. Bower, E. Petersen, W. Metcalfe, H. Curran, M. Füri, G. Bourque, N. Aluri, F. Güthe, Ignition delay time and laminar flame speed calculations for natural gas/hydrogen blends at elevated pressures, in: Proceedings of AMSE TurboExpo, GT2012-69310, American Society of Mechanical Engineers, 2012.
[64] Technical Reference for Hydrogen Compatibility of Materials, Sandia National Labs, https://energy.sandia.gov/transportation-energy/hydrogen/materials-components-compatibility/technical-reference-for-hydrogen-compatibility-of-materials-2/.
[65] J. Goldmeer, Power to Gas: Hydrogen for Power Generation, GE Power, 2019. GEA33861.
[66] T. Komori, S. Shiozaki, N. Yamagami, Y. Kitauchi, W. Akizuki, CO2 emissions reduction method through various gas turbine applications, Mitsubishi Heavy Industries Tech. Rev. 44 (1) (2007).
[67] M. Moliere, N. Hugonnet, Hydrogen Fueled Gas Turbines: Experience and Prospects, Power-Gen Asia, Bangkok, Thailand, 2004.

[68] A. Olbes, J. Pujol, M. Moliere, M. Colas, High Compatibility Between Gas Turbines and Refinery Utilities: The Joint Experience of REPSOL and GE Energy Products, PowerGen Europe, Madrid, Spain, 1997.
[69] K. Payrhber, R. Jones, Scholz, Gas Turbine Flexibility with Carbon Constrained Fuels, 2008. ASME TurboExpo 2008, GT2008-50556.
[70] F. Bonzani, G. Pollarolo, Ansaldo Energia gas turbine operating experience with low BTU fuels, in: Proceedings of the ASME TurboExpo 2004, GT2004-53526, American Society of Mechanical Engineers, 2004.
[71] J. Karg, IGCC Experience and Further Developments to Meet CCS Market Needs, Coal-Gen Europe, PennWell, 2009.
[72] J. Hall, R. Thatcher, S. Koshevets, L. Thomas, R. Jones, Development and Field Validation of a Large-Frame Gas Turbine Power Train for Steel Mill Gases, ASME Turbo Expo, GT2011-45923, Vancouver, Canada, 2011.
[73] R. Jones, J. Goldmeer, B. Monetti, Addressing Gas Turbine Fuel Flexibility, GE Energy, GER4601, revision B, 2011.
[74] J. DiCampli, L. Madrigal, P. Pastecki, J. Schornick, Aeroderivative power generation with coke oven gas, in: Proceedings of the International Mechanical Engineering Congress & Exhibition, IMECE2012-89601, Houston, TX, 2012.
[75] Brown, et al., Siemens gas turbine H2 combustion technology for low carbon IGCC, in: Gasification Technologies Conference, San Francisco, 2007.
[76] J. Goldmeer, J. Bartle, S. Peever, Enabling Ethane as a Primary Gas Turbine Fuel: An Economic Benefit from the Growth of Shale Gas, GE Power, 2015. GEA32198.
[77] J. Goldmeer, W. York, P. Glaser, Fuel and combustion system capabilities of GE's F and HA class gas turbines, in: GT2017-64588, Proceedings of the ASME TurboExpo, American Society of Mechanical Engineers, 2017.
[78] L.H. Cowell, A. Padilla, P. Saxena, Advances in using associated gases in solar turbines DLE industrial gas turbines, in: The Future of Gas Turbine Technology 8th International Gas Turbine Conference, ETN, 2016.
[79] J. Goldmeer, T. Rojas, Burning a Mixture of H2 and Natural Gas, Turbomachinery International, 2012.
[80] A. Bonaldo, M. Andersson, A. Larsson, Engine testing using highly reactive fuels on siemens industrial gas turbines, in: Proceedings of the ASME Turbo Expo 2014, GT2014-26023, American Society of Mechanical Engineers, 2014.
[81] A. Reichart, Technologies for a low carbon society, in: International Gas Turbine Conference, Brussels, Belgium, 2019.
[82] Business Wire, GE's Fuel Flex Technology Helps CEPSA Enable Spain's Oil Refinery to Meet Strict European Emissions Standards, https://www.businesswire.com/news/home/20150518005257/en/GE%E2%80%99s-Fuel-Flex-Technology-Helps-CEPSA-Enable, 2015.
[83] M. Zajadatz, F. Guthe, E. Freitag, T. Providakis, T. Wind, F. Magni, J. Goldmeer, Extended range of fuel capabilty for GT13E2 AEV burner with liquid and gaseous fuels, J. Eng. Gas Turbines Power 141 (2019). American Society of Mechanical Engineers.
[84] S. Stollenwerk, W. Faller, R. Kurz, J. Neeves, Balancing the electric grid with a dual drive centrifugal pipeline compressor, in: 11th Pipeline Technology Conference, Berlin, Germany, 2016.
[85] W. York, W. Ziminsky, E. Yilmaz, Development and testing of a low NOx hydrogen combustion system for heavy-duty gas turbines, J. Eng. Gas Turbines Power 135 (2013). American Society of Mechanical Engineers.

[86] DLN 2.6e Product Technology, GEA33140, GE Power, 2017.

[87] K. Lam, P. Geipel, J. Larfeldt, Hydrogen enriched combustion testing of Siemens industrial SGT-400 at atmospheric conditions, ASME J. Eng. Gas Turbine Power 137 (2015).

[88] J. Larfeldt, M. Bloomstedt, M. Bjorkman, A. Lyckstrom, Hydrogen Co-Firing in Siemens Low NOx Industrial Gas Turbines, Power-Gen International, 2018.

[89] T. Asai, et al., Performance of Multiple-Injection Dry Low NOx Combustor on Hydrogen-Rich Syngas Fuel in an IGCC Pilot Plant, ASME TurboExpo, GT2004-25298, 2014.

[90] S. Dodo, et al., Dry low NOx combustion technology for novel clean coal power generation aiming at the realization of a low carbon society, Mitsubishi Heavy Industries Tech. Rev. 52 (2) (2015).

[91] MPHS successfully tests large-scale high efficiency gas turbine fueled by 30% hydrogen mix, https://www.mhps.com/news/20180119.html.

[92] M. Bothien, et al., Sequential Combustion in Gas Turbines—The Key Technology for Burning High Hydrogen Contents with Low Emissions, (2019)ASME TurboExpo 2019, GT2019–90798.

[93] P. Stuttaford, et al., Flamesheet Combustor Engine and Rig Validation for Operational and Fuel Flexibility With Low Emissions, (2016) ASME TurboExpo 2016. GT2016-56696.

[94] F. Bainier, R. Kurz, Impacts of H2 Blending on Capacity and Efficiency on a Gas Transport Network, 2019. ASME Paper GT2019-90348.

[95] Airlight Alacaes Test Report, retrieved 2019: https://www.aramis.admin.ch/Default.aspx?DocumentID=35281&Load=true.

[96] Airlight Alacaes Status on Twitter: https://twitter.com/alacaes/status/1147088556332146689?s=21.

[97] M. Budt, D. Wolf, R. Span, J. Yan, A review on compressed air energy storage: basic principles, past milestones and recent developments, Appl. Energy 170 (2016) 250–268.

[98] Intermountain CAES Project in Power Engineering: https://www.power-eng.com/articles/print/volume-120/issue-4/features/the-intermountain-energy-project.html.

[99] Dresser-Rand SMARTCAES® Sales Brochure. Dresser-Rand, 2015.

[100] R.B. Schainker, M. Nakhamkin, P. Kulkarni, T. Key, New utility scale CAES technology: performance and benefits, in: Electrical Energy Storage Applications and Technologies Conference 2007 San Francisco, CA, 2007. Available at http://www.energystorageandpower.com/pdf/epri_paper.pdf.

[101] V. De Biasi, Fundamental Analyses to Optimize Adiabatic CAES Plant Efficiencies, Gas Turbine World, 2009.

[102] ADELE-ING Engineering-Vorhaben für die Errichtung der ersten Demonstrationsanlage zur Adiabaten Druckluftspeichertechnik Öffentlicher Schlussbericht, DLR, 2018.

[103] S. Freund, R. Marquardt, P. Moser, ADELE Adiabatic Compressed Air Energy Storage—Status and Perspectives, 2012. VGB Powertech, 5.2013.

[104] S. Zunft, S. Freund, Large-scale electricity storage with adiabatic CAES—the ADELE-ING project, in: Energy Storage Global Conference, 19.-21. Nov. 2014, Paris, France, 2014.

[105] Leithner, et al., Druckluftspeicherkraftwerke zur Netzintegration erneuerbarer Energien-ISACOAST-CC, in: 4. Göttinger Tagung zu aktuellen Fragen zur Entwicklung der Energieversorgungsnetze, 22.–23. März 2012, 2012.

[106] M. Hämmerle, M. Haider, R. Willinger, K. Schwaiger, R. Eisl, K. Schenzel, Saline cavern adiabatic compressed air energy storage using sand as heat storage material, J. Sustain. Dev. Energy Water Environ. Syst. 5 (1) (2017) 32–45. http://www.sdewes.org/jsdewes.

[107] F. Crotogino, K.-U. Mohmeyer, R. Scharf, Huntorf CAES: more than 20 years of successful operation, in: Spring 2001 Meeting, Orlando, FL, 15-18 April 2001, 2001.

[108] J.O. Goodson, History of First U.S. Compressed Air Energy Storage (CAES) Plant (110-MW-26h), Electrical Power Research Institute Report TR-101751, 1992.
[109] BBC, Operating Experience With the Huntorf Air Storage GT Power Station, 1986.
[110] R.H. Schulte, N. Critelli, K. Holst, G. Huff, Lessons From Iowa: Development of a 270 MW Compressed Air Energy Storage Project in Midwest Independent System Operator, Sandia Report SAND2012-0388, 2012.
[111] J.W. Rettberg, M. Holdridge, NYSEG Seneca Compressed Air Energy Storage (CAES) Demonstration Project, Final Report, DOE Award No. DE-OE0000196 and NYSERDA 11052, 2012.
[112] W. Medeiros, et al., Technical Feasibility of Compressed Air Energy Storage Using a Porous Rock Reservoir, California Energy Commission Report CEC-500-2018-029, 2018.
[113] K.D. Wygant, J. Bygrave, W. Bosen, R. Pelton, Tutorial on the application and design of integrally geared compressors, in: Asia Turbomachinery & Pump Symposium, 2016.

Chapter 7

Energy storage services

Sebastian Freund[a], Scott Hume[b], and Joseph Stekli[c]
[a]*Energiefreund Consulting, Munich, Germany,* [b]*Electric Power Research Institute, Charlotte, NC, United States,* [c]*EPRI, Palo Alto, CA, United States*

Chapter outline

This chapter examines how energy storage technologies are used in the electricity markets today and can contribute to power system stability and grid capacity. With the growing share of variable renewables changing the power systems, new market opportunities are emerging and grid service functions, or ancillary services, are becoming increasingly important. Variable energy resources (VER), primarily wind and solar photovoltaics (PV), currently do not often provide ancillary services and their ability to ultimately provide them will be limited by the availability of the resource (i.e., wind or sun). Thus energy storage technologies will become increasingly important in maintaining grid stability for the future power system and for shifting energy from times where VER supply exceeds demand to times of solar and wind scarcity.

7.1 Wholesale energy time-shifting and arbitrage

Energy arbitrage is the classic and most obvious use case for energy storage plants. Energy arbitrage is defined as the purchasing power from the grid at times of lower cost, storing that energy, and then selling it at a later time when prices are higher. However, this business case for energy storage requires relatively wide price spreads between low and high pricing in order to account for the cost of the energy storage asset as well as inefficiencies between the collection and later dispatch of the stored energy. Finding sufficient price spreads has become increasingly difficult of the past few decades, at least in

Thermal, Mechanical, and Hybrid Chemical Energy Storage Systems
https://doi.org/10.1016/B978-0-12-819892-6.00007-1

the United States, due to utility deregulation and price depression caused by overcapacity of generation assets on the grid and falling wholesale prices largely resulting from the addition of VER.

In order to provide energy arbitrage competitively, round-trip efficiency, or the ratio of energy the storage system sends out over the amount of energy it took in (often referred to as the AC–AC efficiency), is very important. The reason for this is that a low round-trip efficiency requires more of the storage media to be used to ultimately deliver a desired amount of energy output. Simply, an energy storage system with a 50% round-trip efficiency, meaning that the storage system only outputs 1 kWh for every 2 kWh it takes in, will require twice as much storage media as a storage system with a 100% round-trip efficiency. This will increase the cost of storage in two ways: (1) Through the additional cost of the material doing the energy storage[a] and (2) Through the additional cost of the energy required to charge the energy storage system. Additionally, the price spread needed to break even by buying low and selling high is inversely proportional to the efficiency, where higher efficiency enables more operating hours and an earlier break-even point over the operating life. Historically, energy arbitrage has been provided by pumped storage hydroelectric plants, of which almost 200 GW have been installed around the world. Most energy storage technologies can be applied to this application.

In most energy markets, energy arbitrage must be completed at energy exchanges where electricity products are typically traded in 15-min or 1-h time and 1 MWh volume increments, with markets for both day-ahead (i.e., market to deliver electricity products the following day) and intraday (i.e., market to deliver electricity products in the next hour or less) time intervals. In the United States, these exchanges are typically operated by Independent System Operators (ISOs) or Regional Transmission Operators (RTOs) such as PJM or the California Independent System Operator (CAISO). In Europe's ENTSO-E region, national exchanges of several countries work together and have effectively created one very large central European market. Examples of the energy prices agreed in day-ahead markets are shown in Figs. 1 and 2 for US and German markets, respectively.

7.1.1 Capacity

Due to the growth in VER and the retirement of relatively firm fossil generation, some electricity markets have created markets or payment schemes that provide fixed payments for assets that can deliver power on an as-demanded basis. These so-called capacity payments, or capacity markets as the case may be,

a. In most energy storage systems, the efficiency is lost either through energy loss caused by having to store the energy for a period of time or through losses caused by the act of sending the energy back to the grid and not by losses in the charging of the energy storage system. Therefore, if 1 kWh of energy is taken in that amount of energy must be stored, even if 0.5 kWh (50% efficiency) or 1 kWh (100% efficiency) is ultimately delivered to the grid.

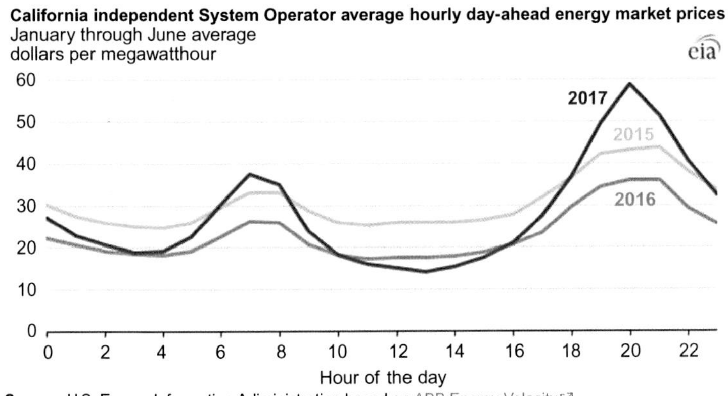

FIG. 1 The "Duck Curve" of hourly electricity prices in California, showing a recent increase in both the depression from growing solar during midday and the demand peak in the evening. *(Source: Greentechmedia.)*

are an effort to provide remuneration for assets that may have difficulty competing in wholesale markets due to higher fuels costs, higher start-up costs, or other causes resulting in relatively high marginal operating cost but can be relied upon to be available when wind or solar PV assets cannot generate electricity. It should be noted that capacity payments differ from the reserve markets or reserve products that exist in many electricity markets in that reserve markets typically operate on a day-ahead or a real-time basis, serving to provide capacity in case of any short-term, unexpected shortfalls in electricity generation or unexpected electricity demand. Instead, capacity auctions typically cover an entire year, or multiple years, and serve to ensure adequate electricity generation capacity is available over the course of that time period.

Capacity markets are being implemented across the globe. The UK established a capacity market in 2014, becoming one of the first European countries to do so. The first auctions in the UK market resulted in capacity pricing of approximately £20/kW-year and the 2018 auction contracted nearly 50GW of capacity, including several GW of storage in the form of existing pumped storage hydropower and new battery energy storage systems. However, UK capacity prices have sharply declined in 2018 and 2019 which has resulted in lesser investment in firm energy generation resources in the UK. France recently has also implemented a capacity market, with clearing prices of about €10/kW-year for contracts in 2022 reported in the 2018 auction. In the United States some ISOs/RTOs have established capacity markets, including MISO, PJM, NYISO, and ISO-NE as of 2019. These ISOs/RTOs assess their capacity requirements based on annual peak load and hold auctions to procure the necessary capacity, with the auctions covering from one year to up to three years in the future. Prices in these auctions have varied widely between the ISOs/RTOs, locational zones

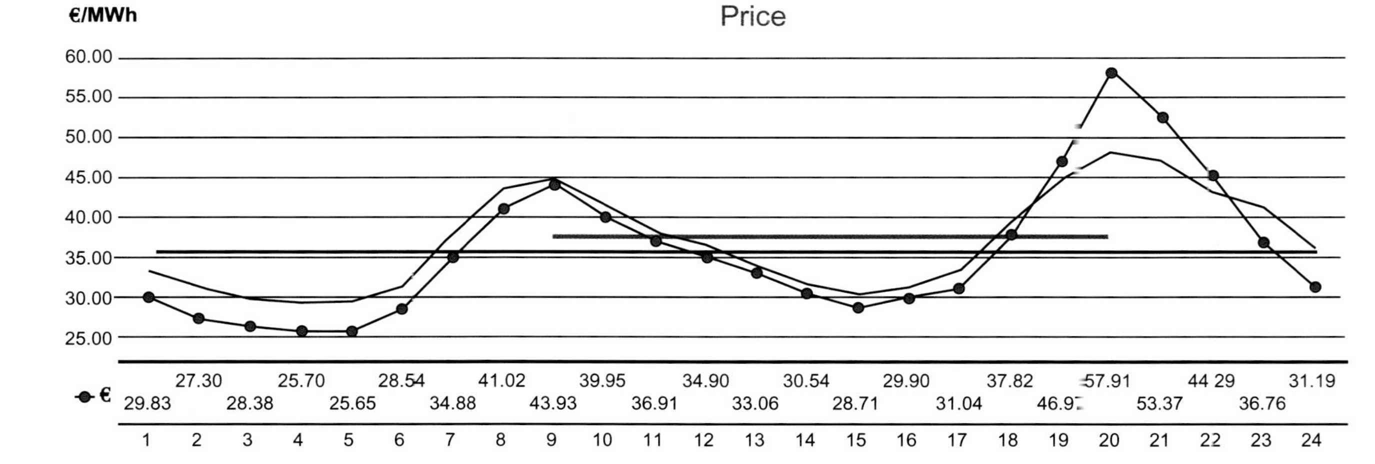

FIG. 2 Hourly electricity prices on a sample day (red dots) and monthly average (gray curve) of the day-ahead spot market in Germany (July 2019). Thick gray line is the average price of the day, thick red line is the average "peak" price of the day. PV during this sunny day causes a price decrease, leading to a similar "Duck" curve as in California even on monthly average hourly prices. *(Source: www.epexspot.com.)*

within the ISOs/RTOs, and by delivery years, with prices ranging from as little as $6/kW-year to $180/kW-year. Participation of storage plants in these markets is dependent upon the duration of storage the plant is capable of and the required duration varies for differing ISOs/RTOs, with NE-ISO requiring only 2 h of duration while PJM requires 10 h of duration as of 2019.

7.1.2 Ancillary services

7.1.2.1 Background

Ancillary services are services used in electric power systems to ensure the operational reliability of the bulk power system. When the United States began regulatory reforms of the wholesale markets in 1996, the Federal Electricity Regulatory Commission (FERC) defined six ancillary services in Order 888 [1]:

(1) *Scheduling, system control, and dispatch*: Service provided by the system operator of a control area to schedule the movement of power through, out of, within, or into a control area.
(2) *Reactive supply and voltage control from generation service*: Producing or absorbing reactive power to maintain transmission voltages on transmission facilities within acceptable limits.
(3) *Regulation and frequency response service*: Generation and nongeneration resources raise and lower their output or demand to maintain power balance (i.e., ensure power supply equals demand on a second to minute basis) in order to maintain frequency at a nominal level (typically 60 Hz in the United States and 50 Hz internationally).
(4) *Energy imbalance service*: Providing a source of energy balance when there is a difference between actual and scheduled delivery of energy. Note that the original FERC definition of balancing error was subhourly.
(5) *Operating reserve—synchronized reserve service*: Power generators held in reserve, but in an manner that would allow them to provide power relatively quickly (typically <10 min) should a system contingency occur (e.g., an online power generator trips and goes offline).
(6) *Operating reserve—supplemental reserve service*: Power generators held in reserve but in a manner that would not allow for immediate response (i.e., they must be started up) but would be available within some defined time period, typically 15 or 30 min.

In order for a grid operator to operate the grid in a stable manner, with minimum frequency and voltage deviations and without interruptions in power service, ancillary services are contracted in addition to energy and capacity from generators. In many markets these services are auctioned within a pool of qualified generators, with participation mandatory in certain territories for certain services, and existing power grid codes ensure power plants comply with the necessary specifications, which include ramp rates and controls. Generators are paid over a contracted period to keep capacity (positive or negative) for these services available on demand by the grid operator.

Historically, ancillary services have been a small component of total electric power costs, with the revenue received by electricity generators in the United States from ancillary services only representing ~3% of total revenue in 2018. The cause of this has been that many of these ancillary services have been provided traditional rotating machinery as part of power delivery, with little or no marginal cost to provide these additional services. However, VERs can generally provide these services by foregoing the opportunity cost of generating electricity, meaning that if the ancillary service is provided, electricity cannot be provided and, due to the fact that solar and wind do not utilize a stored fuel, that electricity generation must be foregone rather than simply delayed as it would be with a coal, gas, or nuclear plant. Therefore as VER penetration increases, ancillary service provision is taking on increasing importance in the electricity system.

Within the organized electricity markets certain ancillary services are procured through several mechanisms such as bid-based auction markets, tariff fixed or formula-based rates (often based upon cost of service), or, less commonly, competitive solicitations. However, certain ancillary services are supplied without market payment, such as Volt/VAR support, due to the inherent properties of electricity generators in operation and a resultant oversupply of the service. However, inverter-based generation options such as wind, solar PV, and battery energy storage systems do not inherently provide some of these ancillary services, meaning that market payments for these services may be required as the penetration of these resources increases.

As VER share on power systems increases, the amount of ancillary services the electricity system requires is likely to increase. This is a result of the increased production variability and forecast uncertainty that VERs have relative to more traditional electricity generator, which ultimately requires more of the services necessary to ensure that electricity supply and demand are matched at all times and system stability is maintained. Although VERs have the potential supply, many ancillary services and several successful demonstrations of these capabilities have been conducted over relatively short time periods (i.e., a few days), government incentives or mandates for VER use have typically only focused on or only reimbursed electricity generation. This has resulted in VER plant design focused solely on maximizing electricity production and has disincentivized any ancillary service delivery from these resources. Therefore it is likely that new connection standards or restructured market mechanisms will be required to drive ancillary service delivery from VERs. Specifically, a consideration of the performance of these technologies and a cost–benefit analysis of VERs providing these services, along with a measure of the confidence system operators have in relying on the availability of VERs to provide the services, will be necessary.

7.1.2.2 Frequency regulation

Frequency regulation, or Primary Reserve as it is typically called in Europe, is a service that serves to smooth out small frequency deviations from supply and demand mismatch over short time periods (seconds to minutes). Participating generators typically must respond within 30 s to 1 min[b] to an automated signal sent out by the grid operator instructing them to ramp their power output up or down in response to relatively small imbalances between power supply and demand. In some markets, frequency regulation up and down have been separated into separate markets, primarily due to the ability of some newer power generators to provide on service more easily or efficiently than another. For example, it is easier for a wind plant to curtail its production and provide frequency regulation down than it is for to increase its power output to provide regulation up due to the limitations of the available wind resource. Similarly, some thermal power plants can and have been designed with a power efficiency curve that allows them to either increase or decrease power output more efficiently from their typical operating point. Most storage technologies are capable of providing frequency regulation, though some like pumped storage hydropower must be already operating at some power output level to provide these services (i.e., they cannot provide frequency regulation from a "cold" start) while others like battery energy storage can respond rapidly enough to participate even if they are not already providing energy to the grid.

Spinning and nonspinning reserves

The second line of defense against changes in grid frequency are spinning and nonspinning, as they are called in the United States, or secondary and minute reserve as they are referred to in Europe. Generators providing these services serve to make up for insufficient electricity production caused by an unexpected outage of an online generator. Spinning or secondary reserves must already be synchronized to the grid and can be called upon at short notice (typically ten minutes) to start producing power at a predetermined rate. Nonspinning reserves serve the same purpose as spinning reserves but are not synchronized to the grid, meaning that they typically take a longer time to get up and running, though they typically must be able to provide power in under 30 min. As spinning reserves must have shorter response time requirements, which historically has required that they are burning some amount of fuel to stay "hot" and ready for operation, they have typically received higher payment than nonspinning reserves. Most energy storage technologies can provide either reserve service, though some such as pumped storage hydropower or thermal energy storage must expend some stored energy while idling to be in a ready enough state to provide spinning reserves.

---

b. In some markets even quicker responses are required and so another ancillary service, called fast frequency response, has been created.

Within the Entso-E electricity markets, the three reserve services are auctioned on a weekly or monthly basis, with positive and negative capacity for secondary and tertiary reserve separately and in daily repeated blocks of several hours. Participation is subject to prequalification with stringent requirements, including proof of 5 min/15 min. from signal to full load ramping requirement for secondary and tertiary reserve, respectively, as shown in Fig. 3. For secondary reserve, the minimum capacity is 5 MWe after a ramp rate of 2% load/min and 4 h of operation capacity is typically required. Further, thermal units are expected to be synchronized and spinning and have a 95% proven availability during a period of min. 50% operational uptime. The latter requirements might be allowed to be accomplished by storage units by pooling or exceptions [2]. Prices for secondary reserve capacity on an annual basis have been on the order of $100 k/MW, but for tertiary, nonspinning reserve capacity, where gas turbine peakers are competing, only on the order of $1 k/MW-year. Prices for negative reserves can be higher depending on the market and on wind penetration.

Primary response and inertia

System inertia is an inherent property of rotational generation equipment where the rotating kinetic energy of the spinning mass (generator, rotor, and turbine elements) is imparted to (or absorbed from) the system when the system frequency changes. This property represents the fastest element of frequency control as the angular momentum of the mass from all synchronous generators connected to the system will be instantaneously slowed as frequency drops. This instant extraction of kinetic energy into the generator acts to inherently counter the frequency excursion, thereby delivering 'strength' to the system. The more inertia on a given system, the slower the rate of change of frequency (RoCoF) that can occur given a system disturbance.

This previously inherent property of the system ensured a sufficient time buffer for primary response action (a controlled parameter) to kick in and counteract the frequency change. Without sufficient inertia, the RoCoF following a

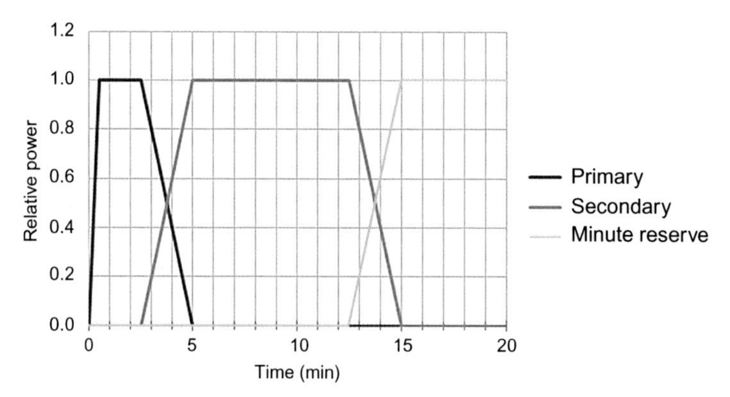

FIG. 3 Principle of reserve operation over time after a frequency drop in Entso-E.

disturbance event could be excessive, resulting in secondary protection mechanisms (underfrequency load shedding) being invoked to protect the system from further frequency excursion. Inertia is measured using units of apparent power and seconds (MVA*s) and is a function of the inertia constant of the synchronous generator (i.e., more mass delivers a higher inertia constant).

In most AC generation systems in the past, this 'system inertia' was taken for granted as all generation was synchronous and as such the generating mass had a direct electromagnetic connection to the prevailing system frequency. With the inclusion of nonsynchronous generation such as solar photovoltaic generators, wind generation and energy storage technologies such as electrochemical batteries and flywheels, the question of system inertia instability is now a serious concern where nonsynchronous generation feed exceeds 60%–70%. Installation of 'synchronous condensers' is typically needed in areas of the grid that are constrained by connection size and the local renewable ratio to ensure system stability is maintained.

Energy storage technologies that feed synchronous generators, such as CAES, LAES, and thermal energy storage systems that rely on a power cycle to generate power, will be advantageous in power systems with increasingly high levels of nonsynchronous renewable generators and batteries.

7.1.2.3 Voltage support

Besides frequency regulation, voltage regulation is required both locally in the distribution grid and regionally in the transmission grid to ensure reliable electricity flow across the power system to every consumer and maintain voltage within an acceptable range. For stabilizing the voltage, both real and reactive power production need to be matched with demand. The variability of the excitation voltage of synchronous generators allows for VAr or power factor adjustment (power electronics of frequency converters may also have this capability) to allow various kinds of storage plants to support voltage regulation. This may be a grid code requirement and there is typically no market mechanism to remunerate such service.

7.1.2.4 Black start

Many power generation devices cannot be brought online without external power being utilized from the grid prior to grid synchronization. If a blackout event has occurred, generation systems that have the ability to be started without external support are classified as being able to deliver 'black-start' service. In the United States, NERC defines black start resources as "generating unit(s) and its associated set of equipment which has the ability to be started without support from the system or is designed to remain energized without connection to the remainder of the system, with the ability to energize a bus, meeting the Transmission Operator's restoration plan needs for real and reactive power capability, frequency and voltage control, and that has been included in the Transmission Operator's restoration plan."

Therefore resources such as generators or energy storage systems must be capable of starting without outside power supply, maintain frequency and voltage under varying load, and maintain rated output for a significant period of time (e.g., 16 h). Though black start service is seldom needed in practice, due to high reliability of the system, it is critical from a contingency perspective to ensure that sufficient resources can be started without any external power sources and these devices will maintain voltage and frequency while load is energized across the system. It is therefore essential that these resources can be relied upon to restore power after a complete or partial black out as quickly as possible. In a generating plant, the characteristics needed for this service favor combustion turbines or hydro generation or combinations of technologies that allow other generating technologies to be black start capable.

7.1.3 Behind the meter and renewable integration

Besides the regulated markets within TSOs and RTOs, further opportunities exist especially for storage plants to provide the right amount of electricity at the right time to reduce cost, improve reliability, and maximize the use of renewables. These include but are not limited to applications such as pairing storage with renewables, using it to shave peak demand or supplying power to islanded grids.

7.1.3.1 Output firming of PV/wind farms for IPPs

For operators of renewable plants to maintain output with a contracted schedule requires energy shifting of renewable generation from more or less random times when it is plentiful/available to times of scheduled output or times that the market demands generation and prices come at a premium. Charging with a fluctuating amount of power as a difference between available power and scheduled output requires good part load capability on charging and fast ramp rates, corresponding to the acceptable tolerance on scheduled output, curtailment options, and rate of change of renewable power from wind gusts, overcasts, or similar events. Lowest part load ($\sim$20%) and fastest ramps (10 s... 30 s) are advantages for operational success. Several large battery installations have been built in the past years together with solar farms; one notable example is the 13 MW/4 h (52 MWh) Li-Ion battery installation of Tesla with KIUC in Kauai, HI. Storage plants should be able to serve this application well within their ramp rate and part load limits.

7.1.3.2 Demand management

Managing peak loads of industrial facilities behind-the-meter to mitigate demand charges or even connection capacity constraints can be a valuable business. Backup batteries, behind-the-meter small generators, and CHP are suitable if they can be switched on a ramp sufficiently fast in anticipation of peak load events. Hence, small storage plants that have capacity available that can be scheduled in sync with large loads can be a good solution.

7.1.3.3 Backup power and micro grids

These applications have the highest demands on flexibility, i.e. start-up times, part load, and ramp rates for micro-grid with frequency stabilization. These are typically served not with storage but with Diesel Gensets, although large battery systems have been used for such applications, especially for short-term backup power with second-fraction start time. For thermal–mechanical power generation, running in islanded mode requires the fastest response rate of input power to generator torque from demand changes, with very tight control of the resulting speed to keep the frequency stable. Unlike backup Diesel generators, many storage plants are not designed for islanded operation and need a stable frequency provided by the grid to sync to. Larger plants that can start and operate without the grid are known as "black-start" capable.

7.2 Examples of energy storage operation on the market

Many pumped hydropower plants were built in a time where markets were not liberalized with the purpose of providing regularly scheduled peak power while enabling higher load of fossil and nuclear plants during off-peak hours. Nowadays, pumped hydro plants may still provide peak power but need to follow the spot market prices from the exchange in addition to any ancillary services they can offer and portfolio optimization the operating company may need. Fundamental to making a margin on the spot market is selling generated electricity at a price that is higher by a factor of 1/efficiency plus operating costs and grid fees than the electricity price paid for pumping. This puts pressure on the capacity factor depending on the magnitude of the price fluctuations and may lead to longer idle periods when peak prices are low, the efficiency is low, or the operation costs and grid fees are high. In many markets during the mid-2010s the margins have been too small to justify new investment in pumped power plants despite their versatility, good grid services, and high efficiency. The following figure shows exemplarily the operation of pump and turbine of 300 MW pumped hydro plant operating on the German day-ahead market in one week in 2013. During price dips on windy nights the pump operates, and during day peaks of ~€55/MWh the turbine operates. The decline of prices over the weekend does not justify turbine operation but pumping at a price $<$ €35/MWh, as shown in Fig. 4.

An application for storage outside the exchange traded market is the provision of firm contractually scheduled renewable power. Some of the largest concentrated solar power plants with 100+ MW and many hours of molten salt storage follow that model and as an independent power producer receive a fixed compensation with a power purchase agreement (PPA) from a local utility. An exemplary recent project was undertaken in 2017 by SolarCity (now Tesla) in Kauai, HI, with a

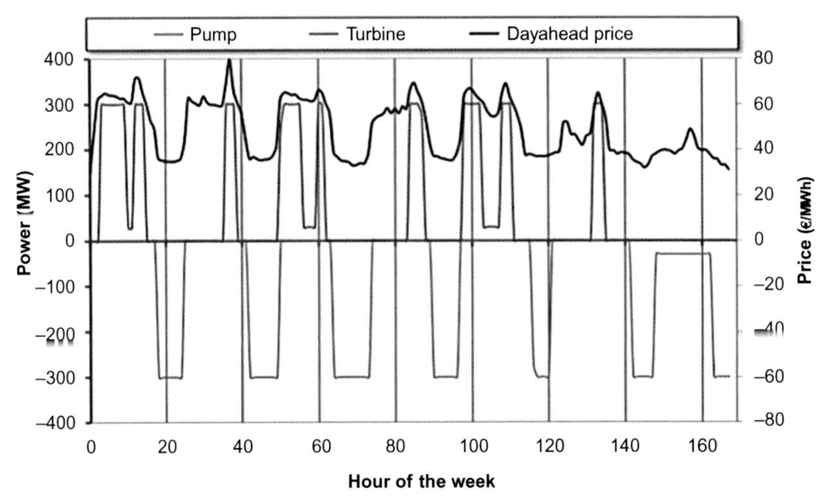

FIG. 4 Operation of a 300 MW pumped hydro plant on the German market with power (left) and day-ahead price (right) over the hours of one week. *(Source: Gutachten zur Rentabilität von Pumpspeicherkraftwerken, Forschungsstelle für E.nergiewirtschaft e.V., München, 2014.)*

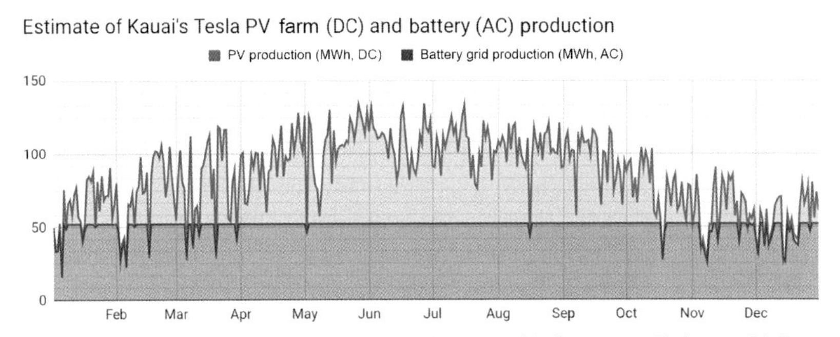

FIG. 5 Battery application for renewable integration to provide firm power. Estimate of daily production of Tesla's Kauai 17 MW PV farm and 52 MWh battery. *(Source: https://www.greencarreports.com/news/1112800_teslas-solar-and-battery-project-in-hawaii-we-do-the-math.)*

17 MW PV farm combined with a 13 MW/52 MWh li-ion battery, as shown in Fig. 5. The battery allows to shift the solar energy away from midday where the market is saturated already with PV production and the system delivers a fixed 52 MWh during certain hours of the evening every day to Kauai Island Utility Cooperative (KIUC) to help meeting their goal of 70% renewables by 2030.

References

[1] Promoting Wholesale Competition through Open Access Non-discriminatory Transmission Services by Public Utilities, FERC Order 888, Final Rule, 75 FERC 61,080 (April 24, 1996).
[2] https://www.regelleistung.net/ext/static/srl.

Chapter 8

Applications of energy storage

Pablo Bueno[a], Craig Turchi[b], Joseph Stekli[c], Hitesh Bindra[d], Brendan Ward[d], James Underwood[e], and David Voss[e]
[a]*Southwest Research Institute, San Antonio, TX, United States,* [b]*NREL, Golden, CO, United States,* [c]*EPRI, Palo Alto, CA, United States,* [d]*Kansas State University, Manhattan, KS, United States* [e]*Solar Turbines Incorporated, San Diego, CA, United States*

Chapter outline

8.1 Introduction

No book on energy storage would be complete without a chapter dedicated to how the concepts explained earlier apply in the real world. At the utility scale, storage has been largely ignored until recently because there was not a real need for it. Power plants could operate in base-load or peak-following mode without storage largely because the economics make sense. Nevertheless, the advent of renewable energy revived the interest of the engineering and scientific

Thermal, Mechanical, and Hybrid Chemical Energy Storage Systems
https://doi.org/10.1016/B978-0-12-819892-6.00008-3

communities to investigate and develop means to store energy to be released on demand.

Most renewable energy sources, such as wind, wave, and solar, are intermittent in nature. Not only that, their intermittency is unpredictable to the degree that power companies require for operational planning. It is in this context that many of the concepts explained earlier in this book have been researched in recent years. However, this is not to say that energy storage cannot be used in more conventional power plants such as fossil fuel, nuclear, or hydroelectric power plants. Indeed, one of the purposes of this chapter is to show that conventional as well as nonconventional power generation technologies can incorporate storage whether in the design phase or as a retrofit.

Fossil fuel plants are still the largest power generators in the world. While the demand for coal has been declining over the last century, it still accounts for about one-third of the worldwide electricity generation. Natural gas has enjoyed a renaissance of sorts since the turn of the century because enhanced recovery techniques such as hydraulic fracturing have made it easier to extract it from shale formations. Shale gas has driven down the price of natural gas which, in turn, has led to an increase in the number of power plants that use it as the primary fuel. Thus fossil fuel power plants still represent the largest market for energy storage.

Nuclear power plants have had a rocky relationship with the general public. While they can be considered a clean energy source, to the extent that they do not emit greenhouse gases, concerns about safety abound among the general population. These concerns are largely unsupported by the safety record of nuclear power plants, but nevertheless they affect public policy and must be addressed. It is in this context that the nuclear power industry has sought to add energy storage to current and future power plants. It should be noted, though, that globally nuclear power has lost market share in recent years.

At the same time, the popularity of renewable energy sources has increased driven by worldwide concerns of climate change. By their very nature, fossil fuel power plants release greenhouse gases such as methane and carbon dioxide into the atmosphere. These gases have been associated with an increase in the global mean surface temperature, ocean acidity, melting of the polar ice caps, variability of global weather patterns, and frequency and intensity of extreme weather events. Thus energy sources that reduce or avoid altogether the emission of greenhouse gases have received much public attention recently. Renewable energy, especially unconventional sources, requires energy storage to be able to supplement fossil fuels and eventually compete with them as base-load power plants. Thus while they are not nearly as ubiquitous as their fossil fuel counterparts, renewable plants can be considered the largest driver of the current push to research and develop storage alternatives.

8.2 Fossil fuel power plants

8.2.1 Description

Fossil fuel power plants are thermal power plants that use fossil fuels such as coal or natural gas to generate electricity. These plants generate heat by burning the fossil fuel and then converting the resulting thermal energy into mechanical energy via a turbine. An electrical generator is then paired with the turbine to generate electricity. Fossil fuel power plants are sometimes called conventional power plants and they account for 85% of the global power generation [1]. Owing to their ubiquity, fossil fuel power plants are good candidates for energy storage.

8.2.2 Principles of operation

Fossil Fuel power plants usually fall in two general categories based on the fluid used in the turbine:

(1) Steam turbine
(2) Gas Turbine

8.2.2.1 Steam turbine power plants

Steam turbine power plants operate on the Rankine cycle. Fuel, typically coal or natural gas, is burned in a furnace from which the combustion gases are drawn into a boiler. At the boiler, the heat of the combustion gases is used to convert water into steam. The steam is then passed through a turbine which converts the thermal energy of the fluid into mechanical energy. This turbine is connected to an electric generator which ultimately produces electricity. Since the steam still contains a significant amount of heat after interacting with the turbine, it circulates through a condenser to further bring down the temperature and pressure. Finally, it is pumped it through the system again. A schematic of the most basic Rankine cycle is shown in Fig. 1.

The overall thermal efficiency of steam turbine power plants is limited by the Carnot efficiency and peaks at about 42%. Additional heating and cooling stages can be added to the process to increase its efficiency.

8.2.2.2 Gas turbine power plants

Gas turbine power plants typically operate on the Brayton cycle. In its most simple inception, natural gas is compressed and burned to increase its enthalpy. The gas then passes through a turbine in which the thermal energy is converted into mechanical energy. An electric generator which is coupled with the turbine ultimately generates electricity. A schematic of the basic Brayton cycle is shown in Fig. 2. The overall thermal efficiency of the Brayton cycle is a strong function of the compressor pressure ratio and peaks at around 55%–60%.

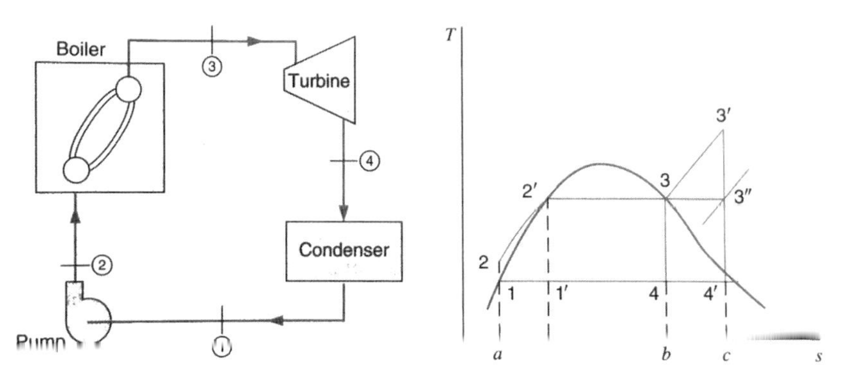

FIG. 1 Simple steam power plant operating on the Rankine cycle [2]. *(From R.E. Sonntag, C. Borgnakke, G.J. Van Wylen, Fundamentals of Thermodynamics, sixth ed., John Wiley and Sons, 2003.)*

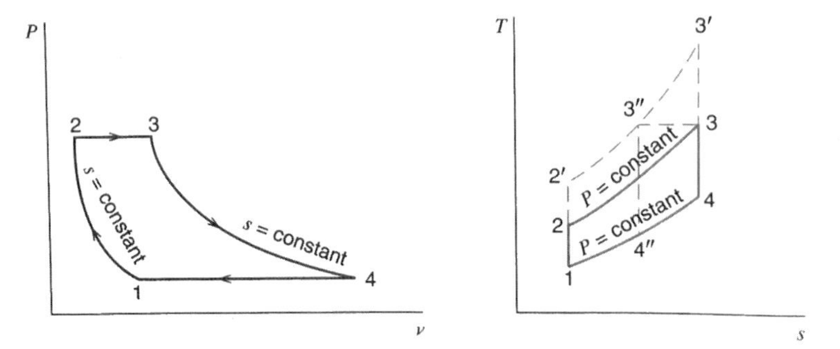

FIG. 2 Air-standard Brayton Cycle [2]. *(From R.E. Sonntag, C. Borgnakke, G.J. Van Wylen, Fundamentals of Thermodynamics, sixth ed., John Wiley and Sons, 2003.)*

Modern gas turbine power plants, however, usually operate as part of a combined cycle power plant. These plants use the waste gas from the turbine and use it to heat water into steam which is subsequently used to spin a steam turbine, thus combining the Brayton cycle with the Rankine cycle. This approach increases the efficiency of the plant over the Carnot limit for a single cycle and can often reach 60%.

8.2.3 Storage options for fossil fuel power plants

Thermal power plants can benefit from energy storage because it gives them greater operational flexibility. For instance, coal-fired power plants are commonly used to generate base load which requires high generating capacity and stability. However, these demands also indicate that coal-fired power plants are not very flexible in their operation. In particular, it is difficult to adjust their load because of long ramp-up and ramp-down times. One way to work around this problem is to add thermal energy storage to enable faster response without causing damage to certain mechanical components [3].

8.2.3.1 Compressed air energy storage

In a conventional gas turbine power station, about two-thirds of the turbine power output are needed to compress the air. In a CAES power plant, there is no compression during turbine operation because the required enthalpy is already contained in the compressed air. This leads to two advantages: first, cheap excess power can be used during low-demand hours to compress the air, and second, all the turbine power output is available for generation [4].

Compressed Air Energy Storage (CAES) for fossil fuel power plants can be broadly classified as diabatic, adiabatic, and isothermal. In general terms, a CAES plant has four basic components: a compressor train, a motor-generator, a gas turbine, and an air storage vessel which is usually an underground cavern. A schematic is shown in Fig. 3.

In adiabatic CAES (ACAES), the air is stored into a constant volume vessel where, as a result of the compression process, its temperature rises. For the process to be truly isentropic, it is necessary to avoid any heat transfer between the air and its surroundings. Thus storage vessels for ACAES applications are usually underground caverns that can hold large volumes of air and are very well insulated from the environment. Additionally, it is necessary to store the heat of compression in a separate medium until the air is needed to generate electricity. At that point, the stored heat is used to preheat the air before it is expanded

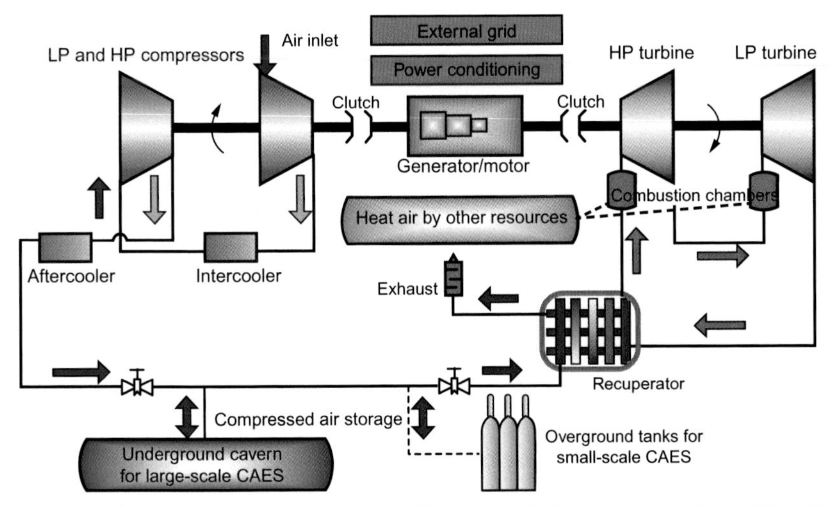

FIG. 3 Schematic of a Generic CAES Power Plant. *(From J. Wang, L. Ma, K. Lu, S. Miao, D. Wang, J. Wang, Current Research and development trend of compressed air energy storage, Syst. Sci. Control Eng. 5(1) (2017) 434–448.)*

through a turbine to extract its energy. The most obvious advantage of this system is that it eliminates the need for an additional heater to increase the temperature of the air before expansion. In theory, the efficiency of such a system should approach 100% but in practice heat losses are unavoidable and the actual achievable efficiency under ideal conditions is around 70%.

A diabatic CAES (DCAES) system differs from ACAES in that the heat of compression is not stored; instead, that heat is released to the atmosphere as waste heat. Since the heat of compression is no longer available to preheat the air, a natural gas burner is usually added before the turbine to bring the air temperature to the required turbine inlet temperature. In an isothermal CAES (ICAES) system, the temperature of the air is maintained constant, or near constant, during compression which allows for storage at higher pressures without temperature variations. This is advantageous because it may eliminate the requirement for an additional fossil fuel heater, much like the ACAES system described previously.

Adiabatic compressed air energy storage

Currently there are no adiabatic compressed air energy storage systems in operation though several have been proposed. For instance, Wojcik and Wang [5] considered the possibility of adding ACAES to a 294-MW combined cycle gas turbine (CCGT) power plant shown in Fig. 4. The power output of the gas turbine is 187.55 MW while the power output of the steam turbine is 105.90 MW. In their study, the ACAES system was allowed to charge for 8 h, hold the energy for 14 h, and then release it for 2 h. Note that the ACAES system uses a large underground cavern for air storage and two above-ground tanks for thermal energy storage (TES) of the heat of compression. The TES system is similar to the molten salt systems that are commonly used in concentrating solar power applications, but it uses Therminol VP1 as the storage medium [5].

The ACAES system is charged during periods of low electricity demand. During air storage, part of the air from the GT compressor is fed into the ACAES side of the plant effectively converting it into the first stage of compression. In this configuration, all other ACAES compressors are also driven by the gas turbine; thus during charging, the entire power output of the gas turbine is used for compression and the electrical output of the GT generator is zero. As this happens, the exhaust gas from the GT is used to keep the HRSG running at a low load. This is advantageous because it eliminates the need to shut down the HRSG during the night. Simultaneously, as the air is being stored, the heat of compression is removed and stored in the TES system [5]. During discharge, the heat stored in the TES system is used to preheat the air before expanding it through two expanders which gives the plant an additional power boost. Other advantages include lower thermal stresses on the HRSG, faster ramping, and less frequent plant shutdowns. The total power output of the plant during the

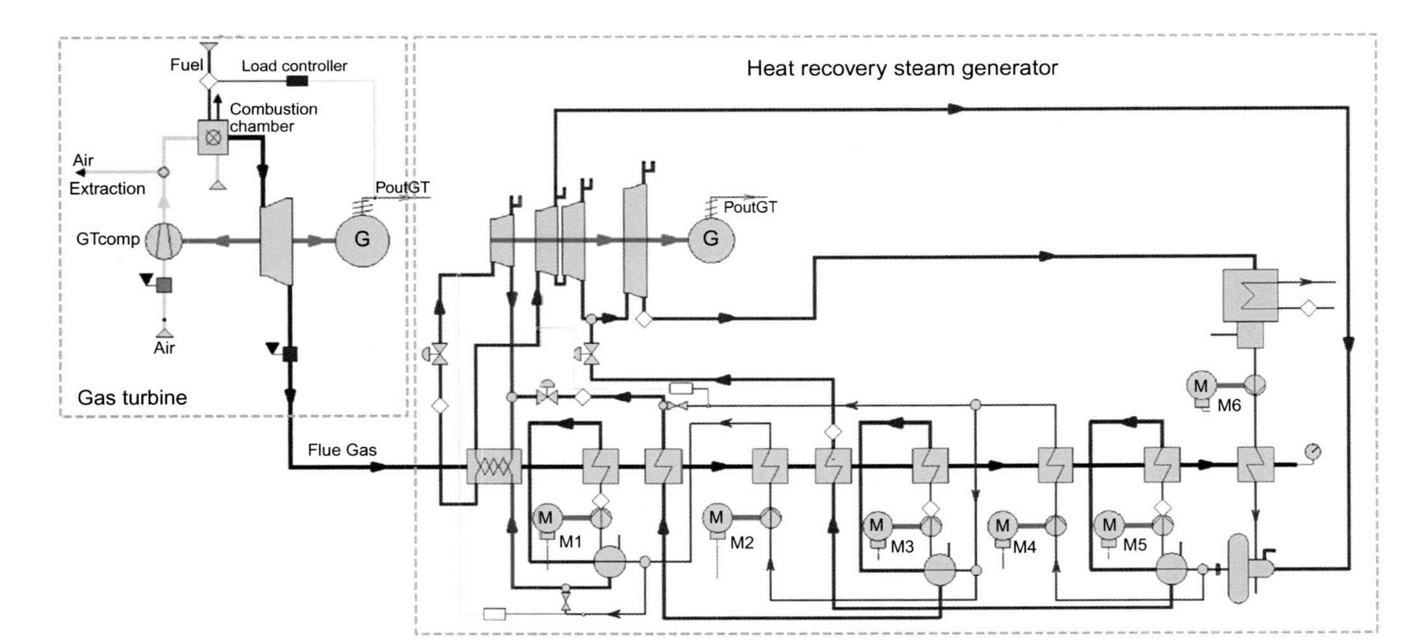

FIG. 4 Schematic of a hybrid 294-MW combined cycle gas turbine power plant with adiabatic compressed air energy storage. *(From J.D. Wojcik, J. Wang, Feasibility study of combined cycle gas turbine (CCGT) power plant integration with adiabatic compressed air energy storage (ACAES), Appl. Energy 221 (2018) 477–489.)*

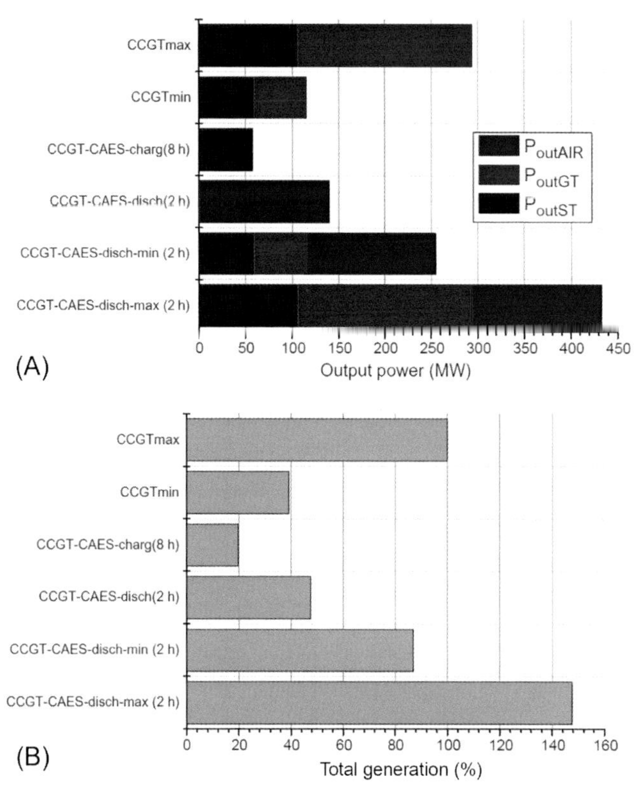

FIG. 5 CCGT-ACAES vs CCGT power output for all considering operating modes (A) in MW and (B) in %. *(From J.D. Wojcik, J. Wang, Feasibility study of combined cycle gas turbine (CCGT) power plant integration with adiabatic compressed air energy storage (ACAES), Appl. Energy 221 (2018) 477–489.)*

2-h discharge process can reach 433 MW; however, the overall efficiency of the hybrid plant is 2% lower than that of the stand-alone CCGT power plant [5]. Comparison of the power output and efficiency of the two power plants is shown in Fig. 5.

Diabatic-compressed air energy storage

Diabatic Compressed Air Energy Storage (DCAES) is a proven technology that has been in operation since the late 1970s. In DCAES systems, the charging process usually takes place during off-peak hours or whenever there is excess supply. At that point, air is compressed via a compressor train into a constant-volume storage vessel. This vessel is usually an underground cavern. When the stored energy is needed, the air is released and heated using a natural gas heater before expanding it through a turbine. The turbine is connected to a generator that ultimately produces electricity. Diabatic systems differ from their adiabatic counterparts primarily in that the heat generated by the compression is dissipated through intercoolers as waste heat. There are currently two power plants in the world that operate DCAES systems: one in Huntorf, Germany (290 MW, commissioned in 1978) and one in McIntosh, AL (110 MW, commissioned in 1991). Their efficiencies are reported to be 42% and 54% respectively [5].

The Huntorf plant has been hailed as an example of success owing to its longevity [4]. Its operational parameters are listed in Table 1. Today the plant is

TABLE 1 Operational parameters of the Huntorf CAES power plant.

Output	
• Turbine operation	290 MW (≤3 h)
• Compressor operation	60 MW (≤12 h)
Air flow rates	
• Turbine operation	417 kg/s
• Compressor operation	198 kg/s
Air mass flow ratio in/out	1/4
Number of air caverns	2
Air cavern volumes (single)	≈140,000 m^3
	≈170,000 m^3
Total cavern volume	≈310,000 m^3
Cavern location	
• Top	≈650 m
• Bottom	≈800 m
Maximum diameter	≈60 m
Well spacing	220 m
Cavern pressures	
• Minimum permissible	1 bar
• Minimum operational (exceptional)	20 bar
• Minimum operational (regular)	43 bar
• Maximum permissible and operational	70 bar
Maximum pressure reduction rate	15 bar/h

From F. Crotogino, K. Mohmeyer, R. Scharf, Huntorf CAES: more than 20 Years of Successful Operation, in: Spring 2001 Meeting, Solution Mining Research Institute, Orlando, FL, April 23–24, 2001.

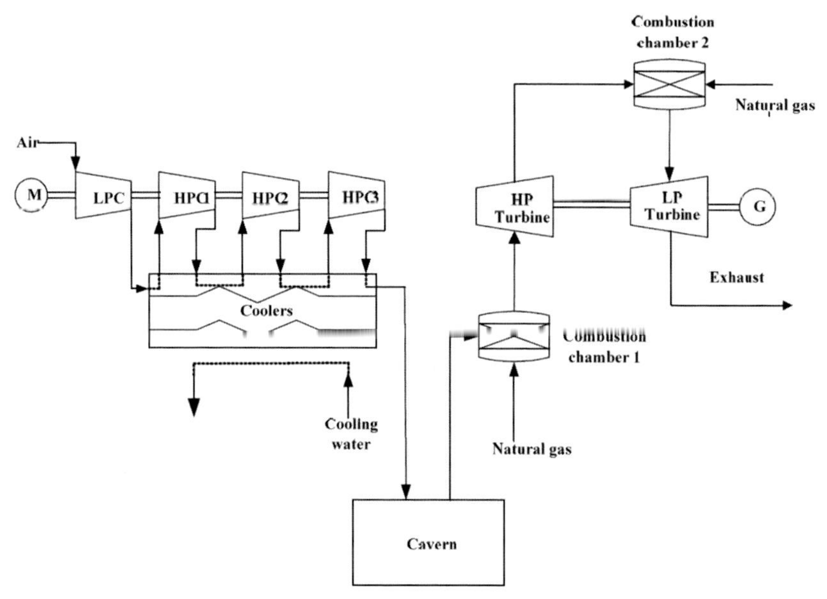

FIG. 6 Schematic of a recuperated CAES power plant. *(From W. Liu, L. Liu, L. Zhou, J. Huang, Y. Zhang, G. Xu, Y. Yang, Analysis and optimization of a compressed air energy storage—combined cycle system. Entropy 16 (2014) 3103–3120.)*

used mostly to generate short-term power while medium-load power stations are ramping up to meet demand (typically around 3 h) and for peak shaving at night after all pumped hydro generation capacity has been exhausted [4].

The McIntosh power plant operates on a recuperated cycle shown in Fig. 6. Here, a recuperator is used to preheat the air from the storage cavern with the exhaust from the LP turbine. To improve the efficiency of this system, Liu et al. [6] proposed adding a bottoming steam cycle. In this system, the waste heat from the LP turbine is used to generate steam through a heat recovery steam generator (HRSG) as in a typical combined cycle power plant. Since the heat of the turbine exhaust is no longer available to preheat the stored air, the recuperator is removed and an additional gas heater is added. Additionally, the heat from the intercoolers used to remove the compression heat is used to preheat the steam before it passes through the high-pressure turbine. The schematic of this system is shown in Fig. 7. The authors report that the addition of the steam cycle increases the efficiency of the plant by 10% [6].

Refs. [7, 8] present thorough overviews of CAES technology applied to fossil fuel power plants and other applications,

8.2.3.2 Thermal energy storage

Thermal energy storage systems can be categorized as sensible, latent, or chemical. In sensible heat storage, the thermal energy is stored directly onto the

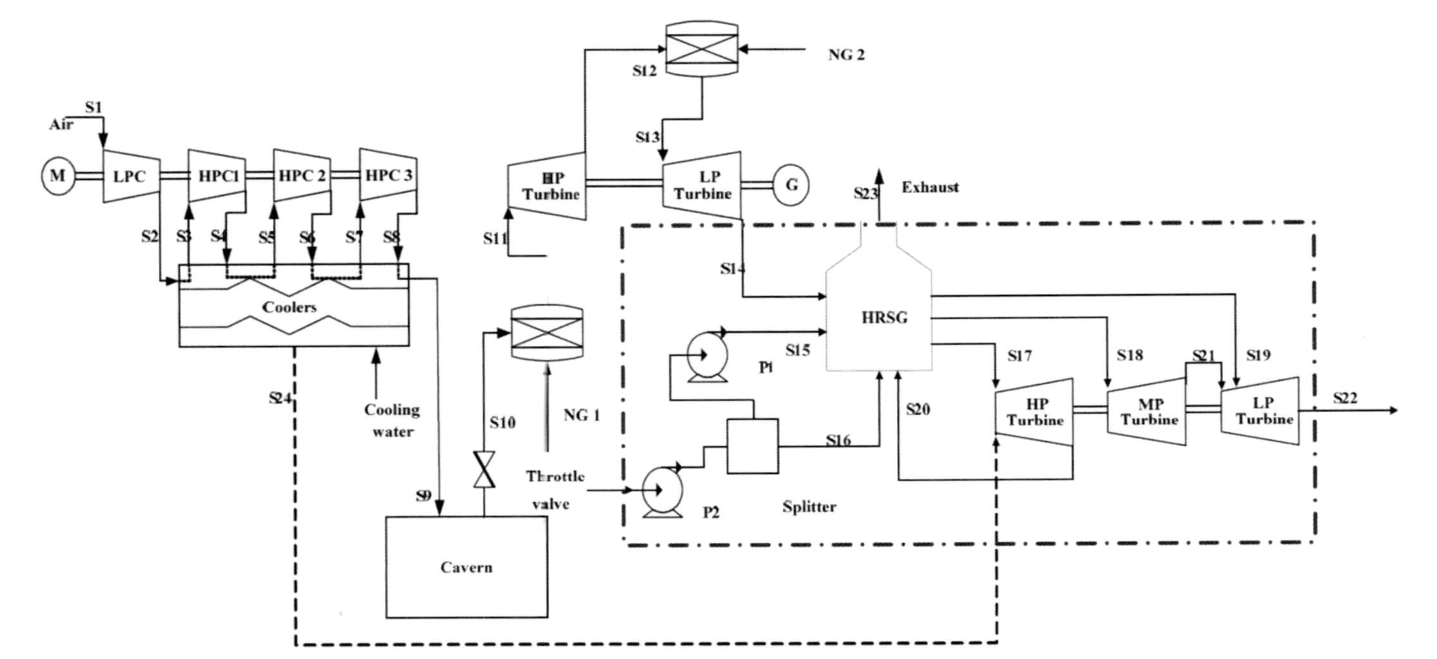

FIG. 7 Schematic of a CAES-CC power plant. *(From W. Liu, L. Liu, L. Zhou, J. Huang, Y. Zhang, G. Xu, Y. Yang, Analysis and optimization of a compressed air energy storage—combined cycle system. Entropy 16 (2014) 3103–3120.)*

storage medium by raising its temperature. The storage medium is usually a solid, such as concrete, rocks, or ceramics, or a liquid, such as molten salt or water. Latent heat storage systems use phase change materials to storage thermal energy in the latent heat. In these systems the phase change is usually solid-solid, solid-liquid, and less commonly liquid-gas. In chemical storage, endothermic reactions are used to store the heat until it is needed, then the direction of the reaction is reversed to release it. Thermal energy storage has been proposed for both conventional as well as combined cycle power plants.

Thermal energy storage for conventional power plants

Several studies have investigated the potential to add thermal energy storage (TES) to conventional power plants operating on the Rankine cycle. The benefits of such a system include decreased ramp times, added flexibility and response, as well as increased capacity.

Wojcik and Wang [9] proposed a generic TES system for a hypothetical 350-MW subcritical oil-fired conventional power plant shown in Fig. 8. In their study, the TES extracts superheated steam upstream of the main control valve, near the boiler reheater, or at the steam taps to the feedwater preheaters and returns it near the deaerator or the condenser as shown in Fig. 9. Their study did not advocate for any particular design or TES technology; rather, it indicated that a TES system retrofitted to a conventional oil-fired power plant could increase its efficiency by as much as 6% at 80% load [9]. The actual type of TES would be up to the plant's designers. Fig. 10 shows the increase in power output while Fig. 11 shows the plant efficiency under various discharge conditions.

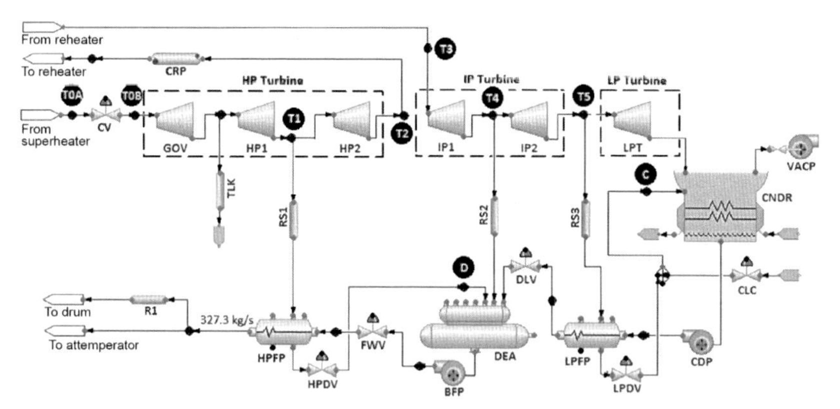

FIG. 8 Schematic of the power block of a generic conventional power plant without thermal energy storage. *(From J.D. Wojcik, J. Wang, Technical feasibility study of thermal energy storage integration into the conventional power plant cycle, Energies 10 (2017) 205.)*

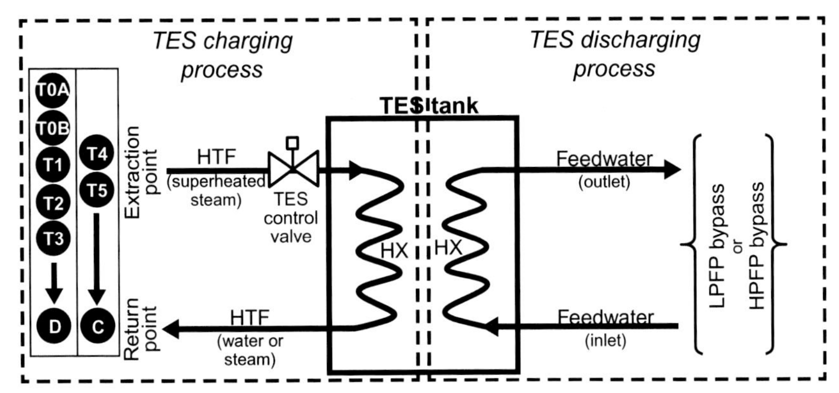

FIG. 9 Generic thermal energy storage system for a fossil fuel power plant. *(From J.D. Wojcik, J. Wang, Technical feasibility study of thermal energy storage integration into the conventional power plant cycle, Energies 10 (2017) 205.)*

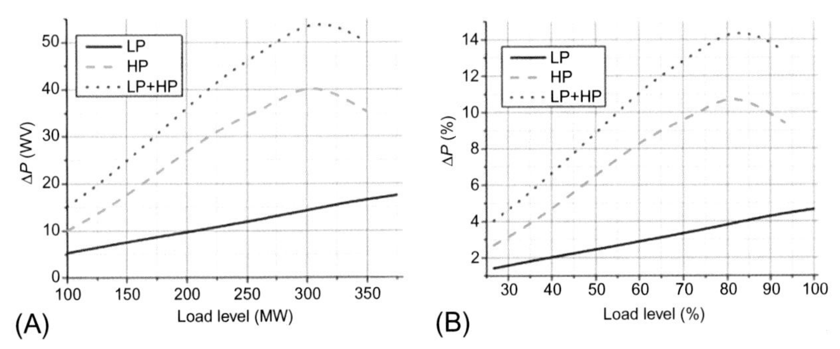

FIG. 10 Increase in plant power output with thermal energy storage under various discharge scenarios (A) in MW and (B) in % of registered capacity where, *LP*, low pressure feedwater preheater bypass; *HP*, high pressure feedwater preheater bypass; *HP + LP*, both LP and HP feedwater preheaters bypass. *(From J.D. Wojcik, J. Wang, Technical feasibility study of thermal energy storage integration into the conventional power plant cycle, Energies 10 (2017) 205.)*

Thermal energy storage for combined cycle power plants

In 1990 the Pacific Northwest Laboratory investigated the possibility of adding TES to an integrated gasification combined cycle (IGCC) power plant [10]. In its study, PNL proposed to add a molten nitrate salt TES that would be charged using the heat from the fuel gas stream and gas turbine exhaust. The stored heat would in turn be used to run a steam bottoming cycle during periods of high demand. This would enable the power plant to not only operate as a base-load power plant but to have peak-following capabilities as well. The IGCC power plant with molten salt TES is shown in Fig. 12.

This concept has both a high-temperature and a low-temperature TES. The high-temperature TES uses a mixture of sodium and potassium nitrate as the

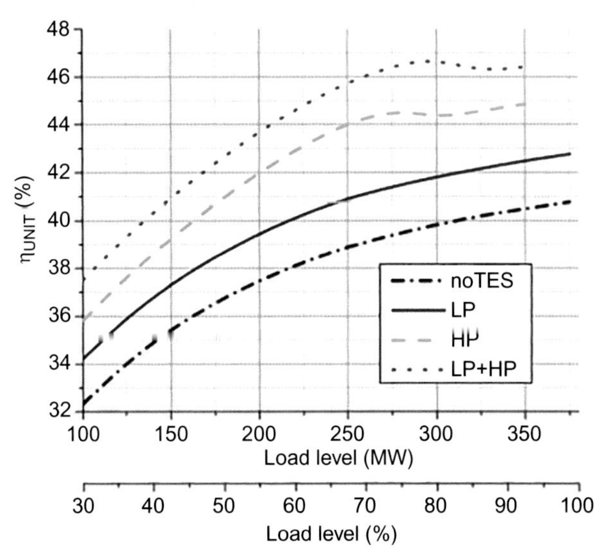

FIG. 11 Plant efficiency with thermal energy storage under various discharge conditions. *(From J.D. Wojcik, J. Wang, Technical feasibility study of thermal energy storage integration into the conventional power plant cycle, Energies 10 (2017) 205.)*

storage medium (60%–40%). The salt in the cold tank is stored at 280°C and pumped continuously through a salt heater, which uses the waste heat from the gas turbine, as well as a convective and a radiant fuel gas cooler to bring it to 566°C. The molten salt then returns to the hot tank until it is needed to produce superheated steam in the steam generator. The superheated steam is used to generate electricity in a conventional Rankine cycle. The low-temperate TES uses rock as the heat storage medium and a heat transfer oil (Caloria HT-43) as the heat transfer fluid in a single storage tank. Hot oil is continuously added or removed from the top while cold oil is continuously added or removed from the bottom. Thus the oil can be separated into hot and cold regions by a thermocline. The cold oil from the tank, which is initially at 121°C, is also heated using the waste heat from the gas turbine to bring it to 288°C before returning to the storage tank. The heat from the low-temperature TES is used to preheat the feedwater of the Rankine cycle before/after it leaves the condenser [10].

In its report PNL considered a power plant that generated base load continuously 24 h a day, from the coal-fired gasification cycle, and intermediate load for 6–16 h from the steam cycle. The peak net power output of the plant was assumed to be 500 MWe. The results of this study indicate that the levelized energy cost (LEC) for the IGCC plant with TES can be as much as 20% lower than the LEC for a conventional plant without TES [10].

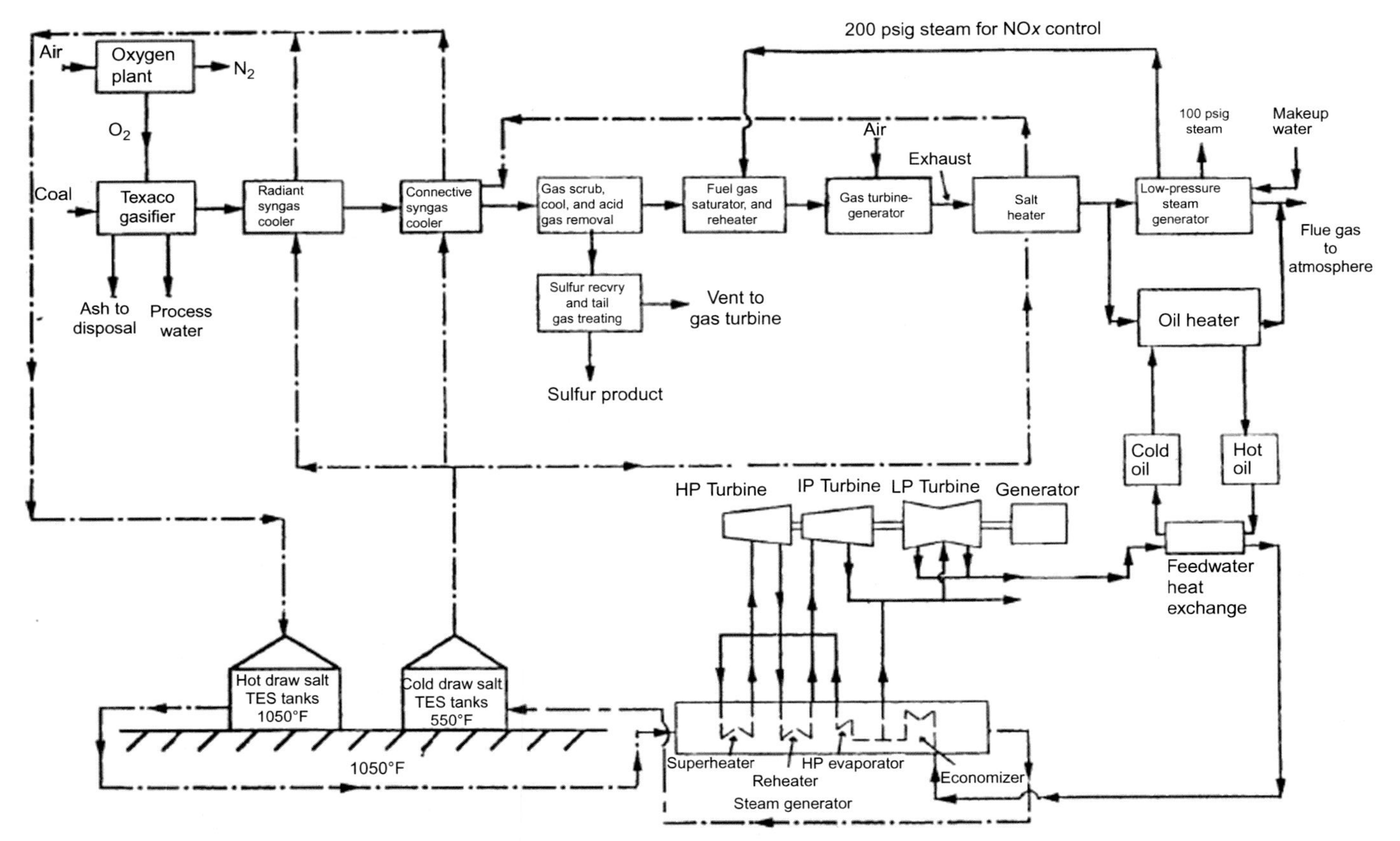

FIG. 12 IGCC power plant with molten salt TES. *(From M.K. Drost, Z.I. Antoniak, D. Brown, S. Somasundaram, Thermal Energy Storage for Integrated Gasification Combined-Cycle Power Plants. Pacific Northwest Laboratory, July 1990.)*

8.3 Thermal energy storage for nuclear energy systems

The main objective of this study is to present thermal energy storage integration concepts suitable for advanced nuclear energy systems and perform thermodynamic analysis of the integrated system. Various energy storage types are considered for different advanced reactor types which use different primary coolants Liquid metals, Molten Salts, and Pressurized Helium gas.

8.3.1 Challenges facing base load producing nuclear power

Currently, stand-alone nuclear power can only be used to meet base load electricity demands (safe operation requires steady reactor power) and consequently, has a lower return in comparison to other power sources. The rising use of renewable energy (wind, solar, etc.) is going to drive-up the grid supply skewness and therefore it may become economically necessary to use nuclear power to meet the peak demands as well. The primary economic challenge for existing Nuclear Power Plants (NPPs) is low revenue caused by the combination of natural gas and renewable energy sources [11]. Low natural gas prices place a ceiling on wholesale electricity prices. Large-scale solar or wind at times drives down the wholesale prices of electricity to near zero. This combination of conditions has resulted in closure of several nuclear plants. In favorable locations, the levelized cost of electricity (LCOE) from nondispatchable utility-scale wind and solar is significantly below nuclear. Operating costs, particularly the fixed Operation and Maintenance costs are lower. This change in technology has disrupted the electricity markets. For example, the wholesale electricity prices on a typical spring day in California have become extremely skewed from 2012 to 2017. In 2012 the marginal price of electricity was set by fossil fuels resulting in a relatively constant wholesale price except at times of peak electricity demand, an environment that favored base-load nuclear power plants. In the following five years, large quantities of photovoltaic (PV) were added that collapses the price at times of high solar output and low demand. However, the prices increase just before and after sunset when dispatchable fossil plants must ramp up and down to meet demand. While subsidies accelerated this trend, the large decreases in utility-scale wind and solar costs changed the market resulting in much more volatile wholesale electric prices. This favors electricity generators that can provide dispatchable electricity. Peak load requirements of nuclear power plants (NPPs) will continue to grow because of increasing skewness in grid power supply due to the increasing supply of intermittent renewable power sources to the grid.

8.3.2 Possible solutions

Under these conditions there are three possible scenarios for nuclear power plants: (1) to operate the plant at reduced load when grid demand is lower,

(2) to use the thermal energy produced by the reactor for any other process applications, (3) store the heat during low demand periods and use it at later times for peak electricity production when demand increases [12].

The first approach can only be advantageous if there are very low prices to produce excess electricity. This is because fuel costs contribute to very small fraction of the electricity production costs in an NPP. Even under negative pricing the use of NPPs to supply peak load demand is complex due to technical constraints associated with reactor behavior. These technical constraints lessen the flexibility of NPPs to follow grid load demand whereas most fossil-fueled plants (NPPs fossil fuel counterparts) can supply peak load demands by adding more fuel and, thus generate far more revenue during those peak hours. The technical challenges of using NPPs to supply peak load demands include the adequate handling of reactivity swings caused by time-varying fuel and moderator temperatures, a higher fuel-failure probability due to thermal-structural cycling, and spatial variations in fission product concentrations. Although there are presently some reactors around the world operating with flexible load-following capabilities, such operation is restricted to slowly varying powers, two to three times a day, and only up to 80% of the power level.

The second option of using nuclear heat for process industry is desirable and has been tried at several facilities in the world but it puts operational constraints on the process integrated as the industrial process plant has to be operated only during load reduction times and should be located near the NPP. For example, in addition to electricity, the Gösgen nuclear power plant has been supplying process heat to the adjacent cardboard factory in Niedergösgen since 1979. Approximately 150 GWh of process steam are extracted annually from the nuclear power plant. Recent attempts of such type have not been successful, and industry experts have suggested that presence of TES systems might improve the NPP integration potential to supply process heat [11]. The process integration can be made techno-economically feasible if energy storage can be leveraged to make nuclear heat truly dispatchable to any desired industry location and at any time of need. In other words, both options 2 and 3 are possible if there exists an efficient and economical mechanism to store heat.

8.3.3 Energy storage methods

Thus a more convenient and effective method to facilitate NPP load-following capabilities is to integrate energy storage with the NPPs. If grid demand is reduced, excess reactor thermal power or plant electrical power is stored in an integrated storage device. This stored energy can be released to the grid when demand is higher than what the NPPs can produce at 100% reactor power. Moreover, the economic prospects offered by energy storage integration are better than a load-following operation, which effectively reduces the reactor power over short time periods and, hence, reduces the total energy produced from NPPs over their lifetime and increases the levelized cost of electricity (LCOE). There are many options for storing energy from an NPP, broadly classified as follows: thermal, mechanical, and electrical energy storage options. Mechanical and electrical storage options include but are not limited to compressed air, pumped hydroelectric, superconducting magnets, batteries, and capacitors. Mechanical energy storage is any kinetic or potential energy stored within a device and generally these systems have very high to and from efficiency (can be close to 98%) but disadvantages include large space requirements and integration limitations. Electrical energy storage options can be easily integrated to an NPP as they do not impact Reactor systems or Balance of Plant, but are economically prohibitive for large-scale (~GWh) energy storage. Thermal energy storage is energy stored in the form of heat in well-insulated solids or fluids, as either sensible heat, stored within a single-phase media, or latent heat, stored within phase change materials. Thermal energy storage options include but are not limited to molten salts, solid media, phase change materials, thermochemical reactions, and steam accumulators. Thermal storage is much more economically competitive for NPPs as compared to electrical or mechanical energy storage options. However, the adoption of a thermal storage and integration with a NPP is largely dependent upon the operating and process conditions of the nuclear heat source and reactor coolant.

Nuclear plants produce heat; thus what is required is a low-cost system to send much of the heat to a storage device at times of low electricity prices and use that steam to generate peak electricity at times of higher prices. However, the adoption of a particular TES storage option is largely dependent upon the operating conditions of the nuclear heat source, reactor coolant, and market. The TES and mechanical energy storage integration can be very complex as compared to EES integration systems. For example, in terms of technology readiness and stand-alone risk associated with storage technologies, mechanical storage such as pumped hydro is one of the best choices; however, its selection puts a constraint on geographical and environmental factors [13].

Options which have been developed or are currently being researched include steam accumulators, sensible heat of a liquid or a solid, latent heat of fusion, reversible thermochemical reactions. The broad classification of these methods can be done into (i) sensible heat, (ii) latent heat, and (iii) thermochemical heat. All of these methods have their own relative advantages or disadvantages dependent upon the requirements. Sensible heat storage solutions have reached higher level of technological readiness than the other options. And large or grid-scale integration requires more robust solutions which imply sensible heat storage options are preferred over other systems. Also, it has been shown by Bindra et al. [14] that sensible heat storage has much higher round-trip efficiency for high energy density storage design as compared to solid-to-liquid phase change heat storage methods. In Fig. 13, various energy storage technologies, their readiness level, and their associated risk levels are described. Packed bed thermal energy storage with ceramic particles such as alumina is one of the low risks and high deployment potential technologies which are ready for commercialization [13, 15].

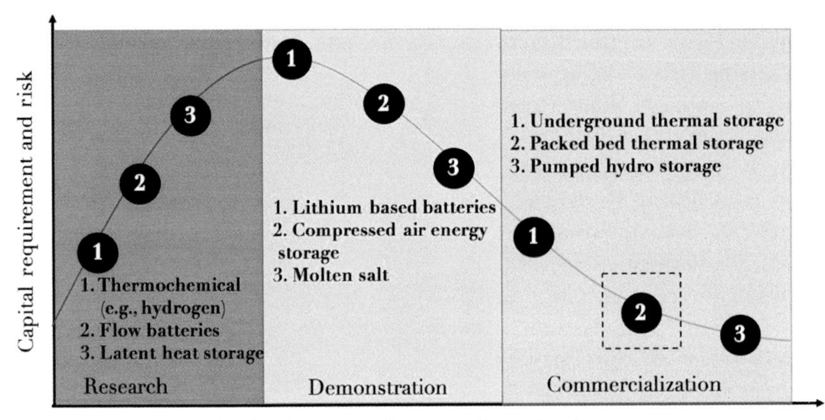

FIG. 13 Different energy storage technologies with their technology readiness and associated economic risk level for integration.

The critical step remains how to select and develop an ideal choice of heat transfer fluid or storage media. Existing NPPs operate using pressurized water and Next Generation Nuclear Power (NGNP) reactors use molten salts, liquid metals, or high-temperature gases as the main coolant. Currently, there are some thermal storage solutions such as molten nitrate salt, also known as solar salt (40% KNO_3 + 60% $NaNO_3$) and packed bed of alumina particles which present very low technological risk and a high deployment potential. These solutions can be good candidates for NGNP high-temperature reactors but have some limitations for integration into existing NPPs. This study evaluates conceptual designs for the storage integration of next generation high-temperature reactor concepts.

8.3.4 Energy storage integration

Thermal energy storage integration to the nuclear reactor system is complex, therefore thermal analysis of integrated system is required for performance and safety evaluation.

In NPPs the reactor coolant (RC) is considered as one of the intermediary layers for radioactivity containment, so for safety measures the coolant is generally not allowed to leave the containment building. This safety philosophy and large volume requirements postulate the TES integration to NPPs via heat exchange between reactor coolant and secondary heat transfer fluid. Thus for nuclear safety requirements the heat exchanger is housed inside the reactor containment building where HTF obtains thermal energy from reactor coolant. The hot HTF is brought out to store energy in a TES system. Nevertheless, the secondary reason for the need of this heat exchange process arises due to the compatibility issues of different TES systems and HTF requirements.

Existing NPPs The most widely implemented nuclear reactors throughout the world are Light Water Reactors (LWRs), which produce saturated steam to operate steam turbines on the Rankine cycle principles. LWRs are further categorized as Boiling-Water Reactors (BWRs), which directly produce saturated steam from heat of the reactor core and Pressurized-Water Reactors (PWRs). The PWRs have intermediate heat exchangers or steam generators which transfer the thermal energy of the pressurized water to the secondary coolant, i.e., water and convert it into steam to operate a steam turbine to produce electricity. Due to thermodynamic and heat transfer limitations, most of the LWRs produce steam at 280°C or less. At these temperatures, thermodynamic efficiency is close to 35% which is considerably lower than thermal power plants powered by fossil fuels.

The previous cited limitations will be less relevant for some future high-temperature reactor designs, which may easily provide the flexibility of adding on-site economical thermal storage.

Advanced Reactor Technologies In recent years the focus for new nuclear power plants has been on higher operating temperatures to improve overall plant efficiency and increase the outlet temperature of the main coolant. The NGNP high-temperature reactors that will be discussed are the Liquid metal (Sodium) Cooled Fast Reactors, Molten Salt Reactors, and the High-Temperature Gas-cooled Reactors, each generating electricity via Rankine or Brayton cycles with different primary coolants. Increasing the outlet temperature allows for other possible applications such as hydrogen production or petroleum refining. The efficiency of thermal energy storage options for next generation plants is also greatly improved, as well as the exergy efficiency by the higher operating temperatures.

Sodium-cooled fast reactors (SFRs) The SFRs use liquid sodium as a primary reactor coolant with normal operating temperatures in the range (350–550°C). SFRs operate at high temperatures and low pressures due to low volatility, meaning higher thermal efficiencies and improved passive safety. The working fluid for SFRs is water/steam which requires an intermediate heat transfer loop to avoid any accidental ingress of water/steam in primary sodium loop and reactor pool. The intermediate loop serves as a primary function of heat exchange between primary, intermediate, and working or secondary loops. Typically, this intermediate loop uses Sodium as the coolant in the design of SFRs.

Molten Salt Reactors (MSRs) These reactors utilize molten salt as the primary coolant circulating through the core which then passes its gained thermal energy to passing through an intermediate heat exchanger. Molten salts as the primary coolants for MSRs are high-temperature salts such as FLiBe or FLiNaK. The primary coolant transfers the heat during normal operation to intermediate loop which are designed with more common molten salts such as solar salt. Intermediate heat transfer fluid transfers the heat to the working cycle which can be Air-Brayton cycle or Water/Steam-Rankine cycle.

*High-Temperature Gas-Cooled Reactors (*HTGRs) are designed with pressurized helium (7 MPa) as the primary coolant flowing through the core and graphite as a moderator. The outlet temperature of the helium can reach temperatures of 850°C and higher, which leads to a high thermal-to-electrical energy efficiency. Thermal energy is transferred from the primary coolant (Helium) to the secondary side using a heat exchanger integrated to either a Helium-Brayton cycle or Water/Steam-Rankine cycle. HTGRs can store excess thermal energy via solid or liquid storage systems by directly extracting heat from the primary coolant or from the working fluid.

Table 2 describes the properties of three different coolants used in three major advanced reactor designs discussed here. These properties and operating conditions are then used for compatibility assessment and conceptual design of storage systems.

8.3.5 Sensible heat storage

As mentioned earlier, there are many options for storing thermal energy from a NPP. Solid and liquid sensible storage solutions have very high Technology Readiness Level (TRL) and both offer advantages for storing thermal energy from Advanced Nuclear Reactors. It has been shown by Bindra et al. [17, 18] that sensible heat storage has much higher exergy efficiency for high energy density storage design. Currently, there are two sensible heat storage methods presenting a very low technological risk and a high deployment potential. These methods are (1) molten nitrate salt, also known as Solar salt (40% KNO_3 + 60% $NaNO_3$) and (2) packed bed of rocks made of materials such as alumina, concrete, etc. Similarly, there are three high-temperature next generation reactor

TABLE 2 Thermal properties of coolants for the three discussed reactor types.

NPP type	SFR	MHTGR	PB-FHR [16]
Reactor coolant	Sodium	Helium	FLiBe
Inlet/outlet temp. (°C)	330/550	490/850	600/700
Density (kg/m^3)	800	2.7902	1889.15
Specific heat (kJ/kg K)	1.3	5.1903	2.4
Thermal conductivity (W/m K)	80	0.413	1
Viscosity (mPa s)	1	0.0526	7.524

concepts (SFR, MSR, and HTGRs) for which sensible heat storage integration is presented here. Specifically,

Two-tank Molten Salt Storage Liquids such as molten salt do not need a tertiary media as they have sufficient thermal capacity to be stored "as it is" when they are hot without any pressurization requirements. In order to keep the salt at hot or high-temperature state separated from the cold state, there are two possible designs—Single tank thermocline or Two-tank design. Although single tank design is more economical it is difficult to maintain thermocline in the tank during flow in or out of the tank. Therefore two-tank design is a preferred system design for the molten salt storage system where one tank is used to store the cold salt and other tank is used to store high-temperature salt. During the storage process, cold salt from the cold storage tank is circulated through the intermediate heat exchangers and then stored in the hot tank, thereby storing the excess thermal energy of nuclear reactor. Similarly, hot salt is circulated through the secondary heat exchanger where the stored energy is rejected to the working fluid during peak demand times and then resulting cold salt is returned back to the cold tank. The temperature difference between cold state and hot state of the liquid determines energy density. However, materials have constraints which limit these temperature differences such as high-temperature molten salts such as FLiBe or FLiNaK have very high melting points which compel the system to have an auxiliary heating mechanism such as electrical heating to ensure the molten salts remain in molten state even when the NPP is not operating or when not storing heat. This auxiliary heating system will lead to higher parasitic losses and can lead to lower efficiency for existing NPPs. However, solar (Nitrate) salts have much lower melting point, i.e., 220°C, and have higher technical feasibility which makes it preferred choice for intermediate loop coolant for MSRs.

Due to the presence of additional resistances in the heat exchanger, the actual storage inlet temperature is lower than the reactor coolant outlet temperature and thus is dependent upon the effectiveness of an indirect heat exchange process or design. Irrespective of the design details of heat exchange equipment, the heat transfer between two streams can be described using heat exchanger effectiveness which in turn can be used to compute the temperature of fluid stream entering the TES system during the storage cycle. Some of the values of heat transfer effectiveness and heat capacity ratio on the basis of primary and intermediate coolant or storage HTF loops are listed in Table 3.

Packed bed of solid rocks. Packed bed systems provide efficient thermal storage due to their large surface area resulting into high heat transfer effectiveness. High-temperature solids such as alumina are good candidates to store energy at very high temperatures (beyond the operating range of advanced reactor coolant temperatures). The major advantage of packed beds over liquid sensible heat storage solutions is that they provide similar or higher energy density and help in maintaining thermoclines by breaking the inertial effects of turbulent convection in thermal transport. This results in single tank system which

TABLE 3 Thermal properties for nitrate molten salt and alumina (at average temperatures for energy storage i.e., 900 K)

Storage type	Nitrate molten salt [19]	Alumina particles
Density (kg/m^3)	1790	3930
Specific heat (kJ/kg K)	1.560	1.195
Thermal conductivity (W/m K)	0.56	9.685
Viscosity(mPa s)	1.02	NA

reduces the capital and operational costs. However, typically they need a tertiary HTF to transfer heat from the RC. But in case of Solar Salt as the intermediate HTF deployed for MSR or SFR designs, the packed beds can be integrated within those systems (Fig. 14).

In this conceptual study a new approach is adopted as mentioned earlier to replace the intermediate loop coolants in SFRs which are typically sodium or sodium-potassium with solar salt. The major advantage of this adoption is that molten salt has better compatibility with storage media such as alumina and do not pose additional safety concerns. In addition, as the nonsolid fraction of the packed beds is occupied with the fluid phase, molten salts have much higher energy storing capacity. This new intermediate loop with packed bed TES integrated in SFR design is shown in Fig. 15.

The packed bed thermal storage performance involves heat exchange between the intermediate loop coolant which serves as a heat transfer fluid (HTF) and the solid storage media. Therefore modeling heat transfer between tertiary HTF and the TES (solids) involves an additional step [18, 20], thus an energy balance must be solved numerically to find the temperature of fluid exiting the packed bed during recovery. This behavior for large bed area and fluid flow passage can be described with one-dimensional thermal transport continuum mechanics approach and has been presented in detail in the previous work [17].

The system operating conditions for this packed bed TES can be described for storage and recovery cycles independently. In the storage cycle, hot HTF from heat exchanger enters from the one end and during the recovery cycle HTF enters at ambient temperature from the other end. More details on the model descriptions can be obtained from previous references [17, 18, 20]. The outlet temperature of HTF during the recovery cycle is then used to compute the exergy efficiency.

Therefore the overall exergy efficiency includes the thermal exergy efficiency, and the effect of parasitic losses in exergy destruction. As most of the parasitic losses such as work done to overcome pressure drop are typically independent of the storage inlet temperature, it can be inferred that higher inlet

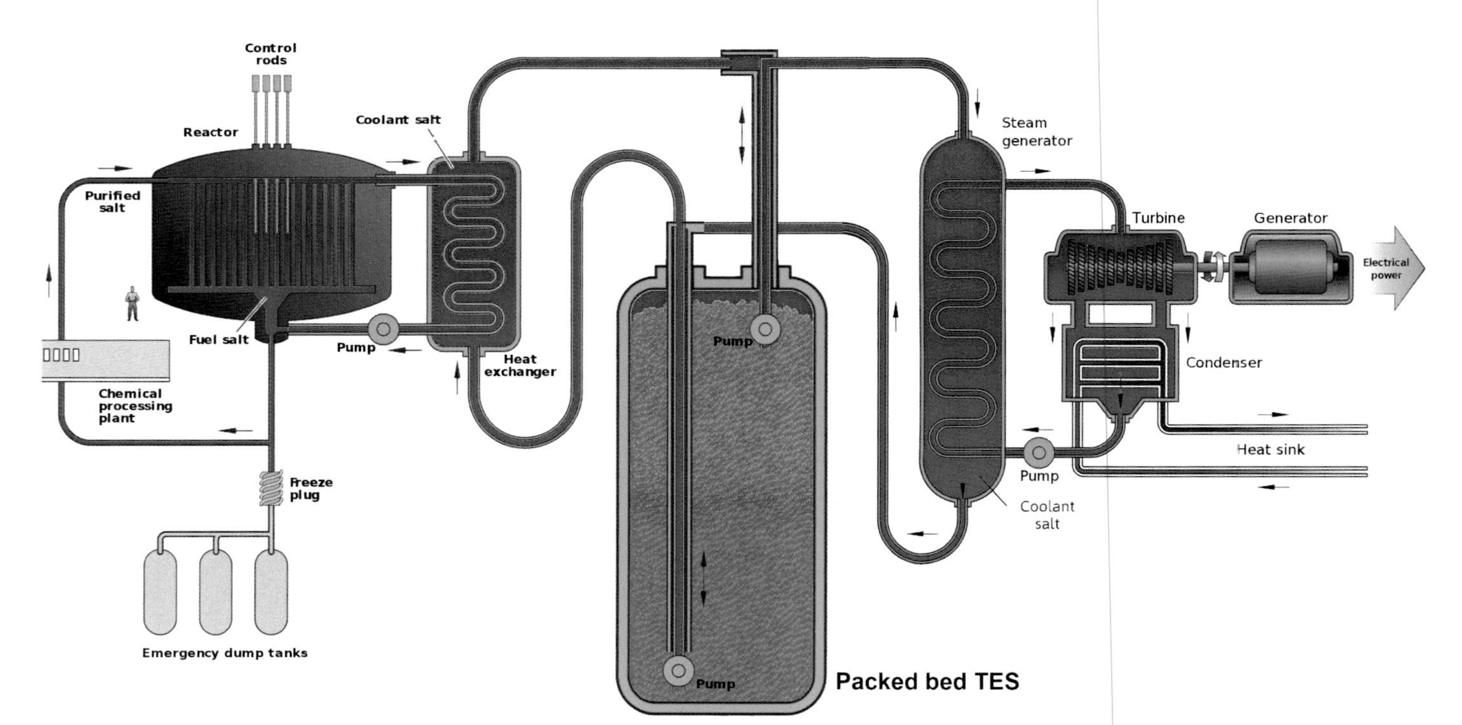

FIG. 14 Packed bed thermal storage system integrated with the MSR by incorporating packed bed storage tank within the intermediate loop.

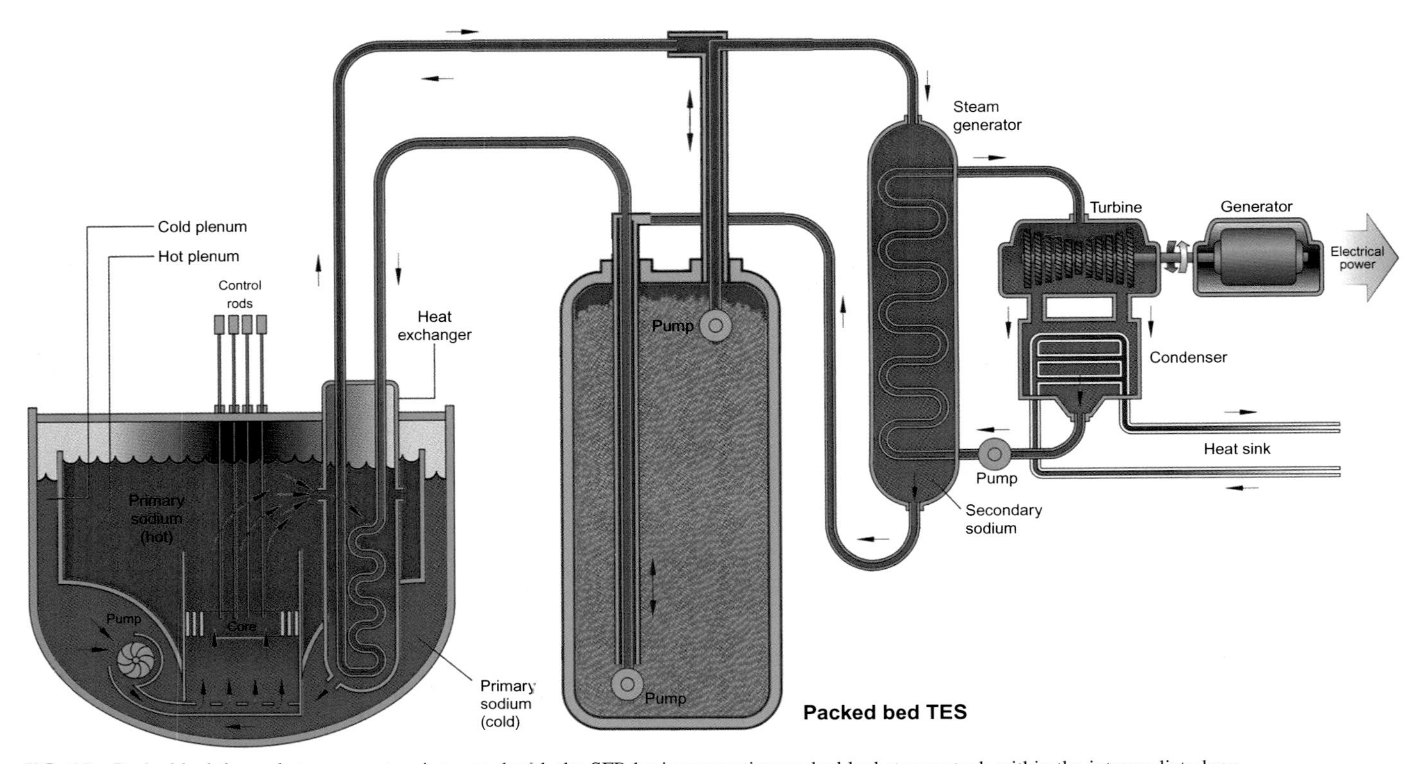

FIG. 15 Packed bed thermal storage system integrated with the SFR by incorporating packed bed storage tank within the intermediate loop.

temperature for storage will lead to lower fractional exergetic loss with the same amount of pressure drop through the storage bed. Therefore it is desirable to operate the system at maximum possible temperature for better efficiency or performance of thermal storage systems.

8.3.6 Exergy recovery and efficiency

For present study of system exergy quantification with storage integration, following assumptions are made: (a) the temperature difference between the RC and HTF streams remains constant throughout the flow path inside the heat exchanger, and (b) the overall heat transfer coefficient is only dependent upon the minimum heat transfer coefficient across the heat exchange equipment. Based on these assumptions and known effectiveness of heat exchange equipment, the inlet temperature for energy storage can be estimated. Therefore introduction of thermal storage reduces the temperature at which energy is produced as storage leads to exergy destruction due to heat transfer resistance. The effect of exergy reduction due to reduction in temperature of energy transfer reduces the energy ultimately going to the power block. In addition to thermal effects, the other effects such as parasitic energy losses lead to exergy destruction which needs to be quantified for efficiency calculations. Examples of such parasitic losses include the heating requirement to maintain molten salt thermal storage always in the liquid state, pumping power to move fluids, and pressure drop costs which are generally significant for packed bed type of thermal storage. The net fractional exergy recovery can thus be expressed as follows

$$\eta_{ex} = \frac{\Xi_{re} - 2W_{pd}}{\Xi_{st}} \tag{1}$$

$$\eta_{ex} = \frac{\Xi_{re}}{\Xi_{st}} - \frac{2W_{pd}}{\Xi_{st}} \tag{2}$$

$$\eta_{ex} = \eta_{th} - \eta_{pd} \tag{3}$$

where Ξ is the exergy of the process fluid and subscript *re* refers to the recovery process and *st* refers to storage process, η_{th} is thermal exergy efficiency and η_{pd} is fractional exergy destroyed due to parasitic losses, mainly pressure drop. In present calculations, W_{pd} is computed as the work required to compensate for the additional pressure drop. In the calculations shown here the fractional exergy destroyed due to pressure drop is approximated to be 6%.

8.3.7 Energy density

If total thermal energy density of the storage system is known it can be used to compute electricity units stored per mass of storage material. The calculations are performed for advanced reactor designs integrated with Alumina or Solar

(Nitrate) Salt storage system. In case of thermal storage, the costs are directly related to the ability of recovered heat to provide useful work. Thus an exergy model which takes into account thermal energy losses, additional work requirement, and entropy generation due to mixing or dispersion effects is developed. Based on the reactor operation temperature, cycle efficiency and exergy efficiency of the storage system can be determined as explained before. If total thermal energy density of the storage system is known, it can be used to compute electricity units stored per mass of storage material.

$$e_{d,th} = C_p(\Delta T_s) \tag{4}$$

where $e_{d,th}$ is the thermal energy density and ΔT_s is the change in the temperature across intermediate loop or storage system. Thermal energy density can be converted to effective energy density using,

$$e_d = e_{d,th}\eta_{exer} \tag{5}$$

where e_d is the effective thermal energy density based on exergy efficiency.

For conceptual analysis the design parameters of different reactor types were obtained from recently published reports. Kairos-FHR is one of the MSR technologies under development and the design parameters from this technology [21, 22] were used to perform thermal storage conceptual design and integration analysis. Similarly, for SFR and HTGR, reports [23, 24] published by DOE on the technologies under development were used. The values for the effectiveness along with the C_r values were set depending on the RC and HTF being integrated together. These values are given in Table 4 and the resulting exergetic efficiency of various thermal storage methods range from 75% to 80% if integrated with NPPs, as seen in Table 5.

The numerical values of storage inlet, outlet, exergy efficiency, and energy density are computed based on the method and model presented in previous

TABLE 4 Heat transfer effectiveness *E* and volumetric thermal capacity ratio (hot to cold fluid) C_r values for the varying RC and HTF proposed combinations.

NPP/TES combination	Effectiveness (*E*)	C_m
HTGR/solar salt—two tank	0.89	0.8
HTGR/alumina (solar salt—HTF)	0.89	0.8
SFR/solar salt—two tank	0.8	1.2
SFR/alumina (solar salt—HTF)	0.8	1.2
MSR/solar salt—two tank	0.6667	1
MSR/alumina (solar salt—HTF)	0.6667	1

TABLE 5 Performance of different thermal storage technologies integrated with corresponding NPPs, detailed earlier in the paper.

Storage type	T_{sin}	T_{sout}	η_{ex}	e_d
Solar (nitrate) salt				
HTGR	573	253	0.72	144
MSR	600	500	0.79	49.4
SFR	500	300	0.78	97.7
Alumina (solar salt HTF)				
HTGR	573	253	0.78	180
MSR	600	500	0.7857	56.79
SFR	500	300	0.8	115.6

Temperatures during the storage cycle are given in °C and thermal energy density e_d in kWh/m^3.

work along with the thermal properties for solar salt and alumina listed in Table 3. The numerical results presented in Table 5 show that in all scenarios the exergy efficiency of integrating storage is similar. HTGR integration with intermediate loop or TES sees a large drop in exergy efficiency as the relative Carnot efficiency drop is significant. While for higher temperature reactors the nitrate molten salt storage media has higher exergy efficiencies but the alumina packed bed storage has higher energy density. As the energy costs are directly related to these functions, alumina is expected to perform significantly better economically and thermodynamically for the HTGR or SFR. Thermal energy density is directly related to the ΔT_s across the TES system and therefore highest energy density results in case of HTGRs as they see a large temperature change within the primary coolant itself.

This work evaluates the exegetic performance of different materials and methods to store thermal energy of forthcoming NPPs. The study reveals that there are various possible options to store thermal energy of next generation NPPs efficiently. Low technological risk sensible heat storage materials such as molten salt, alumina can be integrated with high-temperature NPPs without any technical or economic constraints.

8.4 Concentrating solar power

8.4.1 Thermal energy storage

A very important characteristic of CSP technologies is their ability to provide power even when the sun is not shining. This is possible because most CSP systems can easily and inexpensively incorporate TES. In its simplest form, TES is achieved by storing a CSP plant's hot HTF in a large, insulated tank. Such a system is simple and efficient, but economic viability depends on the thermophysical properties and cost of the HTF.

The current commercial TES option for parabolic trough systems uses the previously discussed solar salt as the storage medium in an *indirect*, 2-tank system (see Fig. 16). Implementation of this TES system into parabolic trough power plants requires an indirect configuration—i.e., different HTF and storage fluids—due to the fact that the oil HTF is too expensive and volatile at high temperatures for extended durations to store directly and there are concerns with solar salt freezing in the solar field due to its high freezing point where it is to be used as the HTF. This indirect configuration has been deployed at dozens of plants in Spain, at the Solana plant in the United States, and at Noor plant in Morocco.

Implementation of 2-tank TES into molten salt power towers is accomplished using *direct* storage of the molten salt HTF (see Fig. 17). The direct configuration eliminates the need for the heat exchanger required with indirect TES, thereby reducing cost and increasing the performance of the TES system. Round-trip efficiency of TES in power towers has been estimated at 98% [25]. Direct TES is currently used in the 20 MW_e Gemasolar power tower in Spain and the 110 MW_e Crescent Dunes power tower in the United States.

The incorporation of TES allows CSP plants to extend or shift energy generation to coincide with peak electricity demand and provides operating flexibility and enhanced dispatchability. TES differs from electricity storage (e.g., grid-scale storage batteries) in several key ways. One difference is that TES is more efficient, when looking at energy round-trip efficiency in the storage system, and better suited to large-scale application than most types of batteries. A second difference is that TES systems have a lower capital cost, when

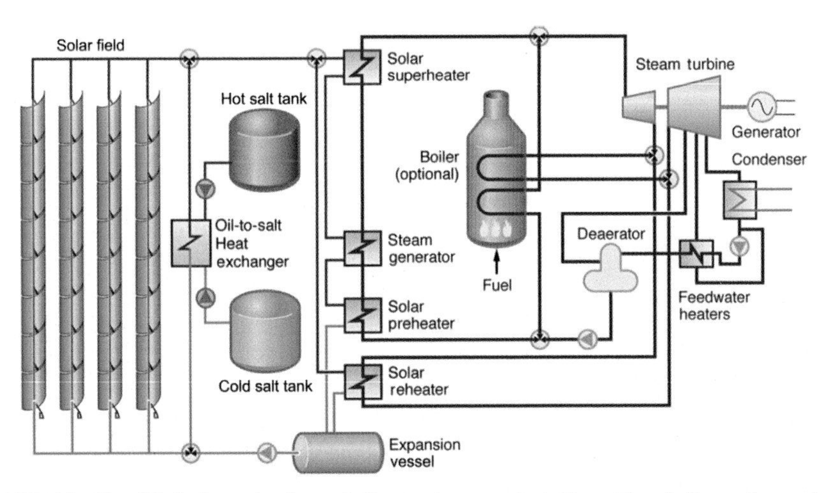

FIG. 16 Simplified schematic of a parabolic trough power plant. *(From Electric Power Research Institute, EPRI, Solar Thermocline Storage Systems: Preliminary Design Study, Electric Power Research Institute, Palo Alto, CA, 2010, p. 1019581.)*

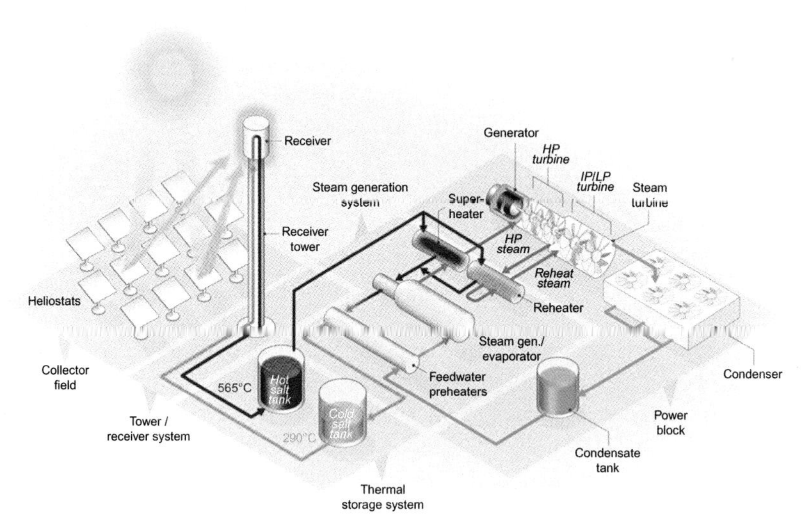

FIG. 17 Schematic of a molten salt power tower. Direct storage of the salt HTF simplifies this design. *(Image credit NREL.)*

integrated with a CSP plant, and have a longer duration, both in terms of the hours of energy that can be stored and the useful life of the storage system, than do batteries. At an estimated subsystem capital cost of $30/kWh$_{th}$ and a thermoelectric conversion efficiency of 41%, the storage cost for the CSP system equates to $73/kWh$_e$ [26], which is nearly an order of magnitude below current battery costs and less than R&D goals for future batteries [ARPA-e GRIDS]. However, a third difference is that a TES system only works because a CSP plant initially produces heat and then converts that heat to electricity. This means that the previously stated benefits for TES only are realized when associated with a CSP plant and these benefits may not be realized if TES were used as a stand-alone grid storage system.

8.4.2 Heat transfer fluids

The conceptual design closest to the current state-of-the-art CSP plant would use an advanced liquid HTF and direct storage of that fluid in a 2-tank TES system (see Fig. 17). Options for this fluid include salts and metals listed in Fig. 18. Another approach would use solid particles in a fashion analogous to a liquid HTF—the particles are heated in a specially designed receiver and stored in hot and cold silos as in a 2-tank system [28].

Development of receivers that directly utilize sCO_2 is also under investigation [28]. If successfully developed, this would allow the power cycle and receiver to share the same fluid. Such receiver designs must accommodate the combination of high temperature and high pressure required by the turbine. Pressurized packed

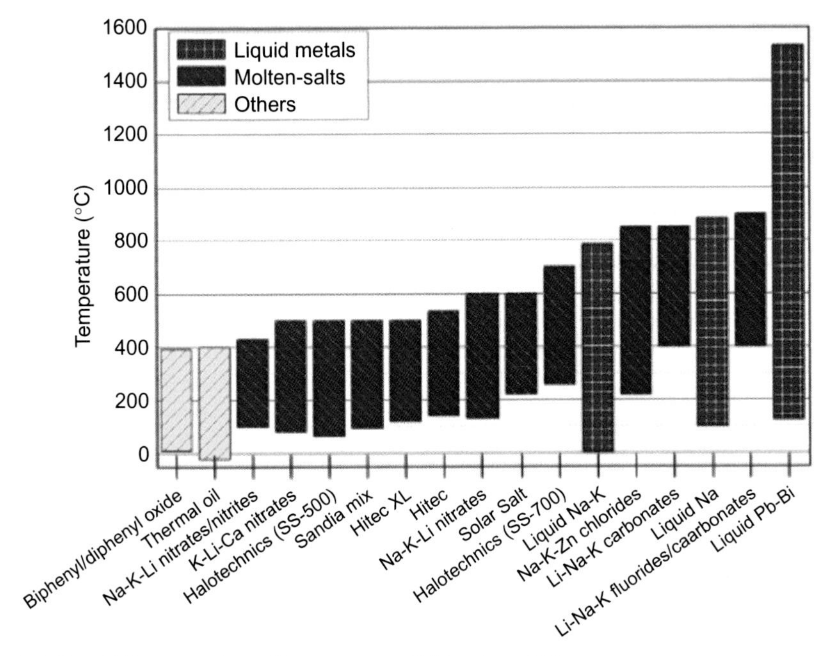

FIG. 18 HTFs proposed for high-temperature CSP systems [27].

bed systems utilizing a solid media, such as alumina, as the storage media have been proposed for use in conjunction with sCO_2 receivers [29].

8.4.3 Thermal energy storage and turbine ΔT

One means of integrating the sCO_2 power cycle into CSP systems is to simply replace the steam-Rankine power cycle of a CSP plant (e.g., Fig. 17) with an sCO_2 power cycle. Such integration could use existing solar salt; however, materials other than nitrate salts would be necessary to operate the system at temperatures beyond about 565°C. New materials under consideration capable of achieving increased operating temperatures in this type of CSP configuration include chloride salts, carbonate salts, and molten metals [27]. Since sCO_2 cycles are highly recuperated and turbine expansion ratio is limited, the temperature window for an optimized recompression cycle is relatively narrow. This works to the detriment of sensible heat storage systems, such as the 2-tank design.

A bottoming cycle may be considered to expand the heat source temperature differential, and thereby more efficiently integrate with 2-tank TES, by absorbing heat from the sCO_2 recuperator exit and lowering the returning temperature back to the receiver. The enlarged temperature difference increases storage energy density and therefore reduces the cost of the sensible energy TES.

Such a combined, or cascaded, cycle is considered by Pratt and Whitney/Rocketdyne in its "sCO_2 above sCO_2 cycle" configuration [30]. Various designs have also been modeled by Argonne National Laboratory [31]; however, all these combined-cycle designs increase power block complexity and capital cost. As an alternative, analysis at NREL suggests the partial-cooling cycle provides a better combination of efficiency and relative simplicity for sCO_2 systems interfacing with a sensible-media TES system [32].

The use of sCO_2 as the HTF in the solar receiver opens the possibility of other types of TES designs. One approach would flow the solar-heated sCO_2 through a bed of ceramic material such as alumina beads, which then acts as a thermocline for thermal energy storage. Internal insulation is suggested as a means to avoid having to provide pressure containment at the high temperature of the solar receiver and turbine inlet. A later variation of this design calls for staging the inlet and outlet flows along the length of the packed vessel to increase exergetic efficiency [29].

As an alternative to the sensible TES system, a phase change TES system that operates over a more narrow temperature window could be utilized. Aluminum and aluminum alloys are promising candidates for this type of TES system, with large heats of fusion in the 580–660°C range. The use of a metallic alloy also eliminates the thermal conductivity limitation found with salt phase change materials (PCMs). Some additional PCM options are the impregnation of a salt PCM within a porous, thermally conductive media such as graphite [33] or encapsulation of small amounts PCM in a thermally stable shell to be placed in a packed bed configuration [34]. Using phase change for thermal energy storage allows one to optimize the power cycle for a narrow temperature differential, thereby allowing integration with a simpler sCO_2 power cycle configuration. However, no commercial PCM systems have been integrated into an operating CSP plant regardless of power cycle.

8.4.4 Potential system designs and current research

8.4.4.1 Solar salt molten salt power tower

Replacement of the steam-Rankine power cycle in the existing design of a solar-salt power tower with the sCO_2 power cycle is conceptually straightforward and requires primarily the development and demonstration of the power system itself. CSP-specific components include the salt-to-CO_2 heat exchanger and air-cooled capability for the power cycle. While this is a low-risk path, the potential benefits versus the venerable steam power option are also low—system efficiency is estimated to be only marginally better by some [35] and worse by others [36]. At these temperatures, the greater benefit may be in power cycle flexibility and capital cost, but operating characteristics and costs for the sCO_2 cycle are still early stage estimates. The potential improvements in cycle efficiency and cost estimated by one team resulted in an estimated 8% reduction in levelized cost of electricity for an sCO_2 Brayton cycle plant operating with solar salt at 600°C versus the state-of-the-art steam Rankine plant at 565°C. The projected lower cost of the power block cost was partially offset by a required larger TES system cost [35]. As noted previously, Cheang et al. [36] concluded that the sCO_2 system was inferior to the current superheated steam-Rankine cycle at a TIT of 550°C. However, in addition to the lower temperature, the Cheang [36] study assumed a smaller capacity system that required a gearbox for the turbine/generator.

8.4.4.2 High-temperature tower with direct TES

Full realization of the power cycle's potential requires development of alternative heat transfer fluids with greater thermal stability than solar salt so the power cycle can realize higher operating temperature and efficiency. The leading candidates for this role include chloride salts [27] and carbonate salts [37]. Chlorides promise lower cost and a freezing point comparable with solar salt. Carbonates are inherently compatible with CO_2 and have thermal properties that are superior to solar salt, albeit with a higher cost and melting point. The corrosivity of new salts is a dominant concern, and the cost of the containment tanks may dominate the cost for high-temperature salt systems if corrosion cannot be managed. A comparison of these salts is shown in Table 6.

TABLE 6 Comparison of commercial solar salt with potential alternatives.

Property/salt	Solar salt	Chloride salt	Chloride salt	Carbonate salt
Typical composition (weight %)	60% $NaNO_3$ 40% KNO_3	68.6% $ZnCl_2$ 23.9% KCl 7.5% NaCl	62.5% KCl 37.5% $MgCl_2$	34% Na_2CO_3 33% K_2CO_3 33% Li_2CO_3
Melting point (°C)	~220	204	426	398
Heat capacity (J/g K)[a]	1.5	0.8	1.1	1.9
Density[a]	1.7	2.4	1.7	2.4
Max temperature (°C)	~585	850	850	~800
Estimated cost ($/kg)	~1	<1	<1	~1
Estimated cost ($/kWh$_{th}$)	14	~15	~10	~10
Reference	SQM	[27]	[38]	[37]

[a]*Typically measured near the salt melting point.*

An alternative direct storage concept being researched involves using flowing particles rather than a high-temperature liquid. Compared to liquids, solid particles can be very low cost and have excellent thermal stability. Conventional corrosion is of little concern, although abrasion and handling issues become important. The primary designs are open receivers with direct illumination of falling particles [28] and sealed receivers that use optically tailored channels to bring solar flux deep into the falling stream of particles [39].

8.4.4.3 High-temperature tower with indirect TES

Introducing a third thermal fluid into the configuration opens a new range of options for receiver design, TES design, and HTF properties. For example, if the HTF does not need to perform double duty as the thermal storage media, one can tolerate a higher specific cost to gain superior thermophysical properties. Molten-metal or gas-phase HTFs may become viable. Metals typically have excellent thermal conductivity and can have low melting points, but tend to be more expensive than salts. Liquid sodium and NaK are probably the best example, although blends based on lead/bismuth and tin have also been explored [27].

8.4.4.4 High-temperature tower with PCM

Transitioning away from sensible-heat TES designs to phase change materials may allow cycle designers to take advantage of the sCO_2 power cycle's preference for a relatively small temperature drop across the turbine and TES system. Aluminum metal (m.p. = 660°C) was an early favorite for this role; however, molten aluminum is notoriously difficult to contain in metal piping as it tends to alloy with most other metals. Work with salt-PCM systems has focused on impregnating graphite foam or other thermally conductive materials with the PCM and encapsulated PCM systems [40, 41].

8.5 Testing

This section describes the different phases and types of testing, current testing standards and procedures, and testing labs that perform ESS testing.

8.5.1 Phases of testing

There are four main phases of testing—research and development testing, production acceptance testing before shipment, commissioning and field acceptance testing, and field performance degradation testing.

Research and Development testing is often performed on system components at a bench or lab scale. These tests are sometimes specific to a given storage technology. As the system moves to higher Technology Readiness Levels (TRLs), more complicated tests are performed, leading up to a pilot and a field demonstration test for first-of-a-kind systems. Field demonstrations are key to determine the ESS performance in real-world operating conditions. The field demo will need to operate in all the different modes that can be experienced by ESS: load following, providing ancillary services, islanding, black starting, performing arbitrage, etc. These tests also demonstrate control system strategies, hybridization with other systems, maintenance requirements [42].

Acceptance testing of production equipment can be performed in the factory or at the customer site or both. System commissioning and controls integration testing will take place prior to final site acceptance testing. Due to the transient and variable nature of ESS, the controls integration is a very important portion of the test.

As is common for most energy systems, performance testing and health monitoring will need to be performed periodically over the life of an ESS. After an ESS is put into commercial service, periodic testing still needs to be performed to gauge performance degradation. Some systems are expected to have very little degradation over their lives, while others will have predictable degradation. These tests will determine whether a given system is meeting expected performance measures, and whether routine maintenance activities and/or component replacement are needed to restore performance levels if performance degradation exceeds a predetermined level. While field testing protocols for BESS are well documented [43], field degradation testing procedures for other systems may need to be developed for a particular system.

8.5.1.1 Testing standards and procedures

ASME Power Test Code (PTC) 53 Mechanical and Thermal Energy Storage Systems is a draft standard for testing mechanical and thermal ESS, and applies to many storage media including phase change, sensible heat, compressed air, gravitational, chemical, kinetic, and electrolytic [44]. PTC 53 describes putting a test boundary around an ESS in order to accurately measure power, energy, and efficiency, as shown in Fig. 19.

Sandia and Pacific Northwest National Labs have jointly developed a uniform test protocol for the DOE titled "Protocol for Uniformly Measuring and Expressing the Performance of Energy Storage Systems" [45]. This protocol is the most comprehensive in describing the different duty cycles that should be applied to EESes during testing. The protocol was developed collaboratively by government and industry over a several year period.

The IEEE (Institute of Electrical and Electronics Engineers) 2030.3 Standard Test Procedures for Electric Energy Storage Equipment and Systems for Electric Power Systems Applications covers many aspects of testing, but is focused toward certain ESS with only electrical inputs and outputs (e.g., no fuel inputs or thermal storage) [46]. IEEE 1679 Recommended Practice for the Characterization and Evaluation of Emerging Energy Storage Technologies in Stationary Applications is intended for new technologies that are still in development [47].

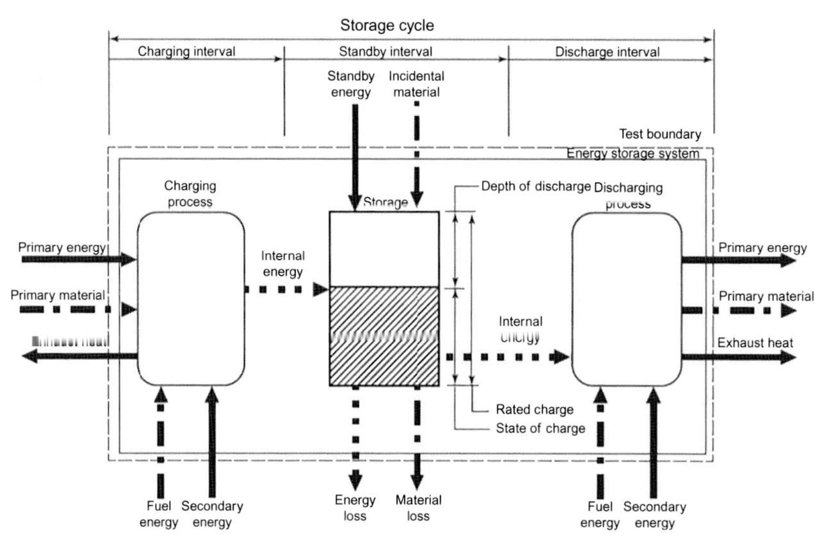

FIG. 19 Generic test boundary [44].

EPRI and the Energy Storage Integration Council (ESIC) have developed detailed test procedures geared toward utilities and other ESS operators in the ESIC Energy Storage Test Manual. This manual was developed around electrical and electrochemical storage. ESIC is a group made up of utilities, ESS developers, government, and research labs [48].

8.5.2 Types of tests and performance metrics

The Sandia/PNNL test protocol defines common performance metrics, which are described as either reference performance tests or duty cycle tests. The Reference Performance metrics include

- Stored energy capacity
- Round-trip energy efficiency
- Response time (see Fig. 20)
- Ramp rate
- Internal resistance
- Standby energy loss rate and self-discharge rate

Self-discharge rate is defined as the "rate at which an energy storage system loses energy when the storage medium is disconnected from all loads, except those required to prohibit it from entering into a state of permanent non-functionality" [45]. Table 7 lists typical self-discharge rates for several different ESS technologies, where self-discharge rate is measured in percent of energy

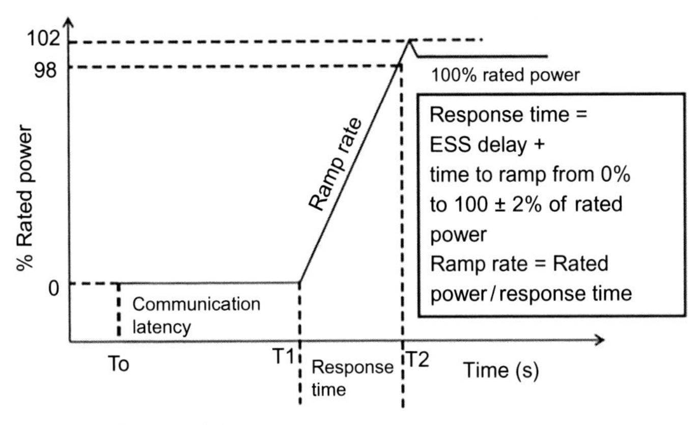

FIG. 20 Response time test [45].

TABLE 7 Self-discharge rate for various energy storage technologies.

ESS technology	Self-discharge rate (% energy lost/day)	Ref.
Pumped hydro storage	0%–0.02%	[49]
Compressed air ES	0%–1%	[49]
NaS battery	0.05%–1%	[50]
Li-ion battery	0.1%–0.3%	[50]
NiCD battery	0.67%–0.6%	[51]
Lead-acid battery	0.033%–0.3%	[51]
Vanadium redox flow battery	0.05%–1%	[50]
Zinc bromine flow battery	8%–33%	[49]
Supercapacitor	20%–40%	[50]
Superconducting magnetic ES	10%–15%	[50]
Thermal ES	0.05%–1%	[52]
Hydrogen	0%–4%	[50]
Synthetic natural gas	Negligible	[50]
Flywheel	20%–100%	[49]

lost per day. Note that standby energy loss is similar to self-discharge rate, but standby energy loss is measured with the main breaker closed and self-discharge rate is measured with the main breaker open.

Since ESS can provide benefits in many different modes of operation, different tests have been developed for each of these use cases, most of which occur under transient conditions. Duty cycles have been developed for these various applications, to measure response vs. system demand. The duty cycles identified are as follows:

- Peak shaving
- Frequency regulation
- Islanded micro-grid
- PV smoothing
- Volt-var support
- Renewables firming
- Power quality
- Frequency control

During the duty cycles, relevant performance metrics are measured, such as:

- duty cycle round-trip efficiency
- reference signal tracking
- State-of-charge excursions (max and min SOC during the duty cycle)
- energy capacity stability (energy capacity at any time in the cycle as a percent of the initial energy capacity)

In addition to grid testing, there are tests that are specific to certain types of storage. For example, thermal storage will include testing of the thermal system heat losses and CAES testing will include air cyclic testing of geological caverns to determine if there is any leakage.

8.5.2.1 Testing facilities

Third-party testing facilities are important for energy storage manufacturers and integrators who may not have the resources to build their own test facility. Independent facilities also offer the advantage of third-party evaluation and the use of common procedures (if they exist). Some test labs can be used both for development testing and for certification of production systems.

ESS testing facilities at independent labs can be used to prove out new technology, provide duration and cycle testing, develop transient controls, develop testing and safety procedures, determine a system's reliability, determine if a system meets performance standards, operate system in different modes (load following vs. island mode). Table 8 contains various system level testing facilities around the world.

TABLE 8 Energy storage system testing laboratory capabilities.

Organization	Facility	Location	Focus area	Max. power	Certification	Development	Ref.
Underwriters Laboratories	Battery and Energy Storage Technology Test Center	Crane, IN	Safety	1 MW	×		[53]
Intertek		Arlington Heights, IL	Batteries		×		[54]
DNV-GL (Det Norske Veritas Germanischer Lloyd)	Energy Storage Performance Test Lab	Chalfont, PA		2 MW	×	×	[55, 56]
Sandia National Lab	Energy Storage Test Pad & Energy Storage Analysis Laboratory	Albuquerque, NM	Grid scale	1 MW		×	[57–59]
National Renewable Energy Lab	Energy Systems Integration Facility	Golden, CO	Integration, Controls, Thermal	1 MW		×	[60]
Pacific Northwest National Lab	Energy Storage Reliability Test Laboratory	Richland, WA	Flow Batteries			×	[61]
Bonneville Power Authority	Energy Storage Test Facility	Vancouver, WA	Large Loads	5 MW		×	[62]
Electric Power Research Institute	Knoxville Test Facility	Knoxville, TN	Distributed	1 MW		×	[63]

Continued

TABLE 8 Energy storage system testing laboratory capabilities—cont'd

Organization	Facility	Location	Focus area	Max. power	Certification	Development	Ref.
Miramar Marine Corps Air Station	Installation Energy Test Bed	San Diego, CA	Microgrid, V2G	3 MW		×	[64, 42]
University of California San Diego	Center for Energy Research	San Diego, CA	Microgrid	1 MW		×	[65]
Alaska Center for Energy and Power	Power Systems Integration Lab	Fairbanks, AL	Microgrid	0.6 MW		×	[66]
Fraunhofer Institute for Solar Energy Systems	Center for Energy Storage Technologies and Systems	Freiburg, Germany	Thermal (−30°C to 550°C)			×	[67]
Institute of Engineering Thermodynamics Chinese Academy of Sciences	Energy Storage R&D Center	Beijing, China	CAES	10 MW		×	[68]
China Electric Power Research Institute	Energy Storage System Integration Test Lab	Zhangbei, China	Grid integration	2.5 MW		×	[69]

8.6 Energy storage codes and standards

Codes and Standards are certainly not considered the most exciting topic in the realm of energy storage. However, it is a necessity in the development, deployment, and operation of an Energy Storage System. Codes are often an afterthought in the development of a product but can provide guidance and reduce development time if embraced early in the process. The accelerated deployment of energy storage systems has increased the need and proliferation of Codes and Standards as they are deployed in new markets and applications. Safe growth as a result becomes highly dependent on codes and standardization for certification. These certifications allow the Authorities Having Jurisdiction (AHJ) to have assurance that the Energy Storage Systems being installed will protect personnel and equipment in a safe manner. The end customer also benefits from knowing that the product will deliver what is expected as the certification requires the product to be fully tested and validated to perform as advertised.

Codes, Standards, Certifications, and Safety Regulations have evolved over time out of necessity as new technologies have been introduced that can pose potential threats to society along with their benefits. Electrical safety codes were born out of the 1893 World's fair where inspector William Merrill was called into deem a large lighting display safe, and shortly thereafter formed Underwriters Laboratories. Due to the success of Underwriters Laboratories multiple organizations across the United States sprung up with different codes making it difficult for electricians to adhere to a single standard. "On November 6, 1896, representatives from each of these organizations came together the organization known as the National Fire Protection Agency" [70]. Shortly after the formation of NFPA the most fair criteria were chosen from the five separate standards and sent to over 1000 reviewers from which the National Electric Code was born. Since then, many organizations have formed internationally increasing the complexity of the regulation environment.

Certification of a system begins at the component level. In the early days of Battery Energy Storage Systems certification was only needed for each of the components such as the inverters and battery racks and were borrowed from other standards for UPS systems until application-specific standards were developed. The next level of certification comes at the assembly level. UL 9540 was one of the first adopted standards for the various types of energy storage systems that we are familiar with today. Additional standards for energy storage have been developed that are specific to the type of facility or application they are installed in and can vary by region.

Codes and Standards must be satisfied to cover the whole lifecycle of energy storage products. The adherence to codes starts in the design phase, followed by construction and testing, on to commissioning and operation, and finally through repair, rehabilitation, and demolishing. They most often being with development at the national level to address a specific new technology, issue,

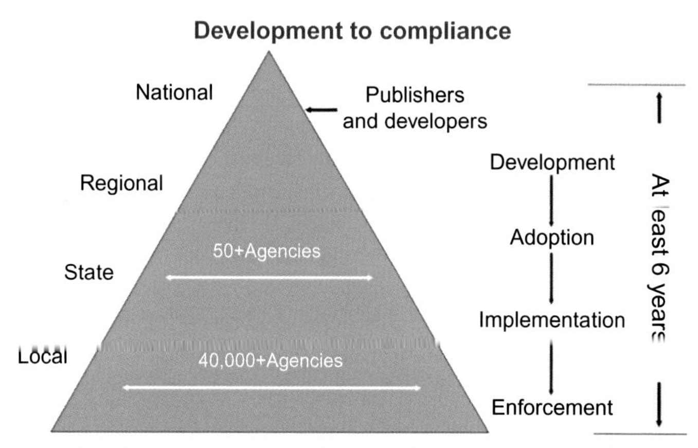

FIG. 21 Overview of development and deployment of Codes and Standards [71]. *(From https://energystorage.pnnl.gov/pdf/PNNL-23578.pdf.)*

or application. Then they are adopted at the regional/state level by different regulating agencies. Examples of this include different ISO/RTOs having slightly different requirements depending on application. The regulations are then implemented and enforced at the local level where AHJs will oversee the installation of an Energy Storage System to insure it is compliance (Fig. 21).

8.6.1 Safety standards and certification by type

Tables 9–14 contain Energy Storage Standards listed by project phase.

Knowing the Energy Storage Standards that must be followed by the technology being pursued and the region in which is installed is only half the battle.

TABLE 9 Energy storage system standards at component level.

Energy storage system components	
Energy storage system components	Standard
Molded-case circuit breakers, molded-case switches, and circuit-breaker enclosures	UL 489
Electrochemical capacitors	UL 810A
Lithium batteries	UL 1642
Inverters, converters, controllers, and interconnection system equipment for use with distributed energy resources	UL 1741
Batteries for use in stationary applications	UL 1973

From D. Rosewater, Energy Storage System Safety—Codes & Standards, EMA Energy Storage Workshop, Singapore, August 2015.

TABLE 10 Energy storage system standards at system level.

Energy storage systems standards	
Energy storage system type	Standard
Stationary energy storage systems with lithium batteries—safety requirements (under development)	IEC 62897
Flow battery systems for stationary applications—part 2-2: safety requirements	IEC 62932-2-2
Recommended practice and requirements for harmonic control in electric power systems	IEEE 519
Standard for interconnecting distributed resources with electric power systems	IEEE 1547
Recommended practice and procedures for unlabeled electrical equipment evaluation	NFPA 791-2014
Energy storage systems	NEC 706
Outline for investigation for safety for energy storage systems and equipment	UL 9540

From D. Rosewater, Energy Storage System Safety—Codes & Standards, EMA Energy Storage Workshop, Singapore, August 2015.

TABLE 11 Energy storage system installation standards.

ES installation standards	
Energy storage installation	Standard
Transportation testing for lithium batteries	UN 38.3
Safety of primary and secondary lithium cells and batteries during transport	IEC 62281
Shipping, receiving, and delivery of ESS and associated components and all materials, systems, products, etc. associated with the ESS installation	DOT regulations
Worker safety	Federal and state OSHA
Competency of third party field evaluation bodies	NFPA 790
Fire and smoke detection	NFPA 1, NFPA 101, NFPA 5000, IBC, IFC, state and local codes

Continued

TABLE 11 Energy storage system installation standards—cont'd

ES installation standards	
Fire suppression	NFPA 1, NFPA 13, NFPA 15, NFPA 101, NFPA 850, NFPA 851, NFPA 853, NFPA 5000, IBC, IFC, state and local codes
Fire and smoke containment	NFPA 1, NFPA 101, NFPA 5000, IBC, IFC, state and local codes
Ventilation, exhaust, thermal management, and mitigation of the generation of hydrogen or other hazardous or combustible gases or fluids	NFPA 1, IEEE/ASHRAE 1635, IMC, UMC, state and local codes
Egress/access/fire department access/ physical security/illumination (operating and emergency)	NFPA 1, NFPA 101, NFPA 5000, IBC, IFC, state and local codes
Working space	OSHA 29 CFR 1910.305(j)(7) and OSHA 29 CFR 1926.441 (if applicable), NFPA 70E, Article 320
Anchoring and seismic protection	NFPA 5000, IBC, state and local codes
Buildings, enclosures, and protection from the elements	IEC 60529, UL 96A, NFPA 5000, IBC, state and local codes
Signage	ANSI Z535, IEEE C-2, NFPA 1, NFPA 70E, NFPA 101, NFPA 5000, IBC, IFC, state and local codes
Emergency shutoff	IEEE C-2, NFPA 1, NFPA 101, NFPA 5000, IBC, IFC, state and local codes
Spill containment, neutralizing, and disposal	NFPA 1, IPC, UPC, IFC, IEEE1578, state and local codes
Electrical safety	IEEE C-2 (National Electrical Safety Code), NFPA 70E, FM Global DS 5-10, DS 5-1, DC 5-19
Communications networks and management systems	IEC 61850

From D. Rosewater, Energy Storage System Safety—Codes & Standards, EMA Energy Storage Workshop, Singapore, August 2015.

Once a design has been completed and built it must still be certified in order for it to be installed and commissioned. At this time there are two primary paths for certification of ESSs.

First, is through CE marking using IECEE CB Scheme, which is an international system for mutual acceptance of test reports and certificates dealing

TABLE 12 Energy storage system commissioning standards.

Commissioning standards	
Energy storage commissioning	Standard
Recommended practice for commissioning of fire Protection and life safety systems	NFPA 3
Building and systems commissioning	ICC 1000

From D. Rosewater, Energy Storage System Safety—Codes & Standards, EMA Energy Storage Workshop, Singapore, August 2015.

TABLE 13 Energy storage system operation and maintenance standards.

ES operation and maintenance	
Energy storage operations and maintenance	Standard
Hazardous materials storage, handling, and use	NFPA 400
Standard on maintenance of electrical equipment	NFPA 70B

From D. Rosewater, Energy Storage System Safety—Codes & Standards, EMA Energy Storage Workshop, Singapore, August 2015.

TABLE 14 Energy storage system incident preparedness standards.

Incident preparedness standards	
Incident preparedness	Standard
Standard for technical rescuer professional qualifications	NFPA 1006
Standard for fire fighter professional qualifications	NFPA 1001
Standard for fire department occupational safety	NFPA 1500
Standard system for the identification of the hazards of materials for emergency response	NFPA 704
Guide for substation fire protection	IEEE 979
Fire fighting	Emergency planning and community right-to-know act (EPCRA)
Fire and explosion investigations	NPFA 921
Fire safety concepts tree	NFPA 550

From D. Rosewater, Energy Storage System Safety—Codes & Standards, EMA Energy Storage Workshop, Singapore, August 2015.

with the safety of electrical and electronic components, equipment, and products [72]. The CB Scheme requires adherence to the applicable IEC Standards, does not including ongoing production inspections, and is a self-certification by the manufacturer [73]. Due to this flexibility in certification strategy this method would most likely receive more scrutiny from AHJs on installation.

A second path is to utilize UL in which there are two options. The first option is the UL ES Certification Program which provides a product level certification and ongoing production inspection subject to UL 9540. This is intended for standardizing product type energy storage systems where there is a repeatable product being produced and installed in a variety of applications. This is a costlier route initially but allows for rapid deployment through reduced project costs. The UL ES Certification Program has the most stringent requirements and less likely to be scrutinized by inspectors. The second option for UL certification is the UL Field Evaluation Program which is intended for one-off project certification. The intent of this program is to allow for systems of low Technology Readiness Levels that are likely to be modified after the initial prototype to be certified after the installation of the initial production. This path still utilizes UL 9540 but can allow the system to be installed, tested, and modified if required before receiving final certification inspection. Testing will often require temporary permit for operation especially if the unit is grid tied.

The best path for certification will be dependent on the technology type, readiness, and repeatability. The Codes and Standards landscape for Energy Storage is constantly evolving, and more standards are on the horizon.

References

[1] BP Statistical Review of World Energy, sixty-eighth ed., 2019.

[2] R.E. Sonntag, C. Borgnakke, G.J. Van Wylen, Fundamentals of Thermodynamics, sixth ed., John Wiley and Sons, 2003.

[3] Integration of Energy Storage in Thermal Power Plants, European Association of Gas and Steam Turbine Manufacturers, March 2017.

[4] F. Crotogino, K. Mohmeyer, R. Scharf, Huntorf CAES: more than 20 Years of Successful Operation, in: Spring 2001 Meeting, Solution Mining Research Institute, Orlando, FL, April 23-24, 2001.

[5] J.D. Wojcik, J. Wang, Feasibility study of Combined Cycle Gas Turbine (CCGT) power plant integration with Adiabatic Compressed Air Energy Storage (ACAES), Appl. Energy 221 (2018) 477–489.

[6] W. Liu, L. Liu, L. Zhou, J. Huang, Y. Zhang, G. Xu, Y. Yang, Analysis and optimization of a compressed air energy storage—combined cycle system, Entropy 16 (2014) 3103–3120.

[7] J. Wang, L. Ma, K. Lu, S. Miao, D. Wang, J. Wang, Current Research and development trend of compressed air energy storage, Syst. Sci. Control Eng. 5 (1) (2017) 434–448.

[8] J. Wang, K. Lu, L. Ma, J. Wang, M. Dooner, S. Miao, J. Li, D. Wang, Overview of compressed air energy storage and technology development, Energies 10 (2017) 991.

[9] J.D. Wojcik, J. Wang, Technical feasibility study of thermal energy storage integration into the conventional power plant cycle, Energies 10 (2017) 205.

[10] M.K. Drost, Z.I. Antoniak, D. Brown, S. Somasundaram, Thermal Energy Storage for Integrated Gasification Combined-Cycle Power Plants, Pacific Northwest Laboratory, July 1990.
[11] C. Forsberg, Light Water Reactor Heat Storage for Peak Power and Increased Revenue—Focused Workshop Report MIT-ANP-TR-170, Technical Report, Massachusetts Institute of Technology, 2017.
[12] M. Shannon, R.B. Bragg-Sitton, C. Rabiti, J.S. Kim, M. McKellar, P. Sabharwall, J. Chen, M.S. Cetiner, T.J. Harrison, A.L. Qualls, Nuclear-Renewable Hybrid Energy Systems: 2016 Technology Development Program Plan, *Idaho Falls (ID): Idaho National Laboratory, Nuclear Science and Technology Division (2016 Mar) Report No.: INL/EXT-16–38165. Contract No.: DE-AC07-05ID14517. Sponsored by the US Department of Energy*, 2016.
[13] Workshop Presentations, Thermal-Mechanical-Chemical Electricity Storage Workshop and Roadmapping Sessions, Technical report, Department of Energy, South-west Research Institute, February 2019.
[14] H. Bindra, et al., Thermal analysis and exergy evaluation of packed bed thermal storage systems, Appl. Therm. Eng. 52 (2) (2013) 255–263.
[15] B. Decourt, R. Debarre, Electricity storage. Leading the energy transition factbook, Schlumberger Business Consulting (SBC) Energy Institute, Gravenhage Google Scholar, 2013.
[16] C. Andreades, et al., Technical Description of the "Mark 1" Pebble-Bed Fluoride-Salt-Cooled High-Temperature Reactor (PB-FHR) Power Plant, UCBTH-14-002, Department of Nuclear Engineering, University of California, Berkeley, 2014.
[17] H. Bindra, P. Bueno, J. Morris, R. Shinnar, Thermal analysis and exergy evaluation of packed bed thermal storage systems, Appl. Therm. Eng. 52 (2013) 255–263.
[18] H. Bindra, P. Bueno, J. Morris, Sliding flow method for exergetically efficient packed bed thermal storage, Appl. Therm. Eng. 54 (2014) 201–208.
[19] M. Sohal, M. Ebner, P. Sabharwall, P. Sharpe, Engineering Database of Liquid Salt Thermophysical and Thermochemical Properties, Idaho National Laboratory, 2010.
[20] J. Edwards, et al., Exergy analysis of thermal energy storage options with nuclear power plants, Ann. Nucl. Energy 96 (2016) 104–111.
[21] M.D.B. Gabriel, S. Huang, M. Hackett, Qualifying Structural Graphite for Kairos Power Fluoride-Salt-Cooled, High-Temperature Reactor, INGSM, Burges, September, 2019.
[22] Kairos, Design Overview for the Kairos Power Fluoride Salt-Cooled, High Temperature Reactor, *NRC Report KP-NRC-1811-002*, 2018.
[23] G. Flanagan, F. Fanning, T. Sofu, Sodium-Cooled Fast Reactor (SFR) Technology and Safety Overview, INL/EXT-06-11057, https://gain.inl.gov, 2015.
[24] H.D. Gougar, C.B. Davis, Reactor Pressure Vessel Temperature Analysis For Prismatic and Pebble-Bed VHTR Designs, Idaho National Laboratory, 2006.
[25] J.E. Pacheco, Final Test and Evaluation Results from the Solar Two Project, SAND2002-0120, Sandia National Laboratories, Albuquerque, NM, USA, January 2002.
[26] System Advisor Model (SAM), Version 2015-06-30, National Renewable Energy Laboratory, https://sam.nrel.gov/.
[27] K. Vignarooban, X. Xu, K. Wang, E.E. Molina, P. Li, D. Gervasio, A.M. Kannan, Vapor pressure and corrosivity of ternary metal-chloride molten-salt based heat transfer fluids for use in concentrating solar power systems, Appl. Energy 159 (2015).
[28] C.K. Ho, B.D. Iverson, Review of high-temperature central receiver designs for concentrating solar power, Renew. Sust. Energy Rev. 29 (2014) 835–846.
[29] H. Bindra, P. Bueno, J.F. Morris, Sliding flow method for exergetically efficient packed bed thermal storage, Appl. Therm. Eng. 64 (2014).

[30] G. Johnson, M. McDowell, Issues associated with coupling supercritical CO_2 power cycles to nuclear, solar and fossil fuel heat sources, in: Presentation in Proceedings of SCCO2 Power Cycle Symposium 2009, RPI, Troy, NY, April 29–30, 2009.
[31] A. Moisseytsev, J.J. Sienicki, Performance Improvement Options for the Supercritical Carbon Dioxide Brayton Cycle, ANL-GenIV-103, Argonne National Laboratory, June 6, 2007.
[32] T. Neises, C. Turchi, A comparison of supercritical carbon dioxide power cycle configurations with an emphasis on CSP applications, Energy Procedia 49 (2014).
[33] W.H. Zhao, D.M. France, W.H. Yu, T. Kim, D. Singh, Phase change material with graphite foam for applications in high-temperature latent heat storage systems of concentrated solar power plants, Renew. Energy 69 (2014) 134–146.
[34] J. Stekli, L. Irwin, R. Pitchumani, Technical challenges and opportunities for concentrating solar power with thermal energy storage, J. Therm. Sci. Eng. Appl. 5 (2013).
[35] C. Turchi, 10 MW Supercritical CO_2 Turbine Test, Final Report under DE-EE0001589, National Renewable Energy Laboratory, January, 2014.
[36] V.T. Cheang, R.A. Hedderwick, C. McGregor, Benchmarking supercritical carbon dioxide cycles against steam Rankine cycles for concentrated solar power, Sol. Energy 113 (2015).
[37] M. Mehos, C. Turchi, J. Vidal, M. Wagner, Z. Ma, C. Ho, W. Kolb, C. Andraka, A. Kruizenga, Concentrating Solar Power Gen3 Demonstration Roadmap, National Renewable Energy Laboratory, NREL/TP-5500-67464 (January 2017).
[38] D.F. Williams, Assessment of Candidate Molten Salt Coolants for the NGNP/NHI Heat-Transfer Loop, ORNL/TM-2006/69, Oak Ridge National Laboratory, June, 2006.
[39] J. Martinek, Z. Ma, Granular flow and heat-transfer study in a near-blackbody enclosed particle receiver, J. Sol. Energy Eng. 137 (5) (2015).
[40] D. Laing, T. Bauer, N. Breidenbach, B. Hachmann, M. Johnson, Development of high temperature phase-change-material storages, Appl. Energy 109 (2013) 497–504.
[41] D. Singh, W. Zhao, W. Yu, D.M. France, T. Kim, Analysis of a graphite foam–NaCl latent heat storage system for supercritical CO_2 power cycles for concentrated solar power, Sol. Energy 118 (2015).
[42] Grid Energy Storage, U.S. Department of Energy, December 2013.
[43] K. Smith, M. Baggu, A. Friedl, T. Bialek, M.R. Schimpe, Performance and health test procedure for grid energy storage systems, 2017 IEEE Power & Energy Society General Meeting, Chicago, IL (July 2017) 16–20.
[44] ASME PTCMechanical and Thermal Energy Storage Systems, American Society of Mechanical Engineers 53 (2018).
[45] D.R. Conover, A.J. Crawford, J. Fuller, S.N. Gourisetti, V. Viswanathan, S.R. Ferreira, D.A. Schoenwald, D.M. Rosewater, Protocol for Uniformly Measuring and Expressing the Performance of Energy Storage Systems, SAND2016-3078R, Sandia National Laboratories, Albuquerque, NM (April 2016).
[46] IEEE Standard Test Procedures for Electric Energy Storage Equipment and Systems for Electric Power Systems Applications, in IEEE Std 2030.3-2016, pp. 1–72, 30 September 2016, https://doi.org/10.1109/IEEESTD.2016.7580998.
[47] 1679–2010—IEEE Recommended Practice for the Characterization and Evaluation of Emerging Energy Storage Technologies in Stationary Applications, Institute of Electrical and Electronics Engineers, 2010.
[48] Energy Storage Integration Council (ESIC) Energy Storage Test Manual 2016, Electric Power Research Institute, 2016.
[49] Electricity Storage and Renewables: Costs and Markets to 2030, International Renewable Energy Agency, 2017.

[50] SBC Energy Institute, Electricity Storage, September 2013, retrieved from https://www.cpuc.ca.gov/WorkArea/DownloadAsset.aspx?id=3170.
[51] D. Mooney, Large-scale energy storage, GCEP Research Symposium, Stanford University, pp. 14–15 (October 2015).
[52] X. Luo, J. Wang, M. Dooner, J. Clarke, Overview of current development in electrical energy storage technologies and the application potential in power system operation, Appl. Energy 137 (2015) 511–536.
[53] UL, Energy Storage Systems Testing and Certification, retrieved from https://www.ul.com/offerings/energy-storage-system-testing-and-certification.
[54] Intertek, Energy Storage/Battery Testing & Advisory Services, retrieved from https://www.intertek.com/uploadedFiles/Intertek/Divisions/Commercial_and_Electrical/Media/PDF/Battery/Energy-Storage-Single-Sheet.pdf.
[55] Det Norske Veritas GL, Energy Storage Performance Testing Solutions, retrieved from https://www.dnvgl.com/services/energy-storage-performance-testing-solutions-7266.
[56] P. Blume, K. Lindenmuth, J. Murray, Power Grid Energy Storage Testing Part 1, Evaluation Engineering, retrieved from https://www.evaluationengineering.com/applications/article/13006571/power-grid-energy-storage-testing-part-1 (1 November 2012).
[57] D. Rosewater, B. Schenkman, Energy storage test bed design issues, EMA Energy Storage Workshop, Singapore (August 2015).
[58] D. Rose, S. Ferreira, Energy Storage Testing and Validation-Independent testing of individual cell level to megawatt-scale electrical energy storage systems, Sandia National Laboratories, retrieved from https://www.energy.gov/sites/prod/files/ESTF.pdf (October 2012).
[59] D. Rosewater, S. Ferrerira, Energy Storage Test Pad and Energy Storage Analysis Laboratory, Sandia National Laboratories, 2012. retrieved from https://energy.sandia.gov/download/21150.
[60] Energy Systems Integration Facility, National Renewable Energy Laboratory, https://www.nrel.gov/esif/facility-tour.html.
[61] Pacific Northwest National Laboratory, Redox Flow Battery Laboratories, retrieved from https://www.pnnl.gov/redox-flow-battery-laboratories.
[62] A.A. Akhil, G. Huff, A.B. Currier, B.C. Kaun, D.M. Rastler, S.B. Chen, A.L. Cotter, D.T. Bradshaw, W.D. Gauntlett, DOE/EPRI Electricity Storage Handbook in Collaboration with NRECA SAND2015-1002, Sandia National Laboratories, Albuquerque, NM, February 2015.
[63] R. Schainker, B. Kaun, EPRI Energy Storage Testing and Demonstrations, EPRI Renewable Council Meeting, (April 2001) pp. 5–6.
[64] E. Wood, Microgrid Knowledge, Miramar Microgrid to Demonstrate One Solution to World's Waste Problem, retrieved from https://microgridknowledge.com/miramar-microgrid-landfill-waste/ (8 February 2019).
[65] W.V. Torre, Energy Storage Research at UC San Diego, CER Seminar, University of California San Diego (25 February 2015).
[66] University of Alaska Fairbanks, Alaska Center for Energy Power, Power Systems Integration Lab, retrieved from http://acep.uaf.edu/facilities/power-systems-integration-lab.aspx
[67] Fraunhofer ISE, Center for Energy Storage Technologies and Systems, retrieved from https://www.ise.fraunhofer.de/en/rd-infrastructure/center/center-for-energy-storage-technologies-and-sytems.html.
[68] Institute of Engineering Thermophysics, Chinese Academy of Sciences, Energy Storage R&D Center, 2018, retrieved from http://english.iet.cas.cn/Institute/6/.

[69] Retrieved from http://www.epri.sgcc.com.cn/html/eprien/col2016201118/2016-12/05/20161205145934234577996_1.html.

[70] E. Beach, The History of the National Electrical Code, retrieved from https://bizfluent.com/about-5062903-history-national-electrical-code.html (26 September 2017).

[71] D.R. Conover, Overview of Development and Deployment of Codes, Standards, and Regulations Affecting Energy Storage System Safety in the United States, Pacific Northwest National Laboratory (August 2014).

[72] IEC System for Conformity Assessment Schemes for Electrotechnical Equipment and Components, CB Scheme, retrieved from https://www.iecee.org/about/cb-scheme/.

[73] D. Rosewater, Energy Storage System Safety—Codes & Standards, EMA Energy Storage Workshop, Singapore (August 2015).

Chapter 9

Path to commercialization

David Voss[a], James Underwood[a], Jason Kerth[b], David K. Bellman[c], Kevin Pykkonen[d], and Kenneth M. Bryden[e]
[a]*Solar Turbines Incorporated, San Diego, CA, United States,* [b]*Siemens Energy, Houston, TX, United States,* [c]*All Energy Consulting, Houston, TX, United States,* [d]*Carbon America, Arvada, CO, United States,* [e]*Iowa State University, Ames, IA, United States*

Chapter Outline

This chapter begins with a description of the market conditions and the role that energy storage can play filling various market needs. Economic considerations and methods to quantify the financial benefits are introduced. Technology considerations that effect the commercialization of an ESS are reviewed. Finally, all of these topics are pulled together into a dispatch model that can be used to show the economic benefit of an ESS, with examples.

9.1 Market place

An engineer may have invented an energy storage device that is second to none but if one cannot sell that concept, then the invention will never see the light of

Thermal, Mechanical, and Hybrid Chemical Energy Storage Systems
https://doi.org/10.1016/B978-0-12-819892-6.00009-5

day. The path to commercialization will likely be full of starts and stops particularly in the electric sector where the markets are fragmented. Fragmentation is abundant from the varieties of potential customers to the rules and regulations which apply to this new invention. Market fragmentation essentially will limit the marketing of the product and the ability to sell it.

9.1.1 Conditions

In general, storage economics involves taking a product (electricity from the grid) when it has low value and then utilizing it at a time when it has more value. The price gap between the time of storage and the time of use is a function of the market construct. The famous duck curve in California (Fig. 1) shows how price can vary throughout the day. These daily variations will only become greater as Variable Renewable Energy (VRE) penetration of the market grows.

Another layer of complexity for the power markets: the cost of power for end customers typically involves some demand/capacity charge. In this cost component, storage can play a much larger role than generation sources. The demand/capacity charge is based on the fact that there is little to no storage in a given electricity market; therefore the generation source must be built to meet the very peak demand. Those who contribute to that peak demand should pay for those added peak units, which are very costly for the amount of energy produced. Those with storage can avoid demand charges by lowering their peak contributions.

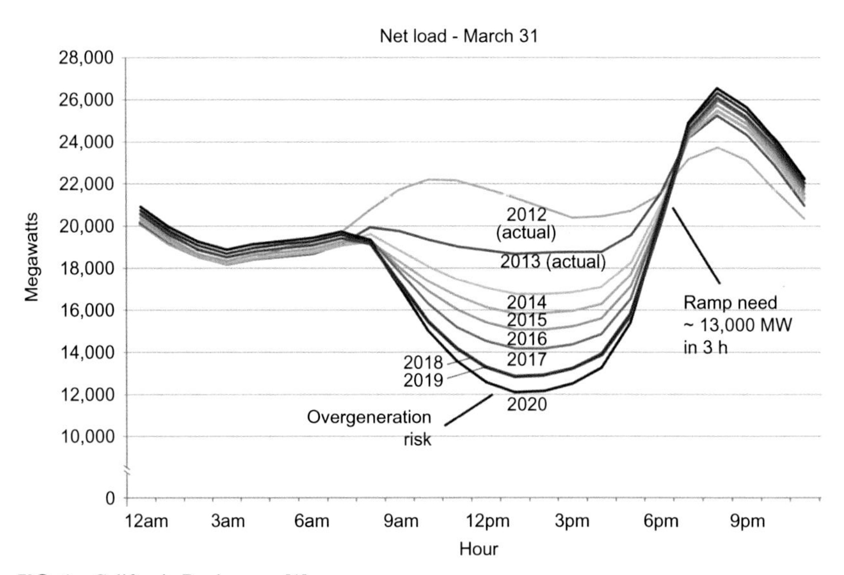

FIG. 1 California Duck curve [1].

9.1.2 Financial benefits

Energy Storage (ES) has many potential benefits to grids and grid-attached generators. In some cases, these benefits can be explicitly monetized through participation in organized market activities, including services such as Capacity or Frequency Regulation. However, most of the posited ES benefits do not have such paths to revenue production for the ES asset owner but may instead be monetized through the mechanisms of contractual arrangements such as Power Purchase Agreements (PPAs) with third parties in the case of merchant ES plants, or authorization to add ES assets to rate bases in the case of regulated rate of return utilities. ES assets can also produce reductions in Operations and Maintenance (O&M) costs for thermal generation plants (i.e., coal, natural gas, nuclear, or geothermal) through reduction or elimination of their need to ramp power output or cycle from generating to shut down states, which benefits accrue to the thermal plant owner but may not be reflected in revenue streams to the ES asset owner.

At the time of this writing, the largest recent ES deployments in the United States have been battery systems under the strong direction of the California Public Utilities Commission. Notably, the justification for these deployments has apparently not been strictly economic: in one instance speed of ES deployment was the paramount concern as mitigation for the impact of the Aliso Canyon gas storage problem, in the other instances there was a directive to PG&E to procure the lowest cost nonfossil fuel solution to local capacity issues. The rate-based utilities were allowed to add the ES assets to their rate bases. The purchase of these assets contributed toward fulfillment of California ES purchase mandates, as well. The relevant observation is that ES assets do not necessarily need to be economically competitive with alternative (fossil) capabilities in order to justify an investment by the eventual asset owner.

The following general discussion describes ES benefits and potential mechanisms for monetizing them to the benefit of an ES asset owner. The discussion is US centric and there may be significant differences in other locations. A more detailed dispatch model is discussed in Section 9.4 of this chapter.

9.1.2.1 Energy arbitrage

Energy Arbitrage is the purchase of ES charging energy at low prices, then its later sale at higher prices. Developers of ES assets that can take advantage of arbitrage opportunities need to closely examine Locational Marginal Prices (LMPs) at their intended point of grid interconnection to understand the economic potential of this revenue stream, perhaps through hypothetical dispatch of their assets against hourly historic prices.

Some grids have seen the surprising advent of negative energy prices at the wholesale level (i.e., the grid will pay loads to take energy) with increasing penetration of intermittent renewable resources. Despite this phenomenon, it appears ES assets often need other revenue streams beyond Energy Arbitrage to justify investment.

9.1.2.2 *Capacity payments*

The provision of "Capacity" is likely the most salient benefit of ES assets. Capacity is the ability to deliver dispatchable energy on demand for a defined time period. With increasing penetration of nondispatchable renewables on worldwide grids, Capacity is increasingly important to compensate for periods of low intermittent generation. Wind and solar energy sources often are the lowest cost sources of electric energy (when they are available). This consequently lowers the energy generation costs and prices during periods of peak generation. These periods offer low charging costs for ES assets, making them more competitive for Capacity services compared to traditional thermal generation assets.

Commitments for grid service such as a Capacity asset are usually over several year periods, meaning the Capacity assets must be available with no notice at any time during the commitment period.

In order for ES assets to provide Capacity services, they must be both capable of providing energy for the minimum requisite period and also be in a charged state at the time they receive a grid signal to generate. This has consequences both for the ES system cost and its operation. Currently, most grids require a generating asset be capable of 4 h of continuous discharge at a given power level to qualify as a Capacity asset, with the notable exception of PJM which recently changed its requirement from 4 to 10 h. From an operational perspective, an ES asset that was seeking revenue from Energy Arbitrage might not be able to also receive Capacity service payments because they would not be available for Capacity service if they were otherwise discharged through Energy Arbitrage activities.

With the exception of ERCOT (Electric Reliability Council of Texas) and CAISO (California Independent System Operator), other organized markets in the United States have periodic Capacity auctions through which resources are selected for participation in Capacity services. ERCOT has no Capacity market at all. In California, Capacity requirements are unfortunately fulfilled through opaque bilateral contracts between the Capacity asset owner and Capacity Adequacy obligees (i.e., the California Investor-Owned Utilities, Municipal Utilities, and Community Choice Aggregators) with the consequence that there is no public disclosure of Capacity pricing levels. Outside of organized markets, there are still technical requirements for Capacity as mandated by NERC (North American Electric Reliability Corporation). Utilities that operate in those markets fulfill their capacity obligations through either self-owned assets or contractual arrangements with asset owners.

9.1.2.3 *Ancillary services*

Ancillary Services are provided at the transmission level. Ancillary Service definitions vary by region, but are generalized here as Frequency Regulation, Spinning Reserve, Nonspinning Reserve, Black Start, and Voltage Support.

9.1.2.4 Frequency regulation

Frequency Regulation (FR) is the short-term modulation of the power output of a generator or ES asset according to a grid-generated dispatch signal which is set according to the momentary balance of gross grid generation to grid load. As the name implies, its objective is to regulate the frequency of the AC power on the grid to remain within acceptable boundaries as defined by NERC. The function of Frequency Regulation is necessarily independent of whether a grid is in an organized market (e.g., CAISO, PJM, Midcontinent Independent System Operator (MISO)) or not (e.g., Western Electricity Coordination Council (WECC)).

Generally in organized markets, qualifying ES assets interconnected at the transmission level (In-Front-of-the-Meter or IFOTM) may participate in FR markets. Some markets such as CAISO split FR services into Regulation-Up and Regulation-Down, with independent dispatch signals. Technical requirements vary by market/region. For instance CAISO rules specify a qualifying FR resource must be able to provide 60 min of service, which has a financial consequence for a participating ES asset both in terms of that asset's storage capacity as well as whether and how the asset is bid into the FR market. There may also be "accuracy" requirements imposed on participating resources, expressed as a measurement of the resource's actual performance compared to the FR dispatch signal.

Note in order for an ES asset to consistently participate in FR service, it is necessary for the FR dispatch signal to be "balanced" between Regulation-Up (i.e., discharging the ES asset) and Regulation-Down (charging the ES asset), else the ES asset will become "stuck" at fully charged or fully discharged states.

There have been discussions of whether and how to allow Behind-the-Meter (BTM) ES assets to participate in FR markets, a complex issue with many technical and market rule implications.

FR is a thin market which is rapidly saturated by ES assets, as was the case in PJM after the introduction of numerous utility-scale battery systems. Estimates of adequate FR service are in the range of 1% of peak grid load.

Spinning reserve

Spinning Reserve is the provision of standby "spinning" generation ready to commence generation within a few minutes of receiving a dispatch signal from the grid. The service definition reflects traditional thermal generating assets (e.g., a coal plant) which can take hours to "heat up," synchronize with the grid power frequency, and begin generation. In recognition of this, spinning reserve generators are compensated through this mechanism to use fuel to be in "hot standby," spinning and ready to quickly synchronize and generate.

In CAISO a Spinning Reserve resource must achieve full generation levels within 10 min of receiving a dispatch signal. Some ES assets may be capable of supplying Spinning Reserve services and market participation.

Nonspinning reserve

Nonspinning Reserve is an identical service to Spinning Reserve except the participating resource (e.g., a combustion turbine) is not already spinning. In CAISO the response time requirements for Nonspinning Reserve and Spinning Reserve are identical. Some ES assets may be capable of supplying Spinning Reserve services and market participation.

Black start

Black Start is a special service for designated generation assets which have been designed to establish an appropriate power signal on the grid after a blackout. Most generators need an established AC power waveform with which to synchronize before beginning generation. Black Start resources are capable of generating such a waveform from scratch, establishing the reference waveform from which other generators can subsequently synchronize to rebuild the entire intact grid. Some ES assets are conceptually capable of Black Start services, although this author is not aware of their use as such on wholesale grids.

Voltage support

Voltage Support is the provision of Reactive Power (Volt-Amps Reactive or "VARs") to the grid. On grids this is necessary to phase align the AC current and voltage waveforms in order to provide a greater proportion of (useful) "Real" power relative to (less/un-useful) "Reactive" power. Traditionally this function has been achieved by large-scale capacitor banks. Some ES assets, in particular those incorporating inverters, may be capable of supplying VARs to the grid and thus participating in these markets.

9.1.2.5 Other grid benefits

ES assets can provide beneficial grid functions beyond traditional generation analogues. In particular, location-specific ES assets can avoid or delay the need for transmission- or distribution-level investments by their owners, or improve the performance of the grid. Some of these applications have been named as "Transmission and Distribution Upgrade Deferral," "Transmission Support," "Reduced Transmission Capacity Requirements," and "Transmission Congestion Relief" [2].

Economic justification for these ES applications is generally by comparison to the next best (non-ES) alternative under the premise that some form of investment is necessary to provide the safe and effective future operation of the grid. Typically, it would be expected that the transmission or distribution asset owner would be the owner of the ES asset in this case, although it is conceivable a contractual arrangement with a separate ES owner could be arranged.

The unique characteristics of ES assets may be a complication for applications of this type. In general, the financial benefits of these applications accrue to the transmission or distribution owner. For competitive fairness reasons, transmission operators must conduct business at arm's length with generation resource owners, even if they are in fact part of the same company. If an ES system is deemed a generation asset, then the transmission owner could in fact not own and operate the ES assets. Likewise, if the ES asset were to be owned and operated by the generation side of the same business, they would be proscribed from coordinating with the transmission owner to operate the ES asset in the manner that would be most beneficial to the transmission or distribution system, defeating the purpose for the application.

9.1.2.6 *Behind-the-meter applications and small-scale grids*

In general, single-family dwelling Behind-the-Meter (BTM) ES applications need public funds such as California's Self-Generation Incentive Program (SGIP) to be economically deployed, perhaps in concert with the adoption of Time-of-Use retail electric rates. Larger Commercial and Industrial (C&I) applications may be more feasible for ES deployments as a mechanism to shift load and avoid onerous peak electric demand charges.

Furthermore, some C&I customers may have process heat requirements and preexisting thermal generation assets that are amenable to nonbattery forms of ES that are the subject of this book. The economic benefits of ES integration in such applications may be substantial but will be highly project specific and may involve consideration of the economic benefits to the C&I business that arise from, for instance, decoupling the production of electricity and the production of process heat and consequent opportunities to change production schedules or operating practices at the facility.

There is growing interest in deployment of (electrically) islanded "microgrids" outside of the United States and in some remote locations in North America. These concepts often include renewables plus ES assets as primary energy sources, backed by a lightly dispatched fossil generator for ultimate reliability. The delivered cost of fossil fuel (usually diesel) to these islanded locations may be sufficiently high to favor conventionally more expensive ES on microgrids, favoring longer duration ES systems to serve lower-probability periods of insufficient renewable energy availability on the grid.

9.1.3 Opportunities and scale

The scale of the ES product will drive the customer segmentation. Can a given ES technology be scaled for utility, industrial, commercial, and/or residential users? What market environment would be most conducive for the ES technology—regulated or deregulated markets? These questions are explored in the following section.

9.1.3.1 Utility scale

Utility-scale markets (5+ MW) would seem likely the easier segment to implement assuming a given ES technology can reach this scale. Many reasons make it perceived to be an easy path. The first is the cost on a per kW basis can be drastically reduced. Fixed costs are spread over a greater basis of energy. Another advantage, many utilities are financially backed by rate payers which would in theory remove some risk from implementing something new. Rate base can also be a disadvantage as it becomes more of a political game versus an economical game. In this paradigm, lobbying may be seen as having a better return on investments than technology investments. Actions from regulated markets are typically skewed toward keeping status quo, not at taking a chance to making the world a better place, given the risk of loss. The utility-scale market will participate in the wholesale market, which allows forward investment hedging. They will already have an active trading desk and corresponding banking relationships. In theory, they could lock in the market value of load shifting solidifying the financial of the project.

9.1.3.2 Industrial scale

Industrial-scale markets (1–500 MW) are likely a very hard sell as the core competency of the industrial customer is on making the widget. A disturbance in a secondary source—nonetheless a cost center—does not bode well in selling an implementation. No matter how compelling your storage technology is you must take into account the industrial plant's production goals. They are continually making capital investments and they have a choice to invest in making more widgets or potentially lowering the cost of widgets. Many times, making more widgets is the strategy chosen. In this mind-set, the lowering of the cost of widgets is no guarantee as other factors going beyond your storage technology could cause prices to rise, e.g., weather, commodity price, outages, etc. The cost of electricity on a widget also may be relatively small compared to the other costs such as feedstock. Financial hedging is somewhat limited as the energy component is likely the only hedgeable component. An active power trading desk is likely limited in scale if it even exists in the company.

The best value to an industrial customer for storage will likely be energy security. Facilities that already invest in backup generation will likely be targets for storage, but at the same time, they typically already have a sunk cost in backup. Rate structures for industrial customers are, in general, more focused on peak capacity than energy which can be a strong selling point for energy storage versus renewable energy.

9.1.3.3 Commercial scale

The Commercial-scale market (1 kW–5 MW) is a fragmented arena, but, in general, you will see a rent versus ownership dynamic. Many commercial customers do not own the infrastructure; therefore they really have no control on these decisions. The property owner often has no incentive/desire to make energy cost savings for their tenants, as most tenants decision to lease are not focused on energy cost, but on locations and amenities to their customer. This breakdown is one of the reasons for government standards. For example, efficiency standards were enacted to make the developers and landowners buy more efficient appliances and devices, causing a trickling down in customer usage; without government programs this may not have occurred for this sector. A storage device at this level will likely have to be marketed like the industrial customer as energy security. There are more varying rate structures for commercial class customers. Some rates will be focused on demand peak charges vs energy. Typically, this customer class sees the highest electricity prices as they do not have a lobbying force comparable to the industrial class lobbyist and the residential-focused consumer advocacy groups. The rate complexity makes it very hard to financially hedge this customer class.

9.1.3.4 Residential scale

The Residential-scale market (<1 MW) is a group where cost can be 3–5 times higher than utility scale as result of fixed cost and small energy usage. You can see this in residential installs of solar where all the various groups have a fixed cost to the customer no matter the size of the install—leading to a high $/kW for installation. In fact, the cost of the panels themselves likely represents less than 20% of the project cost. These costs could be lowered if the customers were adjacent to each other and some sort of aggregation was done. But typically, customers are disparate causing multiple installers to drive to different locations. The rate structure for residential customers is usually not time of use based and mainly energy based—not demand focused. Net metering rules are a detriment to storage implementation as the ability to offset load use at varying times is irrelevant in net metering. The volumes can also be too low to financially hedge the project. Mass marketing by groups such as Residential Electric Providers is needed to aggregate enough volume. An interesting development is being made in this front by Sun Run, which is aggregating thousands of residential ESS to produce a virtual plant to offer to utilities in Oakland, CA [3].

9.2 Economic considerations

In this section the economic aspects of an ESS are explained from different perspectives. There is not always one right way to view the economics of a particular system, so different methods are presented here. First, software-based market modeling is reviewed. Then the Levelized Cost method is presented, using an example. Finally, the Risk and Financing aspects of ES are examined.

9.2.1 Market modeling

The calculation of levelized cost of electricity (LCOE) is a commonly used technique that is described in the next section. However, levelized cost analysis is based on some very high-level assumptions which do not take into account the granularity needed for market timing arbitrages that storage and renewables require. LCOE is an effective apples-to-apples comparison when assumptions can be the same and generalized across the board.

However, each project involves a location which has unique properties which will require unique analysis. We can see that energy storage value comes in its ability to react in seconds or minutes vs hourly. Aggregating the hourly behavior to fit into a model is already requiring a level of assumption. Aggregating up to a yearly period which most LCOE calculations do, can make the analysis too uncertain in its conclusions.

Given our technology advancements in computing hardware and software, running hourly analysis is very achievable. Software companies such as Plexos-Aurora, Promod, etc. offer the ability to easily model hourly simulation of grid data. MS Excel and modern day processors can easily handle 8760 rows with complex logic to simulate the realities of the problem.

Ultimately to bring a technology into market you have to solve a particular problem not a high-level problem. The final investment will be at a precise location with a precise situation. We should use the available data and computing resources to make more accurate models. The dispatch modeling section at the end of this chapter will show this approach in more detail.

9.2.2 Levelized cost methods

The fundamental requirement for any project lifetime value analysis is to resolve the time value of money effect on the cash outflows during the initial project construction and the cash inflows during the operational life of the project. A reasonable treatment can be achieved either by converting all of the future cash flows to the present value or by amortizing the initial investment over the project life so as to determine a net revenue per operational period that accounts for the recovery of the initial investment. The levelized cost methods are basically a form of the latter approach, with the added feature that the costs are normalized by the unit of production rather than simply stated per unit time. For instance, a levelized cost of electricity analysis provides a resultant figure of dollars per MWh. In principle, a user who is familiar with electricity market price dynamics gains some insight into the relative competitiveness of the project based on this figure of merit.

For power generation systems, the Levelized Cost of Electricity (LCOE) is the most common metric that is considered. This metric is also commonly adapted for storage projects, usually under the assumption the metric is stated per unit energy delivered from the storage device. In this case, some treatment is required regarding the amount of electricity that must be absorbed in order to discharge one unit and the effective price of the absorbed electricity.

A similar metric that is further adapted for the characterization of energy storage assets is the Levelized Cost of Storage (LCOS). This partially dissociates the figure of merit from the absolute electricity price assumption, since only the energy losses are priced.

In the simplest constructs of either metric, all investment is assumed to occur in the zeroth operating period of the project and uniform operational costs are assumed for all operating periods in the project life. In this case, the results can be calculated using simple explicit formulae. A generalized LCOE construct which can be used for conventional power generation or energy storage with or without fuel consumption (e.g., diabatic CAES) is developed later. Some exemplary figures are provided along with graphical presentation of results in Table 1 and Fig. 2. Fig. 3 provides further graphical analysis of the components that make up the LCOE.

The Levelized Cost Methods are somewhat helpful to understand competitiveness of a technology with a "rough order of magnitude" level of accuracy and they may be helpful to understand the relative contributions of CAPEX, O&M, and efficiency. Levelized cost methods can be quite inaccurate and even misleading as a comparator between different technologies or as a test to determine if a specific project could be viable. The biggest problem is that for storage assets, the capacity factor could be rather low, and with low capacity factor, the CAPEX component dominates the LCOE result. Also, the value of the CAPEX component becomes quite sensitive to the exact capacity factor. For a storage asset, the capacity factor depends on both the efficiency of the device and the electricity value vs time characteristic of the market where the asset operates. In Fig. 3, a capacity factor of 25% was assumed for all of the technologies other than the combined cycle, but it is highly unlikely that these very different technologies might really operate with identical capacity factor in any market situation. What then can be taken away from the results? Is diabatic CAES really a lower cost technology than Li-Ion batteries? It is impossible to determine, simply given the efficiency characteristics of the storage device, how often such a device may run and what would be the effective buy and sell prices for electricity corresponding to this amount of running in a specific market. Furthermore, the price vs time characteristics of many markets change over the seasons, so even identifying single representative values for the important inputs seems implausible.

A slightly more sophisticated approach to the levelized cost approach is to construct a project cash flow and then determine the levelized cost figure of merit that achieves specified financial goals, such as return on equity investment. In this way, more realistic cash flow features can be considered, including the following:

TABLE 1 LCOE example calculations compared for several energy systems.

$LCOE_{disch}$	$= LC_{fuel} + LC_{electricity} + LC_{O\&M} + LC_{CAPEX}$	The result is total cost per unit discharging energy. It is often helpful to analyze the result component-wise
LC_{fuel}	$= HR_{disch} \cdot P_{fuel}$	Levelized cost component for fuel per unit discharging energy
$LC_{electricity}$	$= ER \cdot P_{electricity}$	Levelized cost component for charging electricity per unit discharging energy
LC_{CAPEX}	$= \frac{Payment}{CF \cdot H}$	Levelized cost component for CAPEX recovery per unit discharging energy
$Payment$	$= \frac{SpecificCAPEX \cdot R}{1-(1+R)^{-N}}$	This is the basic period payment from present value financial formula, for example, available via EXCEL's PMT() function. The specific CAPEX is used, such that the result is period payment per unit of discharging power
$LC_{O\&M}$	$= Cost_{O\&M}$	Cost of Operations and Maintenance per unit discharging energy

Input parameters	Definition	Remark
HR_{disch}	Heat rate	Fuel consumed per unit of discharge energy
ER	Energy ratio	Ratio of electrical energy output per unit electrical energy input. For a storage device without fuel consumption, energy ratio is the inverse of efficiency
P_{fuel}	Price of fuel	
$P_{electricity}$	Price of charging electricity	
CF	Capacity factor for discharging	Ratio of time discharging time to total time. For example, discharging for 2000 h per year equates to capacity factor of 2000/8760 = 0.228
H	Hours per discount period	For example, if discounting is considered per annum, $H = 8760$
R	Discount rate	Rate must correctly correspond to the discount period. A weighted average cost of capital is often considered: where WACC = (%Debt)(Interest Rate for Debt) + (%Equity)(Expected Return on Equity Investment)
N	Number of discount periods in the project life	For example, considering a 20-year project life with annual discounting, $N = 20$
$SpecificCAPEX$	Specific CAPEX	The total project CAPEX (including for charging, discharging, and storage systems) divided by the discharging power rating
$Cost_{O\&M}$	O&M cost	The total project operations and maintenance cost (including for charging, discharging, and storage systems) normalized per unit discharging energy

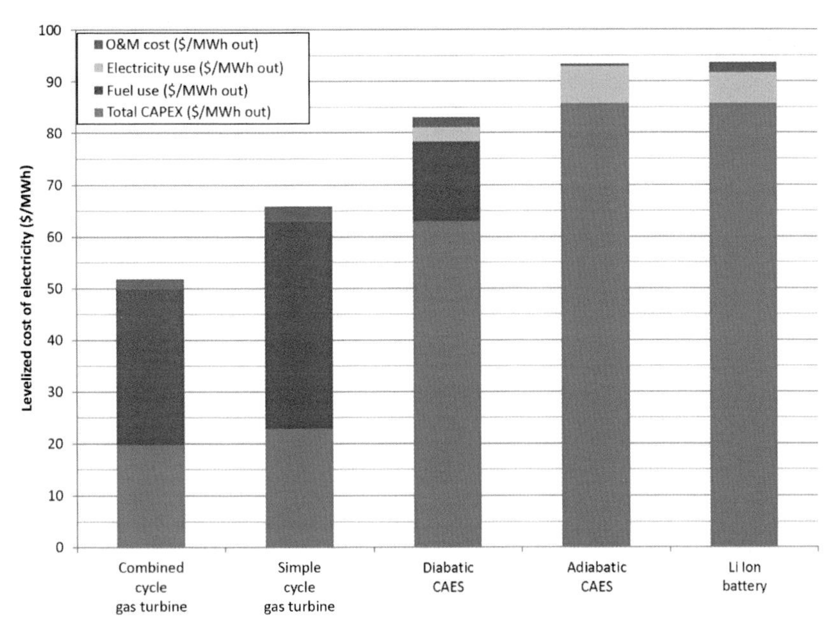

FIG. 2 Levelized cost of electricity.

- Investment can be considered not just at the zeroth period but over a range of periods (this is not illustrated in this example)
- Realistic modeling of asset depreciation and corresponding impact on taxation.
- Operating cost streams could vary with time over the project life, for example cost escalation can be modeled
- Capital reinvestment that might be required during the project life

Fig. 4 illustrates the first few columns of a spreadsheet that carries out this analysis (the spreadsheet extends to the right in years to the project life, but not all columns are shown here) using CAES as an example. In this exercise, we seek a constant number to apply across the 2nd row such that performing an IRR calculation on the cash flow at the last line, we receive the desired investor return (12% in this example). The solution, $97/MWh, is higher than the $83.1/MWh reported in Fig. 2, because Fig. 2 analysis did not take consideration of taxes. In this analysis taxes amount to $15.66/MWh, which explains most of the difference. The remainder of the difference is due to the inclusion of some cost escalation in this model.

Despite the more rigorous treatment, the underlying problems of the levelized cost method remain. Namely, the assumption regarding the capacity factor (this enters the model on the first row) and also the assumption of an average price for electricity purchase.

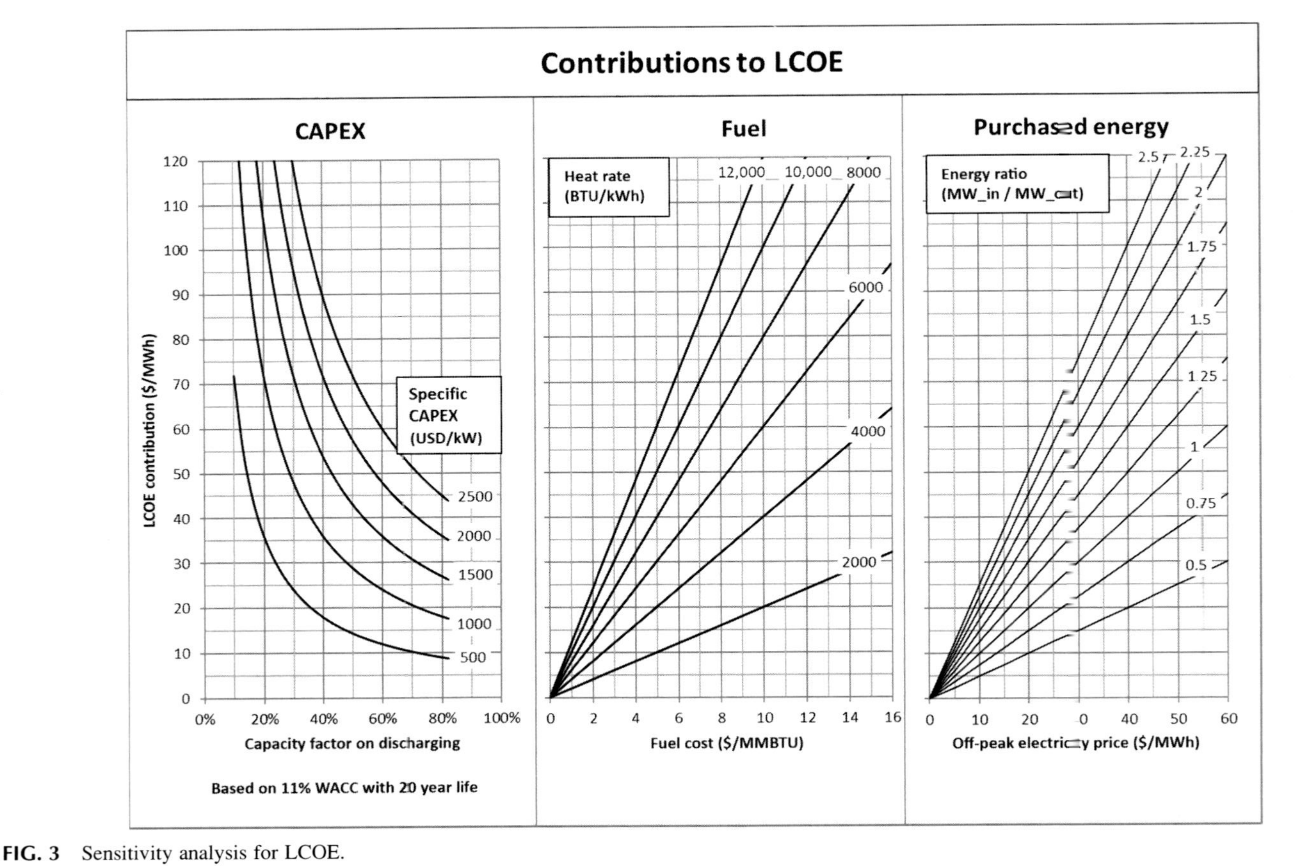

FIG. 3 Sensitivity analysis for LCOE.

		Lifetime totals	Avg ($/MWh) for LCOS	Year 0	1	2	3	4
Total storage discharging	('000 MWh)	4380.00			219.00	219.00	219.00	219.00
Levelized storage cost	($/MWh)				97.00	97.00	97.00	97.00
Total revenues	($M)	424.86	97.00		21.24	21.24	21.24	21.24
Total charging	('000 MWh)	2409.00			120.45	120.45	120.45	120.45
Total charging cost	($M)	(13.39)	(3.06)		(0.61)	(0.61)	(0.62)	(0.63)
Total fuel	('000 MMBTU)	16,941.84			847.09	847.09	847.09	847.09
Fuel cost	($M)	(75.35)	(17.20)		(3.42)	(3.46)	(3.49)	(3.53)
Total O&M	($M)	(8.76)	(2.00)		(0.44)	(0.44)	(0.44)	(0.44)
Total operating costs	($M)	(97.51)	(22.26)		(4.47)	(4.51)	(4.55)	(4.59)
EBITDA	($M)	327.35	74.74		16.77	16.73	16.69	16.65
Dept outstanding - beginning of period	($M)			38.50	38.50	37.75	36.93	36.03
Dept - interest expense	($M)	(45.85)	(10.47)		(3.47)	(3.40)	(3.32)	(3.24)
Dept - principal payment	($M)	(38.50)	(8.79)		(0.75)	(0.82)	(0.89)	(0.97)
Total debt service	($M)	(84.35)	(19.26)		(4.22)	(4.22)	(4.22)	(4.22)
EBITDA	($M)	327.35	74.74		16.77	16.73	16.69	16.65
Depreciation	($M)	(110.00)	(25.11)		(11.00)	(9.90)	(8.91)	(8.02)
Interest expense	($M)	(45.85)	(10.47)		(3.47)	(3.40)	(3.32)	(3.24)
Taxable income	($M)	171.50	39.16		2.31	3.44	4.46	5.39
Tax benefit (Liability)	($M)	(68.60)	(15.66)		(0.92)	(1.37)	(1.78)	(2.16)
After-tax net equity cash flow	($M)	174.40	56.14	(71.50)	11.63	11.14	10.69	10.28
RR for equity investors				12.0%				

FIG. 4 Levelized cost model using rigorous cash flows. Example is diabatic CAES with 100 MW discharging, similar performance figures and financial assumptions as in Fig. 2.

9.2.3 Risk and financing

The ability to finance a project will drive many projects to doom or success. In most cases an individual party does not want or have the ability to wait and see their returns 10–20 years later. Financing a project requires the ability of the banking institute to hedge out some or all of the risk.

9.2.3.1 Capital

Capital markets are highly dependent on interest rates. A low interest rate environment will lead to more capital searching for yields creating a more investment friendly environment. However, this will not make the financing of a storage deal any more likely given the investment options in the market is plentiful in the power sector itself—from corporate bonds to utility stocks to renewable PPA deals.

9.2.3.2 Commodity

The futures market for power is limited in time and locations. At some point a proxy commodity would likely be used to hedge out the commodity risk of a power project. Typically, natural gas contracts would be used as the Henry Hub NYMEX (New York Mercantile Exchange) is very liquid and goes out many years. The best form of hedge used by the market is a utility power purchase agreement (PPA). With a PPA, one can go to the bank and note that utility promises to pay Y and the deal becomes more financeable to the bank. All that feeds the project from materials to feedstock become the risk of the project. The materials used to build an ES technology are a big concern. For example, much is talked about in terms of Lithium and Cobalt commodity risk as EV continue to expand in the market. At some point Li-Ion battery solutions will have to deal with these commodity risks as the market grows. Having a technology not dependent on rare and high cost commodity should make a project more financeable.

Risk not only lies in the commodities, but in the counterparty. As seen in a recent Pacific Gas and Electric (PG&E) case much of the solar deals with PG&E were financed largely as a result of PG&E taking a power purchase agreement with the solar developers. However, now with PG&E in bankruptcy and the ruling to allow PG&E to abandon or restructure the PPA it highlights counterparty risk.

9.2.3.3 Regulatory

There is also regulatory risk which a utility is subject to since certain costs could become disavowed. Case in point is the nuclear project in Georgia Power where the utility company had to take some of the cost increases on the company balance sheet vs being able to pass it all on to the rate payers. Regulatory risk is a real, but less often talked about, risk. Changes in rules and regulations of how a utility can operate and collect money can cause great uncertainty decreasing the ability to finance a project. A history of regulatory consistency and favorable rulings to utilities will likely ease much of the concern to the risk of financing a project.

9.2.3.4 Environmental

Environmental risk becomes another concern for making a project financeable. Using elements or processes which have a history of catastrophic losses will not bode well for financing. The 2019 APS (Arizona Public Service) Li-Ion battery explosion will likely be a very close watched investigation as many Li-Ion storage projects await their financing. The hydrogen storage facility explosion in Norway also is highlighting the concerns of using certain elements.

In order to finance a project, all of the risks highlighted previously need to be acknowledged and at the very least a mitigation plan put in place as the project is pitched to potential buyers.

9.3 Technology considerations

Several attributes of a given ESS technology will have an impact on the suitability for a given application. This section will look at capital and operating costs, efficiency, siting and sizing, and storage duration for various ESS.

9.3.1 Installed (capital) costs

When creating a capital cost model for a project it is important to consider much more than the equipment that needs to be purchased to perform the Energy Storage service. Capital costs are fixed, one-time expenses incurred on the purchase of land, buildings, construction, and equipment used in the creation of an Energy Storage Plant. It is the total cost needed to bring a project to a commercially operable status [4]. All factors must be taken into account to build an accurate Capital Cost model pivotal to the financial success of a project.

There can be many hidden costs within the construction of an ESS that if left out can greatly impact the financial viability of project. When developing the cost model the first thing that comes to mind is the System Hardware cost of the equipment. But is also important to think about applicable sales taxes and tariffs. Neglecting this could give some suppliers an artificial leg up ending in significant impacts to the project bottom line. Similarly, when constructing the plant or system the number of labor hours and average labor rates are factored in but not considering the person-hours per task by labor class may cause the EPC (Engineer Procurement and Construction) labor cost model to be inaccurate. It is also important to consider EPC Soft Costs, which can include SG&A (Selling, General and Administrative) mark up, supply chain costs, and other fees that must be scrutinized when analyzing bids. Ignoring this and viewing only the hourly rates can cause cost overruns in projects. One more consideration is the project developer costs which are involved in the land acquisition, interconnection studies, project origination and acquisition. Taken these factors into account will contribute to a strong Capital Cost model increasing the chances of financial success of the project.

Total Capital cost models have many facets that summate to outputs of Total Equipment Costs, Direct and Indirect Labor, EPC Soft Costs, Developer Direct Costs and Overhead. Using Battery ESS as an example, Fig. 5 shows the bottom-up cost structure model for a BESS unit.

9.3.2 Operational and maintenance (O&M) costs

Many times, focus is placed on the first time capital cost of a system. But the true measure of the project economics is determined by the total lifecycle cost of a system. Some systems may have a very low upfront capital cost but need a larger number of full-time personnel to keep operational or could require regular replacement of costly components. Conversely, systems can be overengineered for the application and require very little maintenance but have a high upfront cost that prices the product out of the market making it economically unviable. It is important when evaluating the energy storage technology for the project to take all of these factors into consideration. The operating and maintenance cost of a system can be a key determining factor on whether the project will be justifiable economically.

In the following section, we will discuss the operational and maintenance costs of different storage technologies. We will look at the common areas of operating costs for all technologies and evaluate the differences in the various types of technologies in operation and maintenance. The objective of this section is not to provide a snapshot of storage costs at the time this book was written but give an approach on how to calculate operational costs and aspects to be aware of for each type of technology.

Every type of storage technology comes with associated operating and maintenance (O&M) costs. These can range from minor monitoring systems with very little upkeep to extensive maintenance procedures which require periodic major overhaul of equipment. Each technology has its particular trade-offs with specific advantages depending on the market they are competing in and servicing.

Previously in the section on levelized cost we quantified the O&M costs. In this section we will quantitatively compare the cost factors for different types of ESS. Though this book is not focused on BESS, they will be used as a baseline to compare other technologies since often it is the advantages or disadvantages compared to BESS that will drive a project to a particular technology.

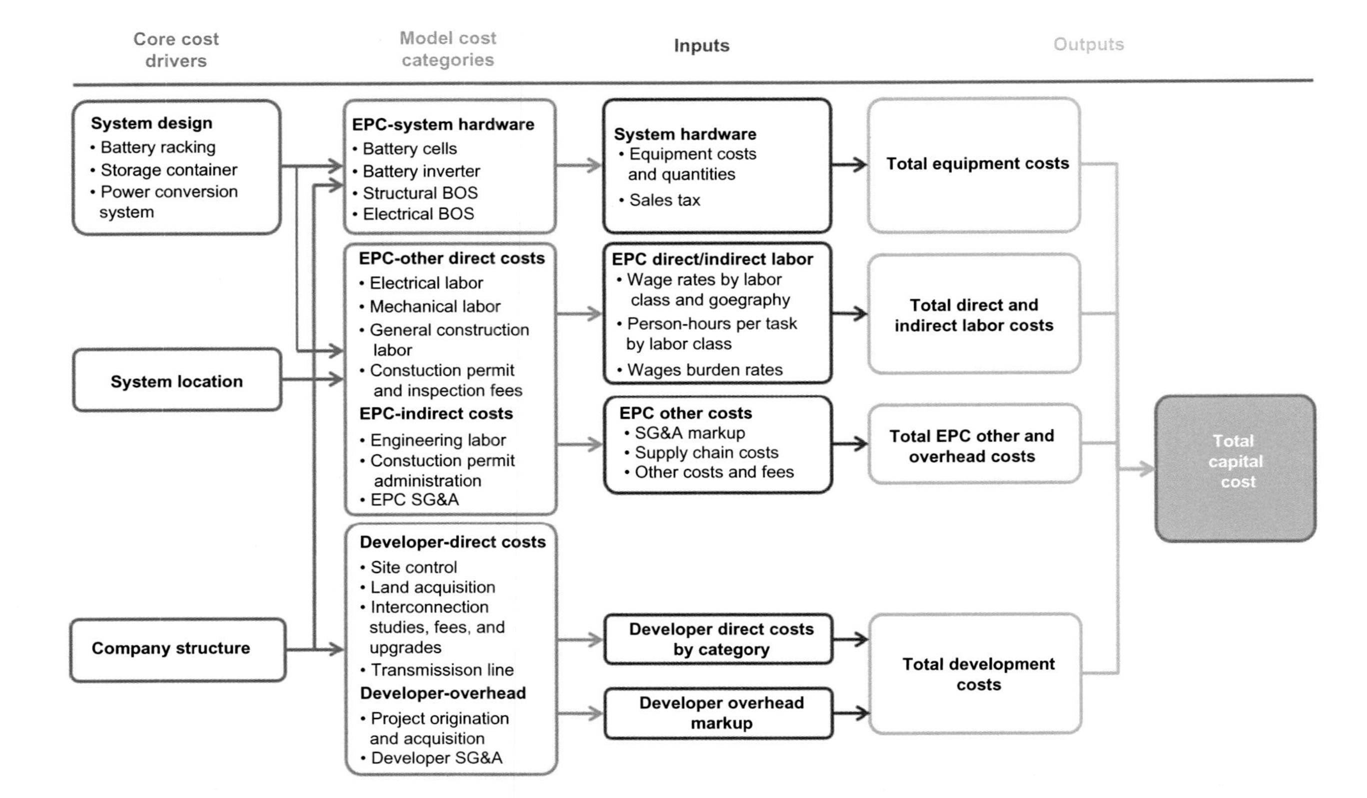

FIG. 5 Structure of the bottom-up cost model for stand-alone storage systems [4].

9.3.2.1 Battery energy storage systems (BESS)

Many types of battery chemistries can be used such as Lead-Acid (VRLA), Nickel-Cadmium (Ni-Cad), Sodium-Nickel, etc. However, the most common type of BESS in today's energy market uses Lithium-Ion batteries and therefore will be the focus of this discussion. The primary operating costs for a Battery Energy Storage System are as follows:

- Heating, Ventilation, and Air Conditioning (HVAC)
- Inverter Cooling
- Parasitic Load Monitoring

The parasitic monitoring will be a constant load and can easily be calculated given the correct information from the supplier. The variable operational costs are the HVAC and Inverter Cooling and will vary with how the system is operated and the ambient conditions where the ESS is sited. The HVAC and Inverter Cooling load will be highly dependent on the following factors:

- Charge Rate—Faster charge = more heat generation
- Charge Duration—Longer charge = more accumulated heat
- Battery Age—more cycles = more heat generation
- Ambient Conditions—hotter or more humid environment = greater nonoperational HVAC load
- Sun Exposure—direct sun = more heat absorption

Much of this information should be available from the BESS supplier. Additional resources may be required to calculate the HVAC loads due to solar flux. Mitigating actors to reduce operational costs can be put into place such as adding a sunshade to reduce HVAC load if the economics are known.

Maintenance of a BESS is rather minimal. Each system has their own unique design features but common maintenance requirements across many systems include:

- HVAC Servicing and Filter Replacement
- Fire Detection and Suppression Servicing
- Inverter Cooling System Servicing
- Battery Replacement for Degradation

By far the most impactful maintenance event is the replacement of batteries due to degradation. This requires the majority of the system cost to be replaced once the batteries have reached end of life which typically occurs around 6000 cycles or about 10 years.

These are key factors that must be taken into account when evaluating the implementation of a Battery Energy Storage System. This approach can be extended to all other energy storage technologies with slight modifications.

9.3.2.2 Pumped hydro

Pumped Hydro technology is rather simple in operation and only requires a pump/turbine and an elevated reservoir. The operational costs for a hydro plant are simply the costs of running the pumps and staffing. The parasitic costs are the monitoring of the critical failure points of the system, which is relatively low when compared to cooling costs of a battery system of comparable size. The largest costs are the maintenance of the system. The reservoir, penstock, and turbine/pump chamber all require periodic monitoring, inspection, and maintenance.

Following is a list of primary considerations for operation cost and maintenance of Pumped Hydro Storage:

- Reservoir
- Penstock
- Pump
- Turbine
- Control Valves

9.3.2.3 Thermal storage

Thermal storage is primarily dominated by molten salt storage through Concentrated Solar Power (CSP) Plants. CSP Plants "fuel" is free but the operation and maintenance is a costly endeavor. Whether a tower style, such as Crescent Dunes Solar Energy Project, or a trough style CSP, such as Solana Generation Station, there are hundreds of miles of mirrors that are only efficient if they retain high reflectivity. This requires a constant cleaning crew of multiple people working every day of the week. This cleaning equipment often consists of spray trucks that also require constant maintenance. The best placement for CSP plants is often in desert climates; however, they typically rely on steam turbines and generators to convert the heat energy into electricity. These steam turbines require large amounts of high-quality water to run for long periods. Even with a recapture system in place the water subsystem of the plant can be a several person operation with overhead costs that can be far greater than anticipated. Additionally, Thermal Storage Plants can require miles of heat tracing buried under feet of insulation to keep the thermal salts from freezing in between charge cycles. The heat tracing utilizes resistive heaters in long strips with a high failure rate. Due to this, the heat tracing can require constant replacement requiring a full-time crew of several people. All of these are important factors when building a robust Thermal Storage cost model.

Following is a list of primary considerations for operation cost and maintenance of Thermal Storage Power Plants:

- Thermal Salt Reservoirs
- Heat Exchangers
- Steam Turbines

Reservoir	Size (MWe)	CPRC ($/kW)	CESC ($/kWh)	ST (h)	TC ($/kWe)
Salt	200	350	1	10	360
Porous media	200	350	0.1	10	351
Hard rock	200	350	30	10	650
Surface piping	20	350	30	3	440

Abbreviations: CPRC, cost for power-related plant components; CESC, cost for the energy storage components; ST, "Typical" hours of storage for a plant; TC, total cost

FIG. 6 Cost of CAES power-related plant and energy storage components [5]

- Water Treatment
- Heat Tracing
- Mirror maintenance

9.3.2.4 Compressed air energy storage

There are several different flavors of Compressed Air Energy Storage (CAES) Systems. The simplest uses an electric motor to run a compressor and a turbo expander to drive a generator. These maintenance costs can be benchmarked against similar turbomachinery applications, to create robust cost models. A unique aspect that must be taken into account in a CAES system is the air reservoir which often is an abandoned mine or salt cavern. In Fig. 6, the capital cost of an air reservoir for various storage media and plant configurations is listed. The cost is related to the types of storage (containers/caverns), the power rating, and the duration of storage. Therefore in Fig. 6, the cost of power-related components such as turbine, expander, etc. is listed as $/unit power and the cost of storage components such as underground caverns and overground cylinders is related to their capacity and listed as $/unit energy stored. [5] The turbomachinery operational and air reservoir capital cost are primary considerations that should be taken into account when developing a CAES cost model.

Following is a list of primary considerations for operation cost and maintenance of Compressed Air Storage Systems:

- Fuel
- Electricity
- Turbomachinery
  - Compressors
  - Expanders
  - Combustor
  - Motor
  - Generator
- Air Reservoir
- Piping Maintenance

9.3.2.5 Flywheel

Flywheel storage technologies consist primarily of mechanical parts that rotate at very high speeds. Since the malfunction of components in the system can have catastrophic failures it is very important to ensure everything is in proper working order. This requires extensive instrumentation and monitoring but has a relatively low O&M cost when compared to the other storage technologies (see Fig. 7).

Following is a list of primary considerations for operation cost and maintenance of Flywheel Energy Storage Systems:

- Rotor
- Motor/Generator
- Bearings
- Power Electronics
- Vacuum Pump

9.3.2.6 *Ultracapacitor*

Due to the simplicity of an Ultracapacitor storage system and it being entirely power electronics with little to no moving parts they have the lowest O&M costs when compared to all storage technologies. This comes at the cost of power density and limited storage duration.

Following is a list of primary considerations for operation cost and maintenance of Ultracapacitor Energy Storage Systems:

- Ultracapacitors
- Power Electronics
- Cooling system

Fig. 7 from Luo [6] compares total cost models for different types of storage technologies. Taking into account the factors listed previously, it can be relatively easy to deduce the cost drivers behind these models.

Cost models make or break a project's financial viability, so it is important that due diligence is taken during the development. Following the approach given previously will provide a good starting basis for a financial model. It is important to look for the hidden factors, to avoid cost overruns and ensure the financial success of a project.

9.3.3 Efficiency

Energy storage devices are not perfect storage devices. Rather, the energy input to them is lost in a number of ways depending on the type of device. For example, friction reduces the energy available from flywheel storage devices. In pumped hydro ES, pumping efficiency, turbine efficiency, and water loss can reduce the amount of energy available relative to the input energy. And diabatic compressed air energy storage loses energy from the heat energy from compression and the subsequent reheat and turbine efficiencies. Accounting for these energy losses is critical in sizing the energy storage system as well as determining the levelized cost of the electricity delivered. As a result, the

Parameter	Sodium-sulfur battery		Li-ion battery		Lead acid		Sodium Metal Halide		Zinc-hybrid cathode		Redox Flow battery	
	2018	2025	2018	2025	2018	2025	2018	2025	2018	2025	2018	2025
Capital cost—Energy Capacity ($/kWh)	400–1000 **661**	(300–675) **(465)**	223–323 **271**	(156–203) **(189)**	120–291 **260**	(102–247) **(220)**	520–1000 **700**	(364–630) **(482)**	265–26[illegible] **265**	(179–199) **(192)**	435–952 **555**	(326–643) **(393)**
Power conversion System (PCS) ($/kW)	230–470 **350**	(184–329) **(211)**	230–470 **288**	(184–329) **(211)**	230–470 **350**	(184–329) **(211)**	230–470 **350**	(184–329) **(211)**	230–47[illegible] **350**	(184–329) **(211)**	230–470 **350**	(184–329) **(211)**
Balance of plant (BOP) ($/kW)	80–120 **100**	(75–115) **(95)**	80–120 **100**	(75–115) **(95)**	80–120 **100**	(75–115) **(95)**	80–120 **100**	(75–115) **(95)**	80–120 **100**	(75–115) **(95)**	80–120 **100**	(75–115) **(95)**
Construction and commissioning ($/kWh)	121–145 **133**	(115–138) **(127)**	92–110 **101**	(87–105) **(96)**	160–192 **176**	(152–182) **(167)**	105–126 **115**	(100–119) **(110)**	157–18[illegible] **173**	(149–179) **(164)**	173–207 **190**	(164–197) **(180)**
Total project cost ($/kW)	2394–5170 **3626**	(1919–3696) **(2674)**	1570–2322 **1876**	(1231–1676) **(1446)**	1430–2522 **2194**	(1275–2160) **(1854)**	2810–5094 **3710**	(2115–3440) **(2674)**	1998–240[illegible] **2202**	(1571–1956) **(1730)**	2742–5226 **3430**	(2219–3804) **(2598)**
Total project cost ($/kWh)	599–1293 **907**	(480–924) **(669)**	393–581 **469**	(308–419) **(362)**	358–631 **549**	(319–540) **(464)**	703–1274 **928**	(529–860) **(669)**	500–60[illegible] **551**	(393–489) **(433)**	686–1307 **858**	(555–951) **(650)**
Q&M fixed ($/kW year)	10	(8)	10	(8)	10	(8)	10	(8)	10	(8)	10	(8)
Q&M variable (cents/kWh)	0.03		0.03		0.03		0.03		0.03		0.03	

Parameter	Pumped storage hydropower[(a)]	Combustion turbine	CAES[(a)]	Flywheel[(b)]	Ultracapacitor[(c)]
Capital cost–energy capacity ($/kW)	1700–3200 **2638**	678–1193 **940**	1050–2544 **1669**	600–2400 **2400**	240–400 **400**
Power conversion system (PCS) ($/kW)	Included in capital cost	N/A	N/A	Included in capital cost	350 (211)
Balance of plant (BOP) ($/kW)					100 (95)
Construction and commissioning ($/kW)				480[(d)]	80[(d)]
Total project cost ($/kW)	1700–3200 **2638**[(f)]	678–1193 **940**	1050–2544 **1669**	1080–2880 **2,880**	**930 (835)**
Total project cost ($/kWh)	106–200 **165**		94–229 **105**	4320–11,520 **11,520**	**74,480 (66,640)**
Q&M fixed ($/kW year)	15.9	13.0	16.7	5.6	1
Q&M variable (cents/kWh)	0.00025	1.05	0.21	0.03	0.03

FIG. 7 Capital and Operating Costs for different ES technology [6].

efficiency of energy storage, commonly referred to as round-trip efficiency, is a primary performance metric.

Round-trip efficiency accounts for the loss of energy between the time that an energy storage device is charged and subsequently delivered (or discharged). The overall energy use of most typical energy generation systems (e.g., solar, wind, conventional coal-fired plants, and gas turbines) can be clearly described by efficiency. In the same way, for most energy storage devices round-trip efficiency is simply

$$\eta_{RT} = \frac{E_{out}}{E_{in}}$$

where E_{in} is typically the electrical energy input (or the charge) into the storage device and E_{out} is the electrical energy delivered (or discharged), and in those cases where electricity is the input and the output, determining round-trip efficiency is straight forward. Table 2 provides typical round-trip efficiencies for several storage technologies.

TABLE 2 Round-trip efficiency of selected storage technologies.

ES technology	Lifetime	Efficiency
Pumped hydro [7]	30–60 years	70%–85%
Flywheel [6]	15–20 years	90%–95%
Compressed air [7]	20–40 years	40%–70%
Molten salt (thermal) [7]	30 years	80%–90%
Li-ion battery [7]	1000–10,000 cycles	85%–95%
Lead-acid battery [7]	6–40 years	80%–90%
Vanadium redox flow battery [6]	5–20 years	65%–80%
Zinc bromine flow battery [6]	5–10 years	65%–80%
Capacitor [6]	1–10 years	60%–70%
Supercapacitor [6]	10–30 years	84%–97%
Superconducting magnetic storage [8]	100,000 cycles	80%–95%
Hydrogen [7]	5–30 years	25%–45%
Synthetic natural gas [8]	30 years	25%–50%
Thermal energy [6]	5–30 years	30%–60%
Liquid air [6]	25+ years	55%–80%

The challenge in applying the concept of round-trip efficiency occurs when the storage mechanism requires heat or fuel in addition to the energy stored (as in the case of diabatic CAES systems).

Consider a CAES system. In a typical system excess electricity from a generation system (e.g., an intermittent source such as wind or a large constant source such as nuclear with intermittent demand) is used to charge the system by compressing and storing air. When electrical demand exceeds the generation capacity, the compressed air is heated with natural gas or oil and expanded through a gas turbine, the complication being that both the stored electrical energy and the fuel are needed to produce the electrical output. Potential metrics for this process include

- heat rate, given as

$$HR = \frac{Q_{fuel}}{E_{delivered}}$$

where Q_{fuel} is the energy content in the fuel and $E_{delivered}$ is the electrical energy delivered;

- charging electrical ratio, given as

$$CER = \frac{E_{delivered}}{E_{in}}$$

where $E_{delivered}$ is the electrical energy delivered and E_{in} is the electrical energy input (or charged) into the storage device; and

- various forms of round-trip efficiency that alternately account for the efficiency of the energy efficiency of the primary energy source, the fuel used to reheat the air, and the efficiency of use of the fuel.

The most straightforward of these definitions of round-trip efficiency for CAES systems can be derived from our earlier definition of round-trip efficiency. This is

$$\eta_{RT} = \frac{E_{out}}{E_{in}} = \frac{E_{delivered}}{E_{elec} + Q_{fuel}}$$

In this case the system efficiency is consistent with the system efficiency for the CAES system. However, there is an argument that can be made that this is not the round-trip efficiency (and that because of the addition of fuel, there is no longer a round-trip) but rather simply the efficiency.

Another approach is to include the efficiency of the primary (i.e., upstream) power generation devices such as nuclear or coal-fired power plants. In this case, the system boundaries are extended to include the primary power generation source, and the E_{in} in the denominator becomes the energy input to the primary power generation source. This is

$$\eta_{PE} = \frac{E_{delivered}}{E_{elec}/\eta_{gen} + Q_{fuel}}$$

where η_{gen} is the efficiency of the primary energy source. The efficiency is no longer referred to as the round-trip efficiency, but recognizing the inclusion of the primary electrical energy source, commonly referred to as the primary energy efficiency, η_{PE} and η_{gen} is the efficiency of the primary power generation source.

One concern with directly including Q_{fuel} in the round-trip efficiency metric for CAES systems is that it is inconsistent with the description of round-trip efficiency for other electrical energy storage devices, i.e., the electrical energy discharged by the device divided by the electrical energy input to the device. There are two approaches to addressing this concern. One approach is to use electrical energy produced by fuel combusted in the gas turbine. This is

$$\eta_{RT} = \frac{E_{delivered}}{E_{elec} + \eta_{turb} Q_{fuel}}$$

where η_{turb} is the efficiency of the gas turbine in the CAES system. Another approach is to remove the energy of the fuel from consideration in round-trip efficiency by subtracting the electrical energy produced by the fuel from the electrical energy delivered by the CAES system. This is

$$\eta_{RT} = \frac{E_{delivered} - \eta_{turb} Q_{fuel}}{E_{elec}}$$

Both approaches arguably put the calculation of round-trip efficiency on the same basis as other electrical energy storage devices, i.e., only electrical energy into the system is included and not the efficiencies of producing the electrical energy. However, it is not clear that one is strongly preferred, and there are yet other approaches that can be taken.

The challenge of how to handle the efficiency metric in energy storage devices and the need for consistency is today an open question that impacts not only CAES devices but has the potential to impact other energy storage devices. For the moment, the metric chosen needs to be driven by the application of the storage system and the need for transparency on the basis of describing the performance of the energy storage system. A clearer description of storage devices in which electrical energy is not the only input (or an input at all) might be to refer to them as hybrid energy storage devices. In this definition, hybrid energy storage devices use both stored energy from primary energy devices and fuel or other conventional energy sources to produce electrical energy. In this case, based on the simple definition of efficiency we could modify the efficiency metric of interest for these hybrid systems to be hybrid energy efficiency, η_{hybrid}, which becomes

$$\eta_{hybrid} = \frac{E_{delivered}}{E_{elec} + Q_{fuel}}$$

However, the use of the term "hybrid energy storage device" is not common or standard. Additionally, the term "hybrid energy efficiency" is not common, and it is not clear from this definition how much energy is stored versus how much energy is produced by the fuel input. This leaves the choice of the best metric to describe the efficiency as the responsibility of the system designer/user based on the application of the energy storage system.

9.3.4 Storage duration

Different ES technologies are sized for different power and energy capacities. This leads to different optimal duration sizes for a given technology. Fig. 8 shows how the various ESS technologies cover a wide range of energy capacities and storage duration times.

As VRE (variable renewable energy) sources increase their share of the energy pie, longer duration times of energy storage are becoming increasingly important. Weekly and even seasonal duration will be needed to offset the peak of weather-dependent VRE. ES applications can be divided into 3 general duration categories: short term, daily, and long term/seasonal as seen in Fig. 9.

The Energy-to-Power (E/P) ratio shows the relationship between energy capacity and power capacity. The E/P ratio is defined by the rated energy storage of the system divided by the rated power of the ESS and is measured in time (typically hours). Note that care must be taken when using E/P ratio for systems

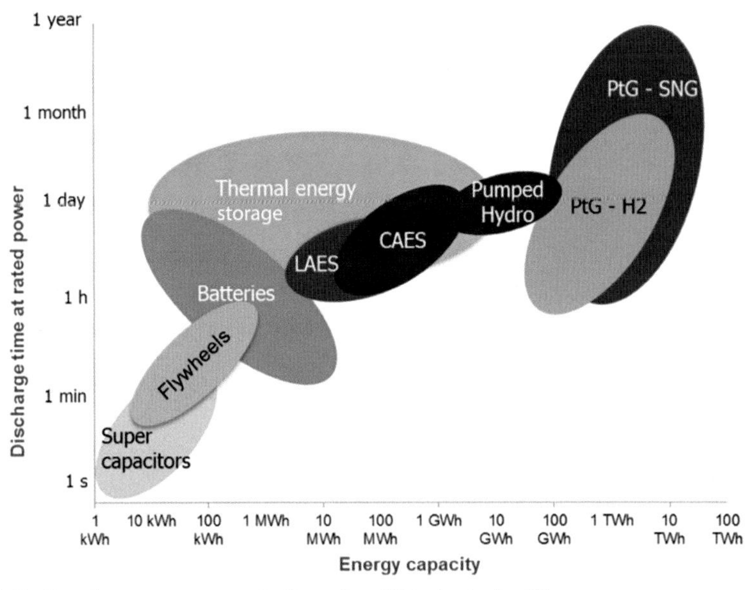

FIG. 8 Duration vs energy capacity for various ES technologies [9].

Duration	Description
Short-term storage	Typically defined as an application where charging and discharging processes last no longer than a few minutes before the power flow changes direction. Because of their very high power capabilities, electricity storage sytems—such as supercapacitators, superconducting coils or mechanical flywheels—are often used in these applications, but many battery storage technologies can also be used
Daily storage	Usually features charge or discharge times of several minutes to a number of hours. Pumped hydro storage, compressed air electricity storage and all types of electro-chemical energy storage systems are suitable for daily storage
Long-term storage or seasonal storage	Usually stores energy over periods of weeks or months. Long-term storage is typically achieved using power-to-gas converters in combination with gas storage systems or large mechanical storage systems such as pumped hydro storage or CAES. Additionally, redox flow batteries and NaS batteries may be able to deliver reasonable weekly storage as their energy-related investment cost declines

FIG. 9 Duration categories for ES applications [10].

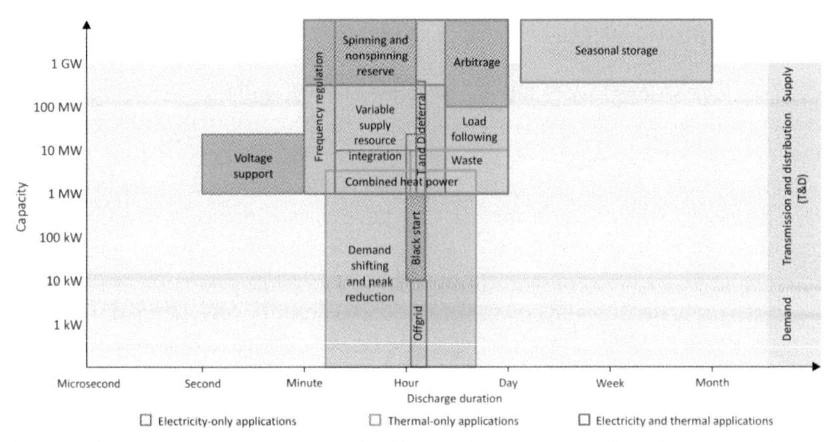

FIG. 10 Duration vs power requirement for different ES applications. *(From International Energy Association, Technology Roadmap—Energy Storage, 2015, p. 14. https://www.iea.org/reports/technology-roadmap-energy-storage.)*

where the charge rates and discharge rates are not equal. In other words, the total cycle time for a full charge and discharge cycle time may be more than 2 times the E/P ratio if the power during discharge and the power during charge are not the same.

Previously we discussed the different economic benefits of ES. Fig. 10 shows how the duration and rated power of an ESS supports the different grid applications.

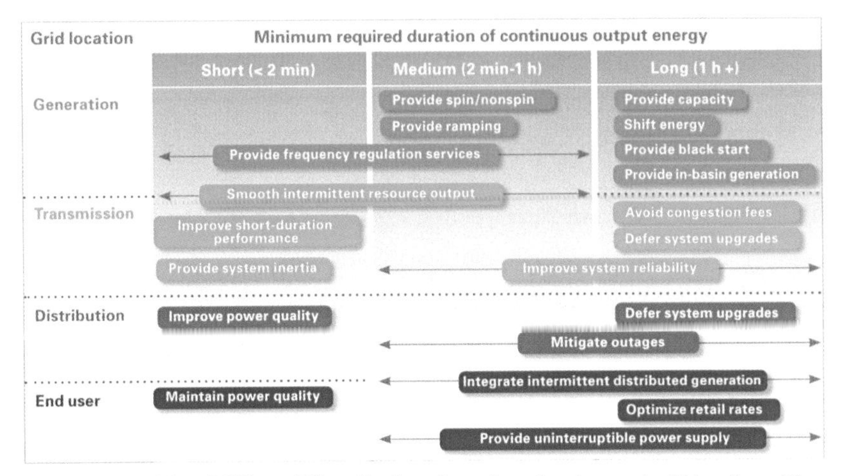

FIG. 11 Suitability of different ES applications for various durations and grid locations. *(From Moving Energy Storage from Concept to Reality, Southern California Edison, 20 May 2011. https://www.edison.com/content/dam/eix/documents/innovation/smart-grids/Energy-Storage-Concept-to-Reality-Edison.pdf.)*

9.3.5 Sizing and siting

The physical location of an energy storage system can have a large impact on the economical size of a system. Systems tied to the transmission grid can have much larger power output than systems that are tied to the distribution systems or that support a microgrid or industrial plant. If a transmission feeder line needs to be installed to the ESS site, this can also have a large impact on the timing and economics of ESS. The duration of a given ESS also effects the suitability of a given technology for various locations as shown in Fig. 11.

ESS energy density and power density will determine how large of a volume or land area needed for a given technology. Power density is typically measured in kW/m^3 and energy density in kWh/m^3. For long-duration storage, energy density is typically more important than power density. For short-duration storage (or portable devices such as vehicles) power density is more important.

Some ESS require certain geological or geographical features, for example, aboveground reservoirs, underground caverns, large bodies of water for submersible underwater tanks, mountainous slopes for railroad tracks. Fig. 12 shows typical land use area requirements for a few ES technologies.

Consideration for microgrid and distribution ESS must look at all factors. By locating an ESS close to the end user, lower efficiencies of a smaller ESS can be offset by reduced cost of the ESS due to modularization and by avoiding the 5%–10% loss penalty in transmission lines.

There are many other factors to consider when locating an ESS, similar to any energy system. If fuel will be combusted, then emissions must be considered. Noise can be an issue, especially if turbomachinery is involved. Safety issues may be involved with the use of hazardous materials such as hydrogen, ammonia, etc.

ES type	Land use factor (m²/MW)
Pumped hydro	1100–4000
Flow batteries	850
Lithium ion batteries	500
Sodium sulfur batteries	200–350
CAES	140

FIG. 12 Land use requirements by ES technology. *(From National Renewable Emery Laboratory, Renewable Electricity Futures Study, vol. 1: Exploration of High-Penetration Renewable Electricity Futures, pp. A-66–A-67. https://www.nrel.gov/docs/fy12osti/52409-1.pdf or https://www.nrel.gov/analysis/re-futures.html.)*

The ambient conditions can also affect the sizing of a system. Some systems may need to be derated from their nominal power rating for temperature effects (e.g., batteries at low temperatures and turbomachinery at high temperatures).

9.3.6 Operational (part load) flexibility

Maintaining grid reliability and satisfactory electricity quality requires modulation of electric power generation in time domains from subsecond to minutes. Traditional generation/energy sources can modulate their power levels at various rates, ranging from minutes to hours for steam thermal plants such as coal, nuclear, and geothermal to subsecond to seconds for combustion turbines to millisecond response times for pure electric devices such as capacitors and batteries. As grids experience higher penetration of intermittent renewables, the need for flexible dispatchable power sources increases dramatically to compensate for the sudden losses of renewable sources for both highly predictable events such as sunset to less predictable ones such as shading of utility-scale solar fields from cloud movements.

For the past hundred plus years, flexibility requirements have been generally well met by traditional generation resources. However, the most economical traditional thermal generators are being forced to modulate power levels and cycle operations in ways for which they were not originally designed with consequent large increases in operating costs and reductions in generating efficiency from running at off-nominal power set points.

Over the past decade, emerging energy storage technologies have demonstrated the ability to likewise meet these requirements. Similar to traditional generation sources, energy storage systems can be characterized by their flexibility or response rate to commands for power modulation. For instance, batteries and flywheels can have millisecond response times, thermal storage systems might have response time characterized by the responsiveness of a steam turbine (but *not* that of a boiler or Heat Recovery Steam Generator), mechanical/gravity storage systems the response time of a nonspinning synchronous generator, and so forth. Response times for select ES technologies were summarized by Luo et al. [6] and excerpted in Table 3 as follows.

TABLE 3 Response time of ES technologies [6].

ES technology	Typical response time
Pumped hydro storage	Minutes
Compressed air energy storage	Minutes
Flywheel	<0.016 s
Lead-acid batteries	<0.25 s
Li-ion batteries	Milliseconds
Capacitors, supercapacitors	Milliseconds
Hydrogen fuel cells	Seconds
Liquid air energy storage	Minutes

From an overall grid perspective, the flexibility of generator response is the sum of individual resource responsiveness (including Demand Response resources) during a given time period. Grid operators/Balancing Authorities face the complex task of scheduling and dispatching resources with adequate flexibility and capacity to meet the anticipated and unanticipated changes in demand and generation balances. As a means to reward resources with quicker response times, some grids (e.g., PJM) have incorporated payment structures for ancillary services with increasing rates according to the fidelity of a resource's actual generation response to the requested modulation as reflected in the dispatch signal. Such methods clearly favor resources such as batteries, flywheels, and capacitors that have very rapid response times.

Recently, there have been first-time deployments of "hybrid" energy storage + thermal generator systems in which battery systems are used in conjunction with a natural gas combined cycle generator but considered as a single dispatchable resource from the perspective of the grid operator. Upon reception of a first dispatch signal, the battery system provides large amounts of power for a short period of time, during which the combined cycle plant can be started, warmed up, synchronized with the grid and start generating. Such a hybrid configuration prevents the need to run the combined cycle plant at a low-efficiency/high specific emissions "minimum generation" set point in anticipation of a request to ramp power, saving operating costs and reducing emissions, and enabling the joint battery + combined cycle plant to qualify as Spinning Reserves without the need to waste natural gas otherwise needed to maintain the combined cycle plant in a "ready to generate" state.

9.4 Dispatch modeling and revenue production

As with any project that requires the deployment of capital, energy storage projects must demonstrate economic merit in order to motivate their implementation. This means that prior to the commitment of capital to construct an energy storage asset, a lifetime value analysis must be established which confirms, with reasonable confidence, that the asset will produce value to its owner in excess of its costs over the extent of its operational life. It is relatively straightforward, albeit by no means trivial, to determine the capital expenditure for an asset: costs for major equipment items and construction efforts are determined and summed, then some means is usually applied to try to manage the inevitable residual uncertainty embodied in the particular methods used. Determining the costs for the ongoing operation as well as the ongoing value production can be more challenging, particularly for an energy storage asset. Quantifying the lifetime cost and value creation entails not only understanding how the asset might operate in the currently prevailing market circumstances, but also forecasting the asset's performance in prevailing market conditions many years into the future.

There are numerous ways an energy storage asset might realize value for its owner, and in most practical cases a particular asset may tap into multiple elements of value creation. The specifics of the value analysis vary from project to project and are heavily dependent on the location of the envisioned project. It is therefore difficult to describe a treatment that is both broadly applicable and comprehensive. The following sections describe various features of value analysis for energy storage assets. An example case of (diabatic) compressed air energy storage (CAES) operating in the open market in Texas (Electric Reliability Council of Texas, ERCOT) is used in the development. The development seeks to determine the economically optimal operation from the point of the energy storage asset owner. This case has some favorable attributes for demonstration purpose. For instance, the diabatic CAES process requires some fuel consumption, so this feature is included in the development. Energy storage schemes which do not require fuel are modeled just as easily by omitting the fuel term. The ERCOT market also features numerous ancillary service products that are cooptimized for hourly dispatch with energy, such that asset owners may make concurrent hourly offers for both energy and ancillary service products.

Rather than jump directly to a fully featured optimization model, the development later explores an array of value modeling techniques, starting with simpler techniques first to make certain points then evolving toward the fully featured approach. It should be noted from the outset that even the fully featured modeling approach described here still has some deficiencies, leading to uncertainty in its outcome which must be effectively managed.

9.4.1 Energy markets and energy price variation

In order to achieve the reliable supply of electricity to serve a group of users, it is at minimum necessary that the installed generating capacity exceeds the maximum anticipated demand. This objective is achieved in most practical cases by defining a balancing consisting of an array of generators and users, and then establishing a balancing authority to ensure that the objective is effectively managed within its respective area. In the case of large, interconnected grids, the balancing inevitably must consider imports to and/or exports from the balancing area, which requires some degree of coordination between neighboring entities. But basically each balancing authority is responsible to match the electrical supply to the demand within its balancing area.

Fig. 13 illustrates the geographic arrangement of balancing authorities in the United States. In some regions, an independent market is established, leading to competition between multiple companies owning generation assets to serve electrical demand. This is called a deregulated wholesale electricity market structure. Examples of this structure include ERCOT, CAISO (California Independent System Operator), NYISO (New York Independent System Operator), and others. In other regions, a single corporate entity is responsible for electricity supply over a designated area. This occurs where a single company owns a sufficient fraction of the generating resources serving an area, such that this company could exercise monopolistic pricing power. In this case, a regulator is required to ensure a fair balance of price to the consumers and profit of the service provider. The regulator is usually a state-level public utility commission. This is the regulated market structure. The southeast and northwest regions of the United States operate in this fashion. The regulated structure does not imply that only one supplier operates in a region. The Public Utilities Regulatory Policies Act of 1978 requires that regulated utilities provide market access

FIG. 13 Electric power markets in the United States [11].

to independent generators. The price for the electricity provided by the independent generator is determined as the cost avoided by the utility in not furnishing the electricity that is instead provided by the independent generator.

The overall service of delivering electricity to consumers consists of essentially three parts: the wholesale electricity market, the transmission delivery, and the retail market. In the wholesale market, electricity prices are often very volatile, as the next section will explain. Furthermore, individual consumers have no practical capability to hedge their exposure to prices in the wholesale market. This is the main service provided by electricity retailers. Retailers manage exposure to the wholesale market for a large group of users and by appropriate hedging of this exposure, offer electricity pricing to end users that is stable and equitable (and also profitable to the retailer). In a given location, the regulated or deregulated concepts can be applied differently over the different scopes. For instance, transmission often follows the regulated model even where the wholesale market is deregulated. In many cases, the retail market is also regulated even while the wholesale market is deregulated.

Whether the wholesale market is regulated or deregulated, one major role of the balancing authority remains essentially the same. The balancing authority determines which units run and at what output level in order to meet the demand. The total electricity demand over a geographical region generally varies over time, for example, as consumers turn on lighting in the evening or run air conditioning when the weather is warm. Since the demand varies in time and supply must match with this demand, the balancing authority must update its decision on which units to run and at what output levels on a continuous basis. The obvious guiding principle to this decision-making process, which is typically enshrined in the regulatory framework governing the balancing authority, is to dispatch units so as to minimize the total cost of serving the load. This objective is achieved in principle by dispatching units in ascending order of their marginal cost until enough units are online with enough output so that the demand is matched. In the typical deregulated market construct, all generators that are dispatched receive the energy price paid to the marginal unit. For instance, if the marginal unit has a bid price of $30/MWh and a particular generator has a bid price of $20/MWh, then that generator is dispatched and paid at a rate of $30/MWh (i.e., $10/MWh profit). Extending the example, a particular generator with a bid price of $40/MWh would not be selected to run and would therefore receive no compensation. The situation is best visualized by considering a graphic such as Fig. 14.

The line in Fig. 14 indicates how much cumulative capacity is available with a marginal cost less than the value indicated by the left axis. At the left side of the curve, the energy bid price of generators generally corresponds to their variable cost. Since fixed costs are "sunk" no matter what the results of the dispatch, any running that returns revenue in excess of the variable cost is positive for generators. At the right of the curve, the bid price might also include a scarcity pricing component. Generators at the right side of the curve do not run

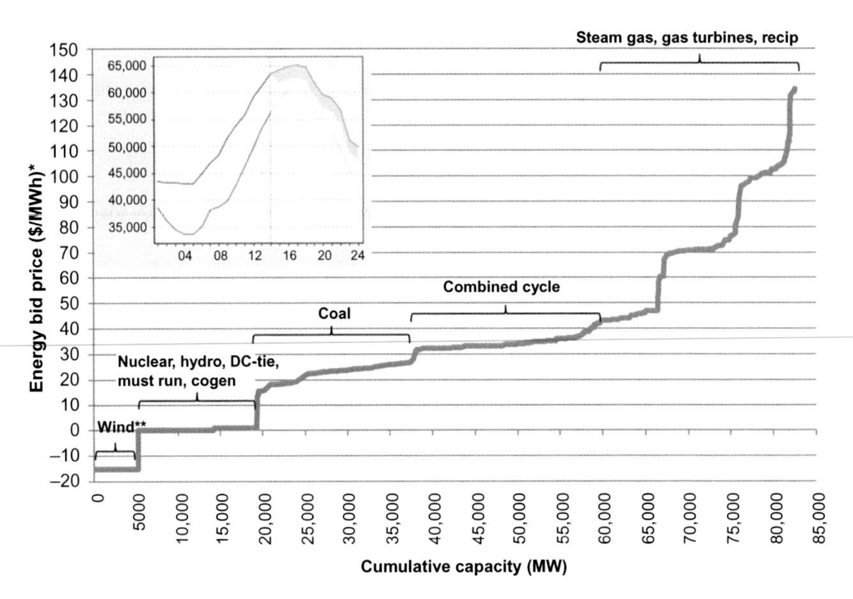

FIG. 14 ERCOT price stack.

often, so they must amortize their initial costs over a relatively smaller number of operating hours. If bid prices for these units only recovered the variable cost, this would not bring enough return to justify investment in these units. The supply curve must therefore become steep at the right side. This is true in this ERCOT example, as ERCOT is an "energy only" market. That is, the economic dispatch of units to serve the energy demand is the only means for generators to earn revenue. Some markets in different regions feature a parallel capacity market concept, where generators receive some compensation by way of simply bidding into the market, i.e., providing capacity, whether or not they are dispatched. The debate over the philosophical merits of alternate market constructs is beyond the scope of this discussion, but in general it hinges upon whether or not the selected construct effectively incentivizes growth of the supply in reasonable correspondence to growth in demand so as to not lead to a supply shortage.

It should be clear from Fig. 14 that as the load on the system moves up or down over time, the price moves correspondingly, basically in accordance with the blue line. The inset chart in the figure shows the variation of the system load with hour of the day. The green and red lines distinguish between day-ahead market results and real-time market results. Further discussion of this detail will follow later. Considering the green line, load is seen to vary between about 34,000 and 61,000 MW over the 24-h period. Reading from the blue line of the main chart, this suggests a price range from about $26/MWh to $44/MWh.

One other point worth emphasizing with regard to Fig. 14 is the importance of fuel price in setting the electricity prices and in turn the profitability of units. The chart notes that the curve is based on $4.29/dt gas price (where dt is decatherm and equals 1 MMBTU). Considering the range of units denoted as "Combined Cycle" in the chart, the price range is roughly between $30/MWh and $40/MWh. Neglecting O&M cost, which is a lesser contributor to the total running cost, and having the corresponding fuel cost, the heat rate of these unit can be estimated. $30/MWh divided by $4.29/dt suggests 6.993 dt/MWh (6993 BTU/kWh in units more familiar to gas turbine engineers). Similar calculation for $40/MWh yields 9.324 dt/MWh. Suppose now that for a particular hour, the market is cleared at $40/MWh. In this hour, the unit offering $30/MWh would be dispatched and would therefore have a profitability of $10/MWh. Now consider these same units having same heat rate, but with a higher gas price of $6/dt. The bid prices in that case would have been $42/MWh and $56/MWh, respectively. In this situation, the profitability of the more efficient unit is $14/MWh. Thus in cases where the marginal unit is gas fired, the profitability of gas-fired units is equal to the heat rate advantage multiplied by the fuel cost. Another interesting outtake is that the price of fuel for the market-clearing unit even effects the profitability of the units not using that fuel. For example, consider the wind turbines, which have a bid price of −$15/MWh (bid price is negative on account that there is no fuel cost and wind turbines enjoy a production tax credit of $15/MWh, so they will break even with any electricity price above −$15/MWh). If the market clears at $40/MWh, as it would with $4.29/dt natural gas, they earn a profit of $55/MWh. In the case of $6/dt gas, they earn a profit of $71/MWh. As a final thought exercise, consider the situation where renewable penetration increases to the extent that the marginal unit that clears the market could be a renewable resource for some or even many hours. What will be the market clearing price under the current market construct? Could it approach −$15/MWh? This profitability scenario would surely deter the growth of renewables to such level, so clearly different market mechanisms are needed in parallel with or in lieu of the current energy market to reach high level of renewables penetration.

This example presents the operation of a deregulated market, but a similar situation occurs in the regulated case. The utility dispatches its lowest cost units first and adds units in ascending order of running cost until the demand is matched. The major difference between regulated and deregulated systems is how profitability is achieved by units. In the regulated situation, the regulator must approve any investment of capital (e.g., additional generation assets) and such investment must be justified by improvement to the cost of electricity payed by the ratepayers. However, once such investment is approved, the costs of recovering the capital are simply passed along to the rate payers, and recovery of this cost is pretty much assured. In the deregulated case, and in particular the energy only case, capital investment costs must be recovered by having a lower operating cost than the market clearing price for a sufficient number of hours, and in the case of "peakers," by including some capital cost recovery component in the energy bid price.

In summary, all electrical grids which experience demand variation over time and are served by an array of generators with different operating costs will have a time-varying value of electricity. The variation in price depends on the variation in demand, the mix of generation assets and their operating cost characteristics, and the prices of different fuels. For completeness and for the purpose of laying conceptual groundwork for later discussions, several other features of electricity market balancing are now described.

9.4.1.1 Day-ahead vs real-time markets

It was mentioned in the earlier discussion that the electricity demand vs time in a balancing area is generally not known in advance. This presents a problem since most electricity generating assets are large machines that do not turn on and off with the simple flip of a switch or even change their output with a simple turn of a knob. In some cases, hours of preparation are required for a generator to come online and produce a specified output. Furthermore, the process by the balancing authority to receive bids from generators, solve the optimal dispatch problem, and issue operating instructions to the fleet of generation assets takes some time. In order to allow sufficient time for effective planning while at the same time maintaining the flexibility to adjust to changing conditions in real time, most balancing authorities operate two distinct markets: a day-ahead market and a real-time market or balancing market. As the name suggests, the day-ahead market is cleared prior to each operating day by solving the dispatch problem against forecasted demand. Dispatch instructions are issued out to all generators participating in the market, so they are able to effectively anticipate the requirements of the next operating day. In most cases, the dispatch is determined on an hourly basis.

The real-time or balancing market is used to correct for difference between the projected supply and demand and the actual supply and demand. Note that with renewable resources contributing to the supply, forecasting uncertainty may exist on the supply side as well as the demand side. In the remainder of the discussion the term "net load" is used to characterize the effective load that must be taken up by dispatchable resources, which is the total load less the variable supply. The real-time market is redispatched on the shortest time interval practical, often 5 min. All generators participating in the market pay or receive a settlement for the difference between their day-ahead (DA) dispatch and what they actually provide real time (RT). A few examples are provided as follows. In these examples, real-time dispatch occurs every 5 min, i.e., $N = 12$ intervals per hour.

Example 1 A generator is dispatched in the day-ahead market to provide 100 MW over a given hour. The day-ahead price is \$40/MWh, so the generator anticipates a \$4000 payment. The generator experiences a mechanical issue such that it is only able to deliver 95 MW over the hour. The real-time price is \$30/MWh for the first 30 min of the hour and \$35/MWh for the remainder of the hour.

The payment to this generator is as follows:

$$\begin{aligned} Payment &= P_{DA}MW_{DA} + \tfrac{1}{N}\sum P_{RT,i}(MW_{RT,i} - MW_{DA}) \\ &= 40 \cdot 100 + \tfrac{1}{12}[6 \cdot 30 \cdot (95 - 100) + 6 \cdot 35 \cdot (95 - 100)] \\ &= 3837.5 \end{aligned}$$

Example 2 A gas turbine generator is dispatched in the day-ahead market to provide 100 MW over a given hour. The day-ahead price is \$40/MWh. Suppose that this generator has a 9 dt/MWh heat rate and gas price is \$4/dt, such that it has an operating cost of \$36/MWh. The generator anticipates a \$4000 payment and \$400 profit in this hour. Suppose the real-time price is \$30/MWh over the entire hour. If the generator operates as planned, it still earns the promised \$400 profit. However, if the generator intentionally drops off-line it earns an even greater payment as follows:

$$\begin{aligned} Payment &= P_{DA}MW_{DA} + \tfrac{1}{N}\sum P_{RT,i}(MW_{RT,i} - MW_{DA}) \\ &= 40 \cdot 100 + \tfrac{1}{12} \cdot 30 \cdot 12 \cdot (0 - 100) \\ &= 1000 \end{aligned}$$

As the fuel cost is not incurred, this payment is all profit. Note that profit could even be improved without completely dropping off-line, but simply reducing load. In the case of a gas turbine it is often favorable to stay online to avoid a start-up, which typically incurs some O&M cost. Assuming a consistent output over the hour, $MW_{RT,\ i} = MW_{RT}$ for all $\boldsymbol{i}$, profit increases linearly between \$400 and \$1000 as actual production ranges from 100 MW (per day ahead dispatch) to 0 MW.

$$\begin{aligned} Profit &= P_{DA}MW_{DA} + \tfrac{1}{N}\sum P_{RT,i}(MW_{RT,i} - MW_{DA}) - P_{Fuel} \cdot HR \cdot MW_{RT,i} \\ &= 40*100 + \tfrac{1}{12} \cdot [30 \cdot 12 \cdot (MW_{RT} - 100) - 4 \cdot 12 \cdot 9 \cdot MW_{RT}] \\ &= 1000 - 6MW_{RT} \end{aligned}$$

This example illustrates how a real-time price lower than the operating cost of a generator induces that generator to reduce its output or drop off and hence how the balancing authority uses real-time market to manage conditions of either surplus generation or shortfall in generation cleared in the day-ahead market.

There are a few important things to understand about the real-time market: First, the magnitude of the real-time market is only the delta between the actual net load and forecasted net load. This is generally a small fraction of the total system demand. Second, in principle the difference between the real-time price and the day-ahead price should be a randomly distributed variable with a mean of zero. In other words, there should be no long-term increase or decrease in profitability for a generator to participate in the real-time market compared to the day-ahead market. In competitive markets, it is often permissible to bid in synthetic load or synthetic generation into the day-ahead market, that is load or generation that has no attachment to physical load or generation and then take a settlement for the difference between the day-ahead and real-time price. This facility in principle mitigates systematic discrepancies between real-time and day-ahead prices to the extent that market participants have equal access to and understanding of the state of the market. Therefore the only remaining price discrepancies ought to be random. The real-time market inevitably shows higher volatility and larger range than the day-ahead market. Imagine if each hourly price found in the day-ahead market had a bell curve of real-time prices centered upon it. The price range defined by the tails of the curves obviously is larger than the price range defined by the means. Since an energy storage device realizes higher profits with higher price volatility, it might seem intriguing to only participate in the real-time market and never participate in the day-ahead market. However, since the real-time market size is generally small, the price sensitivity to market participation is much more pronounced. For instance, charging when the real-time price is low might quickly drive it up. Furthermore, as the relationship to the day-ahead price is in principle random, there is no certainty in any hour that a more favorable price will be available in one market or the other.

9.4.1.2 Frequency regulation

It is a rather unique feature of electricity balancing that the supply and demand must balance in real time. This balance requirement follows from laws of energy conservation and it is not affected at its most basic level by the presence of energy storage systems. Energy storage schemes universally involve conversion of electricity into some other energy form that is more practically storable and then conversion of this stored resource back to electricity again. There is really no storage of the flowing electrons that constitute electricity, at least not over a timeframe that is meaningful to this discussion.

Aside from economic dispatch by the balancing authority in the day-ahead and real-time markets as discussed previously, there are essentially two passive feedback mechanisms and an array of active feedback mechanisms which also factor into this balancing act. In order to understand the working of these feedback mechanisms, it is necessary to realize that the grid frequency is the manifestation of power balance or imbalance in an AC electrical system. The target grid frequency (60 or 50 Hz in most places) is realized if and only if supply and demand are in perfect balance. If load exceeds supply, then the frequency of the entire grid slows down. At lower frequency, less power is delivered to loads, so without any intervention, an overloaded AC system might find an equilibrium at less than the target frequency. Likewise, a system that has too much supply could find an equilibrium at higher than desired frequency.

The first passive control mechanism is therefore the rotational inertia of the connected generators and loads. In a condition of undersupply, the inertia of the system prevents the instantaneous decay of the frequency to a lower equilibrium. Rather, the power imbalance results in a decelerating torque, and the deceleration of the grid is then proportional to the imbalance torque and inversely proportional to the total inertia. Effectively, rotational kinetic energy is feeding the load until a new state with lower kinetic energy is reached.

The second passive, or perhaps more correctly termed "semipassive" feedback control is generator droop setting. Generator droop setting automatically increases the output of a generator's driver in response to deviations in the grid frequency. For example, the generator droop control manipulates the steam turbine's throttle valve or the gas turbine's fuel valve to deliver more power if the grid frequency is low or less power if it is high. This control action is managed by the local controllers of each generator and happens with no intervention by a central balancing authority. In most cases, there is a dead band over which the droop control is inactive. Therefore this control does not respond to small deviations in frequency which occur constantly, but it tries to prevent things from getting too far out of hand in the event of larger excursions.

It is noteworthy that major forms of renewable energy such as wind and solar PV, which feed power to the grid through a solid-state inverter, do not effectively contribute inertia to the system. As such, increased renewables penetration level often comes with an unwanted side effect of lower inherent stability of the electrical system. It is possible through the control of the inverter to realize a sort of "synthetic inertia" in these inverter-connected systems, in order to mitigate this destabilizing effect. However, to the extent that the synthetic inertia is not backed up by actual physical inertia, synthetic inertia is only a manipulation of the response of the system in terms of active power vs time in the event of an observed disturbance of the grid frequency. Implementing droop control or synthetic inertia schemes implies some "holding back" of available output. For combustion units, the equipment is generally capable to deliver higher output, albeit the capability is usually limited to a few percent of design capacity and the incremental output often comes with reduced efficiency. Still, reserve is available and only fuel usage is being held back. Of course, fuel that is not used can still be burned later. For renewables, any hold-back implies waste of the renewable resource. The wind or the sun is not held-back. Any output that is not produced at the moment it is available is lost forever. Thus while droop schemes and synthetic inertia schemes can be implemented in renewable generation devices and are often mandated for system stability reasons, such schemes carry a higher opportunity cost.

The previous sections have discussed the economic means to achieve balance of supply and demand using, as the first means, the day-ahead energy market and then fine-tuning this over the timeframe of minutes using the real-time market. Also, the automatic feedback mechanisms of system inertia and generator droop control have been described. The economic control mechanism is generally quite powerful in terms of magnitude, but the timeframe of minutes is far too slow to achieve the tight control of frequency that is desired. On the other hand, the automatic mechanisms of system inertia and droop control are quite fast-acting but have limited corrective power. Furthermore, the generator droop mechanism somewhat indiscriminately moves generators across the network from their targeted economic dispatch point to some alternate point to maintain system stability. This is acceptable as a very short-term provision, but excessive response by droop controllers is generally not desirable. In order to effectively regulate the system in the 0–5 min interval, a balancing authority must employ one or more other instruments. A typical solution is the use of a frequency regulation service.

Technically, frequency regulation service means that a generator offers a power range over which the balancing authority can adjust the output automatically via a command signal which is updated on very fast interval. This is a sort of centrally operated real-time feedback control. The amount of regulation needed by the balancing authority has to do with how much the load might possibly change within the window of the economic dispatch. Of course, this amount is likely to be a relatively small fraction of the overall system load. An excellent discussion on the ERCOT's operation of its hourly regulation market is provided in [12, 13]. In order to provide this service, note that a generator may have to change its output from the optimal economic point. For example, suppose a 100 MW rated generator offers ± 10 MW of regulation service. In order to provide this service, the generator must operate at 90 MW and have a minimum output capability of 80 MW or less. In cases where the market clearing price exceeds the generator's operational cost, operation at less than its rated capacity implies some economic sacrifice. Therefore some compensation is usually necessary to induce a generator to provide this service.

There is significant variation from place to place regarding how regulation markets operate, in terms of the technical definition of what is offered as well as the compensation scheme. For instance, some places have bidirectional regulation: that is, an offer of 10 MW of regulation means the unit offers to move 10 MW up or 10 MW down. Other places have regulation segregated by direction: a unit may choose to offer to move up 10 MW (but not down) or down 10 MW (but not up). ERCOT, for instance, has separate "UpReg" and "DownReg" products. In some places, a market for regulation is cleared hour by hour (usually the day ahead). In other places, a generator may bid to provide regulation service for a month or longer (this is prevalent in Europe). In some places, regulation service is not an independent and optional market. Rather, any generator that provides energy must also provide a share of the regulation. This is most often the case in regulated markets where there is no commercial advantage or disadvantage incurred by competing entities depending on how the service is provided. The regulated utility owns the assets and simultaneously owns the obligation to balance the system, so the assets can be deployed to this purpose in whatever way minimizes the overall cost. Some places define different regulation products in terms of the speed of response to a command signal. For instance, there could be separate regulation products for response within milliseconds of command vs response within seconds of command vs response within minutes of command. The PJM (Pennsylvania-New Jersey-Maryland Interconnection) market features a unique dynamic regulation product that is supposed to be energy neutral over a period of 15 min. This construct specifically attracts fast-acting but low energy devices, such as batteries. Compensation mechanisms for regulation are similarly diverse. It was illustrated earlier that providing the regulation service generally implies some cost to the generator. In some places this is compensated via a market for this service, similar to but independent from the energy market. Some places provide a "make-whole" payment, such that the energy dispatch determines the profitability of the unit and any economic loss incurred by moving away from the economic dispatch point to provide regulation service is reimbursed. Still other places offer a longer-term capacity like payment to provide regulation service over a longer period.

Frequency regulation service uses real-time control of some generating assets to effectively balance supply and demand within the time horizon of economic dispatch while the system operates normally. The magnitude of frequency regulation service procured by the balancing authority depends on how much variation is anticipated within that time horizon. The frequency regulation resource is not intended to accommodate significant loss of generating capacity in the system, as might occur if a large unit or a significant transmission line trips out of service. In order to ensure overall system reliability, balancing authorities must also make reasonable provisions for such contingency events. Such provision usually entails assigning some generation resources to remain online at lower than their rated capacity or be ready to quickly come online. The magnitude of the reserve provision is usually sized by the most significant one or two events that could remove online capacity within the balancing area. This service may be termed as "spinning reserve" or "synchronized reserve" in the case of online assets or "nonspinning" or "fast-start reserve" in the case of assets standing by but not running. As with the regulation service, operating under turn-down condition or not operating often implies an economic cost which must be compensated to induce generators to offer the service, and as with regulation service, there is significant diversity from place to place in the requirements and compensation schemes to provide these sorts of reserve services. Most places will have some requirement around response time to reach targeted output once the deployment is initiated. Some places will also have a requirement that once initiated the output must be sustained for some minimum duration. After all, it would not be helpful if the reserve provision would deteriorate before the balancing authority could effectively and completely mitigate the initial loss. A minimum sustaining duration prevents energy storage devices with too small of storage durations from participating.

Collectively, regulation service and reserve service are often termed ancillary service, as they arc ancillary to the basic function of providing energy. There are often a few other services in this category, but they usually do not influence bidding decisions of generators or dispatch decisions of balancing authorities and thus are of less importance in the current discussion.

9.4.2 Price array sorting method

A rather instinctive approach to deploying an energy storage asset in a situation of volatile energy prices would be to "buy low and sell high." That is, when energy prices are low, operate the charging part of the storage asset. When prices are high, operate the discharging part of the storage asset. In fact, a reasonable strategy can be developed in this simple way. Suppose a price signal is given as an array of prices. Hourly prices will be presumed for the remainder of this discussion as this is very often the case for day-ahead electricity market dispatch, but the principles described apply just as readily with shorter or longer intervals. The price array can be ordered from the lowest price to the highest price. Operation of the discharging system can be assumed in the hour having the highest price. This 1 h of operation for the discharging system will imply a certain quantity of consumption of the stored resource. Note that this concept is quite general in that the stored resource here might be water in a reservoir, or it could be air in a cavern, or thermal energy in a heat store, or any other form of storable attribute that might form the basis of an energy storage asset. In any case, the consumption of the stored resource over 1 h of discharging operation will necessitate some corresponding amount of charging operation. Depending on the relative size of the charging and discharging devices, the 1 h of discharging might require more or less than 1 h of charging operation. If they are equally sized in terms of rate of resource consumption and production, then 1 h of discharging will require a corresponding 1 h of charging. This corresponding charging energy should be assumed to come during the lowest priced hour. This very first hour of discharging operation will be the most profitable one, having the highest difference between low and high price, and each additional hour of operation will bring lower and lower additional profit. Assuming this first hour is profitable, consideration proceeds down to the next highest price hour (and correspondingly up to the next lowest price hour for charging) and so on through the ordered price array until either all hours are consumed or the calculation at one of the hours suggests that the incremental profit obtained from running in that hour (and thus incurring the corresponding charging cost) is no longer a positive number. This exercise establishes the optimal number of running hours for the asset within the time horizon of the given price array. This optimality can also be characterized by "strike prices" for the charging and discharging systems: the discharging system should run during any hour where the electricity price is higher than this last marginal hour and the charging system should run during any hour where the electricity price is lower than the corresponding low price that was used to calculate the profit for the last marginal hour. Note that some additional care is required to correctly account for the charging costs for partial hours if the size ratio of charging to discharging

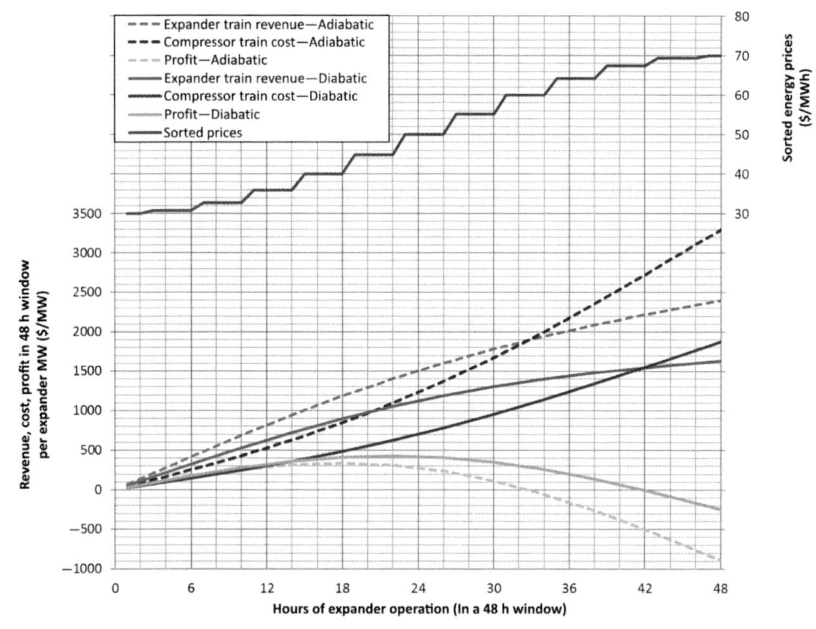

FIG. 15 Energy storage asset dispatch by sorted price array method.

systems in terms of their resource production and consumption is not an integer. Also, in the case of diabatic CAES or other energy storage device that uses fuel or any other commodity in the course of its operation, the costs for this commodity must be duly accounted.

In order to illustrate the method and typical results, consider an electricity price signal that is sinusoidal repeating daily with a minimum of $30/MWh and maximum of $70/MWh. Fig. 15 shows in the upper section the price array values ordered from least to greatest. The presentation appears stepped because in this 48-h window there are 2 cycles of the sinusoid, and prices are given hourly. Therefore the array includes two data points at each of the extrema (max and min) and 4 h having each intermediate value. The lower portion of Fig. 15 shows cumulative costs and revenues for each of two energy storage systems, with parameters defined per Table 4. The terminology in the table refers to CAES examples, but the Adiabatic CAES system serves as a reasonable proxy for almost any unfired storage device having comparable efficiency (note that for an unfired storage device, the inverse of energy ratio is equivalent to the round-trip efficiency). For the diabatic CAES system, a fuel cost of $4/MMBTU is considered, and this cost is subtracted from the discharging revenue line, such that the presented line shows the revenue net of fuel cost. The Operating profiles corresponding to the optimal dispatch are illustrated vs time in Fig. 16.

TABLE 4 Parameters for energy storage devices.

	Diabatic CAES	Adiabatic CAES
Expander flow (lbm/s)	2.5	3.0
Expander power (MW)	1.0	1.0
Compressor flow (lbm/s)	2.5	3.0
Compressor power (MW)	0.782	1.38
Heat rate (BTU/kWh)	3994	0
Specific air consumption (lbm/kWh)	9	10.7
Specific air production (lbm/kWh)	11.5	7.8
Energy ratio (SAC/SAP)	0.783	1.372
Inverse of Energy ratio (SAP/SAC)	1.277	0.729

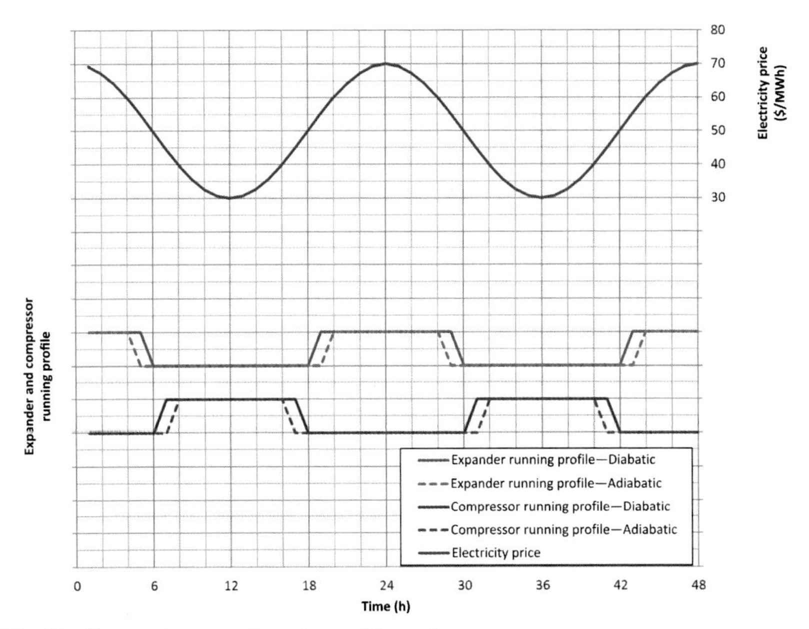

FIG. 16 Energy storage asset running profiles vs time.

The main features of the revenue and cost lines in Fig. 15 are as follows:

- The discharging revenue line has lower and lower slope with each additional hour of operation. This is because the discharging has begun with the most expensive hour in the array and with each additional hour, the electricity price is lower and therefore the revenue less.

- Likewise, the charging cost line has a slope that increases with each additional hour of operation.
- The difference between the revenue line and the cost line represents the profit line (also illustrated). The profit line exhibits a local maximum where the delta between the cumulative revenue line and the cumulative cost line is the greatest. Operating the energy storage asset for the number of hours corresponding to this local maximum is thus the optimal dispatch strategy.
- It is clear that the optimal number of running hours (capacity factor) is not arbitrary and it depends on the characteristics of the storage technology and also the array of prices. This highlights the deficiency of the levelized cost methods.

9.4.3 Stored resource valuation method

For conventional generators, it is a relatively simple process to decide when to run in an environment of variable prices. If the price of electricity exceeds the cost of fuel and O&M that a generator must incur in operation, then operation in that circumstance is profitable. Since all fixed costs are sunk whether or not the generator operates, any operation that earns a profit is better than the alternative of sitting idle and making no profit. Of course, if the price of electricity is less than the cost of fuel and O&M, then operation in that circumstance is unprofitable, which is worse than sitting idle. For a renewable resource that has no fuel cost, any positive electricity price leads to profitable operation. In fact, in cases like wind power in the United States where a production tax credit is offered, energy production is even advantageous when electricity price is negative, so long as the negative price is lower in magnitude than the tax credit. In these examples, real-time decision making by the generator is easy because the consumed resource is clearly priced: in the case of the fossil generator, fuel has a known price and in the case of the wind asset the consumed commodity is available without cost. Unfortunately, this type of logic is not so easily applied for an energy storage asset, because its consumed resource, whether it is water in a reservoir, air in a cavern, or charge in a battery, is generally not priced in any sort of market. The energy storage asset is the sole supplier and consumer of the stored resource. The stored resource clearly has some value in this "market of one." Cost has been invested in equipment to produce the stored resource and energy must also be consumed to produce it. Cost is also invested in the ability to convert the stored resource back into electricity. The stored resource must therefore have some value. If only it could be known, then the energy storage asset could operate based on value decisions in real time just as the conventional generator. The basic strategy would be as follows:

- If the value of electricity is less than the value of resource that can be accumulated by the charging operation, then charging is a profitable decision

- If the value of electricity exceeds the value of the resource that must be released in the discharging operation, then discharging is considered a profitable operation. If fuel or any other resource is used in the operation, as for example in diabatic CAES, this cost must also be accounted.
- If neither charging nor discharging is profitable in accordance with the earlier definitions, then it is best to sit idle.
- If both options are profitable, then operation depends on system capability: If it is possible to operate both the charging and discharging system simultaneously, then this operation is considered profitable, but if simultaneous operation is not possible, then choose the one with the higher profitability. This scenario is likely to be encountered only in diabatic CAES or other energy storage device that similarly has a fuel-burning/power generation feature. If both units operate simultaneously, the CAES device is operating as a conventional gas turbine with the compressor feeding the expander. This situation occurs when electricity price is high enough to merit such operation.

The same price and asset performance data as was used previously is now considered again for the illustration of this method, although only the diabatic CAES asset is considered here. Now the electricity price array can be used to deliver some insight into what the value of the stored commodity might be. Start by considering the minimum price in the array. This point establishes the minimum value of interest for the stored resource. If it is not profitable to accumulate the stored resource even at the cheapest electricity price in the considered time horizon, then no operation is profitable, and the asset should sit idle. Setting profit equal to zero at the minimum electricity price point therefore reveals the minimum endpoint to the range of feasible stored resource value. Likewise, if it is not profitable to discharge even at the highest electricity price, then no operation is profitable. Setting profit equal to zero at the maximum electricity price therefore reveals the maximum endpoint to the range of feasible stored resource value. With the minimum and maximum endpoints established, the interior of this range can be explored by considering an arbitrary array of intermediate points and evaluating in accordance with the concepts previously what the operation and profit will be at each point.

Fig. 17 illustrates the exploration of asset operation over the range of resource valuation considered. Two methods are used to calculate the profit. In the first method, the value of the stored resource is included in the profit. Note however, that this value is only a representation of value by the energy storage asset owner. It could be considered as an asset that has no liquidity as there is no external market for this store of wealth. In the second method, the value of the stored resource is excluded, so that only costs paid into external markets and revenues earned in external markets are included. Finally, the change in inventory over the operating window is included here. This is an

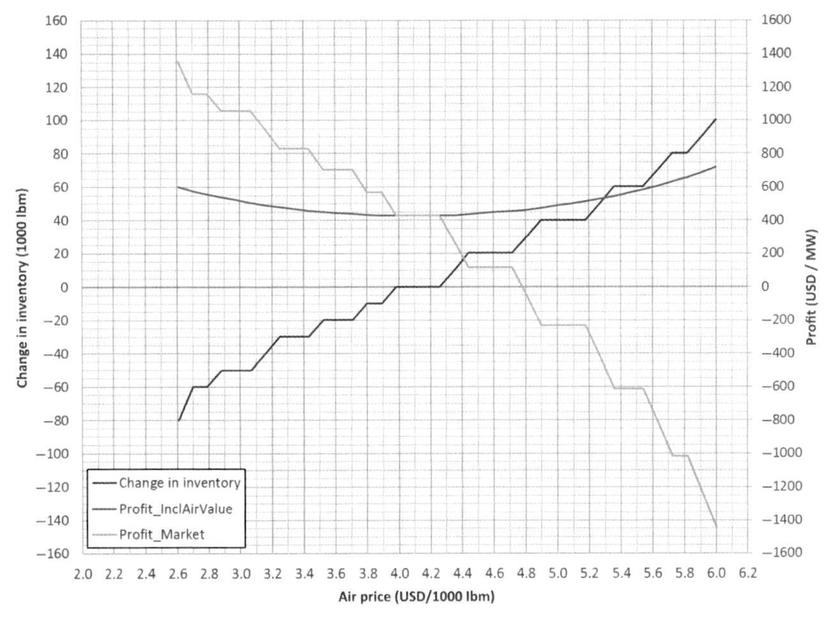

FIG. 17 Profit vs stored resource valuation.

important distinction from the price sorting method given previously, which presumed a balance of charging and discharging such that inventory was restored to its initial condition at the end of the time horizon. In this approach we can see that only one for a very limited range of valuations for the air is the inventory restored to its initial condition. If air is valued less than this amount, then the inventory is diminished at the end of the operating window. Cheap air incentivizes more discharging since during more hours the price of electricity will exceed the costs attributable to the fuel and air and during less hours will the value of the air exceed the cost of electricity needed to compress it. Likewise, more valuable air favors more compression and less expansion over the same window, leading to an increase in the inventory. The profit line for the external market declines monotonically, as high level of expansion on the left side of the chart delivers highest profitability and high level of compression on the right side of the chart delivers heavy losses. Operating in any regime other than a balanced one is of course unsustainable in the long run as it is limited by the size of the storage device storage size. At the unique resource valuation that leads to the inventory being preserved over the window, the internal and external profit functions are identical. At resource valuations where there is inventory depletion (left side of chart) the internal profit is lower than the external profit since the internal figure assigns some value to the lost inventory. Where there is inventory accumulation (right side of the chart), the internal profit is higher than the external profit, again on account of the value assigned to the accumulated resource. It is noteworthy that if the resource valuation is selected so that inventory is preserved, then the dispatch profile of the asset using this method is identical to the profile obtained using price sorting method discussed previously. As such, the profitability at the valuation that balances the inventory is the same as what was shown as a local maximum in the method given previously. It is quite interesting that this point should appear as a local minimum in Fig. 17. This would seem to suggest that more profitable operation is achieved in any regime other than the one that balances the inventory, with profit increasing the further we move from this point in either direction! Perhaps the explanation is that operation away from the balance point can never really deliver additional profit over the long run. If the asset is operated with some depletion of inventory over the current period (i.e., higher profit from external market), it will have to catch up again in the future with offsetting lower profit. Taken together these two likely return profitability to the "equilibrium" point.

9.4.4 Formal mathematical optimization

The methods discussed earlier are helpful to deliver some insight into the problem, and they may even provide dispatch strategies that are quite close to optimal in many circumstances. However, they each have shortcomings in their ability to deliver a truly optimized dispatch strategy under all real-world circumstances. It should be apparent by now that the dispatch of an energy storage asset is a sort of constrained optimization problem. Therefore if the energy storage asset dispatch problem can be represented in the form of a standard optimization problem, then the abundant toolset which has been developed to attack this sort of problem might be brought to bear. The comprehensive representation of all the real-world features of energy storage dispatch problem becomes quite complex. Rather than jumping straight to this end, this section starts off with a basic representation of the problem using the same data and parameters as have been used in the previous sections. In this way, the reader will see useful and interesting connections between the formal optimization and the more simple and insightful methods. Subsequently, additional effects will be added to the model until it is built up into its full form.

The representation of a simple diabatic CAES dispatch problem is given as follows. The objective is to maximize profit, which is the sum over all hours in the selected time horizon of the revenue earned by selling electricity output from the expander train less the cost of fuel consumed by the expander train less the costs of purchased electricity absorbed by the compressor train. The constraint equations account for the mass additions and subtractions to the cavern and specify that the mass at the end of the time horizon should be equal to the mass at the beginning.

Objective function (Maximize)	Remark
$\sum_{t=1}^{T} P_{E,\,t}(MW_{exp,\,t} - MW_{comp,\,t})$ $-P_F MW_{exp,\,t} HR_{exp}$	Energy revenue minus energy cost minus fuel cost

Subject to constraints	Equation count	Remark
$M_{cavern,\,T} = M_{cavern,\,0}$	1	Stored commodity replenishment
$M_{cavern,\,t} = M_{cavern,\,t-1} + SAP_{comp} MW_{comp,\,t} - SAC_{exp} MW_{exp,\,t}$	T	Calculated mass in cavern after each time step. Note that an initial mass must be provided to make the calculation at $t = 1$
Total constraint equations	T + 1	

Decision variables	Eng. unit	Count	Bounds	Remark
$MW_{exp,\,t}$	MW	T	$0 \leq MW_{exp,\,t} \leq HSL_{exp}$	Power generated by expander
$MW_{comp,\,t}$	MW	T	$0 \leq MW_{comp,\,t} \leq HSL_{comp}$	Power absorbed by compressor
$M_{cavern,\,t}$	lbm	T	$M_{min} \leq M_{cavern,\,t} \leq M_{max}$	Mass in cavern (at end of each t period)

Parameters	Eng. unit	Count	Remark
$P_{E,\,t}$	$/MWh	T	Price of energy (buy or sell) for each hour
P_F	$/MMBTU	1	Price of fuel
HR_{exp}	MMBTU/MWh	1	Heat rate of expander
SAP_{comp}	lbm/MWh	1	Specific air production of compressor
SAC_{exp}	lbm/MWh	1	Specific air consumption of expander
HSL_{exp}	MW	1	High sustainable limit of expander
HSL_{comp}	MW	1	High sustainable limit of expander
M_{max}	lbm	1	Maximum allowable air mass in cavern
M_{min}	lbm	1	Minimum allowable air mass in cavern

This problem is a simple linear program which can be solved using the Simplex method [14].[a] If the maximum and minimum allowable air in cavern limits are sufficiently wide that the optimal operation of the system over the time horizon never encroaches on the limits, the solution to the optimization gives the same profitability as found by the methods given previously. Note that there might be minor difference in the operating profile that achieves the optimal profitability. There is often more than one way to achieve the optimal result, and the returned solution depends on some low-level details of the algorithm and its implementation. Solution of the optimization problem yields one other bit of interesting information. In a constrained optimization problem, each constraint in the problem gives rise to a shadow price. The shadow price is synonymous with the Lagrange multiplier concept in optimization. The shadow price has a physical meaning: it is the rate of improvement of the objective with

a. This problem is easily implemented using the solver add-in of Microsoft Excel.

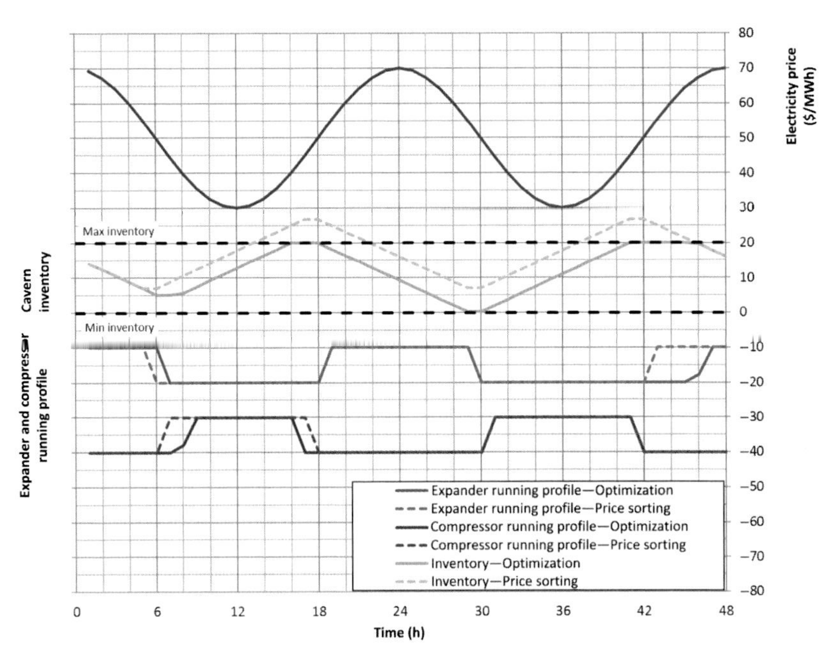

FIG. 18 Dispatch when the storage size is binding.

marginal relaxation of the corresponding constraint. If relaxing a constraint brings no improvement to the objective, then the shadow price of that constraint is zero. Such a constraint is called "nonbinding." In this example, if the cavern size limits are sufficiently wide, these constrains are nonbinding on the solution. However, the constraint that the inventory must be restored at the end of the time horizon will always be binding. As demonstrated previously in 9–17, allowing a lower inventory at the end of the time horizon leads to an increase in profit over the time horizon. It turns out that the shadow price associated with the inventory balance constraint of the optimization problem is precisely the air valuation that leads to inventory balance in the resource valuation method discussed previously.

An important attraction of the optimization method is that it can be used to determine the optimal dispatch even when the cavern size limitation is binding. Fig. 18 illustrates the dispatch of the storage asset when the storage size limits the operation. The optimizer automatically reschedules operation to comply with the storage constraint. Compared to the price sorting scheme, the optimizer scheme expands for one additional hour and it compresses for a shorter duration in order to accommodate the maximum inventory level. In this case the price sorting scheme violates the maximum inventory limitation. Some modification of the logic of the price sorting scheme is needed to prevent it from doing so, but such logic is not so easy to identify. Consider the application of the price sorting scheme, but with operation simply halted as necessary so as not to violate the constraint. If compression is stopped at the point where the inventory limit is reached but then expansion still proceeds "as planned," then this would result in a lower cavern inventory at the end of the time horizon. A reasonable compensating strategy might be to reduce the number of expansion hours in order to restore the balance. Ideally the expanding hours with the lowest profitability should be eliminated and the most lucrative expanding hours should be preserved. However, in this example one of the marginal expanding hours occurs prior to the event of reaching the storage limit, so omitting this expansion hour would in fact allow one more compression hour before the limit is reached. Following this logic, any expansion taken prior to the storage limit being reached mitigates the issue, so rescheduling expansion could be a more profitable approach than eliminating it all together, but how exactly does one make such determinations in the construct of the price sorting method?

Returning to the observation that the shadow price in the optimization scheme represents the marginal value per unit relaxation of the corresponding constraint and the observation that there is a match between the shadow price and the resource valuation that leads to balancing the inventory, it is interesting to consider what has happened to the shadow price in this case where the storage size limit is binding. In this example, the shadow price increases roughly 30%. Returning to Fig. 17 if the higher air price is used to determine the dispatch with no other corrections, the higher air valuation will lead to addition of inventory over the timeframe. In the case where the storage size is binding, the optimization result returns shadow prices not just for the resource balance constraint, but also in each hour where storage limits constrain the operation. It might be possible to realize a correction scheme to the resource valuation dispatch that allows it to achieve resource replenishment over the time horizon, and perhaps such a scheme would see a linkage between the shadow prices for the storage-constrained hours and the identified corrective measures. However, it is clear that the resource valuation method cannot be applied "as is" to achieve a dispatch that optimizes profit and achieves the goal of full resource replenishment over the time period.

This example reveals that even a simple defect can derail the basic price sorting method as well as the stored resource valuation method. Thus while these schemes offer some helpful insight into the storage asset dispatch problem, these methods cannot deliver optimal dispatch in any but the simplest of real-world applications. The optimization mechanism is therefore used as the starting point for development of algorithms that treat additional complicating factors to energy storage asset dispatch.

9.4.4.1 Adding ancillary service cooptimization

The formal mathematical optimization approach can be extended to cooptimize the dispatch for energy and ancillary services. As previously described, technical requirements and compensation methods for ancillary services vary widely from place to place. Furthermore, in most places there are numerous services that might be offered by a storage asset. As a first step, the earlier model will be extended considering a single ancillary service which can be offered by the expander train of a diabatic CAES system. The basic construct of the ancillary service is as follows:

- The service is a basic reserve service. Therefore the expander must be online operating at turndown condition in order to offer the service.
- If the expander train is selected to provide the service, it might be deployed in any level from zero up to the amount offered.
- If the expander is deployed, it receives compensation for the additional energy production due to the deployment. The energy is compensated at the real-time energy price of the deployment. In this example,
- In this example, only hour-by-hour operation is considered.

Objective function (Maximize)	Remark
$\sum_{t=1}^{T} P_{DA,t}(DA_{exp,t} - MW_{comp,t}) + (P_{AS,t} + P_{RT,t}DF_{AS,t})Anc_{exp,t} - P_F MW_{exp,t} HR_{exp}$	Day-Ahead Energy revenue minus Day-Ahead Energy Cost plus Ancillary Service Revenue minus Fuel Cost

Subject to constraints	Equation count	Remark
$M_{cavern,T} = M_{cavern,0}$	1	Stored commodity replenishment
$M_{cavern,t} = M_{cavern,t-1} + SAP_{comp}MW_{comp,t} - SAC_{exp}MW_{exp,t}$	T	Calculated mass in cavern after each time step. Note that an initial mass must be provided to make the calculation at $t = 1$
$DA_{exp,t} + Anc_{exp,t} \leq HSL_{exp}$	T	Expander capability limit
$DA_{exp,t} + Anc_{exp,t}DF_{AS,t} = MW_{exp,t}$	T	Expander real power balance
Total Constraint Equations	3 T + 1	

Decision variables	Eng. unit	Count	Bounds	Remark
$DA_{exp,t}$	MW	T	$0 \leq DA_{exp,t} \leq HSL_{exp}$	Day-ahead energy offered by expander
$Anc_{exp,t}$	MW	T	$0 \leq Anc_{exp,t} \leq HSL_{exp}$	Ancillary service offered by expander
$MW_{exp,t}$	MW	T	$0 \leq MW_{exp,t} \leq HSL_{exp}$	Real power generated by expander
$MW_{comp,t}$	MW	T	$0 \leq MW_{comp,t} \leq HSL_{comp}$	Real power absorbed by compressor
$M_{cavern,t}$	lbm	T	$M_{min} \leq M_{cavern,t} \leq M_{max}$	Mass in cavern (at end of each *t* period)

Parameters	Eng. unit	Count	Remark
$P_{DA,t}$	$/MWh	T	Day-ahead energy price (buy or sell) for each hour
$P_{AS,t}$	$/MWh	T	Price of ancillary service for each hour
$P_{RT,t}$	$/MWh	T	Real-time energy price for each hour

Continued

Parameters	Eng. unit	Count	Remark
$DF_{AS,\ t}$	–	T	Ancillary service deployment factor ranging from 0 to 1
P_F	$/MMBTU	1	Price of fuel
HR_{exp}	MMBTU/MWh	1	Heat rate of expander
SAP_{comp}	lbm/MWh	1	Specific air production of compressor
SAC_{exp}	lbm/MWh	1	Specific air consumption of expander
HSL_{exp}	MW	1	High sustainable limit of expander
HSL_{comp}	MW	1	High sustainable limit of expander
M_{max}	lbm	1	Maximum allowable air mass in cavern
M_{min}	lbm	1	Minimum allowable air mass in cavern

References

[1] NREL, Overgeneration from Solar Energy in California: A Field Guide to the Duck Chart, https://www.nrel.gov/docs/fy16osti/65023.pdf.

[2] J.M. Eyer, Energy Storage Benefits and Market Analysis Handbook, http://dua.jimeyer.net/docs/StorageMarketGuide_DUASandiaReport.pdf, 2004.

[3] Sunrun Has Found a Way to Crack Energy Storage Markets, https://finance-yahoo-com.cdn.ampproject.org/c/s/finance.yahoo.com/amphtml/news/sunrun-found-way-crack-energy-134300856.html.

[4] 2018 U.S. Utility-Scale PhotovoltaicsPlus-Energy Storage System Costs Benchmark, https://www.nrel.gov/docs/fy19osti/71714.pdf

[5] Overview of Compressed Air Energy Storage and Technology Development, https://www.mdpi.com/1996-1073/10/7/991/pdf

[6] J. Luo, et al., Overview of current development in electrical energy storage technologies and the application potential in power system operation, Appl. Energy 137 (2015) 511–536.

[7] Fact Sheet, Energy Storage, EESI, https://www.eesi.org/papers/view/energy-storage-2019, 2019.

[8] K. Bradbury, Energy Storage Technology Review, https://www.kylebradbury.org/docs/papers/Energy-Storage-Technology-Review-Kyle-Bradbury-2010.pdf, 2010.

[9] World Energy Council, E-storage, Shifitng From Cost to Value, https://www.worldenergy.org/assetas/downloads/Resources-E-storage-report-2016.02.04.pdf, 2016.

[10] IRENA: Electricy Storage and Renwabels Costs and Markets to 2030, https://www.irena.org/-/media/Files/IRENA/Agency/Publication/2017/Oct/IRENA_Electricity_Storage_Costs_2017.pdf.

[11] Electric Power Markets National Overview, Federal Energy Regulatory Commission website, https://www.ferc.gov/market-oversight/mkt-electric/overview.asp. (Accessed 23 September 2019).

[12] General Electric International, Inc., Walling R. A. Project Manager, Analysis of Wind Generation Impact on ERCOT Ancillary Services Requirements, Technical report prepared for Electric Reliability Council of Texas, March, 2008. Available at: https://www.nrc.gov/docs/ML0914/ML091420464.pdf.

[13] J. Howard, GE Wind Study Update, Electric Reliability Council of Texas Operations Analysis Dept. internal technical report, August, 2013. Available at: http://www.ercot.com/content/meetings/qmwg/keydocs/2013/1007/GEStudyAnalysis_ERCOTInternalReport.pdf.
[14] Wikipedia contributors, Simplex Algorithm. Wikipedia, The Free Encyclopedia, September 5. 15:14 UTC. Available at:https://en.wikipedia.org/w/index.php?title=Simplex_algorithm&oldid=914162226, 2019. (Accessed 13 September 2019).

Chapter 10

Advanced concepts

Sarah Simons[a], Joshua Schmitt[a], Brittany Tom[a], Huashan Bao[b], Brian Pettinato[c], and Mark Pechulis[c]

[a]*Southwest Research Institute, San Antonio, TX, United States,* [b]*Department of Engineering, Durham University, Durham, United Kingdom,* [c]*Elliott Group, Jeannette, PA, United States*

Chapter outline

10.1 Introduction

Large-scale renewable energy storage is a relatively young technology area that has rapidly grown with an increasing global demand for more energy from sources that reduce the planet's contribution to greenhouse gas emissions. The primary drawback of renewable energy is its dependence on the weather and its inability to store and send power when required. While several forms of energy storage are currently commercially available as discussed in previous chapters, new long-term and short-term storage concepts are continually being developed and improved upon to decrease capital costs and increase energy

Thermal, Mechanical, and Hybrid Chemical Energy Storage Systems
https://doi.org/10.1016/B978-0-12-819892-6.00010-1

conversion efficiencies. Many of the novel ideas described here are based on existing technologies, but taken a step further in technology development or hybridized with other types of energy storage. Other concepts, such as electrical energy storage and chemical storage via electrolytic processes, are unique to this chapter.

This chapter will discuss novel concepts and technology advancements that are under current exploration in the areas of thermal, mechanical, electrical, chemical, and hybrid storage. While not meant to be an exhaustive list of research topics, the goal of this chapter is to give the reader an overview of the current direction of development and its associated level of technology readiness.

10.2 Thermal energy storage

Thermal Energy Storage can be broken into three categories: sensible storage, latent heat, and thermochemical storage. This section will discuss current research regarding sensible and latent heat storage, while thermochemical will be covered separately. The main difference between sensible or latent heat is whether the storage material changes phase. In sensible storage applications, which are the simplest form of thermal energy storage, the storage material stays in a single phase—typically liquids (molten salts, thermal oils, and refrigerants) or solids (rocks, concrete, or metals). On the other hand, in latent heat applications the storage material is referred to as a phase change material (PCM) because it changes phase during the charge and discharge cycles. Example PCMs include water, paraffins, salt hydrates, and fatty acids.

10.2.1 Sensible storage research/facilities

Molten nitrate salt storage is often combined with Concentrated Solar Power (CSP) applications, where the energy received by the receiver is used to heat the salt (Fig. 1). There are already several of these applications in use throughout the world. In the United States, the Department of Energy (DOE) and a consortium of US utilities and industry partners successfully demonstrated this technology with the Solar Two Plant between 1996 and 1999 [1]. The thermal storage system consisted of a cold tank and a hot tank maintained as 290°C and 565°C, respectively. The system had a thermal capacity of 110 MWh and a storage efficiency of 99.5% due to small rate of heat loss compared to the large rate of heat required by the steam generator.

Since then, this technology has become highly commercialized. For example, in 2015 the Crescent Dunes facility near Tonopah, Nevada became operational with net capacity of 110 MWe with 10 h of storage [2].

An example of sensible storage that uses solids as the thermal media completed 20 months of testing in 2017 at the Masdar Institute Solar Platform. This test included several thermal elements in series and parallel to form a

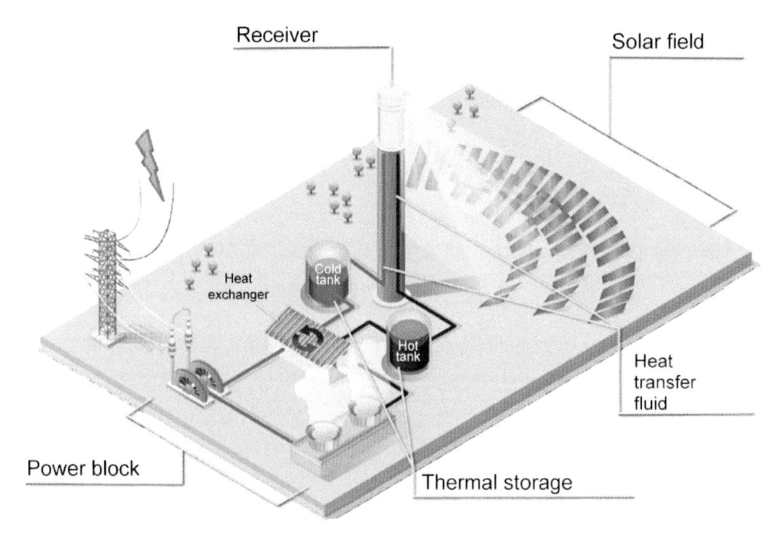

FIG. 1 Thermal energy storage with CSP. *(Source: Shultz, 2019, Presentation at TMCES Workshop San Antonio, TX.)*

heat exchanger. Each thermal element consisted of a steel pipe encased in a concrete-like medium. The heat transfer fluid then flows through the pipes to add or remove energy from the concrete-like block. The test demonstrated operation of a $2 \times 500\ kW_{th}$ pilot plant with peak storage temperatures of 380°C. At the end of the demonstration, there was no degradation in the thermal media [3].

A specific type of thermal energy storage that has received a lot of attention is Pumped Heat Energy Storage (PHES). The basic overview of PHES is electricity in drives a heat pump to charge the system. The electricity is converted to thermal potential stored in hot and cold sensible storage. The thermal potential is stored until it is used to drive a heat engine and deliver electricity out. As discussed in previous chapters, there are two prominent designs—the first using thermoclines of packed bed stores such as gravel, and the second using heat exchanges as shown in Fig. 2.

There are several ongoing efforts studying PHES system. Based on the work of Isentropic UK, Newcastle University is testing a grid-scale PHES system that utilizes two thermocline tanks containing a mineral gravel [4]. The working fluid for the heat pump and heat engine is Argon pressurized to 12 bar and heated up to 500°C. When the Argon is expanded to ambient temperature, it cools to −106°C. In 2018 ARPA-E awarded several efforts to develop PHES systems. Brayton Energy is developing a reversible turbine design. The reversible turbine is proposed to simplify the system and increase durability [5]. Southwest Research Institute (SwRI) is developing a kW-scale proof of concept demonstration focusing on system integration and controls [6]. While SwRI is

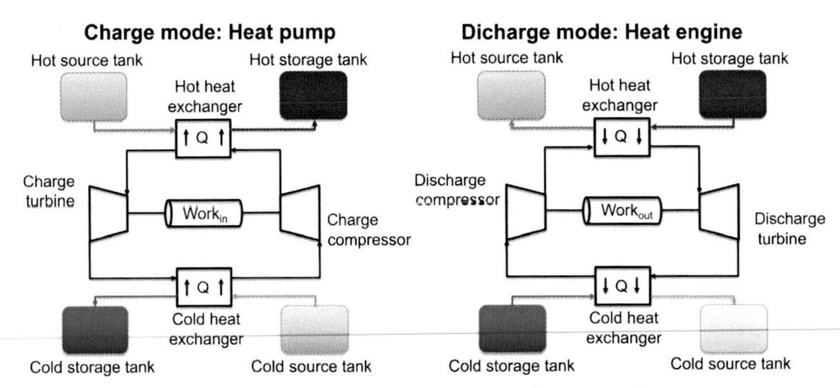

FIG. 2 Schematic of pumped heat energy storage system using heat exchangers.

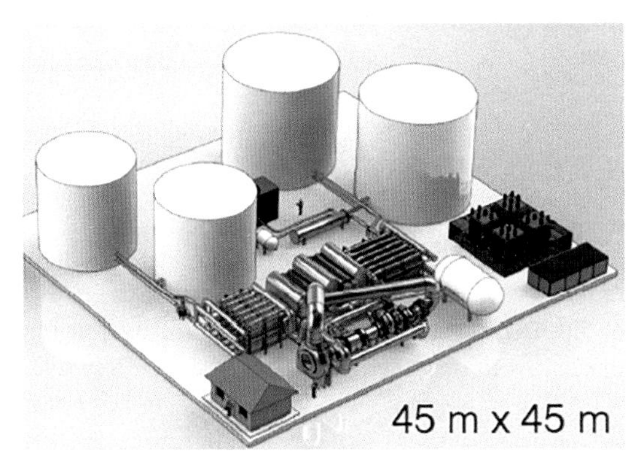

FIG. 3 Model of Malta's 10 MW Pilot System. *(Source: Little, 2019, Presentation at TMCES Workshop, San Antonio, TX.)*

using air for their working fluid, Echogen under their ARPA-E award is developing a PHES system that uses supercritical carbon dioxide [7]. Finally, Malta Inc. is developing a 10 MWe, 10 h storage pilot plant (Fig. 3). Malta will be using molten salt and refrigerant storage with air as the working fluid.

10.2.2 Latent heat research

A commercialized form of latent heat energy load shifts electricity use by creating ice in off-peak hours and then uses the ice during peak hours to cool residential or industrial buildings. Several companies including Ice Energy offer products that work with or in lieu of traditional units. Ice Energy's Ice Bear design has been installed in over 40 utility services since 2005 [8].

Another type of latent heat thermal energy storage uses encapsulation of a PCM. The PCM is encapsulated in shell to increase heat transfer during

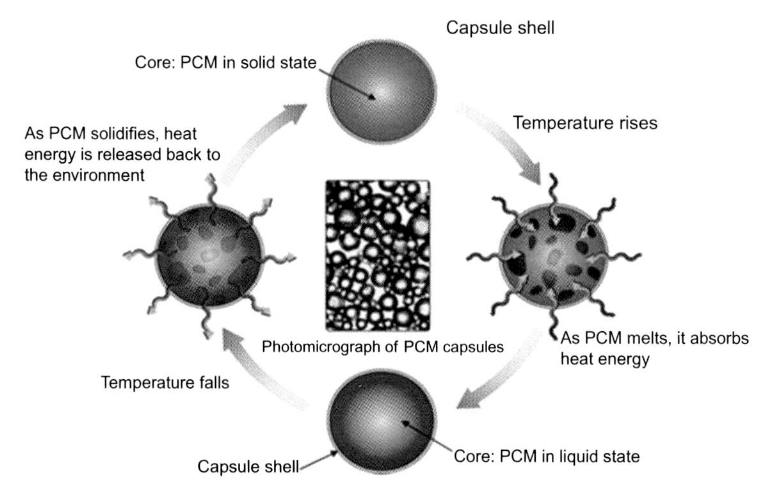

FIG. 4 Encapsulated PCM during charge and discharge modes. *(Source: TMCES Workshop, 2019.)*

discharge. During charge mode, the PCM melts and absorbs heat energy. This heat energy is stored until the PCM solidifies in discharge mode and exchanges the heat back to the environment. This cycle is shown in Fig. 4.

In an effort to decrease molten salt costs in half for CSP plants, Terrafore Technologies, in association with SwRI, developed their TerraCaps design under the US DOE Sunshot Initiative. When molten salt melts, it expands and can rupture the capsule shell. The TerraCaps have been successfully proven to accommodate this change in volume with a void. At temperatures commonly seen in CSP plants with molten salts, encapsulating the PCM can improve the energy density by 52% [9]. Other research by Munoz-Sanchez et al. [10] examined using different encapsulated PCMs throughout a thermocline CSP tank. PCMs capable of 560°C would be placed near the top of the tank, as heat transfer fluid cools thru the thermocline, the encapsulated PCM is replaced by one more suited for operation near 300°C. More effort is required to continue developing and demonstrating efficient and cost-effective encapsulated PCMs.

10.3 Mechanical energy storage

10.3.1 Advanced constant pressure CAES

The concept of a constant pressure (CP-CAES) system is described in detail in Chapter 6. The advantage of such a system is to provide a more consistent inlet pressure to the power train, limiting off-design inefficiencies and losses [11]. Most conventional concepts use an aboveground water reservoir to supply head to the underground cavity. Unfortunately, typical depths of cavities result in a

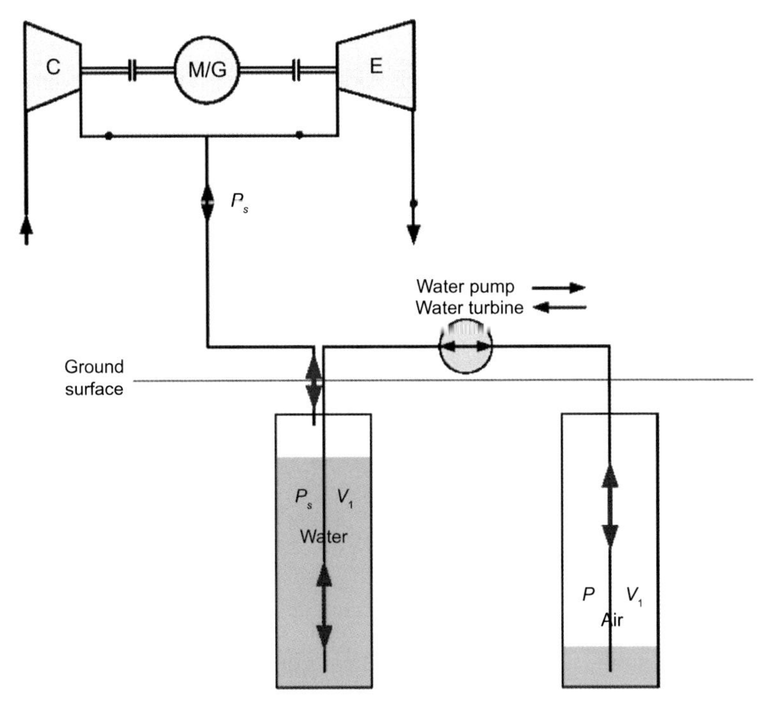

FIG. 5 Schematic diagram of a hybrid CAES and PHES system [12].

water column that can only pressurize the storage to a low pressure, approximately 5 MPa.

An advanced iteration of CP-CAES is shown in Fig. 5. This concept hybridizes CAES with PHES. As with any typical CAES system, air is compressed into an underground cavity. This cavity is filled with water. Once the intended store pressure is reached, a hydraulic machinery, acting as pump, transfers water to a nearby cavity of similar size. This maintains a high pressure in the storage cavity as it fills. The second cavity acts as a hydraulic accumulator at higher pressure than the primary storage cavity.

The air in the secondary cavity is slowly compressed as it is filled with water. The water pump also acts as a load-sharing device in tandem with the primary air compressor. While discharging, the air from the primary cavity is delivered to an expander at constant pressure. In order to maintain this pressure, water must be added to the primary cavity. The accumulated water in the secondary cavity is sent to the hydraulic machinery, acting as a turbine. Both the air expander and the hydraulic turbine are supplying power to the grid.

A study that examined this approach estimated that 23% of the power produced during discharge is in the PHES portion of the system [11]. This study operates the turbine train as a traditional CAES system with supplementary fuel firing. However, this concept could operate with an ACAES power train that stores the heat of compression in a thermal store.

The concept, as demonstrated in Fig. 5, would not require cavities at excessive depths or an aboveground reservoir to operate at constant pressure. Furthermore, the pressures would be such that the energy density could be as high as conventional CAES, on the order of 50 MPa. This is ten times larger than typical pressures with early iterations of CP-CAES.

10.3.2 Poly-generation with CAES

A flexible CAES system could be designed to produce multiple products depending on localized demand. These systems would be sized based on the projected heating, cooling, or electric requirements of the end user, which includes large campuses, industrial sites, and municipalities. The configuration of a poly-generative CAES system is shown in Fig. 6. This system would store compressed air and the heat of compression in hot water. The compressor would run during times of high electric resource, storing compressed air and hot water.

The design of the compression train and the temperature of the water storage would depend on the type of demand being met. Furthermore, the hot water storage could be pressurized and flashed to steam for certain applications. During a time of electric export, heat would be used to set the air expander inlet temperature. If the air entering the expander is not completely reheated, it will exit the expander at a low temperature. This low-temperature air could be used to supplement or replace building air conditioning requirements. If cooled water is required, cool air exiting the expander could pass through a heat exchanger connected to a cooling water system.

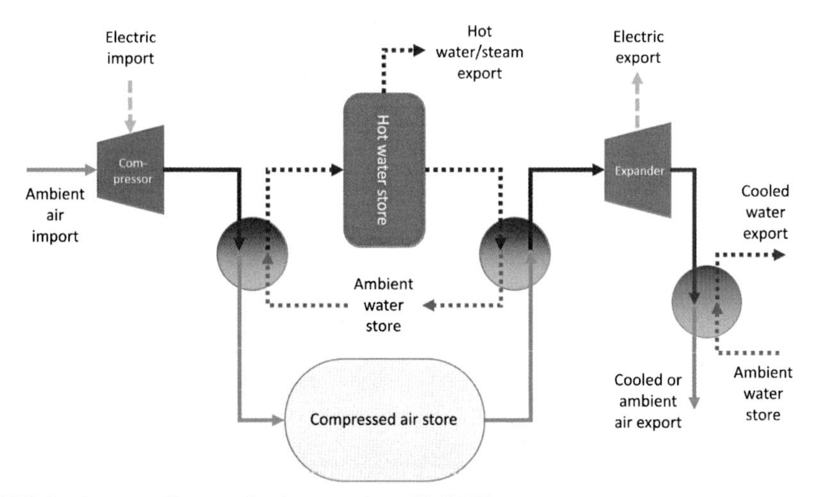

FIG. 6 A system diagram of poly-generation with CAES.

This is similar to adding energy storage to an air cycle system for building cooling, which has shown promising efficiency improvements in previous studies. One key difference between CAES and air cycle cooling is the higher pressures required in CAES for effective energy storage densities [12].

A poly-generative CAES system would provide loads small enough that the compressed air could be stored in conventional, man-made vessels, and is not dependent on being situated near a natural underground cavity. The system could also include supplemental heating, gas or electric, on the hot water storage to provide balance of plant in times of high demand for heating export. The flexibility of available products for export could allow users to flexibly optimize their system to minimize costs [12]. One further advantage to this system is the proposed systems are all commercially available with relatively little technological barriers to adapt to.

10.3.3 Distributed CAES

One approach to CAES could be to decouple the geographical location of the compression train from the geographical location of the storage and power production train, which is known as distributed CAES. As shown in Fig. 7, the compression of air is distributed within the municipalities and is sited near a district heating load [13]. The heat of compression can be used to supply district heating needs while the ambient compressed air is transmitted via pipeline to a nearby storage cavity. The compressed air storage and power generation turbine would

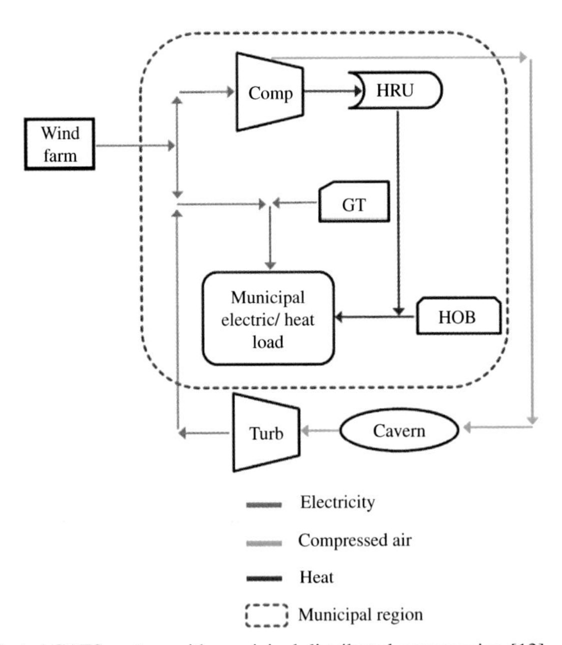

FIG. 7 Distributed CAES system with municipal distributed compression [13].

be a centralized location at a geographic underground cavity that would receive compressed air flow via the municipal pipeline network. A study by Safaei et al. performed a sensitivity analysis that included the length of pipeline and the cost of natural gas for heating. This study concluded that the economic benefit of such a system is only noticeable in areas of high municipal heating needs [13].

A subsequent study by Safaei et al. specifically focused on a case study of distributed CAES in Alberta, Canada. The application examined an installation of a conventional versus a distributed CAES system. The primary driver in economic benefit for distributed CAES is to limit the firing of natural gas for municipal heating. In the Alberta case, a compressor sized to 105 MW, an expander sized to 131 MW, and an average annual heat load of 107 MWth were the basis for the model. The result for the Alberta case is an average annual additional profit of $1.3 million per year, with increasing benefit for shorter compressed air pipelines. While distributed CAES is an attractive alternative to conventional CAES, one limitation is the requirement of a sufficiently sized underground cavity to be geographically near a concentrated heat load so as to minimize the air pipeline length [14].

10.3.4 Supercritical CAES

Supercritical CAES uses the principles of LAES for storage, but achieves the round-trip efficiencies similar to ACAES [15]. A thermodynamic study by Guo, et al. outlines the specific advantages of supercritical CAES. Energy densities can reach as high as 346 MJ/m^3, which is comparable to LAES energy density and 18 times larger than conventional CAES. The round-trip is reported to be as high as 67.41%, a 13% improvement over conventional CAES [16].

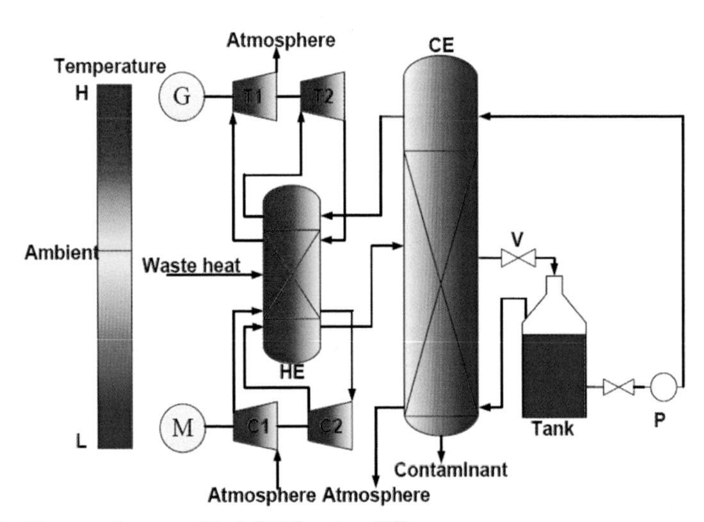

FIG. 8 Diagram of a supercritical CAES system [15].

The configuration of supercritical CAES is shown in Fig. 8. The heat of compression in the compressed air is returned to approximately ambient temperature through the hot exchanger (HE) and thermal store. The fully compressed air is cooled through the cold exchanger (CE) and thermal store. The air is liquefied across a throttle valve and stored in a tank at pressure. During discharge, a cryogenic pump pressurizes the liquid air supercritical pressures before it is vaporized and reheated by the CE and the HE pumps the liquid air. The reheated air is converted into electricity by turbines connected to a generator.

The primary difference in this cycle from LAES is the pressures and of compression and expansion. The thermodynamic study found the highest efficiencies were achieved when the storage pressure was 120 bar and the expansion pressure was 95 bar. Furthermore, it is suggested that a cryogenic liquid expander turbine be used in place of a throttle valve to achieve the highest efficiencies [16].

10.3.5 Porous media CAES with a cushion gas

One alternate approach to standard CAES is to change the cushion gas in unground cavities. In standard, constant-volume CAES, a large portion of the stored air is not used to produce power in order to provide pressure support to the power-producing machinery [17]. This unused volume is referred to as a "cushion gas," which is depicted in Fig. 9.

A study proposes using porous media systems for CAES and replacing the cushion gas with an alternate fluid [17]. If the cavern is kept above the critical pressure of CO_2, the CO_2 will be stored in a dense supercritical state. This dense state will allow for a large mass of stored CO_2 to collect at the bottom of the well. The CO_2 is unique because its compressibility effects would provide near-constant pressure over a wide range of volumes, increasing CAES discharge efficiency. The study outlined the effects of pressurization and the

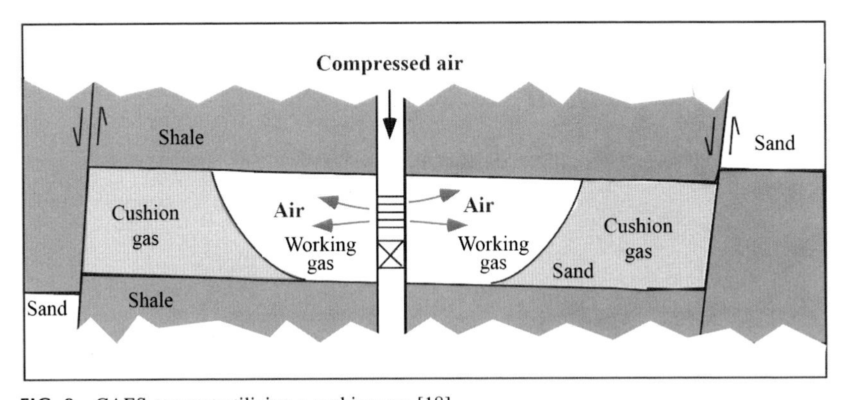

FIG. 9 CAES concept utilizing a cushion gas [18].

gas-gas mixing effects [17]. The results indicate that the air-CO_2 boundary should be as deep as 100 or more meters in the modeled cavity to prevent CO_2 mixing into the emitted air.

CO_2 was the primary gas of interest given its abundance, incentive to be stored underground, and favorable compressibility effects. However, other gasses could be considered as a cushion gas.

10.4 Flywheels

10.4.1 Introduction

Flywheels are devices that store energy in rotational kinetic form. Two major applications for energy storage flywheels include the transportation industry (both automotive and rail) as an alternative to the use of batteries in hybrid vehicles, and in the power industry for applications such as an Uninterrupted Power Supply (UPS), power grid stabilization, and as an immediate, though limited, response to peak demand. This summary concentrates on current flywheel research for power generation applications.

Flywheel energy storage (FES) systems are well suited for short-duration applications. In particular, the systems have been used as UPSs for over 30 years. More recently, FES systems have been used to improve power quality in solar and wind energy projects, maintaining frequency and voltage, and providing steady power during electrical disturbances. This recent usage, along with the development of applicable renewable energy sources, has sparked renewed interest in FES technology.

FES systems have several inherent advantages over other energy storage systems. They can provide energy quickly without needing time to start up. They have an exceptionally long life, in excess of 20 years and can provide hundreds of thousands of discharge cycles, if designed properly. Furthermore, they generate no emissions and do not have the special disposal requirements of a battery. Some inherent disadvantages include low specific energy, leakage losses (windage and bearings), cost, lead time, and a sizeable footprint especially when fabricated out of steel. These disadvantages have been present since the development of flywheels for power generation during the industrial revolution. Considerable progress has been made such that FES has been accepted as a short-duration power supply, but they are currently unsuited for use in medium- to long-duration energy storage applications. Hence, current research and development activities are focused on continued improvement of power loss and increasing the energy density.

10.4.2 Flywheel energy system

A typical modern flywheel energy system (Fig. 10) consists of a rotor mounted in a set of bearings: either rolling element, passive magnetic, active magnetic, or

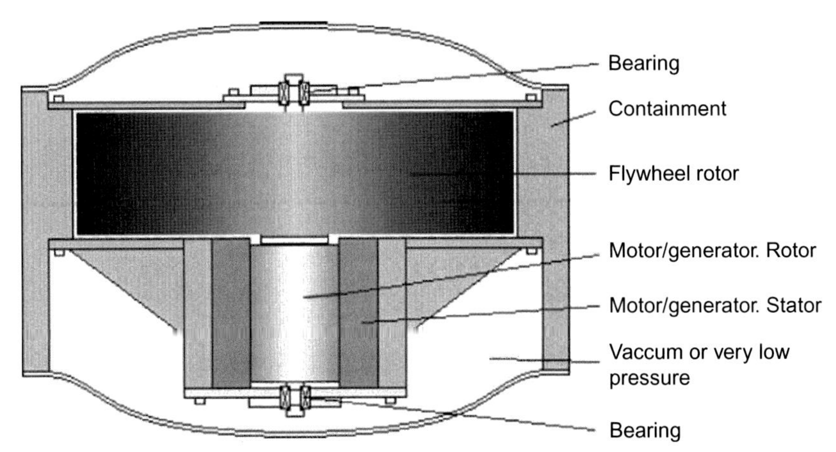

FIG. 10 Basic layout of a flywheel energy storage system. *(From B. Bolund, H. Bernhoff, M. Leijon, Flywheel energy and power storage systems, Renew. Sustain. Energy Rev. 11 (2) (2007) 235–258, https://doi.org/10.1016/j.rser.2005.01.004.)*

some mix of these technologies. The rotor and bearings are assembled into a housing that is typically under vacuum to reduce windage loss. Advanced power electronics and controls (not pictured) are used to provide 60 Hz power at variable speeds while compensating for network power factor.

The rotor itself is comprised of a disk or drum for inertial kinetic energy storage and a motor/generator for charging and discharging. The motor/generator is used as a motor to store energy by increasing the flywheel speed, and as a generator to extract energy thereby decreasing the flywheel speed.

The amount of energy stored is in accordance with Eq. (1)

$$E = \frac{1}{2} \cdot I \cdot \omega^2 \tag{1}$$

where E is the energy stored, I is the mass moment of inertia about the axis of rotation, and ω is the rotational speed in rad/s. Increasing the speed will lead to greater energy storage, but it typically comes at a cost of greater parasitic loss.

The majority of flywheel rotors are fabricated from steel due to its high strength, low cost, and availability. These steel flywheels typically rotate at less than 10,000 RPM and can have a large footprint. Much of the current research into flywheels concentrates on reducing parasitic power loss (bearings and windage), increasing the energy density (higher speeds), and improving the efficiency of the motor/generator and its electronics.

10.4.3 Bearing development

Bearings are selected based on load and speed capacity plus power loss characteristics and cost. This can lead to different solutions such as rolling element bearings, passive magnetic bearings, active magnetic bearings, or some mix of the three.

Bearings are a known source of parasitic power loss. Rolling element bearings have frictional losses, whereas magnetic bearings have essentially no frictional losses. However, they do have hysteretic and eddy current losses. Electric power is required for electromagnetic bearing designs and power for cooling is required for the superconducting magnetic bearing designs. Research is active into improved bearing solutions with reduced power loss characteristics.

Specific areas of research are primarily related to magnetic bearing applications. These include (i) permanent magnet bearings, (ii) active magnetic bearings, and (iii) superconducting magnetic bearings.

Early magnetic bearings used in flywheel applications were strictly passive consisting of permanent magnets with no input or controls needed. It was discovered that the Permanent Magnetic Bearing (PMB) worked well at lifting the shaft but are inherently unstable, thus requiring additional bearings for stabilization. An Active Magnetic Bearing (AMB) operates by sending current through a coil around a conductive material to produce electromagnetism, and relies on feedback and control instrumentation for stability (Fig. 11). A third type of magnetic bearing is the high-temperature superconducting bearing, which provides passive magnetic repulsion in the presence of a magnetic field at temperatures below a characteristic critical temperature.

PMBs are very attractive as a primary bearing due to their relatively low cost and low parasitic losses. Addressing the stability of the PMBs with new flywheel bearing designs is where much research is being concentrated. Researchers are looking at the development of hybrid bearings that combine the passive PMBs with other bearing types that are used to stabilize the shaft. These other bearing types initially included AMBs and High-Temperature Superconductor Bearings (HTSBs). A few examples which discuss designs are presented in the next paragraphs.

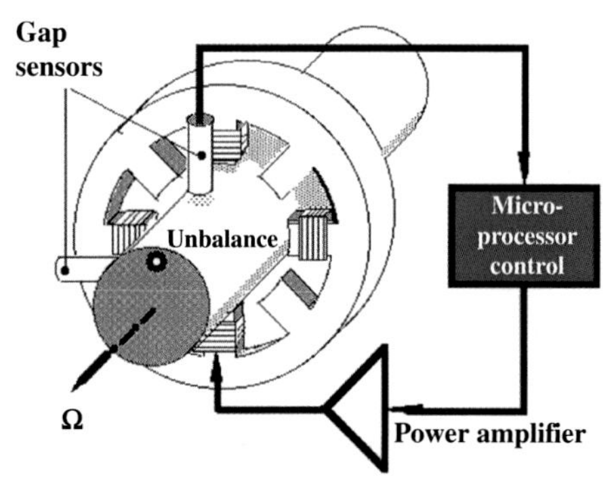

FIG. 11 An active magnetic bearing [2].

Hybrid PMB and AMB bearing designs were discussed in Refs. [1] and [2]. In Ref. [1], the authors looked at different configurations such as air gaps and magnet thicknesses and lengths. The influence of PMB dimensions, the coil parameters of the AMB, and the placement of the PMB and AMB coil placement was studied.

One of the main drawbacks in using an AMB bearing is that energy is required to provide current to the bearing, thus reducing the overall efficiency of the flywheel system. In addition, an active control system with feedback is required during operation. This is why research has progressed from the use of a hybrid system with PMBs and AMBs to those with PMBs and HTSBs. Research on hybrid PMB and HTSB designs involves basic studies on conceptual designs and analysis of forces created by the bearings when lifting and stabilizing the flywheel rotor. Fig. 12 shows a general schematic of the typical setup and Fig. 13 shows a close-up of a high-temperature superconducting bearing.

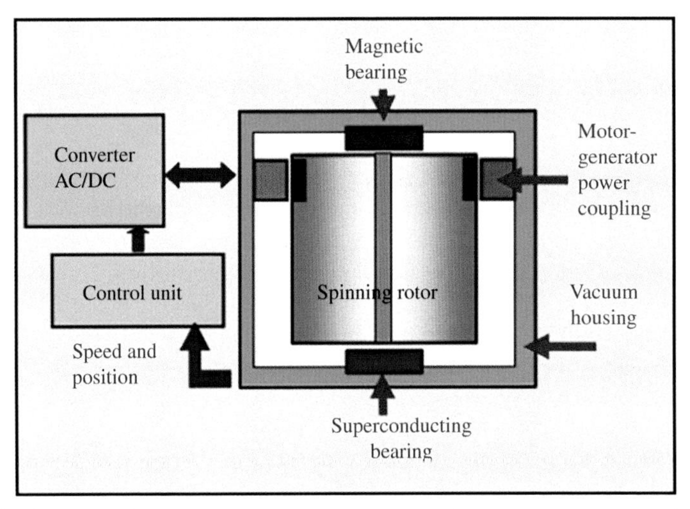

FIG. 12 Typical hybrid bearing with permanent magnet and superconductor bearings [3].

FIG. 13 High-temperature superconducting bearing [19].

One hybrid design [20] combined a PMB with a Superconductor Magnetic Bearing (SMB) to create a passive type of bearing. The SMB uses superconductors that must be cooled below a threshold temperature using liquid nitrogen. A shaft, which was driven by an external motor, was arranged vertically with the PMB at the top and the SMB at the bottom. Static and dynamic forces from the PMB and SMB to lift and stabilize the rotor were measured with each bearing acting separately and together. In addition, resonant speeds were identified. Future testing would involve the flywheel and motor/generator being added to the system.

Another study was performed to evaluate the load-carrying capacity of the bearing in terms of the vertical force on the rotor as a function of its placement [21]. Different arrangements of permanent magnets and superconductors were used to design the optimal configuration (Fig. 14). The rotor was aligned vertically and the superconductor (YBCO) was inside a liquid nitrogen tank.

Other authors discussed the development of a flywheel system with superconducting magnetic bearings made of YBCO and a permanent magnet of Pr-Fe-B [22]. The superconductors were cooled with liquid nitrogen and used to lift the rotor without excessive radial vibration. The rotor was hung vertically and spun with a motor inside a vacuum housing up to 20,000 RPM. The energy storage was measured at 1.4kWh.

One other study investigated the use of a bearing-less motor that could be used for high-speed flywheel applications [23]. These motors use hysteresis

FIG. 14 Hybrid permanent magnetic and superconductor bearing test setup 1: Driving PM discs, 2: Rotor, 3: Lift motor mechanism, 4: Setup motor, 5: Copper disc, 5: HTSB (YBCO) inside a liquid Nitrogen tank [3].

or residual magnetization that would allow them to operate without any bearings. The motors were demonstrated at speeds up to 10,000 RPM. Further development could lead to even higher rotating speeds.

10.4.4 Flywheel material development

Steels and other conventional metal alloys are attractive for flywheel manufacture due to their availability, ease of machining, reduced cost, and simplicity of bearing selection. Disadvantages include lower rotational speed capabilities, and therefore larger size requirement for a given amount of energy storage. Research into low density, high strength materials that can handle higher rotational speeds continues to be active. Composite materials, in particular, have been developed for both flywheel and turbomachinery applications to increase rotational speed. Fig. 15 presents a composite flywheel.

Composite materials can be used to produce a lighter, stronger flywheel that can spin faster and provide the same amount of energy as a steel flywheel, but with a smaller footprint. The main drawback is higher cost to produce the composite rotor flywheel energy system.

Researchers [24] compared a simple one-dimensional flywheel fabricated from steel to another made from a graphite fiber-reinforced epoxy. It was shown that the stored energy of the composite flywheel was much higher than that of steel, as was the energy density (energy storage/unit mass). However, the analysis also showed that the cost of the composite flywheel was 3.66 times higher than that of a comparable steel one. If reduced size were a priority, then the composite flywheel would be the preferred option; whereas, if cost were the biggest consideration, then a traditional steel flywheel would be the better choice.

FIG. 15 Typical composite flywheel.

One other study compared different composite materials to determine an optimal choice to fabricate the flywheel rotor [25]. The authors compared the common boron and graphite epoxy to other improved strength carbon fibers for flywheel application. They calculated the new flywheel design to be 18 times stronger than the common boron and graphite flywheel materials. New advanced materials that will enable even higher speed and a smaller footprint remain a promising and active area of research.

10.4.5 Power quality and hybrid renewable energy systems

Power suppliers are concerned about power quality. Transient disturbances can often affect the quality of the delivered power through voltage sags, swells, and interruptions that can last a few seconds or less. About 85% of all power events are voltage sags lasting less than 2 s and 85% of downtime is caused by disturbances lasting no more than 0.5 s [26]. These events can be mitigated through the use of a large-scale Uninterruptable Power Supply (UPS). Flywheels have excelled in UPS applications, improving the power quality by addressing these short-term disturbance issues by discharging their stored energy when needed.

Much of the supply of energy from renewable sources is affected by power disruptions. Flywheels have been identified as a source of energy to smooth out the energy delivery. There have been a few innovative studies combining renewable energy types with flywheel systems. This can include combining flywheels with wind turbines, solar panels, and supercapacitors to name a few. Two examples are listed as follows.

One study proposes combining flywheels with lead-acid batteries as a hybrid energy storage system to handle power fluctuations from a wind power system [27]. During power increases from the wind turbine the excess energy can be stored in the flywheel/lead-acid battery system and during decreases it can be released from the flywheel/lead–acid battery system. The advantages of the flywheel and battery system are that it takes advantage of the flywheel quick response time (less than 1 s vs 10 s for the battery) and takes advantage of the lead-acid batteries longer discharge duration (up to a few hours for the battery vs up to 15 min for the flywheel).

Another study investigated a hybrid flywheel with a supercapacitor placed inside the rotating flywheel disk [28]. The superconductor energy storage device can increase the energy density of the flywheel system. The superconductor increases the amount of available instantaneous power release. The concept was designed to reduce the size and weight of the flywheel system for the automotive industry, but the technology can be used where size is a limitation for any type of flywheel energy application if the design is developed further.

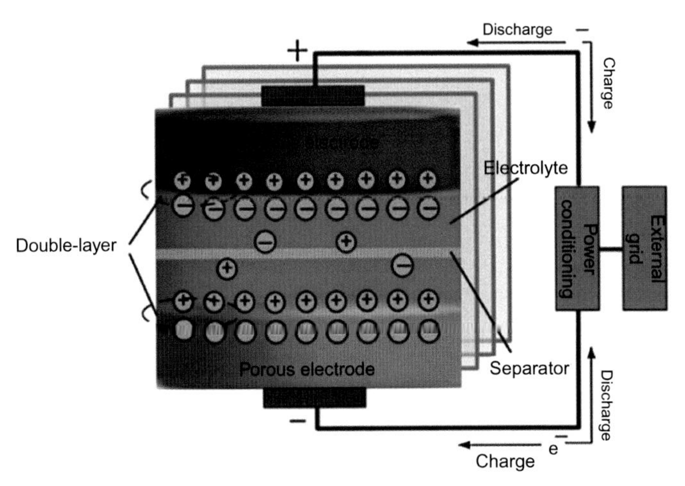

FIG. 16 Schematic diagram of a supercapacitor system [29].

10.5 Electrical energy storage

10.5.1 Supercapacitor

A supercapacitor enhances the characteristics of a capacitor by using two conductor porous carbon (or other high surface area material) electrodes, an electrolyte solution, and a porous membrane separator. The energy is stored as static charge on the surfaces between the electrolyte and the conductor electrodes; since the surface areas of the electrode material are very high and the distance between the plates is very small, this provides a large increase in capacity allowing it to act as an electrochemical battery (see Fig. 16). Supercapacitors to date have primarily been used in short-term storage applications such as pulse power, smoothing out transients, bridging power to equipment, and starting high power dc machines since their most notable features are a relatively high cycle life at over 20,000 cycles, well above batteries, high power density, and good cycle efficiency [30]. The current state-of-the-art supercapacitors still have a low energy density compared to batteries, high self-discharge rate, and high capital cost limiting their application in large-scale and long-term storage but making them useful for short-term power exchanges with large discharge currents and fast responses within large-scale energy storage devices, similar to flywheels. This is a technology currently most useful for energy storage when used in a hybrid combination with long-term storage concepts.

Current research and development is focused on developing electrode materials. Carbon-based materials are the most explored due to the high electrical conductivity, low cost, versatile forms, and chemical stability in a wide range of different solutions. To increase capacitance and electrical properties, the nanoscale structure of the material is manipulated to control and improve the surface area, pore size, and other relevant material properties [31]. Metal oxide materials are also being investigated; however, the ones with the highest specific capacitances, such as Ruthenium oxide, are very expensive and many of these oxides are not chemically stable in various media. Therefore carbon-based composites such as metal oxide/carbon and polymer/carbon are currently being tested to further improve the electrochemical properties of each material. However, the developments to date have focused primarily on increasing charge or capacitance, but have not significantly improved their low energy density which limits their application in large, long-term thermal storage applications.

Energy storage applications that are currently implementing supercapacitors for short-term storage include wind energy and traditional high voltage power supply applications with the aim of improving power quality. Specifically, in wind power applications, the operating point is varying based on environmental conditions; therefore output power predictions are made using fuzzy-rule-based systems which result in some inaccuracies. The addition of supercapacitors can smooth out the output power variations compensating for errors in predictions as well as improve the ability to continuously produce power through transient disturbances and extreme voltage events [32].

Supercapacitors can currently be designed to have a round-trip efficiency up to 95%, and a voltage rating up to 2.7 V with maximum string voltage of 1500 V. To be adapted for large-scale energy storage, the voltage rating must be increased to a significantly higher kV range [33].

10.6 Superconducting magnetic energy storage

Superconducting Magnetic Energy Storage (SMES) systems use energy from an alternating current system and convert and store it as direct current energy in a superconducting magnetic coil through the use of a superconducting coil unit, a power conditioning subsystem (PCS), and a refrigeration and vacuum subsystem (see Fig. 17). The magnet is made of superconducting material such

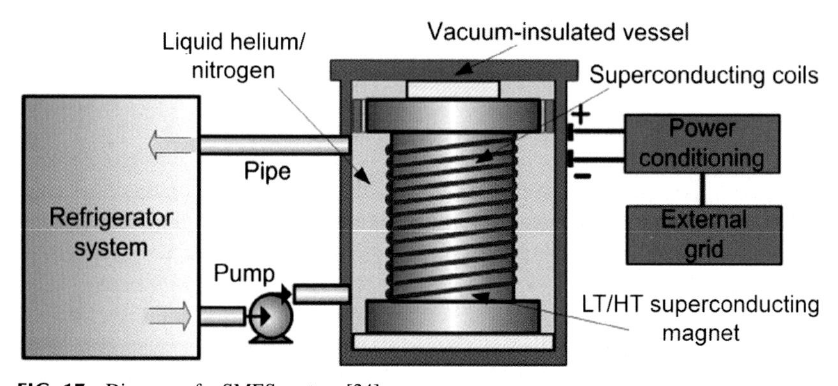

FIG. 17 Diagram of a SMES system [34].

as Niobium-Titanium which, when cryogenically cooled below its superconducting critical temperature, creates an effectively no-loss coil with zero resistance. Once the stored power is required, the DC energy is converted to AC current through the PCS module with an efficiency of over 95%. The conversion process is in the range of milliseconds [34–36].

The operating temperature for a superconducting coil is a balance between the operating requirements and cost. Low-temperature (LT) SMES systems of less than 10 MW are currently available to mitigate power quality issues, load leveling, and spinning reserve [37]. These include the Nosoo power station in Japan; the Upper Wisconsin in the United States operated by American Transmission; and a smaller 20 kW, up to 2 MJ test facility at the University of Houston funded by ARPA-E. Given its short response time, high power density, high efficiency, and long life, developing high temperature (HT), large utility scale (500–1000 MWh) SMES is currently being explored for large power load leveling applications. However, implementation of a structure that would be sufficient to offset the expansion of a large superconducting coil under Lorentz forces is cost prohibitive, not to mention the cost of the coil itself, the amount of daily self-discharge, and the need to minimize the impact of the strong magnetic field it produces.

Current research and development in SMES technology is focused on cost reduction of the coils and material improvements. The most cost effective and feasible utility scale design was first explored in the 1980s which involved placing the coil in an earth trench [36]. It is still too costly to take into a testing and development stage.

10.7 Hybrid energy concepts

10.7.1 Thermal-chemical with mechanical

There are two primary types of heat pumping and refrigeration technology: one is the dominant mechanical vapor compression cycle, and the other is the thermal-driven sorption cycles. This second technology is commonly regarded as a promising alternative to the first one and recognized as the important technology for low carbon future [38, 39]. In fact, these two types have individual distinctive advantages over each other. To combine the merits of different technologies, the hybrid cycles, either sorption-assisted compression cycle [40–43] or compression-assisted sorption cycles [44, 45], have attracted increasing attention in the search to develop more flexible and versatile functionality over a wide range of operational conditions. Sorption cycles, as forging in Chapter 5 especially the solid-gas chemisorption (i.e., thermochemical adsorption), have the unique advantage for energy storage application due to the inherent intermittence. That suggests a hybrid system that integrates a thermal-driven solid-gas sorption cycle with a mechanical-driven vapor compression process

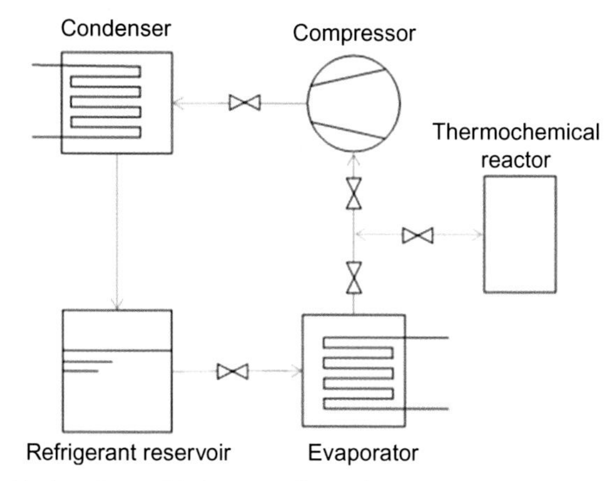

FIG. 18 Hybrid adsorption-assisted compression system.

has the ability of storing thermal and mechanical energy simultaneously in the form of chemical potential.

For a hybrid sorption-assisted compression cycle, the basic configuration (Fig. 18) comprises a condenser, evaporator, refrigerant reservoir, and compressor with an additional thermochemical reactor. The thermochemical sorption cycle shares the same condenser, evaporator, and refrigerant reservoir with the vapor compression cycle [42, 43]. This type of hybrid system prioritizes the continuous cold production by mechanical vapor compression with auxiliary cold storage function simultaneously through thermochemical sorption unit. The compressor helps decrease the desorption temperature, enabling effective utilization of low-temperature heat for desorption and the heat is stored together with mechanical/electrical input through the compressor into the thermochemical reactor. For example, in Ferrucci et al.'s work [43] the solar PV panel and the grid were used to meet electricity requirement of powering the compressor of the vapor compression cycle for cold production. When there was any surplus electricity from the PV panels and no cooling needs, the system switches operation mode and the extra power was used to run the compressor in order to drive the desorption of the $BaCl_2$ ammine contained in the thermochemical reactor. In this instance, the desorption temperature was reduced by up to 20°C as the desorption process occurred with a heat source at 35°C and the electricity was stored in the form of chemical potential energy. The stored energy was later released to produce a cooling effect without the need for the compressor. The system can also operate for heating purpose as the conventional vapor compression heat pump cycle with the thermochemical reactor releases absorption heat at relatively higher temperature.

The compression-assisted sorption cycle mainly consumes thermal energy with relatively smaller amount of supplementary mechanical/electrical energy

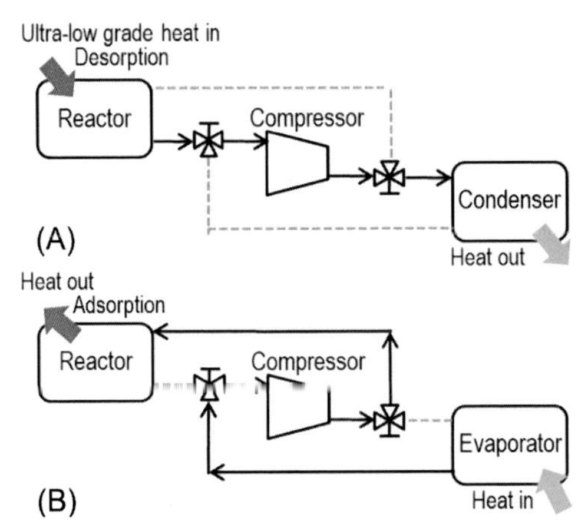

FIG. 19 Hybrid compression-assisted adsorption cycle for thermal and mechanical/electrical energy storage and heat pump. (A) Charging phase: storage of ultra-low grade heat and mechanical/electrical energy. (B) Discharging phase: upgraded heat output or refrigeration.

input. The hybrid concept in Fig. 19 inherits the simple configuration from sorption systems; there are two vessels—one is the thermochemical reactor, the other one is the two-phase heat exchanger (condenser/evaporator)—with one compressor installed in-between two vessels.

In energy charging process (Fig. 19A), low grade heat is supplied for decomposition which releases the working gas while the mechanical/electrical-driven compressor further pressurizes the working gas. Eventually the working gas liquidizes in the condenser at a high pressure and low grade heat and mechanical/electric energy is stored in the form of chemical potential energy. For the energy discharging process (Fig. 19B), it can flexibly perform to meet different requirements: (1) If low grade heat is used again to generate refrigerant vapor in the evaporator and the compressor is also used to pressurize the vapor, the thermochemical reactor will adsorb the high-pressure high-temperature refrigerant vapor exhausted from the compressor while it releases adsorption heat at a much higher temperature than originally stored. (2) Alternatively, for cooling power generation, the evaporation heat of the refrigerant comes from the surroundings while the refrigerant extracts heat from the surroundings and evaporates at a refrigeration temperature and subsequently is adsorbed by the salt adsorbent in the reactor. The refrigerant evaporation produces a refrigeration effect. (3) If there is no mechanical/electric energy available during the energy discharging process, i.e. the compressor is bypassed, the system can operate like the basic thermochemical sorption heat pumping/refrigerating process. In such a combination, the collective effect of thermal-driven and mechanical/electric driven can enhance the performance and capability of heating and cooling because on top of the thermal-driven process, the compressor helps further

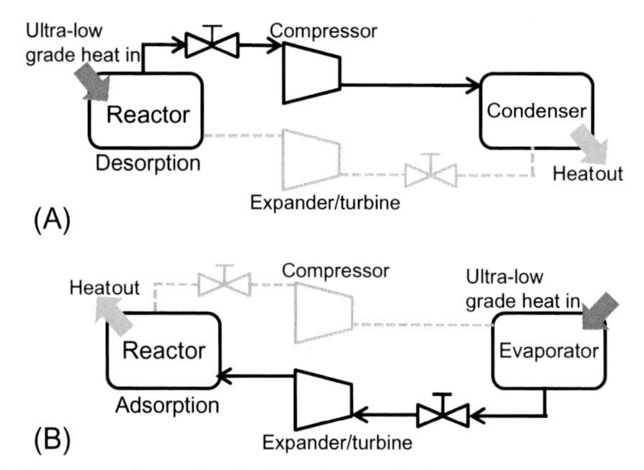

FIG. 20 Hybrid compression-assisted adsorption cycle for thermal and mechanical/electrical energy storage. (A) Charging phase: storing ultra-low grade heat and mechanical/electrical energy. (B) Discharging phase: mechanical/electrical power output.

increase the temperature lift, or further decrease the refrigeration temperature. Additionally, the heating and cooling performance (temperature and capability) can be flexibly adjusted and finely tuned according to the quality, quantity, and availability of heat source/heat sink within a wide temperature range and electric power. When one expander is integrated into the system as illustrated in Fig. 20, it forms the multivector energy storage system, which offers solutions for optimally storing and off-setting simultaneous mechanical energy/electricity and heating or cooling, to reduce the end use energy demands as a whole. The off-peak electricity and ultra-low grade heat can be stored at the same time, then electricity can be generated during the discharging phase. The round-trip efficiency of electricity storage can be larger than 100% due to the input of ultra-low grade heat [46].

To extend the range of applicable temperature for a thermally activated sorption heat pump toward lower temperature of waste heat, van der Pal et al. [47, 48] evaluated a hybrid heat pump system that combined compressor with ammonia-based chemisorption reactors. Without shared condenser and evaporator, the hybrid system was as compact as it could be, consisting of a low-temperature sorbent reactor, high-temperature sorbent reactor (as a resorption cycle [49]), and a compressor. The integration was proven technically feasible to effectively reduce the minimum temperature of the recoverable waste heat and also increase the upgraded heat temperature, although there is the challenge of the mechanical compressor regulation due to the variable mass flow rate from the chemisorption reactor. Bao et al. [46] proposed and investigated the concept of combining compressor and expander with ammonia-based thermochemical resorption processes using manganese chloride ($MnCl_2$), calcium chloride ($CaCl_2$), and sodium bromide (NaBr) as reactant salts. Such an integrated system enables the efficient recovery and storage of low grade heat source from 30°C and 100°C and simultaneously the renewable or cheap electricity with high energy density and high flexibility and versatility. It was concluded that it could achieve temperature lift by between 10°C and 80°C for heat transformation depending on reactant salts used; the round-trip efficiency for electric energy storage could reach 100% and even higher when the heat source temperature was higher than 50°C.

10.7.2 Other hybrid concepts

Generally in the energy storage field, different storage technologies are mostly considered as competitors rather than potential collaborators. Sometimes people still habitually to see the different storage technologies for different types of energies in a competitive mode. In fact, no single technology could easily compete with the energy density of the cheap fossil fuel. It is critical to make these low carbon technologies intelligently work together to achieve better energy management and high integration of renewable energy sources, leading to significantly increased efficiency of the whole process. Researchers and engineers have started developing hybrid energy storage systems (HESS).

The thermochemical sorption-compression hybrid energy system features a high level of interaction between different types of energies that is stored in the same unit through the same process by using one single working medium. This feature renders it the desirable flexibility of switching or adjusting between different energies according to specific requirement, which allows a whole new dimension to process integration. Additionally, the advantage of the minimum energy losses over long-term storage in the form of chemical potential energy promotes its wide range of application. With the ongoing continuous effort on thermochemical technology including material development and reactor optimization, the cyclic performance stability has been proven over more than 10,000 cycles without an evidence of performance degradation [43].

Unlike the thermochemical sorption-compression hybrid concept, most of the hybrid systems studied so far implement different energy storage technologies independently as they operate individually in separated units. The Adiabatic CAES system is one of representative examples of the hybrid concepts. To guarantee a sufficiently high temperature of the air entering the expander to improve CAES efficiency and avoid using fossil fuels, the CAES is combined with thermal energy storage where the heat generated by compression is stored and then returned to the air before expansion [50, 51]. Some researchers proposed hydrogen storage in substitution of high-temperature heat storage as the hydrogen is generated by water electrolysis and later on burns to provide heat for the air expansion [52].

Another typical HESS concept aims at a beneficial coupling of two or more energy storage technologies with complementary operating characteristics, such as energy density and power density, self-discharge rate, ramp rate, efficiency, lifetime, cost, etc. [53, 54]. For example in a common combination, there are typically two energy storages, one energy storage (ES1) that features comparatively faster response and higher efficiency and is dedicated to satisfy "high power" demand, transients and fast load fluctuations. The other energy storage (ES2) is the "high energy" element with a relatively lower discharging rate and lower specific energy installation costs; it is used to cover average power demand. The main advantage of such a HESS is that the configuration and operation is optimized by coupling two energy storages and taking energy and power both into consideration for the optimization, leading to the reduction of the total investment costs and the dynamic stress and losses of the ES2, and the increase of the total system efficiency. Most of the HESS studies so far have focused on mechanical and electrical energy storage in relation with power generation to tackle the issues of fast and strong variations of both load and sources associated with intermittent solar PV and wind power [55]. Particularly, lithium-ion batteries are commonly used as a "high energy" or "high power" element, one of storage devices and their roles depend on their counterpart's operation characteristics. For example, the hybrid flywheels/battery system uses a flywheel to function as the high-power storage that covers peaks, transients, and fast power fluctuations. It helps to avoid battery stress and as a consequence increases battery lifetime [56]. For the hybrid CAES/supercapacitor system, the CAES system is the high power rating, high energy rating and slow ramp rate device; in contrast, the supercapacitor is the fast ramp rate and low power rating, low-energy rating device, but can serve as the "high power" role [57]. In fact, the arrangement of two energy storages for "high energy" and "high power" system can be adopted in thermal energy storage, especially for satisfying the highly dynamic demand of space heating and hot water use, ensuring indoor comfort and device longevity.

Many studies endeavored to established positive synergies between different storage systems. Optimizing design, control, and energy management strategies for HESS at the interface between different types of energies, i.e., mechanical energy, electricity, heat and gas sector, will unfold significant potentials for further improvements of cost, efficiency, and lifetime of low carbon energy systems and increasing integration of renewable energy.

References

[1] J. Pacheco, Final Test and Evaluation Results from the Solar Two Project, Sandia National Labs, 2002. SAN2002-0120.

[2] NREL, Concentrating Solar Power Projects Database, Accessed October 2019, https://solarpaces.nrel.gov/.

[3] N. Hoivik, et al., Long-term performance results of concrete-based modular thermal energy storage system. J. Energy Storage 24 (2019) https://doi.org/10.1016/j.est.2019.04.009.

[4] Newcastle University connects first grid-scale pumped heat energy storage, The Engineer (2019).
[5] J. Keseli, C. Tan, Reversible counter-rotating turbomachine to enable brayton-laughlin cycle, in: Presented at ARPA-E DAYS Kickoff Meeting, March, 2019.
[6] B. Tom, Small-scale PHES demonstration, in: Presented at ARPA-E DAYS Kickoff Meeting, March, 2019.
[7] J. Miller, Low-cost, long-duration electrical energy storage using a CO2-based Electro Thermal Energy Storage (ETES) System, in: Presented at ARPA-E DAYS Kickoff Meeting, March, 2019.
[8] https://www.ice-energy.com/technology/ Accessed October 2019
[9] A. Mathur, Using encapsulated phase change material for thermal energy storage for baseload CSP, in: Presented at DOE CSP Review Meeting, April, 2013.
[10] B. Munoz-Sanchez, et al., Encapsulated high temperature PCM as active filler material in a thermocline-based thermal storage system, Energy Procedia 69 (2015) 937–946.
[11] Y. Kim, Novel Concepts of Compressed Air Energy Storage and Thermo-Electric Energy Storage, EPFL, Lausanne, 2012.
[12] Y. Kim, J. Lee, S. Kim, D. Favrat, Potential and evolution of compressed air energy storage: energy and exergy analyses, Entropy 14 (2012) 1501–1521.
[13] H. Safaei, D. Keith, R. Hugo, Compressed air energy storage (CAES) with compressors distributed at heat loads to enable waste heat utilization, Appl. Energy 103 (2013) 165–179.
[14] H. Safaei, D. Keith, Compressed air energy storage with waste heat export: an Alberta case study, Energy Convers. Manag. 78 (2014) 114–124.
[15] J. Wang, et al., Overview of compressed air energy storage and technology development, Energies 10 (2010) 991.
[16] H. Guo, Y. Xu, H. Chen, X. Zhou, Thermodynamic characteristics of a novel supercritical compressed air energy storage system, Energy Convers. Manag. 115 (2016) 167–177.
[17] C. Oldenburg, L. Pan, Utilization of CO2 as cushion gas for porous media compressed air energy storage, Greenhouse Gases Sci. Technol. 3 (2013) 124–135.
[18] C. Oldenburg, Carbon dioxide (CO2) as cushion gas for compressed air energy storage (CAES), United States Patent 2011/0236134 A1, September 29 (2011).
[19] L.A.R. Corvao, Preliminary Approach Designing and Electromagnetic Bearing for Flywheel Energy Systems, Instituto Superior Tecnico (2008).
[20] G.G. Sotelo, F.S. Rodrigues, J.G. Costa, O.J. de Santiago, R.M. Stephan, Tests With a Hybrid Bearing for a Flywheel Energy Storage System, (2016).
[21] A.C. Cansiz, D.T. McGuiness, A Case Study for a Superconducting Magnet Bearing Optimization, IEEE (2017).
[22] N. Kashima, S. Nagaya, M. Minami, H. Kawashima, Unisuga, Development of a 1 kWh Flywheel Energy Storage System with Superconducting Magnetic Bearing, (2000).
[23] N.W. Stauffer, Designing high-speed motors for energy storage and use, MIT News (2014).
[24] Understanding Flywheel Energy Storage: Does High Speed Really Imply a Better Design?, Active Power White Paper #112, 2008.
[25] M.A. Conteh, E.C. Nsofor, Composite flywheel material design for high speed energy storage, J. Appl. Res. Technol. 14 (3) (2016).
[26] B.D. Miller, Basics of flywheel UPS, Power Magazine (May 2000).
[27] Q. Shao, Y. Zhao, S. Du, Y. Du, A novel hybrid energy storage strategy based on flywheel and lead-acid battery in wind power generation system, Int. J. Control Autom. 8 (7) (2015).
[28] H. Toodeji, A developed flywheel energy storage with built-in rotating supercapacitors, Turk. J. Eng. Comp. Sci. 27 (2019) 213–229.

[29] X. Luo, J. Wang, M. Dooner, J. Clarke, Overview of current development in electrical energy storage technologies and the application potential in power system operation, Appl. Energy J. 137 (2015) 511–536.

[30] H. Chen, T.N. Cong, W. Yang, C. Tan, Y. Li, Y. Ding, Progress in electrical energy storage system: a critical review, Prog. Nat. Sci. 19 (2009) 291–312.

[31] Y. Zhai, Y. Dou, D. Zhao, P. Fulvio, R. Mayes, S. Dai, Carbon material for chemical capacitive energy storage, Adv. Mater. 23 (42) (2011) 4828–4850.

[32] Abbey, C and Geza, J, "Supercapacitor energy storage for wind energy applications, IEEE Trans. Ind. Appl., Vl. 43, No. 3, 2007 pp 769–776. https://ieeexplore.ieee.org/stamp/stamp.jsp?arnumber=4214991.

[33] S. Smith, P.K. Sen, B. Kroposki, Advancement of Energy Storage Devices and Applications in Electrical Power System, IEEE, 2008.

[34] S.C. Smith, P.K. Sen, B. Kroposki, Advancement of energy storage devices and applications in electrical power system, in: 2008 IEEE Power Energy Soc. Gen. Meet. – Convers. Deliv. Electr. Energy 21st Century, IEEE, 2008, pp. 1–8.

[35] Hus, C. and Lee, W., "Superconducting magnetic energy storage for power system applications," IEEE Trans. Ind. Appl., Vo. 29, No. 5, Pp. 990–996, 1992.

[36] Luongo, C., "Superconducting storage systems: an overview," IEEE Trans. Magn., Vo. 32, No. 4, pp. 2214–2223, 1996.

[37] W. Buckles, Hassenzahl, Superconducting magnetic energy storage, IEEE Power Eng. Rev. (2000) 16–20.

[38] F. Ziegler, Novel cycles for power and refrigeration, in: 1st European Conference on Polygeneration, 16–17 October 2007, Tarragona, Spain, 2007.

[39] R.Z. Wang, L.W. Wang, J.Y. Wu, Adsorption Refrigeration Technology: Theory and Application, John Wiley & Sons, Singapore, 2014.

[40] V. Gudjonsdottir, C.I. Ferreira, G. Rexwinkel, A.A. Kiss, Enhanced performance of wet compression-resorption heat pumps by using NH_3-CO_2-H_2O as working fluid, Energy 124 (2017) 531–542.

[41] D.M. van de Bor, C.A. Infante Ferreira, A.A. Kiss, Low grade waste heat recovery using heat pumps and power cycles, Energy 89 (2015) 864–873.

[42] J. Fitó, A. Coronas, S. Mauran, N. Mazet, M. Perier-Muzet, D. Stitou, Hybrid system combining mechanical compression and thermochemical storage of ammonia vapor for cold production, Energy Convers. Manag. 180 (2019) 709–723.

[43] F. Ferrucci, D. Stitou, P. Ortega, F. Lucas, Mechanical compressor-driven thermochemical storage for cooling applications in tropical insular regions. Concept and efficiency analysis, Appl. Energy 219 (2018) 240–255.

[44] G. Angrisani, M. Canelli, C. Roselli, A. Russo, M. Sasso, F. Tariello, A small scale polygeneration system based on compression/absorption heat pump, Appl. Therm. Eng. 114 (2017) 1393–1402.

[45] W. Wu, W.X. Shi, B.L. Wang, X.T. Li, Annual performance investigation and economic analysis of heating systems with a compression-assisted air source absorption heat pump, Energy Convers. Manag. 98 (2015) 290–302.

[46] H.S. Bao, Z.W. Ma, A.P. Roskilly, Integrated chemisorption cycles for ultra-low grade heat recovery and thermo-electric energy storage and exploitation, Appl. Energy 164 (2016) 225–236.

[47] M. van der Pal, R. de Boer, A. Wemmers, S. Smeding, J. Veldhuis, Experimental results and model calculation of a hybrid adsorption-compression heat pump based on a roots compressor and silica gel-water sorption (Chapter 4), in: Thermally Driven Heat Pumps for Cooling, University of TU Berlin, 2013.

[48] M. van der Pal, A. Wemmers, S. Smeding, R. de Boer, Technical and economical feasibility of the hybrid adsorption compression heat pump concept for industrial applications, Appl. Therm. Eng. 61 (2013) 837–840.
[49] H.S. Bao, R.Z. Wang, L.W. Wang, A resorption refrigerator driven by low grade thermal energy, Energy Convers. Manag. 52 (2011) 2339–2344.
[50] E. Barbour, D. Mignard, Y.L. Ding, Y. Li, Adiabatic compressed air energy storage with packed bed thermal energy storage, Appl. Energy 155 (2015) 804–815.
[51] X. Luo, J.H. Wang, C. Krupke, Y. Wang, Y. Sheng, J. Li, Y.J. Xu, D. Wang, S.H. Miao, H.S. Chen, Modelling study, efficiency analysis and optimisation of large-scale adiabatic compressed air energy storage systems with low temperature thermal storage, Appl. Energy 162 (2016) 589–600.
[52] S. Ubertini, A.L. Facci, L. Andreassi, Hybrid hydrogen and mechanical distributed energy storage, Energies 10 (2017) 2035–2050.
[53] A.J. Pimm, S.D. Garvey, B. Kantharaj, Economic analysis of a hybrid energy storage system based on liquid air and compressed air, J. Energy Storage 4 (2015) 24–35.
[54] T. Bocklisch, Hybrid energy storage systems for renewable energy applications. 9th International Renewable Energy Storage Conference, IRES 2015, Energy Procedia 73 (2015) 103–111.
[55] P. Zhao, Y.P. Dai, J.F. Wang, Design and thermodynamic analysis of a hybrid energy storage system based on A-CAES (adiabatic compressed air energy storage) and FESS (flywheel energy storage system) for wind power application, Energy 70 (2014) 674–684.
[56] J. Hou, J. Sun, H. Hofmann, Control development and performance evaluation for battery/flywheel hybrid energy storage solutions to mitigate load fluctuations in all-electric ship propulsion systems, Appl. Energy 212 (15) (2018) 919–930.
[57] S. Lemofouet, A. Rufer, A hybrid energy storage system based on compressed air and supercapacitors with maximum efficiency point tracking (MEPT), IEEE Trans. Ind. Electron. 53 (2006) 1105–1115.

Index

Note: Page numbers followed by *f* indicate figures and *t* indicate tables.

D

E

F

G

H

I

L

N

O

P

T

U

V

W